Lecture Notes in Computer Science 7667

Commenced Publication in 1973
Founding and Former Series Editors:
Gerhard Goos, Juris Hartmanis, and Jan van Leeuwen

T0224128

Tingwen Huang   Zhigang Zeng
Chuandong Li   Chi Sing Leung (Eds.)

Neural
Information Processing

19th International Conference, ICONIP 2012
Doha, Qatar, November 12-15, 2012
Proceedings, Part V

Springer

Volume Editors

Tingwen Huang
Texas A&M University at Qatar, Education City
P.O. Box 23874, Doha, Qatar
E-mail: tingwen.huang@qatar.tamu.edu

Zhigang Zeng
Huazhong University of Science and Technology
Department of Control Science and Engineering
1037 Luoyu Road, Wuhan, Hubei 430074, China
E-mail: zgzeng@gmail.com

Chuandong Li
Chongqing University, College of Computer Science
174 Shazhengjie Street, Chongqing 400044, China
E-mail: licd@cqu.edu.cn

Chi Sing Leung
City University of Hong Kong, Department of Electronic Engineering
83 Tat Chee Avenue, Kowloon, Hong Kong, China
E-mail: eeleungc@cityu.edu.hk

ISSN 0302-9743   e-ISSN 1611-3349
ISBN 978-3-642-34499-2   e-ISBN 978-3-642-34500-5
DOI 10.1007/978-3-642-34500-5
Springer Heidelberg Dordrecht London New York

Library of Congress Control Number: 2012949896

CR Subject Classification (1998): F.1, I.2, I.4-5, H.3-4, G.3, J.3, C.1.3, C.3

LNCS Sublibrary: SL 1 – Theoretical Computer Science and General Issues

Typesetting: Camera-ready by author, data conversion by Scientific Publishing Services, Chennai, India

Printed on acid-free paper

Springer is part of Springer Science+Business Media (www.springer.com)

Preface

This volume is part of the five-volume proceedings of the 19th International Conference on Neural Information Processing (ICONIP 2012), which was held in Doha, Qatar, during November 12–15, 2012. ICONIP is the annual conference of the Asia Pacific Neural Network Assembly (APNNA). This series of conferences has been held annually since 1994 and has become one of the premier international conferences in the areas of neural networks.

Over the past few decades, the neural information processing community has witnessed tremendous efforts and developments from all aspects of neural information processing research. These include theoretical foundations, architectures and network organizations, modeling and simulation, empirical study, as well as a wide range of applications across different domains. Recent developments in science and technology, including neuroscience, computer science, cognitive science, nano-technologies, and engineering design, among others, have provided significant new understandings and technological solutions to move neural information processing research toward the development of complex, large-scale, and networked brain-like intelligent systems. This long-term goal can only be achieved with continuous efforts from the community to seriously investigate different issues of the neural information processing and related fields. To this end, ICONIP 2012 provided a powerful platform for the community to share their latest research results, to discuss critical future research directions, to stimulate innovative research ideas, as well as to facilitate multidisciplinary collaborations worldwide.

ICONIP 2012 received tremendous submissions authored by scholars coming from 60 countries and regions across six continents. Based on a rigorous peer-review process, where each submission was evaluated by at least two reviewers, about 400 high-quality papers were selected for publication in the prestigious series of *Lecture Notes in Computer Science*. These papers cover all major topics of theoretical research, empirical study, and applications of neural information processing research. In addition to the contributed papers, the ICONIP 2012 technical program included 14 keynote and plenary speeches by Majid Ahmadi (University of Windsor, Canada), Shun-ichi Amari (RIKEN Brain Science Institute, Japan), Guanrong Chen (City University of Hong Kong, Hong Kong), Leon Chua (University of California at Berkeley, USA), Robert Desimone (Massachusetts Institute of Technology, USA), Stephen Grossberg (Boston University, USA), Michael I. Jordan (University of California at Berkeley, USA), Nikola Kasabov (Auckland University of Technology, New Zealand), Juergen Kurths (University of Potsdam, Germany), Erkki Oja (Aalto University, Finland), Marios M. Polycarpou (University of Cyprus, Cyprus), Leszek Rutkowski (Technical University of Czestochowa, Poland), Ron Sun (Rensselaer Polytechnic Institute, USA), and Jun Wang (Chinese University of Hong Kong, Hong Kong). The

ICONIP technical program included two panels. One was on "Challenges and Promises in Computational Intelligence" with panelists: Shun-ichi Amari, Leon Chua, Robert Desimone, Stephen Grossberg and Michael I. Jordan; the other one was on "How to Write Better Technical Papers for International Journals in Computational Intelligence" with panelists: Derong Liu (University of Illinois of Chicago, USA), Michel Verleysen (Université catholique de Louvain, Belgium), Deliang Wang (Ohio State University, USA), and Xin Yao (University of Birmingham, UK). The ICONIP 2012 technical program was enriched by 16 special sessions and "The 5th International Workshop on Data Mining and Cybersecurity." We highly appreciate all the organizers of special sessions and workshop for their tremendous efforts and strong support.

Our conference would not have been successful without the generous patronage of our sponsors. We are most grateful to our platinum sponsor: *United Development Company PSC (UDC)*; gold sponsors: *Qatar Petrochemical Company, ExxonMobil* and *Qatar Petroleum*; organizers/sponsors: *Texas A&M University at Qatar* and *Asia Pacific Neural Network Assembly*. We would also like to express our sincere thanks to the IEEE Computational Intelligence Society, International Neural Network Society, European Neural Network Society, and Japanese Neural Network Society for technical sponsorship.

We would also like to sincerely thank Honorary Conference Chair Mark Weichold, Honorary Chair of the Advisory Committee Shun-ichi Amari, the members of the Advisory Committee, the APNNA Governing Board and past presidents for their guidance, the Organizing Chairs Rudolph Lorentz and Khalid Qaraqe, the members of the Organizing Committee, Special Sessions Chairs, Publication Committee and Publicity Chairs, for all their great efforts and time in organizing such an event. We would also like to take this opportunity to express our deepest gratitude to the members of the Program Committee and all reviewers for their professional review of the papers. Their expertise guaranteed the high quality of the technical program of the ICONIP 2012!

We would like to express our special thanks to Web manager Wenwen Shen for her tremendous efforts in maintaining the conference website, the publication team including Gang Bao, Huanqiong Chen, Ling Chen, Dai Yu, Xing He, Junjian Huang, Chaobei Li, Cheng Lian, Jiangtao Qi, Wenwen Shen, Shiping Wen, Ailong Wu, Jian Xiao, Wei Yao, and Wei Zhang for spending much time to check the accepted papers, and the logistics team including Hala El-Dakak, Rob Hinton, Geeta Megchiani, Carol Nader, and Susan Rozario for their strong support in many aspects of the local logistics.

Furthermore, we would also like to thank Springer for publishing the proceedings in the prestigious series of *Lecture Notes in Computer Science*. We would, moreover, like to express our heartfelt appreciation to the keynote, plenary, panel, and invited speakers for their vision and discussions on the latest research developments in the field as well as critical future research directions, opportunities, and challenges. Finally, we would like to thank all the speakers, authors, and participants for their great contribution and support that made ICONIP 2012 a huge success.

November 2012

Tingwen Huang
Zhigang Zeng
Chuandong Li
Chi Sing Leung

Organization

Honorary Conference Chair

Mark Weichold   Texas A&M University at Qatar, Qatar

General Chair

Tingwen Huang   Texas A&M University at Qatar, Qatar

Program Chairs

Andrew Leung   City University of Hong Kong, Hong Kong
Chuandong Li   Chongqing University, China
Zhigang Zeng   Huazhong University of Science and Technology, China

Advisory Committee

Honorary Chair

Shun-ichi Amari   RIKEN Brain Science Institute, Japan

Members

Majid Ahmadi   University of Windsor, Canada
Sabri Arik   Istanbul University, Turkey
Salim Bouzerdoum   University of Wollongong, Australia
Jinde Cao   Southeast University, China
Jonathan H. Chan   King Mongkut's University of Technology, Thailand
Guanrong Chen   City University of Hong Kong, Hong Kong
Tianping Chen   Fudan University, China
Kenji Doya   Okinawa Institute of Science and Technology, Japan
Wlodzislaw Duch   Nicolaus Copernicus University, Poland
Ford Lumban Gaol   Bina Nusantara University, Indonesia
Tom Gedeon   Australian National University, Australia
Stephen Grossberg   Boston University, USA
Haibo He   University of Rhode Island, USA
Akira Hirose   University of Tokyo, Japan
Nikola Kasabov   Auckland University of Technology, New Zealand

Irwin King	The Chinese University of Hong Kong, Hong Kong
James Kwow	Hong Kong University of Science and Technology, Hong Kong
Soo-Young Lee	Advanced Institute of Science and Technology, Korea
Xiaofeng Liao	Chongqing University, China
Chee Peng Lim	Universiti Sains Malaysia, Malaysia
Derong Liu	University of Illinois at Chicago, USA
Bao-Liang Lu	Shanghai Jiao Tong University, China
John MacIntyre	University of Sunderland, UK
Erkki Oja	Helsinki University of Technology, Finland
Nikhil R. Pal	Indian Statistical Institute, India
Marios M. Polycarpou	University of Cyprus, Cyprus
Leszek Rutkowski	Czestochowa University of Technology, Poland
Noboru Ohnishi	Nagoya University, Japan
Ron Sun	Rensselaer Polytechnic Institute, USA
Ko Sakai	University of Tsukuba, Japan
Shiro Usui	RIKEN, Japan
Xin Yao	University of Birmingham, UK
DeLiang Wang	Ohio State University, USA
Jun Wang	Chinese University of Hong Kong, Hong Kong
Li-Po Wang	Nanyang Technological University, Singapore
Rubin Wang	East China University of Science and Technology, China
Zidong Wang	Brunel University, UK
Huaguang Zhang	Northeastern University, China

Organizing Committee

Chairs

| Rudolph Lorentz | Texas A&M University at Qatar, Qatar |
| Khalid Qaraqe | Texas A&M University at Qatar, Qatar |

Members

Hassan Bazzi	Texas A&M University at Qatar, Qatar
Hala El-Dakak	Texas A&M University at Qatar, Qatar
Mohamed Elgindi	Texas A&M University at Qatar, Qatar
Jihad Mohamad Jaam	Qatar University, Qatar
Samia Jones	Texas A&M University at Qatar, Qatar
Uvais Ahmed Qidwai	Qatar University, Qatar
Paul Schumacher	Texas A&M University at Qatar, Qatar

Special Sessions Chairs

Zijian Diao	Ohio University, USA
Hassab Elgawi Osman	The University of Tokyo, Japan
Paul Pang	Unitec Institute of Technology, New Zealand

Publicity Chairs

Mehdi Roopaei   Shiraz University, Iran
Enchin Serpedin   Texas A&M University,USA
Maolin Tang   Queensland University of Technology, Australia

Program Committee Members

Sabri Arik
Emili Balaguer Ballester
Gang Bao
Matthew Casey
Li Chai
Jonathan Chan
Mou Chen
Yangquan Chen
Mingcong Deng
Ji-Xiang Du
El-Sayed El-Alfy
Osman Elgawi
Peter Erdi
Wai-Keung Fung
Yang Gao
Erol Gelenbe
Nistor Grozavu
Ping Guo
Fei Han
Hanlin He
Shan He
Bin He
Jinglu Hu
He Huang
Kaizhu Hunag
Jihad Mohamad Jaam
Minghui Jiang
Hu Junhao
John Keane
Sungshin Kim
Irwin King
Sid Kulkarni
H.K. Kwan
James Kwok
Wk Lai
James Lam
Soo-Young Lee

Chi Sing Leung
Tieshan Li
Bin Li
Yangmin Li
Bo Li
Ruihai Li
Hai Li
Xiaodi Li
Lizhi Liao
Chee-Peng Lim
Ju Liu
Honghai Liu
Jing Liu
C.K. Loo
Luis Martínez López
Wenlian Lu
Yanhong Luo
Jinwen Ma
Mufti Mahmud
Jacek Mańdziuk
Muhammad Naufal Bin Mansor
Yan Meng
Xiaobing Nie
Sid-Ali Ouadfeul
Seiichi Ozawa
Shaoning Paul Pang
Anhhuy Phan
Uvais Qidwai
Ruiyang Qiu
Hendrik Richter
Mehdi Roopaei
Thomas A. Runkler
Miguel Angel Fernández Sanjuán
Ruhul Sarker
Naoyuki Sato
Qiankun Song
Jochen Steil

John Sum
Bing-Yu Sun
Norikazu Takahashi
Kay Chen Tan
Ying Tan
Maolin Tang
Jinshan Tang
Huajin Tang
H. Tang
Ke Tang
Peter Tino
Haifeng Tou
Dat Tran
Michel Verleysen
Dan Wang
Yong Wang
Ning Wang
Zhanshan Wang

Xin Wang
Dianhui Wang
Ailong Wu
Bryant Wysocki
Bjingji Xu
Yingjie Yang
Shengxiang Yang
Wenwu Yu
Wen Yu
Xiao-Jun Zeng
Xiaoqin Zeng
Junping Zhang
Zhong Zhang
Wei Zhang
Jie Zhang
Dongbin Zhao
Hongyong Zhao
Huaqing Zhen

Publications Committee Members

Gang Bao
Guici Chen
Huangqiong Chen
Ling Chen
Shengle Fang
Lizhu Feng
Xing He
Junhao Hu
Junjian Huang
Feng Jiang
Bin Li
Chaobei Li
Yanling Li
Mingzhao Li
Lei Liu
Xiaoyang Liu
Jiangtao Qi
Wenwen Shen
Cheng Wang

Xiaohong Wang
Zhikun Wang
Shiping Wen
Ailong Wu
Yongbo Xia
Jian Xiao
Li Xiao
Weina Yang
Zhanying Yang
Wei Yao
Tianfeng Ye
Hongyan Yin
Dai Yu
Lingfa Zeng
Wei Zhang
Yongchang Zhang
Yongqing Zhao
Song Zhu

Platinum Sponsor

Gold Sponsors

Table of Contents – Part V

Does Social Network always Promote
Entrepreneurial Intentions? Part II: Empirical Analysis

Lu Xiao[*] and Ming Fan

School of management, Jiangsu University,Zhenjiang, Jiangsu 212013, China
hnlulu@126.com, fanming@ujs.edu.cn

Abstract. As a further study of the first part of this paper, the second part is aiming to verify the theoretical model of social network and entrepreneurial intentions. Thus, a structural equation model is developed and several hypotheses are tested based on 157 Chinese College-graduate Village Officials' samples. Results showed that three social network dimensions, which are network size, network heterogeneity, properties of top node, affect entrepreneurial intentions significantly. Specifically, entrepreneurial intention is negatively correlated with properties of top node, and the size of the social network; a significant positive correlation is found between heterogeneity of social network and entrepreneurial intentions. Besides, Entrepreneurial desirability and entrepreneurial feasibility are two mediators between network heterogeneity and entrepreneurial intension. This work has also found that social networks sometimes may become an obstacle for College-graduate Village Official to start a business. The theoretical and practical implications of the study's findings are also discussed.

Keywords: College-graduate Village Official, Social network, Network heterogeneity, Entrepreneurial intentions, Empirical research.

1   Introduction

In the first part of this paper (Does Social Network always promotes Entrepreneurial intentions? Part I: Theoretical model), a theoretical model as well as several hypotheses were developed between social network and entrepreneurial intensions. Three network dimensions are chosen to measure features of social network, which are network size, network heterogeneity, and properties of top node. Two mediators, entrepreneurial desirability and entrepreneurial feasibility are set between social network and entrepreneurial intension. Then, the questionnaire was also developed and revised by our research group. The objective of this work is to launch a questionnaire survey, and to verify the theoretical model with Chinese College-graduate Village Officials samples.

Theoretic model and hypotheses of this paper are as follows.

Hypotheses

H1: College-graduate village official's social networks affect their entrepreneurial intention significantly.

---

[*] Corresponding author.

T. Huang et al. (Eds.): ICONIP 2012, Part V, LNCS 7667, pp. 1–8, 2012.

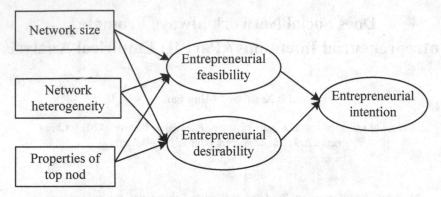

Fig. 1. Theoretic model of social network and entrepreneurial intentions

H1a: Size of social network is positively correlated with entrepreneurial intention.

H1b: Heterogeneity of social network is positively correlated with entrepreneurial intention.

H1c: Properties of top nod are positively correlated with entrepreneurial intention.

H2: Entrepreneurial desirability and entrepreneurial feasibility are two mediators between network heterogeneity and entrepreneurial intension.

2   Date Collection

2.1   Questionnaire Survey

Field surveys and personal interviews are used to collect sample data. During summer holiday in 2011, field surveys with questionnaire were conducted to college-graduate village officials. What's the most difficult for investigators is that amount of samples is constrained because one village has no more than one college-graduate village official; moreover, many villages have none. In all, 163 surveys were distributed and 157 valid responses were received resulting in a valid response rate of 96.3%, which can meet the requirements of research.

2.2   Sample Characteristics

As shown in Table 1, sample date consists of 70 males (44.6%) and 87 females (55.4%), most of them are 22-25 years old (66.9%) with working experience of 1-3 years (90.5%). 73.9% of sample are graduated from normal universities, 19.1% are from "211" universities, 3.2 are from "985" universities. Most of them are majored in economics and management (43.3%), science and engineering (21.0%), Literature and law (14.6%) and agricultural and forestry (10.8%). Sample features consist with research needs.

Table 1. Sample demographics and frequencies

Characteristic	Category	Frequency	Valid percent
Age	Under 21 years old	9	5.7
	22-25 years old	105	66.9
	26-30 years old	37	23.6
	Over 30 years old	6	3.8
Gender	Male	70	44.6
	Female	87	55.4
Tenure	Under 1 year	40	25.5
	1-2 years	62	39.5
	2-3 years	40	25.5
	3-4 years	8	5.1
	over 5 years	7	4.5
Educational background	Senior high school	1	0.6
	Junior College	9	5.7
	College	143	91.1
	Graduate student	4	2.5
University type	"985" universities	5	3.2
	"211" universities	30	19.1
	normal universities	116	73.9
	Junior College	6	3.8
Professional background	Economic and management	68	43.3
	Science and Engineering	33	21.0
	Farming and Forestry	17	10.8
	Literature and law	23	14.6
	Medical Science	3	1.9
	Normal school	5	3.2
	others	8	5.1

3   Analyze and Results

3.1   Exploratory Factor Analysis

In order to verify the questionnaire, exploratory factor analysis was conducted to test the convergent and discriminant validity of the instrument. Convergent validity indicates that all questions intended to measure a construct do reflect that construct. Discriminant validity indicates that a question should not reflect an unintended construct and that constructs are statistically different. Exploratory factor analysis with principal component analysis was used to extract factors in our study. By following the recommended procedures [1], major principal components were extracted as constructs; minor principal components with Eigen value less than 1 were ignored as noise; an item and the intended construct correlation (also known as factor loading) should be greater than 0.5 to satisfy convergent validity; an item and the unintended construct correlation should be less than 0.4 for discriminant validity. We extracted three factors corresponding to the three constructs. The items of "entrepreneurial intentions item4", "Perceived feasibility item1 and item4" were dropped because they did not satisfy the discriminant and convergent requirements. The remaining items showed appropriate validity. Table 2 reports the principal component analysis results with varimax rotation using SPSS17.0.

Table 2. Factor Loadings for the Variables

Construct	Items	factor loading
Entrepreneurial intentions	Estimate the likelihood that you will start your own business in the foreseeable future.	0.997***
	It has crossed my mind to start a business of my own or with my partner.	0.587***
	Do you plan to be self-employed in the foreseeable future?	0.552***
Entrepreneurial Desirability	My personal rating on how desirable it is for me to become an entrepreneur.	0.798***
	My feelings about running my own business.	0.887***
	How tense I would be about running a business.	0.880***
	How enthusiastic I would be about running a business.	0.835***
Entrepreneurial Feasibility	How certain of success are you.	0.743***
	Do you know enough to start a business?	0.704***
	How sure of yourself?	0.838***

3.2   Measurement Model

According to Anderson and Gerbin [2], measurement modeling should be carried out as the first step of structural equation modeling. The purpose of measurement modeling is to ensure instrument quality. Unlike exploratory factor analysis, measurement modeling pre-specifies construct-item correspondences but leaves correlation coefficients (i.e., factor loadings) free to change. The pre-specified construct-item correspondences are then tested for confirmation. Confirmatory factor analysis (CFA) is the conventional statistical method used to specify and test such relationships for the measurement model. With this method, items are expected to be highly correlated with the intended construct only. If an item is not substantially related to the intended construct, or is significantly related to an unintended construct, the pre-specified relationship is invalidated and adjustment of the instrument is required. Therefore, the first requirement of confirmatory factor analysis is that the construct-item correlation be significant [2]. In addition, for an item, the average variance extracted (AVE) by the latent factor should be greater than 0.5: that is, a construct should explain more than 50% of the item variance [3]. Moreover, items of the same construct should be highly correlated. To measure such correlations, two measures, composite factor reliability (CFR) and Cronbach's alpha (a), are required to be greater than 0.7 (Hair and Anderson et al. 1995). If all these criteria (significant correlation, high AVE, high CFR, and Cronbach's α are satisfied, the convergent validity of the items is said to be confirmed. Table 3 reports the result of convergent validity for our sample using statistical software AMOS7.0. All criteria were satisfied.

Table 3. The convergent validity of the measurement model

Construct	Items	AVE	CFR	Cronbach's α
Entrepreneurial intentions	Item 1 Item 2 Item 3	0.548	0.729	0.771
Entrepreneurial Desirability	Item 1 Item 2 Item 3 Item 4	0.723	0.912	0.913
Entrepreneurial Feasibility	Item 1 Item 2 Item 3	0.583	0.801	0.807

One way to confirm discriminant validity is to check that inter-construct correlation is less than the square root of AVE [3]. The underlying rationale is that an item should be better explained by its intended construct than by other constructs. The correlation among constructs is reported in Table 3. Discriminant validity was confirmed in our sample.

In summary, our measurement model confirmed the difference between all the relevance factors used. It also confirmed the internal consistency of the different aspects of the relevance factors as manifested in different questions (items). With that, we could proceed to test the causal relationship among those factors.

3.3　Hypotheses Testing

Hypothesis testing is done by creating a structural equation model in AMOS7.0, which specifies both item-construct correspondence and construct-construct causal relationship excluding the control variables. The coefficients are then solved with maximum likelihood estimation. We followed this procedure and arrived at the results summarized in Figure 2.

Before any conclusions can be drawn for hypothesis testing, the model must fit the data well. A few model fitting indices can be employed here. For example, the chi-square and the degree of freedom ratio (a normalized measure of the "badness of model fit") must be less than 2; the root mean square error of approximation (RMSEA, a measure of the residual) must be less than 0.08; and the goodness of fit index (GFI) must be greater than 0.9 [1, 2, 4]. Our results indicated satisfactory model

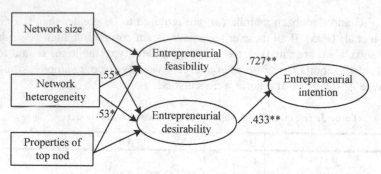

χ2/ df=1.676, NFI=0.920, IFI=0.966, TLI=0.950, CFI=0.965,
RMSEA=0.066, * P< 0. 05, ** P< 0. 01.

Fig. 2. Regional structural equation model

fit. As shown in figure 2, χ 2 / df equates 1.676 (less than 2), NFI, IFI, TLI, CFI are all greater than 0.9, RMSEA of 0.066(less than 0.08). However, the path coefficients between network size, properties of top nod and two mediate varieties are insignificant. Model modification index (MI) show that the direct affects may be better, which is also in line with our theoretical basis, so we modified the model by deleting the indirect effects and increasing the path of direct effect between network size, properties of top nod and entrepreneurial intentions.

Run the modified model in AMOS7.0 and arrived at the results summarized in Figure 3.

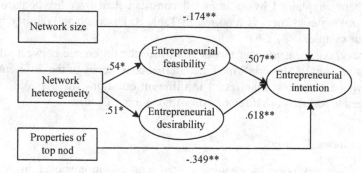

χ2/ df=1.637, NFI=0.926, IFI=0.970, TLI=0.953, CFI=0.969,
RMSEA=0.064, * P< 0. 05, ** P< 0. 01.

Fig. 3. Modified structural equation model

Fitting indices of modified model and original model are compared in table 4. It is shown that after modifying, the fitting indices significantly improved. χ 2 / df equates 1.637, less than the former 1.676, the fitting indices are all more closer to 1 than the former, while RMSEA is 0.064, which is smaller than 0.066 of the former. Therefore, the modified model arrive better results than the original model. So, we use the results of modified model to test the hypothesis.

Table 4. Fit indices comparing of original model and modified model

Models	$\chi 2/$ df	NFI	IFI	TLI	CFI	RMSEA
Original model	1.676	.920	.966	.950	.965	0.066
Modified model	1.637	.926	.970	.953	.969	0.064

According to path coefficients in modified model, the hypotheses testing results are as follows: three social networks dimensions (i.e., network size, network heterogeneity, and properties of top nod) affect entrepreneurial intentions significantly, thus H1 is verified. Specifically, Size of social network and entrepreneurial intentions are negatively associated and H1a is unverified; a significant positive correlation is found between heterogeneity of social network and entrepreneurial intentions and H1b is verified; Properties of top nod and entrepreneurial intentions are negatively associated and H1c is also unverified. Besides, Entrepreneurial desirability and entrepreneurial feasibility are two mediators between network heterogeneity and entrepreneurial intension, but no mediate role found in properties of top nod and network size. Thus, H2 is partly verified.

4　Conclusion

Heterogeneity of social network has a positive effect on college-graduate village officials' entrepreneurial intension through entrepreneurial feasibility and entrepreneurial desirability. College-graduate village officials interact with the members from different industries, different areas, and different sectors can benefit their entrepreneurship, because they can get heterogeneity information and resources about market, business opportunity, investments, policy and so on. Through the complementary advantages, the college-graduate village officials can be more efficient on resource integration, and through analyzing entrepreneurial feasibility and entrepreneurial desirability to generate entrepreneurial intentions.

Size of social network is negatively associated with entrepreneurial intentions, which is contrary to our hypothesis, but an innovation of this study. Usually, we believe that the larger the social network size is, the extensive social relations and information it contained. However, an important aspect was ignored by this point of view. That is, when large size of network concentrated in small amount of different occupations, only can obtain homogeneous information and resource, which is no help to entrepreneurial intentions. In the same way, "Structural holes" theory indicates that only structural holes in network can provide new information and resources. When many network relations concentrate in a few occupations is just like networks with no structural holes, and it can't generate new resources and information [5]. Therefore, in order to gain structural holes with different resources, one should associate with different individuals and groups from different fields.

Properties of top nod are negatively associated with entrepreneurial intentions, which are also contrary to our hypothesis, but an innovation of our study. Preliminary investigation shown college-graduate village officials with more social capital, that is, their network members have high reputation, more rights, high status and more wealth, are not willing to stay in rural area and start a business. We also found that the college-graduate village officials from grassroots and have litter social capital are really rooted in rural areas, hoping to start a business in rural areas through their own efforts. Thus, selecting and training entrepreneurial college-graduate village officials are not only required to examine their capabilities and potential, but also consider their social network characteristics.

5   Limitations and Future Research

Two limitations are apparent in this study. First, because we selected a particular group as our research sample, the amount of them is limited. In China, It is promoted "one college student in one village" system, thus, each village only has no more than one college-graduate village official; second, the independent variables were only confined to three social network dimensions and they are measured by "Chinese New-Year Greeters' Networks", which is Chinese traditional Spring Festival. Thus, the general applicability of the conclusion still needs to be tested. However, Christmas in western countries is just like Spring Festival in China and "Christmas Greeters' Networks" can be studied in future research.

Acknowledgements. This work is supported in part by National Natural Science Foundation of China under grant 71073070 and 71171099, and by the Doctoral Program of Higher Specialized Research Fund of China under g rant 20103227110015.

References

1. Hair, J.F., Anderson, R.E., Tatham, R.L., Black, W.C.: Multivariate Data Analysis with Reading, 4th edn. Prentice Hall, Englewood Cliffs (1995)
2. Anderson, J.C., Gerbing, D.W.: Structure Equation Modeling in Practice: A Review and Recommended Two-Step Approach. Psychological Bulletin 103, 411–423 (1998)
3. Fornell, C., Larcker, D.F.: Structural Equation Models with Unob-Servable Variables and Measurement Error: Algebra and Statistics. J. Market. Research 18, 382–388 (1981)
4. Nunnally, J.C., Bernstein, I.H.: Psychometric Theory. McGraw-Hill, New York (1994)
5. Burt, R.: Structural Holes: The Social Structure of Competition. Harvard University Press, Cambridge (1992)

Rasterization System for Mobile Device

Xuzhi Wang[*], Yangyang Jia, Xiang Feng, Shuai Yu, and Hengyong Jiang

School of Communication and Information Engineering,
Shanghai University, Shanghai 200072, China
wangxzw@shu.edu.cn

Abstract. During the past few years, mobile phones and other handheld devices have gone from only handling dull text-based menu systems to being able to render high-quality three-dimensional graphics at high frame rates. Computer graphics hardware acceleration and rendering techniques have improved significantly in recent years. This paper reviews the state of mobile 3Dgraphics, focusing on the research of raster system. It also discusses technical and practical challenges. Finally, implement the fixed pipeline raster system on the FPGA platform, the results show that the system is able to complete the rendering of complex graphics.

Keywords: Rasterization, FPGA, Hardware acceleration, Triangle filling.

1 Introduction

The popularity of smart phones and tablet PCs, depends the development of mobile GPUs on a large role. Using mobile phones, Internet, shopping, playing the big game, high-definition video, geography navigation, etc., these are inseparable from the role of the mobile GPU.

Traditional mobile phones are aimed at making and receiving telephone calls over a radio link. PDAs are personal organizers that later evolved into devices with advanced units communication, entertainment and wireless capabilities. Smart phones can be seen as a next generation of PDAs since they incorporate all its features but with significant improvements in screen size and resolution, battery life, memory size, graphics processing and processing in general. Tablet personal computers can be described as lightweight notebook computers that have touch sensitive displays for input rather than a keyboard. A mobile internet device is a mobile device that provides wireless internet access and fills the void regarding size between smart phones and tablet personal computers. All these devices are very different in many aspects from each other. Most noticeable differences being screen size, processing speed, storage space and size of the devices itself.

A 3D model from the established to the final display must need some processing, such as lighting, texture mapping, the process consists of three parts: application stage, geometry stage [1], rasterization stage. The fixed rendering pipeline is shown in Fig. 1.

---

[*] Corresponding author.

T. Huang et al. (Eds.): ICONIP 2012, Part V, LNCS 7667, pp. 9–16, 2012.

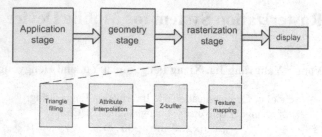

Fig. 1. The fixed function pipeline

The application stage starts and drives the 3D graphics pipeline by feeding 3D models to be rendered according to the information determined in the application stage. The geometry stage mainly complete transform and Lighting. In order to translate the scene from3D to 2D, all the objects of a scene need to be transformed to various spaces. Next in the pipeline is the rasterization stage. The rasterization stage mainly includes four modules, mainly complete texture mapping attribute interpolation.

A 3D model is composed by triangles. Fig. 2 is the rendering process of teapot model composed by 64 triangles. Modeling tool used by the application layer is very commonly used, such as 3Ds MAX. This paper 3Ds MAX model to generate the OBJ files needed by the accelerated graphics pipeline. OBJ file is a model of organizational structure, and contains the vertex coordinates and vertex attributes (color, texture coordinates, normal vector ...). The color of the pixel inside the triangle is obtained by linear interpolation.

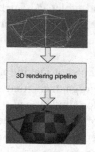

Fig. 2. The rendering process of teapot model

In the field of mobile GPUs, the Lund University of Sweden made a lot of contributions, many algorithms have become the standard of the graphics, such as texture compression algorithm-ETC. South Korea's KAIST University, mainly in architecture design, many high-level articles published in the international arena. The industry is currently in the full three-legged posture. Adreno, which designed by Qualcomm, is widely used in smart phones. Imagination Technologies, launched the MBX / SGX series, adopted Tile-based rendering architecture, reducing the transmission of data between the GPU and memory, lower power consumption; ARM has introduced a have a very strong mobile GPU chip Mali for 3D graphics process, which could process 1080p HD video playback capabilities to the mobile phone and portable players [2].

The rest of the paper is organized as follows. The rasterization algorithm is proposed in section 2. The section 3 describes the implemented of rasterization system. Finally, in section 4, the conclusions are drawn.

2   Algorithm

Rasterization system considerable amount of data processed, it play a decisive role for graphics rendering. Rasterization system has the following parts: triangles filling, attribute interpolation, and the Z-buffer, texture mapping. The texture mapping module, because of a number of division operation for each pixel, is the most time consuming module in the rasterization system, and also is bottlenecks rasterization system, also is the bottlenecks of entire graphics acceleration system.

In this section, some key algorithm of rasterization will be introduced. A novel triangle filling algorithm will be proposed, which reduce the filling time composed with other triangle filling algorithm.

2.1   Triangle Filling

The model is composed by triangle. After the geometric transformation and illumination, we need to show the triangle to the final output device, but the computer screen, cell phone screens and other facilities as a unit to display the pixels on the two-dimensional plane. However, the geometry stage is based on the coordinates in three-dimensional space to represent, so a major task of raster layer is processing the coordinates triangle into pixels, this process is called triangle filling. Triangle filling illustration effect is showed in Fig. 3. The basic principle is that to scan to find out all the pixels inside the triangle. Since the triangle on the three-dimensional space is continuous, the pixel coordinates of the discrete, how valid and correct correspondence is also the focus to complete the triangle to fill.

Because of the importance of triangle filling, many graphics researchers have made a lot of groundbreaking algorithm. Triangle filling algorithm can be divided into two categories: scanning algorithm based bresenham, scanning algorithm based edge function [3, 4, 5, 6].

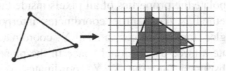

Fig. 3. The illustration of triangle filling

Triangle filling algorithm based edge functions need to scan some of the unwanted pixels, each line needs to scan the unwanted pixels greater than or equal to two. This is because the start of each line scan does not necessarily inside the triangle, if the starting point inside the triangle, you only need to scan two useless pixels, if the starting point in the triangular external, you need to scan more than two non-using the pixel point. The midpoint traversal algorithm proposed in this paper, effective solution of this problem to ensure that each line starting point inside the triangle, so that each line only needs to scan two unwanted pixels, effectively improve the triangle fill efficiency. The midpoint traversal algorithm is shown in Fig. 4.

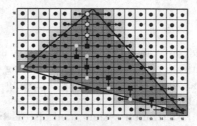

Fig. 4. The midpoint traversal algorithm

2.2   Z-Buffer

Z-buffer algorithm is a very suitable hardware implementation program to eliminate the hided pixels. Use a screen size Z buffer to save the distance from the eye point depth values, and constantly updated in the rendering process. The value of each unit in the Z buffer is the value of the z coordinates of the object as reflected by the corresponding pixel. The initial value of each unit in the Z buffer to take into the great value of z, graphics blanking process to compare the value of the corresponding unit in the z coordinates of the current pixel value and z buffer. Z buffer value is greater than the current pixel, the output of only this pixel to the next module for processing, while the value of the corresponding unit in the z buffer should change the z coordinates of the current pixel value. On the contrary, if the z coordinate value is less than the value in the z buffer, the point discarded.

2.3   Attribute Interpolation

The vertex data is in the property, such as color, depth, texture coordinates. After a triangle filling, the model identified all pixels of the triangle within the triangle inside, but these pixels do not have any attributes, so need according to three vertex's attributes value to interpolate the properties of all pixels inside the triangle.

Commonly used method is barycentric coordinates interpolation algorithm, as show in Fig. 5.Triangles three vertices of X, Y coordinates and interpolation properties of S constitute a three-dimensional space in the plane, to obtain the plane equation, and then substituted into the X, Y coordinates to seek the point after interpolation of the values.

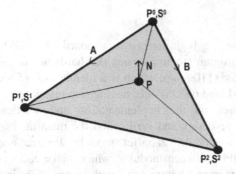

Fig. 5. Barycentric coordinates interpolation algorithm

The interpolation formula is displayed as fellow.

$$\vec{N} \bullet \vec{P} = ax + by + cs + d = 0$$

$$= \left(\left(p_y^1 - p_y^0\right)\left(s^2 - s^0\right) - \left(p_y^2 - p_y^0\right)\left(s^1 - s^0\right)\right) \times \left(p_x - p_x^0\right) +$$
$$\left(\left(p_x^2 - p_x^0\right)\left(s^1 - s^0\right) - \left(p_x^1 - p_x^0\right)\left(s^2 - s^0\right)\right) \times \left(p_y - p_y^0\right) +$$
$$\left(\left(p_x^1 - p_x^0\right)\left(p_y^2 - p_y^0\right) - \left(p_y^1 - p_y^0\right)\left(p_x^2 - p_x^0\right)\right) \times \left(s - s^0\right)$$

$$= \left(\left(p_y^1 - p_y^0\right)\left(p_x - p_x^0\right) - \left(p_x^1 - p_x^0\right)\left(p_y - p_y^0\right)\right) \times \left(s^2 - s^0\right) +$$
$$\left(\left(p_x^2 - p_x^0\right)\left(p_y - p_y^0\right) - \left(p_y^2 - p_y^0\right)\left(p_x - p_x^0\right)\right) \times \left(s^1 - s^0\right) +$$
$$\left(\left(p_x^1 - p_x^0\right)\left(p_y^2 - p_y^0\right) - \left(p_y^1 - p_y^0\right)\left(p_x^2 - p_x^0\right)\right) \times \left(s - s^0\right) = 0 \qquad (1)$$

$$s = \frac{e_0(x, y)}{2S_\Delta}\left(s^2 - s^0\right) + \frac{e_2(x, y)}{2S_\Delta}\left(s^1 - s^0\right) + s^0 \qquad (2)$$

Bilinear interpolation is linear interpolation in two-dimensional linear system, as shown in Fig. 6. The T0~T3 is the sampling points around the four adjacent texel, red point of sampling points. The bi-linear filtering formula shows in Equation 3. I (U', V') represents the color value of T0. I (U'+1, V') represents the color value of T1. I (U', V+1) represents the color value of T2. I (U'+1, V'+1) represents the color value of T3.

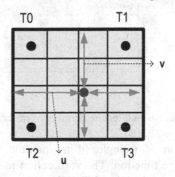

Fig. 6. Bi-linear filtering

$$Filter_{BI}(U,V) = \left(I(U',V') \cdot (1-u) + I(U'+1,V') \cdot (u)\right) * (1-v) +$$
$$\left(I(U',V'+1) \cdot (1-u) + I(U'+1,V'+1) \cdot (u)\right) * (v)$$

$$U' = round(U) \quad V' = round(V) \qquad (3)$$

3   Hardware

In this paper, we use the high-performance board of XUP-LX110T provided by XILINX university program to implement the hardware design. The development board contained a xc5vlx110t chip, which is a member of V5 series, including 17280 slices, 5328Kb BRAM, and 64 DSP48E, having rich resource.

This section focuses on the implementation and verification of rasterization algorithm, and hardware design and verification the main module of the raster system (attribute interpolation module, Z-buffer module, texture mapping module). The introduction of the FIFO between modules, which make each module independent of each other to reduce the impact between the module, one of the innovation of this paper.

The development platform is Virtex-5 LXT FPGA ML505 evaluation board [7]. System architecture diagram is shown in Fig. 7. Programs are stored in DDR2. Exchange of data between the control register is mainly responsible for the CPU and graphics acceleration unit. CPU unit transfer control commands to the graphics acceleration. CPU is Xilinx's soft-core MicroBlaze. The display device settings for the computer display with a resolution of QVGA (320 * 240).

The biggest feature of rasterization stage is that data processing is large, occupy more memory bandwidth. Pixel color information, depth information need to be stored in the external storage device, accessing to the information takes up a lot of bandwidth system. The rasterization stage consumes 70% system power of entire graphics acceleration system. How to reduce bandwidth is becoming the focus of researchers, but also the papers needs to be done the next step.

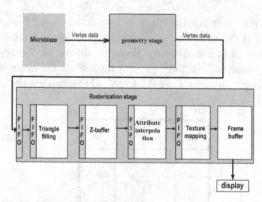

Fig. 7. System architecture diagram of raster system

Fig. 8 is the verification of triangle filling module. The filling algorithm is midpoint traversal based edge function. The verification module is cow composed by 5804 triangle facet. The system clock is 100MHZ.

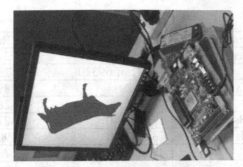

Fig. 8. The verification of triangle filling module

The Fig. 9 is the verification of Z-buffer module. The two triangles have different depth value. From the screen, we can see the pix with greater depth value will be lost.

Fig. 9. The verification of Z-buffer module

The Fig. 10 is the verification of Attribute interpolation module. The verification module also is cow. Compared with Fig. 8, the cow module has rich colors.

Fig. 10. The verification of Attribute interpolation module

The hardware resource consume is shown in table 1. Analysis the table, we can see the BRAM have used by 90%.This is because the Frame buffer is designed in the internal of FPGA chip.

Table 1. Hardware Resource

Device utilization	Number	Ratio
Slice Register	15386 of 69320	22%
Slice LUT	15971 of 69320	23%
DCM_ADVs	1 of 12	8%
BUFG	9 of 32	28%
Bonded IOBs	174 of 640	27%
Block RAM	134 of 148	90%
DSP48Es	45 of 64	70%

4   Conclusion

In this paper, the author introduce key algorithm of rasterization stage in the first, then complete the design of key module, finally, verify the key module in the FPGA board.

With the development of smart phone, tablet PCs, the function of mobile GPU will be more complex. The power consumption is Bottleneck, so low complexity and better performance algorithm is a research priority.

Acknowledgements. This work is supported by the National Natural Science Foundation of China (No.60902086) and the Leading Academic Discipline Project of Shanghai Municipal Education Commission (No.J50104).

References

1. Tomas, A., Jacob, S.: Graphics for the Masses: A Hardware Rasterization Architecture for Mobile Phones. ACM Trans. Graph. (2003)
2. Bresenham, J.E.: Algorithm for Computer Control of a Digital Plotter. IBM Systems J. 4, 25–30 (1965)
3. Sun, C., Tsao, Y., Lok, K., Chien, S.: Universal Rasterizer with Edge Equations and Tile-scan Triangle Traversal Algorithm for Graphics Processing units. In: ICME (2009)
4. Joel, M., Robert, M.: Tiled Polygon Traversal Using Half-Plane Edge Functions. In: Proceedings of the ACM Siggraph/Eurographics Workshop on Graphics Hardware, pp. 15–20 (2000)
5. Jiang, H., Wang, X., Zhu, M., Wan, W., Ma, Y.: A Novel Triangle Rasterization Algorithm Based on Edge Function. In: CSQRWC, pp. 1211–1216 (2011)
6. XILINX. Virtex-5 Family Overview. V5.0 (2009)

Study on Rasterization Algorithm
for Graphics Acceleration System

Xuzhi Wang[*], Wei Xiong, Xiang Feng, Shuai Yu, and Hengyong Jiang

School of Communication and Information Engineering,
Shanghai University, Shanghai 200072, China
wangxzw@shu.edu.cn

Abstract. The 3D devices have been widely applied in people's life, for example, smart phone, internet games, high-definition video, geography navigation, etc. The large scale graphics rendering depends on the computing power of the hardware greatly, the calculation of model rasterization operations need amounts of data. It has become the bottleneck of system performance. This paper proposed the method which can accelerate graphics rasterization procedure, based on the idea of tile to meet the needs of graphical applications on embedded platform. The rules and procedures of algorithm are introduced. By using of the XUP-LX110T, the experiments are carried out. It is verified that the method should been compensated the features for the lack of resources and poor computing power of embedded platforms. It can apply smaller chip area to achieve graphics acceleration with fewer resources.

Keywords: Rasterization, Tile rendering, FPGA, Block_edge_test (HET).

1   Introduction

With the requirement of large scale 3D real-time rendering, we usually need to design a hardware rendering system for the specific 3D scene to make each pixel with the right color, so as to complete realistic model rendering process by transforming two-dimensional vertex into pixels on the screen. 3D graphics rendering depends on the computing power of the hardware greatly, the calculation of model rasterization operations need amounts of data, it becomes the bottleneck of system performance, we should design and optimize high-performance graphics systems for the grating phase of data processing specially. GPU is designed to achieve high-performance 3D graphics and multimedia processing [1], currently GPU uses programmable shader core to compute functions and it supports a common single-precision as well floating-point operations, dynamic flow control, and variable conversion operations. Recently, graphics acceleration module for embedded devices is becoming the focus of the study, specifically a new rendering system has been put forward to resolve the problem how to reduce the memory bandwidth of graphics acceleration in the mobile

---

[*] Corresponding author.

T. Huang et al. (Eds.): ICONIP 2012, Part V, LNCS 7667, pp. 17–24, 2012.

terminal and ultimately reduce system power consumption. However, a large number of areas need to improve in efficiency and power consumption for it.

FPGA (programmable field gate array) is a revolutionary product in recent years, the user can complete the circuit design and through the comprehensive and layout, so the functional verifications can be done quickly on FPGA , high-performance FPGA chip has also reduced the gap between FPGA and ASIC. Apply its field programmable and high performance parallel data processing capability; it makes algorithm simulation and verification be achieved very good results. In this paper, we specially have the hardware graphics rendering unit designed and optimized, and we mainly aim at researching triangle primitive rasterization optimization and its realization.

2   Hardware System Design

Although the full-screen scan conversion uses pipeline optimization and parallel processing, so as to increase the processing speed dramatically, but the increasement of parallel pipelines and water operations leads to consume a large number of registers and memory resources which proportional to the screen resolution (such as the depth buffer, color buffer), as well frequently access to external memory will increase power consumption. In this paper, the method of Tile-based rasterization can be easy to reduce memory consumption to achieve parallelization, and make a better drawing.

Due to the embedded scene is not very complicated generally, as well, the list of triangles for Tile-based rasterization algorithm is not very great and its draw state is limited, for those reasons we can apply the design of multiple channels and multiple parallel processing units to increase rendering efficiency. It can be proved that the power would be more easily to be controlled in the case of using the same resources. This article will focus on the classification of triangular surface, list building, and scan conversion study [2].

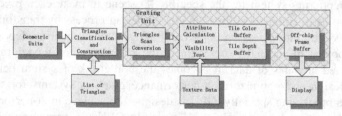

Fig. 1. The block diagram of rasterization based Tile

2.1   Triangular Face Classification

Tile-based graphics drawing needs to split screen into a number of Tiles firstly, then we can deal with each Tile separately; according to the classification method, we can deal with each triangular face list, so that the processing can be drawn in accordance with the list of data on each Tile; then draw all the Tiles separately, and final results are copied to the off-chip frame buffer. These operations can be designed by pipeline and parallel processing approach. In this paper, we propose a bounding box and the edge function-based classification method for triangles classification and build, shown as the Fig. 2.

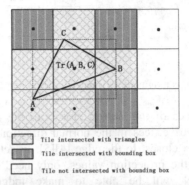

> Tile intersected with triangles
> Tile intersected with bounding box
> Tile not intersected with bounding box

Fig. 2. Classification method based on bounding box and the edge function

Bounding box testing methods can be quite easy to detect whether Tile is covered by the bounding box of triangle surface. But this is an inaccurate detection method, and dark gray Tile would be detected into triangles intersect incorrectly, and in fact, Tile is completely outside of the triangular surface. When the triangular surface is small enough and at the right location, this method would be accurate. Miscarriage of justice will not generate an error rendering results, but classifying the triangle into the list of triangles which is not intersected with the Tile will cause the waste of storage space and bandwidth, and it makes more Tile attempt to draw an area of triangle not in it, which can also cause a decline in the speed of rendering. However, this algorithm will spend shorter time on the classification of triangles. The triangle surface detection method based on the bounding box, in this paper, combined with the flexibility of the bounding box detection and accurate features of edge function. We designed an algorithm based on the bounding box and the edge function of the second detection to improve the detection accuracy and speed effectively. The following will make a systematic exposition of this algorithm [3].

When a point relative to the edge function of the counterclockwise side, the number of its values are less than zero, it turns out that the point is in the internal or at boundary of the triangle, you can also use a similar algorithm to judge whether the Tile is in the internal triangle through this algorithm. When we judge weather a Tile is intersected with the triangular face, you need to determine the relative position of the Tile and side, and that is to judge positional relationship between the four vertices and three sides, this paper presents an optimized approach to avoid detecting each vertex. As show in Fig. 3, actually only need judge the vertex of the closest side. So that we can reduce to 12 times operations of boundary function to 3 times of boundary function on three sides, effectively reduced the amount of computation. If the vertex closest to the medial edge of the triangle was in the lateral side, we can draw a conclusion that the Tile is entirely located in the edge of the outside, otherwise the Tile is located in the medial side or fellowshipped with the side.

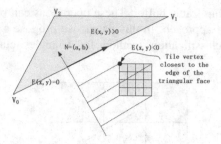

Fig. 3. The relationships judgment between Tile based normal and edge of spatial

To judge which vertices of the Tile is closest to the side of the Tile, we can calculate projection of the vertices in the normal direction of the section edge, the biggest point is the vertex closest to the edge [4][5]. In actual operation, it does not really need to calculate the projection values, only need to determine the normal where the limits of N will be able to make judgments, as the function $e(x, y) = ax + by + c$, so we can get

$$e(x + p, y + q) = e(x, y) + ap + bq \tag{1}$$

Firstly, calculate the Tile vertex in the lower left corner of the boundary function value $e(x, y)$, as the value N, p and q which are not in the limit, $p = \begin{cases} w, a \geq 0 \\ 0, a < 0 \end{cases}$,

$q = \begin{cases} h, b \geq 0 \\ 0, b < 0 \end{cases}$, w and h are the wide and height of Tile, boundary function of Tile

can be represented as $e(x + p, y + q) = e(x, y) + ap + bq$. If the boundary function value of Tile on three sides were greater than zero, then this Tile and triangle are intersected [6][7][8].

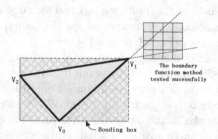

Fig. 4. The Tile which boundary function test was done but not intersected with triangles

In order to avoid using the edge function to detect incorrectly, shown in Fig. 4. The algorithm needs to use the bounding box for a preliminary test to narrow the scope of tests and then using the edge function to conduct further tests; you can achieve better results on the efficiency and accuracy by this method [9].

2.2   Build the List of Triangles

After the intersection test for triangles and Tile, you need to draw the triangles belonging to each Tile. At Tile-based graphics accelerate unit, the entire screen is divided into many blocks, each block will have a copy of the block which needs to draw a list of triangular facets. Before entering the drawing stage, Each triangular face will add the triangle information to the list of covered blocks in accordance with its position on the screen respectively, the list of these triangles will be stored in external memory. This paper particular concerns how to reduce the required memory space of the triangle list and effective data access.

The triangle data contained in each Tile list is unpredictable; the algorithm uses a linked list (Link list) to build the corresponding data structures. Taking into account the storage characteristics of the SDRAM memory (reading and writing while insert wait states according to various circumstances, waiting time will be different), to separate the structure of the data and the list will cause the hardware needing to read the data pointer, after that, then jump to the address of where the data resides, and then jump back to the pointer to read the next position, so that frequent beat is not conducive to improve the efficiency in the use of SDRAM memory, so in this paper, the designation of storage structures is to keep the continuity of continuous reading of the memory address. In addition to the start of each block list pointer, the pointer list of data fields were combined, and apply triangle units to create a list.

2.3   The Level of the Triangle Scan Algorithm

After processing triangles based on Tile lists, you need to scan for the triangular surface of each linked list, so that let the rasterization processor find all pixels in the triangle plane, and calculate the properties of the corresponding pixel. Although using a larger area of the edge function test can first rule out the more pixels, but the accuracy of it is not enough, the test area may still contain many pixels which not belong to the triangle. Smaller area of the edge function test can achieve a more accurate judgment, but it takes more processing time. Choosing what size block to test is the focus of the efficiency. The proposed design level normalized scan algorithm to achieve the levels and scope of scanning [11].

In this block design architecture, length and width of each drawing block are both 32 pixels and the Level 32 is on behalf of them. The edge of the entire drawing block is used for regional function test, if it is determined that the scope of the triangle to be exist, then cut them into four sub-blocks in this range, four sub-block are the 16×16 pixels, that is, Level 16, then do the edge of the four regional function test respectively, also recursive call block_edge_test to itself, each one call Level will be halved until Level down to 1, on behalf of this range size is one pixel, then they directly use the edge function to test whether the pixel inside the triangle needs to output. For the region has been tested and found no a triangle is inside, then it will direct end finding and no longer do a layer of smaller regional test, so in order to avoid unnecessary scanning. The four sub-block processing order is an inverted zigzag, the use of this processing sequence can be more efficient in order to facilitate the use of the memory, the memory configuration and access. Fig. 5 shows the diminishing class relations and regional scanning moving way.

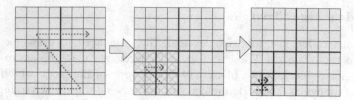

Fig. 5. HET hierarchy scan

Fig. 6 shows system architecture diagram of the algorithm. According to HET algorithm, the each original recursive class needs to cut into four sub-regional areas, and then apply edge function tests to determine whether the sub-block should be processed in the next class. Now we have four operational units, which requires four times arithmetic tests, to be used simultaneously in one step so as to detect which areas covered by the triangle in the four processing units. Then you can skip the unnecessary processed region, handle intersects with the triangle region directly. Next, each sub-region also need to do the recursive computation, so the status of the other sub-regions in the processing hierarchy should be saved, while waiting for a recursive sub-region recursive completed then jumping to the next processing area. We can see that because the increase of the processing unit extended from the original one pixel into 2x2 pixels range, there is less hierarchy depth layer at the end of recursive computation.

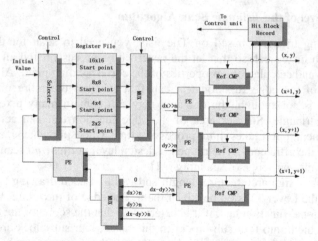

Fig. 6. The HET of four times parallel processing algorithm architecture

3   Experimental Results

Table 1 shows the integrated results of graphics acceleration unit system. The system can run in the range of 67.879 MHZ.

Table 1. Integrated results of graphics acceleration unit system

Device utilization	Number	Ratio
Slice Register	19473 of 69320	28%
Slice LUT	18731 of 69320	27%
DCM_ADVs	1 of 12	8%
BUFG/BUFGCTRLS	9 of 32	28%
Bonded IOBs	239 of 640	37%
Block RAM/FIFO	36 of 148	24%
DSP48Es	21 of 64	32%

According to graphics knowledge, all the models of different complexity can be composed by the basic triangular faces. So the first set of the experimental data is for the triangular face test. Fig. 7 are the rasterization rendering results of graphics acceleration unit based Tile. We can see that the design of the graphics acceleration unit can meet both different geometry structure and different transform in the same geometry.

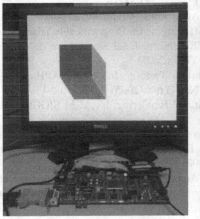

Fig. 7. Rendering cube on the graphics system

4   Conclusion

This paper achieves the accelerated graphics rasterization procedure based on the idea of Tile to meet the needs of graphical applications on embedded platform. we design an algorithm to detect Tile with triangular surface coverage based Tile ideological, and the linked list was designed through the method, so as to compensate the features for the lack of resources and poor computing power of embedded platforms, and apply smaller chip area to achieve graphics acceleration. Level scanning algorithm was designed for application in the scan conversion of triangles in the linked list, so as to reduce the probability of the redundant pixels and improve the efficiency of rasterization. From the comprehensive results, it is seen that this design has achieved graphics rasterization with fewer resources.

Acknowledgements. This work is supported by the National Natural Science Foundation of China (No.60902086) and the Leading Academic Discipline Project of Shanghai Municipal Education Commission (No.J50104).

References

1. Hans, H.: Embedded 3D Graphics Core for FPGA-based System-on-chip Applications. In: FPGA World Conference, pp. 8–13 (2005)
2. Fredrik, E.: A Tile-based Triangle Rasterizer in Hardware (2009)
3. Kim, D., Kim, L.: Area Efficient Pixel Rasterization and Texture Coordinate Interpolation. Comput. Graphics (2008)
4. Juan, P.: A Parallel Algorithm for Polygon Rasterization. In: Proceedings of the 15th Annual Conference on Computer Graphics and Interactive Techniques, pp. 17–20 (1988)
5. Michael, D.M., Chris, W., Kevin, M.: Incremental and Hierarchical Hilbert Order Edge Equation Polygon Rasterization. In: Proceedings of the ACM SIGGRAPH/EUROGRAPHICS Workshop on Graphics Hardware, pp. 65–72 (2001)
6. Oberman, S.F., Siu, M.Y.: A High Performance Area Efficient Multifunction Interpolator. In: Symposium on Computer Arithmetic, pp. 272–275 (2005)
7. Voicu, P., Paul, R.: Forward Rasterization. ACM Trans. Graph. 62, 375–411 (2006)
8. Kyusik, C., Kim, D., Kim, L.: A 3-way SIMD Engine for Programmable Triangle Setup in Embedded 3D Graphics Hardware. In: IEEE International Symposium on Circuits and Systems, pp. 4546–4549. IEEE Press, New York (2005)
9. Tomas, A.: Fast 3D Triangle-Box Overlap Testing. In: SIGGRAPH 2005 ACM SIGGRAPH 2005 Courses, pp. 1–4. ACM, New York (2005)
10. Steven, M.: A Sorting Classification of Parallel Rendering. In: ACM SIGGRAPH ASIA 2008 Courses. ACM, New York (2008)
11. Joel, M., Robert, M.: Tiled Polygon Traversal Using Half-plane Edge Functions. In: Proceedings of the ACM SIGGRAPH EUROGRAPHICS Workshop on Graphics Hardware, pp. 15–21. ACM, New York (2000)
12. Nguyen, H.T.: An Efficient Data-Parallel Architecture for Volume Rendering. In: IEEE Region 10's Ninth Annual International Conference, pp. 664–671. IEEE Press, New York (1994)

Simultaneous Learning of Several Bayesian and Mahalanobis Discriminant Functions by a Neural Network with Memory Nodes

Yoshifusa Ito[1,*], Hiroyuki Izumi[2], and Cidambi Srinivasan[3]

[1] Aichi Medical University, 1 Yasagokarimata, Nagakute, 480-1195 Japan
[2] Aichigakuin University, 12 Araike, Iwasaki, Nisshin, 470-0195 Japan
[3] University of Kentucky, Lexington, 40506, USA
ito@aichi-med-u.ac.jp, hizumi@psis.agu.ac.jp, srini@ms.uky.edu

Abstract. We construct a one-hidden-layer neural network capable of learning simultaneously several Bayesian discriminant functions and converting them to the corresponding Mahalanobis discriminant functions in the two-category, normal-distribution case. The Bayesian discriminant functions correspond to the respective situations on which the priors and means depend. The algorithm is characterized by the use of the inner potential of the output unit and additional several memory nodes. It is remarkably simpler when compared with our previous algorithm.

Keywords: Simultaneous learning, Discriminant function, Bayesian, Mahalanobis.

1 Introduction

It is well known that a neural network can approximate a Bayesian discriminant function which is actually a posterior probability [1–5, 7–14]. We have noticed that, in the two-category, normal-distribution case, the logit transform (the inverse logistic transform) of the Bayesian discriminant function is a Mahalanobis discriminant functions if shifted by a constant [5]. Our tool for the logit transform is the network with the output unit having the logistic activation function. The inner potential of the output unit of the network is inevitably the logit transform of the output [5].

On the other hand, we have shown that if a neural network is capable of learning a Bayesian discriminant function, it can learn several Bayesian discriminant functions simultaneously when equipped with several linear nodes [4].

Furthermore, we have proposed a method to enable the network to determine the constant for the conversion of the Bayesian discriminant function to a Mahalanobis discriminant function [9] and later established that the method is compatible with the algorithm for simultaneous learning of several Bayesian discriminant functions [10].

---

* Corresponding author.

T. Huang et al. (Eds.): ICONIP 2012, Part V, LNCS 7667, pp. 25–33, 2012.
© Springer-Verlag Berlin Heidelberg 2012

As a result we have obtained an algorithm for a neural network to learn several Bayesian discriminant functions and corresponding Mahalanobis discriminant functions simultaneously. However, this algorithm was complicated, in the sense that the network must be trained twice and, for the first training, all the teacher signals must be shifted by the respective sample means beforehand [10].

We have recently found an alternative algorithm for obtaining the constant [11] to be used for the conversion of the Bayesian discriminant function. The accuracy of this method is comparable with the previous one [11]. In the case of the new method, the network does not need to be trained twice. The goal of this paper is to show that this method is also compatible with the algorithm for the simultaneous learning of several Bayesian discriminant functions and simplifies our previous algorithm in [10]. This method may be applied in the case where the positions (the means) of the distributions depend on the situations.

2   Discriminant Functions

We treat the case where the number of the categories is two and the state-conditional probability distributions are normal. We suppose that there are m situations γ_j, $j = 1, ..., m$, on which the prior probabilities $P_j(\theta_i), i = 1, 2$, and the mean vectors $\mu_{ij}, i = 1, 2$, depend, where θ_1 and θ_2 are the categories. We set $\Theta = \{\theta_1, \theta_2\}$ and $\Gamma = \{\gamma_1, ..., \gamma_m\}$. The covariance matrices Σ_1 and Σ_2 are supposed not to depend on the situations. Hence, the normal distributions are denoted by $N(\mu_{1j}, \Sigma_1)$ and $N(\mu_{2j}, \Sigma_2)$.

Let $x \in \mathbf{R}^d$ be signals to be classified and denote by $p_j(x|\theta_i)$, $i = 1, 2$ the state-conditional probability distributions in the situation γ_j. The posterior probabilities $P_j(\theta_i|x)$ are defined by

$$P_j(\theta_i|x) = \frac{P_j(\theta_i, x)}{P_j(\theta_1, x) + P_j(\theta_2, x)} = \frac{P_j(\theta_i)p_j(x|\theta_i)}{p_j(x)}, \quad i = 1, 2, \ j = 1, ..., m, \quad (1)$$

where $p_j(x) = P_j(\theta_1)p_j(x|\theta_1) + P_j(\theta_2)p_j(x|\theta_2)$ is the probability density function in the situation γ_j. The posterior probability (1) can be used as a Bayesian discriminant function [15]. We set $\psi_{Bj}(x) = P_j(\theta_i|x)$. If $\psi_{Bj}(x) > \frac{1}{2}$, x is allocated to the category θ_1. Let σ be the logistic function: $\sigma(t) = (1 + e^{-t})^{-1}$. Since σ and its inverse, the logit function σ^{-1}, are monotone, the logit transform of $P_1(\theta_i|x)$:

$$q_{Bj}(x) = \sigma^{-1}(P_j(\theta_1|x)) = \sigma^{-1}(\psi_{Bj}(x)) = \log \frac{P_j(\theta_1|x)}{P_j(\theta_2|x)} \quad (2)$$

is also a Bayesian discriminant function [15]. If $q_{Bj}(x) > 0$, x is allocated to the category θ_1.

If the difference $q_{Bj_1}(x) - q_{Bj_2}(x)$ is linear in x for any j_1, j_2, then there exists a function q_0 such that

$$q_{Bj}(x) = q_0(x) + u_{Bj}(x), \quad j = 1, \cdots, m, \quad (3)$$

where u_{Bj}'s are linear. This decomposition is not unique up to a linear function.

When the state-conditional probability distributions are normal $N(\mu_{ij}, \Sigma_i)$, the Bayesian discriminant function in the situation γ_j is

$$q_{Bj}(x) = \log \frac{P_j(\theta_1)}{P_j(\theta_2)} - \frac{1}{2} \log \frac{|\Sigma_1|}{|\Sigma_2|} + \frac{1}{2} \sum_{i=1,2} (-1)^i (x - \mu_{ij})^t \Sigma_i^{-1} (x - \mu_{ij}) \quad (4)$$

We suppose, for simplicity, that the variance-covariance matrices Σ_i are not degenerate. The function (4) can be decomposed into the sums of the form (3):

$$q_0(x) = -\frac{1}{2} x^t (\Sigma_1^{-1} - \Sigma_2^{-1}) x, \quad u_{Bj}(x) = q_{Bj}(x) - q_0(x). \quad (5)$$

Though the decomposition is not unique, we do not need to fix its form because our neural network decomposes the function by itself while learning.

For the normal distribution $N(\mu_{ij}, \Sigma_i)$, the Mahalanobis distance between x and μ_{ij} is defined by

$$d_{ij}(x, \mu_{ij}) = |(x - \mu_{ij})^t \Sigma_i^{-1} (x - \mu_{ij})|^{1/2}, \; i = 1, 2.$$

Hence, the difference multiplied by $-1/2$

$$q_{Mj}(x) = -\frac{1}{2} \{ (x - \mu_{1j})^t \Sigma_1^{-1} (x - \mu_{1j}) - (x - \mu_{2j})^t \Sigma_2^{-1} (x - \mu_{2j}) \} \quad (6)$$

can be used as a Mahalanobis discriminant function in the situation γ_j. We set $\psi_{Mj} = \sigma(q_{Mj})$. When $\psi_{Mj}(x) > \frac{1}{2}$ the signal x is allocated to the category θ_1 in the case of Mahalanobis discriminant analysis. If the difference of any two of the Mahalanobis discriminant functions q_{Mj} are linear, they can be decomposed similarly to (3): $q_{Mj}(x) = q_0(x) + u_{Mj}(x)$, where q_0 can be the same as in (5).

Observe that the difference between $q_{Bj}(x)$ and $q_{Mj}(x)$ is a constant:

$$C_j = q_{Bj}(x) - q_{Mj}(x) = u_{Bj}(x) - u_{Mj}(x) = \log \frac{P_j(\theta_1)}{P_j(\theta_2)} - \frac{1}{2} \log \frac{|\Sigma_1|}{|\Sigma_2|}. \quad (7)$$

We use this fact for constructing the neural network which is capable of converting a Bayesian discriminant function to a Mahalanobis discriminant function.

In the algorithm of this paper, the constants C_j are estimated by the neural network after training. By (4) and (7), we have

$$q_{Bj}(\mu_{1j}) = C_j + \frac{1}{2} (\mu_{1j} - \mu_{2j})^t \Sigma_2^{-1} (\mu_{1j} - \mu_{2j}),$$

$$q_{Bj}(\mu_{2j}) = C_j - \frac{1}{2} (\mu_{1j} - \mu_{2j})^t \Sigma_1^{-1} (\mu_{1j} - \mu_{2j}),$$

$$q_{Bj}\left(\frac{\mu_{1j} + \mu_{2j}}{2}\right) = C_j - \frac{1}{8} (\mu_{1j} - \mu_{2j})^t (\Sigma_1^{-1} - \Sigma_2^{-1})(\mu_{1j} - \mu_{2j}).$$

Hence,

$$C_j = 2q_{Bj}\left(\frac{\mu_{1j} + \mu_{2j}}{2}\right) - \frac{1}{2} (q_{Bj}(\mu_{1j}) + q_{Bj}(\mu_{2j})). \quad (8)$$

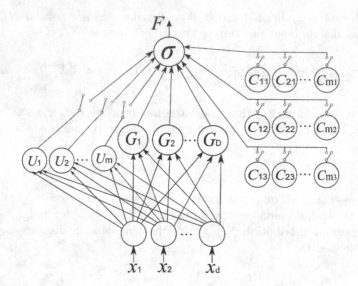

Fig. 1. A one-hidden-layer neural network with m linear nodes U's and D non-linear units G's and $3m$ additional memory nodes C's ($D = \frac{1}{2}d(d+1)$). The activation function of the output unit is the logistic function σ. The linear and memory nodes can be individually turned on and off.

Of course, in the simulation the means $\hat{\mu}_{ij}$ of the teacher signals and the inner potential $\hat{q}_{Bj}$ of the output unit are used instead of μ_{ij} and q_{Bj} to estimate C_j.

We use a neural network of the structure illustrated in Fig. 1.

This network is based on a proposition in [6]. It has more hidden layer units than the minimal requirement but its training is easier. There are unit vectors $v_k, k = 1, ..., \frac{1}{2}d(d+1)$, for which squares of the inner products $(v_k \cdot x)^2$ are linearly independent. Reflecting this fact, the network has $\frac{1}{2}d(d+1)$ non-linear units on the hidden layer. The activation function of the non-linear units G_k is a shifted logistic function $g(t) = \sigma(t+2)$. This shift is necessary to avoid $g^{(2)}(0) = 0$. Without the memory nodes C's, the inner potential of the output unit can realize a formula

$$\Sigma_{k=1}^{\frac{1}{2}d(d+1)} a_k g(\delta v_k \cdot x) + w_j \cdot x + c_j, \tag{9}$$

when the linear node U_j is connected, where v_k are the unit vectors fixed beforehand, and the connection weights a_k and coefficient δ can be optimized. The linear part $w_j \cdot x + c_j$ is from the U_j, and the vector w_j and constant c_j can also be optimized. The formula (9) can approximate any quadratic form with any accuracy [6]. Hence, the inner potential and output of the network approximate $q_{Bj}(x)$ and $\psi_{Bj}(x) = \sigma(q_{Bj}(x))$ respectively, when the network is trained successfully.

The teacher signals are $\{(x^{(t)}, \theta^{(t)}, \gamma^{(t)})\}_{t=1}^n, x^{(t)} \in \mathbf{R}^d, \theta^{(t)} \in \Theta, \gamma^{(t)} \in \Gamma$. Let $F(x, \gamma, w)$ be the output of the network, where w is the weight vector and $\gamma \in \Gamma$. Let $\xi(x, \theta, \gamma)$ be a function on $\mathbf{R}^d \times \Theta \times \Gamma$, and $E_j[\cdot|x]$ and $V_j[\cdot|x]$ be the conditional expectation and variance in the situation γ_j. Training of the network is based on the proposition below, which is a modification of the one in [13].

Proposition (Ruch et al.).   *Set*

$$\mathcal{E}(w) = \sum_{j=1}^{m} P(\gamma_j) \int_{\mathbf{R}^d} \sum_{i=1}^{2} (F(x,w,\gamma_j) - \xi(x,\theta_i,\gamma_j))^2 P_j(\theta_i) p_j(x|\theta_i) dx. \qquad (10)$$

Then,

$$\mathcal{E}(w) = \sum_{j=1}^{m} P(\gamma_j)\{ \int_{\mathbf{R}^d} (F(x,\gamma_j,w) - E[\xi(x,\cdot,\gamma_j)|x])^2 p_j(x) dx$$

$$+ \int_{\mathbf{R}^d} V[\xi(x,\cdot,\gamma_j)|x] p_j(x) dx\}. \qquad (11)$$

If $\xi(x,\theta_1,\gamma_j) = 1$ and $\xi(x,\theta_2,\gamma_j) = 0$, $E[\xi(x,\cdot,\gamma_j)|x] = P_j(\theta_1|x)$. Hence, the training is carried out by minimizing

$$\mathcal{E}_n(w) = \frac{1}{n} \sum_{t=1}^{n} (F(x^{(t)},\gamma^{(t)},w) - \xi(x^{(t)},\theta^{(t)},\gamma^{(t)}))^2$$

$$= \frac{1}{n} \sum_{j=1}^{m} \sum_{\gamma^{(t)}=\gamma_j} (F(x^{(t)},\gamma^{(t)},w) - \xi(x^{(t)},\theta^{(t)},\gamma^{(t)}))^2. \qquad (12)$$

This method of training is used extensively in the literature [1–5, 7–14].

3   Simulations

We illustrate here the result of a simulation with $m = 3$. Listed in Table 1 are the probabilistic parameters used in the simulation. The shapes of the state-conditional probability distributions and the Bayesian and Mahalanobis discriminant functions in the three situations are shown in Fig. 2. The teacher signals $\{(x^{(t)},\theta^{(t)},\gamma^{(t)})\}_{t=1}^{n_1}$ and the test signals $\{(x^{(t)},\gamma^{(t)})\}_{t=1}^{n_2}$ are independently constructed with the parameters in Table 1. Here, we illustrate the result of simulation with $n_1 = 5000$ and $n_2 = 1000$ though we have tried various numbers.

Table 1.   The probabilities of the situations, prior probabilities and the state-conditional probability distributions. The covariance matrices depend only on the category.

Situation	$P(\gamma_j)$	$P_j(\theta_1)$	$P_j(\theta_2)$	Σ_1	Σ_2	μ_{1j}	μ_{2j}
γ_1	0.3	0.4	0.6	$\begin{pmatrix} 2.0 & 1.0 \\ 1.0 & 2.0 \end{pmatrix}$	$\begin{pmatrix} 1.0 & -0.1 \\ -0.1 & 1.0 \end{pmatrix}$	(1, 0)	(-1, 1)
γ_2	0.4	0.7	0.3	$\begin{pmatrix} 2.0 & 1.0 \\ 1.0 & 2.0 \end{pmatrix}$	$\begin{pmatrix} 1.0 & -0.1 \\ -0.1 & 1.0 \end{pmatrix}$	(0, 0)	(0, 0)
γ_3	0.3	0.5	0.5	$\begin{pmatrix} 2.0 & 1.0 \\ 1.0 & 2.0 \end{pmatrix}$	$\begin{pmatrix} 1.0 & -0.1 \\ -0.1 & 1.0 \end{pmatrix}$	(0, -1)	(0, 0)

While learning, the memory nodes are turned off and, if the teacher signal generated in the situation γ_j comes, the linear node U_j is connected and others are disconnected. After training, the network outputs the simulated Bayesian discriminant function $\hat{\psi}_{Bj}$ when the linear node U_j is connected. Hence, if the estimated mean $\hat{\mu}_{1j}$ is then fed to the network, the output is $\hat{\psi}_{Bj}(\hat{\mu}_{1j})$, implying that the inner potential of the output unit is $\hat{q}_B(\hat{\mu}_{j1})$, an approximation of $q_{Bj}(\mu_{j1})$. Then, the memory node C_{j1} is connected and the potential is memorized in it. Following this, $\hat{\mu}_{2j}$ and $(\hat{\mu}_{1j}+\hat{\mu}_{1j})/2$ are fed in turn, and the respective inner potentials are memorized in C_{j2} and C_{j3}. This is repeated for all j.

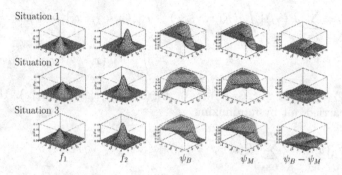

Fig. 2. The probability distributions, Bayesian and Mahalanobis discriminant functions based on Table 1 and the differences of the two discriminant functions

Consequently, if the linear node U_j is connected, and the memory nodes C_{j1}, C_{j2} and C_{j3} are connected with weights -1/2, -1/2 and 2 after training, the output is $\hat{\psi}_{Mj}(x) = \sigma(\hat{q}_{Mj}(x)) = \sigma(\hat{q}_{Bj}(x) - \hat{C}_j)$, where $\hat{C}_j = 2\hat{q}_{Bj}(\frac{\hat{\mu}_{j1}+\hat{\mu}_{j2}}{2}) - \frac{1}{2}(\hat{q}_{Bj}(\hat{\mu}_{j1}) + \hat{q}_{Bj}(\hat{\mu}_{j2}))$. By (4), (6), (7), (8), $\hat{\psi}_{Mj}(x)$ is an approximation of the Mahalanobis discriminant function $\psi_{Mj}(x)$. Thus, the network can approximate both the Bayesian and Mahalanobis discriminant functions in all the situations if the linear and memory nodes are chosen.

The outputs of the network for the initial parameters, the simulated Bayesian and Mahalanobis discriminant functions $\hat{\psi}_{Bj}, \hat{\psi}_{Mj}$ and their differences from the theoretical Bayesian and Mahalanobis discriminant functions are illustrated in Fig. 3. In the situation γ_1, the difference is a little large but smaller than that between the theoretical Bayesian and Mahalanobis discriminant functions.

The allocation results of 1000 test signals by the theoretical discriminant functions, $\psi_{Bj}(x)$ and $\psi_{Mj}(x)$, simulated discriminant functions, $\hat{\psi}_{Bj}(x)$ and $\hat{\psi}_{Mj}(x)$, in each situations are listed in Table 2. Among the 1000, 318, 406 and 276 test signals are generated from the situations γ_1, γ_2 and γ_3. Hence, the discriminant functions were tested respectively with these numbers of test signals.

Each of the four discriminant functions correctly classified only about 75% or a little more of the test signals. This result, not so impressive, is caused by the overlapping of the probability distributions of the two categories in the respective situations. However, the allocations by ψ_{Bj}'s and $\hat{\psi}_{Bj}$'s, coincided at 977 test

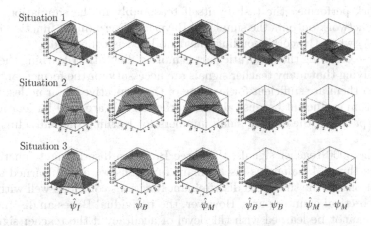

$$\hat{\psi}_I \qquad \hat{\psi}_B \qquad \hat{\psi}_M \qquad \hat{\psi}_B - \psi_B \qquad \hat{\psi}_M - \psi_M$$

Fig. 3. The outputs for the initial parameters, the Bayesian and Mahalanobis discriminant functions learned by the network and their deviations from their theoretical counter parts.

Table 2. The allocation results of the 1000 test signals. The numbers of signals from each category in the respective situations are in the column "Sgl". The results by the respective discriminant functions are in other columns. The lowest row shows the numbers of correctly allocated signals by the respective discriminant functions.

Situation	Category	Sgl	ψ_B	ψ_M	$\hat{\psi}_B$	$\hat{\psi}_M$
γ_1	θ_1	129	118	153	119	151
	θ_2	189	200	165	199	167
γ_2	θ_1	301	404	346	400	352
	θ_2	105	2	60	6	54
γ_3	θ_1	136	127	191	111	181
	θ_2	140	149	85	165	95
Total	θ_1	566	649	690	657	684
	θ_2	434	351	310	343	316
Correct			789	752	782	762

signals, and those by ψ_{Mj}'s and $\hat{\psi}_{Mj}$'s at 964. These imply that training of the network was successful.

4   Discussions

Thus, we have shown that our new method in [11] for estimating the constant for converting a Bayesian discriminant function to a Mahalanobis discriminant function is compatible with the algorithm for the simultaneous learning of several Bayesian discriminant functions.

The main part of the network learns the quadratic part which is common through the situations and the linear nodes U_j on the hidden layer learn the respective linear parts. Though the decomposition (3) is not unique as stated, the network performed the task by itself reasonably in the simulation. As its proof, the network outputs the simulated Bayesian discriminant function in each situation γ_j when the corresponding linear node U_j is connected.

Of course, learning the quadratic part is more difficult than learning the linear parts, implying that many teacher signals are necessary for the former, and small numbers of them are sufficient for learning of the respective linear nodes. Hence, our algorithm is efficient because all the teacher signals are used for learning the quadratic part and the divided small numbers for learning the individual linear parts.

With the 5000 teacher signals, the three Bayesian discriminant functions as well as the three Mahalanobis discriminant functions were well learned so that the allocations by the simulated discriminant functions coincided well with those by the theoretical counter parts. However, the individual Bayesian discriminant functions cannot be learned with this level of accuracy if the teacher signals in the respective situations are separately used.

Though the simulation is restricted to the two-category, normal state conditional distribution case, the algorithm can be extended to multi-category cases if the categories are appropriately paired [8].

Acknowledgements. This work was supported by a Grant-in Aid for scientific research (22500213) from the Ministry of Education, Culture, Sports, Science and Technology of Japan.

References

1. Funahashi, K.: Multilayer Neural Networks and Bayes Decision Theory. Neural Networks 11, 209–213 (1998)
2. Ito, Y., Srinivasan, C.: Multicategory Bayesian Decision Using a Three-layer Neural Network. In: Kaynak, O., Alpaydın, E., Oja, E., Xu, L. (eds.) ICANN 2003 and ICONIP 2003. LNCS, vol. 2714, pp. 253–261. Springer, Heidelberg (2003)
3. Ito, Y., Srinivasan, C.: Bayesian Decision Theory on Three-layer neural networks. Neurocomput. 63, 209–228 (2005)
4. Ito, Y., Srinivasan, C., Izumi, H.: Bayesian Learning of Neural Networks Adapted to Changes of Prior Probabilities. In: Duch, W., Kacprzyk, J., Oja, E., Zadrożny, S. (eds.) ICANN 2005. LNCS, vol. 3697, pp. 253–259. Springer, Heidelberg (2005)
5. Ito, Y., Srinivasan, C., Izumi, H.: Discriminant Analysis by a Neural Network with Mahalanobis Distance. In: Kollias, S.D., Stafylopatis, A., Duch, W., Oja, E. (eds.) ICANN 2006. LNCS, vol. 4132, pp. 350–360. Springer, Heidelberg (2006)
6. Ito, Y.: Simultaneous Approximations of Polynomials and Derivatives and Their Applications to Neural Networks. Neural Comput. 20, 2757–2791 (2008)
7. Ito, Y., Srinivasan, C., Izumi, H.: Learning of Bayesian Discriminant Functions by a Layered Neural Network. In: Ishikawa, M., Doya, K., Miyamoto, H., Yamakawa, T. (eds.) ICONIP 2007, Part I. LNCS, vol. 4984, pp. 238–247. Springer, Heidelberg (2008)
8. Ito, Y., Srinivasan, C., Izumi, H.: Multi-category Bayesian Decision by Neural Networks. In: Kůrková, V., Neruda, R., Koutník, J. (eds.) ICANN 2008, Part I. LNCS, vol. 5163, pp. 21–30. Springer, Heidelberg (2008)

9. Ito, Y., Izumi, H., Srinivasan, C.: Learning of Mahalanobis Discriminant Functions by a Neural Network. In: Leung, C.S., Lee, M., Chan, J.H. (eds.) ICONIP 2009, Part I. LNCS, vol. 5863, pp. 417–424. Springer, Heidelberg (2009)
10. Ito, Y., Izumi, H., Srinivasan, C.: Simultaneous Learning of Several Bayesian and Mahalanobis Discriminant Functions by a Neural Network with Additional Nodes. Australian J. Intell. Inform. Process. Syst. 11, 1–7 (2010); Proceedings of ICONIP 2010
11. Ito, Y., Izumi, H., Srinivasan, C.: Learning of Mahalanobis Discriminant Functions by a Neural Network (submitted, preparation)
12. Richard, M.D., Lipmann, R.P.: Neural Network Classifiers Estimate Bayesian a Posteriori Probabilities. Neural Comput. 3, 461–483 (1991)
13. Ruck, M.D., Rogers, S., Kabrisky, M., Oxley, H., Sutter, B.: The Multilayer Perceptron as Approximator to a Bayes Optimal Discriminant Function. IEEE Trans. Neural Networks 1, 296–298 (1990)
14. White, H.: Learning in Artificial Neural Networks: A Statistical Perspective. Neural Comput. 1, 425–464 (1989)
15. Duda, R.O., Hart, P.E.: Pattern Classification and Scene Analysis. Joh Wiley & Sons, New York (1973)

Reinforcement of Keypoint Matching by Co-segmentation in Object Retrieval: Face Recognition Case Study

Andrzej Śluzek[1,*], Mariusz Paradowski[2], and Duanduan Yang[3]

[1] Khalifa University, Abu Dhabi Campus, UAE
andrzej.sluzek@kustar.ac.ae
http://www.kustar.ac.ae
[2] Wroclaw University of Technology, Poland
[3] Motorola(China) Electronics LTD, Shanghai, China

Abstract. The paper investigates a certain group of problems in visual detection and identification of near-identical *the-same-class* objects. We focus on difficult problems for which: (1) the intra-class visual differences are comparable to inter-class differences, (2) views of the objects are distorted both photometrically and geometrically, and (3) objects are randomly placed in images of unpredictable contents. Since detection of *the-same-person* faces in complex images is one of such problems, we use it as the case study for the proposed approach. The approach combines a relatively inexpensive technique of near-duplicate fragment detection with a novel co-segmentation algorithm. Thus, the initial pool of matching candidates can be found quickly (however, with limited *precision*, i.e. many false positives can be detected). It is shown that the subsequent co-segmentation can effectively reject false positives and accurately extract the matching objects from random backgrounds.

Keywords: Object matching, Near-duplicates, Co-segmentation, Image mapping, Face identification.

1   Introduction

Detection of similarly looking (near-duplicate) image fragments is one of the fundamental operations in CBVIR (content-based visual information retrieval), e.g. [1,2,3]. In most cases, the ultimate objective of this operation is to detect *the-same-class* objects present in matched images. Unfortunately, certain practical limitations exist in this (superficially) straightforward problem. Typically, *the-same-class* objects (even if almost identical) may be distorted both physically and photometrically, and often only small visually similar patches can be found within their outlines. Moreover, if classes of objects are visually similar, near-duplicate patches may exist in *different-class* objects, and eventually it might be difficult to differentiate between intra-class and inter-class visual characteristics.

---

* Corresponding author.

T. Huang et al. (Eds.): ICONIP 2012, Part V, LNCS 7667, pp. 34–41, 2012.

Therefore, *precision* and *recall* (i.e. credibility and completeness of the returned results) have to be compromised in the retrieval of such objects.

Several approaches have been proposed to handle this 'conflict of interests'. The most typical method of improving *recall* is to use topology instead of geometry (e.g. [6]) or to apply non-linear mappings (e.g. [7]) for determining image correspondences. To improve *precision* of the retrieval, image co-segmentation seems to be a promising technique. It has been proposed to identify identical, but visually difficult objects, [8], and also to match highly flexible objects (e.g. animals, [9]).

This paper discusses problems with such difficulties. We assume that multiple classes of objects are defined (where each class consists of almost identical objects though some photometric distortions and minor physical deformations within the class may exist). However, objects from different classes can also be generally similar, i.e. the inter-class visual differences are almost at the same level as the intra-class differences.

It can be argued that visual identification of human faces is an example of such a scenario. An individual face is a well-defined unique class, but visual variations due to hairstyles, glasses, expressions, makeup, hats, illumination conditions, etc. are unavoidable. At the same time, human faces are generally similar so that is it possible that under certain conditions images of different faces may look almost as similar as images of the same face.

Therefore, the problem of face *detection-and-matching* in random images has been selected as the case study for the proposed methodology. However, we **do not** propose a dedicated system for face recognition. In particular, we do not use any knowledge (or learning process) related to the human face structure or anatomy. Faces are consider just pieces of visual data embedded into wider contexts of images. Thus, without any changes, the same technique is suitable for any other application where a need exists to distinguish between generally similar yet uniquely defined classes of objects. Identification of animals, botanical specimens, cars, etc. can be mentioned as other examples of similar problems.

The proposed method consists of two steps. First, we apply a relatively inexpensive algorithm (see [3]) detecting near-duplicate patches in matched images. Overview of this algorithms and its limitations are presented in Subsection 2.1.

In the second step, detected in the first step fragments containing near-duplicate patches are segmented using a novel co-segmentation algorithm. The algorithm is briefly introduced in Subsection 2.2 (its full description is available in [10]). Experimental results are discussed in Section 3. To evaluate performances of the method, we primarily use face images from a popular publicly available[1] Caltech101 databases. Section 4 concludes the paper.

2   Prior Contributions

2.1   Detection of Near-Duplicate Patches

Identification of similar objects randomly located on diversified backgrounds is often implemented by detecting near-duplicate image patches (where for locally

---

[1] http://www.vision.caltech.edu/Image_Datasets/Caltech101

planar objects near-duplicity is typically modeled by affine mappings). Thus, similarly looking fragments of faces can be prospectively detected as affine-covariant patches in matched images, i.e. only purely visual data are used and no structural or anatomic model of faces is employed (e.g. [4,5]). Note than any other visually similar contents would be detected in the same manner, and face retrieval is just an example rather than a specialized application.

In this paper, we use the algorithm published in [3] which detects near-duplicate patches as convex hulls of keypoint clusters for which a consensus exists between visual similarities and geometric configurations. Exemplary pairs of extracted near-duplicate patches (including patches actually representing human faces) are shown in Fig. 1.

Fig. 1. Exemplary pairs of near-duplicate patches detected in various images

This straightforward approach has achieved relatively good performances (as reported in [5]) in face *detection-and-identification*. Using face images of Caltech101 as a benchmark dataset, 0.95 *precision* and 0.78 *recall* have been reported. However, significant limitations of the method have been highlighted as well. It indiscriminately detects near-duplicate fragments of various sizes both in *the-same-face* images and *different-face* images (see examples in Fig. 2).

Fig. 2. Near-duplicate patches found in Caltech101. Examples of *the-same-face* patches are deliberately mixed up with *different-face* (sometimes of the opposite sex) patches.

In particular, the reported performances have been obtained under the assumption that only near-duplicate patches covering a significant percentage of the face outlines are accepted (i.e. at least one of the matched images must be from a preprocessed database with outlined face areas). The optimum threshold coverage is approx 40%. Without this assumption, *recall* is even higher (i.e. 0.88), but *precision* drops to below 0.4. Thus, we can predict that in unprocessed random images containing human faces (frontal views of reasonable quality) the majority of *the-same-face* cases would be retrieved by near-duplicate patch detection, but more than a half of the returned matches are false positives (presumably indicating humans with similarly looking noses, eyes, chins, etc.).

2.2   Principles of the Co-Segmentation Algorithm

Co-segmentation is a simultaneous segmentation of the same object(s) from a pair of images. Thus, in the second step the detected near-duplicate patches are considered seeds of same objects and fragments around such patches are co-segmented using the algorithm reported in [10].

Typical current co-segmentation methods apply basically the same graph-cut framework solved by minimizing a Markov Random Field energy function through a min-cut/max-flow algorithm (e.g. [11]). In general, the MRF energy E_{MRF} of two co-segmented images I_1 and I_2 is expressed as a sum of the *image* energy E_{Im} and the *foreground* (object) energy E_F.

$$E_{MRF}(I_1, I_2) = E_{Im}(I_1, I_2) + E_F(I_1, I_2). \tag{1}$$

The algorithm we use follows the standard approaches (e.g. [11,12]) regarding the image energy (which consists of the *deviation penalty* and *separation penalty* functions). However, a novel foreground energy E_F is proposed. First, using the keypoint correspondences established in the patch detection as the *control points*, two non-linear mappings between the matched images are built (i.e. $f_{12} : I_1 \rightarrow I_2$ and $f_{21} : I_2 \rightarrow I_1$). The principles of building these mappings are detailed in [13]; in general, they are based on TPS (thin plate splines) warping (see [14]). Subsequently, the foreground energy of matched images is defined as follows:

$$E_F(I_1, I_2) = \sum_{p_i \in I_1} \sum_{p_j \in I_2} z_{ij} s_{ij}. \tag{2}$$

where $z_{ij} = 1$ if both pixels p_i and p_j are labeled *foreground* (otherwise it is equal to 0). The value of s_{ij} is:

$$s_{ij} = -1 \ if \ f_{12}(p_i) = p_j \ or \ f_{21}(p_j) = p_i \ (and \ 0 \ otherwise). \tag{3}$$

In [10], this co-segmentation algorithm has been benchmarked against [11], and found superior both in terms of performances and timing, in particular in cases of photometric distortions (exemplary results are compared in Fig. 3).

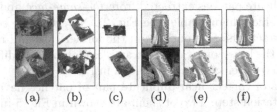

(a)   (b)   (c)   (d)   (e)   (f)

Fig. 3. Exemplary pairs of images (a,d) and the co-segmentation results using the [11] algorithm (b,e) and the results for the applied algorithm (c,f)

Performances of near-duplicate patch detection followed by the proposed co-segmentation are evaluated in the following section.

3   Experimental Verification

3.1   Methodology

The case study experiments have been conducted using exactly the same low-level tools and the same data as in [5] so that the results can be objectively compared. Keypoints used to detect near-duplicate patches are extracted by Harris-Affine detector [15] and represented by SIFT descriptors [16]. However, any other keypoint matching schemes and/or other techniques of detecting near-duplicate patches can be alternatively employed without any changes to the proposed method.

285 images of human faces (19 faces, each shown on 15 images with diversified backgrounds) from Caltech101 dataset have been compared against each other, i.e. 40,470 image pairs have been matched altogether. 1,995 image pairs are the *ground-truth* correct matches. The images are not preprocessed (i.e. the outlines of human faces are not provided) so that near-duplicate patches of any size are preliminarily considered *the-same-face* objects. Image pairs with near-duplicate background patches only (there are a few such cases) are not counted.

3.2   Analysis of the Results

The summary of the case study results is provided in Table 1.

Table 1. Results of the case study experiment (compared to the results published in [5])

Measure	Proposed algorithm	[5] algorithm
Returned image pairs	1,811	4,464
Correct image pairs	1,750	1,754
Precision	0.97	0.39
Recall	0.88	0.88

The most important conclusion from Table 1 is that *the-same-face* retrieval using the algorithm reinforced by co-segmentation is superior in terms of *precision* (with practically the same *recall*). This is somehow a surprising effect because near-duplicate patches extracted from *the-same-face* and *different-faces* images often look indistinguishable (see Fig. 2). Nevertheless, co-segmentation almost always extracts the face outlines in the former case and returns no foreground objects in the latter one. Several examples (confusingly difficult examples have been deliberately selected) are provided in Fig. 4.

Only four cases of correct near-duplicate pair of patches have been rejected by the co-segmentation step. Two of them are shown in Fig. 5 in the context of whole images to show that they might be confusing even for a human observer. The other two cases are also for the same subjects.

Compared to the technique presented in [5], the number of false positives is very small (61 *versus* 2,710) so that *precision* is very high (97%). Three examples of false positives are given in Fig. 6. It can be seen that the actual face outlines are only partially incorporated into the returned forgrounds.

(a)   (b)   (c)

Fig. 4. Near-duplicate patches extracted from *the-same-face* images (a) and the corresponding foregrounds (b) returned by co-segmentation (scale distortion is due to the size normalization of co-segmented images). In case of near-duplicate patches extracted from *different-face* images (c) co-segmentation returns no foregrounds.

(b)   (c)

Fig. 5. Examples of image pairs with correct detection of near-duplicate patches for which co-segmentation was unsuccessful

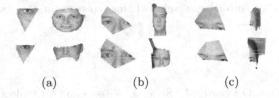

(a)   (b)   (c)

Fig. 6. Three examples of successful co-segmentation over near-duplicate pairs from *different-faces* images

4   Concluding Remarks

A method combining keypoint-based image matching and co-segmentation has been proposed. It usefulness in *the-same-class* object detection in problems where intra-class and inter-class visual differences might be of similar levels has been demonstrated using *the-same-face* image retrieval as a case study.

Co-segmentation has been used in a novel manner. Unlike [8], where co-segmentation is used to preliminary identify the image correspondences, we use it to verify the identified correspondences. Also, in contrast to most co-segmentation procedures (e.g. [9,11]) we replace the color-based foreground energy function by the correspondence-based function. Therefore, the method can robustly handle photometric distortion but instead of returning the whole foreground objects, it returns only the objects (or their fragments) which have similar geometric configurations. Nevertheless, non-linear distortions of the configurations can be tolerated.

Fig. 7. Co-segmentation results of selected photos

We have also preliminarily tested the method using images of animals, birds and cars from various publicly available databases. The method seems to well identify only actually identical fragments of individual objects rather than returning their whole outlines (which is a typical output from other co-segmentation methods). Some positive examples of those tests are provided in Fig. 7. Negative examples, where car fragments are preliminary selected (based on patch retrieval) for co-segmentation but eventually rejected, are shown in Fig. 8.

Fig. 8. Two pairs of sub-images selected for co-segmentation which yield no foregrounds

References

1. Philbin, J., Chum, O., Isard, M., Sivic, J., Zisserman, A.: Object Retrieval with Large Vocabularies and Fast Spatial Matching. In: IEEE Conference on Computer Vision and Pattern Recognition (CVPR 2007), pp. 1–8 (2007)
2. Zhao, W.-L., Ngo, C.-W.: Scale-rotation Invariant Pattern Entropy for Keypoint-based Near-duplicate Detection. IEEE Trans. Image Process. 2, 412–423 (2009)
3. Paradowski, M., Śluzek, A.: Local Keypoints and Global Affine Geometry: Triangles and Ellipses for Image Fragment Matching. In: Kwaśnicka, H., Jain, L.C. (eds.) Innovations in Intelligent Image Analysis. SCI, vol. 339, pp. 195–224. Springer, Heidelberg (2011)
4. Chum, O., Perdoch, M., Matas, J.: Geometric min-Hashing: Finding a (Thick) Needle in a Haystack. In: IEEE Conference on Computer Vision and Pattern Recognition (CVPR 2009), pp. 17–24 (2009)
5. Śluzek, A., Paradowski, M.: Visual Similarity Issues in Face Recognition. Int. J. Biometrics 4, 22–37 (2012)
6. Tell, D., Carlsson, S.: Combining Appearance and Topology for Wide Baseline Matching. In: Heyden, A., Sparr, G., Nielsen, M., Johansen, P. (eds.) ECCV 2002, Part I. LNCS, vol. 2350, pp. 68–81. Springer, Heidelberg (2002)
7. Li, H., Kim, E., Huang, X., He, L.: Object Matching with a Locally Affine-Invariant Constraint. In: IEEE Confernce on Computer Vision and Pattern Recognition (CVPR 2010), pp. 1641–1648 (2010)
8. Cech, J., Matas, J., Perdoch, M.: Efficient Sequential Correspondence Selection by Cosegmentation. IEEE Trans. PAMI 32, 1568–1581 (2010)

9. Vicente, S., Rother, C., Kolmogorov, V.: Object Cosegmentation. In: IEEE Confernce on Computer Vision and Pattern Recognition (CVPR 2011), pp. 2217–2224 (2011)
10. Yang, D., Śluzek, A.: Co-segmentation by Keypoint Matching: Incorporating Pixel-to-pixel Mapping into MRF. Nanyang Technological University, SCE - unpublished report, Singapore (2010)
11. Hochbaum, D.S., Singh, V.: An Efficient Algorithm for Co-segmentation. In: 12th IEEE International Conference on Computer Vision (ICCV 2009), pp. 269–276 (2009)
12. Mukherjee, L., Singh, V., Dyer, C.R.: Half-integrality based algorithms for cosegmentation of images. In: IEEE Conference on Computer Vision and Pattern Recognition (CVPR 2009), pp. 2028–2035 (2009)
13. Yang, D., Śluzek, A.: A Low-dimensional Local Descriptor Incorporating TPS Warping for Image Matching. Image Vision Comput. 28, 1184–1195 (2010)
14. Bookstein, F.L.: Principle Warps: Thin Plate Splines and the Decomposition of Deformations. IEEE Trans. PAMI 16, 460–468 (1989)
15. Mikolajczyk, K., Schmid, C.: Scale and Affine Invariant Interest Point Detectors. Int. J. Comput. Vision 60, 63–86 (2004)
16. Lowe, D.G.: Distinctive Image Features from Scale-invariant Keypoints. Int. J. Comput. Vision 60, 91–110 (2004)

Effect of Luminance Gradients in Measurement of Differential Limen

Hiroaki Kudo*, Takuya Kume, and Noboru Ohnishi

Graduate School of Information Science, Nagoya University,
Furo-cho, Chikusa-ku, Nagoya 464-8603, Japan
{kudo,kume,ohnishi}@ohnishi.m.is.nagoya-u.ac.jp

Abstract. Objects in the visual field affect the appearance in the neighborhood. One of explanations to show a degree of its effect is a concept of a field of perception. It forms a contour map that is measured by upper and lower limen around a visual object. To clarify factors forming a field, we measured upper and lower limen for distribution of luminance according to a linear expression equation and a quadratic polynominal equation. The results show that upper and lower limen are affected by luminance itself, magnitude of luminance gradients. We conjectured a field of perception is shaped mainly by the maginitude of luminance gradients and its trend from the results.

Keywords: Luminance gradients, Differential limen, Field of perception.

1 Introduction

To examine figure's appearance effect, visual illusions have been studied. It is known well since about 150 year ago that Mach band shows a difference between psychologically observed brightness pattern and physical luminance pattern. Recently, the method which localize Mach edge using the 3rd derivative, is proposed from the view of spatial filtering [1].

As a localization on edge only but a whole pattern of a visual target, some theories have been proposed as explanations of phenomenon on the visual illusions [2]. As one of such theories, there is an approach which is based on field theories. Fig. 1 shows a concept image of a field of perception. Each filled figure (left: rectangle and right: triangle) is a visual target. And the thin lines, which looks like a contour map, show a degree of effect into surrounding region (background).

In order to measure a degree of effect, some methods are proposed. In a psychological aspect, the differential limen is measured. Fig. 2 shows the field by such a measurement. Fig. 2(a) shows an arrangement of a visual target. This example is consisted of an object (a low luminance circle) and a background (high luminance). A small spot light is displayed near the edge of an object,

---

* Corresponding author.

T. Huang et al. (Eds.): ICONIP 2012, Part V, LNCS 7667, pp. 42–49, 2012.
© Springer-Verlag Berlin Heidelberg 2012

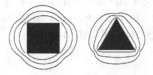

Fig. 1. A concept image of a field

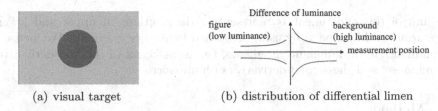

(a) visual target   (b) distribution of differential limen

Fig. 2. Measurement of a field of perception[3]. (a):visual target (interest region is a neighborhood of a right part of a arc) (b):differential limen around an interest arc.

and a subject gaze at a point of a small spot light. Initially, luminance of a spot is set a higher value than luminance of background in the visual target. The luminance of a spot is decreased until the subject can not identify the difference between small light luminance and surrounding region's luminance. The position and the luminance are recorded. This procedure is equal to the measurement of upper limen at a measurement point. After measuring at a various positions of the visual target, we can obtain the field as a top-right curved line in Fig. 2(b). Measurement at inside of an object (a circle) is also possible, the field is plotted as top-left curved line in Fig. 2(b). Then, two lower curved lines are obtained by measuring for the lower limen as a similar procedure[3]. In the traditional paper, the top-right line is proposed only, that means measuring upper limen on a background only.

Then, we may consider this measurement is same to a measurement of upper limen around the Mach band. If a curved line is caused by perceived brightness (sensation) based on the lateral inhibition in vision[4], the brighter region may be perceived at the position near the luminance edge, then upper limen may increase. The curved line matches its hypothesis. However, it is not clear why lower limen decrease giving it is true.

Table 1. Properties of distribution of visual target

condition	equation	average luminance	gradient of luminance	second derivative of luminance	trend of magnitude of second derivative
(a)	linear	low	constant	zero	
(b)	linear	high	constant	zero	
(c)	quadratic	low	non-constant	negative	increase
(d)	quadratic	high	non-constant	positive	decrease

In this paper, we focused on the distribution of luminance gradients and its trend. Upper limen and lower limen were measured for visual stimuli which was composed with distribution of smoothed luminance. It doesn't cause a percept of Mach band.

2   Experiment I: Effect on Changes of Luminance Gradients

The aim of this experiment is representing distributions of upper and lower limen around a Mach band by using smoothed luminance gradients. We focused on luminance itself, luminance gradients, i.e., a coefficient of the first derivative of luminance, and the second derivative of luminance.

2.1   Method

Upper and lower limen are measured according to a procedure of method of limits. A small test target at a measurement position was displayed. The luminance of a target was changed by the descent/ascent series for measurement of upper/lower limen. The limen is calculated by a subtraction of the luminance which is supposed distributed function from the measured luminance. Luminance is changed by one step (gray scale steps for a display) per 0.8 second. A test target is used a rectangle of 12 $[h] \times 1[w][pixels]$ Five trials were repeated for each measurement position. One subject (21 years, men) is participate with this experiment.

We used the LCD display (EIZO ColorEdgeCG210, size $432[mm] \times 324[mm]$, resolution $1,600 \times 1,200[pixels]$, dot pitch$0.270[mm] \times 0.27[mm]$, R/G/B 256 $[steps]$ for each color, which is calibrated from hardware specs 10 bits. We set a gamma value 2.2/1.0, maximum luminance 170/171, minimum luminance 0.6/0.5 for the low/high (average) luminance condition. A subject observed a visual target with a chin support in a darkroom ($\leq 1[lx]$). Visual distance between a subject and a display was $0.54[m]$.

2.2   Visual Target

In this experiment, we set following visual targets. Distributions of luminance of a visual target are according to a linear expression or a quadratic polynominal. Low luminance and high luminance conditions are set for both distribution. From the point of view of local luminance of measuring position, the conditions are classified in Table 1.

The luminances are according to the following equations for condition (a), (b), (c) and (d).

$$(a) \quad L = \frac{(14.6 - 32.3)}{40}x + 23.45, \qquad (b) \quad L = \frac{(104 - 167)}{40}x + 135.5,$$

$$(c) \quad L = -0.0132x^2 - 0.444x + 28.8, \qquad (d) \quad L = 0.0381x^2 - 1.568x + 120.5.$$

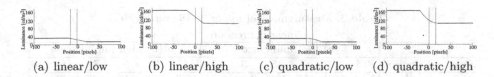

(a) linear/low   (b) linear/high   (c) quadratic/low   (d) quadratic/high

Fig. 3. Luminance distribution of visual target

Table 2. Luminance at measurement positions

linear expression distribution

	(a) low luminance	(b) high luminance
position $[px]$	-16 -11 -7 -5 -3 -1	1 3 5 7 11 16
luminance$[cd/m^2]$	30.1 27.9 26.3 25.2 24.7 23.7	134 131 127 125 118 111

quadratic polynomial distribution

	(c) low luminance	(d)high luminance
position $[px]$	-16 -11 -7 -5 -3 -1	1 3 5 7 11 16
luminance$[cd/m^2]$	32.4 31.8 31.2 30.6 29.5 29.0	119 116 113 111 107 105

L and x means luminance $[cd/m^2]$ and measurement point$[pixels]$. The origin of measurement position is the center of visual target. It coordinates the positive to the right direction. These distributions of luminance gradients are displayed at the center of regions (40$[pixels]$ which correspond to 1.08$[arcdeg]$) on a LCD monitor. Outside regions of its range are displayed by a constant luminance. The left side is set at high luminance (32.3$[cd/m^2]$(a)(c), 167$[cd/m^2]$(b)(d)), and right is set at low luminance (14.6, 104$[cd/m^2]$). The profiles of each distribution are shown in Fig. 3. (The origin is located at the position 640 [pixels].) The vertical two dashed lines indicate the ends of a measurement range.

We measured upper and lower limen at separate positions (1, 3, 5, 7, 11, and 16$[pixels]$). At the position of 1 and 16$[pixels]$, the visual angle forms 0.028$[arcdeg]$ and 0.458$[arcdeg]$. The luminance values according to the equation at each measurement positions are shown in Table 2.

2.3   Results and Discussion

Results of measurements on upper and lower limen for five trials at each position are shown in Table 3 and Fig. 4. Means and standard deviations (SD) are calculated.

In Fig. 4, the values are calculated by subtracting the value of an equation at the measurement positions from averaged value of measurements. The horizontal axis shows measurement positions and the vertical axis shows the difference of luminance. A filled circle in a legend means results of upper limen, and an open circle means ones of lower limen. A tick on the each plot shows the range of ±1-SD.

As a validation of the effect on distribution of luminance, by two-sided t−test, whether the luminance differences are same at the ends of the measurement

Table 3. Measurement of upper and lower limen

linear expression

limen	position[px]	(a) low luminance						(b) high luminance					
	position[px]	-16	-11	-7	-5	-3	-1	1	3	5	7	11	16
upper	mean[cd/m^2]	1.04	1.10	0.75	1.05	0.62	0.91	2.67	3.07	3.07	2.27	2.40	2.67
	SD[cd/m^2]	0.43	0.35	0.26	0.33	0.21	0.38	0.60	0.68	0.53	0.68	0.53	0.60
lower	mean[cd/m^2]	-0.68	-0.76	-0.94	-0.82	-0.91	-0.69	-3.34	-3.87	-3.74	-3.87	-3.74	-3.34
	SD[cd/m^2]	0.22	0.26	0.21	0.52	0.37	0.24	0.73	1.15	0.68	0.65	0.68	0.73

quadratic polynomial distribution

limen	position[px]	(c) low luminance						(d) high luminance					
	position[px]	-16	-11	-7	-5	-3	-1	1	3	5	7	11	16
upper	mean[cd/m^2]	0.59	0.70	0.70	0.69	0.90	1.12	4.54	3.47	3.74	2.67	3.74	2.67
	SD[cd/m^2]	0.00	0.24	0.43	0.43	0.28	0.36	0.65	0.65	0.53	0.84	1.16	0.84
lower	mean[cd/m^2]	-0.47	-0.58	-0.80	-0.68	-0.99	-0.98	-4.01	-2.14	-2.80	-2.00	-1.34	-1.74
	SD[cd/m^2]	0.43	0.00	0.28	0.22	0.22	0.40	1.12	0.27	1.29	0.60	0.42	0.68

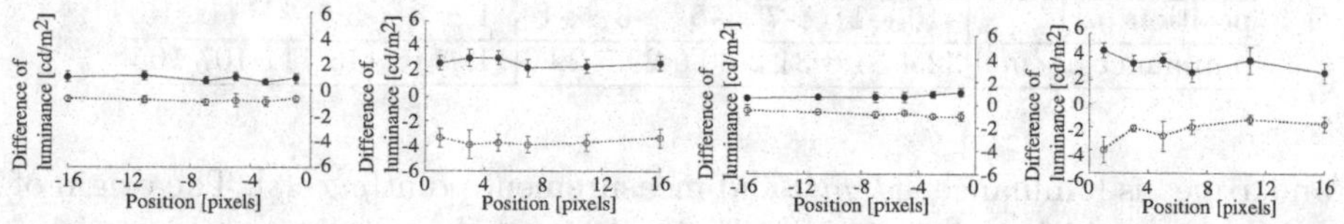

(a) linear/low   (b) linear/high   (c) quadratic/low   (d) quadratic/high

Fig. 4. Distributions of upper and lower limen for luminance gradients

positions (i.e., a pair of -1 and -16, and 1 and 16[pixels]) or not, is tested. The p-values and the significant level are shown in Table 4. One asterisk and two asterisks show a significance level at 5% and 1%, respectively. Therefore, marked condition means the luminance differences at the ends are not same.

From Table 4, upper and lower limen are not changed at the ends of measurement points for linear expression distribution of luminance gradients. In Fig. 4 (a) and (b), we can conjecture that plotted lines hold almost constant values.

On the other hand, luminance differences of upper limen are not same for the quadratic polynomial distribution, especially, for high luminance condition (d). The significance level is 1%. In the Fig.4 (d), we can identify upper limen at the position 1[$pixel$] takes a high value than one at the position 16[$pixels$]. The trends of distribution seem that the value takes higher as the measurement position locates nearer the origin. The value at 1 [$pixel$] takes low value than one at 16[$pixels$], for lower limen, In Fig. 4(c), we can see the same trend for distributions of Fig. 4(d).

The properties of these distributions are similar to a field of perception (Figs. 1 and 2). Here, we can see that the absolute value of limen takes a high value as an average luminance becomes high. It is natural that differential limen are depend on the averaged luminance according to the power law of perception. However, we must note that the luminance is lower at the position -1 [$pixels$] than one at the position -16 [$pixel$] as a distribution in Fig. 3(c). From this

Table 4. Results of t–test on the luminance differences of limen

equation	limen	low luminance	high luminance
linear	upper	0.675	1.000
	lower	0.933	1.000
quadratic	upper	0.042 *	0.009 **
	lower	0.121	0.011 *

point of view, we supposed this property was caused by the change of luminance gradients, i.e., the second derivative of luminance.

In Fig 4(c), the absolute values of the second derivative luminance increase, as the position locates more larger value along a horizontal axis. On the contrary, the absolute values of second derivative decrease in Fig. 4(d).

We conjectured that both upper and lower limen were affected on the change of luminance gradients.

3   Experiment II: Effect on Luminance Gradients

3.1   Method

In this experiment, we set distribution of visual target according to a linear expression equation. We assumed to express the combinations of linear expression equations as a quadratic polynomial equation for a luminance distribution.

Two conditions of averaged luminance at a measurement position, are set (low 32.8 $[cd/m^2]$ and high 106 $[cd/m^2]$). For low luminance condition, the gradient is set by 0.0, 0.4, 0.6, or 0.8$[cd/m^2/pixel]$. For high luminance condition, 0.0, 0.5, 1.0, or 1.5 $[cd/m^2/pixel]$ is set. A displayed range of luminance gradients is limited from -20 $[pixels]$ to 20 $[pixels]$. Measurement position is set at 0 $[pixel]$. Fig. 5 shows profiles of visual targets. The gamma of a display setting is set 2.2/1.0 for low/high luminance condition. The maximum and minimum luminance are set by 140 and 0.5 $[cd/m^2]$ in both conditions. A similar procedure of measurement was done as one of Experiment I.

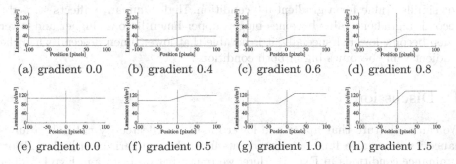

   (a) gradient 0.0   (b) gradient 0.4   (c) gradient 0.6   (d) gradient 0.8

   (e) gradient 0.0   (f) gradient 0.5   (g) gradient 1.0   (h) gradient 1.5

Fig. 5. Profiles of visual targets ((a)-(d)low luminance (e)-(h)high luminance)

Table 5. Measurements of upper limen and lower limen

limen		gradient $[cd/m^2/pixel]$							
		low				high			
		0.0	0.4	0.6	0.8	0.0	0.5	1.0	1.5
upper	mean$[cd/m^2]$	0.33	0.76	0.98	1.43	3.17	3.27	3.93	4.15
	SD$[cd/m^2]$	0.27	0.27	0.41	0.56	0.53	0.69	0.41	0.44
lower	mean$[cd/m^2]$	-0.64	-0.86	-0.98	-1.39	-1.86	-2.51	-2.62	-3.06
	SD$[cd/m^2]$	0.52	0.26	0.41	0.26	0.27	0.27	0.41	0.27

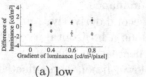

(a) low

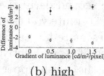

(b) high

Fig. 6. Distributions of upper limen and lower limen for gradietial luminance

Table 6. Results of $t-$test on the luminance difference of limen

limen	low luminance		high luminance	
upper	0.014	.*	0.022	*
lower	0.044	*	0.000	**

3.2 Results and Discussion

Measurement results of upper and lower limen for five trials are shown in Table 5 and Fig. 6. A two-sided $t-$test is applied to the values of gradient 0.0 and 0.8 for a low luminance condition, and the values of gradient 0.0 and 1.5 for a high condition. The $p-$value and significance level's marks are shown in Table 6.

Table 5 and Fig. 6 show that the upper limen become higher and the lower limen become lower when a gradient value becomes bigger. Table 6 shows that there is a significance difference for all conditions. First, low or high luminance condition affects upper and lower limen, because the values of (a) and (b) are not simillar value for a gradient 0.0 condition. High luminance affects strongly. Second, a gradient value becomes bigger, upper limen become higher and lower limen become lower. Especially, it is notable that lower limen descend when the gradient value becomes big in both conditions.

4 Discussion

We can conjecture that upper and lower limen are affected by not only a luminance itself but also luminance gradients from the comparison of high and low luminance conditions in Exp. II. Here, we remember on results in Exp.I. Upper and lower limen are not changed so much for a distribution according to a linear expression equation. The gradient is constant for each meaurement position, then a luminance factor only works. This implies that a luminance factor does not affect so much. The low luminance result for a distribution according to a quadratic polynominal equation, is important. Although the second derivative of luminance is negative, a magnitude of gradient becomes big. Then, luminance becomes lower and the magnitude of a gradient becomes bigger for going to the positive direction of horizontal axis in Fig.4 (c),(d). In this case, the upper limen ascend. It implies there is the case that a gradient factor is affected stronger than luminance factor. Furthermore, these distributions of upper and lower limen are alike curved lines of a left side in Fig.2(b). The sensation neighborhood of an edge at the darker region of Mach band becomes more darker is well known. It is similar to a situation of condition (c). On the other hand, a high luminance result is natural. Two factors are not conflict. The luminance becomes lower and the magnitude of luminance gradients decrease. Then, upper limen descend and lower limen ascend. This distributions look alike curved lines of a right side in Fig.2(b). That is a situation at the brighter region of Mach band. And, its sensation looks alike a distribution in condition (d).

5   Conclusions

Visual objects affect the appearance of neighborhood objects. One of the explanations is a concept of a field of perception. It forms a contour map around a visual object by measuring upper and lower limen. To clarify factors forming a field, we measured upper and lower limen for distribution of luminance according a linear expression equation and a quadratic polynomial equation. The results shows that both limen are affected by luminance, magnitude of luminance gradients, and its trend. We conjecture that these results imply a field of perception is formed by the sensation of Mach band from a distribution's similarity.

References

1. Wallis, S.A., Georgeson, M.A.: Mach Edges: Local Features Predicted by 3rd Derivative Spatial Filtering. Vision Res. 49, 1886–1893 (2009)
2. Eriksson, E.S.: A Field Theory of Visual Illusions. Br. J. Psychol. 61, 451–466 (1970)
3. Ando, R., Kudo, H., Ohnishi, N.: An Examination of the Distribution of Induction Field in Vision for Negative or Positive Type Images. Vision: The Journal of the Vision Society of Japan 18, supplement for ACV 2006, 140 (2006)
4. Bekesy, G.V.: Mach- and Hering-Type Lateral Inhibition in Vision. Vision Res. 8, 1483–1499 (1968)

Robust Controller for Flexible Specifications Using Difference Signals and Competitive Associative Nets

Weicheng Huang, Shuichi Kurogi*, and Takeshi Nishida

Kyusyu Institute of technology, Tobata, Kitakyushu, Fukuoka 804-8550, Japan
{ko@kurolab2.,kuro@,nishida@}cntl.kyutech.ac.jp
http://kurolab2.cntl.kyutech.ac.jp/

Abstract. This paper describes a robust controller using difference signals and multiple competitive associative nets (CAN2s) and flexibly applicable to different control specifications or objectives. Using difference signals of a plant to be controlled, the CAN2 is capable of leaning piecewise Jacobian matrices of nonlinear dynamics of the plant as well as the control objective. By means of employing the GPC (generalized predictive controller), a robust control method to switch multiple CAN2s to cope with plant parameter change is introduced. We show the effectiveness of the present method via numerical experiments of a crane system.

Keywords: Robust controller, Flexible control specifications, Switching of multiple CAN2, Difference signals, Jacobian matrix, Control of nonlinear plant.

1 Introduction

This paper describes a robust controller using difference signals and competitive associative nets (CAN2s) and flexibly applicable to different control specifications or objectives. Here, the CAN2 is an artificial neural net for learning an efficient piecewise linear approximation of nonlinear functions by means of competitive and associative schemes[1,2]. The effectiveness is shown in several areas such as plant control, function approximation, rainfall estimation, time series prediction, and so on [3,4,5]. As an application of the CAN2, we are developing control methods using CAN2 for learning and utilizing piecewise linear models of nonlinear and time varying plant dynamics, and we have presented a robust control method using multiple CAN2s for learning and controlling a nonlinear plant whose parameter values may change [5]. We have clarified the method from the point of view of Jacobian matrices of the nonlinear plant, and shown the effectiveness through numerical experiments of the control of an overhead traveling crane model.

In this paper, we show an improvement of the method to flexibly cope with the following practical cases. Namely, for the crane control system, we sometimes want to reduce overshoot with acceptable settling time, while we sometimes want to reduce settling time with acceptable overshoot, and we would like to change the control objectives

---

* Corresponding author.

T. Huang et al. (Eds.): ICONIP 2012, Part V, LNCS 7667, pp. 50–58, 2012.
© Springer-Verlag Berlin Heidelberg 2012

case by case. Although this task can be done by the conventional robust controllers after tuning the control parameters for each specification, the tuning process for all possible changes of plant parameters may be complicated.

In the next section, we show the formulation of the present controller using difference signals and multiple CAN2s, where an improvement from the previous method [5] is shown in Sect. 2.4 for obtaining the CAN2s for practical specifications. And then, in Sect. 3, we examine the effectiveness of the method through numerical experiments applied to a nonlinear crane system involving changeable parameter values.

2 Predictive Controller Using Difference Signals and CAN2s

We formulate the controller using difference signals and multiple CAN2s to cope with the parameter change and the specification change.

2.1 Plant Model Using Difference Signals

Suppose a plant to be controlled at a discrete time $j = 1, 2, \cdots$ has the input $u_j^{[p]}$ and the output $y_j^{[p]}$. Here, the superscript "[p]" indicates the variable related to the plant for distinguishing the position of the load, (x, y), shown below. Furthermore, we suppose the dynamics of the plant is given by

$$y_j^{[p]} = f(\boldsymbol{x}_j^{[p]}) + d_j^{[p]} , \tag{1}$$

where $f(\cdot)$ is a nonlinear function which may change slowly in time and $d_j^{[p]}$ represents zero-mean noise with the variance σ_d^2. The input vector $\boldsymbol{x}_j^{[p]}$ of the function consists of the input and output sequences of the plant as $\boldsymbol{x}_j^{[p]} \triangleq \left(y_{j-1}^{[p]}, \cdots, y_{j-k_y}^{[p]}, u_{j-1}^{[p]}, \cdots, u_{j-k_u}^{[p]} \right)^T$, where k_y and k_u are positive integers, and the dimension of $\boldsymbol{x}_j^{[p]}$ is $k = k_y + k_u$. Then, for the difference signals $\Delta y_j^{[p]} \triangleq y_j^{[p]} - y_{j-1}^{[p]}$, $\Delta u_j^{[p]} \triangleq u_j^{[p]} - u_{j-1}^{[p]}$, and $\Delta \boldsymbol{x}_j^{[p]} \triangleq \boldsymbol{x}_j^{[p]} - \boldsymbol{x}_{j-1}^{[p]}$, we have the relationship $\Delta y_j^{[p]} \simeq \boldsymbol{J}_f \Delta \boldsymbol{x}_j^{[p]}$ for small $\| \Delta \boldsymbol{x}_j^{[p]} \|$, where $\boldsymbol{J}_f = \partial f(x)/\partial x \big|_{x=\boldsymbol{x}_{j-1}^{[p]}}$ indicates the Jacobian matrix (row vector). If $\boldsymbol{J}_f$ does not change for a while after the time j, then we can predict $\Delta y_{j+l}^{[p]}$ by

$$\widehat{\Delta y}_{j+l}^{[p]} = \boldsymbol{J}_f \widetilde{\Delta \boldsymbol{x}}_{j+l}^{[p]} \tag{2}$$

for $l = 1, 2, \cdots$, recursively. Here, the elements of $\widetilde{\Delta \boldsymbol{x}}_{j+l}^{[p]} = (\widetilde{\Delta y}_{j+l-1}^{[p]}, \cdots, \widetilde{\Delta y}_{j+l-k_y}^{[p]}, \widetilde{\Delta u}_{j+l-1}^{[p]}, \cdots, \widetilde{\Delta y}_{j+l-k_u}^{[p]})^T$ are the past and the predictive input and output given by

$$\widetilde{\Delta y}_{j+m}^{[p]} = \begin{cases} \Delta y_{j+m}^{[p]} & \text{for } m < 1 \\ \widehat{\Delta y}_{j+m}^{[p]} & \text{for } m \geq 1 \end{cases} \quad \text{and} \quad \widetilde{\Delta u}_{j+m}^{[p]} = \begin{cases} \Delta u_{j+m}^{[p]} & \text{for } m < 0 \\ \widehat{\Delta u}_{j+m}^{[p]} & \text{for } m \geq 0. \end{cases} \tag{3}$$

Here, see Sect. 2.3 for the predictive input $\widehat{\Delta u}_{j+m}^{[p]}$ $(m \geq 0)$. Then, we have the prediction of the plant output from the predictive difference signals as

$$\widehat{y}_{j+l}^{[p]} = y_j^{[p]} + \sum_{m=1}^{l} \widehat{\Delta y}_{j+m}^{[p]}. \tag{4}$$

2.2 CAN2 for Learning Nonlinear Plant Involving Parameter Change

A CAN2 has N units. The ith unit has a weight vector $\boldsymbol{w}_i \triangleq (w_{i1}, \cdots, w_{ik})^T \in \mathbb{R}^{k \times 1}$ and an associative matrix (row vector) $\boldsymbol{M}_i \triangleq (M_{i1}, \cdots, M_{ik}) \in \mathbb{R}^{1 \times k}$ for $i \in I = \{1, 2, \cdots, N\}$ (see Fig. 1(a)).

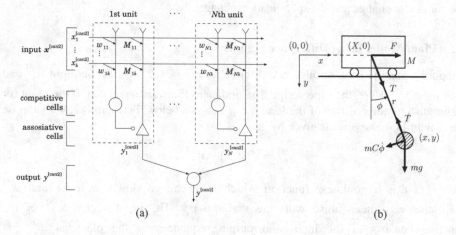

(a)   (b)

Fig. 1. Schematic diagram of (a) CAN2 and (b) overhead traveling crane system

In order for the CAN2 to learn a given dataset $D^n = \{(\boldsymbol{x}_j^{[p]}, y_j^{[p]}) \mid j = 1, 2, \cdots, n\}$ obtained from the plant, we feed the input and output of the CAN2 as $(\boldsymbol{x}^{[can2]}, y^{[can2]}) = (\Delta \boldsymbol{x}_j^{[p]}, \Delta y_j^{[p]})$. Then, the CAN2 after learning (see [4] for the learning method) is supposed to approximate the output $\Delta y_j^{[p]} = f(\Delta \boldsymbol{x}_j^{[p]})$ by

$$\widehat{\Delta y}_j^{[p]} = \boldsymbol{M}_c \Delta \boldsymbol{x}_j^{[p]}, \tag{5}$$

where the index of the unit, c, is selected by

$$c = \underset{i \in I}{\operatorname{argmin}} \| \Delta \boldsymbol{x}_j^{[p]} - \boldsymbol{w}_i \|^2. \tag{6}$$

Now, let us examine (5) and (6). From (2) and (5), we can see that the associative matrix approximates the Jacobian matrix, or $\boldsymbol{M}_c \simeq \boldsymbol{J}_f$. Furthermore, (5) and (6) indicate an assumption that the associative or the Jacobian matrix can be identified by $\Delta \boldsymbol{x}_j^{[p]}$, although it depends on the original signal $\boldsymbol{x}_{j-1}^{[p]}$ from the definition

$J_f = \partial f(x)/\partial x \big|_{x=x_{j-1}^{[p]}}$. However, when J_f does not change for a while and $\Delta x_{j-m}^{[p]}$ $(m = 1, 2, \cdots)$ involves k $(= \dim \Delta x_j^{[p]})$ linearly independent vectors, we can derive $J_f = \left[\Delta x_{j-1}^{[p]}, \cdots, \Delta x_{j-k}^{[p]}\right]^{-1} \left[\Delta y_{j-1}^{[p]}, \cdots, \Delta y_{j-k}^{[p]}\right]$. This indicates that J_f can be identified by $2k$-dimensional vector, $\Delta x_j'^{[p]} = (\Delta y_{j-1}^{[p]}, \cdots, \Delta y_{j-k-k_y}^{[p]}, \Delta u_{j-1}^{[p]}, \cdots, \Delta u_{j-k-k_u}^{[p]})$ which involves all elements in $\left[\Delta x_{j-1}^{[p]}, \cdots, \Delta x_{j-k}^{[p]}\right]$. Thus, the relation from $\Delta x_j'^{[p]}$ to $\Delta y_j^{[p]}$ becomes a single valued function $\Delta y_j^{[p]} = J_f \Delta x_j^{[p]} = g(\Delta x_j'^{[p]})$. Then, considering the linear approximation $\Delta y_j^{[p]} = g(\Delta x_j'^{[p]}) \simeq J_g \Delta x_j'^{[p]}$, the relation from $\Delta x_j'^{[p]}$ to J_g is a vector-valued function. Therefore, the Jacobian J_f for the enlarged vector $\Delta x_j^{[p]} = (\Delta y_{j-1}^{[p]}, \cdots, \Delta y_{j-k_y'}^{[p]}, \Delta u_{j-1}^{[p]}, \cdots, \Delta u_{j-k_u'}^{[p]})$ with $k_y' = k+k_y$ and $k_u' = k+k_u$ is a vector-valued function of $\Delta x_j^{[p]}$, or $J_f = J_f(\Delta x_j^{[p]})$. Thus, let's use this enlarged vector $\Delta x_j^{[p]}$ below from here, so that the control using $\Delta y_{j+l}^{[p]}$ predicted from $\Delta x_j^{[p]}$ is expected to be robust to the parameter change of the plant because parameter values are reflected by J_f which can be identified by $\Delta x_j^{[p]}$. Furthermore, note that $\Delta x_j^{[p]}$ involves the differential control trajectory $\Delta y_{j-l}^{[p]}$ and $\Delta u_{j-l}^{[p]}$ for $l = 1, 2, \cdots$, and then J_f is supposed to be able to memorize the control trajectory for a certain control objective (see Sect. 3.3 for details).

2.3   GPC for Difference Signals

The GPC (Generalized Predictive Control) is an efficient method for obtaining the predictive input $\hat{u}_j^{[p]}$ which minimizes the control performance index [6]:

$$J = \sum_{l=1}^{N_y} (r_{j+l}^{[p]} - \hat{y}_{j+l}^{[p]})^2 + \lambda_u \sum_{l=1}^{N_u} (\widehat{\Delta u}_{j+l-1}^{[p]})^2, \qquad (7)$$

where $r_{j+l}^{[p]}$ and $\hat{y}_{j+l}^{[p]}$ are desired and predictive output, respectively. The parameters N_y, N_u and λ_u are constants to be designed for the control performance. We obtain $\hat{u}_j^{[p]}$ by means of the GPC method as follows; the CAN2 at a discrete time j can predict $\Delta y_{j+l}^{[p]}$ by (2) and then $\hat{y}_{j+l}^{[p]}$ by (4). Then, owing to the linearity of these equations, the above performance index is written as

$$J = \|r^{[p]} - G\Delta u^{[p]} - \bar{y}^{[p]}\|^2 + \lambda_u \|\widehat{\Delta u}\|^2 \qquad (8)$$

where $r^{[p]} = \left(r_{j+1}^{[p]}, \cdots, r_{j+N_y}^{[p]}\right)^T$ and $\widehat{\Delta u}^{[p]} = \left(\widehat{\Delta u}_j^{[p]}, \cdots, \widehat{\Delta u}_{j+N_u-1}^{[p]}\right)^T$. Furthermore, $\bar{y}^{[p]} = \left(\bar{y}_{j+1}^{[p]}, \cdots, \bar{y}_{j+N_y}^{[p]}\right)^T$ and $\bar{y}_{j+l}^{[p]}$ is the natural response $\hat{y}_{j+l}^{[p]}$ of the system (1) for the null incremental input $\widehat{\Delta u}_{j+l}^{[p]} = 0$ for $l \geq 0$. Here, we actually have $\bar{y}_{j+l}^{[p]} = y_j^{[p]} + \sum_{m=1}^{l} \overline{\Delta y}_{j+m}^{[p]}$ from (4), where $\overline{\Delta y}_{j+l}^{[p]}$ denotes the natural response of the difference system of (2) with J_f replaced by M_c. The ith column and the jth row of the matrix G is given by $G_{ij} = g_{i-j+N_1}$, where g_l for $l = \cdots, -2, -1, 0, 1, 2, \cdots$ is the unit step response $y_{j+l}^{[p]}$ of (4) for $\widehat{y}_{j+l}^{[p]} = \widehat{u}_{j+l}^{[p]} = 0 \; (l < 0)$ and $\widehat{u}_{j+l}^{[p]} = 1(l \geq 0)$. It is easy to derive that the unit response g_l of (4) is obtained as the impulse response of (2). Then, we have $\widehat{\Delta u}^{[p]}$ which minimizes J by $\widehat{\Delta u}^{[p]} = (G^T G + \lambda_u I)^{-1} G^T (r^{[p]} - \overline{y}^{[p]})$, and then we have $\widehat{u}_j^{[p]} = u_{j-1}^{[p]} + \widehat{\Delta u}_j^{[p]}$.

2.4   Control and Training Iterations to Obtain CAN2 for Multiple Specifications

To improve the control performance, we execute iterations of the following phases.

(i) **control phase:** Control the plant by a default control schedule at the first iteration, and by the above GPC using the CAN2 obtained at the previous iteration otherwise.
(ii) **training phase:** Using the dataset $D^n = \{(x_j^{[p]}, y_j^{[p]} | j = 1, 2, \cdots, n)\}$ obtained from the control phase, apply the batch learning method to the CAN2 [4].

The control performance at an iteration depends on the CAN2 obtained at the previous iteration. So, for the actual control of the plant, we use the best CAN2 obtained through a number of iterations. Here, we would like to cope with the following practical cases. Namely, in the crane control situations, we sometimes want to reduce overshoot with acceptable settling time, while we sometimes want to reduce settling time with acceptable overshoot. For such cases, we store and selectively use the best CAN2 for the specification of each case.

2.5   Switching Multiple CAN2s For Robustness to Parameter Change

To cope with parameter change of the plant, we employ the following steps using multiple CAN2s, each of which, we denote $\text{CAN2}^{[\theta_s]}$, is the best CAN2 obtained through the control and training iterations for parameter θ_s and $s \in S = \{1, 2, \cdots\}$.

step 1: At the time j in the control phase, select the unit for each $\text{CAN2}^{[\theta_s]}$ by (6), or $c^{[s]} = \underset{i \in I}{\operatorname{argmin}} \| \Delta x_j^{[p]} - w_i^{[s]} \|^2$, where $w_i^{[s]} \; (i \in I)$ are the weight vectors of $\text{CAN2}^{[\theta_s]}$.

step 2: Select the s^*th CAN2 which has the minimum MSE (mean square prediction error) for the recent N_e outputs,

$$s^* = \underset{s \in S}{\operatorname{argmin}} \frac{1}{N_e} \sum_{l=0}^{N_e-1} \| \Delta y_{j-l}^{[p]} - M_{c^{[s]}} \Delta x_{j-l}^{[p]} \|^2 , \tag{9}$$

where $M_{c^{[s]}}$ is the $c^{[s]}$th associative matrix of $\text{CAN2}^{[\theta_s]}$.

Note that **step 2** is necessary because $c^{[s]}$ obtained by **step 1** indicates the optimal unit only for the Voronoi partition of the sth CAN2 and different CAN2 has different Voronoi partition. For evaluate the fitness to the recent data, the above MSE seems to be a reasonable criterion.

3 Numerical Experiments of Crane System

In order to examine the effectiveness of the present method, we execute numerical experiments of the following crane system.

3.1 Overhead Traveling Crane System

We consider the overhead traveling crane system shown in Fig. 1(b). From the figure, we have the position and motion equations given by

$$(x, y) = (X + r \sin \phi, r \cos \phi) \tag{10}$$

$$m(\ddot{x}, \ddot{y}) = (-T \sin \phi - mC\dot{\phi} \cos \phi, mg - T \cos \phi - mC\dot{\phi} \sin \phi) \tag{11}$$

$$M\ddot{X} = F + T \sin \phi \tag{12}$$

where (x, y) and m are the position and the weight of the suspended load, $(X, 0)$, M and F are the position, weight and driving force of the trolley, r and ϕ are the length and the angle of the rope, T is the tension of the rope, and C is the viscous damping coefficient. From (10) and (11), we have the nonlinear second order differential equation of ϕ given by $r\ddot{\phi} + (C + 2\dot{r})\dot{\phi} + g \sin \phi + \ddot{X} \cos \phi = 0$. Thus, with (12), the transition of the state $x = \left(\phi, \dot{\phi}, X, \dot{X}\right)^T$ is given by

$$\dot{x} = h(x) = \begin{bmatrix} \dot{\phi} \\ -\dfrac{C + 2\dot{r}}{r}\dot{\phi} - \dfrac{g}{r} \sin \phi - \dfrac{F + T \sin \phi}{rM} \cos \phi \\ \dot{X} \\ \dfrac{F + T \sin x_1}{M} \end{bmatrix}, \tag{13}$$

where $T = m\sqrt{(\ddot{x} + C\dot{\phi} \cos \phi)^2 + (\ddot{y} - g + C\dot{\phi} \sin \phi)^2}$ is also a function of x. The control objective is to move the horizontal position of the load, $x = X + r \sin \phi$, to a destination position x_d by means of operating F.

3.2 Parameter Settings

The parameter values of the crane system are set as follows: the trolley weight $M = 100$kg, the damping coefficient $C = 0.5$m/s, the maximum driving force $F_{\max} = 10$N, and we have examined the robustness to the rope length r for $2, 3, \cdots, 10$ [m], and the load weight m for $10, 20, \cdots, 100$ [kg]. We obtain the discrete signals by $u_j^{[p]} = F(jT_v)$ and $y_j^{[p]} = x(jT_v)$ with $T_v = 0.5$s. Here, we use the virtual sampling method shown in [3], where the discrete model is obtained with the virtual sampling period T_v while the observation and operation are executed with shorter actual sampling period $T_a = 0.01$s. We use $k_y' = k_u' = 4$ for the enlarged input vector $\Delta x_j^{[p]}$, and $N_y = 20$, $N_u = 1$ and $\lambda_u = 0.01$ for the GPC. We used $N_e = 8$ samples for switching multiple CAN2s. Let crane$^{[\theta]}$ denote the crane with the parameter θ. Especially, let

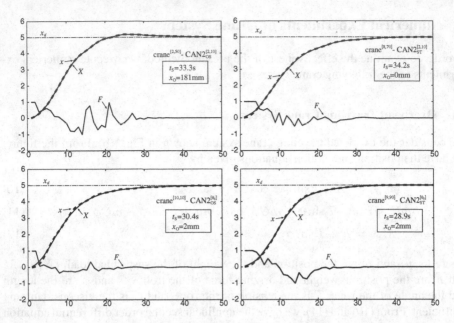

Fig. 2. Examples of resultant time course of x[m], X[m] and F[10N] for the control using $CAN2_{OS}^{[2,10]}$ (top left), $CAN2_{ST}^{[2,10]}$ (top right), $CAN2_{OS}^{[\theta_S]}$ (bottom left) and $CAN2_{ST}^{[\theta_S]}$ (bottom right). These show the worst results from the point of view of their control objectives to reduce overshoot (left side) and settling time (right side), respectively, and see Table 1 for details.

$\theta = \theta_s$ for $s = 1, 2, 3, 4$ denote $\theta = (r, m) = (2, 10), (2, 100), (10, 10)$ and $(10, 100)$, respectively, which are used for training CAN2s. Let $CAN2_{OS}^{[\theta_s]}$ and $CAN2_{ST}^{[\theta_s]}$ denote the CAN2 with the smallest overshoot and settling time, respectively, obtained through 20 iterations of control and training for crane$^{[\theta_s]}$. We have used $N = 20$ units for each CAN2. Let $CAN2_{OS}^{[\theta_s]}$ and $CAN2_{ST}^{[\theta_s]}$, respectively, denote the set of all $CAN2_{OS}^{[\theta_s]}$ and $CAN2_{ST}^{[\theta_s]}$ for $s \in S = \{1, 2, 3, 4\}$ used for the switching controller.

3.3 Results and Remarks

Examples of resultant time courses of x, X and F are shown in Fig. 2, where we can see that the controller using $CAN2_{OS}^{[2,10]}$ and $CAN2_{ST}^{[2,10]}$ show worse results than the switching controller using $CAN2_{OS}^{[\theta_s]}$ and $CAN2_{ST}^{[\theta_s]}$ from the point of view of their control objective to reduce overshoot and settling time, respectively..

A statistical summary of the resultant overshoot and settling time are shown in Table 1. From the column of "trained θ_i" of "overshoot", we can see that single $CAN2_{OS}^{[\theta_i]}$ ($i = 1, 2, 3, 4$) has achieved the control with 0mm overshoot. Similarly, from the column corresponding to "trained θ_i" of "settling time", we can see that $CAN2_{ST}^{[\theta_i]}$ ($i = 1, 2, 3, 4$) has achieved smaller settling time than the corresponding $CAN2_{OS}^{[\theta_i]}$. From the columns of "mean", "min", "max" and "std" of "test", we can see that the statistical properties of the control result for the cranes with 90 combinations

Table 1. Experimental result of overshoot and settling time. The columns of "trained θ_i" indicate the result by the controller using $\mathrm{CAN2}_{\mathrm{OS}}^{[\theta_i]}$ and $\mathrm{CAN2}_{\mathrm{ST}}^{[\theta_i]}$ applied to the crane with the corresponding parameters $\theta_1 = (2, 10)$, $\theta_2 = (10, 10)$, $\theta_3 = (2, 100)$ and $\theta_4 = (10, 100)$, respectively. The columns of "mean", "min", "max" and "std" of "test" for "overshoot" and "settling time" indicate the minimum, maximum and standard deviation for the crane control of all 90 combinations of parameter values (r, m) for $r = 2, 3, \cdots, 10$[m] and $m = 10, 20, \cdots, 100$[kg]. The rows of $\mathrm{CAN2}_{\mathrm{OS}}^{\theta_S}$ and $\mathrm{CAN2}_{\mathrm{ST}}^{\theta_S}$ indicate the result of the switching controller for $\theta_S = \{\theta_1, \theta_2, \theta_3, \theta_4\}$. The thick and the italic figures indicate the best (smallest) and the worst (biggest) results in each column block.

CAN2 used for the controller	overshoot [mm]					settling time [s]				
	trained θ_i	test				trained θ_i	test			
		mean	min	max	std		mean	min	max	std
$\mathrm{CAN2}_{\mathrm{OS}}^{[\theta_1]}$	0	2.3	0	181	19.0	**31.0**	30.7	20.8	*35.6*	3.8
$\mathrm{CAN2}_{\mathrm{OS}}^{[\theta_2]}$	0	0.4	0	30	3.1	42.6	*47.6*	27.9	*88.8*	*8.3*
$\mathrm{CAN2}_{\mathrm{OS}}^{[\theta_3]}$	0	*285.7*	0	*1043*	*255.4*	31.2	33.2	**16.6**	54.9	7.6
$\mathrm{CAN2}_{\mathrm{OS}}^{[\theta_4]}$	0	26.6	0	97	28.5	36.5	37.2	27.9	52.0	3.4
$\mathrm{CAN2}_{\mathrm{OS}}^{[\theta_S]}$	—	**0.02**	0	**2**	**0.2**	—	32.0	25.0	37.7	**2.4**
$\mathrm{CAN2}_{\mathrm{ST}}^{[\theta_1]}$	7	16.6	0	182	32.8	19.7	26.0	**17.1**	34.2	4.7
$\mathrm{CAN2}_{\mathrm{ST}}^{[\theta_2]}$	5	*86.3*	5	*246*	*55.8*	22.5	26.6	20.2	*47.0*	*7.1*
$\mathrm{CAN2}_{\mathrm{ST}}^{[\theta_3]}$	8	39.2	0	179	38.7	**18.9**	**19.6**	17.3	**26.6**	1.7
$\mathrm{CAN2}_{\mathrm{ST}}^{[\theta_4]}$	2	12.5	1	53	10.7	26.4	26.6	22.4	32.7	2.1
$\mathrm{CAN2}_{\mathrm{ST}}^{[\theta_S]}$	—	**8.0**	0	47	7.9	—	23.9	21.0	28.9	**1.3**

of parameter values. From the control objective, the "max" overshoot for $\mathrm{CAN2}_{\mathrm{OS}}^{[\theta_i]}$ and the "max" settling time for $\mathrm{CAN2}_{\mathrm{ST}}^{[\theta_i]}$ should be small as much as possible, but they do not always have small values. On the other hand, the result of the switching controller using $\mathrm{CAN2}_{\mathrm{OS}}^{[\theta_S]}$ and $\mathrm{CAN2}_{\mathrm{ST}}^{[\theta_S]}$ shows better properties. Namely, the "max" overshoot for $\mathrm{CAN2}_{\mathrm{OS}}^{[\theta_S]}$ is 2mm with "std" being 0.2 and the "max" settling time for $\mathrm{CAN2}_{\mathrm{ST}}^{[\theta_S]}$ is 28.9mm with "std" being 1.3, which indicates better property than the result for the above single CAN2s. Furthermore, the overshoot for $\mathrm{CAN2}_{\mathrm{ST}}^{[\theta_S]}$ and the settling time for $\mathrm{CAN2}_{\mathrm{OS}}^{[\theta_S]}$ have smaller "std" values than the corresponding single CAN2s, which is a desirable property.

The above robust and stable properties of the switching controller for the crane with 90 wide-range test parameter values is considered to be owing to the switching method shown in Sect. 2.5 which selects the best $M_{c[s*]} \simeq J_f$ for predicting $\Delta y_{j+l}^{[p]}$ at each time j, which is capable because J_f is a function of $\Delta x_j^{[p]}$ depending on θ as shown in Sect. 2.2. Furthermore, for different control specifications, we can apply the same controller by selecting CAN2s for each control objective with the same predictive control parameters of (7). This is considered to be possible because $\Delta x_j^{[p]}$ involves the control trajectory depending on the control objective, and then the trained and selected $M_{c[s*]}$ reflects not only the Jacobian of the plant but also the control objective. However, this may not be adequate for linear plants, so that the analysis of the applicability of the present method is remained for our future research studies.

4   Conclusion

We have presented a robust controller using difference signals and multiple CAN2s and flexibly applicable to different specifications. The CAN2 using difference signals of a plant is shown to be able to learn Jacobian matrices of nonlinear dynamics of a plant as well as the control trajectory. A robust control method using GPC is constructed to switch multiple CAN2s to cope with plant parameter change. Via numerical experiments of a crane system, we have shown the effectiveness of the present method.

Acknowledgements. This work was partially supported by the Grant-in Aid for Scientific Research (C) 24500276 of the Japanese Ministry of Education, Science, Sports and Culture.

References

1. Kohonen, T.: Associative Memory. Springer (1977)
2. Ahalt, A.C., Krishnamurthy, A.K., Chen, P., Melton, D.E.: Competitive Learning Algorithms for Vector Quantization. Neural Networks 3, 277–290 (1990)
3. Kurogi, S., Nishida, T., Sakamoto, T., Itoh, K., Mimata, M.: A simplified Competitive Associative Net and a Model-Switching Predictive Controller for Temperature Control of Chemical Solutions. In: Proceedings of International Conference on Neural Information Processing, pp. 791–796 (2000)
4. Kurogi, S., Sawa, M., Ueno, T., Fuchikawa, Y.: A Batch Learning Method for Competitive Associative Net and Its Application to Function Approximation. In: Proceedings of SCI 2004, vol. 5, pp. 24–28 (2004)
5. Kurogi, S., Yuno, H., Nishida, T., Huang, W.: Robust Control of Nonlinear System Using Difference Signals and Multiple Competitive Associative Nets. In: Lu, B.-L., Zhang, L., Kwok, J. (eds.) ICONIP 2011, Part III. LNCS, vol. 7064, pp. 9–17. Springer, Heidelberg (2011)
6. Clarki, D.W., Mohtadi, C.: Properties of Generalized Predictive Control. Automatica 256, 859–875 (1989)

Moments of Predictive Deviations as Ensemble Diversity Measures to Estimate the Performance of Time Series Prediction

Kohei Ono, Shuichi Kurogi*, and Takeshi Nishida

Kyushu Institute of technology, Tobit, Kitakyushu, Fukuoka 804-8550, Japan
{ko@kurolab2.,kuro@,nishida@}cntl.kyutech.ac.jp
http://kurolab2.cntl.kyutech.ac.jp/

Abstract. This paper presents an analysis of moments of predictive deviations as measures of ensemble diversity to estimate the performance of time series prediction. As an extension of the ambiguity decomposition of bagging ensemble, we decompose the fourth power of ensemble prediction error and clarify the effect of the moments of predictive deviations of ensemble members to the ensemble prediction error. We utilize this analysis for estimating the performance of time sires prediction, and show the effectiveness by means of numerical experiments.

Keywords: Moments of predictive deviations, Ensemble diversity measures, Performance estimation, Time series predictive.

1 Introduction

This paper presents an analysis of moments of predictive deviations of ensemble members as measures of ensemble diversity to estimate the performance of time series prediction. Here, the diversity representing the degree of disagreement involed in the ensemble has been well analyzed to apply to ensemble learning algorithms such as NC (negative-correlation) learning, where for training dataset with known target values, an appropreate tuning of the trade-off between the minimization of each prediction error and the maximization of the covariance of prediction errors within the ensemble is shown to give a better performance (see [1,2]). As an extension of the above analysis based on the ambiguity decomposition of the square error of ensemble prediction, we first examine the fourth power of ensemble prediction error for the dataset with unknown target values and clarify the effect of the moments of predictive deviations for estimating the prediction error.

Next, we try to apply the above analysis for estimating the performance of multistep prediction of time series whose target values are unknown. Here, note that the performance estimation of single step prediction is the same as that of regression problem, for which we have already analyzed to show that the out-of-bag estimate is effective[3]. Although we can imagine that the model parameter of learning machines for single- and multi-step prediction is different, its quantitative difference has not been clarified yet.

---

* Corresponding author.

T. Huang et al. (Eds.): ICONIP 2012, Part V, LNCS 7667, pp. 59–66, 2012.
© Springer-Verlag Berlin Heidelberg 2012

By means of numerical experiments in this paper, we examine the quantitative effect of the moments of predictive deviations to multistep prediction, and apply it to model selection for good prediction of time series. As a result of comparison with the conventional model selection method based on the holdout estimate for time series, we show that the present method has better performance.

In the next section, we show the notation of bagging, followed by the introduction of error decompsition of bagging ensemble prediction error and the moments of predictive deviations, and then describes multistep prediction of time series. In **3**, we show the results of numerical experiments and the effectiveness of the present analysis.

2   Bagging, Diversity and Time Series Prediction

2.1   Bagging for Regression Problem

Let $D^n \triangleq \{(x_i, y_i) | i \in I^n\}$ be a training data set, where $x_i \triangleq (x_{i1}, x_{i2}, \cdots, x_{Ike})^T$ and y_i denote an input vector and the target value, respectively, and $I^n \triangleq \{1, 2, \cdots, n\}$. We suppose the relationship given by

$$y_i \triangleq r_i + e_i = r(x_i) + e_i \ , \tag{1}$$

where $r_i \triangleq r(x_i)$ is a nonlinear target function of x_i, and e_i represents zero-mean noise with the variance σ_e^2.

We introduce the bagging ensemble as follows; let $D^{n\alpha^\sharp, j} = \{(x_i, y_i) | i \in I^{n\alpha^\sharp, j})\}$ be the jth bag (multiset, or bootstrap sample set) involving $n\alpha$ elements, where the elements in $D^{n\alpha^\sharp, j}$ are resampled randomly with replacement from the training dataset D^n. Here, α (> 0) indicates the bag size ratio to the given dataset, and $j \in J^{\text{bag}} \triangleq \{1, 2, \cdots, b\}$. Here, note that $\alpha = 1$ is used in many applications (see [4,5]), but we use variable α for improving generalization performance (see [3] for the effectiveness and the validity). With multiple learning machines θ^j ($\in \Theta^{\text{bag}} \triangleq \{\theta^j | j \in J^{\text{bag}}\}$) which have learned $D^{n\alpha^\sharp, j}$, the bagging for estimating the target value $r_i = r(x_i)$ is done by

$$\hat{y}_i^{\text{bag}} \triangleq \hat{y}^{\text{bag}}(x_i) \triangleq \frac{1}{b} \sum_{j \in J^{\text{bag}}} \hat{y}_i^j \equiv \left\langle \hat{y}_i^j \right\rangle_{j \in J^{\text{bag}}} \ , \tag{2}$$

where $\hat{y}_i^j \triangleq \hat{y}^j(x_i)$ denotes the prediction by the jth machine θ^j. The angle brackets $\langle \cdot \rangle$ indicate the mean, and the subscript $j \in J^{\text{bag}}$ indicates the range of the mean. For simple expression, we sometimes use $\langle \cdot \rangle_j$ instead of $\langle \cdot \rangle_{j \in J^{\text{bag}}}$ in the following.

2.2   Error Decomposition and Moments of Predictive Deviations

To analyze the error of bagging ensemble prediction, we have examined the bias-variance decomposition and the ambiguity decomposition [1], and we here show a slightly different formulation in order to deal with unknown target values as follows. Firstly, we denote $y_i^j = r_i + \beta_i + \epsilon_i^j$, where $\beta_i = \langle y_i^j \rangle_j - r_i$ represents the bias, and

$\epsilon_i^j = y_i^j - \beta_i$ the predictive deviation. Then, we have the mean square error of the predictions for all bags to the training target value as

$$\left\langle (\hat{y}_i^j - y_i)^2 \right\rangle_j = \left\langle (\beta_i + \epsilon_i^j - e_i)^2 \right\rangle_j = (\beta_i)^2 + \left\langle (\epsilon_i^j)^2 \right\rangle_j - 2\beta_i e_i + e_i^2 , \quad (3)$$

and the square error of the bagging prediction to the true target value, which we sometimes call generalization error in the following, as

$$(\hat{y}_i^{\text{bag}} - r_i)^2 = (\beta_i)^2 = \left\langle (\hat{y}_i^j - y_i)^2 \right\rangle_j - \left\langle (\epsilon_i^j)^2 \right\rangle_j + 2\beta_i e_i - e_i^2 . \quad (4)$$

Here, (3) corresponds to the bias-variance decomposition and (4) the ambiguity decomposition, where the variance term $\langle (\epsilon_i^j)^2 \rangle_j$ is called ambiguity as a measure of diversity. Differently from the decompositions shown in [1], the above decompositions show the effect of overfitting $\beta_i e_i$ which should be taken into account for reducing the generalization error although it is hard to be estimated. Intuitively from the ambiguity decomposition, larger variance is supposed to reduce the generalization error much more. Furthermore, since we can obtain only the variance term when predicting unknown y_i, we expect that the variance is useful to estimate the generalization error. However, the variance has no relationship with the generalization error in our experiments shown below, which is of course owing that the first term may cancel the effect of the variance and it is well known that the variance becomes bigger as the complexity of the learning model increases.

So, we decompose the fourth power of the error as follows:

$$(\hat{y}_i^{\text{bag}} - r_i)^4 = C - \left\langle (\epsilon_i^j)^4 \right\rangle_j - 4(\hat{y}_i^{\text{bag}} - y_i) \left\langle (\epsilon_i^j)^3 \right\rangle_j - 6(\hat{y}_i^{\text{bag}} - y_i)^2 \left\langle (\epsilon_i^j)^2 \right\rangle_j , \quad (5)$$

where C represents the sum of the terms which do not explicitly involve ϵ_i^j, and note that $\hat{y}_i^{\text{bag}} - y_i = \beta_i - e_i$ involves unknown y_i. Then, in order to reduce the right hand side for a constant C, both $\langle (\epsilon_i^j)^4 \rangle_j$ and $\langle (\epsilon_i^j)^2 \rangle_j$ should be larger while $|\langle (\epsilon_i^j)^3 \rangle_j|$ should be smaller and larger for the corresponding terms being negative and positive, respectively. Now, to evaluate the relationship to the generalization error without any dependency among these terms, we examine the following moments of predictive deviations, i.e. the skew S_i and the kurtosis K_i as well as the variance V_i,

$$V_i \triangleq \sigma_i^2 \triangleq \langle (\epsilon_i^j)^2 \rangle_j , \quad S_i \triangleq \frac{\langle (\epsilon_i^j)^3 \rangle_j}{\sigma_i^3} , \quad K_i \triangleq \frac{\langle (\epsilon_i^j)^4 \rangle_j}{\sigma_i^4} . \quad (6)$$

For all test data to be predicted, we use mean variance (MV) $\langle V_i \rangle_i B!(B$ mean absolute skew (MAS) $\langle |S_i| \rangle_i$ and mean kurtosis (MK) $\langle K_i \rangle_i$, where not $\langle S_i \rangle_i$ but $\langle |S_i| \rangle_i$ is used to adapt unknown polarity of $(\hat{y}_i^{\text{bag}} - y_i)\langle (\epsilon_i^j)^3 \rangle_i$.

2.3   Time Series Prediction

As explained before, we have developed the above analysis for estimating the performance of multistep prediction of time series, which is formalized as follows. Suppose

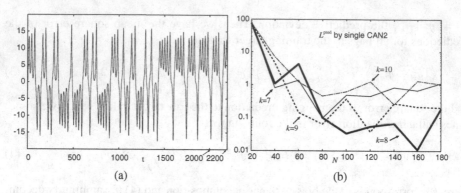

Fig. 1. (a) Lorenz time series and (b) holdout L vs. N for determining k

for a given time series of real values $y(t)$ ($\in \mathbb{R}$) for $t \in T_{0:t_{\text{given}}} \triangleq \{0, 1, 2, \cdots, t_{\text{given}} - 1\}$, we should predict the successive time series $y(t)$ for $t \in T_{t_{\text{given}}:t_{\text{pred}}} \triangleq \{t_{\text{given}}, t_{\text{given}} + 1, \cdots, t_{\text{given}} + t_{\text{pred}} - 1\}$. To solve the problem, we substitute $y_t := y(t)$ and $x_t := (y(t-1), y(t-2), \cdots, y(t-k))^T$ and use $y_t = r(x_t) + e_t$ in (1)), where we regard k as the embedding dimension (see the theory of Chaotic time series [7] for details). Then, the learning and the prediction can be formulated as a regression problem by the bagging ensemble described above, where we execute the multistep prediction $\hat{y}_t = \hat{y}^{\text{bag}}(\hat{x}_t)$ for $t = t_{\text{given}}, t_{\text{given}} + 1, \cdots$ successively with $\hat{x}_t = (x_{t1}, x_{t2}, \cdots, x_{tk})$ involving $x_{tj} = y_{t-j}$ $(t - j < t_{\text{given}})$ and $x_{tj} = \hat{y}_{t-j}$ $(t - j \geq t_{\text{given}})$.

3   Numerical Experiment

For the Lorenz time series (see [7] and Fig. 1(a)) with $t_{\text{given}} = 2000$ given data and $t_{\text{pred}} = 100$ prediction data, or

$$T_{0:2000} = \{0, 1, 2, \cdots, 1999\} \text{ and } T_{2000:100} = \{2000, 2001, \cdots, 2099\},$$

we use CAN2 (see **A.3** and [3] for details) to solve this problem. In this experiment, the signal $y(t)$ does not involve observation error $e(t)$ so that $y(t) = r(t)$ and we evaluate the performance of the prediction $\hat{y}(t)$ by the mean square error,

$$L^{\text{pred}} \triangleq \left\langle (\hat{y}(t) - y(t))^2 \right\rangle_{t \in T^{pred}}, \tag{7}$$

which we call prediction loss in the following.

3.1   Embedding Dimension k

In the following, we show the result for the embedding dimension $k = 8$ determined from Fig. 1(b) or the result of holdout method (see **A.1**) with $t_{\text{train}} = 1900$ training data and $t_{\text{holdout}} = 100$ holdout prediction data. Here, we examined $k = 7, 8, 9, 10$ and $N = 20, 40, \cdots, 180$ by means of single CAN2 because bagging CAN2 is time

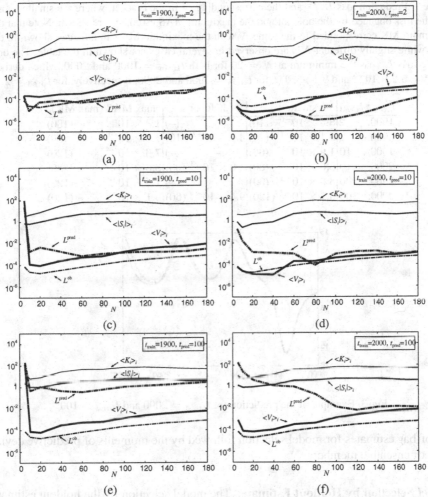

Fig. 2. Experimental results for $t_{\text{train}} = 1900$ (left) and 2000 (right) training data, or $T_{0:1900}$ and $T_{0:2000}$, and the prediction horizons $t_{\text{pred}} = 2$ (top), 10 (middle) and 100 (bottom)

consuming and k to be used is supposed not to depend on learning machines, where N is the number of units of the CAN2 which represents the complexity of the learning model. Furthermore, from Takens' theorem, we may use any of $k > 2d = 6$ for the dimension $d = 3$ of the original Lorentz dynamic system, but to reduce the computational cost we had better use smaller k which achieves an acceptable prediction performance.

3.2 Analysis of Prediction Performance

For $t_{\text{train}} = 1900$ and 2000 training datasets involving $y(t)$ for $t \in T_{0:1900}$ and $T_{0:2000}$, respectively, we have trained the bagging CAN2s with $b = 200$, respectively, and made predictions. We show the prediction performance for three prediction horizons $t_{\text{pred}} = 2, 10, 100$ in Fig. 2. In the following, we examine the conventional holdout and

Table 1. Prediction loss L^{pred} and the corresponding N in the bracket, where the smallest loss, and the loss obtained by the holdout and the maximum MAS estimates are shown. Note that the maximum MK corresponded to the same N with the maximum MAS in all cases shown in the table while slightly different N's are observed in several cases not shown in the table. The out-of-bag loss L^{ob} has the minimum at $N = 20$ for both $t_{\mathrm{train}} = 1900$ and 2000, which derives $L^{\mathrm{pred}} = 5.2 \times 10^{-5}$ and 9.1×10^{-7} for $t_{\mathrm{train}} = 1900$ and 2000, respectively, for $t_{\mathrm{pred}} = 2$.

t_{train}: t_{pred}	smallest		holdout		max MAS (max MK)	
1900:	2	8.5×10^{-6} (10)	–	(–)	1.8×10^{-4}	(100)
2000:	2	9.1×10^{-7} (20)	4.2×10^{-6} (10)		1.0×10^{-6}	(80)
1900:	10	1.9×10^{-4} (60)	–	(–)	7.6×10^{-4}	(120)
2000:	10	8.6×10^{-5} (80)	1.0×10^{-3} (60)		8.6×10^{-5}	(80)
1900:	100	3.6×10^{-1} (60)	–	(–)	1.4×10^{0}	(120)
2000:	100	6.1×10^{-3} (120)	9.4×10^{-1} (60)		6.1×10^{-3}	(120)

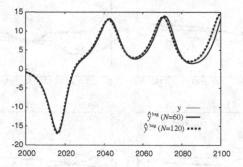

Fig. 3. Example of the prediction for $t_{\mathrm{train}} = 2000$ and $t_{\mathrm{pred}} = 100$

out-of-bag estimates for model selection followed by the moments of predictive deviations of ensemble members.

Model Selection by Holdout Estimate. The model selection by the holdout estimate for this experiment is done by selecting N which achieves the minimum L^{pred} for $T_{1900:t_{\mathrm{pred}}}$ and use it to predict $y(t)$ for $t \in T_{2000:t_{\mathrm{pred}}}$. We show the prediction error obtained by the holdout estimate in Table 1, and we can see that it did not work better than the "max MSE" estimate explained below especially for longer term prediction for $t_{\mathrm{pred}} = 10$ and 100.

Out-Of-Bag Estimate for Short Term Prediction. As explained before, the out-of-bag estimate (see A.2 and [3] for details) has been shown effective in regression problem or single step prediction. From Fig. 2(a) and (b) (see also the caption of Table 1), this estimate seems work for small horizon $t_{\mathrm{pred}} = 2$, but does not work for larger $t_{\mathrm{pred}} = 10$ and 100.

Moments of Predictive Deviations of Ensemble Members. First, from Fig. 2, we can see that the mean variance (MV) $\langle V_i \rangle_i$ has the maximum at the largest $N = 180$

in (a), (b), (d) and (f), and the smallest $N = 5$ in (c) and (e), thus MV cannot be used for estimating the smallest L^{pred}. Next, by means of taking a deep look at L^{pred} and the mean absolute skew (MAS) $\langle |S_i| \rangle_i$, we can see that L^{pred} vs. N has two local minima in Fig. 2(a), (b), (c) and (e), and they have almost the same N with the minimum and the maximum MAS, respectively. Furthermore, in Fig. 2 (d) and (f), the minimum L^{pred} and the maximum MAS seems to have almost the same N. This observation suggests a model selection method using the maximum MAS for large t_{pred}, and the result is shown in Table 1. The result indicates a better performance than the hold out method for all cases for $t_{\text{train}} = 2000$, where the case for $t_{\text{train}} = 2000$ and $t_{\text{pred}} = 2$ does also show a better performance because the maximum MAS corresponds to the second smallest local minima of L although N is very different. In Fig. 3, we show the prediction for $t_{\text{train}} = 2000$ and $t_{\text{pred}} = 100$ obtained by two model selection methods.

As described in Table 1, the maximum of the mean kurtosis (MK) $\langle K_i \rangle_i$ has almost the same N with the maximum MAS. So, we may be able to use the maximum MK for the model selection.

4   Conclusion

We have analyzed to show that the moments of predictive deviations as ensemble diversity measures can be used for estimating the performance of time series prediction. From the fourth power of bagging ensemble prediction error, we have clarified the effect of the moments of predictive deviations of ensemble members to the ensemble prediction error. We have utilized the result of this analysis for estimating the performance of time series prediction, and shown the effectiveness through numerical experiments. However, the relationship between the model complexity N and the moments of predictive deviations has not been clarified enough. Furthermore, the performance of the present method may dend on the problems to be solved, learning machines to be used, and so on, so we would like to analyze the present method much more by means of applying it to many other problems in our future research studies.

Acknowledgements. This work was partially supported by the Grant-in Aid for Scientific Research (C) 24500276 of the Japanese Ministry of Education, Science, Sports and Culture.

References

1. Brown, G., Wyatt, J., Tino, P.: Managing Diversity in Regression Ensembles. J. Mach. Learn. Res. 6, 1621–1650 (2005)
2. Chen, H.: Diversity and Regularization in Neural Network Ensembles. PHD thesis, University of Birmingham (2008)
3. Kurogi, S.: Improving Generalization Performance via out-of-Bag Estimate Using Variable Size of Bags. J. Japan. Neural Network Society 16, 81–92 (2009)
4. Kohavi, R.: A Study of Cross-Validation and Bootstrap for Accuracy Estimation and Model Selection. In: Proceedings of the Fourteenth International Conference 18 on Artificial Intelligence (IJCAI), pp. 1137–1143 (1995)

5. Efron, B., Tbshirani, R.: Improvements on Cross-Validation: the.632+ bootstrap method. J. American Stats. Associ. 92, 548–560 (1997)
6. Breiman, L.: Bagging Predictors. Mach. Learn. 26, 123–140 (1996)
7. Aihara, K.: Theories and Applications of Chaotic Time Series Analysis, Sangyo Tosho, Tokyo (2000)

A   Appendix: Prediction Tools

A.1   Holdout Method for Time Series Prediction

This method is often used to estimate the performance of prediction for unknown time series because it requires only training dataset as follows. For a given set of $y(t)$ for $t \in T_{0:t_{given}}$, a set of $y(t)$ for $t \in T_{t_{train}:t_{holdout}} = \{t_{train}, t_{train} + 1, \cdots, t_{given} = t_{train} + t_{holdout} - 1\}$ are hold out for evaluating the prediction performance while the remaining $y(t)$ for $t \in T_{0:t_{train}} = \{0, 1, 2, \cdots, t_{train} - 1\}$ are used for training.

A.2   Out-of-Bag Estimate for Generalization Error

Out-of-bag estimate is used to estimate the generalization error of bagging ensemble as follows: For the given dataset $D^n = \{(x_i, y_i)|i \in I^n\}$, the out-of-bag prediction error is defined by $L^{ob} \triangleq \langle (\hat{y}_i^{ob} - y_i)^2 \rangle_{i \in I^n}$, where $\hat{y}_i^{ob} \triangleq \langle \hat{y}_i^j \rangle_{j \in J^{ob}}$ is the mean of the out-of-bag predictions $\hat{y}_i^j = \hat{y}^j(x_i)$ by the ensemble members which have learned the bags not involving the ith data (x_i, y_i).

A.3   CAN2

The CAN2 (competitive associative net 2) is a neural net for learning efficient piece-wise linear approximation of nonlinear function by means of the following schemes (See [3] for details): A single CAN2 has N units. The jth unit has a weight vector $w_j \triangleq (w_{j1}, \cdots, w_{jk})^T \in \mathbb{R}^{k \times 1}$ and an associative matrix (or a row vector) $M_j \triangleq (M_{j0}, M_{j1}, \cdots, M_{jk}) \in \mathbb{R}^{1 \times (k+1)}$ for $j \in I^N \triangleq \{1, 2, \cdots, N\}$. The CAN2 after learning the training dataset $D^n = \{(x_i, y_i)|y_i = r(x_i) + e_i, i \in I^n\}$ approximates the target function $r(x_i)$ by $\hat{y}_i = \tilde{y}_{c(i)} = M_{c(i)}\tilde{x}_i$, where $\tilde{x}_i \triangleq (1, x_i^T)^T \in \mathbb{R}^{(k+1) \times 1}$ denotes the (extended) input vector to the CAN2, and $\tilde{y}_{c(i)} = M_{c(i)}\tilde{x}_i$ is the output value of the $c(i)$th unit of the CAN2. The index $c(i)$ indicates the unit who has the weight vector $w_{c(i)}$ closest to the input vector x_i, or $c(i) \triangleq \underset{j \in I^N}{\text{Armin}} \|x_i - w_j\|$. The above function approximation partitions the input space $V \in \mathbb{R}^k$ into the Voronoi (or Dirichlet) regions $V_j \triangleq \{x \mid j = \underset{i \in I^N}{\text{Armin}} \|x - w_i\|\}$ for $j \in I^N$, and performs piece-wise linear prediction for the function $r(x)$. The bagging CAN2 is the bagging version of the CAN2.

Self-Organized Three Dimensional Feature Extraction of MRI and CT

Satoru Morita

Faculty of Engineering, Yamaguchi University,
2-16-1 Tokiwadai, Ube, Japan
smorita@yamaguchi-u.ac.jp

Abstract. We can observes a section of the body using MRI and CT . CT is suitable for the blood flow and the diagnosis of the wrong point of the bone by the computed tomograph, and MRI is suitable for the diagnosis of the cerebral brain infarction and the brain tumor. Because different nature is observed to so same the observation object, a, doctor, uses CT and an MRI image complementary, and sees a patient. The feature which appears in both images remarkably is extracted using the CT image and the MRI image by this paper. Various three-dimensional filters are generated using the ICA base in the self-histionic target from the characteristic image for that image, and how to extract a remarkable feature from the feature image which could get is proposed by this research. A remarkable feature is extracted from the CT image and the MRI image of the patient which actually has a tumor , and its effectiveness is shown.

Keywords: MRI, CT, Self-Organized 3D Feature Extraction, ICA.

1 Introduction

CT and MRI are used so that generally a doctor may observe the inside of the patient's body. CT uses X-ray photography to cut a body into circular sections, and it takes an image by computer processing. Magnetism is used, and MRI looks at the section of the body. Water, which occupies $60\% - 70\%$ of the human body is easily resonant with the magnetism used. It is suitable for blood flow and diagnosis of the wrong point of the bone by the computed tomograph, and MRI is suitable for the diagnosis of the cerebral brain infarction and the brain tumor. Because different nature are observed of the same observation object, a doctor uses CT and an MRI image complementarily, and sees a patient. The feature which appears in both images remarkably is extracted by using the CT image and the MRI image by this research. The calculation model of the initial vision which extracts a feature from the bottom-up is being discussed[1]. Koch and Ullman proposed the general idea by using various filters concerning the calculation theory to compute the saliency map which shows a saliency in a two-dimensional plane from more than one feature map of every feature that bottom-up generated[2]. Itti and Koch applied a calculation model of the form which can

T. Huang et al. (Eds.): ICONIP 2012, Part V, LNCS 7667, pp. 67–74, 2012.

be applied to the image analysis[3]. A filter here is the feature extraction of the multiplex resolution, and it is constructed using general Fourier transformation and a rotation filter[4]. Because these use a more general filter, it is necessary for the data to be arranged in a mesh image uniformly. Fourier transformation and a Wavelet change are proposed to extract a feature in the bottom-up from the image. The information related to the frequency by crossing a sin wave in the original signal is achieved using the Fourier transformation. A base function becomes a sine wave here. The information related to the frequency by crossing a mother base function in the original signal can get it with Wavelet. It is the technique which the base function defined is used. On the other hand, it was thinking about how to extract a feature signal by estimating an ICA base[5]. An adaptation example applied to the mesh image arranged uniformly though the example which adapts itself is introduced corresponding to the signal and the two-dimensional image. The receptive field of the human vision has the arrangement which isn't especially uniform mesh. but the self-organizing filter generated from only the observation of the image is used. Therefore the general idea of the ICA base estimation is introduced in the model which generates a saliency map. The receptive field of the human vision has the arrangement which isn't especially uniform mesh. For example, the density of the receptive field is high in the center, and is low in the circumference with a foveated vision. The filter which can be computed is generated in the self-histionic target from the observation to the receptive field which has such arrangement which isn't uniform mesh[15][14]. It is specially presumed that a doctor observes a CT image and a MRI image complementary and it is diagnosed by this research. The general idea of Saliency Map is that it is usually facing a two-dimensional image and is expanded to analyze a three-dimensional image because a CT image and an MRI image have three-dimensional data. A method to apply to base function estimation and a method to apply to the receptive field of the arrangement is mentioned in the section 2. We extend it to the method applied to the three dimensional information such as MRI and CT. It is actually applied to the MRI image and the CT image in the section 3, and based on the ICA base function estimation, and the effectiveness of the initial vision model to extract a remarkable feature.

2   Base Function Estimation in the Observation of the Three Dimensional Image

It is thinking about the method which an ICA base function is estimated for the image and the signal. The self-organization of the feature extracting cell of the color and the motion are modeled based on the ICA base function estimation in this paper[15][14]. The feature is detected such that the data projected to the subspace is independent and distributed. When the data that $X_l = \{x_{l1}, x_{l2}, \cdots, x_{ln}\}$ is centerized, Y_l is defined in the following as

$$Y_l = W_l X_l, \tag{1}$$

where $Y_l = \{y_{l1}, y_{l2}, \cdots, y_{ln}\}$, where $W_l = \{w_{l1}, w_{l2}, \cdots, w_{ln}\}$. The saliency map and the ICA base function is estimated for the lth data X_l. $x_{l}i$ is i th input data related to CT and MRI on the three dimensional receptive field X_l in this paper. The independence $J0_l$ of Y_l is evaluated as

$$J0_l = \Sigma_{i,j(i \neq j)} E\{y_{li} \cdot y_{lj}\}. \tag{2}$$

The degree of the distribution $J1_l$ of Y_l is evaluated as

$$J1_l = \frac{1}{\Pi_i E\{y_{li} \cdot y_{li}\}}, \tag{3}$$

where,

$$|w_l i| = 1 (l = 1, \cdots, m, i = 1, \cdots, n). \tag{4}$$

The general evaluation J_l is defined as

$$J_l = k_0 \cdot J0_l + k_1 \cdot J1_l, \tag{5}$$

where $J_l(l = 1, \cdots, m)$ which is saliency map value here. E is the expected value of probability variables. It returns in the problem which W_l is decided as so that J_l may become the smallest here. It becomes the base function which W_l is estimated at. It is solved by the method of steepest descent.

When the input data that $X_l = \{x_{l1}, x_{l2}, \cdots, x_{ln}\}$ is done as to the centration are taken. They are $Y_l = \{y_{l1}, y_{l2}, \cdots, y_{ln}\}$ and $W = \{w_{l1}, w_{l2}, \cdots, w_{ln}\}$. The feature $Y_l = \{y_{l1}, y_{l2}, \cdots, y_{ln}\}$ is extracted from n input data of using equation (1).

The salience map is defined based on the feature value $Sa_l = max\{y_{l0}, \cdots, y_{ln}\}$ and the feature number $Sb_l = \Sigma_{i=0}^{n} f(y_{li})$, where a function f used step function. A saliency map based on the feature value is generated by using $Sa_l(l = 1, \cdots, m)$. A saliency map based on the number of features is generated by using $Sb_l(l = 1, \cdots, m)$. The parameter k_0 and k_1 are defined as $k_0 = 1$ and $k_1 = 0$ simply in the experiment.

It is known that the arrangement of the receptive field is the circle and the rectangle and so on. Generally it is seldom arranged by the receptive field of the biological system uniformly. The receptive field is arranged according to the probability distribution which is dense in the center and is sparse in the periphery. Figure 1(a) shows the three dimensional probability distribution of the receptive field of the foveated vision in $Z = 0$ surface, $X = 0$ surface and $Y = 0$ surface. Figure 1(b) shows the position of recetptors generated based on the three dimensional probability distribution. Receptive field is arranged to the three dimension space. The receptive field which it could usually get here becomes input. If the number of the receptive field is a 10 , n in $X = \{x_1, x_2, \cdots, x_n\}$ becomes 10. When it thinks about a cell to react to the MRI image and the CT image to mention it in the next knot usually. n becomes $10 \cdot 2 = 20$ when the input which varies in each receptive field is received.

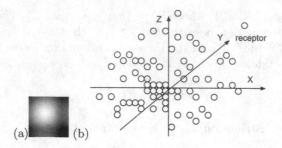

Fig. 1. (a)The three dimensional probability distribution of the receptive field of the foveated vision in $Z = 0$ surface, $X = 0$ surface and $Y = 0$ surface. (b)The position of recetptors generated based on the three dimensional probability distribution.

3   The Feature Extracting Cell Which Reacts to the CT Image and the MRI Image.

The CT image of the patient of the brain tumor and an MRI image were analyzed. Both resolution and size vary according to the MRI image in the CT image. The three dimensioal CT image composed of the 119 pieces two dimensional images sliced in the 1.2mm width. The three dimensional MRI image is 106 pieces two dimensional images sliced in the 2.0mm width. The size of the image is a $256 * 256$ pixel, and the image view of the CT image is 218mm, and the image view of the MRI image is 232mm. The voxel number of CT image is $106 * 256 * 256$, and the voxel number of MRI image is $119 * 256 * 256$. The size of CT image is $232 * 232 * 212 mm^3$ and the size of MRI image is $218 * 218 * 143 mm^3$. The three dimensional image of $256 * 256 * 256$ is generated over again here to align the pixel of the MRI image and the CT image. The histogram of the x direction, the y direction and the z direction is made, and the position of the top, the left side and the right side of the head is detected to position an MRI image and a CT image. The object except a human body here is removed. When an image is actually confirmed, the revision of the minute position is added. The histogram of the MRI image and the CT image is made, and the maximum of the histogram and a minimum are set up between 255 from 0. The space of $256 \times 256 \times 256$ of the same size is made in common with the CT image and the MRI image in the same position from the part it is reflected and which it wants to observe.

A figure 1 shows the probability distribution of the receptive field of the foveated vision by the glay level. Fifteen input is taken as the photoreceptor, the probability of that arrangement that receptive field exists in the center is high. The probability of that arrangement that receptive field exists in the center is high, and probability in the periphery lowers. Figure 1 shows the probability distribution of the receptive field of the foveated vision by the glay level. A figure 2 shows flow from the photoreceptor which reacts to the CT image and the MRI image to the primary visual cortex.

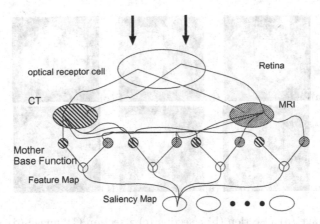

optical receptor cell

Retina

CT

MRI

Mother
Base Function

Feature Map

Saliency Map

Fig. 2. Flow from the photoreceptor which reacts to the CT image and the MRI image to the primary visual cortex

- A photoreceptor receives the intensity information of the MRI image and the CT image.
- It gets a feature by multiplying the base function computed in the intensity information which a photoreceptor received.
- Saliency map value is calculated from the feature value and the number of features.
- When there is two and more base function, a saliency map value that J value is minimum in the all is used as a pixel.

Figures 3 (a) (b) and (c) show a MRI image, and a figures 3 (d) (e) show (f) are a CT image. Figures 3 (a) (b) and (c) show a x section, a y section and a z section, and figures 3 (d) (e) (f) are a x section, a y section and a z section respectively. The size of the figure is $256 \times 256 \times 256$. The method which the next point of view was decided as simply in the probability from the whole of the image was used here. Figures 4 are eight feature images which could get it. Figures 4 (a) (b) and (c) show the x section, the y section and the z section of feature images respectively. The input of the receptive cell takes 4 input to receive the input of the MRI image and the CT image. Figures 5 are eight base functions of the receptive cell which the MRI image which could get it is taken as. Figures 5 (a) (b) and (c) show on the x section, the y section and the z section respectively. Figures 5(d)(e) and (f) show eight base functions of the receptive cell which the CT image which could get it is taken as. Figures 5 (d) (e) and (f) show the x section, the y section and the z section respectively. It is seen that the color is black in the center, and white in the periphery. On the other hand, it is seen that the color is white in the center, and black in the periphery. It is seen that the color is white only in a certain direction. There are relations between the four images in the neighbour, and it is seen that the base function is the reverse relation with a certain base function. There is relationship between these cells and a cell to react to the stimulus to move to the on-center off-surround form cell seen

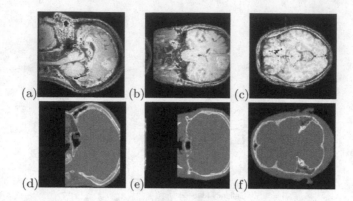

Fig. 3. MRI image (a) x section (b) y section (c) z section CT image (d) x section (e) y section (f) z section

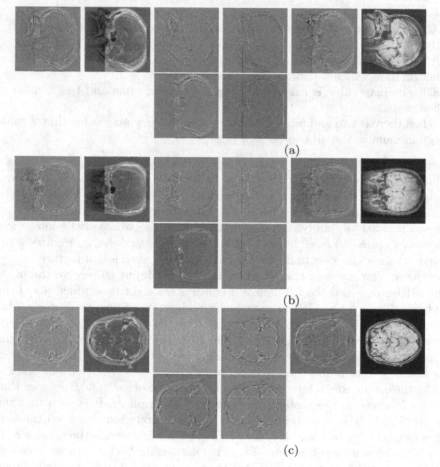

Fig. 4. Eight feature images (a) (the x section) (b) (the y section) (c) (the z section)

(a)
(b)
(c)
(d)
(e)
(f)

Fig. 5. Eight base functions of the receptive cell which a MRI image is taken as. (a) (the x section) (b) (the y section) (c) (the z section) Eight base functions of the receptive cell which CT is taken as. (d) (the x section) (e) (the y section) (f) (the z section).

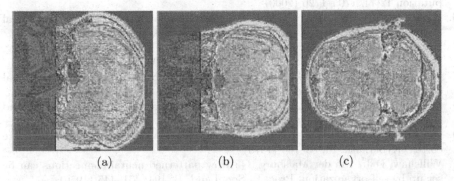

(a)   (b)   (c)

Fig. 6. (a) A saliency map of the surface based on the feature number in x section. (b) A saliency map of the surface based on the feature number in y section. (c) A saliency map of the surface based on the feature number in z section.

by the receptive field of the actual gangliocyte and the primary visual cortex, a off-center on-surround form cell and the cell of the orientation sensitivity and the specifical direction which it has strongly. A more remarkable base function becomes clear by increasing the number of the receptive cell. It is meaningful that the self-organizing cell which is common with the actual receptive field is generated. A figure 6 is a saliency map based on the number of features which it could get against a receptive cell to react to the CT image and the MRI image. Figures 6 (a) (b) and (c) show on the x section, the y section and the z section respectively. It is understood that a point is concentrated on the part of the remarkable feature.

4   Conclusion

The self-organization of the feature extracting cell which reacts to the CT image and the MRI image was modeled based on the base function estimation. The input of the receptive field of the arrangement which isn't uniform mesh is taken. The self-organizing cells such as the various cells confirmed in the brain are generated, and an effectiveness as a model is high from being a simple method.

References

1. Marr, D.: VISION. WHF Freeman and Company (1982)
2. Koch, C., Ullman, S.: Shifts in selective visual-attention towards the underlying neural circuitry. Hum. Neurobiol. 4, 219–227 (1985)
3. Itti, L., Koch, C.: Feature combination strategies for saliency-based visual attention systems. J. Electron Imaging 10(1), 161–169 (2001)
4. Mallet, S.G.: A theory for multiresolution signal decomposition: The wavelet representation. IEEE Tran. on PAMI 11, 674–693 (1989)
5. Hyvarinen, A., Hoyer, P.O.: Emergence of phase and shift invariant features by decomposition of natural images into independent feature subspace. Neural Computation 12(7), 1705–1720 (2000)
6. Hubel, D.H., Wiesel, T.N.: Receptive fields, binocular interaction and functional architecture in the cat's visual cortex. J. Physiol. 160, 106–154 (1962)
7. Hubel, D.H., Wiesel, T.N.: Receptive fields of cells in striate cortex of very young, visually inexperienced kittens. J. Neurophysiol. 26, 994–1002 (1963)
8. Blakemore, C., Cooper, G.F.: Development of the brain depends on the visual environment. Nature 228, 477–478 (1970)
9. Blakemore, C., Mitchell, D.E.: Environmental modification of the visual cortex and the neural basis of learning and memory. Nature 241, 467–468 (1973)
10. von der Malsburg, C.: Self-organization of orientation sensitive cells in the striate cortex. Kybernetik 14, 85–100 (1973)
11. Willshaw, D.J., von der Malsburg, C.: How patterned neural connections can be set up by self-organization. Proc. R. Soc. Land. B. 194, 431–445 (1976)
12. Kohonen, T.: Self-organized formation of topographically corerct feature maps. Biol. Cybern. 43, 59–69 (1982)
13. Horn, B.K., Schunck, B.G.: Determining optical flow. Artif. Intell. 17, 185–203 (1981)
14. Morita, S.: Generating Saliency Map Related to Motion Based on Self-organized Feature Extracting. In: Köppen, M., Kasabov, N., Coghill, G. (eds.) ICONIP 2008, Part II. LNCS, vol. 5507, pp. 784–791. Springer, Heidelberg (2008)
15. Morita, S.: Generating Self-organized Saliency Map Based on Color and Motion. In: Leung, C.S., Lee, M., Chan, J.H. (eds.) ICONIP 2009, Part II. LNCS, vol. 5864, pp. 28–37. Springer, Heidelberg (2009)

OMP or BP? A Comparison Study of Image Fusion Based on Joint Sparse Representation

Yao Yao, Xin Xin, and Ping Guo

School of Computer Science & Technology,
Beijing Institute of Technology, Beijing 100081, China
{bityaoyao,xxin}@bit.edu.cn, pguo@ieee.org

Abstract. In image fusion techniques based on joint sparse representation (JSR), the composite image is calculated from the fusion of features, which are represented with sparse coefficients. Orthogonal matching pursuit (OMP) and basis pursuit (BP) are the main candidates to estimate the coefficients. Previously OMP is utilized for the advantage of low complexity. However, noticeable errors occur when the dictionary of JSR cannot ensure the coefficients are sparse enough. Alternatively, BP is more robust than OMP in such cases (though suffered from larger complexity). Unfortunately, it has never been studied in image fusion tasks. In this paper, we investigate JSR based on BP for image fusion. The target is to verify that 1) to what extent can BP outperform OMP; and 2) what is the trade-off between BP and OMP. Finally, we conclude, in some cases, fusion with BP obviously outperforms the one with OMP under an affordable computational complexity.

Keywords: Joint sparse representation, Image fusion, Sparse coefficient approximation, Basis pursuit, Orthogonal matching pursuit.

1 Introduction

Image fusion is to integrate the complementary information from two or more images of the same scene into a single composite image. Consequently, the image is more suitable for human visual perception, as well as computer-processing tasks such as image segmentation, feature extraction, object recognition, etc. Images of a specific scene can be acquired by using different sensor modalities or the same camera with different settings. Different images have their own characteristics and also share some common information. A fusion system recognizes these properties and combines the information into the final composite image. Image fusion has been widely utilized among digital imaging, remote sensing, medical image processing, pre-processing for computer vision problems, etc [1-3].

Recently the approach of joint sparse representation (JSR) [4, 5] has achieved a great success in solving the image fusion task. In JSR, image patches are denoted as sparse linear combinations of a set of prototype atoms. The set of atoms is also called the over-complete dictionary, in which the number of basis atoms exceeds the number of image patch dimension. The feature of an image patch can be represented by the

---

T. Huang et al. (Eds.): ICONIP 2012, Part V, LNCS 7667, pp. 75–82, 2012.

sparse coefficients. In JSR based image fusion, features of an image patch include the common features shared in both image patches, and innovation features to the individual of each image patch. Intuitively, the sparse coefficients of an image patch are the combination of the two. The composite image is calculated by the fusion of all the features from the two input images.

There are two typical algorithms to calculate the sparse coefficients: orthogonal matching pursuit (OMP) [7] and basis pursuit (BP) [6]. OMP computes the sparse coefficients in a greedy stepwise way. And BP finds the sparse coefficients with a relaxation from ℓ^0-norm to ℓ^1-norm. According to Donoho *et al* [8], on the condition that the ℓ^0-norm of sparse coefficients is smaller than a constant decided by the dictionary, both methods can be guaranteed to have bounded approximation solutions of sparse coefficients estimation. Previously, OMP is mainly utilized in image fusion tasks for the advantage of low computational complexity.

Although OMP method has achieved success, in some cases of real applications, it would fail to perform well. The main reason is that the condition to guarantee the bound cannot be always fulfilled. As a result, the approximation solutions of sparse coefficients have approximation error larger than the expected bound. So the performance of JSR based image fusion with OMP is limited.

To solve the limitation of OMP, BP is an attractable alternative approach. From Elad's book [9], BP works more robustly and has smaller approximation error than OMP as the cardinality of the solution getting larger, even though the constraint in [8] is not satisfied. Unfortunately, this method has not been investigated in solving image fusion tasks. One reason is that the computational complexity of BP is larger than OMP. However, in some cases, it is worthwhile to spend more time for better performance.

In this paper, we investigate JSR based on BP for image fusion. We aim at answering the following two questions: 1) To what extent can BP outperform OMP in JSR-based image fusion tasks? 2) What is the trade-off between the OMP's advantage of computational complexity and BP's advantage of better performance? Through experimental verification, we demonstrate that when the error bound is demanded to be relatively small, the BP approach could outperform OMP approach significantly. Moreover, we further demonstrate that in some cases, fusion with BP obviously outperforms the one with OMP under an affordable computation complexity.

For convenience, we denote the JSR-based image fusion with OMP as JSR-OMP, and call JSR-based image fusion with BP as JSR-BP.

2   Joint Sparse Representation

2.1   Fusion Framework Based on JSR

Assuming there are two source images I_1, I_2, we adopt a framework similar to that of Yin and Li [4]. Firstly, the source images are divided into image patches using a sliding window strategy [12]. The window size $k \times k$ and the window moving step are two important factors. A small moving step avoids blocking artifacts in results. So the moving step is set to be one pixel. When the sliding window strategy is applied on a source image with size of $M \times N$, there will be $(M-k+1) \times (N-k+1)$ image patches which are ordered to form vectors of size $n = k \times k$.

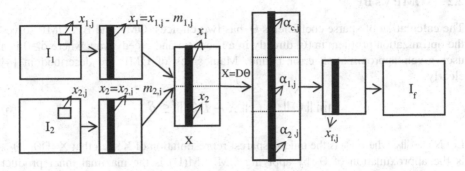

Fig. 1. Framework of image fusion based on JSR

In Figure 1, $x_{1,j}$ and $x_{2,j}$ are vectors of size n, formed from the jth pair of corresponding image patches from two source images, respectively. $m_{1,j}$ and $m_{2,j}$ are mean values of $x_{1,j}$ and $x_{2,j}$. x_1 and x_2 are left part of image patches subtracted by its mean value. Secondly, $x_{1,j}$ and $x_{2,j}$ are fused. The weights for mean values $m_{1,j}$, $m_{2,j}$ of two image patches are set to be 0.5. So $x_{f,j}$ is obtained by (1), where $\alpha_{c,j}$ is the common sparse coefficients shared by x_1 and x_2, and $\alpha_{1,j}$ and $\alpha_{2,j}$ are innovation sparse coefficients of x_1 and x_2, respectively. And finally the fused images are constructed from the fused image patches and the overlapping pixels are averaged.

$$x_{f,j} = \Phi(\alpha_{c,j} + \alpha_{1,j} + \alpha_{2,j}) + 0.5(m_{1,j} + m_{2,j}) \tag{1}$$

To calculate $\alpha_{c,j}$, $\alpha_{1,j}$ and $\alpha_{2,j}$, we use JSR to model the joint sparsity [10] of x_1 and x_2 on the same over-complete dictionary $\Phi \in \mathbb{R}^{n \times m}$. Then $x_1 = \Phi\alpha_{c,j} + \Phi\alpha_{1,j}$, and $x_2 = \Phi\alpha_{c,j} + \Phi_{2,j}$. The JSR model can be expressed as

$$X = D\Theta \tag{2}$$

$$X = \left[x_1^T, x_2^T\right]^T \in \mathbb{R}^{2n} \tag{3}$$

$$D = \begin{bmatrix} \Phi & \Phi & 0 \\ \Phi & 0 & \Phi \end{bmatrix} \in \mathbb{R}^{2n \times 3m} \tag{4}$$

$$\Theta = \left[\alpha_c^T, \alpha_{1,j}^T, \alpha_{2,j}^T\right]^T \in \mathbb{R}^{3m} \tag{5}$$

X is a large vector piled with two input vectors in the group. The over-complete dictionary Φ can be carefully chosen or trained to ensure the vectors in the group can be grossly sparsely represented. D is large dictionary concatenated with Φ. Θ is the sparse coefficients of JSR model. The shared common sparse coefficients α_c and individual innovation sparse coefficients $\alpha_{1,j}$ and $\alpha_{2,j}$ can be obtained directly from Θ.

2.2   OMP v.s BP

The calculation of sparse coefficients Θ has two choices: OMP and BP. OMP solves the optimization problem in (6) directly in a heuristic and greedy way, where $\delta \geq 0$ is a user given approximation error bound. Main steps of OMP are described in [4] clearly.

$$\min_{\Theta} \| \Theta \|_0 \text{ s.t. } \| X - D\Theta \|_2^2 \leq \delta \tag{6}$$

Let $N = \|\Theta_0\|_0$, where Θ_0 is the unique sparest representation of X such that $X = D\Theta_0$. $\Theta_{0,\delta}$ is the approximation of Θ_0 by applying OMP. M(D) is the maximal inner product between columns of D which are normalized to unit ℓ^2-norm. If $N < (1/M(D)+1)/2$, then the following is guaranteed [8].

$$\| \Theta_{0,\delta} - \Theta_0 \|_2^2 \leq \frac{\delta}{1 - M(D)(2N-1)} \tag{7}$$

In real applications, the dictionary doesn't always satisfy the above constraint unless one chooses or trains it carefully. Here the dictionary Φ for image patch representation is trained with some learning algorithms. Denote N_1 and N_2 as the ℓ^0-norm of the unique sparest representation of x_1 and x_2 under Φ, respectively. Suppose Φ ensures the above constraint, i.e. $N_1 < (1/M(\Phi)+1)/2$ and $N_2 < (1/M(\Phi)+1)/2$. According to (4), after normalization M(D) still equals to M(Φ). And from (5), N is larger than either N_1 or N_2. That means the dictionary D in JSR probably cannot ensure the above constraint. As a result, the approximation solutions of sparse coefficients may have approximation error larger than the expected bound in (7). So the performance of JSR-OMP is limited.

Although BP has the same problem of the dictionary, the approximation error of BP becomes much smaller than that of OMP as the sparse coefficients become denser [9]. So JSR-BP is expected to outperform JSR-OMP in such cases. BP computes the sparse coefficients by relaxing the ℓ^0-norm in (6) to ℓ^1-norm and then solves a convex optimization problem in (8). And here the convex optimization problem in (8) is solved with Least Angle Regression Stagewise (LARS) [11, 16] algorithm.

$$\min_{\Theta} \| \Theta \|_1 \text{ s.t. } \| X - D\Theta \|_2^2 \leq \delta \tag{8}$$

3   Experiments

3.1   Experimental Setup

Following the configurations in [12], the sliding window size of an image patch is set to be 8 × 8 and the moving step of the window is set one pixel to avoid blocking artifacts and to ensure the performance of fusion. The dictionary Φ for image patch representation is trained from the image patches of input source images by an online learning algorithm [13]. The size of the dictionary Φ is set to be 64×256. So the size of concatenated dictionary D for JSR is 128×768. Different values of the approximation error bound δ are tested, including 0.05, 0.01, 0.005, and 0.001. The $Q^{AB/F}$ measurement [15] is chosen as the evaluation metric. The better the algorithm performs the larger the $Q^{AB/F}$ value would be. The experiments are conducted on a variety of image pairs obtained from the Online Resource for Research in Image Fusion [14]. We select three as representative to show the experiment result. The three examples are chosen randomly without any preference. The image size is 512×512. And all the experiments are implemented in Matlab 7.13 on an Intel Core i3 3.07-GHz & 3.06-GHz PC with 2-GB RAM.

3.2   Improvements of JSR-BP over JSR-OMP

Figure 2 shows the fusion results of both JSR-OMP and JSR-BP. When we set the approximation error bound δ to be a small value, the results of both methods are almost the same good by human visual perception. Table 1, Table 2 and Table 3 show the $Q^{AB/F}$ measurements of JSR-OMP and JSR-BP with different approximation bound δ on three different image pairs. From the results, it can be observed that as the value of δ becomes smaller, the performance measurements of both methods become larger. This means that when the approximation error bound is lower, the fusion result is better.

In general, JSR-BP outperforms JSR-OMP. In Table 1, when δ is 0.001, JSR-BP outperforms JSR-OMP by 17%. In Table 2, when δ is 0.005, the performance increases by 32.9%. And in Table 3, when δ is 0.001, the performance increases by 59.6%. This is because when δ is small enough to affect the approximation of sparse coefficients, BP is more robust to give an approximation solution having a much smaller approximation error than OMP under the concatenated dictionary of JSR.

3.3   Trade-off Analysis

In Table 4, the running time is averaged over different pairs of images. As the approximation error bound becomes smaller, the average running time of JSR-OMP changes a little. But the average running time of JSR-BP increases when δ becomes smaller. It increases to an unacceptable time when δ is 0.001. This is because that BP is more complicated for approximation of sparse coefficients than OMP.

But by analyzing all the tables together, we find that there are some cases where the running time of JSR-BP is acceptable and the fusion performance of JSR-BP is much better than that of JSR-OMP. For example, when δ is 0.01, the averge running time of JSR-BP is about 15 minutes and the fusion performance of JSR-BP is much better than that of JSR-OMP in Table 1, Table 2 and Table 3. The increment percentage of performance measurements are 10.3%, 28.2% and 48%, respectively. From the experiment results, when δ is relatively small we found that JSR-BP has high performance and high computational complexity, and that JSR-OMP always has low computational complexity and low performance compared with JSR-BP. It is reasonable to have a trade-off where JSR-BP outperforms JSR-OMP greatly under an acceptable computational complexity. For example, in the experiments, δ from 0.005 to 0.01 can be such trade-off places.

Table 1. QAB/F measurements on Figures 2(a) and 2(b)

δ	0.05	0.01	0.005	0.001
JSR-OMP	0.5398	0.5597	0.5616	0.5683
JSR-BP	0.4830	0.6172	0.6382	0.6651
Increment(%)	-10.5	10.3	13.6	17.0

Table 2. QAB/F measurements on Figures 2(e) and 2(f)

δ	0.05	0.01	0.005	0.001
JSR-OMP	0.3760	0.3949	0.3953	0.3953
JSR-BP	0.4140	0.5061	0.5255	0.5137
Increment(%)	10.1	28.2	32.9	29.9

Table 3. QAB/F measurements on Figures 2(i) and 2(j)

δ	0.05	0.01	0.005	0.001
JSR-OMP	0.3400	0.3666	0.3699	0.3715
JSR-BP	0.3699	0.5426	0.5695	0.5930
Increment(%)	8.7	48.0	54.0	59.6

Table 4. Average running time (in seconds) over different pairs of images

δ	0.05	0.01	0.005	0.001
JSR-OMP	244	282	243	287
JSR-BP	296	1081	2377	6621

4   Related Work

Image fusion algorithms are generally categorized into two classes: spatial domain and transform domain methods. Spatial domain methods use the pixels of source images directly as image feature to fuse. Carper et al [17] combined panchromatic image pixels and multi-spectral image pixels directly with a weighted averaging criterion. Transform domain methods regard the coefficients in transform domain as image feature to fuse. Multi-scale transforms are widely used, including pyramid decompositions [18], discrete wavelet transform [1], ridgelet [19], curvelet [20], and contourlet [21]. Considering sparse representation (SR) is able to extract stable and meaningful representation of images, Yang and Li [12] adopted SR with OMP to compute sparse coefficients which are regarded as image features. Later Yang and Li [22] improved SR-based fusion with simultaneous orthogonal matching pursuit (SOMP). Recently Yin and Li [4] proposed image fusion based on JSR. And Yu et al [5] adopted a similar JSR-based fusion method but with different fusion rules.

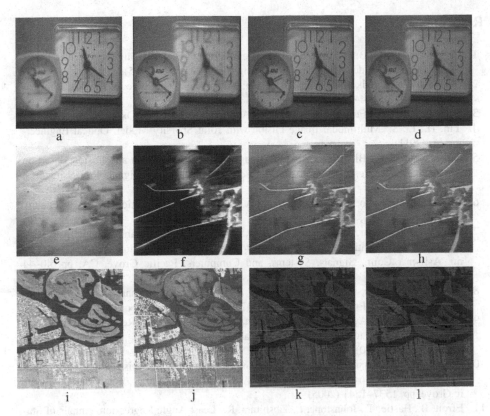

Fig. 2. Source image pairs and fusion results: (a) and (b) are multi-focus source images. (e) and (f) are multi-sensor source images. (i) and (j) are multi-spectral source images. (c), (g) and (k) are results of JSR-OMP with δ=0.001. (d), (h) and (l) are results of JSR-BP with δ=0.001.

5   Conclusion

In this paper, we have made a comparison study of two algorithms for image fusion, namely OMP and BP, under the framework of JSR. Through experimental verification on real world data, we demonstrate that BP algorithm outperforms OMP algorithm significantly when the error bound is relatively small. Moreover, we further show the trade-off of the two methods, from the performance view, as well as the computational complexity view. It can be concluded that in some cases, fusion with BP obviously outperforms the one with OMP under an affordable computation complexity.

Acknowledgments. This is supported by the grants from the National Natural Science Foundation of China (Project Nos. 90820010, 60911130513, 60805004). Prof. Ping Guo is the author to whom all correspondence should be addressed, his email is pguo@ieee.org.

References

1. Li, H., Manjunath, B.S., Mitra, S.K.: Multisensor Image Fusion Using the Wavelet Transform. Graphical Models and Image Processing 57(3), 235–245 (1995)
2. Goshtasby, A.A., Nikolov, S.: Image Fusion: Advances in The State of The Art. Information Fusion 8(2), 114–118 (2007)
3. Stathaki, T.: Image Fusion: Algorithms and Applications. Elsevier, Oxford (2008)
4. Yin, H., Li, S.: Multimodal Image Fusion with Joint Sparsity Model. Optical Engineering (6) (2011)
5. Yu, N., Qiu, T.S., Bi, F., Wang, A.Q.: Image Features Extraction and Fusion Based on Joint Sparse Representation. IEEE J. Selected Topics in Signal Processing 5(5), 1074–1082 (2011)
6. Chen, S.S., Donoho, D.L., Saunders, M.A.: Atomic Decomposition by Basis Pursuit. SIAM Review 43(1), 129–159 (2001)
7. Pati, Y.C., Rezaiifar, R., Krishnaprasad, P.S.: Orthogonal Matching Pursuit: Recursive Function Approximation with Applications to Wavelet Decomposition. In: Proc. 27th Annu. Asilomar Conf., Signals, Systems, and Computers, Pacific Grove, CA, pp. 40–44 (1993)
8. Donoho, D.L., Elad, M., Temlyakov, V.N.: Stable Recovery of Spare Overcomplete Representations in the Presence of Noise. IEEE Trans. Information Theory 52(1) (2006)
9. Elad, M.: Towards Average Performance Analysis. In: Sparse and Redundant Representations: From Theory to Applications in Signal and Image Processing, pp. 137–152. Springer, New York (2010)
10. Duarte, M., Sarvotham, S., Baron, D., Wakin, M., Baraniuk, R.: Distributed Compressed Sensing of Jointly Sparse Signals. In: Proc. Asilomar Conf., Signals, Syst. Comput., Pacific Grove, pp. 1537–1541 (2005)
11. Efron, B., Hastie, T., Johnstone, I., Tibshirani, R.: Least Angle Regression. Annals of Statistics 32(2), 407–499 (2004)
12. Yang, B., Li, S.: Multifocus Image Fusion and Restoration With Sparse Representation. IEEE Trans. Instrum. Meas. 59(4) (2010)
13. Marial, J., Ponce, J., Sapiro, G.: Online Dictionary Learning for Sparse Coding. In: Proc. International Conference on Machine Learning, Montreal, Canada, pp. 689–696 (2009)
14. Online Resource for Research in Image Fusion, http://www.ImageFusion.org
15. Xydeas, C.S., Petrovic, V.: Objective Image Fusion Performance Measure. Electronics Letters 36(4), 308–309 (2000)
16. Sparse Modeling Software, http://spams-devel.gforge.inria.fr
17. Carper, T.W., Lillesand, T.M., Kiefer, R.W.: The Use of Intensity-Hue-Saturation Transformations for Merging SPOT Panchromatic and Multispectral Image Data. Photogrammetric Engineering and Remote Sensing 56, 459–467 (1990)
18. Akerman, A.: Pyramid Techniques for Multisensor Fusion. In: Proceedings of SPIE, vol. 1828, pp. 124–131 (1992)
19. Chen, T., Zhang, J., Zhang, Y.: Remote Sensing Image Fusion Based on Ridgelet Transform. In: Proc. Geo. Remote Sens. IEEE Int. Symp., pp. 1150–1153 (2005)
20. Nencini, F., Garzelli, A., Baronti, S., Alparone, L.: Remote Sensing Image Fusion Using The Curvelet Transform. Information Fusion 8(2), 143–156 (2007)
21. Do, M.N., Vetterli, M.: The Contourlet Transform: An Efficient Directional Multiresolution Image Representation. IEEE Trans. on Image Processing 14(12), 2091–2106 (2005)
22. Yang, B., Li, S.: Pixel-Level Image Fusion with Simultaneous Orthogonal Matching Pursuit. Information Fusion 13(1), 10–19 (2012)

An Improved Approach to Super Resolution Based on PET Imaging

P.M. Yan, Meng Yang, Hui Huang, and J.F. Li

School of Communication and Information Engineering,
Shanghai University, Shanghai, China
pmyan@mail.shu.edu.cn, yangmeng412721@126.com

Abstract. The low spatial resolution of Positron Emission Tomography imaging (PET) is due to the width of detector and some physical parameters (such as scattering fraction, counting statistics, positron range and patient's motion). To overcome this problem and improve the resolution of PET image, a high effective sub-pixel registration algorithm based on Keren's method is proposed, and a new iteration algorithm of registration is introduced to improve the registration accuracy. Compared with Keren's method, this method can improve the registration accuracy highly. This new registration algorithm is applied into super-resolution PET imaging. What's more, this new super-resolution approach will be demonstrated in this paper.

Keywords: PET imaging, registration, Keren's method, super-resolution, sinogram, images.

1   Introduction

Positron emission tomography (PET) is recognized as one of the most advanced medical equipment in the field of nuclear medicine. PET imaging with 18F-fluorodeoxyglucose (18F-FDG) has been recently used in clinical to diagnose status of metabolism of patients with cancer and examine the extent of diseases. What's more, PET imaging with 18F-fluorodeoxyglucose (18F-FDG) has also been recognized to be a highly precision diagnosis without pain [1]. As all known, high resolution (HR) medical images are very helpful for a doctor to make a correct diagnosis. However, one of the problems of PET imaging is its relatively poor spatial resolution. The resolution of PET image is limited by physical parameters such as the positron range, non-colinearity, counting statistics and detector width.

To improve the resolution of PET images, super-resolution (SR) techniques are introduced into PET imaging in recent years. Increasing the sample frequency of the axial can acquire sets of low resolution images from different points of view [2]. SR techniques can combine multiple low resolution images to generate one or more high-resolution images. Kennedy et al. [3] introduced a super-resolution method which acquired the low-resolution images by shifting and rotating the detector. Chang et al. [4] optimized Kennedy's algorithm. In order to reduce the processing time and

T. Huang et al. (Eds.): ICONIP 2012, Part V, LNCS 7667, pp. 83–90, 2012.
© Springer-Verlag Berlin Heidelberg 2012

memory storage, Chang et al. proposed a method which just uses parts of low-resolution images for super-resolution reconstruction. Additionally, Chang et al. [5] also proposed an approach that acquired multiple low-resolution images by shifting and rotating the pixel grids instead of Kennedy's algorithm (acquire low-resolution images from different point of view). Zhi Yuan et al. [6] introduced a high accuracy registration approach-genetic algorithm which is used to filter out all the mismatches. Kye Young Jeong et al. [7] obtained over-sampled data from the wobbling motion and proposed a novel and efficient super-resolution scheme for the sinograms. The registration algorithm of super-resolution directly affects the quality of super-resolution image. To reach the sub-pixel accuracy, we apply an improved registration method based on Keren's method to optimize the super-resolution PET imaging. This paper is organized as follows: in section 2, we propose an improved registration for the super-resolution PET imaging, the principle on how to achieve the super-resolution PET imaging and the steps of proposed algorithm. In section 3, simulation environment and results are introduced. Conclusions are described in section 4.

2   Methods

SR is a technology which is mainly used to reconstruct high-resolution images from several low-resolution images [8]. In PET imaging, multiple low-resolution images with sub-pixel shift can be obtained by wobbling motion. In this section, an improved registration algorithm based on Keren's method is proposed for super-resolution PET imaging.

2.1   GAPY (GAussian PYramid Iterative Reconstruction Based on Keren's Method)——An Improved Algorithm Based on Keren's Method

Image registration is a significant part in SR imaging. In Keren's method, a Gaussian pyramid structure is proposed to increase computing speed and improve robustness [9,10].

In Keren's method, the original image (size of $N{\times}N$) is filtered by a Gaussian filter, and then down-sampled to an image with size of $N/2{\times}N/2$. This scheme is repeated until the smallest image reached. Great motion translations can be small at this reduction level. Firstly a small image (usually 64×64) as the top image is used to computing the motion parameters. This process is iterated for a few times to optimize the motion parameters. At last, a larger image is corrected by the computed parameters. It can be called "Push". This process continues until the original image is similarly reached. In Keren's method, the relation between two images f and g can be expressed by a (horizontal shift), b (vertical shift) and θ (rotation angle) [9]:

$$g(x, y) = f(x\cos\theta - y\sin\theta + a, y\cos\theta - x\sin\theta + b) \tag{1}$$

The $cos\theta$ and $sin\theta$ can be expended by Taylor series for the first two terms:

$$g(x,y) \approx f(x+a-y\theta-x\theta^2/2, y+b+x\theta-y\theta^2/2)$$
$$\approx f(x,y) + (a-y\theta)\frac{\partial f}{\partial x} + (b+x\theta)\frac{\partial f}{\partial y} \tag{2}$$

$E\,(a, b, \theta)$ is the error function between images f and g:

$$E(a,b,\theta) = \sum[(a-y\theta)\frac{\partial f}{\partial x} + (b+x\theta)\frac{\partial f}{\partial y} + f - g]^2$$
$$= \sum[a\frac{\partial f}{\partial x} + b\frac{\partial f}{\partial y} + \theta(x\frac{\partial f}{\partial y} - y\frac{\partial f}{\partial x}) + f - g]^2 \tag{3}$$

When the value of $E\,(a, b, \theta)$ is minimum, the value of a, b and θ are needed. So it should be computed its derivatives by a, b and θ respectively, and make sure its derivatives equations are equal to zero:

$$\frac{\partial E}{\partial a} = \frac{\partial E}{\partial b} = \frac{\partial E}{\partial \theta} = 0 \tag{4}$$

Equation (4) can be expressed as (5):

$$a\sum\left(\frac{\partial f}{\partial x}\right)^2 + b\sum\frac{\partial f}{\partial x}\frac{\partial f}{\partial y} + \theta\sum R\frac{\partial f}{\partial x} = \sum\frac{\partial f}{\partial x}(f-g)$$
$$a\sum\frac{\partial f}{\partial x}\frac{\partial f}{\partial y} + b\sum\left(\frac{\partial f}{\partial y}\right)^2 + \theta\sum R\frac{\partial f}{\partial y} = \sum\frac{\partial f}{\partial y}(f-g) \tag{5}$$
$$a\sum R\frac{\partial f}{\partial x} + b\sum R\frac{\partial f}{\partial y} + \theta\sum R^2 = \sum R(f-g)$$

Which R represents $x\frac{\partial f}{\partial y} - y\frac{\partial f}{\partial x}$. So the above matrix can be expressed as (6):

$$\begin{bmatrix} \sum\left(\frac{\partial f}{\partial x}\right)^2 & \sum\frac{\partial f}{\partial x}\frac{\partial f}{\partial y} & \sum R\frac{\partial f}{\partial x} \\ \sum\frac{\partial f}{\partial x}\frac{\partial f}{\partial y} & \sum\left(\frac{\partial f}{\partial y}\right)^2 & \sum R\frac{\partial f}{\partial y} \\ \sum R\frac{\partial f}{\partial x} & \sum R\frac{\partial f}{\partial y} & \sum R^2 \end{bmatrix} \times \begin{bmatrix} a \\ b \\ \theta \end{bmatrix} = \begin{bmatrix} \sum\frac{\partial f}{\partial x}(f-g) \\ \sum\frac{\partial f}{\partial y}(f-g) \\ \sum R(f-g) \end{bmatrix} \tag{6}$$

In this case, we can use C, U and V to simplify matrix equation (6):

$$
C = \begin{bmatrix} \sum\left(\dfrac{\partial f}{\partial x}\right)^2 & \sum\dfrac{\partial f}{\partial x}\dfrac{\partial f}{\partial y} & \sum R\dfrac{\partial f}{\partial x} \\[2ex] \sum\dfrac{\partial f}{\partial x}\dfrac{\partial f}{\partial y} & \sum\left(\dfrac{\partial f}{\partial y}\right)^2 & \sum R\dfrac{\partial f}{\partial y} \\[2ex] \sum R\dfrac{\partial f}{\partial x} & \sum R\dfrac{\partial f}{\partial y} & \sum R^2 \end{bmatrix} \quad U = \begin{bmatrix} \sum\dfrac{\partial f}{\partial x}(f-g) \\[2ex] \sum\dfrac{\partial f}{\partial y}(f-g) \\[2ex] \sum R(f-g) \end{bmatrix} \quad V = \begin{bmatrix} a & b & \theta \end{bmatrix}^T \tag{7}
$$

So equation (6) can be expressed as:

$$
CV = U \tag{8}
$$

In Keren's method, Gaussian pyramid is introduced to increase computing speed and expand the application scope of motion parameters (even big translation is small at this reduction level). The original image is sub-sampled by Gaussian pyramid to form multi-level images with different size. In every level, the images use registration by iterations, motion parameters matrix V are used for correct guess image g, and the motion parameters V also can be correction by registration once again. When differences between f and the guess image g are reached to zero or the number of iterations reaches a certain value, motion parameters are used to "push" the next level. This process continues until the bottom images are reached [9].

Keren's method can improve the registration accuracy and reconstruction efficiency. But the disadvantage of Keren's method is the iteration algorithm. The first level image are sub-sampled from the original images, even though differences between f and g are reached to zero, there still exist error in the first level. Motion parameters are still inaccurate compared with real value. Actually, image in the top is the result of sub-sampling from the original image. The error exists between the top levels and then the computed motion parameters are directly applied into the next layer images, registration accuracy will be adversely affected. The "push" doesn't exactly represent the real movement of the original image, so a certain error will exist in Keren's method.

In this paper, an improved registration algorithm (GAPY) is used. The motion parameter is firstly used to "push" the bottom image. And then starts to re-sample from the bottom layer image by Gaussian pyramid to obtain the operation image, so it can offset big movement, correct the bottom image and the top image directly. It can effectively avoid the loss of information. Compared with Keren's method, the resolution of the reconstructed image can be improved obviously. The steps are shown as follows:

1. The Gaussian pyramid is used for sub-sampling the original image. Here, using Keren's method, the original image (usually 256×256) is sub-sampled to the image of size 128×128, and then the next level is 64×64. In this paper, Gaussian pyramid to form 3 level images can be used, so the size of the smallest image is 64×64.
2. The top image is used for registration just once, the calculated motion parameters are directly used to "push" the bottom image. After the movement of bottom

image, the bottom image is sub-sampled again by Gaussian pyramid to optimize every level image.

3. The new top image is used for registration again. Then step 1 to 2 is repeated until motion parameters are close to zero or a maximum number of iterations is reached.

4. When the step 3 is finished, the next level image is repeat step 1 to 3, this process is continues until the original image is reached.

This method is used to directly "push" the bottom image by computing motion parameter of the top images and sub-sampled to correct the top image through more iteration. In this paper, the algorithm proposed can improve registration accuracy than Keren's method, and then this algorithm will be applied into super-resolution of PET imaging.

2.2   Super-Resolution Implement to PET Imaging

High resolution images can be reconstructed from multiple low-resolution images through super-resolution techniques. When the low-resolution images acquired, it will be need the multiple low-resolution images with sub-pixel shift among each other [11,12]. When acquire multiple low-resolution images, the observation model [1] can be represented as follow:

$$y_k = D_k B_k M_k x + n_k = W_k x + n_k, \quad for \quad 1 \le k \le p \qquad (9)$$

Where D_k is sub-sampling matrix, B_k represents blurring matrix, M_k is a warping matrix, x is the high-resolution image, and n_k represents a noise vector, y_k is low-resolution image which has been observed, then x will be obtained by super-resolution techniques.

In PET scanner, the geometric transformations are a combination of rotation and translation within the transaxial plane, and wobbling motion can increase the rate of spatial sampling and sets of image data during the acquisition. The wobbling motion will be described in the next section.

In the axial direction each acquisition is moved through wobbling motion to provide a set of four low-resolution slices, and the four different points of view are shifted with the sub-pixel shift and rotation with a small angle. In this paper, we use GAPY registration algorithm for PET super-resolution imaging, the initial high-resolution sinogram can be obtained by interpolating a low-resolution sonogram.

3   Experiments

3.1   Simulation Environment

Wobbling motion is introduced to acquire sets of low-resolution images, sinograms of phantom are obtained in the transaxial plane, the elements of geometry are considered to be at rest during each time-step [13]. Multiple sinogram sets are obtained through wobbling motion between every time-step.

Wobbling motion means there is always exist one point to do radial motion along the diameter direction instead of following the trajectory of circular motion. When the wobbling motion rotates a circle, the trajectory of a point is a circle, as show in Fig.1.

Through the Wobbling motion, not only the resolution of transaxial plane will be improved, but also multiple low-resolution images of PET may be acquired. In this paper, wobbling motion is used to increase the numbers of sinogram samples points.

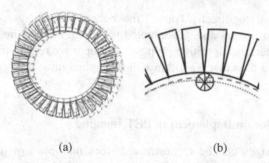

(a)   (b)

Fig. 1. (a) Wobbling motion with sub-pixel shift. (b) Wobbling motion along one point to do radial motion.

The wobbling motion can be described as:

$$dM(t) = A\sin(2\pi f t + phi) \tag{10}$$

Where $dM(t)$ is the translation vector at time t, A is the maximum displacement vector, f is the movement frequency, phi is the phase at $t=0$, and t is the time value [8]. For simulation, we adopt a phantom with spheres of different diameters (respectively 40, 32, 18, and 4mm), and the different spheres are arranged in the center of the field of view (FOV) of the scanner, as show in Fig.2, and all data were acquired in 2D mode. The data may be acquired four times by moving the center of the system. Uploading the data into matlab, and this software [15,16] can register the four low-resolution images and then reconstruct a HR images.

Fig. 2. Resolution phantom for physical simulation with spheres of different diameters (respectively 40, 32, 18, and 4mm)

3.2   Simulation Results

The image reconstructed from HR sinogram that using GAPY is compared with the image reconstructed from original sinogram. Ordered subsets expectation maximum

(OSEM) algorithm may be adopt as the reconstructed method. The reconstruct images can be seen in Fig.3 (a) and (b). From the figure, a conclusion that the proposed registration method can improve the spatial resolution of PET image effectively will be obtained.

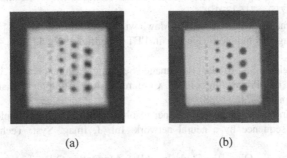

(a)   (b)

Fig. 3. (a) PET imaging without super- resolution.   (b) PET imaging with the novel registration method.

4   Conclusions

In this paper, a novel registration algorithm GAPY is introduced to optimize super-resolution reconstruction of PET imaging and then the registration accuracy is highly improved. The wobbling motion can get low-resolution images with sub-pixel shift and a small angle rotation. This GAPY algorithm will be applied into PET imaging and authenticated by computer simulation. From the simulation results, the conclusion can be drawn as follows: the proposed registration algorithm can improve the resolution of PET images effectively.

Acknowledgments. This research is supported by Leading Academic Discipline Project of Shanghai Municipal Education Committee (No. J50104).

References

1. Park, S.C., Park, M.K., Kang, M.K.: Super-resolution image reconstruction: A technical overview. IEEE Signal Process. 20(3), 21–36 (2003)
2. Kennedy, J.A., Israel, O., Frenkel, A., Bar-Shalom, R., Azhari, H.: Improved image fusion in PET/CT using hybrid image reconstruction and super-resolution. Int. J. Biomed. Imag. 46, 846 (2007)
3. Kennedy, J.A., Israel, O., Frenkel, A., Bar-Shalom, R., Azhari, H.: Super-resolution in PET imaging. IEEE Trans. Med. Imag. 25(2), 137–147 (2006)
4. Chang, G.P., Pan, T., Qiao, F., Clark Jr., J.W., Mawlawi, O.R.: Comparison between two super-resolution implementations in PET imaging. J. Nucl. Med. 49, 63 (2008)
5. Chang, G.P., M.S.: A new implementation of Super Resolution technique in PET imaging. Rice University, AAT 1455224 (2008)

6. Zhi, Y., Alireza, A., Yan, P.M., Kamata, S.: Image registration based on genetic algorithm and weighted feature correspondences. In: International Symposium on Consumer Electronics, pp. 42–46 (2009)
7. Kye, Y.J., Kyuha, C., Woo, H.N., Ji, H.K., Jong, B.R.: Image resolution improvement based on sinogram super-resolution in PET. Engineering in Medicine and Biology Society (EMBC), pp. 5712–5715 (2010)
8. Chang, G., Pan, T., Clark Jr., J.W., Mawlawi, O.R.: Optimization of super-resolution processing using incomplete image sets in PET imaging. J. Nucl. Med. 49(suppl. 1), 393 (2008)
9. Keren, D., Peleg, S., Brada, R.: Image Sequence Enhancement Using Sub-Pixel Displacement. In: Proceedings IEEE Conference on Computer Vision and Pattern Recognition, pp. 742–746 (1988)
10. Lu, Y., Inamura, M., Valdes, M.: Super-resolution of the undersampled and subpixel shifted image sequence by a neural network. Int. J. Imag. Syst. Technol. 14(1), 8–15 (2004)
11. Chang, G., Pan, T., Qiao, F., Clark Jr., J.W., Mawlawi, O.R.: Improving PET image resolution and SNR using super resolution postprocessing. J. Nucl. Med. 48, 411 (2007)
12. Chang, G., Pan, T., Qiao, F., Clark Jr., J.W., Mawlawi, O.R.: Reducing PET scan duration by improving SNR using super-resolution techniques. Med. Phys. 34(6), 2354 (2007)
13. GATE Users Guide, http://www.opengatecollaboration.org
14. Zhi, Y., Yan, P.M., Li, S.: Super resolution based on scale invariant feature transform. In: 7-9th International Conference on Audio, Language and Image Processing, pp. 1550–1554 (2008)
15. Fessler, J.A.: Image reconstruction toolbox, http://www.eecs.umich.edu/fessler/code
16. Matlab toolbox superresolution_v_2.0, http://lcavwww.epfl.ch/software/superresolution

Pedestrian Analysis and Counting System with Videos

Zhi-Bin Wang[1], Hong-Wei Hao[2], Yan Li[1], Xu-Cheng Yin[1,*], and Shu Tian[1]

[1] School of Computer and Communication Engineering,
University of Science and Technology Beijing, Beijing 100083, China
[2] Institute of Automation, Chinese Academy of Sciences, Beijing 100090, China
wzb1818@yahoo.cn, hongwei.hao@ia.ac.cn, toffer@yeah.net,
xuchengyin@ustb.edu.cn, tshu23@gmail.com

Abstract. Reliable estimation of number of pedestrians has played an important role in the management of public places. However, how to accurately count pedestrians with abnormal behavior noises is one challenge in such surveillance systems. To deal with this problem, we propose a new and efficient framework for pedestrian analysis and counting, which consists of two main steps. Firstly, a rule induction classifier with optical-flow feature is designed to recognize the abnormal behaviors. Then, a linear regression model is used to learn the relationship between the number of pixels and the number of pedestrians. Consequently, our system can count pedestrians precisely in general scenes without the influence of abnormal behaviors. Experimental results on the videos of different scenes show that our system has achieved an accuracy of 98.59% and 96.04% for the abnormal behavior recognition and pedestrian counting respectively. Furthermore, it is robust against the variation of lighting and noise.

Keywords: Video surveillance, Behavior recognition, Pedestrian counting.

1 Introduction

Along with the rapid development of the worldwide urbanization, the crowded phenomenon has become more and more frequently. It is timely to realize intelligent crowd surveillance and management though computer vision and pattern recognition techniques [1, 2]. Therefore, video surveillance system that provides automated services for pedestrian analysis and counting has played an important role in the management of public places relating to safety and security [3, 4].

Estimating the number of pedestrians automatically is a practical task to video surveillance system. An accurate and real-time estimated results can provide valuable information for managers to make decisions. In general, there are two categories of methods for pedestrian counting, i.e. (1) feature regression-based methods and (2) detection or tracking-based methods. In the first category, a regression model is used to learn the relationship between features and the number of pedestrians [5, 6]. The feature usually includes pixel, texture, shape or silhouette. And, the regression method

---

* Corresponding author.

T. Huang et al. (Eds.): ICONIP 2012, Part V, LNCS 7667, pp. 91–99, 2012.

may be linear regression, neural network, Gaussian process regression or discrete classifier. These methods have one remarkable advantage, which is that pedestrian's privacy can be well preserved [4]. But, the accuracy of pedestrian counting is heavily affected by camera perspective and overlapping among pedestrians. The second category tries to detect and track individuals in video sequences with feature trajectories, and then the number of pedestrians in scenes can be counted [7, 8]. Such methods have good counting performance when pedestrian density is very low. But, as the density increases, they are hardly used for real-time applications due to their high computational complexity and instability.

To our knowledge, most of pedestrian counting methods have been achieved in general scenes without the influence of abnormal behavior. In fact, abnormal behavior classification is an important factor for improving the reliability of pedestrian counting system. In the literature, various approaches have been proposed [3]. Among these methods, the analysis of feature trajectories is a kind of promising method to abnormal behavior recognition [9, 10]. It comprises tracking each object in the scene, and learning models for the resulting object tracks. Because of abnormalities definition varying with applications, it is quite difficult to compare these solutions.

Since the abnormal behavior recognition is necessary for improving the accuracy of pedestrian counting, this paper presents a new and efficient framework for pedestrian analysis and counting to deal with this situation. The proposed system consists of two processing steps, abnormal behavior recognition and pedestrian counting. In the first stage, a rule induction classifier with optical-flow feature is designed to recognize the abnormal behaviors which include three categories, cluster, across and reverse. Then, in the second stage, a linear regression model is used to estimate the number of pedestrians in general scenes without the influence of abnormal behaviors. Therefore, this system not only can count the pedestrians, but also deal with the occurred abnormal events.

The rest of this paper is organized as follows. In section 2, we provide a brief of system architecture. Section 3 describes detailedly the methods to abnormal behavior recognition and pedestrian counting. We present the experimental results on different video clips in Section 4. Finally, Section 5 concludes this work.

2   System Overview

The proposed system is specially used in the public places relating to security, such as Olympic park, World Expo Park, etc. It has two tasks, abnormal behavior recognition and pedestrian counting. The relationship between them is that the output of abnormal behavior recognition is the input of pedestrian counting. To accomplish these tasks, this system was divided into three parts, feature extraction, behavior classification and pedestrian counting, as shown in Fig.1.

In the first part, we extracted the optical-flow feature to distinguish different behaviors. What's more, we adopted an improved Lucas-Kanade (L-K) algorithm to compute the optical flow. This improved algorithm based on the pyramid of image sequences can overcome the drawback of the original L-K method. Besides, the Harris corner was used in this method to strengthen the robustness against noise and reduce the computational complexity.

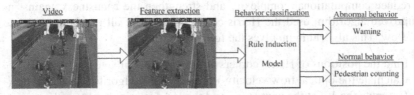

Fig. 1. Architecture of pedestrian analysis and counting systems

Then, in the second part, a rule induction classifier with optical-flow feature was designed to recognize the abnormal behaviors. They were defined as three categories in this paper, including cluster, across and reverse. Due to the stable motion characteristics, the rule induction classifier had achieved a better classification results. That is to say, if there are abnormal events happing, this system can display warning information to managers, or else count the pedestrians.

Finally, in the third part, we adopted a linear regression model to estimate the pedestrians in general scenes without the influence of abnormal behaviors. Besides, we took three measures to improve the accuracy of pedestrian counting which will be discussed detailedly in section 3.

3   Methods of Behavior Classification and Pedestrian Counting

3.1   Feature Extraction

In most cases, pedestrians do not move randomly but follow specific motion patterns. So, the optical-flow feature can be used to distinguish different behaviors.

Optical flow is a commonly used method to estimate velocity in sequential images, and various techniques have been proposed. The L-K algorithm is a well known gradient-based method for optical flow estimation [11]. It assumes that the optical flow is essentially constant in a local neighborhood, and solves the basic optical flow equations for all pixels in that region by the least squares criterion. But this method can be only used when the velocity vector between two frames is small. For a variable frame rate, we adopted an improved L-K method based on the Gaussian pyramid of image sequences [12]. Its working schematic is shown in Fig.2.

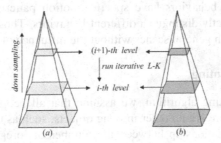

Fig. 2. Schematic of the pyramid L-K algorithm: (*a*) the current frame; (*b*) the next frame

To reduce computational complexity and strengthen the robustness against noise, this improved algorithm used the Harris corners instead of all pixels to calculate the optical flow. This algorithm involves the follows steps.

(1) Find the positions of Harris corners at the top of the image pyramid;
(2) Calculate the optical flow velocity $V_{(i+1)}$ using L-K algorithm at the level $(i+1)$;
(3) Compute the $V_{(i)}$ at the next layer, and the $V_{(i+1)}$ is initial value of $V_{(i)}$;
(4) Return to step (2) until $i=0$.

Because optical flow is calculated from the highest level of image pyramid, the error will be enlarged in the next layer. Therefore, the original image can not be decomposed too much. In general, the number of level is within the scope of 2 to 6.

3.2   Behavior Classification Based on Rule Induction

Rule induction is an important method for classification. Like its name, these methods use a set of if-then rules to describe the classification process [13]. In general, a rule induction classifier is comprised of two parts, rules set and classifier structure. The rules which can be written as a collection of consecutive conditional statements are used to describe classified patterns. And all the rules are organized to recognize test sample by classifier structure. Due to the accuracy for special patterns, a rule induction classifier is adopted to recognize the abnormal behavior.

With the optical-flow feature and pixels variation, the rule induction classifier is described as follows.

Rule 1: if the direction of optical-flow is opposite to the setting value, then output the reverse abnormality;

Rule 2: if the variation of pixels and velocities are below a threshold respectively, then output the cluster abnormality;

Rule 3: if the changes of pixels in special area exceed a threshold, then output the across abnormality;

Rule 4: if there are no abnormal behaviors, then estimate the number of pedestrians.

Because the abnormal behaviors have specific motion patterns, the rule induction classifier is able to exactly distinguish different behaviors. Thus, we can estimate the number of pedestrians in general scenes without the influence of abnormal behaviors.

3.3   Pedestrian Counting

In the pedestrian counting algorithm, we assume that all people move on a known ground plane where there are no other moving objects, such as cars or animals. So, if we have known the relationship between the number of foreground pixels and the number of pedestrians, then we can obtain an approximate estimation. However, such method suffers from three disadvantages, (1) abnormal behavior occurring, (2) lighting and noise influence, and (3) perspective changes.

In this paper, we employed three measures to solve these issues. First, the rule induction classifier was used to recognize abnormal behaviors, making sure that pedestrian counting was in general scenes. Second, an adaptive threshold method was utilized to segment pedestrians from background. Besides, this method was robust against the variation of lighting and noise. Finally, a fixed line region was used to gather the pixels from pedestrians so that all people in images had same changes.

Segmentation results of the adaptive threshold method for an image can be illustrated in Fig.3 (b). In this method, an image is equally divided into fixed number sub-blocks. Taken a $m \times n$ pixels sub-block for example, the averaging pixels can be calculated as follows.

$$T = \left(\sum_{x=0}^{m} \sum_{y=0}^{n} d(x, y) \Big/ (m \times n) \right) - p \qquad (1)$$

Here, $d(x,y)$ is a difference image of three frames. The constant $P > 0$ can adjust the segmentation results. If $d(x,y) \geq T$, the brightness of a pixel is 255, or else it is 0.

In Fig.3, the region $\overline{abcd}$ selected on the road is a region-of-interest (ROI) which is used to pedestrian counting, and the red line $\overline{ef}$ is employed to gather pixels from pedestrians when they walk across this interesting line.

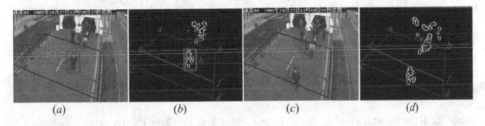

(a)   (b)   (c)   (d)

Fig. 3. Pixel sampling: (a) the reference pedestrian at the front of the red line, (c) at the end; (b) and (d) the threshold images, which show the process of pixel sampling when the pedestrian walks across the red line

Suppose a pedestrian with height h is walking across the red line from Fig.3 (a) to Fig.3 (c), we can acquire the number of pixels of this pedestrian according to Fig.3 (b) and Fig.3 (d). Using this method, we had acquired ten groups of average pixels with different number of pedestrians in videos. The corresponding relationship between the number of average pixels and the number of pedestrians is displayed in table 1.

Table 1. The relationship between the number of average pixels and the number of pedestrians

Num of pedestrians	1	2	3	4	5	6	7	8	9	10
Num of videos	22	19	20	21	20	18	16	18	16	15
Average pixels	904	1815	2867	3733	4660	5653	6373	7335	8222	9186

In table 1, we establish ten groups of average pixels with different number of pedestrians. And each group contains the same number of pedestrians in different scenes. For example, in the first group data, there are 22 videos, and each includes one person. We gather the pixels from these videos respectively. Thus, we can acquire 22 different pixel sets. Because the pixels varied with different pedestrians, we adopted the average pixels to represent one pedestrian.

Using the ten groups of data, the relationship between the number of average pixels and the number of pedestrian can be learned by the least squares criterion. The fitted linear equation is as follows.

$$y = ax + b \tag{2}$$

Here, x, y is the number of pedestrians and pixels respectively, and $a = 912.97$, $b = 54$.

For different applications, the parameters of a, b should be re-learned using the method discussed in this paper.

4   Experiments

The performance of the exploited system was tested on the CASIA Pedestrian Counting Dataset [14]. This dataset consisted of two parts, Image Data and Video Data. In experiment, we chose the videos in scene 5, which were captured from an outdoor entrance with safe guard check in. The frame size is 352×288 pixels and sampling rate is 25 *fps*. The pedestrians range from 20×40 pixels to 30×60 pixels in the video frames. To evaluate the performance of our system, we split the videos into two categories, without abnormal behaviors and with abnormal behaviors, which are displayed in list 1 of table 2.

Table 2. Experimental results: list 1 is test video set; list 2 is the recognition results of abnormal behavior; list 3 is the pedestrian counting results

List 1	Total	Without abnormal	With abnormal
Test set	103	33	70
List 2	Actual	Prediction	*Err_rate*
Total	142	142	0.0141
Cluster	32	32	0
Reverse	50	51	0.02
Across	60	59	0.02
List 3	Actual	Prediction	*Err_rate*
Total	2119	2091	0.0396

Based on the motion characteristics, the abnormal behaviors of pedestrians were classified into three categories, cluster, across and reverse, as shown in Fig.4. The error rate for abnormal behavior recognition is calculated as follows.

$$Err\_rate = (\sum_{i=1}^{n} |d_i - t_i|) \Big/ \sum_{i=1}^{n} t_i \qquad (3)$$

Here, d_i, t_i is the number of pedestrians of prediction and actual respectively.

In table 2, the list 2 presents detail results of abnormal behavior recognition. The total number of abnormal behaviors occurring is 142 in 70 videos. And 103 videos are used to evaluate the proposed system.

It is worth noting that an accuracy of 98.59% is achieved for abnormal behavior recognition. There are two reasons for that. Firstly, pedestrians do not move randomly but follow specific motion patterns. So, the rule induction classifier can exactly distinguish different abnormal behaviors. Secondly, there are no other moving objects, such as cars or animals so that reduce the influence factor for recognition.

When there are not abnormal events happing, this system estimates pedestrians using the formulate (2). The list 3 in table 2 has shown the counting results. We have made an accuracy of 96.04% for the pedestrians counting in 103 videos.

The Fig.4 illustrates the application results of our pedestrian analysis and counting system. The accurate and real-time pedestrian counting results can provide valuable information for managers to make decisions. For example, when the number of pedestrians exceeded the maximum set value, this system will show the warning information of crowd to alter the managers.

| (a) | (b) | (c) |

Fig. 4. Application results. The yellow region and red text are warning information when the abnormal events occurred. The number of pedestrians is displayed in lower right corner of image. (*a*) the cluster abnormal behavior, (*b*) the across abnormal behavior, (*c*) the reverse abnormal behavior.

5   Conclusion

This paper describes a pedestrian analysis and counting system. We divided this system into three parts, feature extraction, behavior classification and pedestrian counting. Experimental results on 103 videos in the CASIA Pedestrian Counting

Dataset have demonstrated the system's accuracy and robustness. Besides, we simply discuss the relationship between abnormal behavior recognition and pedestrian counting. And we regard the abnormal behavior as the influence factor for improving the reliability of pedestrian counting system. Due to application requirement, the proposed system is specially used in the public places relating to security, such as Olympic park, World Expo Park, etc. Therefore, the robustness of our system has not been verified in other datasets. In the further work, we will study a machine-learning based model to accurately express the relationship between abnormal behavior recognition and pedestrians counting on different datasets.

Acknowledgments. The research was partly supported by National Natural Science Foundation of China (61105018, 61175020), and the R&D Special Fund for Public Welfare Industry (Meteorology) of China (GYHY201106039, GYHY201106047).

References

1. Zhan, B.B., Monekosso, D.N., Remaqnino, P., Velastin, S.A., Xu, L.Q.: Crowd Analysis: A Survey. Machine Vision and Applications 19(5-6), 345–357 (2008)
2. Haritaoglu, I., Harwood, D., Davis, L.S.: W^4: Real-time Surveillance of People and Their Activities. IEEE Trans. Pattern Anal. Mach. Intell. 22(8), 809–830 (2000)
3. Xiong, G.G., Cheng, J., Wu, X.Y., Chen, Y.L., Ou, Y.S., Xu, Y.S.: An Energy Model Approach to People Counting for Abnormal Crowd Behavior Detection. Neurocomputing 83, 121–135 (2012)
4. Zhang, J.P., Tan, B., Sha, F., He, L.: Predicting Pedestrian Counts in Crowded Scenes with Rich and High-dimensional Features. IEEE Trans. Intell. Transp. Syst. 12(4), 1037–1046 (2011)
5. Chan, A.B., Vasconcelos, N.: Counting People with Low-level Features and Bayesian Regression. IEEE Trans. Image Process. 21(4), 2160–2177 (2012)
6. Liu, J., Wang, J., Lu, H.: Adaptive Model for Robust Pedestrian Counting. In: Lee, K.-T., Tsai, W.-H., Liao, H.-Y.M., Chen, T., Hsieh, J.-W., Tseng, C.-C. (eds.) MMM 2011 Part I. LNCS, vol. 6523, pp. 481–491. Springer, Heidelberg (2011)
7. Viola, P., Jones, M., Snow, D.: Detecting Pedestrians Using Patterns of Motion and Appearance. Int. J. Comput. Vis. 63(2), 153–161 (2005)
8. Morris, B.T., Trivedi, M.M.: A Survey of Vision-based Trajectory Learning and Analysis for Surveillance. IEEE Trans. Circuits. Syst. Video. Technol. 18(8), 1114–1127 (2008)
9. Mahadevan, V., Li, W.X., Bhalodia, V., Vasconcelos, N.: Anomaly Detection in Crowded Scenes. In: 2010 IEEE Computer Society Conference on Computer Vision and Pattern Recognition, pp. 1975–1981. IEEE Press, San Francisco (2010)
10. Jeong, H., Chang, H.J., Choi, J.Y.: Modeling of Moving Object Trajectory by Spatio-Temporal Learning for Abnormal Behavior Detection. In: 8th IEEE International Conference on Advanced Video and Signal Based Surveillance, pp. 119–123. IEEE Press, Klagenfurt (2011)
11. Lucas, B., Kanade, T.: An Iterative Image Registration Technique with an Application to Stereo Vision. In: 7th International Joint Conference on Artificial Intelligence, pp. 674–679. IJCAI Press, Vancouver (1981)

12. Bouguet, J.Y.: Pyramidal implementation of the Lucas Kanade feature tracker description of the algorithm (2004),
    http://robots.stanford.edu/cs223b04/algo_tracking.pdf
13. Michalski, R.S.: A theory and methodology of inductive learning. Artificial Intelligence 20(2), 111–161 (1983)
14. Li, J.W., Huang, L., Liu, C.P.: Robust People Counting in Video Surveillance: Dataset and System. In: 8th International Conference on Advanced Video and Signal Based Surveillance, pp. 54–59. IEEE Press, Klagenfurt (2011)

Vehicle License Plate Localization and License Number Recognition Using Unit-Linking Pulse Coupled Neural Network

Ya Zhao and Xiaodong Gu[*]

Department of Electronic Engineering, Fudan University,
220 Handan Road, Shanghai, 200433, P.R. China
{10210720050,xdgu}@fudan.edu.cn

Abstract. In this paper, Unit-linking Pulse Coupled Neural Network (U-PCNN) is applied in vehicle license Plate Localization (PL) and license Number Recognition (NR) being part of optical character recognition(OCR). PL and NR are cores of License Plate Recognition (LPR) system. In PL, firstly, the proposed algorithm based on U-PCNN edge detection highlights plate regions, and then using those results obtains plate locations. In NR, employing U-PCNN extracts features of license number. The experimental results show that the license plates properly extracted were 224 over 233 input images (96.137%) and the NR accuracy is 96.67%.

Keywords: Unit-linking Pulse Coupled Neural Networks (U-PCNN), Plate Localization(PL), Intelligent Transportation System (ITS), Optical Character Recognition (OCR).

1   Introduction

Vehicle License Plate Recognition (LPR) system is an important part in Intelligent Transportation System (ITS) for LPR has been widely applied in transportation control and surveillance systems. The LPR system consists of various technologies such as image processing, pattern recognition, neural networks and so on[1]. So lots of people just dedicated on one aspect. [2, 3] did Plate Localization (PL), [4, 5] did the plate Characters Segmentation (CS) and [6, 7, 8] did Character Recognition (CR).

However, the LPR system is made up of 1).PL, extraction of a license plate region; 2).CS, segmentation of the plate characters; and 3).CR, recognition of each character. These three steps are serial and the later one impacted by the former one, any robust step can remedy others. But the lack of coherence is obvious by using different algorithms from different researchers because the requirements of inputs and outputs cannot match unhindered [9].Some studies made the whole system [10], but each step needs individual codes so that the speed slowed down and need more resources.

---

[*] Corresponding author.

T. Huang et al. (Eds.): ICONIP 2012, Part V, LNCS 7667, pp. 100–108, 2012.

In this paper, a new LPR algorithm mainly focused on PL and CR based on U-PCNN is proposed. These two parts are considered the core techniques of LPR. And the second step was conducted by a robust algorithm called projection [5] which is well matched with the proposed LP outputs and CR inputs. U-PCNN is the simplified traditional PCNN. U-PCNN is a further modified model avoiding parameters selecting [11, 12], and [11, 13] detail the steps to do thinning, edge detection and other image processing application. The firing condition of U-PCNN can be easily controlled. And PCNN is easy to implement in hardware by the same structure units.

LP algorithm in this paper is based on U-PCNN edge detection [11]. This algorithm does not use the edge features directly, while let widened edges highlight plate regions. How to extract plate regions is illustrated in section 2. Section 3 presents the algorithm of NR by matching time sequence features, followed the conclusion in section 4.

2   Plate Localization Based on U-PCNN

PL technique will directly determine the LPR and its accuracy as the first step of the LPR system. The survey [9] had summarized existed LP methods and their accuracies, so this article will not repeat. The features of plate region itself are the cores for all these algorithms like humans did this task. But humans can perform much effort. These features include color information, high contrast between characters and background, textures and edges, brightness, symmetry, rectangle, ratio and so on. No matter what features are employed, it is not as easy as humans accomplished this task. PCNN exhibits synchronous pulse bursts in cat and monkey visual cortexes. With this biological background, it is reasonable to expect PCNN to perform better.

2.1   Edge Detection by U-PCNN

Traditional PCNN which has five equations is somehow complicated [11]. These equations represent three parts of a Pulse Coupled Neuron (PCN), the input part, the modulation field and the pulse generator. In image processing application, the input part consists of $F_{ij}(n) = I_{ij}$ and $L_{ij}(n)$ channel, where I_{ij} is the intensity value of the image pixel. $L_{ij}(n)$ channel takes neighbor pixels' information in and makes firing condition uncontrollable in traditional PCNN. Let $L_{ij}(n)$ be a simply one de-

$$\text{rived U-PCNN: } L_{ij}(n) = \begin{cases} 1 \text{ if } \sum_{k \in N_{ij}} Y_k(n-1) > 0 \\ 0 \text{ else} \end{cases} \tag{1}$$

Other equations are the same as the Traditional PCNN for image processing. This U-PCNN makes firing condition controllable. Edge detection in this section and binary image thinning used in next section are both based on this less-parameter PCNN.

U-PCNN based edge detection technique has two advantages. One is the width of the edge is controlled by setting iteration times N [13]. N time-iteration leads to (N-1) pixel-width edge. The other one is fast speed which is very necessary for LPR.

There are several edge detection techniques such as sobel operator, canny operator and morphology method. It is hard to compare which is better due to different applications. Most acknowledged LP algorithm is edge detection integrated with morphology [2].The operator in morphology is based on the distance between object and camera. However, the proposed LP method can highlight plate region without any prior knowledge. So this method can be used in more conditions in theory.

2.2   Highlight Plate Region by U-PCNN

Different iteration times N are set in the following experiment. The result is shown by the figures in Fig 1.

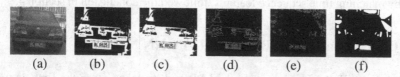

(a)	(b)	(c)	(d)	(e)	(f)

Fig. 1. (a) original image, (b) U-PCNN edge detection of 15 pixel width (N=16),(c) U-PCNN edge detection of 254 pixel width(n=255),(d) U-PCNN edge detection of 1 pixel width (N=2),(e) SOBEL vertical mask edge detection, (f) Morphological processing

The original image is not in good illumination. And in this experiment, just the second channel of RGB image is chosen to be processed other than changing RGB into gray by any rules. It can shorten the time. This should be a small revolution.

Carefully compare image (b) and (c) of Fig.1. No matter how wide the edge is, license plate regions are obvious and changeless while other regions merge together or disappear. Although the firing condition of U-PCNN is easier to be analyzed than other PCNN, it still seems interesting and not easy to figure out why the characters in plate did not merge together. To find the fact, look at (d) which is one-pixel width edge by setting N equals 2.and (e). Image (e) is sobel edge detection. As the width of the edges increased, the edges merged together because the width of characters' edges increased.The edges of neighboring characters overlapped each other. But the firing pulses are disabled to jump across the strokes of each character and edges increased out of the first edges. So the merged edges make the background of plate region. As a result, plate region is highlighted by those serried characters.

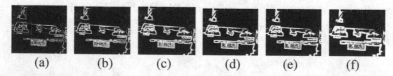

(a)	(b)	(c)	(d)	(e)	(f)

Fig. 2.   (a) N=4, 3-pixel width edge, (b)N=6, 5-pixel width edge, (c)N=8, 7-pixel width edge,(d) N=10, 9-pixel width edge, (e)N=12, 11-pixel width edge, (f)N=14, 13-pixel width edge

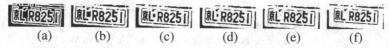

(a)	(b)	(c)	(d)	(e)	(f)

Fig. 3. Focus on Plate Region corresponding to Fig.2. from (a) to (f)

Fig.2 and Fig.3 show how edges merge together without change of license characters' stroke. Look at the regions in red rectangle of (a) in Fig.2. (like the car front glass region), the edges on the front glass merged in to one region as the edge widened. Some background regions merge as well. Not like license characters, they are stable as edge added beyond the first contour, only background changed and the fired pixels cannot cross the strokes of characters. Unbroken expanse of the plate background and independent characters make the Plate Region easy to extract. The width of this algorithm is not strict, for this exampled image, N can be14, 15 or bigger. N is 16 in the following experiments. Bigger N means more iteration, it will be time consuming. 16 times iteration only needs 0.033sec for this 421×502 pixels image.

In conclusion, the width-controlled edge detection method lessens texture of less textural regions like car front glass and background, and highlights some features of plate regions, 1) full of edges, 2) high contrast between characters and background, and 3) the almost uniform thickness of character strokes. The more features used in localization, the higher accuracy it should be. The result of experiments proves it.

2.3   Further Processing of Localization

The width-controlled edge detection based on U-PCNN algorithm highlights the Plate Region. So the posterior steps are not complicated. First, reverse the binary images of 15-pixel width edge by NOT operator. Then the characters are 1 and the back of plate is 0 (Fig.4. (a)). Step II, remove big and small regions which cannot be region of interest (ROI) (Fig.4. (b)). Step III, count change times in one row of each block after segment the whole image into 4 blocks like Fig.4 (c). Step IV, find out the row of the highest change density (no matter in which block), e.g. the row number is "y". Step V, compare the change times of neighbor rows to the whole "y" row's change times and yield the difference of change times. Those rows whose difference are not bigger than two third of the y's, it is regarded as rows of Plate Region. Step VI, use those rows including Plate to calculate the angle of avertence of the plate by "radon" operator in matlab, e.g. theta. Finally, rotate the whole image by theta and segment the same rows as Step VI.

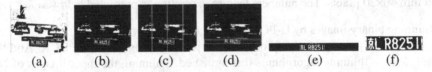

(a)   (b)   (c)   (d)   (e)   (f)

Fig. 4.   (a) Step I, (b)Step II, (c)Step III,(d)Step IV & V, (e)Step VII, (f) is the result

These seven steps only extract one feature of plate region that is full of changes. Other features are embodied in the "Highlight" step. The result is like (f) in Fig.4.

2.4   Comparisons between Proposed Algorithm and Morphology Based Method

The algorithm in ref.[2] was mainly based on modified sobel operator vertical edge detection and morphology. (b) and (c) in Fig.5 is the results. (b) has some features like license which change 7 times like plate region. (c) is the plate region, but two front characters are missed because the morphology operator is not suitable and there is a big margin between second and third characters which made the result a mess.

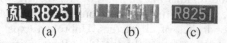

(a)   (b)   (c)

Fig. 5. (a)is the result of U-PCNN based, (b) and (c) are both the results of ref.[2]

The images of vehicles are from the website of a commercial corporation called "Green Intelligence Technology".[1] 199 images from the website and other 34 images taken by self are the experimental image set. The accuracy rate of ref.[2] is only 157 over 233. Algorithms in other papers are also not well performed for our dataset.

3   Number Recognition Based on U-PCNN

The outputs of proposed PL algorithm are binary images with all seven characters in the plate region. There are margins on top, bottom and two sides (see Fig.5 (a)) are not wanted for the NR algorithm. Projection method in [4] can cut out all those margins and outputs every character. The outputs are binary images.

The main idea of NR algorithm is feature matching. The features are gotten by U-PCNN fired images. U-PCNN can segment images by firing pixels with similar intensities [12]. However, to the binary images, the pixel whose linking domain includes more than one fired pixels will fire in next iteration so long as it never fired before.

3.1   Feature Extraction of License Plate Character

The proposed algorithm just takes numbers of license for test to verify the efficiency of the feature matching algorithm. Two features are extracted from input images. All images are from the proposed PL algorithm outputs by projection algorithm and resized into 40×20 pixels. The numbers feature extraction is presented below.

1. Thinning binary images by U-PCNN

Thinning is always the preprocessing of pattern recognition like OCR. And the morphology thinning algorithm is distinguished. Thinning the input images of NR by U-PCNN has two reasons. One is that the thinning method just changes two elements of U-PCNN for edge detection, [13] for reference. So it is fluent to the whole LPR system. Another is that this method has good anti-noise property.

2. Feature extraction

(a) Process with U-PCNN

After thinning based on U-PCNN, the width of strokes is one or two pixels. To get time sequences by U-PCNN, the object is "1" and background is "0" in binary images. U-PCNN need no parameter set. The matrix $Y_{ij}(n)$ records the firing condition of the $(n-1)_{th}$ time. Three times iteration yielded fired images $Y_{ij}(1)$, $Y_{ij}(2)$ and $Y_{ij}(3)$.

(b) Get features

Divide the image in two methods, as shown in Fig.6. Method 1 divided images into eight blocks and method 2 into four. And for iteration results do processing below

Fig. 6. Method 1 in the left and **Method 2** in the right

    (i) Calculate the total number of fired pixels, i.e. sum $Y_{ij}(1)$, $Y_{ij}(2)$ and $Y_{ij}(3)$.

    (ii) Calculate the number of fired pixels in each block under division method 1 and method 2 for $Y_{ij}(1)$, $Y_{ij}(2)$ and $Y_{ij}(3)$.

From (i), each image gets one three-demission feature. This feature only contains time sequence information, so named it as Time Sequence Characteristic (TSC).

From (ii), each image has three 12-demission features. These features contain block time signals corresponding to 12 blocks. The block time signal is named Block Time Characteristic (BTC) which reflects the structural information of time sequence.

TSC includes the perimeter information of the objects because thinning method extracts the skeleton. And three fired images from iteration includes outside corner information. But only TSC cannot reflect structural information inside the characters. So BTC is needed. BTC likes grid feature, but it is from time sequence fired images. These features are easy to get. The preprocessing thinning method based on U-PCNN is insensitive to noise, so the TSC and BTC features are also anti-noise.

3.2   Number Recognition rules

The complete NR algorithm is described as follows.

    Step I, For the input images, extract TSC and BTC features based on U-PCNN.

    Step II, For each feature, compare them to those features of templates. And different voting weights are given to corresponding index.

    (a) Calculate the difference between it and those features of template images. The signal of feature difference is gotten, named it as DIF[i] ("i" is from 0 to 9, which means the difference signal between input image to ten templates.).

    (b) Calculate the absolute sum of the all components of DIF[i], find the minimum one, e.g. the index of the minimum is 5. Then, get index of the maximum of all components in DIF[i] is the minimum. If it is the same as the former index, some weight is voted.

For the TSC, voting weight is 1. For BTC, voting weights are set as (9-n), where n is the iteration time. For $Y_{ij}(1)$, n is 1 and the voting weight is 8.

Step III, Make the final decision according to the vote result. The index which has the most votes should be the result of the input images. NR finished.

The whole procedure above can be considered as a number recognizer. It is unlike normal matching algorithms which calculate the Euler distance between input images to templates. It has less features and less computational complexity.

3.3   Experiment Result and Analysis

The recognition method can be generalized to letters and Chinese characters of plate. Because the license character set is finite. In the experiment, only numbers 0 to 9 are chosen. These number characters are all from real scene which may suffer distortion, noise and uneven illumination. And the binary results are affected (Fig.7 left).

Fig. 7. One group numbers from real scene v.s.   The template numbers

There are ninety experimental numbers, each number has nine images. The tested numbers are not large enough so some noise was added in them to verify the validity of the algorithm. The results are in Tbl.1. Tbl.1 verifies that this algorithm is not sensitive to pepper and salt noise but very sensitive to Gaussian noise when use the real scene images. There are some shortages in this compare for that the real scene image already has some noise (Fig.7 left). To verify its anti-noise characteristic, real scene characters are replaced by template numbers. The results are in Tbl.2.

Table 1. Original real scene characters with different noise

	Original Real scene Characters (ORC)	ORC with 0.01 pepper & salt noise	ORC with 0.02 pepper & salt noise	ORC with 0.1 pepper & salt noise	ORC with 0.2 pepper & salt noise	ORC with 0.01 Gaussian noise
Accuracy rate	87/90	82/90	73/90	73/90	52/90	65/90

Table 2. Templat copies with different noise

	template Characters (TC)	TC with 0.01 pepper & salt noise	TC with 0.02 pepper & salt noise	TC with 0.1 pepper & salt noise	TC with 0.2 pepper & salt noise	TC with 0.01 Gaussian noise
Accuracy rate	90/90	85/90	82/90	79/90	69/90	51/90

Compare Tbl.1 with Tbl.2, it is easy to figure out that this algorithm is sensitive to Gaussian noise but can overcome very strong pepper and salt noise which is more common in digital cameras. Meanwhile, another conclusion by comparing Tbl.1 and Tbl.2 is that the real scene characters include very less Gaussian noise.

4   Conclusion

Two core techniques using U-PCNN in this paper can efficiently be used in LPR, with both the Localization and NR accuracies over 96%. The work also verified the prominence of image processing using U-PCNN. In PL step, U-PCNN based algorithm performs better than morphology based one. And in NR step, the proposed method shows good anti-noise performance by comparison, especially in tolerating pepper and salt noise.

Acknowledgements. This work was supported in part by Shanghai National Natural Science Foundation under grant 12ZR1402500 and National Natural Science Foundation of China under grant 61170207.

References

1. Anagnostopoulos, C.N.E., Anagnostopoulos, I.E., Loumos, V., Kayafas, E.: A License Plate-Recognition Algorithm for Intelligent Transportation System Applications. IEEE Transactions on Intelligent Transportation Systems 7(3), 377–392 (2006)
2. Qiu, Y.J., Sun, M., Zhou, W.L.: License Plate Extraction Based on Vertical Edge Detection and Mathematical Morphology. In: IEEE Conference on Computational Intelligence and Software Engineering, pp. 1–5. IEEE Press, Wuhan (2010)
3. Mai, V.D., Miao, D.Q., Wang, R.Z., Zhang, H.Y.: An Improved Method for Vietnam License Plate Location. In: 2011 International Conference on Multimedia Technology (ICMT), pp. 2942–2946 (2011)
4. Li, H., Zhang, H.H.: Improved Projection Algorithm for License Plate Characters Segmentation. In: 2009 Second International Conference on Intelligent Computation Technology and Automation, pp. 534–537 (2009)
5. Xia, H.D., Liao, D.C.: The Study of License Plate Character Segmentation Algorithm based on Vetical Projection. In: 2011 International Conference onConsumer Electronics, Communications and Networks (CECNet), pp. 4583–4586 (2011)
6. Wu, J., Xiao, Z.T.: New License Plate Character Recognition Algorithm Based on ICM. In: 2011 International Conference on Control, Automation and Systems Engineering (CASE), pp. 1–4 (2011)
7. Zhu, Y.Q., Li, C.H.: A Recognition Method of Car License Plate Characters Based on Template Matching Using Modified Hausdorff Distance. In: 2010 International Conference on Computer, Mechatronics, Control and Electronic Engineering (CMCE), vol. 6, pp. 25–28 (2010)
8. Jin, Q., Quan, S.H., Shi, Y., Xue, Z.H.: A Fast License Plate Segmentation and Recognition Method Based on the Modified Template Matching. In: 2nd International Congress on Image and Signal Processing, CISP 2009, pp.1–6 (2009)
9. Anagnostopoulos, C.-N.E., Anagnostopoulos, I.E., Psoroulas, I.D., Loumos, V., Kayafas, E.: License Plate Recognition From Still Images and Video Sequences: A Survey. IEEE Transactions on Intelligent Transportation Systems 9(3), 377–391 (2008)
10. Wen, Y., Lu, Y., Yan, J.Q., Zhou, Z.Y.: An Algorithm for License Plate Recognition Applied to Intelligent Transportation System. IEEE Transactions on Intelligent Transportation Systems 12(3), 830–845 (2011)

11. Gu, X.D., Zhang, L.M., Yu, D.H.: General Design Approach to Unit-linking PCNN for Image Processing. In: Proceedings in Int. Joint Conf. Neural Networks, Montreal, Canada, pp. 1836–1841 (2005)
12. Gu, X.D., Guo, S.D., Yu, D.H.: A New Approach for Automated Image Segmentation Based on Unit-linking PCNN. In: Proceedings in IEEE International Conference on Machine learning and Cybernetics, Beijing, China, pp. 175–178 (2002)
13. Gu, X.D., Yu, D.H., Zhang, L.M.: Image Thinning Using Pulse Coupled Neural Network. Pattern Recognition Letters 25, 1075–1084 (2004)

ROI-HOG and LBP Based Human Detection via Shape Part-Templates Matching

Shenghui Zhou, Qing Liu, Jianming Guo, and Yuanyuan Jiang

School of Automation, Wuhan University of Technology, Wuhan, China
shenghuizhou2009@126.com, {qliu2000,jmGuo62}@163.com,
enxin_402@hotmail.com

Abstract. Currently, Histogram of Oriented Gradient (HOG) descriptor serves as the predominant method when it comes to human detection. To further improving its detection accuracy and decrease its large dimensions of feature vectors, we introduce an improved method in which HOG is extracted in the Region of Interest (ROI) of human body with a combined Local Binary Pattern (LBP) feature. Via establishing human shape part-templates tree, a template matching approach is employed to improve detection results and segment human edges. The experimental results on INRIA database and images from practical campus video surveillance demonstrate the effectiveness of our method.

Keywords: Pedestrian detection, ROI-HOG descriptor, LBP feature, Shape part-template matching.

1 Introduction

In 2005, Dalal[1] introduced Histogram of Oriented Gradient (HOG) descriptor for human detection. Due to its excellent performance, HOG has been widely accepted as one of the best features to capture edge information. Over the last several years, among its continuous improvement, combing HOG with additional features is an effective way. Xiaoyu Wang[2] proposed an approach capable of handling partial occlusion by combing HOG and Local Binary Pattern (LBP)[3] feature; Qing Liu[4] integrated HOG with color frequent and skin color feature to describe pedestrian, adopted a two-stage detection method, and trained each stage classifier separately. Although these methods improve detection accuracy to some degree, they also lead to certain feature dimension redundancy which means extra computation burden.

As for obtaining human targets from video sequences, researchers take various approaches to enhance real-time property of HOG. Changjiang Li[5] combined HOG and Kalman Filter for video-based human detection and tracking to make the relationship that detection did not only serve as the prerequisite of tracking, but also benefited from tracking. In spite of accelerated detection speed, however, the algorithm above can hardly deal with the occasions of human body blocks in real situation.

On the basis of previous works and in the hope of solving the problems mentioned above, we propose an improved feature extraction method in which, with added LBP feature, HOG is extracted in the Region of Interest (ROI) of human body edge

T. Huang et al. (Eds.): ICONIP 2012, Part V, LNCS 7667, pp. 109–115, 2012.

repeatedly for the purpose of strengthening the collection of contour information. We use simple linear SVM as the baseline classifier in comparison with Dalal's result. Using the Hierarchical Part-Template Matching method introduced by Zhe Lin[6] for reference, we create a human body shape part-template tree. When it comes to human detection, we combine part-template matching with our descriptor to precisely present different targets. Experimental results on INRIA dataset and campus video surveillance validate the effectiveness of our newly developed HOG method.

2   HOG Descriptor in ROI

HOG has exhibited great effectiveness and robustness when representing human body's local edge and shape information. However, for each scanning window, it is doubtless that human's shape and edge information are not evenly distributed, which implies the possibility of setting ROI for the extraction of HOG. By analyzing the positive samples of INRIA dataset, we find that human head and torso's gestures and locations are comparatively fixed, leg's take second place, and arm's information is the most variable and complicated. Meanwhile, the shape and edge information of background might confuse human detection.

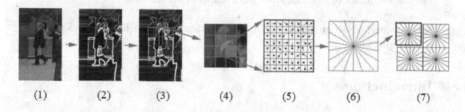

(1)   (2)   (3)   (4)   (5)   (6)   (7)

Fig. 1. Human Body Edge ROI and HOG Feature Extraction

In order to further analyze the distribution of human body HOG feature, as shown in Fig.1-(2), we use Sobel operator to extract the shape and edge information for graying positive sample images in INRIA dataset. Fig.1-(3) shows the four ROI rectangles (head-torso part in red, right and left arm parts in green and legs part in blue) obtained by qualitative analysis. After setting ROI, the dimension of a scanning window's ROI-HOG descriptor is reduced to 1764. The location, size and feature vector information of each ROI are shown in Table1.

For each pixel in a ROI rectangle, the gradient computation process of Dalal's approach is fully transplanted. To be concrete, as our algorithm involves the quantization for gradient extraction ROIs and the human body templates matching, fixed-size feature extraction blocks and cells are now used. Each detection window is divided into cells of size 8×8 pixels (Fig.1-(5)) and each group of 2×2 cells (Fig.1-(4)) is integrated into a block in a sliding fashion, so blocks overlap with each other. Each cell consists of a 9-bin (Fig.1-(6)) HOG between the interval [-π/2, π/2] and each pixel contributes a gradient magnitude weighted vote to the bins. Each block is thus represented by a 36-D vector (Fig.1-(7)) that is normalized to L1 unit length.

Table 1. Details of Human Body Edge ROI

Categories	Initial Point Position (X-axis, Y-axis)	Size Width×Height (pixels)	Feature Vector Dimensions
Head-Torso	(16, 12)	32×32	324
Right and Left arms	(4, 24) / (36, 24)	24×40/24×40	288/288
Legs	(12, 64)	40×56	864

3 Extraction of Cell-Structured LBP Feature

LBP feature is a texture descriptor that can filter out noisy edges using the concept of uniform pattern. Because HOG performs poorly when the background is cluttered with noisy edges, the combination of HOG and LBP makes up the deficiency.

Simply, the LBP code of a pixel is obtained by comparing its intensity with those of its neighbors. To be more specific, let LBP_N^R denote the LBP feature that takes N sampling points on a circle of radius R (N=8, R=1 in our experiment). For each pixel p_c in a scanning window, its surrounding points are denoted as p_i (i=0,1,...,7), then LBP code of pixel p_c is defined as:

$$LBP_{p_c} = \sum_{i=0}^{7} 2^i \times z(p_i - p_c) \quad , z(a) = \begin{cases} 1, a > 0 \\ 0, a \leq 0 \end{cases} \tag{1}$$

In this way, we compute LBP feature for each pixel in scanning window from top left to bottom right and build LBP histogram for the window by utilizing function (2):

$$H_j = \sum_{p_c} J\{LBP_{p_c} = j\} \ (j = 0, ..., m - 1) \quad , J(A) = \begin{cases} 1, \ A \ is \ true \\ 0, A \ is \ false \end{cases} \tag{2}$$

where m is the number of LBP patterns (the number of LBP_8^1 patterns is $2^8 = 256$).

One typical type of LBP is the uniform pattern (denoted as $LBP_{8,1}^2$), which means the number of 0-1 (or 1-0) transitions of a 8-bit binary code is no more than 2. In the computation of $LBP_{8,1}^2$ histogram, different uniform patterns contribute unweighted votes to corresponding bins and all non-uniform patterns vote into one bin, which is:

$$LBP_{p_c} = \begin{cases} \sum_{i=0}^{7} 2^i \times z(p_i - p_c), u \leq 2 \\ 85, u > 2 \end{cases} \tag{3}$$

where u is the number of transitions. We use decimal number 85 (transformed from 01010101) to denote the bin into which all of the non-uniform patterns are counted.

Via utilizing $LBP_{8,1}^2$ operator, the number of LBP patterns can be reduced from 256 to 59. Meanwhile, in our experiments on INRIA dataset, we find that for each positive sample image, the pixels owing uniform pattern LBP features account for more than 85%, which means $LBP_{8,1}^2$ operator can effectively represent the most useful visual texture information in a scanning window.

Being similar to HOG extraction, we divide each scanning window into several non-overlapping LBP-Cells of size 16×16 pixels. For each LBP-Cell, we build histogram according to 59 patterns and normalize feature vector using L1-normaliztion. As for a 64×128 pixels detection window, 1888-D LBP feature vector is obtained.

4   Shape Template Matching

Using the approach introduced by Zhc Lin[6] for reference, we also create human shape part-templates for further analysis of the distribution of HOG feature in each scanning window. Due to their complicated gestures, human arms are excluded when establishing the shape templates tree. Table 2 shows the parts of human body used in our experiment with their definitions and parameters of the degree of freedom.

Table 2. Details of Mimic Human Body Parts

Categories	Definition of Degree of Freedom	Numbers of Degree of Freedom
Head	Position relative to Torso	3 (Left, Middle, Right)
Torso	Width	2 (Wide, Narrow)
Upper Legs	Position relative to Torso	3 (Vertical, 30°Left, 30°Right)
Lower Legs	Position relative to Upper Legs	3 (Vertical, 30°Left, 30°Right)

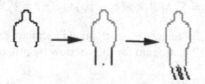

Fig. 2. An Example of Shape Templates Tree Matching

As shown in Fig.2, the whole shape templates tree includes three layers, which are head-torso (HT) layer, upper-legs (UL) layer and lower legs (LL) layer respectively. We adopt top-down sequence when doing the matching and the matching result of the upper layer will decide the matching choices in the lower layer directly. Following this rule and based on the freedom parameters of each body parts, we calculate that the HT layer consists of 3×2=6 different templates, the UL layer includes 3×3×2=18 templates and the LL layer owns 3×3×18=162 templates. For a whole matching process, we have to make 6+9+9=24 matching choices.

In each layer, we compute and compare the matching score for each part template candidate using an approach similar to the extraction of HOG feature. In the gray scanning windows, we calculate gradient magnitudes and edge orientations using differential opcrator (-1, 0, 1) for each pixel both in horizontal and vertical directions. In each 8×8 pixels non-overlapping cell, edge orientations are also quantized into 9 bins (each pixel contributes a gradient magnitude weighted vote to the bins). Differently, we establish feature histogram without normalization for the cells instead of the blocks when extracting HOG feature. If the pixels in the part templates match the pixels locating in human body edges in the scanning window, it is obvious that the feature histogram of the cells nearby will display a high gradient value.

Let G(t) denote the edge orientation at pixel t, its corresponding orientation bin index B(t) is computed as:

$$B(t) = [G(t)/(\pi/9)] \tag{4}$$

where operator [x] denotes the maximum integer less than or equal to x.

Because of the non-overlapping cells, we can identify the cell where pixel t locates in. Using $H = \{h_i\}$ (bin index i=0, 1, ..., 8) representing the unnormalized orientation histogram, then the matching score of pixel t can be computed as:

$$f(t) = \sum_{b=-\sigma}^{\sigma} \omega(b)h_{B(t)+b} \tag{5}$$

where σ denotes a neighborhood range and $\omega(b)$ is a symmetric parameter. To be concrete, in our experiment, $\sigma=1$, $\omega(1) = \omega(-1) = 0.25$, $\omega(0) = 0.5$.

Given a part template T which has N sampling points, the matching score is computed as the average pixel score along the template:

$$f(T) = \frac{1}{N}\sum_{t\in T} f(t) \tag{6}$$

As mentioned above, the whole template tree includes three layers. Following the whole top-to-down sequence, we compute the matching scores for all of the part templates in one layer and choose the one owning the biggest total score to represent the human edge and shape information in the scanning window; the chosen template determines the next layer's part template indexes based on their relative position in the template tree. At last, we get three part templates from the three different layers to constitute a whole human body edge.

5   Experimental Results

To evaluate our algorithm, we conduct two main verification experiments: 1) Detection Error Tradeoff (DET) curves; 2) human detection and silhouette segment.

We conduct DET curve plotting experiments on INRIA dataset which contains 2,416 positive and 1,218 negative training images plus 1,132 positive and 453 negative testing images. Due to its variable pose articulations, occlusions, clutters, INRIA dataset is deemed to be a very challenging and generally acknowledged dataset.

As shown in Fig.3, we compare our results with Dalal's DET curves obtained by using the same Linear SVM. Obviously, using the same Linear SVM classifier, our curve keeps below Dalal's throughout, which means we achieve better performance in terms of detection accuracy. Besides, in practical projects, FalseNeg detection windows should be effectively avoided. Our algorithm shows remarkable superiority in the aspect of limiting missing rate. Numerically, when missing rate is less than 0.01, we achieve a false positives per-window rate which is less than 0.01.

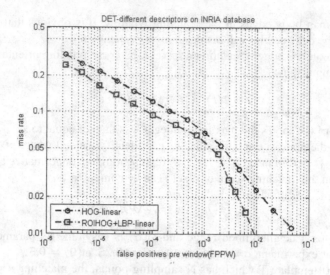

Fig. 3. Experimental Detection Error Trade off (DET) curves

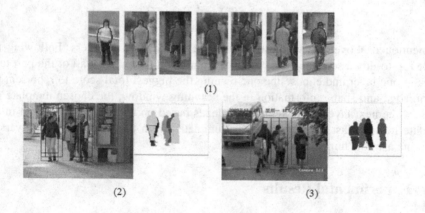

Fig. 4. Human Target Detection and Segment

Using the ROI-HOG&LBP human detector, we also evaluate our part-template matching approach applying on both INRIA person dataset (Fig.4-(1), (2)) and the practical campus video surveillance (Fig.4-(3)). All the detection and edge matching results are carried out without any background extraction. For each image, all of the positive detection windows are displayed and the human edges are subjected to an empty image for the purpose of evaluating the accuracy of our segment approach.

As shown in Fig.4, we get quite clear human detection results for integral images via utilizing ROI-HOG&LBP classifier, which is the further evidence of our algorithm's accuracy. For the positive samples of INRIA dataset (Fig.4-(1)), part-template tree matching based silhouette segment performs very well. When it comes to integral images (Fig.4-(2), (3)), although our current numerical threshold method for part template matching cannot always deal with the occlusions and clutters between different human targets accurately and steadily, we can obtain body silhouettes with a small number of redundant data, which we believe can be diminished in future works.

6   Conclusions

The main work completed in this paper is summarized as follows: firstly, on the basis of HOG human detector, we introduce a ROI-HOG&LBP human descriptor and experimentally verify its effectiveness and efficiency on INRIA dataset; secondly, via establishing human shape part templates tree, a template matching approach is employed to improve detection results and segment human edges. In the future, we will use quantitative method to analyze the distribution of HOG feature in order to gain ROIs more exactly. The threshold method of part template matching process also needs further research.

References

1. Dalal, N., Triggs, B.: Histograms of Oriented Gradients for Human Detection. In: IEEE Computer Society Conference on Computer Vision and Pattern Recognition, vol. 1, pp. 886–893. IEEE Press, New York (2005)
2. Wang, X.Y., Han, T.X., Yan, S.C.: An HOG-LBP Human Detector with Partial Occlusion Handling. In: 12th IEEE International Conference on Computer Vision, pp. 32–39. IEEE Press, New York (2009)
3. Ojala, T., Pietikinen, M., Maenpaa, T.: Multiresolution Gray-Scale and Rotation Invariant Texture Classification with Local Binary Pattern. In: IEEE Trans on Pattern Analysis and Machine Intelligence, vol. 24, pp. 971–987. IEEE Press, New York (2002)
4. Liu, Q., Qu, Y.: HOG and Color Based Adaboost Pedestrian Detection. In: 7th IEEE International Conference on Natural Computation, vol. 1, pp. 584–587. IEEE Press, New York (2011)
5. Li, C.Y., Guo, L.J., Hu, L.C.: A New Method Combining HOG and Kalman Filter for Video-Based Human Detection and Tracking. In: 3rd International Congress on Image and Signal Processing, vol. 1, pp. 290–293. IEEE Press, New York (2010)
6. Lin, Z., Davis, L.S.: Shape-Based Human Detection and Segmentation via Hierarchical Part-Template Matching. In: IEEE Transactions on Pattern Analysis and Machine Intelligence, vol. 32, pp. 604–618. IEEE Press, New York (2010)

Matrix Pseudoinversion
for Image Neural Processing

Rossella Cancelliere[1], Mario Gai[2], Thierry Artières[3], and Patrick Gallinari[3]

[1] Universitá di Torino, Turin, Italy
rossella.cancelliere@unito.it
[2] National Institute of Astrophysics, Turin, Italy
[3] LIP6, Université Pierre et Marie Curie, Paris, France

Abstract. Recently some novel strategies have been proposed for training of Single Hidden Layer Feedforward Networks, that set randomly the
weights from input to hidden layer, while weights from hidden to output
layer are analytically determined by Moore-Penrose generalised inverse.
Such non-iterative strategies are appealing since they allow fast learning,
but some care may be required to achieve good results, mainly concerning the procedure used for matrix pseudoinversion. This paper proposes a
novel approach based on original determination of the initialization interval for input weights, a careful choice of hidden layer activation functions
and on critical use of generalised inverse to determine output weights.
We show that this key step suffers from numerical problems related to
matrix invertibility, and we propose a heuristic procedure for bringing
more robustness to the method. We report results on a difficult astronomical image analysis problem of chromaticity diagnosis to illustrate
the various points under study.

Keywords: Pseudoinverse matrix, Weights initialization, Supervised
learning.

1 Introduction

In the past two decades, single hidden layer feedforward neural networks (SLFNs)
have been one of the most important subject of study and discussion among neural
researchers. Methods based on gradient descent have mainly been used, although
defining different learning algorithms; among them there is the large family of techniques based on backpropagation, widely studied in its variations [1]. The start-up
of these techniques assigns random values to the weights connecting input, hidden
and output nodes, these weights are then iteratively adjusted.

In any case, gradient descent-based learning methods are typically slow, frequently require small learning steps, and are subject to convergence to local
minima. Therefore, many iterations may be required by such algorithms in order to achieve an adequate learning performance, and many trials are required
to avoid poor local minima.

Some non iterative procedures based on the evaluation of generalized pseudoinverse matrices have been recently proposed as novel learning algorithms for

T. Huang et al. (Eds.): ICONIP 2012, Part V, LNCS 7667, pp. 116–125, 2012.

SLFNs: among them the method to improve performance of multilayer percep-tron by Halawa [2], the extreme learning machine (elm) [3] and some studies more application oriented [4,5,6]. Usually, input weights (linking input and hidden lay-ers) are randomly chosen, and output weights (linking hidden and output layers) are analytically determined by the Moore-Penrose (MP) generalized inverse (or pseudoinverse). These theoretically appealing methods have many interesting features among which is their non iterative nature, that makes them very fast, but some care may be required because of the known numerical instability of the process of pseudoinverse determination (see [7]).

This paper presents a deep investigation on a few major issues of pseudo-inversion-based learning of SLFN's and on related numerical problems; we put in light that this technique can not be used without a careful analysis, because of the eventual presence of almost singular matricies whose pseudoinversion can cause instability and consequent error growth. We also suggest a method based on threshold tuning to give greater stability to solutions.

We first recall the main ideas on SLFN learning through matrix pseudoin-version in section 2. In section 3 we present the main methods to compute out-put weights and their possible issues. In section 4 we describe the astronomical problem of chromaticity diagnosis through image analysis. Finally in section 5 we propose a new initialization criterion for input weights, we compare various ways of evaluating pseudoinverse and discuss results and performance trends.

2   How to Train Weights by Pseudoinversion

A standard SLFN with P inputs, M hidden neurons, Q output neurons, non-linear activation functions ϕ on the hidden layer and linear activation functions otherwise, computes an output vector $o = (o_1, ..., o_Q)$ from an input vector $x = (x_1, ..., x_P)$ according to:

$$o_k = b_k^O + \sum_{i=1}^{M} w_{k,i}\phi(\mathbf{c}_i \cdot \mathbf{x} + b_i^H) \qquad k = 1, \cdots Q \tag{1}$$

where the vector $\mathbf{c}_i$ denotes the weights connecting input units to hidden neuron i, $w_{k,i}$ denotes the weight linking the hidden neuron i to the output neurons k, b_k^O and b_i^H denote biases for output and hidden neurons.

Considering a dataset of N distinct training samples of (input, output) pairs $(\mathbf{x}_j, \mathbf{t}_j)$, where $\mathbf{x}_j \in \Re^P$ and $\mathbf{t}_j \in \Re^Q$, learning a SLFN aims at producing the desired output $\mathbf{t}_j$ when $\mathbf{x}_j$ is presented as input, i.e. at determining weights w, c, and biases b such that:

$$t_j^k = b_k^O + \sum_{i=1}^{M} w_{ki}\phi(\mathbf{c}_i \cdot \mathbf{x}_j + b_i^H) = \sum_{i=1}^{M+1} w_{ki}\phi(\mathbf{c}_i \cdot \mathbf{x}_j + b_i^H) \quad k=1, \cdots Q \ j=1, \cdots N \tag{2}$$

We made the assumptions $\phi(\mathbf{c}_{M+1} \cdot \mathbf{x}_j + b_{M+1}^H) = 1$ and $w_{k,M+1} = b_k^O$ to include the bias terms in the sum; in the following we will deal with a generic number M

of hidden neurons to simplify notation. The above equations can thus be written compactly in matrix form as

$$T = Hw, \qquad (3)$$

where:

$$
w = \begin{vmatrix} w_1^T \\ \vdots \\ w_M^T \end{vmatrix}_{M \times Q} \quad
T = \begin{vmatrix} t_1^T \\ \vdots \\ t_N^T \end{vmatrix}_{N \times Q} \quad
H = \begin{vmatrix} \phi(c_1 \cdot x_1 + b_1) & \cdots & \phi(c_M \cdot x_1 + b_M) \\ \vdots & \ddots & \vdots \\ \phi(c_1 \cdot x_N + b_1) & \cdots & \phi(c_M \cdot x_N + b_M) \end{vmatrix}_{N \times M} \qquad (4)
$$

H is the hidden layer output matrix of the neural network; the i-th column of H is the i-th hidden node output with respect to inputs $x_1, x_2, \cdots x_N$. In most cases of interest, the number of hidden nodes is much lower than the number of distinct training samples, i.e. $M \ll N$, so that H is a non-square matrix; in this case a *least-squares solution* is searched, determining c, b, w such that the following cost functional is minimized:

$$
E_D = \sum_{j=1}^{N} \sum_{k=1}^{Q} \left(t_j^k - \sum_{i=1}^{M} w_{ki} \phi(c_i \cdot x_j + b_i) \right)^2 = \|Hw - T\|^2. \qquad (5)
$$

As stated above, gradient-based iterative learning algorithms, that require to adjust input weights and hidden layer biases, are generally used to search the minimum of $\|Hw - T\|^2$.

The least-squares solution w^* of the linear system is then (see [8])

$$w^* = H^+ T, \qquad (6)$$

where H^+ is the Moore-Penrose pseudoinverse of matrix H.

The solution w^* has some important properties, in fact it is one of the least-squares solutions of the general linear system (3), hence it reaches the smallest training error. Besides, it has the smallest norm among all least-squares solutions and it is unique.

Huang et al. [3] proved that, contrarily to what happens in traditional training algorithms, input weights and hidden layer biases of a SLFN do not need to be adjusted, but they can be randomly assigned. In doing so, SLFN output weights can be simply considered as the solution of a linear system and analytically determined through generalized inversion (or pseudoinversion) of the hidden layer output matrix, assuming that hidden neurons activation functions are infinitely differentiable. These ideas gave rise to an algorithm, which is extremely fast because it works in a single pass. Besides, pseudoinversion provides the smallest weights norm solution: this point is quite relevant, since, according to Bartlett's theory [9] the smaller is the norm of weights, the better generalization performance is usually achieved by the network.

3   Pseudoinverse Computation

Hereafter we address the main critical issue which concerns learning of output layer weights by pseudoinversion.

Several methods are available, (see e.g. [10]), to evaluate the Moore-Penrose pseudoinverse matrix: in particular, the orthogonal projection (OP) method can be used when $\mathbf{H}^T\mathbf{H}$ is nonsingular, so that

$$\mathbf{H}^+ = (\mathbf{H}^T\mathbf{H})^{-1}\mathbf{H}^T \tag{7}$$

but it is known that this method is affected by severe limitations when the matrix $\mathbf{H}^T\mathbf{H}$ is almost singular. In this case, the computation of the pseudoinverse is highly unstable, and consequently the product $\mathbf{H}^+\mathbf{H}$ is potentially much different from the unit matrix. A possible solution consists in the addition of a regularization term to the cost functional (5)

$$E = E_D + \lambda E_W \tag{8}$$

where λ is the regularization coefficient that controls the relative strength of the data-dependent error E_D and the regularization term E_W. For L_2 regularization the term E_W usually takes the form $E_W = \frac{1}{2}\mathbf{w}^T\mathbf{w}$, so that the cost functional (5) becomes:

$$E = E_D + \lambda E_W = \frac{1}{2}\sum_{j=1}^{N}\sum_{k=1}^{Q}\left(\left(t_j^k - \sum_{i=1}^{M} w_{ki}\phi(\mathbf{c}_i \cdot \mathbf{x}_j + b_i)\right)^2 + \frac{\lambda}{2}\sum_{i=1}^{M}|w_{ki}|^2\right) \tag{9}$$

With this approach, i.e. regularised OP (ROP), the solution (6) becomes:

$$\mathbf{w}^* = (\mathbf{H}^T\mathbf{H} + \lambda\mathbf{I})^{-1}\mathbf{H}^T\mathbf{T}. \tag{10}$$

A different approach consists in evaluating the singular value decomposition (SVD) of $\mathbf{H}$, $\mathbf{H} = \mathbf{U}\mathbf{S}\mathbf{V}^T$ (see for instance [11]). $\mathbf{U}^{N\times N}, \mathbf{V}^{M\times M}$ are unitary matrices and $\mathbf{S}^{N\times M}$ has elements $\sigma_{ij} = 0$ for $i \neq j$ and $\sigma_{ii} = \sigma_i$ for $i = j$, with $\sigma_1 \geq \sigma_2 \geq \cdots \geq \sigma_p \geq 0$, $p = \min\{N, M\}$; the elements σ_i are the singular values of the decomposition.

We can also define the pseudoinverse matrix $\mathbf{H}^+$ through the SVD of $\mathbf{H}$ (see [7] for more details) as:

$$\mathbf{H}^+ = \mathbf{V}\mathbf{\Sigma}^+\mathbf{U}^T \tag{11}$$

where $\mathbf{\Sigma}^{+M\times N}$ has elements $\sigma_{ij}^+ = 0$ for $i \neq j$ and $\sigma_{ii}^+ = 1/\sigma_i$ for $i = j$. If $\sigma_1 \geq \sigma_2 \geq \cdots \geq \sigma_k > \sigma_{k+1} = \cdots = \sigma_p = 0$, the rank of matrix $\mathbf{H}$ is k and the inverses of the $p - k$ zero elements are replaced by zeros.

Also, this method may exhibits some problems in the presence of almost singular matrices because of the presence of very small singular values σ_i, whose numerical inversion may cause instability in the algorithm.

In section 5 we will therefore propose a solution through the introduction of a cut-off threshold, acting as a sort of 'regularisation' that stabilises the algorithm.

4   The Astronomical Problem

Astrometry, i.e. the precise determination of stellar positions, distance and motion, is largely based on imaging instrumentation fed by telescopes operating in the visible range. The measured image profile of a star however depends on its spectral type, i.e. the emitted light distribution as a function of frequency, so that its measured position appears affected by an error called chromaticity. Chromaticity was identified for the first time [12] in the data analysis of the space mission *Hipparcos* of the European Space Agency (ESA).

The chromaticity issue becomes even more important for the current European Space Agency mission Gaia [13] for global astrometry, aiming at much higher precision and approved for launch in 2013.

The detection and correction of chromaticity in different conditions has been addressed in recent years [14,15]; to this purpose, a single-hidden layer feedforward neural network (SLFN), trained by a classical BP algorithm, was used to solve this diagnosis task.

Each image is first reduced to a one-dimensional signal $s(x)$ along the single measurement direction of Gaia, by integration on the orthogonal direction, also to reduce the data volume and telemetry rate.

With respect to our previous studies, we also adopt as input to the neural network processing a more convenient set of statistical moments M_k for image encoding:

$$M_k = \sum_n (x_n - x_{COG})^k \cdot s(x_n) \cdot s_A(x_n) , \tag{12}$$

where $s(x_n)$ is the above mentioned signal, $s_A(x_n)$ is the signal from an ideal instrument, and x_{COG} is the signal barycenter.

Chromaticity can be more conveniently defined using the concept of blue (effective temperature $T = 30,000\,K$) and red ($T = 3,000\,K$) stars, in particular it is the barycenter displacement between them, and this is the neural network target. Correct diagnostics allows effective correction of this kind of error, therefore a good approximation by the neural network results in a small *residual chromaticity* after neural processing.

The 11 neural network inputs are the red barycenter and the moments of red and blue stars from Eq. 12 ($k = 1, \cdots, 5$), computed for each instance.

We have a total of 13000 instances, built according to the above prescriptions, and split in a training set of 10000 instances and a test set with 3000 instances. The data set was provided by the Astrophysical Observatory of Turin of the Italian National Institute for Astrophysics.

5   Experimental Results on Chromaticity Diagnosis

We investigate in this section on the one hand the influence of activation functions choice and on the other hand the importance of procedures used for evaluating the pseudoinverse. As an important result, we further put in evidence a numerical problem encountered dealing with singular value decomposition, that may lead to poor performance and for which we propose a solution that increases the robustness of the method, allowing it also to reach better performance with respect to classical backpropagation.

We investigate neural networks with 11 input neurons corresponding to the 11 statistical moments describing each image, 1 output neuron and a number of hidden neurons varying from 50 to 600. For each initial setup, the number of hidden nodes has been gradually increased adding one node each time and average results are computed over 10 simulation trials for each selected size of SLFN. All the simulations are carried out in the Matlab 7.3 environment.

The use of sigmoidal activation functions has recently been subject to debate because they seem to be more easily driven towards saturation due to their non-zero mean value [16]; hyperbolic tangent seems to be less sensitive to this problem, therefore we utilize this one. Besides, in the perspective of mitigating saturation issues, we propose a uniform random choice of input weights in the range $(-1/\sqrt{M}, 1/\sqrt{M})$, where M is the number of hidden nodes. This links the size of input weights, and therefore of input values to hidden neurons, to the network architecture, thus forcing the use of the almost linear central part of the hyperbolic activation function when exploring the performance as a function of an increasing number of nodes. Because for the hyperbolic tangent this part is centred on zero the presence of this factor contributes also to decrease hidden neurons outputs.

The technique we propose, named Hidden Space Related Pseudoinversion (HSR-Pinv), combines together this choice of initialization intervals and activation functions with the pseudoinverse evaluation method described by eq. (11). It is compared with the commonly used technique that initialize weights uniformly in the interval $(-1, 1)$, and employs sigmoidal activation functions (named σ SVD).

For the sake of comparison, we evaluate the pseudoinverse matrix using the different methods of evaluation presented in section 3, i.e. OP method, described by eq.(7), and ROP method described by eq.(10).

The performance reached by a standard SLFN neural network, with hyperbolic tangent as hidden neuron activation function trained using the backpropagation algorithm is the reference result. We initialized NN weights randomly according to a uniform distribution, looking for the minimum value of RMSE when the number of hidden nodes is gradually increased from 10 to 200. This range was selected according to the following three main reasons. On one side, previous investigations [3] show that the limiting performance of SLFN trained by backpropagation is achieved with a lower number of hidden neurons with respect to pseudoinversion methods. Besides, we trained some different models by varying the learning parameter η in the range (0.1, 0.9), as shown for a few cases in Fig. 1 (left), and the resulting RMSE level appears to reach an asymptotic value for all values of η quickly, using only a few ten hidden neurons. Finally, the iterative nature of backpropagation results in a heavy computational load, which also depends on the SLFN size.

Since we used a sufficiently large dataset for training and no overfitting arose, best results are achieved without the need for a regularization term, and learning consists then in minimizing the cost functional defined by eq. (5). With such a setting the best RMSE obtained is 3.81, with 90 hidden neurons.

Results are reported in Table 1. We record in the first row the best performance i.e. the best mean value over 10 simulation trials for all methods. For each tabulated value, the corresponding number of hidden neurons M_* used in the simulation is reported on the nearby column. The corresponding standard deviations are recorded in the second row and in the third row there is the optimal performance, i.e. the minimum value over 10 simulation trials.

Table 1. Comparison of performance on the test set

Method:	HSR-Pinv	M_*	σ-SVD	M_*	ROP	M_*	OP	M_*
Mean RMSE	1.77	573	8.52	88	7.69	90	9.43	60
St. Dev.	0.73	573	2.54	88	2.28	90	3.91	60
Min. RMSE	0.81	348	1.51	448	4.83	90	6.45	60

We remark that the best performance, in terms of both mean and minimum RMSE, is achieved by our proposed HSR-Pinv technique (first column). We also verified that the difference between HSR-Pinv mean RMSE value and the value obtained with backpropagation is significant within the 99% confidence interval and that the difference between HSR-Pinv min. RMSE value and σ-SVD min. RMSE value is significant within the 80% confidence interval.

Besides, our proposed HSR-Pinv technique achieves smaller variance, meaning a better behaviour with respect to over-fitting and it appears to be more able to take advantage of a larger hidden layer.

Fig. 1 (right) gives more insights on the evolution of the performance with the number of hidden units, showing the mean RMSE trend vs. number of hidden nodes (solid line); it is interesting to note that there is an error peak for a number of hidden neurons approximately equal to 250.

This error peak is related to the presence of singular values close to zero in matrix **S**, as discussed in section 3. This correlation is put in evidence by plotting also the ratio of the minimum singular value and the Matlab default

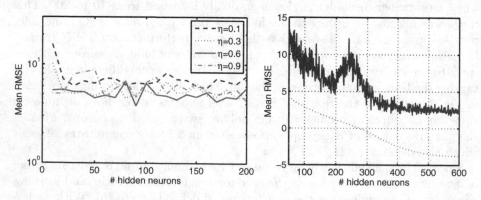

Fig. 1. Left: RMSE vs. number of hidden neurons for backpropagation training. Right: Mean RMSE vs. number of hidden neurons, and threshold ratio (logarithmic units).

threshold, below which singular values are considered too small (dotted line, using logarithmic units). The ratio approaches unity in the peak region.

This fact has potentially dramatic effects on performance, and must therefore be taken into account when SVD is used to evaluate $\mathbf{H}^+$, mainly because, as it happens in this case, the best performance can be reached with a number of hidden neurons larger than this critical dimension; the existence of this phenomenon has therefore to be known to avoid to early stop training in presence of error growth.

As a novel strategy to treat this problem we suggest a careful tuning of the threshold. We selected as a new threshold the mean value of the smallest singular values obtained over the 10 trials when using a hidden layer with 180 hidden neurons, because this is approximately the dimension of hidden neuron space when RMSE starts to increase.

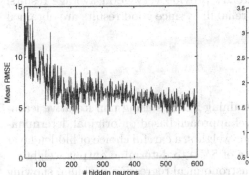 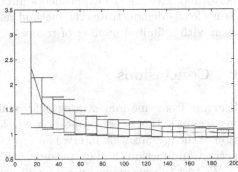

Fig. 2. Left: Mean RMSE vs. number of hidden neurons with threshold tuning. Right: Statistical analysis of parameter space

The effect of such a cut-off in fact is a sort of regularisation that provides a significant improvement in the method robustness, as can be seen from the relevant reduction of error peak, at the expense of a slight increase of RMSE values, as shown in Fig. 2 (left). Since the performance (mean RMSE = 4.12, St. Dev. = 0.83, min RMSE = 1.15) remains quite attractive, threshold tuning appears to be an interesting option. The above results show a clear advantage of the proposed strategy over standard training for SLFN. We provide some hints for a better understanding of such a phenomenon.

The actual model learning includes many random initialisations, at fixed network size, followed by the determination of corresponding output weights; a possible viewpoint is to keep the one that yields the best results, i.e. the optimal value recorded in the last row of Table 1.

Such a training phase, based on multiple random choices of weights, can be supposed to explore more extensively the parameter space, with respect to the trajectories followed by backpropagation algorithm, that develop continuously from a single random starting point. Yet this approach is reasonable provided that the number of trials, i.e. the number of different random extractions and output weights evaluations, remains limited.

We verified the presence in the parameter space of many *good solutions* that can be reached also with a relatively small number of initial random choices. We first performed a set of 1000 different training trials. Then for a given subset size s, ranging from 10 to 200, we randomly built 100 subsets each containing s trials among the initial 1000 cases, in order to achieve statistical information. We then selected in each subset the minimum value of test error, and evaluated mean values and standard deviations of the distributions of such minima over the 100 subset instances. The test results are summarised in Fig.2 (right).

As expected, the mean value of the error distribution decreases with increasing number of trials; also, the range of variability decreases quickly, meaning that a limited number of trials is indeed required to attain good results, i.e. close to the best case. In our experiments, 100 trials are enough to reach an almost optimal performance, while 10 experiments already provide very good average results. This clearly demonstrates the method reliability, since good results are obtained even with a limited number of trials.

6   Conclusions

Starting from emerging strategies for training SLFN's using non iterative learning schemes, this paper proposes a novel approach based on original determination of the initialization interval for input weights, a careful choice of hidden layer activation functions and on prudent use of SVD to determine output weights.

We test this approach on a difficult astronomical regression problem, showing that we reach the best performance with respect to pseudoinversion by orthogonal projection, with or without regularization, and classical backpropagation.

We also put in evidence the presence of error peaks with increasing number of hidden units. We validate the hypothesis that this trend is due to numerical instability and we propose the adoption of threshold tuning in the SVD context in order to reduce the effects of singular values related peaks, to obtain a more robust learning scheme while still reaching very good results.

Acknowledgements. The activity was supported by Fondazione CRT, Torino, and Università degli Studi di Torino in the context of the World Wide Style Project.

References

1. LeCun, Y.A., Bottou, L., Orr, G.B., Müller, K.-R.: Efficient BackProp. In: Orr, G.B., Müller, K.-R. (eds.) NIPS-WS 1996. LNCS, vol. 1524, pp. 9–50. Springer, Heidelberg (1998)
2. Halawa, K.: A method to improve the performance of multilayer perceptron by utilizing various activation functions in the last hidden layer and the least squares method. Neural Processing Letters 34, 293–303 (2011)

3. Huang, G.-B., Zhu, Q.-Y., Siew, C.-K.: Extreme Learning Machine: Theory and applications. Neurocomputing 70, 489–501 (2006)
4. Nguyen, T.D., Pham, H.T.B., Dang, V.H.: An efficient Pseudo Inverse matrix-based solution for secure auditing. In: IEEE International Conference on Computing and Communication Technologies, Research, Innovation, and Vision for the Future (2010)
5. Kohno, K., Kawamoto, M., Inouye, Y.: A Matrix Pseudoinversion Lemma and its Application to Block-Based Adaptive Blind Deconvolution for MIMO Systems. IEEE Transactions on Circuits and Systems I 57(7), 1449–1462 (2010)
6. Ajorloo, H., Manzuri-Shalmani, M.T., Lakdashti, A.: Restoration of Damaged Slices in Images Using Matrix Pseudo Inversion. In: 22th International Symposium on Computer and Information Sciences, Ankara, pp. 98–104 (2007)
7. Bini, D., Capovani, M., Menchi, O.: Metodi numerici per l'algebra lineare. Ed. Zanichelli, Bologna (1988)
8. Bishop, C. M.: Pattern Recognition and Machine Learning. Ed. Springer, Berlin (2006)
9. Bartlett, P.L.: The sample complexity of pattern classification with neural networks: the size of the weights is more important that the size of the network. IEEE Trans. Inf. Theory 44(2), 525–536 (1998)
10. Ortega, J.M.: Matrix Theory. Plenum Press, New York (1987)
11. Golub, G., van Loan, C.: Matrix computations. The Johns Hopkins University Press, London (1996)
12. Le Gall, J.Y., Saisse, M.: Chromatic Aberration of an All-Reflective Telescope. In: Instrumentation in Astronomy V, London. SPIE, vol. 445, pp. 497–504 (1983)
13. Perryman, M.A.C., et al.: GAIA - Composition, formation and evolution of the galaxy. Concept and technology study, Rep. and Exec. Summary. In: European Space Agency, ESA-SCI, Munich, Germany, vol. 4 (2000)
14. Gai, M., Cancelliere, R.: Neural network correction of astrometric chromaticity. MNRAS 362(4), 1483–1488 (2005)
15. Cancelliere, R., Gai, M.: Efficient computation and Neural Processing of Astrometric Images. Computing and Informatics 28, 711–727 (2009)
16. Glorot, X., Bengio, Y.: Understanding the difficulty of training deep feedforward neural networks. In: 13th International Conference on Artificial Intelligence and Statistics, AISTATS 2010, Chia, Italy (2010)

Learn to Swing Up and Balance a Real Pole Based on Raw Visual Input Data

Jan Mattner*, Sascha Lange, and Martin Riedmiller

Machine Learning Lab,
Department of Computer Science,
University of Freiburg,
79110, Freiburg, Germany
{mattnerj,slange,riedmiller}@informatik.uni-freiburg.de
http://ml.informatik.uni-freiburg.de

Abstract. For the challenging pole balancing task we propose a system which uses raw visual input data for reinforcement learning to evolve a control strategy. Therefore we use a neural network – a deep autoencoder – to encode the camera images and thus the system states in a low dimensional feature space. The system is compared to controllers that work directly on the motor sensor data. We show that the performances of both systems are settled in the same order of magnitude.

Keywords: Neural network, Pole balancing, Deep autoencoder, Reinforcement learning, Visual input.

1 Introduction

One of the main applications of reinforcement learning (RL) is controlling a dynamic system. In the ideal case the RL controller adapts automatically to a new system without the need to incorporate additional domain specific knowledge. Thus part of the setup has to be learned as well as the controlling task itself. A promising way is to use the raw and high dimensional image data of a camera, which monitors all relevant parts of the dynamic system, in order to learn a low dimensional feature space, that can be used for classical RL methods [1]. However when applying this to a real dynamic system, there appear three major problems. Firstly we have a delay. The camera information is not instantaneously available and the applied actions will not immediately have an effect. This results in observation and action delay, which cannot be handled by standard RL methods. Secondly, an image is just a snapshot of the dynamic system, so we have to add motion information of the moving parts of the system. Therefore a correct velocity estimation is needed. Thirdly, by creating a low dimensional map (feature space) of a manifold some information is always lost or obscured. This noise in the feature space may disturb the actual control task.

We analyze these problems and apply the proposed solutions to the well-known pole balancing task (see figure 1). Here the controller has to apply actions

---

* Corresponding author.

T. Huang et al. (Eds.): ICONIP 2012, Part V, LNCS 7667, pp. 126–133, 2012.

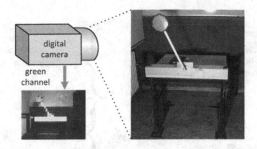

Fig. 1. The pole mounted on a table. The controller has to learn to swing up and balance the pole using only the downsampled visual data from a camera in front of the pole.

via a motor to a pole with a weight at its end in order to swing up and balance the pole in an upright position. The actions are so weak such that from the start position a constant action doesn't bring the pole into the goal area. So the controller has to learn to really swing the pole in both directions to increase it's momentum and beeing able to reach the top. For reducing the dimensionality of the raw image data and creating the feature space a deep encoder neural network is used. The actual controller for the pole balancing task uses a kernel based approximate dynamic programming method [2].

2   Related Work

The pole balancing task and its dynamics are well known [3]. It is also known as 'inverted pendulum', 'rotary pendulum' or 'cart-pole task' with an additional cart which has to be moved instead of direct motor interaction with the pole. It exists as a simulation [4] or real task [5]. In early research [6] a neural network was used to control the cart with small input images based on the simulation data. Visual input was used for a fixed control strategy too [7,8], however the main issue was to explicitly extract and calculate the position of the cart and the angle of the pole by handcrafted methods. We instead use deep neural networks [9] to autonomously encode the system state and learn a control strategy by RL methods. The deep networks were quite successful [10] and have proven to be able to provide a useful feature space for RL methods [1]. In our work we want to use this approach for the challenging pole balancing task and compare its performance with strategies which are learned directly on motor sensor data instead of the feature space.

3   The Feature Space

In order to apply RL methods to control the system we have to reduce the dimensionality of the input image data. Instead of manually extracting features, which describe the distinct system states, we are using here a deep autoencoder

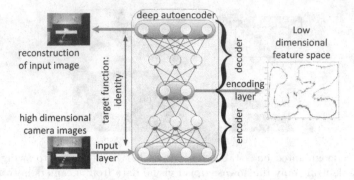

Fig. 2. Overview of an autoencoder

[9]. This is a neural network that can be divided into two separate networks: the encoder and decoder (see figure 2). The input layer neurons of the encoder correspond to the pixel values of the camera images. After multiple hidden layers the encoder reduces the information of the image to a small number of output neurons, which build the encoding layer. The topology of the decoder is mirrored, so these encoding neurons are the input layer of the decoder and its (overall) output layer is again an image, ideally the same image that has been used as the encoder's input. Hence the autoencoder is actually trained to yield the identity function of the camera images and therefore pushing all information through the bottleneck of the encoding layer, which then contains all necessary features to describe the visually perceptible system state. For a detailed description see [1]. The architecture of the used autoencoders is described in 5.1

This feature space has the property that visually close system states are represented by points in the feature space which are close together as well. Thus if there are only small changes, the evolution of the system states over time is reflected as a line of sample points through the feature space. In the case of the pole balancing task the pole can only turn around one axis. This circle topology causes the feature points to form a closed strap (see figure 3). Due to the used gradient descent update method the autoencoder may get stuck in a bad local optimum. For a 2D feature space this could mean dense agglomerations and many crossings of the strap. This can result in two completely distinct system states that cannot be distinguished only by their feature points. This problem is discussed in the next section.

4   Technical Details

In a real dynamic system there are always delays, be it due to sensor information capturing, transport and processing or due to inertia of the physical objects which move within the system. [11] introduces Deterministic Delayed Markov Decision Processes (DDMDP). In the pole balancing task we have deterministic observation and action delay. In [11] is also shown, that a DDMDP can be reduced to a simple MDP, which is needed for RL methods, by augmenting the state space. Thus for observation delay o and action delay a the state vector has to be augmented by the last $o + a$ actions. So we just have to determine the delays.

The observation delay is quite easy to measure. This is the time from the moment when the camera acquires the image until the point right before the application of the action, for which the image has been used for the first time to choose an action. This of course heavily depends on the synchronisation of sensor and motor.

Determining the action delay is more involved. This is the time from the action application until the action can be uniquely distinguished in the observation data from all other actions. Therefore we design an experiment with two action sequences, which are identical up to time step t, when one sequence applies a different action. Having the velocity data obtained by the motor sensor in several sample runs, we can determine, e.g. with the Welch's t-test (which suits well, as we have continuous data with possibly different variances), for a certain α-level at which time step the velocity sample values are drawn from different distributions.

A correct velocity estimation is crucial for controlling a dynamic system. As the feature space learned by the autoencoder is based on still camera images, we have to manually add velocity information. Adding action history as described above does not suffice unless all past actions are used, which is of course no option since this would drastically increase the dimensionality. Instead we have to estimate the velocity by adding either a past feature point (implicit) or the difference of the current and a past feature point (explicit) to the augmented state representation.

Estimating the velocity in such a way is of course problematic because the distance between the feature points usually does not correspond to the actual changes in the real world. But although the distance and direction in feature space as velocity introduces a systematic error, it works fairly well in practice. This is most likely because in general the velocities in the feature space are locally similar, i.e. states with similar position and velocity in the real world fall into the same area in the feature space and have similar predecessor feature points. Thus the controller can implicitly learn these local similarities. This problem does not inhibit the controller from learning a successful policy and therefore can be neglected.

The automatic mapping of the high dimensional images to the feature space does not work perfectly in all cases. One phenomenon is ambiguities of feature points. Figure 3 shows a two dimensional feature space with a crossing of the strap in horizontal and vertical direction. The reconstruction in (v2) of the vertical point at the crossing is wrong. Apparently the autoencoder preferred here the horizontal hypothesis since the vertical reconstruction corresponds to the horizontal one in (h2). Only knowing a feature point near the crossing, it is impossible to decide if the pole is pointing up or down. This would be an ambiguous state. A simple solution is to add the information of where the pole was

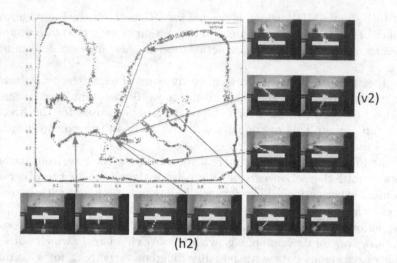

Fig. 3. A feature space with a crossing and each 3 corresponding example images. The example images consist of the original (left) and the reconstruction by the autoencoder (right).

one step before. If we know the previous feature point we can clearly tell that (v2) actually is on the vertical line. However this information is already given if we have added a correct velocity estimation.

5   Results

5.1   Experiment Set-Up

The empirical evaluation was done for the pole balancing task (see figure 1) with camera input (CAM) and direct motor sensor input (SEN). The controller has to learn to swing up and keep the pole upright in the goal area, which is defined to be at angle 0 with a buffer of 0.5 in both directions. The motor can apply actions from −100% (counterclockwise) to +100% (clockwise). However the controller is restricted to three actions: −40%, 0% (brake), +40%. This way it is not possible to reach the goal area in one go from the start of an episode by just applying constantly the same action. So the controller needs at least one swing to one side and then move down away from the goal and swing up to the other side. The goal is not terminal so it is a continuous time problem. The discounting factor is set to 0.95. For each time step the controller gets a reward of 0 if the pole is in the goal area and −1 for any other state. Each episode consists of 100 steps (or: cycles) but when the camera and motor run out of sync some of these transitions have to be discarded for learning.

The camera runs at $60Hz$, however camera and motor are synchronized such that a full cycle with observation acquisition and action application lasts $100ms$. This comprises 6 images but only the first and fourth image are processed and encoded to 2-dimensional feature vectors f_1 and f_4. We use explicit velocity estimation, so the state vector at time step t is $s_t = (f_1, f_1 - f_4)$. For a better velocity estimation we use the latest available image instead of the last state. The motor applies the action $83ms$ after the first image of a cycle is taken, which is the observation delay. The Welch's t-test of two action sequences with 30 sample runs each and an α-level of 0.001 has revealed that there is no action delay. This can be explained by the design of one motor cycle. Internally, the motor applies an action at the beginning of its cycle and returns its sensor information at the end of this cycle, after $33ms$. Thus the true action delay is less or equal to $33ms$, however due to the cycle design the actions can be uniquely distinguished within the same cycle of action application. All in all with the total delay of $83ms$ and the overall cycle time of $100ms$ we have to augment the state space by 1 past action. For explicit velocity estimation for the camera experiment the augmented state vector at time t results in the 5-dimensional vector $h_t = (s_t, a_{t-1})$ with $a_i \in \{-40, 0, +40\}$. The sensor experiment uses the same augmented state space, but here the state consists of the two scalars 'pole angle' and 'angular velocity' which leads to a 3-dimensional augmented state vector.

For each camera based controller a separate autoencoder was trained offline on 2000 independent training images of size 40x30, which took about $30min$ on an 8-core CPU. The input layer therefore has $40x30 = 1200$ neurons, followed by 2 convolutional hidden layers of size 40x30 and 20x15 with convolutional kernels of size 5x5, which is important for tracking the 3x3 pixel large green marker at the end of the pole. Then fully connected layers of size 150, 75, 37 and 18 up to the encoding layer of 2 neurons complete the encoder. The decoder's architecture is exactly mirrored up to the output layer of size 40x30, which should reconstruct the input image.

5.2 Empirical Evaluation

For the CAM and SEN experiments 5 independent controllers were each trained for 35000 transitions. A test run tests a controller of a given number of transitions for 100 steps. Thus the mean reward per step (MRPS), which serves as performance measurement, of a test run is in the range of the maximum (0) and minimum reward (-1). Therefore an MRPS of e.g. -0.21 means that the pole was in 21 steps not in the goal area. Since a random controller never reached the goal area (MRPS: -1), the beginning of the learning phase was supported by a fixed exploration policy until the controller collected one valid transition into the goal area. Then an ϵ-greedy exploration strategy was applied.

Figure 4 shows the mean learning curves with standard deviation of the 5 independent CAM and SEN controllers with 5 test runs each. Although the CAM controllers worked on higher dimensional input vectors and had to learn the relations of the feature space, they only needed about 2 to 3 times more transitions to come up with a competitive strategy.

The box plot with minimum and maximum in Figure 5 shows the final performance of the best controllers for both experiments and for a CAM controller with a hand-picked autoencoder (well unfolded strap, only one crossing). Each was tested 20 times. The best SEN controller learned a perfect strategy, so in

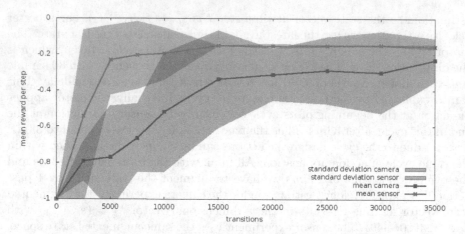

Fig. 4. Mean learning curves with standard deviation

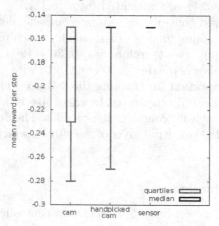

Fig. 5. Best final controllers of the camera and sensor experiments and of a camera controller with hand-picked autoencoder

all tests the pole needed 15 steps into the goal area and stayed there. In 13 out of 20 tests the best CAM controller performs similarly well $(-0.15, -0.16)$. Although the CAM controllers are in general less stable, in practice this means that the pole reaches the goal area just some steps later than in the mean case or drops out of the goal area for one downswing and is then immediately balanced again. The special controller with hand-picked autoencoder is almost as stable as the best SEN controller, only in 1 out of 20 tests the pole drops out of the goal area. Indeed the quality of the feature space is crucial for the performance. Considering the 1200 dimensional input for the overall system, even the lower stability at such a high performance level is a remarkable result. Another CAM controller outside these experiments with a hand-picked autoencoder was able to run in infinite loop for several minutes without any such dropout. An example video is available online at http://youtu.be/E0wZcYcoh-g.

6   Conclusions

We have seen that the automatic feature extraction by a deep autoencoder can be successfully applied to the inherently instable pole balancing task. The control strategies can be easily obtained by RL methods and can compete with strategies trained on direct motor sensor data instead of the feature space. Both controller types show performances which are settled in the same order of magnitude. A shortcoming of the used autoencoder approach is that the moving parts have to be clearly visible, which is why a big green marker had to be attached to the pole. However the autoencoder turned out to be robust towards small lighting changes, which resulted in a small shift in the feature space, though not to position or background changes.

References

1. Riedmiller, M., Lange, S., Voigtlaender, A.: Autonomous Reinforcement Learning on Raw Visual Input Data in a Real World Application. In: International Joint Conference on Neural Networks (2012)
2. Ormoneit, D., Sen, Ś.: Kernel-Based Reinforcement Learning. Mach. Learn. 49, 161–178 (2002)
3. Sutton, R.S., Barto, A.G.: Reinforcement Learning: An Introduction. MIT Press, Cambridge (1998)
4. Riedmiller, M.: Neural Fitted Q Iteration - First Experiences with a Data Efficient Neural Reinforcement Learning Method. In: Gama, J., Camacho, R., Brazdil, P.B., Jorge, A.M., Torgo, L. (eds.) ECML 2005. LNCS (LNAI), vol. 3720, pp. 317–328. Springer, Heidelberg (2005)
5. Riedmiller, M.: Neural Reinforcement Learning to Swing-Up and Balance a Real Pole. In: IEEE International Conference on Systems, Man and Cybernetics, pp. 3191–3196. IEEE Press, New York (2005)
6. Tolat, V.V., Widrow, B.: An Adaptive 'Broom Balancer' with Visual Inputs. In: IEEE International Conference on Neural Networks, pp. 641–647 (1988)
7. Wenzel, L., Vazquez, N., Nair, D., Jamal, R.: Computer Vision Based Inverted Pendulum. In: Proceedings of the 17th IEEE Instrumentation and Measurement Technology Conference, pp. 1319–1323 (2000)
8. Wang, H., Chamroo, A., Vasseur, C., Koncar, V.: Hybrid Control for Vision Based Cart-Inverted Pendulum System. In: American Control Conference, pp. 3845–3850 (2008)
9. Hinton, G.E., Salakhutdinov, R.R.: Reducing the Dimensionality of Data with Neural Networks. Science 313, 504–507 (2006)
10. Ciresan, D.C., Meier, U., Gambardella, L.M., Schmidhuber, J.: Deep Big Simple Neural Nets Excel on Handwritten Digit Recognition. Neural Comput. 22, 3207–3220 (2010)
11. Katsikopoulos, K.V., Engelbrecht, S.E.: Markov Decision Processes with Delays and Asynchronous Cost Collection. IEEE Trans. Autom. Control 48, 568–574 (2003)

GPU-Based Biclustering
for Neural Information Processing

Alan W.Y. Lo, Benben Liu, and Ray C.C. Cheung

Department of Electronic Engineering,
City University of Hong Kong, Hong Kong
{wingyulo5,benbenliu2@student.cityu.edu.hk}, r.cheung@cityu.edu.hk

Abstract. This paper presents an efficient mapping of geometric biclustering (GBC) algorithm for neural information processing on Graphical Processing Unit (GPU). The proposed designs consist of five different versions which extensively study the use of memory components on the GPU board for mapping the GBC algorithm. GBC algorithm is used to find any maximal biclusters, which are common patterns in each column in the neural processing and gene microarray data. A microarray commonly involves a huge number of data, such as thousands of rows by thousands of columns so that finding the maximal biclusters involves intensive computation. The advantage of GPU is its ability of parallel computing which means that for those independent procedures, they can be carried out at the same time. Experimental results show that the GPU-based GBC could reduce the processing time largely due to the parallel computing of GPU, and its scalability. As an example, GBC algorithm involves a large number of AND operations which utilize the parallel GPU computations, that can be further practically used for other neural processing algorithms.

Keywords: High Performance Computing (HPC), Biclustering, Graphics Processing Unit (GPU).

1   Introduction

Neural data analysis has become one of the most important topics in neural information processing and biomedical research. For instance, biologists correlate the gene information with diseases; a cancer class can be discovered based on tissue classification [1]. Genes that have similar expression patterns may have common biological functions. There can be tens of thousands of genes involved in a microarray experiment, so efficient data analysis methods are indispensable for finding similar data expression patterns. Biclustering is an important method to analyze large neural information [2], and biological datasets [3], which is a process that partitions the data samples based on a number of similar criteria so that the existing patterns in the data can be analyzed. The biclustering process can be formulated in multidimensional data space for analysis [3].

Geometric biclustering (GBC) algorithm [4] identifies the linearities of microarray in a high dimensional space of biclustering algorithm. It provides higher

T. Huang et al. (Eds.): ICONIP 2012, Part V, LNCS 7667, pp. 134–141, 2012.

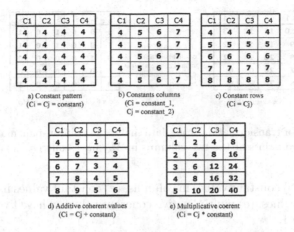

C1	C2	C3	C4
4	4	4	4
4	4	4	4
4	4	4	4
4	4	4	4
4	4	4	4

a) Constant pattern
(Ci = Cj = constant)

C1	C2	C3	C4
4	5	6	7
4	5	6	7
4	5	6	7
4	5	6	7
4	5	6	7

b) Constants columns
(Ci = constant_1,
Cj = constant_2)

C1	C2	C3	C4
4	4	4	4
5	5	5	5
6	6	6	6
7	7	7	7
8	8	8	8

c) Constant rows
(Ci = Cj)

C1	C2	C3	C4
4	5	1	2
5	6	2	3
6	7	3	4
7	8	4	5
8	9	5	6

d) Additive coherent values
(Ci = Cj + constant)

C1	C2	C3	C4
1	2	4	8
2	4	8	16
3	6	12	24
4	8	16	32
5	10	20	40

e) Multiplicative coerent
(Ci = Cj * constant)

Fig. 1. The five types of patterns detected in biclustering

accuracy and reduces the computational complexity for coherent pattern detection compared to other methods. A hypergraph partitioning method [5] is proposed to further optimize the performance of the GBC algorithm. This method reduces the size of matrix in each operation in order to speed up the GBC algorithm using a software partition tool called hMetis [6]. However, it is still time consuming to identify patterns in a large-scale microarray data. In this paper, we explore the parallelizing possibility of the GBC algorithm running on the GPU platform. There exists related work comparing the speed and power efficiency of different computing platforms such as FPGAs, CPUs and GPUs [7]. The contributions of this paper are:

1. analyzing the parallelism feature of the geometric biclustering algorithm;
2. efficiently mapping the geometric biclustering algorithm on GPU platform;
3. performance optimization of different memory usage for GPU-based GBC.

This paper is organized as follows. Section 2 describes the background of this work; Section 3 introduces the geometric biclustering algorithm and its parallelizing possibility; Section 4 discusses the efficient mapping on GPU device; Section 5 presents our results of the implementation; Section 6 and Section 7 discusses the design limitation and summarizes the paper.

2   Background

Data analysis requires the computation of a microarray data matrix [8], in which the response pattern in the matrix can identify disease subphenotypes, predict disease progression and activities of new compounds [1]. Biclustering is a simultaneous clustering technique on the row and column dimensions of the data matrix [9]. The biclustering can be formulated in multidimensional data space [3], which can detect five coherent patterns [10] including (a) constant value in the

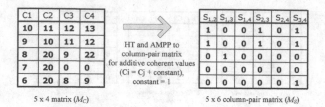

C1	C2	C3	C4
10	11	12	13
9	10	11	12
8	20	9	22
7	20	0	0
6	20	8	9

5 x 4 matrix (M_C)

HT and AMPP to
column-pair matrix
for additive coherent values
(Ci = Cj + constant),
constant = 1

$S_{1,2}$	$S_{1,3}$	$S_{1,4}$	$S_{2,3}$	$S_{2,4}$	$S_{3,4}$
1	0	0	1	0	1
1	0	0	1	0	1
0	1	0	0	0	0
0	0	0	0	0	0
0	0	0	0	0	1

5 x 6 column-pair matrix (M_S)

Fig. 2. Example of transformation from data matrix to column-pair matrix. Set bit in M_S represents two points at the two columns in M_C have coherent relationship.

entire pattern, (b) constant values in columns, (c) constant values in rows, (d) additive coherent values, (e) multiplicative coherent values. These five patterns are shown in Figure 1.

As it is very slow to compare over tens of thousands of data patterns, it is necessary to explore the use of high-performance computing platform such as GPUs and FPGAs to speedup GBC. Because of its intrinsic parallelisum feature, FPGAs are capable of accelerating many applications in various areas. However, developing efficient FPGA design involves tedious concurrent timing analysis and hardware debugging.

General-purpose computing on graphics processing units (GPGPU) is the means of using a GPU, which typically handles computation only for computer graphics, to perform computation in applications traditionally handled by the CPU. GPUs have a parallel throughput architecture that emphasizes executing many concurrent threads slowly, rather than executing a single thread very quickly. Compute Unified Device Architecture (CUDA) is a parallel computing architecture developed by NVIDIA, which is accessible to software developers through variants of industry standard languages like C/C++. Recent work on accelerating sequence analysis applications using CUDA includes MUMmerGPU [11] and Smith-Waterman [12]. Our work is novel because there exists no previous acceleration on GBC algorithm.

3   Geometric Biclustering

Geometric biclustering (GBC) is important in gene data analysis. In this section, we introduce the work flow of the GBC algorithm, identify the most computation intensive part of GBC and discuss the parallel feature of the algorithm. Therefore the algorithm can be implemented on GPU efficiently.

The microarrary data is converted to 2-D column-pair space sub-biclusters by using the HT and AMPP [4]. The column-pair biclusters are combined to maximal biclusters, and the number of genes in the merged biclusters is reduced in this process. In the evaluation of the maximal biclusters, the biclusters are filtered out if the number of conditions is fewer than the given parameter. Finally the valid maximal biclusters are the target search patterns.

GBC algorithm is used to find the maximum common patterns within a microarray as shown in Algorithm 1. Assume we have an m by n data matrix and

---

Algorithm 1. Column-pair combination

Input: column-pair matrix M_S, $S_{i,j} \in M_S$, where $1 \leq i \leq n, 1 \leq j \leq n$
Input: $TH1$, a threshold to accept combined column-pairs
Output: Combined matrix M'_S

1: //Remove the column-pair with total set bit less than $TH1$
2: **for all** $S_{i,j} \in M_S$ **do**
3:    **if** Total set bit of $S_{i,j} \geq TH1$ **then**
4:       $M'_S \leftarrow S_{i,j}$
5:    **end if**
6: **end for**
7: //Compare each column-pair in M'_S
8: **for all** $S1 \in M'_S$ **do**
9:    **for all** $S2 \in M'_S$, column in $S2$ not subset of $S1$ and column in $S1$ not subset of $S2$ **do**
10:       **if** At least one column in $S1$ and $S2$ is in common **then**
11:          **if** At least one column in $S1$ and $S2$ is different **then**
12:             $S_{combine} = S1 \& S2$
13:             **if** Total set bit of $S_{combine} \geq TH1$ **then**
14:                $M_{temp} \leftarrow S_{combine}$
15:             **end if**
16:          **end if**
17:       **end if**
18:    **end for**
19:    $M'_S = M'_S \cup M_{temp}$
20: **end for**

$(n)(n\text{-}1)/2$ column-pairs in total. Fig. 2 illustrates an example of transforming a *5* by *4* data matrix (M_C) to a *5* by *6* column-pair matrix (M_S) in the GBC work flow. The target is finding an additive coherent pattern in the data matrix, where the value of the next column is equal to the value of the current column plus a constant (Ci=Cj+constant). The input matrix M_C is transformed to M_S by using HT and AMPP for additive coherent value. The set bit (logic '1') in $S_{1,2}$ indicates there is an additive coherent relationship of that position in C1 and C2. All the column-pairs in M_S are combined in the combination process.

4   Mapping of the 5 GPU Versions

Table 1 shows the five different versions of mapping the GBC on the GPU device. As an example, a sample microarray is of 4950 rows and 100 columns. In order to find the maximal biclusters, we need to perform eliminations with the parallel GPU computations. After reading the data file, a ColTag structure is created, consisting of 4950 ColTag Nodes. After those rows with the number of 1s less than Threshold (threshold is 10 in this case) are removed, there are still 463 ColTag Nodes remained. The number of ColTag Nodes expands ($463 \rightarrow 2460 \rightarrow 4650$) and then shrinks ($4650 \rightarrow 3820 \rightarrow 2560 \rightarrow 1300 \rightarrow 460 \rightarrow 100 \rightarrow 10$). Finally there are 10 ColTag Nodes remained, which means there are 10 maximal biclusters found from the input data. The pattern of the maximal bicluster is printed followed by the corresponding indices. We can see that the size of the ColTag structure reaches its maximum at loop 2. After looping 7 times, there are 10 ColTag Nodes remained in the ColTag structure, hence the maximal biclusters are finally obtained.

Table 1. Summary of the GPU implementations

GPU version	Description	Feature
1	Initial baseline	straightforward implementation
2	Based on version-1	Smaller number of cudamemcpy() function calls
3	Using shared memory	Reusing small-size data
4	Based on version-2	Reducing data transfer
5	Index processing on GPU	Minimizing the index processing time

The execution time of CPU version is around 100 minutes. With different index representation in programming, the runtime will vary. In the CPU version, the indices of each row are represented by a fixed size integer array, then the time spent on comparing the indices is much longer. The length of the linked list is at most 10 and the length of the fixed integer array is 100, so it is reasonable for the execution time to be more than 10 times longer.

Both the GPU-based and the CPU-based GBC programs produce the same results. The size of the ColTag structure is first expanded and then shrinks as in CPU version, and we obtain ten maximal biclusters. The execution time of GPU-version 1, 2, 3, and 4 are around 14.95, 8.76, 9.19, and 8.06 minutes respectively. Fig. 3 shows the program flow of the GPU-version 4.

The maximal biclusters are identified as those found by the CPU version. Although it is about 22 minutes faster than the CPU version, it takes about 78.73 miniutes. This GPU-version demonstrates the parallel processing power of the GPU. In this version, the index and data manipulations are completely in GPU, rather than the index manipulation in CPU and the data manipulation in GPU as in the previous GPU-versions. Due to the power of the parallel computing of GPU, it runs much faster than the CPU version.

The execution time of GPU-version 5 is the longest because of the time-consuming index processing execution in the GPU. Among all the GPU-versions, version 4 has the shortest execution time. The time reduction of version 2 from version 1 is due to the reduction of the number of times of calling the function *cudamemcpy()*. This function is time consuming and costs overhead of bus arbitration before transferring data between main memory and global memory. The execution time of version 3 is smaller than version 2 because of using shared memory. The access time of shared memory is much shorter than that of global memory. However, the result shows that the execution time is a little longer. The reason for this would be the number of times of data reuse is not large, actually 10 times of reuse in this case. Version 4 further reduces the execution time because the data is only sent back to the main memory after the maximal biclusters are found instead of sending to main memory for storing the ColTag structure.

5   Implementation Results

5.1   Evaluation Platform and Benchmarks

In this section, we implement the designs on NVIDIA GeForce GT430 with 96 CUDA cores with Processor Clock of 1400 MHz. In order to evaluate our

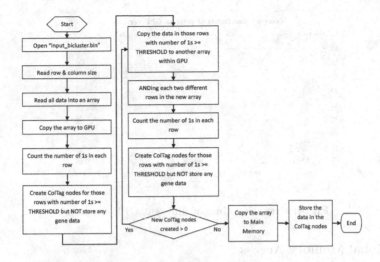

Fig. 3. Program Flow of GPU-version 4

design, we use both simulated data and real data as benchmarks to evaluate the GBC accelerator on a GPU. The implementation result of the simulated data is shown below for further explanation as the other implementations generate similar results.

5.2 Execution Time

All of the implementations give the same results of maximal biclustering. Fig. 4 shows the execution time of different versions of GPU implementation. Version 4 gives the best result as less as 8.06 minutes. From version 1 to version 4, the index processing is executed in CPU and the combining processing is executed in GPU. Compared with version 1, version 2 is able to minimize the number of data transfer between CPU and GPU, therefore achieves less time than version 1 since the data transfer between host and device is quite time consuming. In version 3, although shared memory is used for fast access, the number of times of reuse data is not large, actually 10 times of reuse in this case. Therefore the execution time is even a little longer than version 2. Version 4 reduces the execution time by further reducing the data transfer between CPU memory and GPU global memory. Different from version 1 to 4, the index processing is also executed on GPU instead of CPU. Since there are too many sequential operations during index processing, it is better to execute this process on CPU taking data transfer time into consideration. Therefore version 5 has the longest execution time.

6 Discussions

During GPU programming many memory and thread manipulations need to be taken care of which have a huge effect on the overall performance. Below lists some limitation and important considerations in our work.

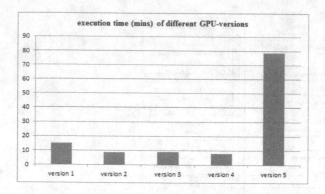

Fig. 4. Execution time of different GPU-versions

6.1   Global Memory Access

Global memory is good for its huge size but bad for slow access. There is no cache for GPU global memory and the access delay to global memory is usually 400 to 600 clock cycles. All the accesses to global memory are executed sequentially. Consequently the access to global memory is often the bottleneck of the overall performance. Coalesced global memory access is that the access to global memory of threads in a half-warp can be executed in parallel under specific conditions.

6.2   Control Flow

In some cases, tasks need to be assigned to threads according to Thread ID which will result in divergence. Divergence is the case that threads in a same warp have different instruction paths. When divergence happens, threads in a same warp on different paths will execute in sequential order instead of concurrently.

7   Conclusions and Future Works

In this paper, different versions of GPU-based biclustering methods are designed. Version 4 achieves the shortest execution time, around 8.06 minutes. The parallel computing ability of GPU can significantly improve the efficiency of the program. Since the AND-operations of each 2 different rows are independent, they map perfectly and run concurrently on the GPU-cores. Those computation intensive parts are being processed in GPU. However, the main drawback that limits the execution time of the program is the inefficient data transfer between main memory and global memory. From the results of GPU-version 1 and GPU-version 2, reducing the number of calls of the data transfer function can halve the execution time. The efficiency can be greatly improved if this bottleneck can be resolved. In order to produce an efficient computation, the combination of CPU and GPU should be carefully designed. Future work includes the extension of other neural processing algorithms and the mapping on hybrid computing platforms such as FPGAs.

Acknowledgement. This work was supported by a grant from the Research Grant Council of the Hong Kong Special Administrative Region, China (Project No. CityU 9041799).

References

1. Golub, T.R., Slonim, D.K., Tamayo, P., Huard, C., Gaasenbeek, M., Mesirov, J.P., Coller, H., Loh, M.L., Downing, J.R., Caligiuri, M.A., Bloomfield, C.D., Lander, E.S.: Molecular Classification of Cancer: Class Discovery and Class Prediction by Gene Expression Monitoring. Science 286(5439), 531–537 (1999)
2. Dhillon, I.S., Mallela, S., Modha, D.S.: Information-Theoretic Co-clustering, pp. 89–98 (2003)
3. Gan, X., Liew, A.W.C., Yan, H.: Discovering Biclusters in Gene Expression Data based on High-Dimensional Linear Geometries. BMC Bioinformatics 9(1), 209 (2008)
4. Zhao, H., Liew, A.W.C., Xie, X., Yan, H.: A New Geometric Biclustering Algorithm based on the Hough Transform for Analysis of Large-scale Microarray Data. Journal of Theoretical Biology 251(3), 264–274 (2008)
5. Wang, D.Z., Yan, H.: Geometric Biclustering Analysis of DNA Microarray Data based on Hypergraph Partitioning. In: IDASB Workshop on BIBM 2010, pp. 246–251 (2010)
6. Karypis, G., Kumar, V.: Multilevel K-Way Hypergraph Partitioning
7. Thomas, D.B., Howes, L.W., Luk, W.: A Comparison of CPUs, GPUs, FPGAs, and Massively Parallel Processor Arrays for Random Number Generation. In: Proc. FPGA, pp. 63–72 (2009)
8. Labiod, L., Grozavu, N., Bennani, Y.: Clustering Categorical Data Using an Extended Modularity Measure, pp. 310–320 (2010)
9. Cheng, Y., Church, G.M.: Biclustering of Expression Data. In: The Eighth International Conference on Intelligent Systems for Molecular Biology, pp. 93–103 (2000)
10. Madeira, S.C., Oliveira, A.L.: Biclustering Algorithms for Biological Data Analysis: A Survey. IEEE/ACM Transactions on Computational Biology and Bioinformatic. 1, 24–45 (2004)
11. Schatz, M., Trapnell, C., Delcher, A., Varshney, A.: High-Throughput Sequence Alignment Using Graphics Processing Units. BMC Bioinformatics 8(1), 474 (2007)
12. Manavski, S., Valle, G.: CUDA Compatible GPU Cards as Efficient Hardware Accelerators for Smith-Waterman Sequence Alignment. BMC Bioinformatics 9(suppl. 2), S10 (2008)

Color Image Segmentation Based on Regional Saliency

Haifeng Sima[1,2], Lixiong Liu[1], and Ping Guo[1]

[1] School of Computer Science&Technology,
Beijing Institute of Technology, Beijing,10081, China
[2] School of Computer Science&Technology,
Henan Polytechnic University, jiaozuo, 454000, China
smhf@hpu.edu.cn, lxliu@bit.edu.cn, pguo@ieee.org

Abstract. In this paper, we propose a novel segmentation model integrated the salient regional features into mean shift (MS) clustering segmentation as fusion matrixes. Firstly, a regional visual saliency map of the given image is obtained based on quantification image in HSV color space. Then saliency factors are extracted from salience map from each channel in L*a*b space in two steps: region saliency(S-R) and pixels-region (P-R). Fuse the salient factors derived from former salient features with original components of the image as new input features, who are involved in the mean-shift procedure for segmentation. This paper takes advantage of regional salience to guide the MS vectors moving to accurate modes, and decreases premature and ill convergence at local area. The introduction of salient factors enhances the accuracy of the pixels clustering for region segment. Experiment results carried on Berkeley database and comparison with human segmentation results demonstrated that our algorithm has better performance on nature color images segmentation.

Keywords: Color quantification, Region saliency, Salient feature fusion, Mean shift, Color segmentation.

1 Introduction

Regional saliency is an efficient technique defined to capture essential attributes for region detection. Because of the purpose of segmentation is to find the homogeneous and semantic regions and to extract correct targets of the image. So extraction of the region saliency is helpful for segmentation. Almost all of image segmentation approaches are data-driven models, and can be divided into four techniques: pixel clustering, regional driven, boundary- driven and model-driven [1]. Recently, many of the images saliency approaches are applied to image target detection, and various types of image analysis. Saliency detection plays an important role in content-based applications: such as object detection [2], object extraction and segmentation [3,4], image compression and editing [5,6], image retrieval and annotation [7,8].There are many methods for saliency map calculation: some approaches are based on pixels, some are based on analyzing the spectral features of the angle, and some regional saliencies are based on quantization of color image [9]. Byoung Chul Ko et al. [10] presents a seg-

T. Huang et al. (Eds.): ICONIP 2012, Part V, LNCS 7667, pp. 142–150, 2012.
© Springer-Verlag Berlin Heidelberg 2012

mentation algorithm used visual saliency model and semantic clustering. This method first split the image area, at the same time, create attention window (AW). On the basis of the regional and pixel saliency, use support vector machine to select the outstanding areas, then merge regions together again into a semantic object by region merging and clustering. Shyjan Mahamud et al. [11] has developed a segmentation method, which identified the closed boundary with calculation of space close to the salient features. Yin-zhu Xue et al. [12] proposed a new saliency model by computing local/global color difference, orientation difference, and spatial distribution by distance-weighted color similarity between each pixel and each region, and the original regions are obtained with mean shift. Saliency calculation methods can be divided into the calculation of the global contrast or local contrast. The former saliency model are on the basis of the average of whole image, such as literature [3]; the latter are getting the approximate area at first and then assigns corresponding saliency values for the regions, such as the method in [13]. In [14], it computes the saliency map with the statistical properties on the Fourier spectrum of the natural background. It proposed inverse Fourier transform of the results spectrum of the residuals as a saliency map. Mean shift method is looking for local density extremums in the given three-dimensional color space and the coordinates space. The regional saliency of image represents the real homogeneity regions of the color image. The regions salient feature was introduced in the MS search process and lead pixels to rapid and accurate convergence to its own category.

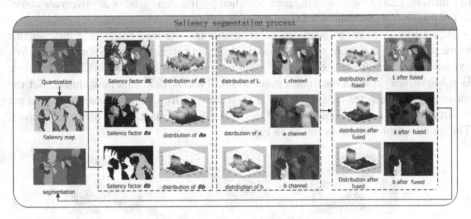

Fig. 1. Saliency segmentation process

This paper proposes a regional saliency model based on the color quantification. It defines region saliency factor for each channel of the given image in Lab color space. Saliency factor and original image channels are fused as new features, who are involved in image segmentation by MS method. The experiments show that saliency factors improve the local convergence in the MS algorithm well. In addition, this measurement smoothen the homogeneous regions with salient feature, and enlarges regional differences between inhomogeneous regions. The rest of the paper is organized as follows. Section 2 presents the region saliency computational method and segmentation procedure. In order to demonstrate the effectiveness of our method, experimental results and comparison with popular methods are presented in Section 4. Sections 5 draw the conclusions and future work.

2   Regional Saliency Model and Segmentation

Generally, quantization methods can be divided into two categories. One is to cut the color space to sub-space. The other is to build the most adaptive palette by pixel clustering with statistical density distribution. In this paper, we cut the color space at first and get adaptive palette by K-means clustering.

2.1   Quantization in HSV Color Space

The quantization includes two steps: Firstly, split HSV color space into color bins on the three channels. The HSV components values are quantified to the following ranges: 0–360° for hue and 0–1 for saturation and value. After examining the hue circle, the intervals of major colors can be identified to be homogenous is 15°. So, the range size can be divided by 15°. The range size of saturation and value can be divided by 0.2. So all colors are mapped to (24×5×5 = 300) color bins. The same or similar color is mapped to a corresponding region of space. Three-dimensional pixel histogram distribution statistics bins, reflecting the size of the area occupied by the same kind or approximate colors in the image. The formula of 3-D histogram as follows:

According to peaks principle of color histogram, the bins from the cube contains more pixels can represent concentrated area of the corresponding color visually. We extract the maximum bin form 26 adjacent bins for each bin in the HSV color space. B = Max (bins) contain maximum pixel should be classified to a new cluster. Set the mean color of selected bins to be initial class centers for clustering later. Next, to get the image color quantization results Q_{image} by the K nearest neighbor means clustering. The quantization process is shown in Fig. 2.

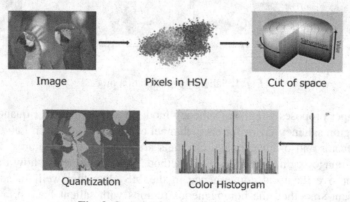

Image   Pixels in HSV   Cut of space

Quantization   Color Histogram

Fig. 2. Quantization process scheme

2.2 Region Saliency Calculation

Pixel saliency is defined by color contrast with all colors in the image. In order to make the saliency factor to express the mapping between the pixel and the region accurately, the calculation of saliency includes two parts: The first part is region saliency defined as R-Saliency. It indicates the saliency between quantized regions to global image, which are denoted by S_R.

$$S_R(I(x,y)) = \sum_{\forall \sigma(x,y) \in I} (1 + \exp(Q(x,y) - Q_{others})^{-1})^{-1} \tag{1}$$

The second part is pixel-region saliency, which is defined as difference between original channel value of the image and the quantized value. This type of saliency are denoted by $S_{p/L}$, $S_{p/a}$ and $S_{p/b}$, which are called P-R saliency. The P-R saliency is calculated by color contrast between pixels and mean value of region. The saliencies of pixels in the same region have the same value. Here we used the sigmoid function as the fusion operator on original image and saliency, and get normalize saliency value as follows :

$$S_{p/L}(L(x,y)) = \sum_{\forall L(x,y) \in I} \left(1 + \exp\left((L(x,y) - L_{reg\_mean})\right)^{-1}\right)^{-1} \tag{2}$$

$$S_{p/a}(a(x,y)) = \sum_{\forall a(x,y) \in I} \frac{1 - \exp\left((a(x,y) - a_{reg\_mean})\right)^{-1}}{1 + \exp\left((a(x,y) - a_{reg\_mean})\right)^{-1}} \tag{3}$$

$$S_{p/b}(b(x,y)) = \sum_{\forall b(x,y) \in I} \frac{1 - \exp\left((b(x,y) - b_{reg\_mean})\right)^{-1}}{1 + \exp\left((b(x,y) - b_{reg\_mean})\right)^{-1}} \tag{4}$$

2.3 Segmentation Algorithm

To fuse color information and saliency features information of three channels and compose 6-dimensional input matrixes: $x = (X_c, X_{r+s})$. X_c denotes coordinates of the points, and X_{r+s} denote color and regional saliency matrixes features. Subscript r denotes three color channel, and s denote saliency of three channels. The kernels function of probability density estimation:

$$K_{hc,hr+s} = \frac{C}{hc, hr + s} k\left(\left\|\frac{x^s}{h^c}\right\|^2\right) k\left(\left\|\frac{x^{r+s}}{h_{r+s}}\right\|^2\right) \tag{5}$$

Because of computational complexity, it is a better choice to take regional saliency as the smooth factor than as redundant matrixes in the MS procedure. The factors of the input matrixes in MS are denoted by $\theta = S_R * P_R$. Each component is calculated as follows.

$$L'(x_i, y_i) = L(x_i, y_i) \times \theta_L$$
$$a'(x_i, y_i) = a(x_i, y_i) \times \theta_a \tag{6}$$
$$b'(x_i, y_i) = b(x_i, y_i) \times \theta_b$$

The new mean shift vector with mixed features is calculated by following Eq.7. Where $G = -K$. The Fig.3 displays the segmentation procedure on detail of the image.

$$M_{h,G}(x) = \frac{\sum_1^n G\left(\frac{L_i'-L\prime}{h}, \frac{a_i'-a\prime}{h}, \frac{b_i'-b\prime}{h}\right)x_i}{\sum_1^n G\left(\frac{L_i'-L\prime}{h}, \frac{a_i'-a\prime}{h}, \frac{b_i'-b\prime}{h}\right)} - x \tag{7}$$

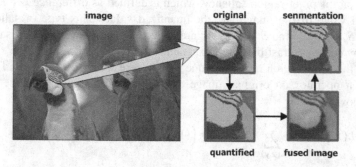

Fig. 3. Segmentation illustration of detail

3   Analysis of Experimental Results

In this section, we test the proposed algorithm on the Berkeley segmentation datasets (BSD) [15], which have unified resolution 321 * 481. Select 200 color images of natural scenes from BSD included different types of natural images, such as natural scenery, animals, people and buildings etc. The experiments results indicate that the proposed algorithm enlarged region convergence in MS and get ideal segmentation results. Comparison with MS segmentation results are shown in Fig.4 and Table 1. The parameters h_r and h_s are generated by experience for global adaptability in the database. It is easy to find that our approach effective mitigate over-segmentation and misclassification phenomenon of the MS segmentation algorithm under the same color and spatial resolution.

From the Table1 and Fig 4, we can see that our method got fewer and proper regions than MS with same bandwidth parameters. In order to verify the segmentation efficiency, we compare the proposed approach with most popular boundary detection segmentation method (LESBR) [16] and region-based segmentation method (SRDS) [17]. In Fig.5 displays four images segment results comparison with two former algorithms and human segmentation results.

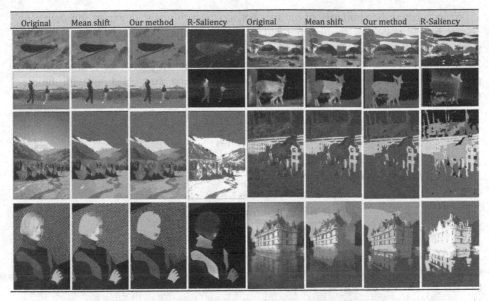

Fig. 4. Saliency segmentation results and MS results

Table 1. Regions comparison of two segmentation algorithms

	MS	Ours	h_s, h_r	Min area		MS	Ours	h_s, h_r	Min area
143090.jpg	74	21	16, 9	50 pixels	113009.jpg	184	118	16, 7	50 pixels
296059.jpg	67	18	16, 9	50 pixels	46076.jpg	101	57	16, 9	50 pixels
35070.jpg	174	49	16, 9	50 pixels	148026.jpg	350	175	16, 9	50 pixels
286092.jpg	61	24	16, 9	20 pixels	102061.jpg	106	78	16, 9	50 pixels

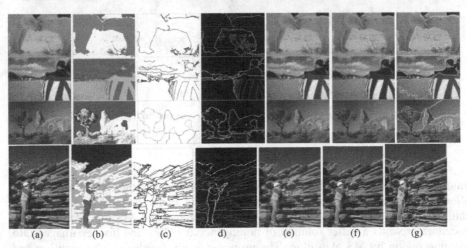

(a) (b) (c) d) (e) (f) (g)

Fig. 5. Column (a) is the original images, column (b) is the regional saliency of our method, column (c) is segment results of our approach and column (d) is human segmentation results. The columns (e-g) are segmentation results of three algorithms (ours, LESBR and SRDS).

We compared the similarity of outlines for four algorithms with expert segmentation results by the matching strategy defined in [18]. It defines precision and recall to be proportional to the total number of unmatched pixels between two segmentations S1 and S2. Unmatched pixels are those for which a suitable match cannot be found within a particular distance threshold τ_d. In this paper the matching we carried out with the distance threshold $\tau_d = 3$. The precision/recall measures are defined as follows:

$$Precision(S_1, S_2) = \frac{Matched(S_1, S_2)}{|S_1|} \tag{8}$$

$$Recall(S_1, S_2) = \frac{Matched(S_1, S_2)}{|S_2|} \tag{9}$$

$$F_\beta = \frac{(1 + \beta^2)Precision \times \text{Recall}}{\beta^2 \times Precision + \text{Recall}} \tag{10}$$

The histograms in Fig.6 describe the segmentation precisions respectively under the distance threshold τ_d=3, which is reasonable for resolution of given images. In the Eq. (9), the β=0.3. It is easy to see the proposed approach bring better precision. Upon the chosen comparison strategy, our algorithm has provided much approximate contours with manual expert segmentation results. Moreover, excess segmentation contour is relatively less than LESBR and SRDS by contrast.

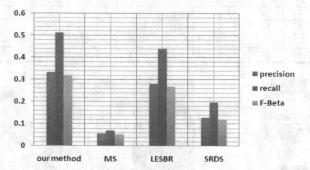

Fig. 6. Precision comparison of four given algorithms (our method, MS, LESBR, SRDS)

4   Conclusion

This paper provides a novel color image segmentation algorithm based on regional saliency. A visual saliency map is obtained by color image quantization. The fused features base on S-R and P-R saliency are incorporated into MS input feature. The procedure presents a better solution of homogeneous colors and divides images into semantic region by MS algorithm. The proposed method reduces over-partition and wrong partition in the spatial clustering. As the future work, to investigate efficient segmentation algorithms incorporate novel saliency computation approach is important. At the same time, it could be beneficial to incorporate high level factors like texture, symmetry into saliency maps for segmentation.

Acknowledgement. This work was supported by the grants from the National Natural Science Foundation of China (Project No. 60911130513, 60805004). Prof. Ping Guo and Lixiong Liu are the authors to whom all correspondence should be addressed.

References

1. Gunther, H.: Region saliency as a measure for colour segmentation stability. Image Vision Computing 26(2), 211–227 (2008)
2. Achanta, R., Hemami, S., Estrada, F., Süusstrunk, S.: Frequency-tuned salient region detection. In: Proceedings of IEEE Conference on Computer Vision and Pattern Recognition (CVPR), Miami, Florida, USA, pp. 1597–1604 (2009)
3. Han, J., Ngan, K., Li, M., Zhang, H.: Unsupervised extraction of visual attention objects in color images. IEEE Transactions on Circuits and Systems for Video Technology, IEEE Consumer Electronics Society 16(1), 141–145 (2006)
4. Ko, B., Nam, J.: Object-of-interest image segmentation based on human attention and semantic region clustering. Journal Optical Society of America 23(10), 2462–2470 (2006)
5. Christopoulos, C., Skodras, A., Ebrahimi, T.: The JPEG2000 still image coding system: an overview. IEEE Transaction on Consumer Electronics 46(4), 1103–1127 (2002)
6. Achanta, R., Susstrunk, S.: Saliency Detection for Content-aware Image Resizing. In: Proceedings of International Conference on Information Processing (ICIP), Cairo EGYPT, pp. 1005–1008 (2009)
7. Feng, S.H., Xu, D., Yang, X.: Attention-driven salient edges and regions extraction with application to CBIR. Signal Processing 90, 1–15 (2010)
8. Hare, J.S., Lewis, P.H.: Saliency-based Models of Image Content and their Application to Auto-Annotation by Semantic Propagation. In: Proceedings of Multimedia and the Semantic Web/European Semantic Web Conference (2005)
9. Huang, Z.Y., He, F.Z., Cai, X.T., Zou, Z.Q., Liu, J., Liang, M.M., Chen, X.: Efficient random saliency map detection. Information Sciences 54(6), 1207–1217 (2011)
10. Ko, B.C., Nam, J.Y.: Object-of-interest image segmentation basedon human attention and semantic region clustering. Journal Opt. Soc. Am. A 23(10), 2462–2470 (2006)
11. Shyjan, M., Lance, R.W., Karvel, K.T., Xu, K.L.: Segmentation of Multiple Salient Closed Contours from Real Images. IEEE Transactions on Pattern Analysis and Machine Intelligence 25(4), 433–445 (2003)
12. Xue, Y.Z., Liu, Z., Shi, R.: Saliency detection using multiple region-based features. Optical Engineering 50(5), 1–9 (2011)
13. Cheng, M.M., Zhang, G.X., Mitra, N.J., Huang, X., Hu, S.M.: Global contrast based salient region detection. In: Proceedings of Conference on Computer Vision and Pattern Recognition, Colorado, USA, pp. 21–23 (2011)
14. Hou, X.D., Zhang, L.Q.: Saliency detection: a spectral residual approach. In: Proceedings of IEEE Conference on Computer Vision and Pattern Recognition, Minneapolis, Minnesota, USA, pp. 1–8 (2007)
15. Martin, D., Fowlkes, C.: The Berkeley segmentation database and benchmark. Computer Science Department, Berkeley University,
http://www.eecs.berkeley.edu/Research/Projects/CS/vision/bsds

16. Donoser, M., Riemenschneider, H., Bischof, H.: Linked Edges as Stable Region Boundaries. In: Proceedings of International Conference on Computer Vision and Pattern Recognition, San Francisco, USA, pp. 1665–1672 (2010)
17. Achanta, R., Estrada, F.J., Wils, P., Süsstrunk, S.: Salient Region Detection and Segmentation. In: Gasteratos, A., Vincze, M., Tsotsos, J.K. (eds.) ICVS 2008. LNCS, vol. 5008, pp. 66–75. Springer, Heidelberg (2008)
18. Estrada, F.J., Jepson, A.D.: Quantitative Evaluation of a Novel Image Segmentation Algorithm. In: IEEE International Conference on Computer Vision and Pattern Recognition, San Diego, USA, pp. 1132–1139 (2005)

Mass Classification in Digitized Mammograms Using Texture Features and Artificial Neural Network

Man To Wong, Xiangjian He, Hung Nguyen, and Wei-Chang Yeh

Faculty of Engineering and Information Technology,
University of Technology Sydney, Broadway, NSW 2007, Australia
eemtwong@gmail.com,
{Xiangjian.He,Hung.Nguyen,Wei-Chang.Yeh}@uts.edu.au

Abstract. A technique is proposed to classify regions of interests (ROIs) of digitized mammograms into mass and non-mass regions using texture features and artificial neural network (ANN). Fifty ROIs were extracted from the MIAS MiniMammographic Database, with 25 ROIs containing masses and 25 ROIs containing normal breast tissue only. Twelve texture features were derived from the gray level co-occurrence matrix (GLCM) of each region. The sequential forward selection technique was used to select four significant features from the twelve features. These significant features were used in the ANN to classify the ROI into either mass or non-mass region. By using leave-one-out method on the 50 images using the four significant features, classification accuracy of 86% was achieved for ANN. The test result using the four significant features is better than the full set of twelve features. The proposed method is compared with some existing works and promising results are obtained.

Keywords: Mammogram, Mass classification, Artificial neural network.

1 Introduction

Breast cancer is the leading cause of death of women in the U.S [1]. Currently the most effective method for early detection of breast cancers is mammography [2]. Mammography is currently the only widely accepted imaging method used for routine breast cancer screening [3]. Masses are one of the important signs of breast cancers [4]. It is difficult to detect masses because their features can be obscured or can be similar to normal breast parenchyma. Reading mammograms is a demanding job for radiologists. A computer aided detection (CAD) system can provide a consistent second opinion to a radiologist and greatly improve the detection accuracy.

In most of the mass detection CAD schemes, image preprocessing is first used to preprocess the digitized mammogram to suppress noise and improve the contrast of the image. Then image segmentation is used to locate the suspicious regions, called regions of interest (ROIs) in this paper [5]. The ROIs can contain a mass or normal tissue. In this paper, a method is reported for the classification of ROIs as mass or non-mass regions. This will reduce the number of false positives in mass detection.

T. Huang et al. (Eds.): ICONIP 2012, Part V, LNCS 7667, pp. 151–158, 2012.

This paper describes a method to extract texture features from ROIs, select significant features using the sequential forward selection (SFS) [6] technique, and classify ROIs into mass or non-mass regions using an ANN and the selected significant features. One advantage of using texture features is that computers are better than human observers in analyzing second-order statistical features. This paper shows that significant texture features derived from GLCM can successfully classify ROIs into mass or non-mass regions using an ANN.

Many previous research papers in mass classification of ROIs using texture features have been reported. Christoyianni et al. [7] used the Radial-Basis-Function (RBF) and Multilayer perceptron (MLP) neural networks to classify ROIs into abnormal or normal breast tissue by using two types of texture features: gray level histogram moments (GLHM) and ten texture features derived from GLCM. In his paper, for GLCM, three angles were used: 0, 45 and -45 degrees. However, the three GLCMs for the three angles were not used together and the test result was reported separately for each of the three GLCMs and GLHM. In his paper, for MLP, best result was obtained with 2 hidden layers and the best total classification accuracy was 84.03%. Petrosian et al. [8] used texture features from GLCM and a modified decision-tree classifier to perform mass classification. The classification capabilities of the features were analyzed by their correlation coefficients and class distances. Groups of three features were selected and put into a decision-tree classification scheme. With the leave-one-out (LOO) method [6], the test result was about 76% sensitivity and 64% specificity. Chan et al. [9] used 8 GLCM based texture features of each ROI for classification of masses and normal tissue. The importance of each feature in distinguishing masses from normal tissue was determined by stepwise linear discriminant analysis. It was reported that five of the texture features were important for classification. The classifier achieved an average area under the ROC curve (A_Z) of 0.84 during training and 0.82 during testing.

The purpose of this paper is to classify ROIs into mass or normal breast tissue using texture features derived from gray level co-occurrence matrix (GLCM), and using an ANN as a classifier. The sequential forward selection (SFS) technique is then used to select the significant features. The objective of this paper is to show that a small number of significant features derived from GLCM can have better or comparable performance in classification accuracy when compared to the full set of features in mass classification using the ANN.

2   Texture Features and Feature Selection

GLCM matrix is used to measure the texture-context information. The texture-context information is specified by the matrix of relative frequencies $P(i, j, d, \theta)$ with which two neighboring pixels separated by distance d and along direction θ occur on the image, one pixel with gray level i and the other with gray level j [5] [10]. To simplify the computational complexity, θ is often given as 0, 45, 90 and 135 degrees. When θ is 0, the element $P(i, j, d, \theta)$ can be expressed as [5] :

$$P(i,j,d,0) = \#\{((x_1,y_1),(x_2,y_2)), |x_2 - x_1| = d, y_2 - y_1 = 0\}. \tag{1}$$

$I(x, y)$ is the intensity value of the pixel at the position (x, y), $I(x_1, y_1)$ equals i, $I(x_2, y_2)$ equals j, and $\# S$ is the number of elements in the set S. After the number of neighboring pixel pairs R used in computing a particular GLCM matrix is obtained, the matrix is normalized by dividing each entry by R, the normalizing constant [10].

In this paper, for each ROI, six texture features were derived from each GLCM. These features are angular second moment (ASM), inertia, entropy, inverse difference moment (IDM), sum of squares: variance (abbreviated as variance) and correlation [8] [10]. In the equations below, the notation $p(i,j)$ is used to represent the $(i, j)th$ entry in a normalized GLCM matrix and $p(i,j)$ is obtained by dividing each entry of the matrix $P(i, j)$ by R, the normalizing constant [10]. Also $\Sigma_{i,j}$ represents $\Sigma_{i=0}^{n-1}\Sigma_{j=0}^{n-1}$ where n is the number of gray levels per pixel in the image.

$$ASM = \sum_{i,j} p(i,j)^2. \tag{2}$$

$$Inertia = \sum_{i,j} (i-j)^2 p(i,j). \tag{3}$$

$$Entropy = -\sum_{i,j} p(i,j)\log(p(i,j)). \tag{4}$$

$$IDM = \sum_{i,j} \frac{1}{1+(i-j)^2} p(i,j). \tag{5}$$

$$Variance = \sum_{i,j} (i-\mu)^2 p(i,j). \tag{6}$$

$$Correlation = \frac{\Sigma_{i,j}(i-\mu_x)(j-\mu_y)p(i,j)}{\sigma_x \sigma_y}. \tag{7}$$

$$\mu_x = \sum_i i \sum_j p(i,j). \tag{8}$$

$$\mu_y = \sum_j j \sum_i p(i,j). \tag{9}$$

$$\sigma_x^2 = \sum_i (i-\mu_x)^2 \sum_j p(i,j). \tag{10}$$

$$\sigma_y^2 = \sum_j (j-\mu_y)^2 \sum_i p(i,j). \tag{11}$$

In this paper, in finding the GLCM, d was set to 2. Two directions were used for θ : 0 and 45 degrees. Hence, a total of 12 texture features are generated for each ROI. Sequential forward selection (SFS) [6] is then used to select the most significant features from the original 12 features. SFS involves the following steps:

- Use K Nearest Neighbor (K-NN) as the classifier, and the leave-one-out (LOO) method as the evaluation technique.
- Select the first feature that has the highest classification accuracy among all the features using K-NN and LOO method.
- Select the next feature, among all unselected features, together with the previous selected features that gives the highest classification accuracy.
- Repeat the previous step until enough features have been selected, or until the classification accuracy stops increasing.

3   Artificial Neural Network (ANN) Classifiers

ANNs have been widely used in the field of pattern recognition. The advantages of ANNs include their capability of self-learning, and their suitability to solve problems that are too complex for conventional techniques, or hard to find algorithmic solutions [5]. There are two common types of ANN classifiers for mammogram masses: the multilayer perceptron with backpropagation (MLP) and the radial basis function (RBF) network [5]. Christoyianni et al. [7] has reported that the MLP classifier performed better than RBF in mass classification accuracy but RBF networks have the advantage of fast learning rates. The main drawback of the MLP using backpropagation is the long training time. The objective of this paper is to show that a small number of significant texture features derived from GLCM can perform better or have comparable performance in classification using MLP when compared to the full set of features. With a smaller number of features, the network size can also become smaller and this helps to reduce the learning time.

In this paper, one hidden layer is used for the MLP.   The number of input units of MLP is equal to the number of texture features. As the output of MLP is either mass or non-mass, the number of output units is one. The number of hidden neurons is chosen according to the suggestion of [11]. The approach is to start with a very small number of neurons, train and test the network while recording its performance. Then the number of hidden neurons is slightly increased, and the network is trained and tested again. Repeat this procedure until no significant improvement is noted or the classification accuracy is acceptable. In this paper, the number of case samples used was small. To maximize the utility of the available cases for training, the leave-one-out (LOO) method was used in testing [6] [8]. Using the above procedure, it was found that the four significant features have comparable performance as the full set of 12 features. Moreover, using the same dataset and LOO method, MLP has better classification accuracy than the K-NN classifier.

4   Experimental Results

4.1   The MIAS MiniMammographic Database

The MIAS MiniMammographic Database is provided by the Mammographic Image Analysis Society, UK [12]. The mammograms are digitized at 200 micron pixel edge and have a resolution 1024 x 1024. The MIAS database provides ground truth for each abnormality in the form of circles, with an approximation of the centre and the radius of each abnormality. The types of abnormality in the database include calcification, masses, architectural distortion and asymmetry. Mammograms which do not contain any abnormality (classified as normal) are also provided. Three types of background tissues are given: fatty, fatty-glandular and dense-glandular tissue.

Fig. 1. An example of ROI with mass at the centre

Two types of masses are used in testing: spiculated masses and circumscribed masses. Fifty regions of interests (ROIs) are selected from the MIAS database, 25 of them containing masses and 25 containing normal tissue only. As the ground truth data of the database includes the position of the mass centre, a square window of 160 x 160 pixels is selected with the mass centre as the centre of the square. For a normal mammogram, the square window is selected randomly inside the breast region.

4.2   Feature Selection and Mass Classification Using K-NN

For each ROI, 12 texture features are derived from two GLCMs, one at $d = 2$ and $\theta = 0$ and the other at $d = 2$ and $\theta = 45$ degrees. Sequential forward selection (SFS) [6] is used to select the significant features. K-NN is used as the classifier (with K=3) and leave-one-out method [6] [11] is used. The four significant features found by SFS are correlation (0), angular second moment (0), inverse difference moment (0) and correlation (45). The number inside the bracket represents the value of θ.

The following evaluation criteria are used:

$$\text{Sensitivity} = n_m \, / \, N_m \, . \tag{12}$$

$$\text{Specificity} = n_n \, / \, N_n . \tag{13}$$

$$\text{Overall Classification accuracy} = (n_m + n_n) \, / \, (N_m + N_n) \, . \tag{14}$$

n_m is the number of correctly classified ROIs with masses, n_n is the number of correctly classified ROIs which contain normal tissues only (no mass), and N_m and N_n are the total number of ROIs with masses and without masses respectively.

The overall classification accuracy defined above is used to determine performance of the K-NN classifier. The overall classification accuracy of K-NN (using K=3 and LOO method) of the four significant features found by SFS is 80% while the overall accuracy using all the 12 features is 74% (see Table 1). In Tables 1, 2 and 3, overall classification accuracy defined above is abbreviated as overall.

Table 1. Performance of K-NN using significant features and all features (K=3, LOO is used)

Number of features	Sensitivity	Specificity	Overall
4 significant features	72%	88%	80%
All 12 features	68%	80%	74%

4.3   Mass Classification Using ANN

The four significant features found in Section 4.2 are used to classify the ROIs to regions with mass or without mass. As discussed in Section 3, the number of input units is 4, the number of output unit is 1, and the number of hidden neurons in the hidden layer changes from 2 to 12 units. Leave-one-out method (LOO) [6] is used.

Table 2. Performance of ANN using 4 significant features & LOO testing

No. of hidden neurons	Sensitivity (%)	Specificity (%)	Overall (%)
2	76	84	80
3	76	84	80
4	72	76	74
5	72	88	80
6	80	84	82
7	76	84	80
8	76	96	86
10	72	92	82
12	80	88	84

From Table 2, best overall classification accuracy of the ANN for the four significant features is 86% and is obtained when the number of hidden neurons is 8. In Table 3, all of the 12 features are used in the ANN. The numbers of input and output units are 12 and 1 respectively. The best overall classification accuracy obtained is 84% when the number of hidden neurons is 6 or 9.

By comparing Tables 1, 2 and 3, for ANN, the four significant features have slightly better overall classification accuracy than the full set of 12 features (86% vs 84%). The best overall classification accuracy for K-NN is 80% while the best overall accuracy for ANN is 86% (for 4 significant features) and 84% (for full set of 12 features). ANN has better mass classification performance than K-NN. Also by considering the performance in sensitivity, ANN performs much better than K-NN. When compared to existing methods in [7] and [8], the test result using the four significant features in this paper is promising in terms of overall classification accuracy, sensitivity or specificity. The best result of Christoyianni et al. [7] reported an overall accuracy of 84.03% using MLP with 2 hidden layers, using the MIAS Database. The best test result by Petrosian et al. [8] had 76% sensitivity and 64% specificity, using the mammograms from University of Michigan hospitals.

Table 3. Performance of ANN using all 12 features and LOO testing

No. of hidden neurons	Sensitivity (%)	Specificity (%)	Overall (%)
3	76	80	78
4	72	80	76
5	76	88	82
6	80	88	84
7	76	84	80
8	80	84	82
9	84	84	84
10	76	88	82
11	80	88	80
12	80	72	76
13	80	84	82

5   Conclusion

An effective technique to classify ROIs of mammograms into mass or non-mass regions has been proposed. Twelve texture features are derived from the GLCMs of each ROI. By using the sequential forward selection, four significant features can be selected from the twelve features. These four significant features are then used to classify the ROIs into mass or non-mass regions using an ANN. Test results using leave-one-out (LOO) method indicate that the four significant features have better classification accuracy than the full set of features using an ANN. Also performance evaluation between ANN and K-NN using LOO indicates that ANN has better classification accuracy than K-NN. By using the four significant features in mass classification using ANN and LOO method on 50 ROIs, a classification accuracy of 86% can be achieved. The result is promising when compared to techniques reported in [7-8].

References

1. Constantinidis, A.S., Fairhurst, M.C., Deravi, F., Hanson, M., Wells, C.P., Chapman-Jones, C.: Evaluating Classification Strategies for Detection of Circumscribed Masses in Digital Mammograms. In: Proceedings of 7th Int'l Conference on Image Processing and Its Applications, pp. 435–439 (1999)
2. Bovis, K., Singh, S., Fieldsend, J., Pinder, C.: Identification of Masses in Digital Mammograms with MLP and RBF Nets. In: Proceedings of the IEEE-INNS-ENNS International joint Conference in Neural Networks, pp. 342–347 (2000)
3. Tang, J., Rangayyan, R.M., Xu, J., Naqa, I.E., Yang, Y.: Computer-Aided Detection and Diagnosis of Breast Cnacer with Mammography: Recent Advances. IEEE Trans. on Information Technology in Biomedicine 13(2), 236–251 (2009)
4. Cheng, H., Cai, X.P., Chen, X.W., Hu, X.L., Lou, X.L.: Computer Aided Detection and Classification of Microcalcifications in Mammograms: a Survey. Pattern Recognition 36, 2967–2991 (2003)
5. Cheng, H., Shi, X., Min, R., Hu, L., Cai, X., Du, H.: Approaches for Automated Detection & Classification of Masses in Mammograms. Pattern Recognition 39, 664–668 (2006)

6. Whitney, A.W.: A Direct Method of Nonparametric Measurement Selection. IEEE Trans. on Computers C-20(9), 1100–1103 (1971)
7. Christoyianni, I., Dermatas, E., Kokkinakis, G.: Neural Classification of Abnormal Tissue in Digital mammography Using Statistical Features of the Texture. In: Proc. of the 6th IEEE Int'l Conf. on Electronics, Circuits & Systems, vol. 1, pp. 117–120 (1999)
8. Petrosian, A., Chan, H., Helvie, M., Goodsitt, M., Adler, D.: Computer-aided Diagnosis in Mammography: Classification of Mass and Normal Tissue by Texture Analysis. Physics in Medicine and Biology 39(12), 2273–2288 (1994)
9. Chan, H., Wei, D., Helvie, M., Sahiner, B., Adler, D., Goodsitt, M., Petrick, N.: Computer-aided Classification of Mammographic Masses and Normal Tissue: Linear Discriminant Analysis in Texture Feature Space. Physics in Med. & Biol. 40, 857–876 (1995)
10. Haralick, R.M., Shanmugam, K., Dinstein, I.: Texture Features for Image Classification. IEEE Trans. Syst. Man Cybernet. 3(6), 610–621 (1973)
11. Masters, T.: Practical Neural Network Recipes in C++. UK edn. Academic Press (1993)
12. Suckling, J., Parker, J., Dance, D., Astley, S., Hutt, I., Boggis, C., Ricketts, I., Stamatakis, E., Cerneaz, N., Kok, S., Taylor, P., Betal, D., Savage, J.: The Mammographic Image Analysis Society digital mammogram database. In: Proceedings of the 2nd International Workshop on Digital Mammography, vol. 1069, pp. 375–378 (1994)

Image Dehazing Algorithm Based on Atmosphere Scatters Approximation Model

Zhongyi Hu[1,2], Qing Liu[1,3,*], Shenghui Zhou[3],
Mingjing Huang[3], and Fei Teng[3]

[1] Intelligent Transport Systems Research Center,
Wuhan University of Technology,
Wuhan, Hubei, P.R. China 430070
[2] Intelligent Information Systems Institute,
Wenzhou University,
Wenzhou, Zhejiang, P.R. China 325035
[3] School of Automation,
Wuhan University of Technology,
Wuhan, Hubei, P.R. China 430070
hujunyi@163.com

Abstract. Due to the scattered light of suspended particles in the atmosphere, the images taken in the foggy day become gray and are lack of visibility. In order to unveil the clear images structures and colors, the author propose an algorithm based on atmosphere scatters approximation model, which adopts the extended Jones Matrix and Stokes Law to calculate approximate the transmission of light in the atmosphere, so as to eliminate some of the scattered light. Both the light intensity in the atmosphere and haze concentration are obtained by means of Dark Channel Prior, afterward the extinction function for light transmission is used for calculation to restore the foggy images. The experimental results show that the algorithm can not only effectively improve scenery visual effect under different condition of haze, and provide clear pictures for machine vision applications in the foggy day.

Keywords: Scattering Approximate Model, Atmospheric Optical, Dehazing Algorithm, Expanded Jones Matrix, Stokes Laws.

1 Introduction

The atmospheric particles can be divided into two kinds: Water vapor condensation and suspended matter, both of which absorb the moisture from the air and then gradually extend to the formation of fog. In meteorology, according to the atmospheric conditions, the visibility is divided into 10 grades[1], from dense fog to thick fog to moderate fog to slight haze etc which is corresponding with grade 0, grade 1 and grade 2 etc. The atmospheric condition less than or equal to Grade 6 will influence the visual effects of the images taken outdoor.

---

* Corresponding author.

T. Huang et al. (Eds.): ICONIP 2012, Part V, LNCS 7667, pp. 159–168, 2012.

The atmospheric light scattering increases and the direct light of objects diminishes, which results in worse blur images and leads to abnormal operation for those common video system. As the blur image taken in fog seriously affects our daily life, transportation, outdoor surveillance, industry, agriculture and aviation shooting, it is urgent to go on with the research of defogging algorithm for both real life and theory research.

As for the defogging of a single image, the depth-of-field information is needed. While the depth-of-field information is unknown, it is very difficult to obtain a clear and realistic scene from a foggy image. Fortunately, in recent years, there are more and more researchers who devote themselves to the field of image defogging and get a lot of achievements. Fattal [2] can estimate reflectivity of the scene only by processing foggy images. He introduced a defogging method based on the method of independent component analysis (Independent Component Analysis, ICA). Besides, the method also can't deal with gray images[3].

Tan [4] made a comparison of the images taken in the foggy weather and sunny weather under the same scene, and found that the contrast of the image without fog is higher than the foggy one. Therefore he used an image local contrast maximization method to restore the blur images and realize image defogging. The result obtained by the method is credible visually. It can recover the utmost details and structures of the foggy images[5]. References [6-9] use the polarization characteristics of scattering light in foggy weather to recover the image. But the shooting conditions are very strict because two or more different polarization images are required, it can hardly come into use in real life. He [10] introduced a defogging method based on dark channel prior. This method holds physical effectiveness, when most of the regions of a scene target are essentially similar to the atmospheric light, or when there is no shadow being projected onto objects, not only the statistical knowledge of dark primary colors prior may be a useless, but also the restoration transmission chart might become limited and too complicated.

Based on the advantages and disadvantages of the algorithms mentioned above, this paper presents a new improved defogging method, by using atmospheric scattering approximation model to eliminate the scattering light caused by haze and obtaining atmospheric light and light extinction transmission function through the dark primary colors, we can restore the blur image and get a clear one.

2   Atmospheric Optical Transmission Approximate Model

2.1   Bouguer Lambert Beer Law

The theory of atmosphere light transmission was formed in the early 80s. Light waves are applicable to Bouguer Lambert Beer Law in the linear range spread[11], [12], as equation (1) shown:

$$dI_r(v) = -I_r t(v, z)dz \tag{1}$$

In this equation, $I_r(v)$ means the energy density where the light source shined to target on the scene, dz means the distance between light source and target, it is a function of the frequency of light wave, $t(v, z)$ called Extinction function or Attenuation Function, and the value of it is related to the light frequency v and distance z.

2.2   Atmospheric Light Scattering Characteristics

In defogging research field, the researchers only need to consider the visible light propagation properties in the atmosphere but not wave frequency. Suppose the object is in the sunny weather meteorologically, namely the sky is without any fog or haze, therefore the light into the camera is only related to the light of target objects themselves, equation (1) is similar to equation (2) [6], light intensity is a function of the distance between the camera and the target object.

$$I(x) = L_{object}t(x) = L_{object}\exp(-\beta x) \tag{2}$$

In this equation, x means the distance between the camera and target object, $I(x)$ means the intensity of the light into the camera, L_{object} stands for the light strength of the target object, $t(x)$ means light transmission function in the atmosphere, namely the extinction function, β means the extinction parameters.

Actually because there exists industrial pollution and fog haze, meteorologically sunny weather is very scarce. In foggy weather, the light into camera is not only relevant with target object's own light source, but also relevant with atmospheric light scattering, equation (2) can be rewritten as equation (3).

$$I(x) = L_{object}t(x) + A(x) \tag{3}$$

Among them, $A(x)$ is the atmosphere scatter light received by camera, it can be expressed as $A(x) = A_\infty(1 - t(x))$ [13]then equation 3 can be written as

$$I(x) = L_{object}t(x) + A_\infty(1 - t(x)) \tag{4}$$

Where A_∞ means the intensity of the atmosphere scattering light.

As we all know, the nature is generated by a number of different atoms and molecules radiation, therefore the natural light is a light source without polarization properties [14]. From the point of probability and statistics, the amplitude of the natural light in any direction are almost the same, so the natural light can be expressed by two mutually perpendicular polarized light with equal value and no connection, so light intensity can also be expressed as the equation(5).

$$I = I_\equiv + I_\perp \tag{5}$$

Where I means the total of natural light, and $I_\equiv$ means the horizontal polarization natural light equivalent, and $I_\perp$ means vertical polarization equivalent natural light.

Schematic reveals that the nature exists scattering light in the fog weather, which cause the light amplitude change in all directions and light intensity increases, reduces in some direction, therefore light intensity can also be expressed as the equation(6).

$$I = I_{\equiv} + I_{\perp} + I_{other} \tag{6}$$

Among them, I_{other} means light intensity except for the horizontal polarization and vertical polarization equivalent.

2.3   Use Extended Jones Matrix and Eliminate Scattering Light

Polarization is a very important concept in physical optics, we can use various ways to describe polarized light and polarization devices, such as matrix method, the index function method and bond with the ball method, especially, the way to use the Jones matrix (Jones) and Mueller matrix (Mueller) represent the polarization devices has a very good effect. Jones matrix and Mueller matrix have both similarities and differences, the two matrix between the light waves superposition and the phase information of the operation is different, the former does not keep phase operation information, lighting waves of coherent superposition, while the latter keep phase operation information, coherent light waves of superposition, therefore we choose Mueller matrix to participate in eliminating the scattering operations. Jones matrix includes horizontal line up partial device, vertical line up partial device, +45°C lines up partial device, -45°C lines up partial device, 1/4 wave plate (vertical fast axis), 1/4 wave plate (level fast axis), dextral rounded partial device and left-lateral rounded partial device.

Due to the mist in the image, white balance can be simplified as close to the average image. For some complex image, the color of the image changes, white balance is equal to image's local average approximately [15]. To offset the scattering, this paper takes up the horizontal and vertical polarizer in the matrix operations.

$$M_{\equiv} = \begin{vmatrix} 1 & 0 \\ 0 & 0 \end{vmatrix} \tag{7}$$

$$M_{\perp} = \begin{vmatrix} 0 & 0 \\ 0 & 1 \end{vmatrix} \tag{8}$$

Equation (7) $M_{\equiv}$ means the horizontal polarization components, equation (8) $M_{\perp}$ means vertical polarization components, and expands the matrix effectively.

$$M_{\equiv} = \begin{vmatrix} 1 & -s \\ 0 & 0 \end{vmatrix} \tag{9}$$

$$M_{\perp} = \begin{vmatrix} 0 & 0 \\ -s & 1 \end{vmatrix} \tag{10}$$

In equation (9) and (10) ,s means scattering coefficient, and the value of it is decided by the scattering intensity, that is, by the atmosphere mist concentration, experiments showed that, s ranges from 0 to 0.1, distinguish level is less than

0.01. Stokes Law [16] (Stokes Law, 1845) can be applied to visible light intensity, if let natural light go through the polarized elements, its light intensity is:

$$I = I_0 M \tag{11}$$

Among them, I_0 stands for the incident light intensity, while I stands for the intensity of light after polarized.

2.4 Atmospheric Scattering Approximation Model

In equation (6), horizontal polarization of natural light and vertical polarization are represented respectively by stokes vectors, and we choose expanded Jones matrix to mathematical operation, the following approximation equation exists:

$$I = I_{\equiv} + I_{\perp} + I_{other} \approx IM_{\equiv} + IM_{\perp} \tag{12}$$

Among them, I means an approximate value of light intensity into the camera, which is a key element to the defogging algorithm discussed in this paper. It eliminates part of light scattering participated in equation (4), and improve the method's effectiveness. So equation (4) can also be expressed by equation (13).

$$I(x)M_{\equiv} + I(x)M_{\perp} = L_{object}t(x) + A_{\infty}(1 - t(x)) \tag{13}$$

Through the equation (13), we can obtain atmosphere scattering approximation model, and calculated imaging light intensity of the object, and obtain the defogging algorithm based on the atmosphere scattering model, simplify the parameters, so we can get equation (14).

$$L_{object} = \frac{I(x)M_{\equiv} + I(x)M_{\perp} - A_{\infty}(1 - t)}{t} \tag{14}$$

In equation (14), $I(x)$ stands for the strength of the light into the camera, which is a known quantity, so we just calculate the scattering eliminating parameters s of s of A_{∞}, t, $M_{\equiv}$, $M_{\perp}$, which are introduced in the third and fourth quarter respectively.

3   Airlight and Air Extinction Coefficient Estimate

Previously, most of the dehazed algorithms acquire air radiance by pixel estimates in the most hazy density area. The specific methods is to choose the highest intensity of the current image as airlight [4], however, in the actual scene, the most intensity is possibly corresponding to a pixel of white destination objects, so this method may deviate a lot. Another method is put forward by literature [10], by using dark channel prior to select the brightest 0.1% of pixels in a hazy image. Through filtering, the brightest pixel is considered as airlight. This paper is consistent with the latter that the robustness of airlight based on simple method of dark channel is stronger than "the brightest pixel" method. Literature [10] will have the hazy image partitioned into several regions, every region is marked $\omega(x)$, and Equation (4) is acted on every little area $\omega(x)$, and the minimum value is obtain by the two sides of equation.

$$\min_{y \in \Omega(x)} (\frac{I^c(y)}{A^c}) = t(x) \min_{y \in \Omega(x)} (\frac{L^c_{object}(y)}{A^c}) + (1 - t(x)) \tag{15}$$

Since the RGB Image's three channels are independent, so for each color channel the minimum value can be operated independently, and then the minimum value can be obtained from the three minimum values again, equation (16) will be.

$$\min_c(\min_{y \in \Omega(x)} (\frac{I^c(y)}{A^c})) = t(x) \min_c(\min_{y \in \Omega(x)} (\frac{L^c_{object}(y)}{A^c})) + (1 - t(x)) \tag{16}$$

According to dark channel prior, the dark channel J^{dark} of fogless radiation intensity J tends to be 0, and Ac is always positive, then equation (17) will be.

$$\min_c(\min_{y \in \Omega(x)} (\frac{L^c_{object}(y)}{A^c})) = 0 \tag{17}$$

Plugequation (17) into equation (16), we get equation (18).

$$t(x) = 1 - \min_c(\min_{y \in \Omega(x)} (\frac{I^c(y)}{A^c})) \tag{18}$$

Because of the worsening of global environment, even through in the sunny weather, the air will always have some scattering particles. Therefore, when we observe the distance object, the haze will always exist, and influence people's visual habit. The haze is an important clue for human eyes to judge the depth of field [6, 17]. This phenomenon is called air perspective. Therefore, in order to meet people visual demand, the algorithm of this paper is not completely to dehaze, but selectively reserve a small amount of haze for distant scene, the dehazed image after this will look more natural. In equation (18) a constant type parameter $\in (0, 1)$ is adopted, so Equation (18) can be rewritten for equation (19).

$$t(x) = 1 - \omega \min_c(\min_{y \in \Omega(x)} (\frac{I^c(y)}{A^c})) \tag{19}$$

After adjustment, the proposed algorithm can reserve a small amount of haze, the value of ω depends on a specific application. In the experiment of this paper, ω is 0.95.

4   The Experiment and Evaluation

4.1   The Analysis and Contrast of Dehazed Effect

In the paper, the density of the fog is used to estimate the value of the Jones matrixes parameters s through Equation (9) and (10), and get results through

Equation (14). In many dehazed image blocks without the sky, at least, one color channel has very low intensity in some pixel. Due to additional airlight, a hazy image compared with its corresponding dehazed image is more brightly in the communication function value lower area, the hazy image of dark channel has more higher intensity in the heavy hazy area [10], [17]. The density of haze can be estimated by dark channel intensity or intensity specific value, and then determine the horizontal polarization matrix and vertical polarization matrix undetermined parameter s. In order to verify the effectiveness of the method presented in this paper, we take a group of photoes for the experiment, and output the histogram comparison results before and after dehazed. In figure 1, the

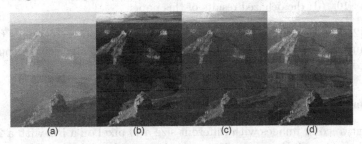

Fig. 1. To Compare The Dehazed Results: (a) input original image, (b)reference [2], (c)reference [10], (d) the dehazed results of this paper

sky takes a small scale of the whole image. From the restored image , the restore of sky in this paper is more credible than reference [2] and [10], in reference [2] the color of sky distorte obviously, and is over-saturated, reference [10] is more close to the real, but I think the sky won't be so blue in the hazy weather. so we make a comparison of the dehaze results in figure 1 using gray histogram, as is shown in figure 2. From figure 2, we can see that the gray histogram of original image is between 40 and 235, apparently because of haze, the goal of scene is hazy and white, and in reference [2] and [10] the gray histogram is more widely than original image, range from 0 to 235, but obviously there is a problem: concentrate on the range of 0 and 50, as a result, the image turn darker. The gray histogram of dehaze results in this paper is more widely than that of

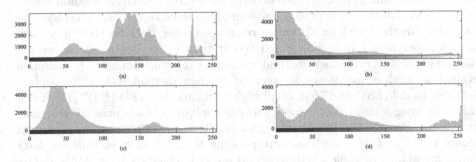

Fig. 2. The Gray Histogram Comparison of The Dehaze Results: in order from (a) to (d) as the input, references [2], reference [10] and the results in this paper

Fig. 3. To Compare The Dehazed Results: (a) input original image, (b)reference [2], (c)reference [10], (d) the dehazed results of this paper

reference [2] and [10], and distribution from 0 and 255 also corresponds to visual rules. Figure 3 shows, the result demonstrates that the algorithm proposed in this paper is suitable for the image defogging with different sky.

4.2   Operation Speed Comparison

We choose two color images with different size and pixel on a PC with a 2.0 GHz Intel Pentium Core i3 Processor, the operation speeds of three different dehazed algorithm is given by table 1. The experiment shows that operation times of reference [2] and [10] are similar, but the algorithm in this paper provides three times operational speed. And the dehazed results from figure 1 to figure 3 also shows the algorithm in this paper is better.

Table 1. Different dehazed methods to operation time comparison

image resolution (w*h)	Fattal's	He's	Our
Fig. 1 450*600	16	15.6	5.5
Fig. 3 210*278	3.4	3.2	1.1

4.3   Laplace Gradient Model Evaluation

To evaluate the image quality, currently the subjective evaluation method taking people as the observer's is predominate and also some researchers use objective evaluation method with or without a reference image. Because the experiment in this paper uses a hazy image and the original image doesn't exist, we use the image quality evaluation method without a reference image. The common evaluation methos are "grayscale average gradient method (GMG)", "Laplace gradient model (LS)" and "half of all high width method (FWHM)" [18]. Typically, the images more clear, more clear-cut contours of each pixel attachments gray value changes the greater the greater the value of the Laplace gradient mode. Fig. 1 and fig. 3 are the corresponding figure of fourteen different hazy and dehazed images, table 2 is the comparisons of Laplace gradient model, the results show that the algorithm to dehaze is used hazy images acquired by different sence and density of hazy weather, the LS model of its dehazed image is better

Table 2. Laplace gradient normal comparison

image	original image	Fattal's	He's	Our
Fig. 1	19.4003	33.4775	33.7446	37.1853
Fig. 3	25.7211	39.6062	48.7993	58.7263

than it of original image,and has different degrees of improvement compared with other defogging methods. In the paper,2590 hazy images are downloaded from Baidu Library, then manual trims and deletes unnatural picture elements such as Logo in images, uese the defogging method provided in paper to recover the images, and calculate Laplace gradient model of both dehazed image and original imagethe result is over 95% percent of dehazed image's LS model is better than original image, experiments proved that the algorithm based on atmosphere scatters approximation model is suitable for restore operation of most foggy images.

5   Summary and Discussion

The purpose of this paper is to defog a image, therefore a very simple but effective dehaze algorithm is proposed, which is called atmosphere scattering approximation model. The model is based on the transmission of light in the atmosphere and the particle scattering theory principle, and smart use of the expansion of the Jones matrix in stokos law mathematical operations. The experimental results in this paper used more than 2500 hazy images shows that, acquiring the density of image through the dark primary colors and ensuring the expansion parameters of the Jones matrix, can make most dehazed effect of images more ideal, but can not apply all the hazy images, such as image mutation in hazy area. In addition, if the dehazed algorithm needs to be used in monitoring, intelligent equipment, remote sensing intelligent equipment and so real places, they must be conducted effective optimization, increased processing speed. Above the existing problems, in the next stage the author team will study more perfect model to adapt to different environment hazy image.

Acknowledgments. The authors acknowledge the financial supported by the Fundamental Research Funds for the Central Universities and Zhejiang Provincial Natural Science Foundation of China(project No.: LY12F02015). The author is grateful to the anonymous referee for a careful checking of the details and for helpful comments that improved this paper.

References

1. Liu, J.B., Wu, J., Yang, C.P.: Theory of Optical Transmission in the Atmosphere. Beijing University of Posts and Telecommunications Press, Beijing (2005)
2. Fattal, R.: Single Image Dehazing. ACM Transactions on Graphics 27 (2008)

3. Pedone, M., Heikkila, J.: Robust Airlight Estimation for Haze Removal from a Single Image. In: 2011 IEEE Computer Society Conference on Computer Vision and Pattern Recognition Workshops (CVPRW), pp. 90–96. Colorado Springs, CO., United states (2011)
4. Tan, R.T.: Visibility in Bad Weather from a Single Image. In: 26th IEEE Conference on Computer Vision and Pattern Recognition (CVPR), Anchorage, AK, United states, pp. 1–8 (2008)
5. Fan, G., Cai, Z.X., Bin, X., et al.: Review and Prospect of Image Dehazing Techniques. Journal of Computer Applications 2417–2421 (2010)
6. Schechner, Y.Y., Narasimhan, S.G., Nayar, S.K.: Instant Dehazing of Images Using Polarization. In: 2001 IEEE Computer Society Conference on Computer Vision and Pattern Recognition (CVPR), Kauai, HI, United states, pp. 325–332 (2001)
7. Schechner, Y.Y., Narasimhan, S.G., Nayar, S.K.: Polarization-Based Vision Through Haze. In: ACM SIGGRAPH 2009 Courses, SIGGRAPH 2009, New Orleans, LA, United states (2009)
8. Shwartz, S., Namer, E., Schechner, Y.Y.: Blind Haze Separation. In: 2006 IEEE Computer Society Conference on Computer Vision and Pattern Recognition (CVPR), New York, NY, United States, pp. 1984–1991 (2006)
9. Narasimhan, S.G., Nayar, S.K.: Chromatic Framework for Vision in Bad Weather. In: 2000 IEEE Conference on Computer Vision and Pattern Recognition (CVPR), Hilton Head Island, SC, USA, pp. 598–605 (2000)
10. He, K., Sun, J., Tang, X.: Single Image Haze Removal Using Dark Channel Prior. IEEE Transactions on Pattern Analysis and Machine Intelligence 33, 2341–2353 (2011)
11. Ishimaru, A. (ed.): Wave Propagation in Random Media and Scattering. Science Press, Beijing (1986)
12. Cartney, M.: Optics of the Atmosphere. Wiley, New York (1983)
13. Narasimhan, S.G., Nayar, S.K.: Vision and the atmosphere. In: ACM SIGGRAPH 2009 Courses, SIGGRAPH 2009, New Orleans, LA, United states (2009)
14. Chen, L.: Physical Optics. Hefei University Press, Heifei (2007)
15. Tarel, J., Hautiere, N.: Fast Visibility Restoration from a Single Color or Gray Level Image. In: 2009 IEEE 12th International Conference on Computer Vision, Kyoto, Japan, pp. 2201–2208 (2009)
16. Wolf, M., Born, E.: Optics. Electronic Industry Press, Beijing (2005)
17. Ancuti, C.O., Ancuti, C., Hermans, C., Bekaert, P.: A Fast Semi-inverse Approach to Detect and Remove the Haze from a Single Image. In: Kimmel, R., Klette, R., Sugimoto, A. (eds.) ACCV 2010, Part II. LNCS, vol. 6493, pp. 501–514. Springer, Heidelberg (2011)
18. Xun, G.Z., Chen, B., Wang, Z.G., et al.: Adaptive Optics Image Restoration Theory and Methods. Science Press, Beijing (2010)

Design of a Data-Oriented PID Controller for Nonlinear Systems

Shin Wakitani, Takuya Nawachi, and Toru Yamamoto

Hiroshima University,
1-4-1 Kagamiyama, Higashihiroshima, Hiroshima, Japan
wakitani-shin@hiroshima-u.ac.jp

Abstract. It is difficult to maintain the desired control performance by applying a fixed PID controller because most process systems have nonlinearity. The CMAC-PID controller which uses Cerebellar Model Articulation Controller (CMAC) has been proposed as the effective method to these systems. However, some problems of CMAC related to the learning time or the huge memory needed are pointed out due to its structure. The issue of the learning time has been solved by the authors' method with the Fictitious Reference Iterative Tuning (FRIT) approach in recent years. In this paper, the method to deal with the problem of the required computer memory is considered by using Group Method of Data Handling (GMDH) approach. According to this method, even though the proposed GMDH-PID tuner keeps the same performance as the CMAC-PID tuner, the required computer memory is drastically reduced. The effectiveness of the proposed method is examined by a simulation.

Keywords: Nonlinear control, PID control, CMAC, FRIT, GMDH, Off-line learning.

1 Introduction

PID control [1,2,3] is still applied in many process systems. Most control objects have nonlinearity and it is difficult to maintain a desired control performance by using a fixed PID controller. To solve this problem, one of on-line PID parameter tuners is proposed as the CMAC-PID controller [4] using Cerebellar Model Articulation Controller (CMAC) [5] that is a type of Neural Network and its effectiveness is validated. However the CMAC-PID controller has some problems as follows:

- Many experiments are necessary to learn the CMAC.
- Large memory is required to realize the controller on a computer.

And, these problems become negative effects for a practical realization of the CMAC-PID controller. In recent years, the off-line learning method which uses the Fictitious Reference Iterative Tuning (FRIT) [6], [7] is proposed [8] to reduce a computational cost. According to the method in [8], an iterative learning is possible by using a pair of operating data.

T. Huang et al. (Eds.): ICONIP 2012, Part V, LNCS 7667, pp. 169–176, 2012.
© Springer-Verlag Berlin Heidelberg 2012

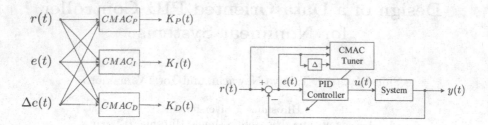

Fig. 1. Structure of a CMAC tuner   **Fig. 2.** Block diagram of a CMAC-PID controller

In this research, the scheme that transforms the tuner from CMAC to GMDH [9], [10] is proposed. According to the proposed method, the GMDH tuner can obtain the same performance of the CMAC tuner by combination of few N-Adalines. Moreover, it only decides weights of N-Adalines in the GMDH tuner, thus the required computer memory is drastically reduced compared with the CMAC-PID controller. In this paper the method which transforms the tuner from CMAC to GMDH is first explained then, the effectiveness of the prosed method is numerically examined by a simulation.

2   Design of a CMAC-PID Controller Using Operating Data

2.1   PID Controller Design

The following PID controller is first designed

$$u(t) = u(t-1) + K_P(t)\Delta e(t) + K_I(t)e(t) + K_D(t)\Delta^2 e(t), \tag{1}$$

where $u(t)$ is the control input and $e(t)$ is the control error which is defined by

$$e(t) := r(t) - y(t). \tag{2}$$

In (2), $r(t)$ and $y(t)$ are the reference signal and the system output respectively. Moreover, Δ is the differencing operator which is given by $\Delta := 1 - z^{-1}$, where z^{-1} is the back-ward operator which denotes $z^{-1}e(t) := e(t-1)$. In the CMAC-PID controller, PID gains $K_P(t), K_I(t)$ and $K_D(t)$ are on-line computed by independent CMACs shown in Fig.1. The block diagram of the CMAC-PID controller is shown in Fig.2.

2.2   Off-Line Learning by CMAC-FRIT

The CMACs are learned by the FRIT method using operating data (from here on, the method is called CMAC-FRIT). The block diagram of the off-line learning scheme is shown in Fig.3. In Fig.3, $r_0(t), e_0(t), u_0(t)$ and $y_0(t)$ mean operating

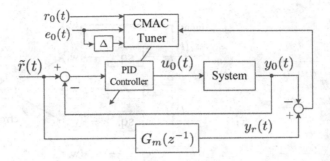

Fig. 3. Block diagram of a CMAC-FRIT

data which are obtained in advance by using a fixed PID controller. Moreover, $\tilde{r}(t)$ is the fictitious reference signal and denotes as follows [6]:

$$\tilde{r}(t) = C^{-1}(z^{-1})u_0(t) + y_0(t). \tag{3}$$

Where, $C(z^{-1})$ is the polynomial equation of the PID controller. In the CMAC-FRIT, weights in CMACs to calculate PID gains at t are updated by iterating following adjustments [8],

$$\left. \begin{array}{l} K_P^{new}(t) = K_P^{old}(t) - \eta_P \dfrac{\partial J(t+1)}{\partial K_P^{old}(t)} \dfrac{1}{N} \\[3mm] K_I^{new}(t) = K_I^{old}(t) - \eta_I \dfrac{\partial J(t+1)}{\partial K_I^{old}(t)} \dfrac{1}{N} \\[3mm] K_D^{new}(t) = K_D^{old}(t) - \eta_D \dfrac{\partial J(t+1)}{\partial K_D^{old}(t)} \dfrac{1}{N} \end{array} \right\}. \tag{4}$$

Here, η_P, η_I and η_D are learning rate, N is the number of weight tables of the CMAC and J is the criterion which is denoted as

$$J(t) = \{y_0(t) - y_r(t)\}^2. \tag{5}$$

In (5), $y_r(t)$ is the output of $G_m(z^{-1})$ which is a desired response model and is given by a user.

3   GMDH-PID Controller Design

In this section, the method to realize GMDH-PID tuner which has the same performance as the CMAC-PID tuner constructed in Section 2 is considered. The structure of GMDH tuner is shown in Fig.4. GMDH is a kind of multi-layered networks constructed by a suitable combination of N-Adalines which is illustrated in Fig.5. In Fig.5, 'sq.' and '×' denote the square and the multiplication,

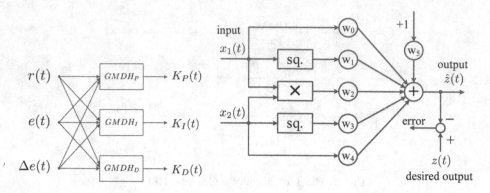

Fig. 4. Structure of a GMDH tuner

Fig. 5. Structure of an N-Adaline

respectively. Furthermore, a N-Adaline is composed by an unit which consists of 2-inputs and 1-output. When $x_1(t)$ and $x_2(t)$ are chosen as the input variables, the output variable $z(t)$ is expressed as follows:

$$\hat{z}(t) = w_0 x_1(t) + w_1 x_1^2(t) + w_2 x_1(t) x_2(t) + w_3 x_2^2(t) + w_4 x_2(t) + w_5, \quad (6)$$

where w_i are weights, and are determined by the least squares method so that the output of a N-Adaline and the CMAC tuner are in accord when the same signals $(r_0(t), e_0(t), \Delta e_0(t))$ are inputted. Moreover, the schematic diagram of a GMDH network is shown in Fig.6. On determining the network structure, a measure is adopted in order to select the N-Adalines which describe the CMAC tuner as strictly as possible. The measure is expressed as

$$R^2 = \sum_{t=1}^{N}(\hat{z}(t) - \bar{\hat{z}})^2 / \sum_{t=1}^{N}(z(t) - \bar{z})^2. \quad (7)$$

In (7), $\bar{\hat{z}}(t)$ and $\bar{z}(t)$ are expressed a average of $\hat{z}$ and z respectively. In this research, the number of GMDH layer is beforehand decided in consideration of the memory capacity of an intended computer and the optimal network that R^2 of the last layer becomes the largest is chosen.

Finally, the procedure of the design method of the proposed GMDH-PID controller is summarized as follows.

1. The first operating data are obtained by using a fixed-PID controller.
2. The CMAC-PID controller is constructed and the weights in the CMAC tuner are learned by FRIT using the operating data.
3. To construct the GMDH tuner, weights of a N-Adaline are decided by least squares method so that the outputs of a N-Adaline to the operating data are comparable to the CMAC tuner outputs.
4. The GMDH network whose R^2 of a last N-Adaline is the largest in all combinations is chosen as a GMDH tuner.

The block diagram of the proposed GMDH-PID controller is shown in Fig. 7.

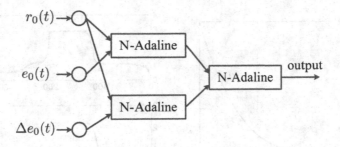

Fig. 6. Block diagram of a GMDH network

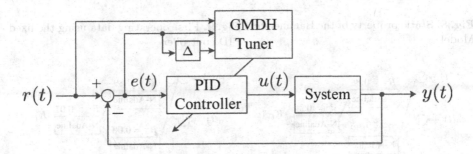

Fig. 7. Brock diagram of the proposed method

4   Simulation Example

To evaluate the effectiveness of the proposed method, Hammerstein model [11] which is given by (8) was applied:

$$y(t) = 0.6y(t-1) - 0.1y(t-2) + 1.2x(t-1) - 0.1x(t-2) + \xi(t) \\ x(t) = 1.5u(t) - 1.5u^2(t) + 0.5u^3(t) \Bigg\} . \quad (8)$$

Where $\xi(t)$ is white Gaussian noise with zero mean and variance 1.0×10^{-4}. The static property of the model is as shown in Fig.8. Moreover, the reference signal values for each instant of time are set as shown in (9):

$$r(t) = \begin{cases} 1.0 \ (0 \le t < 100) \\ 3.0 \ (100 \le t < 200) \\ 0.5 \ (200 \le t < 300) \\ 1.5 \ (300 \le t < 400) \end{cases} . \quad (9)$$

The fixed PID controller is first employed in order to get the initial operating data, whose PID gains were determined by using the CHR method [2], and these PID gains are set as follows:

$$K_P = 0.049, \quad K_I = 0.054, \quad K_D = 0.025. \quad (10)$$

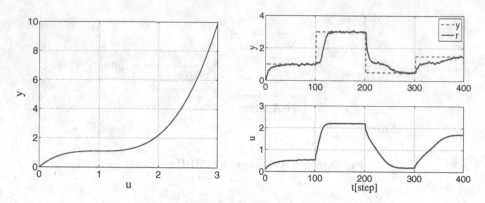

Fig. 8. Static property of the Hammerstein Model

Fig. 9. First operating data using the fixed PID controller

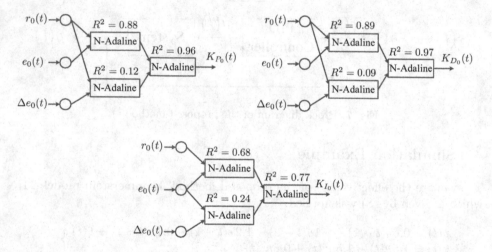

Fig. 10. Constructed GMDH tuners to calculate each PID gains

The control result is as shown in Fig.9. The control performance, especially the result between 200 and 400[step], is not good because the PID gains are too small. Next, the FRIT method utilizes the available data to facilitate the CMAC tuner to learn its weights. Afterwards, the GMDH tuners are constructed by the proposed method. As a result, the GMDH networks like Fig.10 are constructed to output each PID gains. Note, the number of the layer is set as 2 in advance.

The control result using proposed GMDH-PID controller is shown in Fig.11 and the trajectory of PID gains are shown in Fig.12. These figures show that the PID gains are adjusted appropriately at each balance points, thus the good control performance are maintained. Furthermore, a control result and a trajectory of PID gains when the CMAC-PID controller is applied to the system are shown in Fig.13 and Fig.14. From Fig.12 and Fig.14, approximately-same PID gains are outputted by the CMAC tuner and the GMDH tuner when the transitions

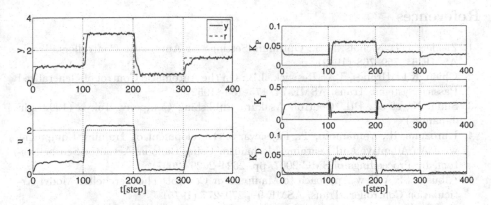

Fig. 11. Control result by using the proposed method

Fig. 12. Trajectory of the PID parameters corresponding to Fig. 11

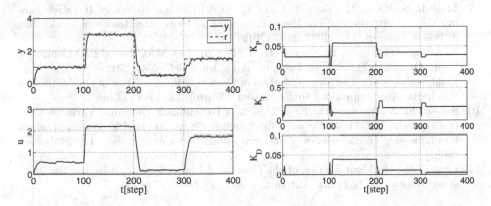

Fig. 13. Control result by using the CMAC-PID controller

Fig. 14. Trajectory of the PID parameters corresponding to Fig. 13

of PID gains are compared, thus the same control performance can be obtained by each controllers.

5 Conclusions

In this paper, the design method of the GMDH-PID controller using operating data is proposed. In the proposed method, the CMAC tuner is learned by FRIT using operating data. After that, the GMDH tuner which has the same performance as the CMAC tuner is constructed. As a result, the required computer memory can be reduced drastically. The control performance is evaluated through simulation. In the simulation, it is confirmed that the trajectory of outputted PID gains from CMAC and GMDH tuners have virtually the same performance. The proposed method is going to be applied to a real system as a future work.

References

1. Ziegler, J.G., Nichols, N.B.: Optimum Settings for Automatic Controllers. Trans. ASME 64, 759–768 (1942)
2. Chien, K.L., Hrones, J.A., Reswick, J.B.: On the Automatic Control of Generalized Passice Systems. Trans. ASME 74, 175–185 (1952)
3. Suda, N., et al.: PID Control. Asakura Publishing Company, Tokyo (1992) (in Japanese)
4. Kurozumi, R., Yamamoto, T., Fujisawa, S.: Development of Training Equipment with An Adaptive and Learning Mechanism Using Balloon Actuator-Sensor System. In: Proceedings of SMC 2007, pp. 2624–2629 (2007)
5. Albus, J.S.: A New Approach to Manipulator Control: The Cerebellar Model Articulation Controller. Trans. ASME 97, 270–277 (1975)
6. Kaneko, O., Souma, S., Fujii, T.: Fictitious Reference Iterative Tuning in the Two-Degree of Freedom Control Scheme and Its Application to a Facile Closed Loop System Identification. Trans. SICE 42, 17–25 (2006)
7. Masuda, S., Kano, M., Yasuda, Y., Li, G.D.: A Fictitious Reference Iterative Tuning Method with Simulations Delay Parameter Tuning of the Reference Model. Int. J. Innov. Comput. I. 6, 2927–2939 (2010)
8. Wakitani, S., Ohnishi, Y., Yamamoto, T.: Design of a CMAC-Based PID Controller Using Operating Data. Dist. Comput. Artif. Intell. 545–552 (2012)
9. Ivakhnenko, A.G.: The Group Method of Data Handling, a Rival of the Method of Stochastic Approximation. Soviet Autom. Control. 13, 43–55 (1968)
10. Sakaguchi, A., Yamamoto, T.: A Design of Generalized Minimum Variance Controllers Using a GMDH Network for Nonlinear Systems. IEICE Transactions on Fundamentals of Electronics, Communications and Comput. Sci. 11, 2901–2907 (2001)
11. Lang, Z.Q.: On Identification of the Controlled Plants Described by the Hammerstein System. IEEE Trans. Automat. Contr. 39, 569–573 (1994)

Improving the Robustness of Single-View-Based Ear Recognition When Rotated in Depth

Daishi Watabe[1,*], Takanari Minamidani[1], Hideyasu Sai[1],
Katsuhiro Sakai[1], and Osamu Nakamura[2]

[1] Saitama Institute of Technology, 1690 Fusaiji, Fukaya, Saitama, Japan
dw@sit.ac.jp
[2] Kogakuin University, 1-24-2 Nishishinjyuku, Shinjuku, Tokyo, Japan

Abstract. An algorithm is proposed that improves the robustness of ear biometric systems with the aim of developing a surveillance system based on ear biometrics. To deal with pose variations that are rotated in field depth, the Gabor jets of different poses are estimated and used as training data for a discriminant analysis-based classifier. Experimental evaluations show the effectiveness of the proposed algorithm, and the potential for improving the robustness of a single-view-based ear-biometric surveillance system.

Keywords: Ear biometrics, Gabor filter, Robustness.

1   Introduction

1.1   Background

In 1964, Iannarelli [1] showed that the shape of an ear has sufficient information to uniquely identify individuals and does not change with age. Based on this research, tracings of a human ear left on walls (ear prints) have been used in forensic science, much like fingerprints, over the last 40 years in the United States.

The advantage in human ear shapes is their larger size compared with fingerprints. They are also easier to identify at a distance. Hence, unlike fingerprints, human ear shapes can possibly be used for criminal surveillance using images taken at a distance from a surveillance camera. Hence, ear identification is a possible candidate to enhance the reliability of commonly used face-based criminal surveillance systems, which can be compromised by wearing make-up, spectacles, beards, and moustaches or through plastic surgery.

In a surveillance scenario, we cannot generally expect the cooperation of the person being identified. Hence, pose variations between the input and registered images are of considerable concern. For an effective ear-biometric surveillance system, robustness to pose variations is also very important. Improving this robustness and providing an effective surveillance system using the prominent features of an ear is the main theme of this study.

T. Huang et al. (Eds.): ICONIP 2012, Part V, LNCS 7667, pp. 177–187, 2012.

1.2   Related Studies

Most robust ear recognition systems [2], [3] are based on 3D data obtained using laser-range-finders both for registered data and input data. Hence, to identify a suspected person, they must have been previously registered using 3D laser-range-finders. As long as this is the case, as in gate-control scenarios, 3D-based systems are reliable. However, this is a problem in common criminal surveillance scenarios, where mug shots (including profiles), previous green-card ID photos (as in M-378) or images from 2D surveillance video cameras are used as registered images. Because laser-range-finders are still expensive, it may take some time before these mug shots, ID photos and surveillance cameras are replaced by 3D laser-range-finders.

Furthermore, existing 2D-based ear recognition systems [4], [5] have been developed for scenarios where the cooperation of the person to be identified is available. Consequently, robustness to the relatively large pose variations required for surveillance is not necessarily considered.

Given the above studies, the authors developed a 2D-based robust ear recognition algorithm [6], [7], [11], and a 99.4 % Rank-1 recognition rate was reported on the XM2VTS database with flowing pose variations of a maximum of 38.0 %, an average of 15.6 % difference in size, and a maximum of 31.1° with an average of 13.1° in-plane rotation. Although the method is promising, the robustness examined in that study was limited to in-plane pose variations. Robustness against off-angle pose variations is also necessary for a more reliable surveillance system.

1.3   Aim of This Study

The aim of this study is to improve the robustness against off-angle rotation by extending the methods presented in [6], [7]. There, it was reported that robustness was improved by using Gabor features for various poses, subjected to independent component analysis (ICA) plus linear discriminant analysis (LDA) training.

This method is reliable when the Gabor features for various poses are obtained beforehand. This requirement can be hard to achieve in real-world surveillance scenarios. The possibility could arise that only one criminal mug shot is available in registration data. In such cases, construction of images corresponding to various poses produced from the image can be useful. Such constructions from in-plane rotation can be reasonably obtained by rotating the image (though ignoring the change in shading). However, constructions of off-angle pose variations are not easy to obtain. Accepting that full constructions are difficult, we tried to produce only local constructions near each feature point. Estimations of the Gabor features corresponding to off-angle pose variations were created [10] and subjected to variants of discriminant analysis-based training. Empirical evaluations demonstrating robust improvements with this proposed method are provided.

2   Proposed Method

In this section, the Gabor jet, the linear jet transformation, and the discriminant analysis that constitute the system are outlined. The development of the proposed algorithm is provided.

2.1   Gabor Jet

Let $\mathbf{x} = (x, y)$ be a point in a plane. A 2D plane-wave defined by wavevector $\mathbf{k} = (k_x, k_y)$ and restricted by a Gaussian function determines the so-called Gabor function:

$$\psi(\mathbf{x}) = \frac{|\mathbf{k}|^2}{\sigma^2} \exp\left(-\frac{|\mathbf{k}|^2 |\mathbf{x}|^2}{2\sigma^2}\right)\left[\exp(i\mathbf{k} \cdot \mathbf{x}) - \exp\left(-\frac{\sigma^2}{2}\right)\right] \tag{1}$$

Here σ denotes the width of this function determined by the Gaussian function. The factor $\exp(-\sigma^2/2)$ is a compensating term that eliminates averages; this condition is required from wavelet theory, but if σ is large enough, this term can be neglected.

Gabor functions are characterized as localized wavy shapes in various directions induced by plane waves. Gabor filters, the convolutions with these Gabor functions, extract directions and wavelengths of the localized wavy shapes of an image near the point under consideration.

Ear cartilage, with its wavy shapes in various directions, is very distinctive. Endpoints, junctions, and protuberances of ridges of the ear cartilage are obvious feature points (Fig. 1). Surface measurements of the wavy shapes near these feature points are coded using a Gabor filter.

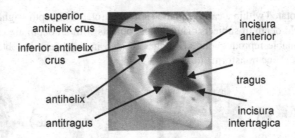

Fig. 1. Feature points of an ear

Five wave lengths $4, 4\sqrt{2}, 8, 8\sqrt{2}, 16$ are employed as parameter values of the Gabor filters, to cover the various widths of the ridges of ear cartilage appearing in the experimental data. Furthermore, to cover all directions evenly, eight directions for every octant are employed. Using this bank of Gabor filters, Gabor jets are sampled at the selected feature points.

To localize the chosen feature points, a modified version of the graph matching method introduced in [11] is used. The algorithm is outlined as follows. Shapes of grids are averaged over the registered images. Raster scanning of this averaged graph is performed on an input image and Gabor jets at each of the scanning points are sampled. For each sampling, the Gabor jets are projected onto the principal subspace of the registered Gabor jets. The similarity between an input Gabor jet and its projection onto the principal component analysis (PCA) subspace is called a jet space similarity [11]. Candidate ears can be detected as matches where jet space similarities are higher than a given threshold. With these matches of candidate ears, by distorting the grids elastically, feature points are finally located on the bases of the highest scores of jet space similarities.

2.2   Linear Jet Transformation

It is known that Gabor jets are robust against pose variations. We attempted to improve this robustness by increasing the number of registration data corresponding to the various poses estimated from a single registration image.

First, the issue arising from reproducing the pose variation using image processing is examined. Pose variations within a camera plane can be estimated by image-rotation processing (Fig. 2 left). This reproduction is not necessarily accurate, because the change in shading induced by the pose variation is not considered. However, when shading is relatively small under adequate lighting, it is a relatively minor issue compared with the difficulty of reproducing images rotated in depth (Fig. right).

Fig. 2. A subject rotated within a camera plane (left) and rotated in depth (right) The difficulty originates in the field depth of subjects. Thus, plane objects without depth are easier to process (Fig. 3 left). Reasonable reproductions of images rotated in depth can be obtained using the image processing of affine transformations.

Fig. 3. A planar object rotated in depth (left) and a tangent plane approximation of a subject rotated in depth (right)

Locally, near the feature points, the subject is approximated by a tangent plane. The tangent plane does not have a depth. Hence, the image of this plane rotated in depth can be estimated (Fig. 3 right). This estimated image reflects local features under pose variations near the feature points. Similar to the tangent plane, Gabor jets only represent local features. Motivated by the above, Gabor jets of the subjects rotated in depth will be estimated. This reproduction method, estimating Gabor jets of subjects with different poses [10], is outlined in the following.

Let the x-y coordinates map the camera plane and the z-axis be perpendicular to the camera plane. Suppose that a planar object, initially placed parallel to the camera plane, is rotated ϕ around its x-axis and then θ around its y-axis. By observing the transformations of unit vectors, a point on the planar object initially at $\mathbf{u} = (x, y)$ is transformed to x given by

$$\mathbf{x} = \mathbf{Au}, \qquad \mathbf{A} = \begin{pmatrix} \cos\phi & \sin\theta\sin\phi \\ 0 & \cos\theta \end{pmatrix}.$$

If this planar object is initially placed at (ϕ_1, θ_1) and not parallel to the camera plane, the above transformation is

$$\mathbf{x} = \mathbf{A}(\phi_2, \theta_2)\mathbf{A}(\phi_1, \theta_1)^{-1}\mathbf{u}.$$

Under this transformation, the transformation of the Gabor jets corresponding to the pose change can be estimated. In the following, $\mathbf{A}(\phi_2, \theta_2)\mathbf{A}(\phi_1, \theta_1)^{-1}$ is denoted as $\mathbf{A}$ for simplicity. Components of the Gabor jets after the transformation are obtained by the convolution of the Gabor function and the transformed image $I(\mathbf{A}^{-1}\mathbf{x})$. Using $\mathbf{x} = \mathbf{Au}$, $\mathbf{x}' = \mathbf{Au}'$, this is

$$j'_k(\mathbf{x}) = \int I(\mathbf{A}^{-1}\mathbf{x}')\Psi_k(\mathbf{x} - \mathbf{x}')dx'$$
$$= \int I(\mathbf{u} - \mathbf{u}')\Psi_k(\mathbf{Au}')|\mathbf{A}|du$$

Assuming the approximation

$$\Psi_k(\mathbf{Au}')|\mathbf{A}| \approx \sum_{k'} c_{kk'}(\mathbf{A})\Psi_{k'}(\mathbf{u}') \tag{2}$$

the transformation of the Gabor jets is simply written as

$$j'_k(\mathbf{x}) \approx \sum_{k'} c_{kk'}(\mathbf{A})j_{k'}(\mathbf{u})$$

Matrix $\mathbf{c}^{(A)} = (c_{kk'}(\mathbf{A}))$ is obtained by multiplying both sides of Eq. (2) by $\overline{\Psi_k(\mathbf{u}')}$ and by integrating both sides (see Appendix).

This estimation algorithm assumes that the normal vectors of the tangent plane are stabilized in advance so that the angles (ϕ, θ) between these normal vectors and the vector normal to the camera plane can be established. It is not easy to establish these angles for each individual ear image in advance in a real scenario. Hence, model normal vectors at each feature point are established in advance based on an exhaustive search of θ and ϕ, with which better recognition rates are obtained for each feature point.

2.3   Discriminant Analysis

In addition to the real registration data, the estimated Gabor jets for other poses obtained through the above procedure are used as training data, combined into a class of information for each individual. Using this class information, we try to improve the robustness against pose variations.

For the training algorithm, multiple discriminant analysis [11] is employed. This algorithm provides coordinate transformations into the coordinates where class separations are easier. The matrix w that performs this coordinate change is obtained by maximizing the following function defined by

$$J(\mathbf{W}) = \frac{\left|\mathbf{W}^t \mathbf{S}_b \mathbf{W}\right|}{\left|\mathbf{W}^t \mathbf{S}_w \mathbf{W}\right|}, \tag{3}$$

where $\mathbf{S}_b$ is the between-class scatter and $\mathbf{S}_w$ is the within-class scatter. The column vectors ω_i of the matrix $\mathbf{W}$ are obtained by solving the following generalized eigenvalue problem

$$\mathbf{S}_b \omega_i = \lambda_i \mathbf{S}_w \omega_i, \tag{4}$$

where the number of samples is not large enough compared with the dimensions of the data (in our setting, the dimension of the Gabor jets), the within-class scatter $\mathbf{S}_w$ is degenerate and not all vectors ω_i are necessarily obtained accurately. To solve this problem, the within-class scatter $\mathbf{S}_w$ is replaced by the total scatter $\mathbf{S}_T$ in [12] and dimensional reduction with PCA is used before discriminant analysis in [9]. Combining the above two in the experiments, dimensional reduction using PCA is first applied and the total scatter $\mathbf{S}_T$ is used instead of the within-class scatter $\mathbf{S}_w$.

2.4   Outline, Novelty, and Expected Effect of the Proposed Method

The proposed method is based on the three algorithms mentioned above. First, the transformation that enhances robustness against pose variations is obtained. The transformation matrix is calculated based on multiple discriminant analysis, where not only the Gabor jets from the registration image, but also the estimated Gabor jets of different poses obtained through the linear jet transformation are used for training data. With this transformation, the Gabor jets of the registered images are transformed to the coordinates where class (individual) separations are easier. With these coordinates, robustness is improved because pose variations are taken into consideration within each class. Given an input image, feature points are then detected using jet space similarity and the Gabor jets are sampled from the feature points. These Gabor jets are also transformed using the transformation matrix enhancing the robustness mentioned above. Finally, the correlation between the Gabor jet of the registered images and the input image are computed to determine similarity.

The novelty of this method is the use of the estimated local features from a single image as training-data for discriminant analysis. Discriminant analysis does not function with one image per person. Hence, the lack of data should be compensated for, using the estimated training data from this single image. The estimated local features for other poses are subjected to training together with these single registration data.

Similar to the discriminant analysis trained using real data, the discriminant analysis trained by these estimated data is expected to improve the robustness against pose variation.

3   Experiments

3.1   Images for the Experiments

There are three types of pose variations expected in the acquisition of ear images, namely yawing, rolling and pitching (Fig. 4). For surveillance, the subject's pose may not only be rotated within the camera plane, as in pitching, but also rotated outside the camera plane, as in rolling and yawing. Hence, recognition experiments that included images of subjects with variations in pose were performed using the HOIP database [13].

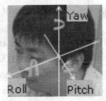

Fig. 4.   Rotations of an ear caused by pose variations

The HOIP database is a database of 300 subjects photographed from 511 directions. These images are obtained by rotating a turn-table every five degrees. In these facial images, the size of the ear approximately fits within a 90×70 window.

In the following, a yaw angle of 0° corresponds to a frontal face view, 90° to a true left profile, and 180° for a back view. The feature points of the ear are located using jet space similarities [11] (Fig. 5 left). Where detection has not been successful (false negatives and positives as in Fig. 5. right), errors are manually corrected.

Fig. 5. Example of ear detection using jet space similarity

3.2   Experimental Method

To examine robustness against yaw-angle pose variations, recognition and verification experiments are performed using ear images taken from the true left profile as a registration image. Input images are taken from yaw angles spanning from 30° to 120°, every 10°, except 90° corresponding to the registration image (Fig. 6).

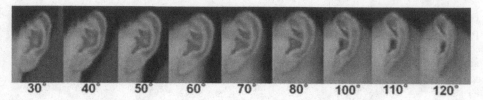

Fig. 6.   Example of input images
(yaw angles 0°, 90° and 180° corresponding to frontal face, true left, and back views)

To examine the effect of our proposed method using estimated data for pose variations as training data for discriminant analysis, robustness in the following three cases is examined and compared.

- No training data for other poses are used.
- Estimated data for other poses are used for training.
- Real data for other poses are used for training.

Without training data from other poses, discriminant analysis cannot be used and is excluded from the process. Using training data from other poses, estimated or real data with yaw angles of 75° and 105° are included in the training data for discriminant analysis besides the registration data taken from 90°.

For subjects with more than four (out of the seven) visible feature points at all viewing angles as input, registration and training data were selected. Hence, the number of subjects slightly depends on the yaw angle of the input data (Table 1).

Table 1. Number of subjects at each yaw angle

Angle(°)	30	40	50	60	70	80	100	110
Number of Subjects	162	159	166	165	168	163	169	168

Accuracy is evaluated using the Rank-1 recognition rate for 1:N recognition and by using the Equal Error Rate (EER) for 1:1 verification as defined in [14]. Precisely speaking, the Rank-1 recognition rate and EER depend on the number of registered subjects. Hence, the comparison between different angles may not be very accurate. However, comparisons between the three methods are reasonable.

4   Experimental Results

The Rank-1 recognition rates and EER of the three methods are displayed as input images with various yaw angles (Fig. 7). Here, since 90° is the angle for registration, at 90°, formally, the Rank-1 recognition rate is set to 1 and EER is set to 0. Fig. 7 clearly shows that using the estimated data for other poses improves the accuracy of most of the input images with yaw angles between 30° and 120°. Although the proposed algorithm using estimated data does not surpass the accuracy of the algorithm using real data, trends indicating the improvement in robustness are clearly similar. Hence, the proposed method based on using estimated data as training data improves the robustness of the algorithm without training data over a wide range.

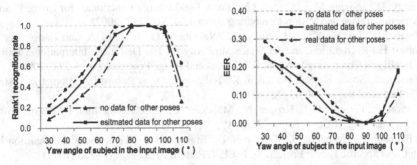

Fig. 7. Robustness against yaw angle variation
(1:N recognition, Rank-1 recognition rate (left), 1:1 verification, EER(right))

5   Summary

An algorithm was proposed for improving the robustness of an ear biometric system with the aim of developing a surveillance system based on ear biometrics. Constructions of images corresponding to off-angle pose variations are not easy to obtain when the subject has a field depth. Locally, near the feature points, the surface can be approximated by a tangent plane. Because the tangent plane does not have a depth, the image of this plane rotated in depth can be estimated. Similar to the tangent plane, Gabor jets only represent local features. Motivated by the analogy with these planes, Gabor jets of each ear feature rotated in depth were estimated. These estimated Gabor features were subjected to variants of discriminant analysis-based training. Experimental evaluations showed the effectiveness of the proposed algorithm, and the potential for improving the robustness of a single-view-based ear surveillance-system.

Acknowledgments. The facial data in this paper are used by permission of the Softopia Japan Foundation. It is strictly prohibited to copy, re-use, or distribute the facial data without permission. This work was supported by KAKENHI 22700219. An earlier report (in Japanese) of this study was presented in [15].

References

1. Iannarelli, A.: Ear Identification. Forensic Identification Series. Paramount Publishing Company, Fremont (1989)
2. Chen, H., Bhanu, B.: Human Ear Recognition in 3D. IEEE Transactions on Pattern Analysis and Machine Intelligence 29, 718–737 (2007)
3. Yan, P., Bowyer, K.: Biometric Recognition Using 3D Ear Shape. IEEE Transactions on Pattern Analysis and Machine Intelligence 29, 1297–1308 (2007)
4. Yuizono, T., Wang, Y., Satoh, K., Nakayama, S.: Study on Individual Recognition for Ear Images by Using Genetic Local search. In: Proceedings of the 2002 Congress on Evolutionary Computation, pp. 237–242 (2002)

5. Hurley, D.J., Nixon, M.S., Carter, J.N.: Force Field Energy Functionals for Image Feature Extraction. Image and Vision Computing Journal 20, 311–317 (2002)
6. Watabe, D., Sai, H., Ueda, T., Sakai, K., Nakamura, O.: ICA, LDA, and Gabor Jets for Robust Ear Recognition, and Jet Space Similarity for Ear Detection. International Journal of Intelligent Computing In Medical Sciences and Image Processing 3, 9–29 (2009)
7. Watabe, D., Soma, K., Sai, H., Sakai, K., Nakamura, O.: A Robust Ear Biometric System. ITE Technical Report 34, 69–72 (2010)
8. Wiskott, L., Fellous, J.M., Kruger, N., Malsburg, C.: Face Recognition by Elastic Bunch Graph Matching. IEEE T. Pattern Anal. 19, 775–779 (1997)
9. Belhumeur, P., Hespanha, J., Kriegman, D.: Eigenfaces v.s. Fisherfaces: Recognition Using Class Specific Linear Projection. IEEE T. Pattern Anal. 19, 711–720 (1997)
10. Maurer, T., Malsburg, C.: Single-View Based Recognition of Faces Rotated in Depth. In: International Workshop on Automatic Face and Gesture Recognition (1995)
11. Watabe, D., Sai, H., Sakai, K., Nakamura, O.: Ear Biometrics Using Jet Space Similarity. In: Canadian Conference on Electrical and Computer Engineering, pp. 1259–1263 (2008)
12. Fukunaga, K.: Statisitical Pattern Recognition. Academic Press (1990)
13. Yamamoto, K., Niwa, Y.: Human and Object Interaction Processing (HOIP) Project. In: Joho Shori Gakkai Shinpojiumu Ronbunshu (in Japanease)
14. ISO/IEC, Information technology—Biometric performance testing and reporting—Part 1:Principles and framework, ISO/IEC 19795-1 (2006)
15. Watabe, D., Zichong, H., Souma, K., Sai, H., Sakai, K., Nakamura, O.: Improving Robustness of Single-view-based Recognition of Ears Rotated in Depth. Journal of The Institute of Image Information and Television Engineers 65, 1016–1023 (2011) (in Japanese)

Appendix

Multiplying both sides of Eq. (2) by $\overline{\Psi_{k'}}(u')$ and integrating both sides results in

$$\sum_{k'} c_{kk'}(A) \int \Psi_{k'}(u') \overline{\Psi_{k'}}(u') du' \approx \int \Psi_k(Au') \overline{\Psi_{k'}}(u') \,|\,A\,|\, du'$$

$$= \int \Psi_k(x') \overline{\Psi_{k'}}(A^{-1}x') dx'$$

Using the following notation,

$$\langle k'|k''\rangle = \int \Psi_{k'}(u') \overline{\Psi_{k'}}(u') du'$$

$$\langle k''(A^{-1})|k\rangle = \int \Psi_k(x') \overline{\Psi_{k'}}(A^{-1}x') dx' \tag{6}$$

we have

$$\sum_{k'} \langle k''|k'\rangle c_{kk'} = \langle k''(A^{-1})|k\rangle$$

which in matrix form becomes

$$\begin{bmatrix} \langle k_1|k_1\rangle & \cdots & \langle k_1|k_N\rangle \\ \langle k_2|k_1\rangle & & \vdots \\ \vdots & & \vdots \\ \langle k_N|k_1\rangle & \cdots & \langle k_N|k_N\rangle \end{bmatrix} \begin{bmatrix} c_{kk_1} \\ c_{kk_2} \\ \vdots \\ c_{kk_N} \end{bmatrix} = \begin{bmatrix} \langle k_1(A^{-1})|k\rangle \\ \langle k_2(A^{-1})|k\rangle \\ \vdots \\ \langle k_N(A^{-1})|k\rangle \end{bmatrix}$$

Denoting the matrix on the left hand side as $\mathbf{T}$ and rearranging yields

$$T\begin{bmatrix} c_{\mathbf{k}_1\mathbf{k}_1} & & c_{\mathbf{k}_N\mathbf{k}_1} \\ c_{\mathbf{k}_1\mathbf{k}_2} & & \\ & & \\ c_{\mathbf{k}_1\mathbf{k}_N} & & c_{\mathbf{k}_N\mathbf{k}_N} \end{bmatrix} = \begin{bmatrix} \langle \mathbf{k}_1(\mathbf{A}^{-1})|\mathbf{k}_1\rangle & & \langle \mathbf{k}_1(\mathbf{A}^{-1})|\mathbf{k}_N\rangle \\ \langle \mathbf{k}_2(\mathbf{A}^{-1})|\mathbf{k}_1\rangle & & \\ & & \\ \langle \mathbf{k}_N(\mathbf{A}^{-1})|\mathbf{k}_1\rangle & & \langle \mathbf{k}_N(\mathbf{A}^{-1})|\mathbf{k}_N\rangle \end{bmatrix}$$

Writing the above as $\mathbf{T} = \mathbf{C}^{\mathbf{T}}\mathbf{S}$, the transformation matrix $\mathbf{C}$ for Eq. (3) is given by

$$\mathbf{C} = \mathbf{S}^{\mathbf{T}}\mathbf{T}^{-1}.$$

Since $\mathbf{T}, \mathbf{S}$ are determined from Eq. (6), integration of Eq. (6) follows. It is enough to integrate $\langle \mathbf{k}(\mathbf{M})|\mathbf{k}'\rangle$, for an arbitrary 2×2 matrix $\mathbf{M} = (\mathbf{a_{ij}})$. Integration of Eq. (6) is analytically obtained using the following equality:

$$\iint \exp\left(-ax^2 - by^2 - cxy - igx - ihy\right)dxdy = \frac{2\pi}{\sqrt{4ab-c^2}}\exp\left(-\frac{ah^2 - cgh + bg^2}{4ab-c^2}\right).$$

Implementation of Face Selective Attention Model on an Embedded System

Bumhwi Kim, Hyung-Min Son, Yun-Jung Lee, and Minho Lee[*]

School of Electrical Engineering and Computer Science, Kyungpook National University,
1370 Sangyeok-Dong, Buk-Gu, Daegu 702-701, South Korea
{bhkim,miniim24,yjlee}@ee.knu.ac.kr, mholee@knu.ac.kr

Abstract. This paper proposes a new embedded system which can selectively detect human faces with fast speed. The embedded system is developed by using OMAP 3530 application processor which has DSP and ARM core. Since the embedded system has the limited performance of CPU and memory, we propose a hybrid system combined the YCbCr based bottom-up selective attention with the conventional Adaboost algorithm. The proposed method using the bottom-up selective attention model can reduce not only the false positive error ratio of the Adaboost based face detection algorithm but also the time complexity by finding the candidate regions of the foreground and reducing the regions of interest (ROI) in the image. The experimental results show that the implemented embedded system can successfully work for localizing human faces in real time.

Keywords: bottom-up selective attention, face detection, embedded, modified census transform, Adaboost.

1 Introduction

The human visual system can effortlessly detect an interesting object within natural or cluttered scenes through a selective attention mechanism. This mechanism allows the human vision system to effectively process visual scenes with a higher level of complexity. There have been several studies about the selective attention which process stimuli from the retina to the visual cortex. Itti, Koch, and Niebur [1] introduced a brain-like model in order to generate a saliency map (SM). Koike and Saiki [2] proposed that a stochastic winner take all (WTA) enables the saliency-based search model to change search efficiency by varying the relative saliency, due to stochastic shifts of attention. Kadir and Brady [3] proposed an attention model integrating saliency, scale selection, and a content description, thus contrasting with many other approaches. Ramström and Christensen [4] calculated saliency with respect to a given task by using a multi-scale pyramid and multiple cues. Their saliency computations were based on game theory concepts. And Jeong et al. [5] introduced a dynamic saliency model which considers temporal dynamics of saliency degrees changing through time

---

[*] Corresponding author.

T. Huang et al. (Eds.): ICONIP 2012, Part V, LNCS 7667, pp. 188–195, 2012.

at each salient point. This model is based on a modified static saliency model which additionally considers symmetry information. Ban et al. [6] proposed an affective saliency model which considers psychological distance as well as visual features.

On the other hand, face detection is still active research area in machine vision. Among the several face detection algorithms, the Haar-like feature based Adaboost algorithm proposed by Viola & Jones [7] is widely used. But Haar-like feature based face detector is weak at handling illumination change, because haar-like features are calculated by differentiating 2 or more image regions. Another Adaboost based face detection algorithm is MCT feature based face detector. The MCT feature based face detector is robust to illumination change because the MCT feature is considers the relativeness of the intensity. Thus, in this paper, MCT feature based Adaboost algorithm is used for face detection together with a bottom-up selective attention model. There have been previous approaches to hybrid the bottom-up selective attention model with the Adaboost based face detection. Cerf et al. [8] and Kim et al. [9] proposed face detection models which use bottom-up selective attention model as a pre-filter of an image. In our research, we present an embedded platform using OMAP3530 application processor that can detect human face with faster speed and reliable performance in real world environment.

This paper is organized as follows. Section 2 describes the proposed face detection model based on bottom-up selective attention model and MCT feature based Adaboost. Experimental results will be represented in Section 3. Section 4 presents conclusions and further works.

2   Proposed System

Figure 1 depicts the proposed model. The input images are obtained from camera and are transferred to embedded system. In the embedded system, there are two main functions; the bottom-up selective attention model and MCT feature based Adaboost face detection. The bottom-up selective attention model extracts the face candidate regions. And then MCT feature based Adaboost face detector operates within the face candidate regions which can alleviate the false positive error of the conventional Adaboost method as well as reduce the computational time. Finally, the detected facial images are displayed on LCD screen.

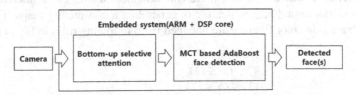

Fig. 1. The proposed system

2.1   Bottom-Up Selective Attention Model

In human visual processing, the intensity, edge and color features are extracted in the retina. These features are transmitted to the visual cortex through the lateral geniculate nucleus (LGN). While transmitting those features to the visual cortex, intensity, edge, and color feature maps are constructed using the on-set and off-surround mechanism of the LGN and the visual cortex. And those feature maps make a bottom-up selective attention model in the lateral intral-parietal cortex (LIP) [10].

In order to implement a human-like visual attention function, we consider the simplified bottom-up selective attention model proposed in [11]. The selective attention model reflects the functions of the retina cells, the LGN and the visual cortex. Since the retina cells can extract edge and intensity information as well as color opponency, we use these factors as the basic features of the selective attention model [11].

Human brain processes the visual features in an RGB color space. Thus the previous selective attention model used the RGB color space as a raw feature. But the proposed hardware system acquires YUV (YCbCr) color space image. The conversion module between YCbCr and RGB color spaces should be considered, but the computational cost is pretty high for the conversion. Thus, in this paper, YCbCr color space based selective attention model is proposed. Fig. 2 shows the flow chart of the modified selective attention model. The edge information (E) can be obtained by applying the Sobel operator on the Y image.

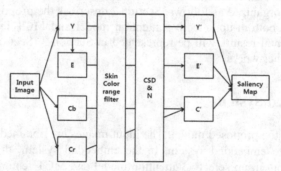

Fig. 2. Simplified diagram of saliency processing

In order to provide the proposed model with face color preference property, the skin color filtered intensity scale image is considered together with the original intensity scale image. According to a given task to be conducted, those three intensity scale images are differently biased. For face preferable attention, a skin color filtered intensity scale image works for a dominant feature in generating a conspicuity maps. The ranges of Y, Cb, Cr for skin color filtering are obtained from following rules in Eq. (1) [11].

$$
\begin{aligned}
& Y > 80 \\
& 77 \leq Cb \leq 127 \\
& 133 \leq Cr \leq 173 \\
& \text{where } Y, Cb, Cr = [0, 255]
\end{aligned}
\tag{1}
$$

According to our experiments, the filtered Y, Cb and Cr images show dominant contribution for face color plausible filtering. In the Center-Surround Difference and Normalization (CSD&N) block, computes the on-center, off-surround algorithm which are important function in LGN of human brain. For the 4 Gaussian pyramid images, Y, E, Cb and Cr, each with 5 different scales, a total of 20 feature maps are computed [5]. The feature maps corresponding to Y, E, Cb and Cr are combined into 3 conspicuity maps. Y', E' and C' using a CSD&N process [5] are obtained by the following equations.

$$Y' = \bigoplus_{c=0}^{1} \bigoplus_{s=c+2}^{c+3} N(Y(c,s)) \tag{2}$$

$$E' = \bigoplus_{c=0}^{1} \bigoplus_{s=c+2}^{c+3} N(E(c,s)) \tag{3}$$

$$C' = \bigoplus_{c=0}^{1} \bigoplus_{s=c+2}^{c+3} N(Cb(c,s) + Cr(c,s)) \tag{4}$$

where Y', E' and C' stand for intensity, edge and color conspicuity maps, respectively. And, c and s represent scale indices for fine scale (center) and coarse scale (surround), respectively. Each conspicuity map represents the degree of relative pop-out at every local area, which is generated by center-surround difference operation mimicking on-center-off-surround mechanism of biological system. These are obtained through a cross-scale addition "$\oplus$"[14]. $N(\bullet)$ is a normalization function so that every pixel value lies in the range 0 to 255. And the final selective attention model is obtained by pixel by pixel adding of three conspicuity maps.

2.2   Face Detection Using MCT Feature Based Adaboost

We adopted an Adaboost approach using MCT features for the face detection algorithm to correctly localize human faces in the candidate regions obtained by the bottom-up selective attention model. The MCT feature is a simple vector with values 0 or 1. A 3x3 moving window in an image generates the MCT features. The MCT features represent the relativeness between global average intensity and the intensity values in current window of an image [8].

There are two data sets for face feature extraction and learning for the Adaboost model. One is called a positive dataset in which every image has a face and the other is a negative dataset with non-facial images. For the two data sets, MCT features are extracted in order to select the proper features and train the Adaboost face detection model. The Adaboost learning algorithm is used to boost the classification performance of a simple learning algorithm [8]. It works by combining a collection of weak classifiers to form a stronger classifier [8].

For each feature, the weak learner determines the optimal threshold classification function, such that a minimum number of examples are misclassified. According to our experiments, a face detection model using Adaboost algorithm based on MCT features shows wrong face detection results in some cases, which can be avoided by considering face color preferable selective attention as a preprocessor. In addition, the Adaboost algorithm takes longer processing time for some scenes with complex orientation features and produces inaccurate face detection results for the scenes with complex background. In contrast, the proposed model can enhance the face detection performance by reducing the computation load as well as increasing face detection accuracy especially for a natural scene with complex orientation information.

2.3   Hardware Implementation

In this paper, a portable embedded system which can handle complex operations such as convolution, floating point calculations for real-environment is implemented. The embedded system has disadvantages such as slow operation time, limited memory and resources. To overcome those problems, supplement co-processor using FPGAs has been designed to conduct the repetitive and parallel operation as a modular distributed system. This embedded system has been designed with bigger goal such as human augmented cognition system. This paper represents the first step towards the bigger goal as face detection in embedded system.

Image data from the camera interface are given as input to the embedded system. Then the OMAP 3530 performs the face detection with the proposed model. Finally, the detected face image by these processes can be shown on the LCD display.

Fig. 3. Developed embedded control board

Figure 3 shows the developed embedded system. The developed embedded system mainly consists of OMAP 3530 with specifications such as 520MHz CPU, Texas Instruments; TI [15] TMS320C64x+ DSP processor based on ARM [16] Cortex-A8, and modular distributed 4 channel FPGAs (EP2C672C8) which has 700,000 logic elements. In addition, 2Gb NAND flash memory, 2048Mb LP-DRAM, camera interface, LCD interface, audio interface, LAN interface are included on the control board. And ARM Linux kernel 2.6.28 release candidate 9 is used as embedded hardware operating system.

To improve the performance of the algorithm in embedded system, the proposed hardware system utilizes co-processing between DSP and ARM core. The MCT feature based Adaboost face detection algorithm processes a moving window operation to detect faces of various sizes. Thus, it burdens the ARM processor with high computational cost. So we consider the DSP processor for parallel processing.

3   Experimental Results

TI supports the ARM with DSP co-processing system (C6EZRuLib[17]) which can access device through the Linux system. C6EZRun is a free, open-source development tool from TI that allows the user to seamlessly use the DSP on heterogeneous SoC processors form TI, specifically targeting ARM and DSP devices. For the implementation, we prepared C-language based selective attention model creation, candidate area detection algorithm and Adaboost based face detection algorithm. Then we generate ARM and DSP co-processing application by using C6EZRunLib.

Table 1. Time comparisons between ARM only and ARM with DSP

	Average processing time (ms)	
Single	ARM only	ARM with DSP
vs.		
Co-processor	3223.5.	208.8

Table 1 shows the comparison of the execution times for the implementation of the proposed face detection algorithm to be run on ARM only and when the same function is executed on the DSP, but called from the ARM (using c6runlib[17]). From the result, it can be observed that the execution times taken by ARM and DSP co-processing are 15 times less than the ARM core alone. This code was executed on the proposed hardware platform.

Table 2. Time comparisons between Adaboost only and Adaboost with selective attention

	Average processing time (ms)	
Adaboost	Adaboost only	Selective attention with Adaboost
vs.		
proposed		(Adaboost)
model	273.5	208.8 (95.3)

Table 2 represents the comparisons between Adaboost only and proposed model. According to the result, the Adaboost execution time is reduced almost 3 times. The proposed model's execution time is about 1.3 times faster than Adaboost only method, even though there is additional execution time in selective attention model. Only compare the Adaboost's execution time, the proposed model is about 3 times faster than Adaboost only method.

Fig. 4 represents an example of embedded system processing results. Fig. 4 (a), (b), (c) and (d) show the original image, detected face images by using Adaboost only, bottom-up saliency map and detected face from the original image by using the proposed model, respectively. According to the result, proposed model successfully reject unnatural facial image.

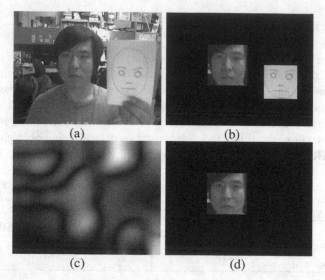

Fig. 4. Example embedded system face detection result on LCD display; (a) original image, (b) detected face image in an embedded platform using Adaboost only, (c) bottom-up saliency map image, (c) detected face image in an embedded platform using proposed platform

4   Conclusion and Future work

In this paper, the proposed face detection model mimics the human early visual mechanisms in order to localize the human face areas by combining the MCT based Adaboost face detector. We also developed an embedded hardware system for operating the proposed system. The experimental results show that the face detection model is successfully ported to embedded system. And also we modified RGB based bottom-up selective attention model to YCbCr based one. For detecting face candidate area from an input image, YCbCr based skin color range filter is applied as a human facial preferable information. But, the skin color rage filter is weak at illumination change. In the future, the top-down model which can bias human facial information in color and form domains is applied instead of static skin color range filter.

Compared to the DSP core, FPGA is shows plausible results by the parallel processing of the data. It reduces the processing time in CSD&N of bottom-up selective attention model which is calculated by convolution operation. In the future, we would like to develop an embedded system with 4 channel FPGAs to enhance the performance of the top-down model which can localize human faces. The final goal of this research is to construct a human augmented cognition system which not only detects and recognizes human face but also handle human voice information such as conversion between voice and text in an embedded system.

Acknowledgements. This research was supported by the Converging Research Center Program funded by the Ministry of Education, Science and Technology (2012K001342).

References

1. Itti, L., Koch, C., Neibur, E.: A Model of Saliency–Based Visual Attention for Rapid Scene Analysis. IEEE Transactions on Pattern Analysis and Machine Intelligence 20(11), 1254–1259 (1998)
2. Koike, T., Saiki, J.: Stochastic Guided Search Model for Search Asymmetries in Visual Search Tasks. In: Bülthoff, H.H., Lee, S.-W., Poggio, T., Wallraven, C. (eds.) BMCV 2002. LNCS, vol. 2525, pp. 408–417. Springer, Heidelberg (2002); Goldstein, E.B.: Sensation and perception, 4th edn. An international Thomson publishing company, USA (1996)
3. Kadir, T., Brady, M.: Scale, Saliency and Image Description. Int. J. Comput. Vis. 45, 83–105 (2001); Goldstein, E.B.: Sensation and perception, 4th edn. An international Thomson publishing company, USA (1996)
4. Ramström, O., Christensen, H.I.: Visual Attention Using Game Theory. In: Bülthoff, H.H., Lee, S.-W., Poggio, T., Wallraven, C. (eds.) BMCV 2002. LNCS, vol. 2525, pp. 462–471. Springer, Heidelberg (2002)
5. Jeong, S., Ban, S.W., Lee, M.: Stereo Saliency Map Considering Affective Factors and Selective Motion Analysis in a Dynamic Environment. Neural Netw. 21, 1420–1430 (2008)
6. Ban, S.W., Jang, Y.M., Lee, M.: Affective Saliency Map Considering Psychological Distance. Neurocomputing 74, 1916–1925 (2011)
7. Viola, P., Jones, M.J.: Robust Real-Time Face Detection. International Journal of Computer Vision 57(2), 137–154 (2004)
8. Cerf, M., Harel, J., Einhäuser, W., Koch, C.: Predicting Human Gaze Using Low-Level Saliency Combined with Face Detection. In: Proceedings of the Twenty-First Annual Conference on Neural Information Processing Systems, NIPS 2007 (2007)
9. Kim, B., Ban, S. W., Lee, M.: Improving AdaBoost Based Face Detection Using Face-Color Preferable Selective Attention. In: Fyfe, C., Kim, D., Lee, S.-Y., Yin, H. (eds.) IDEAL 2008. LNCS, vol. 5326, pp. 88–95. Springer, Heidelberg (2008)
10. Goldstein, E.B.: Sensation and Perception, 4th edn. International Thomson Publishing Company, USA (1996)
11. Park, S.J., An, K.H., Lee, M.: Saliency Map Model with Adaptive Masking Based on Independent Component Analysis. Neurocomputing 49, 417–422 (2002)
12. Mahmoud, T.M.: A New Fast Skin Color Detection Technique. World Academy of Science. Engineering and Technology 43, 501–505 (2008)
13. Bell, A.J., Sejnowski, T.J.: Edges are the Independent Components of Natural Scenes. In: NIPS, pp. 831–837 (1996)
14. Fröba, B., Ernst, A.: Face Detection with the Modified Census Transform. In: Proceedings of the Sixth IEEE International Conference on Automatic Face and Gesture Recognition (FGR 2004), pp. 91–96 (2004)
15. Texas Instruments, http://www.ti.com
16. ARM, http://www.arm.com/
17. C6EZRun Software Development Tool for TI DSP+ARM Devices

From Image Annotation to Image Description

Ankush Gupta and Prashanth Mannem

International Institute of Information Technology, Hyderabad - 500032, India
{ankush.gupta,prashanth}@research.iiit.ac.in

Abstract. In this paper, we address the problem of automatically generating a description of an image from its annotation. Previous approaches either use computer vision techniques to first determine the labels or exploit available descriptions of the training images to either transfer or compose a new description for the test image. However, none of them report results on the effect of incorrect label detection on the quality of the final descriptions generated. With this motivation, we present an approach to generate image descriptions from image annotation and show that with accurate object and attribute detection, human-like descriptions can be generated. Unlike any previous work, we perform an extensive task-based evaluation to analyze our results.

Keywords: Natural Language Processing, Knowledge-based Information Systems, Information Retrieval, Natural Language Generation.

1 Introduction

Image Annotation is the task of assigning labels to an image that describe the content of the image. Automatic image annotation [1][2] is useful in various applications like image indexing, image retrieval, search engine optimization and increasing accessibility to visually impaired users. But a list of labels is often ambiguous. For example, an image annotated with labels {*green, airport, jet*} does not convey the information whether attribute *green* is associated with *airport* or *jet* and whether *'jet is parked at airport'* or *'jet is taking off from airport'*. Whereas, in a sentence *'A green jet is parket at the airport'*, the relations between labels are explicit. Hence, there is an urgent need for automatic conversion of image annotations to natural language, which have a stronger semantic content (Figure 1).

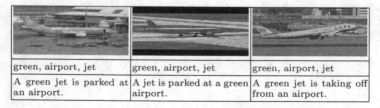

green, airport, jet	green, airport, jet	green, airport, jet
A green jet is parked at an airport.	A jet is parked at a green airport.	A green jet is taking off from an airport.

Fig. 1. Importance of Image Descriptions. A description can differentiate images, but independent labels cannot. All the images have same annotation (second row) but different semantics (third row).

T. Huang et al. (Eds.): ICONIP 2012, Part V, LNCS 7667, pp. 196–204, 2012.
© Springer-Verlag Berlin Heidelberg 2012

Mitchell et al. (2012) states that *"In computer vision, object detections form the basis of action/pose, attribute, and spatial relationship detections."* However, none of the previous approaches evaluate the effect of incorrect object detection on the accuracy of the generated descriptions. Our focus is to show that with accurate image labels (objects and attributes), high-quality human-like descriptions can be generated. With this motivation, we present a novel approach to generate image descriptions from annotated images.

To summarize, our contributions are: (i) A novel approach to generate descriptions from image annotations. (ii) Studying the effect of incorrect label detection on the quality of descriptions generated by comparing with earlier approaches. (iii) Extensive evaluation of all the stages involved in the approach. (iv) Task-Based evaluation of the results (to our best knowledge, this is the first attempt to perform task-based evaluation to access the quality of generated image descriptions).

2   Related Work

Generating descriptions for images is a relatively new field of research, with majority of the work done in past 2-3 years. Most of the approaches use some object detectors, classifiers to predict image content (e.g. objects, attributes, scenes, stuff, actions, etc) and then generate descriptions from scratch [4][5][6][9][10] or transfer available descriptions from visually similar training images [7][8]. Gupta et al. (2012) do not rely on any detectors and makes use of only the available descriptions of the similar images to compose a description for an unseen image. Kuznetsova et al. (2012) makes use of both (visual recognition and available descriptions) to produce an image specific description. Feng et al. (2010) use the output of a probabilistic image annotation model and a relevant text document to generate captions for news images. Aker et al. (2010) assume that relevant text for an image is provided. However, none of these approaches study the effect of erroneous detection of image content on the accuracy of generated descriptions. In this paper, we assume that the input image is annotated with basic content information (objects, attributes) and then generate the descriptions by extracting other relevant information (verb, preposition, determiner) from the descriptions of training images.

3   Our Approach

A simple description of an image (*'A white dog is lying on the bed'*) contains following information : (a) Objects present in the image : *<bed>,<dog>* (b) Attribute : *<white>* (c) Attribute-Object pair : *<white, dog>* (d) Subject : *<dog>* (e) Verb : *<lie>* (f) Preposition : *<on>*

Given an image and its annotation ((a),(b)), our task is to generate an image description. As most of the previous approaches have used PASCAL[1] sentence

---

[1] http://vision.cs.uiuc.edu/pascal-sentences/

INPUT	OUTPUT
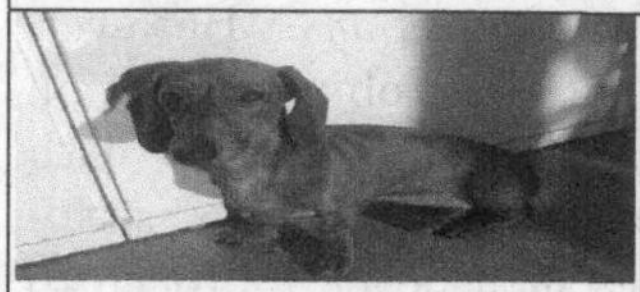 $< dog, wall, white, brown >$	**Attr-Obj** : $< brown, dog > < white, wall >$ **Subject** : dog **Verb** : sit **Preposition** : $next\_to$ **Description** : $A\ brown\ dog\ is\ sitting\ next\ to\ a\ white$ $wall$.

Fig. 2. Overview of our approach to automatically generate description from an annotated image.

Table 1. Phrases extracted from a sample image description. Words are replaced by their most frequently used form (determined using WordNet).

Description	Phrases w/ Synonym
A white Swiss airplane approaches the runway.	$(aeroplane_{subject})$, $(airport_{object})$, $(white_{attribute}, aeroplane_{subject})$, $(swiss_{attribute}, aeroplane_{subject})$, $(aeroplane_{subject}, approach_{verb}, airport_{object})$

dataset, we also report our results using the same. PASCAL dataset consists of 1000 images each containing 5 human-generated descriptions. We first annotate each image by extracting 'attribute' and 'object' information from these descriptions automatically. Then, we associate each 'attribute' to corresponding 'object', determine the 'subject' of the sentence, predict 'verb', 'preposition' and 'determiner' from the available descriptions of training images to get the quadruplet $((det_1, attribute_1, subject), verb, prep, (det_2, attribute_2, object))$, which is used for generating descriptions[2] (Figure 2).

3.1   Annotating Images Using Available Descriptions

To apply our approach on a dataset of images and their corresponding descriptions, we first need to annotate the image with objects and attributes present in the image. We extract this information automatically by processing the available descriptions using the Stanford CoreNLP toolkit[3]. We follow a similar approach mentioned in [3] and extract 14 (unlike 9 in [3]) distinct types of phrases from the available image descriptions : $(subject)$, $(object)$, $(subject, verb)$, $(object, verb)$, $(subject, verb, object)$, $(subject, prep, object)$, $(verb, prep, object)$, $(object, prep, object)$, $(attribute, subject)$, $(attribute, object)$, $(det, attribute, subject)$, $(det, attribute, object)$, $(det, subject)$, $(det, object)$. For generalization, each subject and object in above phrases is expanded up to at most 3 hyponym levels (determined using WordNet synsets). Table 1 shows phrases extracted from sample image descriptions. The phrases used to annotate the test image are : $(subject)$, $(object)$, $(attribute, subject)$, $(attribute, object)$.

---

[2] Each image in PASCAL dataset is annotated with at most two objects and attributes.

[3] http://nlp.stanford.edu/software/corenlp.shtml

For PASCAL dataset, we annotate the image with two objects appearing in maximum number of descriptions. So, if an object is mentioned in all the 5 descriptions, there are high chances that it is important to describe the image. For each object, we take one attribute. Hence, for each image we get the annotation labels ($attribute_1$, $attribute_2$, $object_1$, $object_2$). We also report our results on IAPR TC-12[4] dataset. As the descriptions of IAPR consist of 3-5 sentences, we extract 4 objects (unlike 2 for PASCAL) getting the following annotation : ($attribute_1$, $attribute_2$, $attribute_3$, $attribute_4$, $object_1$, $object_2$, $object_3$, $object_4$).

3.2 Attribute-Object Association

Image Annotation does not convey any information about the relationship between labels. Hence, we consider the labels ($attribute_1$, $attribute_2$, $object_1$, $object_2$) as independent. We link each attribute and object by using the phrases ($attribute$, $subject$) and ($attribute$, $object$) extracted in Section 3.1. If $attribute_1$ has been associated to $object_1$ more number of times than $object_2$ in the descriptions of training images, then $attribute_1$ is assigned to $object_1$ ($attribute_2$ to $object_2$) and vice-versa.

3.3 Determining Subject of Description

We need to determine out of $object_1$ and $object_2$, which one is more likely to be the subject of a sentence. If the phrase ($<subject=object_1>$) appears more number of times in the descriptions than ($<subject-object_2>$), $object_1$ is selected as the subject of the sentence ($object_2$ becomes the object) and vice-versa. Example : In the sentence 'John is eating apple', 'John' is the subject and 'apple' is the object[5].

3.4 Predicting Verb

The verb depends on both the subject and object of a sentence. Example : If subject is 'aeroplane' and object is 'sky', the most likely verb is 'fly' but if the object is 'airport', the verb most frequently used is 'park'. Once the subject and object of the description are determined (Section 3.3), we look for phrases of type ($subject$, $verb$, $object$) in the training image descriptions and extract the most frequently used verb (transitive verb). In case, no phrases of such type are there, verb is predicted by multiplying the counts of ($subject$, $verb$) and ($verb$, $prep$, $object$)[6] for all the verbs and using the one with the maximum value. If the count is 0, we do not use verb while generation. Example : 'There is a white cow with the big ear in the picture.'

---

[4] http://www.imageclef.org/photodata
[5] For image annotation, both 'John' and 'apple' are objects.
[6] prep can take any value.

3.5 Predicting Determiner

The selection of the determiner depends on both the adjective (attribute) and noun (subject/object). The most frequently used determiners (including cases where no determiner has been used) are selected for subject and object depending on the count of phrases $(det, subject)$, $(det, object)$, $(det, attribute, subject)$ and $(det, attribute, object)$.

3.6 Predicting Preposition

In case the detected verb in Section 3.4 is a transitive verb, we do not need a preposition for generation. Otherwise, we predict the preposition having maximum count of $(verb, prep, object)$. If no verb is detected, we estimate the preposition using counts of phrase $(subject, prep, object)$.

3.7 Sentence Generation

Once we get a quadruplet of the form $((det_1, attribute_1, subject), verb, prep, (det_2, attribute_2, object))$, the task of generation is to map this into a grammatical sentence. We use SimpleNLG [14] for this purpose, which is a surface realizer with significant coverage of English syntax and morphology. SimpleNLG allows us to initialise the basic constituents of a sentence (subject, verb, object) with appropriate lexical items. Attributes are assigned to subject and object by means of the modifier function. In case of intransitive verbs, object is realised as a prepositional phrase. Various features of the constituents can be set. Example : tense (present) and aspect (progressive) of the verb, voice (active, passive), number (plural, singular), position of the attribute (prenominal, postnominal). Various advantages of using SimpleNLG for the same problem have been discussed in [3].

For the PASCAL dataset, we generate a single sentence of one of the following forms :

(a)$<(det_1, attribute_1, subject), verb, prep, (det_2, attribute_2, object)>$
(b)$<(det_1, attribute_1, subject), verb, (det_2, attribute_2, object)>$ (Transitive Verbs)
(c)$<(det_1, attribute_1, subject), prep, (det_2, attribute_2, object)>$ (No Verb Detected)

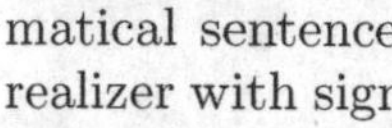 green, white, car, saloon	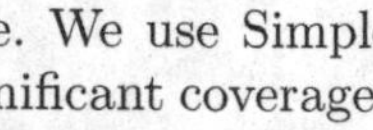 loaded, chair, shelf, plastic	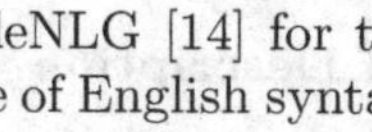 train, green, yellow, building	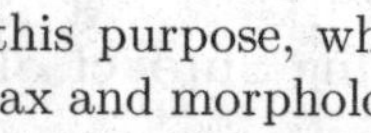 boat, beach, water, person, sandy	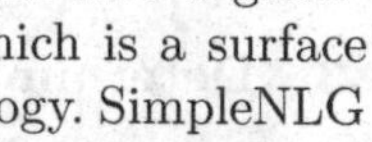 white, airport, engine, building, aeroplane
A green car is parked in front of a white saloon.	A plastic chair is sitting next to a loaded shelf.	A yellow train is passing a green building.	A person is trailing the water. There is a boat in a sandy beach.	A white building is standing on an airport. There is the engine of an aeroplane.

Fig. 3. Examples of good descriptions generated from the PASCAL (first three images) and IAPR TC-12 (last two images) dataset. Second row is the annotation of the image.

For IAPR TC-12 dataset, we generate two sentences of the form :

Sentence 1 : $<(det_1,attribute_1,subject_1),verb,prep_1,(det_2, attribute_2,object_1)>$

Sentence 2 : $<(det_3,attribute_3,subject_2), prep_2, (det_4, attribute_4, object_2)>$

Figure 3 shows some good examples of image descriptions generated by our approach on PASCAL and IAPR TC-12 datasets.

4 Experiments

4.1 Datasets

We use the UIUC PASCAL Sentence dataset [15] and the IAPR TC-12 Benchmark [16] to test the performance of our approach. PASCAL dataset contains 1000 images each having 5 human-written descriptions. The IAPR TC-12 benchmark comprises of 20,000 images each containing a description of up to 5 sentences. Unlike PASCAL dataset, it has complex images with complicated descriptions, hence making it a challenging dataset for automatic generation of image descriptions.

4.2 Experimental Details

We partition each dataset into 90% training and 10% testing. This is repeated to generate results for the entire dataset. For each image in testing dataset, we extract object and attribute information from its description (Section 3.1). Training dataset is used as a corpus to compute the counts of various types of phrases.[7]

4.3 Evaluation

Automatic Evaluation. Automatic evaluation of Natural Language Generation is a challenging task [17]. Most of the previous approaches have used BLEU [18] and Rouge [19] metrics for automatic comparison of generated descriptions against human-written descriptions and we also report our scores on same metrics. For PASCAL dataset, we also compute the average BLEU and Rouge-1 Precision scores between each human-written description to the set of others over entire dataset. The results are shown in Table 2.

Table 2. Our automatic evaluation results for generating image descriptions. Higher score means better performance. B-n means n-gram BLEU score using exact match for each word. B-n-s means n-gram BLEU score by matching synonyms. Rouge-1 means Rouge-1 Precision score without considering synonyms. Rouge-1-s match synonyms.

Dataset	B-1	B-1-s	B-2	B-2-s	B-3	B-3-s	Rouge-1	Rouge-1-s
PASCAL	0.74	**0.79**	0.55	**0.61**	0.35	**0.42**	0.55	**0.60**
PASCAL Human (std.)	0.64	0.66	0.42	0.44	0.24	0.26	0.50	0.52
IAPR TC-12	0.33	**0.36**	0.18	**0.21**	0.07	**0.09**	0.38	**0.41**

---

[7] In case descriptions are not available, google counts can be used.

Table 3. Precision of Object Extraction, Attribute-Object Association, Verb and Preposition Prediction

Dataset	Object	Attrbute-Object	Verb	Preposition
PASCAL	98.13	94.54	74.95	86.42
IAPR TC-12	85.7	87.5	58.14	64.32

Incorrect subject	Incorrect attr-obj	Incorrect verb	Incorrect prep	Object Repitition
A white table is sitting on a buddha statue.	A brown person is milking a young cow.	A young person is cutting a cake.	A wooden table is sitting on a wooden chair.	There is a patio set and the wet patio in the picture.

Fig. 4. Examples of bad results as judged by human evaluators. Second row shows the error in the description. Third row shows the generated description.

Human Evaluation. To get a better insight of the performance, we manually evaluate the approach at various stages. We provide human evaluators with a set of images and objects extracted from its description, predicted attribute-object pairs, verb and preposition. The results on 500 images from each dataset are shown in Table 3. We design two task-based experiments to more accurately evaluate the quality of the generated descriptions. In the first experiment, given an image description (generated by our approach) and three images, the user is asked to select the image which best describes the description. The images are selected from the same class group as the test image. So, if the description shown is of a test image of 'bird' class, the remaining two images are also of 'bird'. We ask human evaluators to evaluate 200 images from each dataset. Accuracy of 89% and 73% is obtained on PASCAL and IAPR TC-12 datasets respectively. In the second experiment, we add the description generated by our approach to the available 5 human-written descriptions of PASCAL dataset and show these six descriptions for each image to human evaluators. The order of descriptions is random, so that evaluators cannot differentiate between human-written and system-generated output. Evaluators are asked to select the description which least fits the given image. User is also provided an option of selecting *'Unable to Decide'*, in case he finds each description apt. Bad cases are those when user selects the description generated by our approach. This experiment is performed on 200 images from PASCAL dataset. We achieved an accuracy of 82%. In 71% of the cases, user is unable to decide which description is least apt for the image, suggesting that our approach generates descriptions close to human-written descriptions. Some of the results rated as 'bad' by human evaluators are shown in Figure 4.

4.4 Comparison with Previous Approaches

Though we cannot compare our approach directly with previous approaches as none of them assume an annotated image as input and each approach use a tuple

Table 4. Comparison of our approach with *BabyTalk* [6], *CorpusGuided* [5] and *Linguistics-over-Vision* [3] each reporting their best results on PASCAL dataset. While we consider annotated image as input for generation, other approaches generate descriptions from an unannotated image. Bleu-n-s means n-gram BLEU score by matching synonyms. Rouge-1-s means Rouge-1 Precision score considering synonyms.

Approach	Bleu-1-s	Bleu-2-s	Bleu-3-s	Rouge-1-s
BabyTalk	0.30	-	-	-
CorpusGuided	0.41	0.13	0.03	0.31
Linguistics-over-Vision	0.54	0.23	0.07	0.41
Ours	**0.79**	**0.61**	**0.42**	**0.60**

of a different form and consider their own sets of attributes, objects, verbs and prepositions. But still we report our scores (using Automatic Evaluation) for future research and to emphasize the effect of attribute and object detection on generation accuracy (Table 4). This is an encouraging result as it highlights the importance of the correct object and attribute detection for the image description generation task.

5 Conclusion

We proposed a novel approach for generating high-quality image descriptions from an annotated image without using any computer vision techniques. Experimental results demonstrate that the quality of the generated descriptions is highly sensitive to the retrieval precision of objects and attributes present in an image. Extensive evaluation of each stage shows that predicting verb has the least accuracy. One direction of future work would be to use some vision based inputs in addition to corpus statistics to predict the verb.

References

1. Guillaumin, M., Mensink, T., Verbeek, J., Schmid, C.: TagProp: Discriminative Metric Learning in Nearest Neighbor Models for Image Auto-Annotation. In: ICCV (2009)
2. Makadia, A., Pavlovic, V., Kumar, S.: Baselines for image annotation. In: IJCV, pp. 88–105 (2010)
3. Gupta, A., Verma, Y., Jawahar, C.V.: Choosing Linguistics over Vision to Describe Images. In: AAAI (2012)
4. Li, S., Kulkarni, G., Berg, T.L., Berg, A.C., Choi, Y.: Composing simple image descriptions using web-scale n-grams. In: CONLL, pp. 220–228 (2011)
5. Yang, Y., Teo, C.L., Daume III, H., Aloimonos, Y.: Corpus-guided sentence generation of natural images. In: EMNLP, pp. 444–454 (2011)
6. Kulkarni, G., Premraj, V., Dhar, S., Li, S., Choi, Y., Berg, A.C., Berg, T.L.: Baby Talk: Understanding and generating simple image desciptions. In: CVPR (2011)
7. Ordonez, V., Kulkarni, G., Berg, T.L.: Im2Text: Describing images using 1 million captioned photographs. In: NIPS (2011)

8. Farhadi, A., Hejrati, M., Sadeghi, M.A., Young, P., Rashtchian, C., Hockenmaier, J., Forsyth, D.: Every Picture Tells a Story: Generating Sentences from Images. In: Daniilidis, K., Maragos, P., Paragios, N. (eds.) ECCV 2010, Part IV. LNCS, vol. 6314, pp. 15–29. Springer, Heidelberg (2010)
9. Mitchell, M., Dodge, J., Goyal, A., Yamaguchi, K., Stratos, K., Han, X., Mensh, A., Berg, A., Berg, T., Daume III, H.: Midge: Generating Image Descriptions From Computer Vision Detections. In: EACL (2012)
10. Yao, B.Z., Yang, X., Lin, L., Lee, M.W., Zhu, S.-C.: I2T: Image Parsing to Text Description. IEEE (2010)
11. Kuznetsova, P., Ordonez, V., Berg, A.C., Berg, T.L., Choi, Y.: Collective Generation of Natural Image Descriptions. In: ACL (2012)
12. Feng, F., Lapata, M.: How many words is a picture worth? Automatic caption generation for news images. In: ACL, pp. 1239–1249 (2010)
13. Aker, A., Gaizauskas, R.: Generating image descriptions using dependency relational patterns. In: ACL (2010)
14. Gatt, A., Reiter, E.: SimpleNLG: A realisation engine for practical applications. In: ENLG, pp. 91–93 (2009)
15. Rashtchian, C., Young, P., Hodosh, M., Hockenmaier, J.: Collecting Image Annotations Using Amazon's Mechanical Turk. In: NAACLHLT Workshop on Creating Speech and Language Data with Amazon's Mechanical Turk (2010)
16. Grubinger, M., Clough, P.D., Muller, H., Thomas, D.: The IAPR TC-12 benchmark: A new evaluation resource for visual information systems. In: LREC (2006)
17. Belz, A., Reiter, E.: Comparing Automatic and Human Evaluation of NLG Systems. In: EACL (2006)
18. Papineni, K., Roukos, S., Ward, T., Zhu, W.: BLEU: A method for automatic evaluation of machine translation. In: ACL, pp. 311–318 (2002)
19. Lin, C.Y., Hovy, E.: Automatic evaluation of summaries using n-gram co-occurrence statistics. In: NAACLHLT, pp. 71–78 (2003)

Computer Aided Writing –
A Framework Supporting Research Tasks,
Topic Recommendations and Text Readability

André Klahold, Mareike Dornhöfer, and Madjid Fathi

University of Siegen, 57068 Siegen, Germany
m.dornhoefer@uni-siegen.de, {aklahold,fathi}@informatik.uni-siegen.de
http://www.uni-siegen.de/fb12/ws

Abstract. Although the concept of computer aided writing and word processing is already about 50 years old, there are only few features supporting text creation for authors in today's standard word processing tools. This work presents a Computer Aided Writing (CAW) framework, which supports the user not only during research tasks, but also in creating a text and varieties of this initial text. A feature for proof reading of the text highlights paragraphs which might cause readability problems. The CAW framework bases on methods from the research fields of Knowledge Discovery from Text (KDT) and Recommender Systems.

Keywords: Computer Aided Writing, Research Tools, Recommender Systems, Natural Entity Recognition, Readability.

1 Introduction

The creation of a text basically consists of the three stages: *Planing, Translating* and *Reviewing*. A theory about the cognitive process behind these stages is given by Hayes and Flower [1]. McCutchen defines another theory about the aspects of writing expertise and how a computer may support a human during the writing process [2]. There are different possible motivations for writing, like writing a new article, a book, a scientific text, newspaper or journal articles, e-mails, instruction manuals etc.. During the *Planing* or concept phase of all these different types of text, an author creates/outlines ideas for his future work. He organizes the text structure, which encompasses the succession of different facts. Afterwards, the *Translating* phase allows the processing and transformation of the outlined ideas into written words. For the *Reviewing* and checking of the text, the author proof reads and corrects parts of the text as well as the text as a whole. If there is a general problem in the structure of the text or the text amount varies from a given requirement, a cyclic repetition of the phases *Planing, Translating* and *Reviewing* is always possible [1]. A computer is able to support the human user during all three phases, independent of the type of text. The research of Computer Aided Writing is discussed in the literature already for the last 50 years. The following sections will give an overview, which methods

T. Huang et al. (Eds.): ICONIP 2012, Part V, LNCS 7667, pp. 205–212, 2012.

and possibilities are available and how they may be combined into a *Computer Aided Writing (CAW) framework*.

2 Planing, Translating and Reviewing – Status Quo

2.1 Planing

The *Planing* process [1] integrates different knowledge bases. The process is mainly built on the author's long term memory, but is also heavily influenced by current research tasks of the author. According to Machill et.al. [3], who executed a study about research behaviour in professional editorial environments, computers are the most used research instrument today. Machill et.al. determined the following numbers [3]: Computer based research tools are applied with a percentage of 47%. The time during research where search engines are applied amounts to 4.1%. Insignificantly less used than search engines, are specifically chosen web sites. In most cases Google is the dominating search engine (especially in Germany with 99.3%) [3], [4]. Regarding the specific web sites, most authors use 10 web sites for about 40% of their research [3], [4]. Having a look at the page rank algorithm of Google it becomes apparent that prominent web sites are prioritized [4]. If the author, who executes a research task via Google only takes the first results into consideration, the numbers above show how limited the web based view of the world really is [4]. Only a relative small number of sources built the foundation of computer based research [4]. The question for the proposed *CAW framework* is, how to improve the research of an author and make more sources available to him?

2.2 Translating and Reviewing

Word processing consists of a *Translating* and a *Reviewing* phase [1] and arose with the replacement of typewriters with computers. Already in 1965 Magnuson [5] described a prototype, which supported the separation of texts and sentences in the sense of typesetting. A recommendable overview about the history of computer aided writing is given by Haigh [6].

3 Computer Aided Writing Framework (CAW)

Today it is of course possible to support the different stages of text creation with the help of computers. The proposed *Computer Aided Writing framework (CAW framework)* is basically built on two primary building blocks: a research component and a recommendation component. The *research* component supports the author during the outlining, while the *recommendation* component supports him during the creation and subsequent proof reading of the text. Both components are described in detail in the following subsections.

3.1   Research Component

The research component bases on a corpus of articles from different sources. An indexing function continually gathers the source articles and organizes them into the corpus. Contrary to a generic search engine, the focus of the *CAW research component* is a topic research. This way the component supports the author in defining topics very easily. A prototypic interface allows him to enter a topic word and receive similar or related words in the form of recommendations. The recommended word associations are determined by a previous developed algorithm, which resembles human word associations [7].

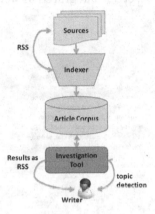

Fig. 1. CAW research workflow

Figure 2 depicts an excerpt of a prototype interface for a topic search based on the word entered on the right hand side of the figure, in this case *ipod*. This way the author is able to define different topics or associations. The sum of the available and future texts, which contain a related topic are offered as a RSS feed to the authors. The calculation of the related texts also bases on a previous developed module, an extended *CRIC* method [8].

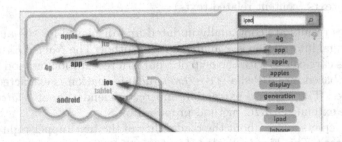

Fig. 2. Topic definition with the help of word associations

3.2   Recommendation Component

The *CAW framework* recommendation component is integrated into an open source web editor, which already offers an integrated spell checker. The recommendation component consists of different singular functions, which are all compliant to the requirements for a generic recommender system. A recommender system is a system, which actively offers a subset of *beneficial* elements to a user with a specific context (translated definition based on [9]). Exceptions to this definition are the functions for the optimization of the text representation and the text modification. Next to visually highlighting the current sentence and the connected references in the rest of the text, a support function for a correct usage of grammatically aspects like brackets, quotation marks, etc. is integrated as well.

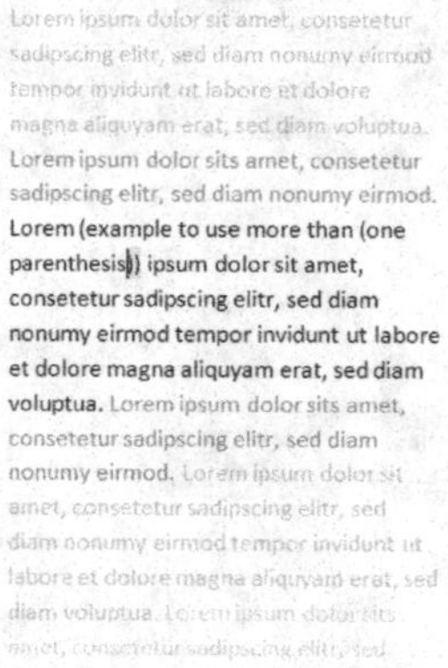

Fig. 3. Highlighting of the current writing context and grammar aid

During the creation of the text a simultaneous analysis process is running. This process analyses:

+ Named Entity Recognition (people, places etc., which are part of the text)
+ Temporal information (dates)
+ Keywords (the central words of the text from a statistical point of view) and
+ Related texts (content related texts)

The analyses results are continually updated and displayed for the author. Per click contextual information is available for the author. The *Named Entity Recognition* [10] is carried out with the help of a pattern recognition method as well as a dictonary based on DBPedia [11]. *Temporal* information is extracted via rule based queries. The *keyword* extraction and recommendation of *related* articles built on the extended *CRIC* method proposed in [8].

Figure 4 depicts a hint about the readability of the text (upper right corner of the text editor). This hint is displayed in form of traffic lights (green, yellow or red). Next to the readability of the overall text, the readability for the singular paragraphs is calculated as well. The visualization options for paragraph based

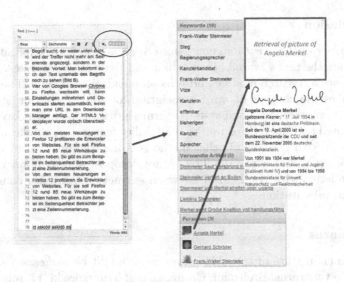

Fig. 4. CAW recommendation component; content of A. Merkel extracted via DBPedia [14]

readability analyses are discussed by Karmakar and Zhu [12]. The readability for the *CAW framework* is calculated with a newly developed readability index based on different former readability indices [13]. The following criteria are part of the calculation process:

+ Average length of sentences
+ Average number of syllables per word
+ Percentage of *Simple words*
+ Sentences per paragraph
+ Passive constructs
+ Filler words

Simple words are calculated with regards to a frequency analysis of the relevant corpus of existing articles. Words, which occur often, are classified as *simple*. If the readability index is only average or in a lower range, the author has got the possibility to see the paragraphs, which cause a reduced readability index. The paragraphs are annotated with the reasons for the reduced readability. Subsequent adaptations of the text allow a dynamic calculation of a new readability index. In this context one study about readability has been executed by Crossley, Allen and McNamera [15].

4   Conclusion and Potential Features for Extending the CAW Framework

4.1   Conclusion

The given paper outlines a *Computer Aided Writing framework* for supporting authors during the text *Planing, Translating* and *Reviewing* processes. The core components are a research module and a recommendation module. The research module applies the concept of word association for topic definition steps. The recommendation module encompasses Natural Entity Recognition, keyword extraction, related text analysis as well as readability features. A similar approach, but carried out with the help of other methods is the computer assisted writing system proposed by Liu et.al. [16]. The authors propose a system for writing an assisted love letter with the help of keyword extraction, sentence construction and synonym substitution.

4.2   Thesaurus

A potential new feature of the *CAW framework* might be a *thesaurus*. For the definition of a thesaurus, Breitman, Casanova and Truszkowski [17] refer in their work to the definition of the ANSI/NISO Thesaurus standard [18], which specifies a thesaurus as *"a controlled vocabulary arranged in a known order and structured so that equivalence, homographic, hierarchical, and associative relationships among terms are displayed clearly and identified by standardized relationship indicators that are employed reciprocally."* Thesauri are especially helpful for writers, who often create texts about the same topic (e.g. sports journalists) to find new synonyms and to consequently vary their texts. A thesaurus is therefore a potential feature for extending the recommendation component, as well as the research component.

4.3   Automatic Text Summarization

Automatic Text Summarization (ATS) is a field, which has been extensively researched in the last years. Essentially *"Automatic Text Summarization is the process of automatically creating a compressed version of the given text."*[21]. The way how this process is designed varies between different approaches [19]. In the following three different approaches, which are only a small amount of the existing thesis for ATS, are shortly characterized. The approach of Geng et.al. [19] applies the technique of *"term co-occurence ... and linkage information of different subjects"* as well as statistical calculations for the summarization process, which allows a dynamically decidable summary size. The process is three-fold, consisting of subject determination, subject terms extraction and summary generation [19]. Zhang and Li [20] propose a different approach for ATS, where they apply sentence clustering and extraction of sentences based on their priority. Therefore they make calculations about word form, word order, sentence and word semantic similarity. Manne et.al. [21] propose a concept consisting of *"text preprocessing, term weight determination, term relationship exploration, stance ranking and summary generation"*. They use techniques like part of speech tagging and Hidden Markov models. Concerning the *CAW framework*, ATS would allow an author so see a summarized version of his article. This way he would be able to determine, if his core intention and message are relayed in a shortened version of said text.

4.4   Sentiment Analysis

Two closely linked research areas are those of opinion mining and sentiment analysis for texts. In their survey about opinion mining, sentiment analysis and subjectivity analysis, Pang and Lee [22] approach the terms opinion mining and sentiment analysis *"more or less interchangeably"*. Sentiment analysis is a field of research, where the sentiment or intention of a given text is detected. The most prominent example for these forms of analyses are the detection of a positive or negative sentiment of a text. Regarding the *CAW framework* the feature of sentiment analysis would be favorable in addition to the readability feature, as the author would directly have a feedback not only about the readability of his text, but also about the impression his text would have on a reader.

References

1. Hayes, J.R., Flower, L.: A Cognitive Process Theory of Writing. College Composition and Communication 32(4), 365–387 (1981)
2. McCutchen, D.: From Novice to Expert: Implications of Language Skills and Writing. Relevant Knowledge for Memory during the Development of Writing Skill. Journal of Writing Research 3(1), 51–68 (2011)
3. Machilli, M., Beiler, M., Zenker, M.: Journalistische Recherche im Internet. Bestandsaufnahme Journalistischer Arbeitsweisen in Zeitungen, Hörfunk, Fernsehen und Online. Schriftenreihe Medienforschung, Landesanstalt für Medien NRW 60, 164–304 (2008)
4. Klahold, A.: Computer Aided Journalism. In: Medienkonvergenz - Transdisziplinär (Medienkonvergenz / Media Convergence), De Gruyter (2012)
5. Magnuson, R.A.: Automated Documentation. Research Analysis Corporation McLean, Virginia, USA, pp. 1–13 (1965)
6. Haigh, T.: Remembering the office of the future: The origins of word processing and office automation. IEEE Annals of the History of Computing 28(4), 6–31 (2006)
7. Klahold, A., Uhr, P., Fathi, M.: Imitation of the Human Ability of Word Association as a Basis for Topic Detection. IEEE Transactions on Knowledge and Data Engineering, Status in Review (2012)
8. Klahold, A.: CRIC: Kontextbasierte Empfehlung unstrukturierter Texte in Echtzeitumgebungen. Dissertation, University of Siegen (2006)
9. Klahold, A.: Empfehlungssysteme. Grundlagen, Konzepte und Systeme, pp. 1–188. Vieweg + Teubner, Wiesbaden (2009)
10. Nadeau, D., Sekine, S.: A survey of named entity recognition and classification. Linguisticae Investigations 30, 3–26 (2007)

11. Auer, S., Bizer, C., Kobilarov, G., Lehmann, J., Cyganiak, R., Ives, Z.G.: DBpedia: A Nucleus for a Web of Open Data. In: Aberer, K., Choi, K.-S., Noy, N., Allemang, D., Lee, K.-I., Nixon, L.J.B., Golbeck, J., Mika, P., Maynard, D., Mizoguchi, R., Schreiber, G., Cudré-Mauroux, P. (eds.) ASWC 2007 and ISWC 2007. LNCS, vol. 4825, pp. 722–735. Springer, Heidelberg (2007)
12. Karmakar, S., Zhu, Y.: Visualizing Multiple Text Readability Indexes. In: International Conference on Education and Management Technology (ICEMT), pp. 133–137 (2010)
13. DuBay, W.H.: Smart language: Readers, Readability and the Grading of Text. Costa Mesa: Impact Information (2006)
14. Content A. Merkel (including signature (labeled as public domain)) (2012), http://de.wikipedia.org/wiki/Angela_Merkel, retrieved via DBPedia http://de.dbpedia.org/page/Angela_Merkel
15. Crossley, S., Allen, D.B., McNamara, D.S.: Text readability and intuitive simplification: A comparison of readability formulas. Reading in a Foreign Language 23(1), 84–102 (2011)
16. Liu, C., Lee, C., Yu, S., Chen, C.: Computer assisted writing system. Expert Systems with Applications 38(1), 804–811 (2011)
17. Breitman, K.K., Casanova, M.A., Truszkowski, W.: Semantic Web. Concepts, Technologies and Applications. Springer, New York (2007)
18. ANSI/NISO: Guidelines for the Construction, Format, and Maintenance of Monolingual Thesauri (1993)
19. Geng, H., Zhao, P., Chen, E., Cai, Q.: A Novel Automatic Text Summarization Study Based on Term Co-Occurrence. In: 5th IEEE International Conference on Cognitive Informatics (ICCI), pp. 601–606 (2006)
20. Zhang, P., Li, C.: Automatic text summarization based on sentences clustering and extraction. In: 2nd IEEE International Conference on Computer Science and Information Technology (ICCSIT), pp. 167–170 (2009)
21. Manne, S., Pervez, S.M.Z., Fatima, S.S.: A novel automatic text summarization system with feature terms identification. In: Annual IEEE India Conference (INDICON), pp. 1–6 (2011)
22. Pang, B., Lee, L.: Opinion mining and sentiment analysis. Foundations and Trends in Infromation Retrieval 2(1-2), 1–135 (2008)

Aspect-Oriented Design and Implementation of Secure Agent Communication System

Ozgur Koray Sahingoz and Emin Kugu

Turkish Air Force Academy, Computer Engineering Department, 34149, Istanbul, Turkey
{sahingoz,e.kugu}@hho.edu.tr

Abstract. Programming distributed and large systems requires dividing the extensive programming codes into smaller entities such as; modules, objects, agents, etc. Especially multi-agent systems (MAS) are emerged as a promising paradigm for constructing complex distributed software-intensive systems. For successfully and effectively achieving a common goal in MAS, agents need to interact with each other according to a coordination protocol. In order to use the agent technology in real applications, it is inevitable to ensure security of this coordination protocol, which crosscuts with different software modules in MAS by hindering modularity and independence. Therefore, the coordination concern should be separated from computation modules, and should be designed by security experts specifically. Aspect-oriented programming has been demonstrated to be an attractive technology for separation of concerns. It can also help the development of secure communication mechanism of MASs. This paper presents an aspect-oriented approach for separation of secure communication mechanisms from agents' functional components. By this approach, it is aimed to enable system developers to implement the MAS's main application codes and security codes separately.

Keywords: Aspect Oriented Programming, MAS, Agent Communication, Security.

1 Introduction

A software agent is a program that represents a user or another program and works towards its goals with autonomously in a dynamic environment. To extend the application areas of software agents, *mobility* feature is added to agent's abilities. With this feature, an agent can move from one host to another according to its dynamically gathered information (or in a predetermined route) to perform a task, which is given by its owner. The spreading use of multi-agent systems (MAS) in large-scale software systems has revealed new software quality requirements such as modularity, mobility, distribution, security and adaptability.

Most of the current MASs typically encapsulate core agent functionalities by defining single thread of objects as agents, but they provide little support for these agents (and hosts) security and confidential agent communication. Especially with increasing use of mobile agents in heterogeneous and open agent environments, MAS

T. Huang et al. (Eds.): ICONIP 2012, Part V, LNCS 7667, pp. 213–220, 2012.

environments are generally created and hosted by different organizations. Therefore, a special attention must be paid to different types of security mechanisms in MAS development and maintenance processes.

It should be assumed that there can always be some malicious agents on local or remote hosts, and these agents try to get some private information by interacting with stationary or incoming mobile agents. Therefore, adding some additional security mechanisms at the application layer is a good choice for enabling secure agent communication. This can be achieved either by using message encryption, which enables a security mechanism against monitoring by malicious agents, or by signing that assures the message integrity. These security mechanisms will help use of MAS in real applications in a modular and extensible form.

As the complexity of MAS increases, the functional and nonfunctional concerns of agents cannot be modularized based only on object-oriented abstractions. These concerns spread across different system components, and consistently cut across the several agent elements, such as goals, actions, and plans. These crosscutting concerns in MAS can be exemplified as learning, mobility, error handling, and security [1].

Aspect-Oriented Programming (AOP) approach is an emerging software development paradigm that is a suitable solution for these crosscutting concerns, and it enables improved separation of concerns by modularizing these concerns in a unit, called as *aspect*. AOP improves the system modularity, decreases its complexity and enables the reusability of these modular parts of the software. By this approach, structure of the MAS will be more modular, and it improves the efficiency of the development and maintenance of the MAS.

MASs generally employ a number of agents that interacts with each other for achieving a common goal. These interactions are done via an Agent Communication Language (ACL), and used in negotiation, cooperation and coordination operations between agents. For achieving secure agent communication, it is possible to encrypt the whole ACL Message. However, it can be more usable encrypting only required part of the ACL Message without changing its structure.

There are some works [7-10] in the literature, which use AOP concept with distributed computing and security. In this paper, an aspect oriented security mechanism is implemented for setting a   secure MAS, which enables confidential agent communication without changing the structure of ACL Messages. As a non-functional concern, secure communication is a key criterion in determining the success of a MAS. It can be separated by AOP concepts from the functional part of the system. By this way, secure communication module can be addressed not only by security experts, but also it can be upgraded in any part of the software development lifecycle.

This paper is structured as follows: In the next section, necessary background information is detailed. Section 3 depicts frameworks and design issues of the aspect oriented secure agent communication system. Experimental study and performance study is shown in Section 4. Finally, conclusions are explained in Section 5.

2   Background Information

Aspect Oriented Programming: Object oriented programming concept was emerged as a revolutionary programming approach to the modular programming. However, it also poses new problems. In large-scale object oriented systems, some concerns such as security, logging and exception handling are implemented in each object separately. This leads crosscutting concerns by using repetitive code and loss of centralized modularity management.

Aspect-oriented programming was emerged as new programming paradigm, in which it grouped related functionalities into aspects (as a single module) and made a modular approach to these crosscutting concerns. This approach improves modularity, decreases complexity, and enhances the reusability of the system.

At the programming level, an aspect is designed as a modular unit to implement a concern, which is scattered in the system. An aspect contains two main components: advices and pointcuts. An advice is the behavior of the aspect, and it contains the code that must be executed when this behavior is called. A pointcut is a specification, and it expresses when, where and how to invoke the advice. These aspects connect to the main system by join points, which are well-defined places in the structure of the program. Finally, an aspect weaver composes core functionality of system with aspects, thereby yielding an evolved woven system.

Agent Security: As in most distributed applications, security is an extremely important criterion in Multi Agent Systems design. With the extending use of mobile agents in the MAS, not only the security of mobile agents, but also protection of hosts is challenging tasks in system development. In NIST's special publication [2], security requirements of Agent Frameworks are listed on five main topics: Confidentiality, Integrity, Accountability, Availability, and Anonymity. In this paper, it is focused on confidential communication that is particularly difficult to achieve, as the agent platform is considered as untrustworthy. In this subsection, therefore, mechanisms for secure communication are explained.

Encryption is an essential strategy for implementing confidential messaging in MASs. There are mainly two types of encryption; symmetric key encryption and asymmetric key encryption. Although asymmetric key encryption is more secure than symmetric one, its encryption time is longer and keeping all agents public key needs too much storage. Therefore, some distributed applications prefer usage of both encryptions together; firstly, asymmetric key encryption is used for setting up a session key. After that, messages are encrypted by using the new temporary session key. Because the symmetric key encryption algorithms do not consume more resources and time. As soon as the communication is finished, the session key is invalidated.

Agent Communication and Fipa ACL: Agents communicate with each other (and with users) via an Agent Communication Language. Firstly, KQML[3] was developed as an ACL but later FIPA ACL[4] was developed by FIPA, which is a non-profit association and proposes open standards for agents' behavior and interactions. The FIPA-ACL is grounded in speech act theory in which every message describes an action or a communicative act that is intended to be carried out. Although the FIPA ACL protocols have become a de facto standard, it lacks security functionality. Therefore, for enabling a secure ACL communication, application layer modifications are needed. To enable a confidential communication in MAS, it is possible to encrypt the whole ACL Message. However, it can be a beneficial way to encrypt only required part of the ACL Message without changing the structure of it.

JADE: Agent platforms provide a framework for development and execution of distributed agents. Java Agent DEvelopment Framework (JADE) [5] is probably one of the most used platforms to build a multi agent system. JADE is fully implemented in Java language, and its platform has only one main container, which controls necessary agent management functionalities. If there is a need for other containers, which are launched on different hosts, they have to register themselves with this main container at start up. JADE Agents can send a message to another agent based on the standardized agent communication language, FIPA-ACL. .If this communication is between agents in the same JADE container, then Java event-passing mechanism is used, otherwise if agents are in different containers then Java RMI is used [6].

3   Design and Implementation of System

In this paper, an aspect oriented secure agent communication system is developed. To develop this system, JADE Agent Development framework was chosen because of its widespread usage, its support on FIPA specifications, and its simplified APIs to access agent management services.

In most of the previously developed multi agent systems, security is not the main concern in the system development. However, to use these systems in real application areas, some security requirements should be met. Security is a crosscutting concern in large-scale distributed systems. Therefore, AOP is a good choice for grouping these functionalities in a single unit called as aspect. And it is also an essential topic to make these update without changing the original MAS. By this way, new modular unit can be implemented by security experts and there is no need to understand the framework and workflow of the main system.

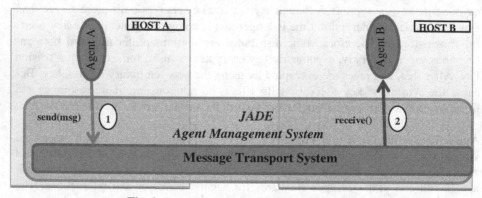

Fig. 1. Agent Communication in JADE System

This system was originally implemented in Java and JADE framework, and it is restructured with AspectJ, which is a mature AOP implementation for Java. In this paper, it is aimed to show that AspectJ is useful for implementing secure communication concerns in a MAS. In JADE systems agents communicates by using the Message Transport System of the JADE as depicted in Figure 1.

Agents communicate with each other by using the *send* method of the Agent class. A sample messaging code can be exemplified as following program segment part a. In a similar way, the other agent gets the incoming message by using the "receive' method of the Agent class (as shown in the following code (part b)).

```
*****
ACLMessage msg = new ACLMes-
sage(ACLMessage.INFORM);
msg.setLanguage("fipa-sl");
msg.setOntology("brokerage ");
msg.setContent( "........" );
msg.addReceiver( new AID(
"agent2", AID.ISLOCALNAME) );
send(msg);
****
```

part a

```
                        *****
ACLMessage msg= receive();
// "receive" method is described
in "Agent" class
if (msg!=null) {
   // Necessary Operations
}
block();
}
*****
```

part b

JADE agents use FIPA-ACL agent communication language according to the FIPA standard [4] for message transfers. The FIPA-ACL is based on speech act theory in which communication between agents is thought of as similar to conversations between humans.

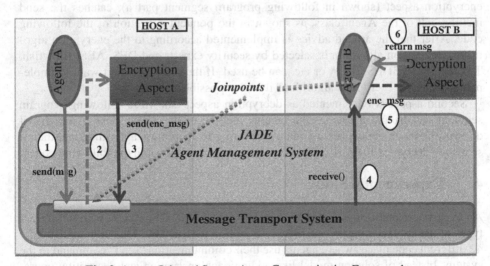

Fig. 2. Aspect Oriented Secure Agent Communication Framework

As depicted in previous sections there are two ways for sending encrypted message. In one hand the whole FIPA-ACL message can be encrypted, on the other hand, only the necessary part of the message (the content part) is encrypted. In this work, second way is chosen for not changing the structure of the FIPA-ACL message and continues to use JADE message transport system

For secure agent communication, the first naive approach is sending the encrypted message directly to the communicating partner. By this way, messages cannot to be accessible to any other parts (even the agent platform), and at no time during communication a message appears in clear text. Unfortunately, applying encryption effectively can be quite complicated. Agent communication codes are also often spread out in the MAS. Therefore, Encryption and decryption, as well as other security concerns, will also be spread out in the MAS, and this will make implementation difficult. This crosscutting nature of secure communication makes it a potentially ideal candidate for making implementation by using AOP.

With usage of aspect-oriented approach, encryption could be added to MAS without the application programmers having prepared for. Secure communication mechanism and encryption methodologies are collected in a single aspect and it has a clear advantage from a security-engineering point of view by this way. This approach allows a better separation of concern in MAS, and therefore, a better division of labor between application developers and security engineers. In addition, it will be easy to change the security policy at compile-time, because all necessary codes are collected in a single module, and it is not in a crosscutting manner.

The main requirement for this aspect-based update is the specifying the necessary joinpoints for both message sending and message receiving operations. After defining these joint points, the necessary aspects can be added to MAS as shown in Figure 2.

Two aspects have been developed for this secure communication MAS. First aspect, encryption aspect (shown in following program segment part a), catches the send method call of the Agent class, as shown in the pointcut definition of the following code. After that the around advice is implemented according to the encryption algorithm. Encryption algorithm is selected by security experts and DES, AES, Blowfish, Twofish, RC4, 3DES, IDEA or etc. can be used. If the security expert want to implements a specific encryption algorithms it is also possible.

Second aspect is implemented as decryption aspect (shown in following program segment part b) and it catches the receive method of the Agent class. After that it decrypt the incoming cyphered content according to encryption algorithm, and then it sends the decrypted FIPA ACL message to the agent as shown in the following codes

4   Experimental Study

An experiment platform is tested for a small scale multi agent system with 9 different types of agents executing on this MAS. According to structure of the system, there are 42 different code places where agents use their communication "send" command in the system. Instead of solution model like adding each encryption and decryption operations to these code places one by one, by using AOP approach, developing two different aspects is sufficient and will be more secure/robust than the first solution approach.

Test platform is developed on a PC, which is configured as in Table 1.

Table 1. Test Platform Properties

Properties	Values
CPU	Intel (R) Core(TM) i7-2630QM CPU @2.00 GHz.
Operating System	64 bit Windows 7 Ultimate Edition
RAM	6.00 Gbyte
Message Size	1575-3575 Bytes
Independent Runs	50,000 times
AspectJ version	1.6.12

Table 2 shows the composing an ACL message with and without AOP. Undoubtedly, adding these aspects slightly decreases the performance of the system with comparison to adding manually. Because, while developing these aspects, all exceptional situations are considered in the aspect codes, which cannot be executed in most of the joinpoints.

Table 2. Performance Comparison

	Without Aspects (10⁻⁶ sec)	With Aspects (10⁻⁶ sec)
ACL Creation Time	329	342*

* Encryption time should be added to this time. It differs according to used encryption algorithm.

AOP is typically used in large scale and complex software development processes for meeting the some nonfunctional software requirements like modularization and quality of services. By using AOP, a large scale MAS could easily be converted to a securely communication in a modular way.

5   Conclusion

Security is a vital and challenging task in Multi Agent Systems. There are many MASs, which do not give sufficient importance to security related issues, and in most of them, agents communicate with each other insecurely over Internet. To enable a confidential communication mechanism, previous system codes should be changed, and appropriate encryption/decryption mechanisms should be adapted carefully.

In this paper, how to set up a secure agent-to-agent communication with Aspect Oriented Programming methodology is described without changing previously written codes. The greatest advantage of AOP is that it provides full separation of concerns between main application and security logics by separating crosscutting concerns, such as secure communication, into modules. It also increases some software engineering quality factors such as; ease of evolution, readability, understandability, maintainability, extendibility, and modularity. With this modularization, the MAS developers can respond more flexibly to requests for updates of system, and it is possible to make upgrades without changing the previously written codes. For example, if some security policies of MAS change later on, these updates can be easily handled by separate modification of these security aspects without changing the existing code of the MAS. Hence, this approach reduces the complexity of the system, and provides better modularity to design a MAS with secure communication.

As Agent Communication Language, FIPA ACL is chosen to formulate messages, and by this way, different vendors' agents can communicate with each other. A Case Study is presented as extensions to JADE environment. Implementing application-level encryption/decryption using AOP approach is promising for system upgrades. This approach eliminates the problem of tangling and scattering in large scale MAS. This study also shows, with the usage of aspects, adding new functionalities to a previously developed system are also possible and effective.

References

1. Garcia, A., Kulesza, U., Sant'Anna, C., Chavez, C., Lucena, C.: Aspects in Agent-Oriented Software Engineering: Lessons Learned. In: Workshop on Agent-Oriented Software Engineering (held with AAMAS 2005) (2005)
2. Jansen, W., Karygiannis, T.: Mobile Agent Security - NIST Special Publication 800-19 (2000), http://csrc.nist.gov/publications/nistpubs/800-19/sp800-19.pdf
3. Finin, T., Fritzson, R., McKay, D., McEntire, R.: KQML as an agent communication language. In: Proceedings of the Third International Conference on Information and Knowledge Management (CIKM 1994), pp. 456–463. ACM (1994)
4. Foundation for Intelligent Physical Agents (FIPA). Fipa2000 Agent Specification, http://www.fipa.org
5. JADE (Java Agent DEvelopment Framework), http://jade.tilab.com/
6. Vitaglione, G., Quarta, F., Cortese, E.: Scalability and Performance of JADE Message Transport System. In: Proceedings of AAMAS Workshop on AgentCities, Bologna, Italy (2002)
7. Sahingoz, O.K.: Secure Communication with Aspect Oriented Approach in Distributed System Programming. In: Academic IT Conference 2012 - Usak, Turkey (2012) (in Turkish)
8. Bedi, P., Agarwal, S.K.: Managing Security in Aspect Oriented Recommender System. In: 2011 International Conference on Communication Systems and Network Technologies (CSNT), pp. 709–713 (2011)
9. Bedi, P., Agarwal, S.K.: Aspect-Oriented Mobility-Aware Recommender System. In: World Congress on Information and Communication Technologies (WICT), pp. 191–196 (2011)
10. Szpryngier, P., Matuszek, M.: Selected Security Aspects of Agent-Based Computing. In: Proceedings of the 2010 International Multiconference on Computer Science and Information Technology (IMCSIT), pp. 205–208 (2010)

Texture Segmentation
Based on Neuronal Activation Degree of Visual Model

Jin Ma, Fuqing Duan, and Ping Guo[*]

Image Processing and Pattern Recognition Laboratory,
Beijing Normal University, Beijing 100875, China
dragonfly_2003@163.com, fqduan@bnu.edu.cn, pguo@ieee.org

Abstract. In the study of object recognition, image texture segmentation has being a hot and difficult aspect in computer vision. Feature extraction and texture segmentation algorithm are two key steps in texture segmentation. An effective texture description is the important factor of texture segmentation. In this paper, the neuronal activation degree (NAD) of visual model is exploited as the texture description of image patches. By processing the length and direction of NAD, we develop an effective segmentation strategy. First, the length of the NAD are used to partition blank area and non-blank area, then the mark index of neuron is used, which is maximally activated to identify the label of each segment unit to get an initial segmentation. Finally, region merging steps is exerted to get a desired result.

Keywords: Texture Segmentation, Visual Model, Neuronal Activation Degree, Region Merging.

1   Introduction

Natural image segmentation is an important and difficult problem in computer vision. The quality of segmentation directly influences the outcome and quality of image analysis and understanding. As an important method for image segmentation, texture segmentation has been a hot research topic, and an effective texture description method will have a great impact on the segmentation.

In 2009, Karklin *et al* [1] propose that, to produce invariant representations of images, higher-level visual neurons encode statistical variations that characterize local image regions, and a visual perception model can be trained in which neural activity encodes the probability distribution most consistent with a given image [3]. In this work, we propose to use the neuron activation degree (NAD) vector of the visual perception model as a texture description.

This feature is used previously by Guo *et al* in the field of image classification and annotation, and a better result can be obtained on certain kinds of image [4]. Shao *et al* [5] use the feature in image segmentation and blocks with similar NAD feature which is clustered by K-means algorithm. It attains better result than N-cuts

---

[*] Corresponding author.

T. Huang et al. (Eds.): ICONIP 2012, Part V, LNCS 7667, pp. 221–228, 2012.

segmentation in certain kind of image. However, for NAD vector it has to be calculated on large enough blocks, the segmentation result is too coarse. Ma *et al* [6] improve the result by sliding sampling which assign the NAD vector to the 3×3 pixel segmentation unit centered at each patch on which the feature is calculated. Although it can get better result, there are also two main problems: First, the number of categories is manually determined by users, which requires users have some prior knowledge of the image. Second, as the training set for visual model only contains non-blank patches, the NAD vectors of blank patches will not sparse enough and the length of NAD is much longer than that of non-blank patches.

To solve the first problem, instead of clustering the NAD vectors, we assign each pixel the mask number of y which is maximally active (argmax |y|) which is elaborated as "winner maps" [1]. However, a shortcoming is that more fragment and noise will be introduced. To solve this problem, small regions eliminating and merging steps are taken in this work.

Liu *et al* [10] proposed an unsupervised image segmentation by region merging strategy, which is implemented by using a novel dissimilarity measure on considering the impact of color difference, area factor and adjacency degree. A binary partition tree is generated to record the whole merging sequence and based on the analysis of binary partition tree (BPT), and an appropriate subset of nodes is selected from the BPT to represent a meaningful segmentation result. BPT was introduced in [2] to systematically represent the hierarchical segmentation of an image in an efficient way.

To solve the second problem, we analyze the NAD vector of blank and non-blank area, and partition them by clustering algorithm. Then further step is taken to segment different kind of non-blank area.

By testing on the Berkeley Segmentation [7], it demonstrated that the feature we exert is effective feature, and algorithm described in this paper is an effective texture segmentation strategy based on this feature.

This paper is organized as follows: in Section 2 the length and direction of NAD vector is analyzed and a strategy for the partition of blank and non-blank area is proposed; In Section 3, we elaborate the process of small region elimination and region merging strategy; In Section 4, the experiment results are demonstrated, and the comparison of the segmentation result is presented. Finally, conclusions are given.

2   Initial Segmentation

As is described in [6], we get NAD vector of each pixel of an image by sliding sampling and calculating the visual model, and take the NAD vector as a description of texture feature. The next step is how to determine the category of all segmentation units. An intuitive approach is to cluster all NAD, thus forming the initial segmentation of the image as described in [6], which can be depicted in Fig. 1. The area "blank area" is named, within which the variance of gray value change very little.

Through observation and analysis, we found that the clustering segmentation has the following two drawbacks. First, blocks in blank area are clustered into more than one catogroy; second, the number of category is determined by users, which requires a *priori* knowledge.

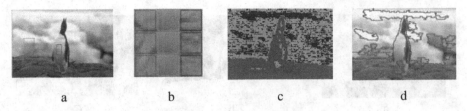

a   b   c   d

Fig. 1. a). Sampling from 3 distinct regions; b). Sampled blocks with frame color correspond to Fig. 1-a; c). K-means clustering; d). Segmentation result after small region elimination

2.1 Analysis of NAD and the Isolation of Blank Area

To figure out the reason of the first drawback, several blocks in the three distinct area are sampled which is shown in Fig. 2-a, and the NAD vectors of each block are calculated. The length value *len_NAD* and direction vector *dir_NAD* of NAD is calculated using equations (1) and (2).

$$len\_NAD = \sqrt{\sum (NAD\ (i))^2} \tag{1}$$

$$dir\_NAD = NAD\ /\ len\_NAD \tag{2}$$

As it can be seen in Fig. 2-b, the average value of *len_NAD* of blank area is significantly greater than non-blank area and it is necessary for us to separate blank area and non-blank area, which is implemented by K-means clustering algorithm, with K=2 in our experiment. The partition result is shown in Fig. 3-b.

However, it does not easily distinguish the two kind of non-blank area just using *len_NAD*, and thus further study of *dir_NAD* is needed. It can be seen in Fig. 2-c, using *dir_NAD* can separate the two kinds of non-blank area without influencing blank area.

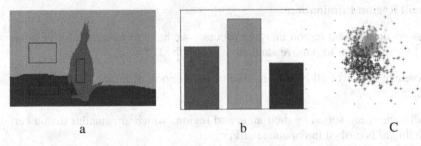

a   b   C

Fig. 2. a). Sample from three distinct area; b). Average value of *len_NAD*; c). Distribution of *dir_NAD* after PCA transformed into two dimention.

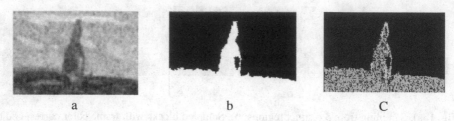

           a   b   C

Fig. 3. a). *len_NAD* value of each pixel; b). Partition image into blank and non-blank area by K-means; c).Category of pixel in non-blank area determined by maximally activated neuron

2.2   Initial Segmentation of Non-blank Area

Another drawback is that it's necessity of determine the number of categories. In Ref. [3], Karklin proposed "winner map" which determine the category of each segment unit by the neuron that is maximally activated. Similarly, after removal of the blank area, we determine the category of one unit by the dimension with max value in NAD vector and with which the initial segmentation is formed.

3   Region Merging Strategy

With the method described above, an initial segmentation is established. Region merging process algorithm is presented to get a desirable result of segmentation in this Section. In [10], Liu *et al* elaborates an effective region merging strategy by exerting a novel dissimilarity measure on considering the impact of color difference, area factor and adjacency degree of each pair of adjacent regions.

    As the process of region merging needs to calculate the similarity between every pairs of adjacent regions, the large number of fragments will increase the computation time. Therefore, small region elimination is taken in advance.

3.1   Small Region Elimination

In order to get an efficient region merging process, we have to remove a lot of noise data. The following steps to remove small area are applied.

Step 1: Average vector of all NAD direction of each region of initial segmentation is calculated;

Step 2: All regions are sorted by their area, and regions which are smaller than a certain threshold *thd* is pushed into a queue *eliVec*.

Step 3: Pop the first region R_c in the head of *eliVec*. Calculate feature distance between the popped region and all its adjacent regions $R_{a1}, R_{a2}, ..., R_{an}$. Then merge the popped region with region R_a with smallest distance to R_c.

$$R_a = \arg \min_{i=1\ldots n} \left\| NAD\ (R_{ai}) - NAD\ (R_c) \right\| \tag{3}$$

If the area of new region R_c' ($R_c' = R_c \cup R_a$) is smaller than *thd*, we push it into the tail of the queue *eliVec*.

Step 4: If queue *eliVec* is empty, terminated. Otherwise, return to **Step 3**.

With above algorithm, we can get segmentation with relatively smaller number of regions. Fig. 4-b shows the small regions elimination result of initial segmentation with *thd* = 50. As it can be seen, the process decreases the region number of segmentation reasonably without degrade the category information.

3.2   Region Merging Process

As is analyzed in [6], the area threshold should not be too large. Generally speaking, the result of small regions elimination still has excessive number of regions. Thus, region merging process is applied as in following steps:

1. Determine the segmentation number *NR*;
2. Calculate the average NAD direction vector y_i and its area a_i of each region. Calculate the adjacent degree of each pair of adjacent regions by the following formula:

$$r_{ij} = \max(\frac{p_{ij}}{p_i}, \frac{p_{ij}}{p_j})$$ (4)

The p_i, p_j and p_{ij} represent the boundary length of region i, boundary length of region j and the length of common boundary of region i and region j, respectively.

3. Calculate distance between two adjacent regions using the following two formulas:

$$dist\ (i, j) = (1 - r) \times (a_i \cdot \|y_i - y_{mrg}\| + a_j \|y_j - y_{mrg}\|)$$ (5)

$$y_{mrg} = \frac{(a_i \cdot y_i + a_j \cdot y_j)}{a_i + a_j}$$ (6)

4. Pairs of region with smallest distance are merged. If the number of remain regions is smaller than *NR*, return to step 2; otherwise, output the result.

a   b   c   d

Fig. 4. a). Initial segmentation; b). Segmentation result after small region elimination; c). Similarity of adjacent regions (red means small value and blue means large value); d). Segmentation result after region merging process

By executing the merge steps of Fig. 4-b given the *NR* number equal to 10, we can get the segmentation result as shown in Fig. 4-d. It can be seen, the region merging process can reliably merge pairs of adjacent regions with smaller area, similar feature and lower adjacent degree.

4   Experiments

4.1   Experiment Description

In the experiments, 40 grayscale images have been selected to form a training set. The number of covariance coefficients (y) is set to 60 and the number of image features used for describing distributions (b_k) is set to 250.   A number of 10×10 pixels patches randomly extracted from the training set is used to train the Visual Model. The process has been elaborated in [3].

With the trained model, several images selected from Berkeley Segmentation Database [7] are used to test our algorithm. For each test image, we use sliding sampling method to extract a number of 10×10 pixels patches with step value equal to 3, and assign the texture value of each patch to its central segmentation unit with scale of 3×3 pixels.After the process of initial segmentation, blank area partition, small region elimination and region merging process, a segmentation result with remaining region number equal to 10 is obtained.

Finally, the result is evaluated by following four methods, Boundary Displacement Error (BDE) [11], Probabilistic Rand Index (PRI) [12], Variation of Information (VoI) [9], and Global Consistency Error (GCE) [7]. We also compare the segmentation result with multiscale N-cuts (m-Ncuts) with the same region number remained [8], and the result is shown in Fig. 5.

4.2   Experiment Results Analysis

For most test images as shown in Fig. 5, our method can obtain a better segmentation result except the second image. We find that the stone in top right corner of the image cannot be distinguished from its environment, and the reason is that the stone and its environment has been filtered as blank area due to its ambiguity. This is a shortcoming of our algorithm which require further improvement. Another shortcoming of our method is that in region merging step the region number also need to be determinated by users which is makes it unconvenient to some extent. Further analysis need to be done to determine a proper segmentation number automatically for the algorithm.

5   Summary

In this paper, we propose an algorithm to segment texture image based on neuronal activation degree of visual model. By analyzing the defect of clustering segmentation using NAD vector as texture description, the length and direction vector of NAD, the image is partitioned into blank area and non-blank area.   Then, initial segmentation is obtained on considering the maximally activated neuron of NAD vector. Applying region merging process, the small regions are eliminated to get final segmentation with given region number. Using Berkeley Segmentation Database to test the proposed method, it is found that a better performance is obtained compared with multiscale N-cuts algorithm with the same segmentation parameter.

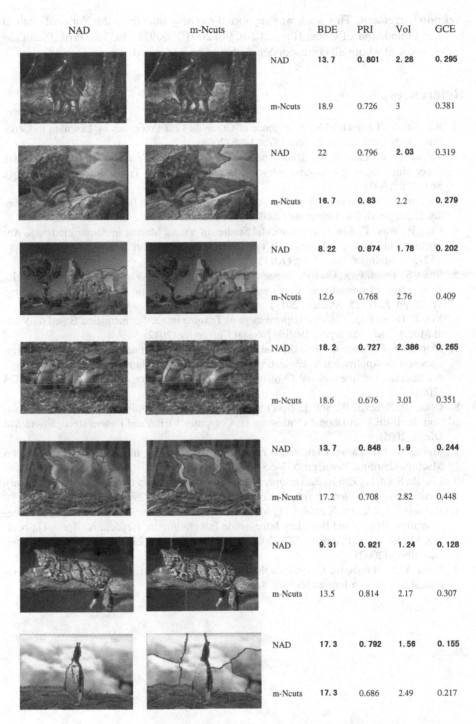

NAD	m-Ncuts		BDE	PRI	VoI	GCE
		NAD	13.7	0.801	2.28	0.295
		m-Ncuts	18.9	0.726	3	0.381
		NAD	22	0.796	2.03	0.319
		m-Ncuts	16.7	0.83	2.2	0.279
		NAD	8.22	0.874	1.78	0.202
		m-Ncuts	12.6	0.768	2.76	0.409
		NAD	18.2	0.727	2.386	0.265
		m-Ncuts	18.6	0.676	3.01	0.351
		NAD	13.7	0.848	1.9	0.244
		m-Ncuts	17.2	0.708	2.82	0.448
		NAD	9.31	0.921	1.24	0.128
		m-Ncuts	13.5	0.814	2.17	0.307
		NAD	17.3	0.792	1.56	0.155
		m-Ncuts	17.3	0.686	2.49	0.217

Fig. 5. Segmentation result and comparison with m-Ncuts

Acknowledgement. This work was supported by the grants from the National Natural Science Foundation of China (Project No. 90820010, 60911130513). Prof. Ping Guo is the author to whom all correspondence should be addressed.

References

1. Karklin, Y., Levicki, M.S.: Emergence of Complex Cell Properties by Learning to Generalize in Natural Scenes. Nature 457, 83–86 (2009)
2. Salembier, P., Garrido, L.: Binary partition tree as an efficient representation for image processing, segmentation, and information retrieval. IEEE Trans. Image Process. 9(4), 561–576 (2000)
3. Karklin, Y.: Hierarchical Statistical Models of Computation in the Visual Cortex. PhD thesis, Carnegie Mellon University (2007)
4. Guo, P., Wan, T., Ma, J.: Experimental Studies of Visual Models in Automatic Image Annotation. In: Jacko, J.A. (ed.) HCI International 2011, Part I. LNCS, vol. 6761, pp. 562–570. Springer, Heidelberg (2011)
5. Shao, S., Duan, F.Q., Guo, P.: Image Segmentation Based on Visual Perception Model. In: Proceedings of International Conference on Computer Application and System Modeling, vol. 2, pp. 723–725. Xiamen (2011)
6. Ma, J., Duan, F.Q., Guo, P.: Improvement of Texture Image Segmentation Based on Visual Model. Technical report, Beijing Normal University (2012)
7. Martin, D., Fowlkes, C., Tal, D., Malik, J.: A Database of Human Segmented Natural Images and its Application to Evaluating Segmentation Algorithms and Measuring Ecological Statistic. In: International Conference on Computer Vision, Vancouver, pp. 416–423 (2001)
8. Cour, T., Benezit, F., Shi, J.: Spectral Segmentation with Multiscale Graph Decomposition. In: IEEE International Conference on Computer Vision and Pattern Recognition, San Diego (2005)
9. Meila, M.: Comparing Clusterings: An Axiomatic View. In: International Conference on Machine Learning, Bonn, pp. 577–584 (2005)
10. Liu, Z., Shen, L., Zhang, Z.: Unsupervised image segmentation based on analysis of binary partition tree for salient object extraction. Signal Process. 91, 290–299 (2011)
11. Freixenet, J., Muñoz, X., Raba, D., Martí, J., Cufí, X.: Yet Another Survey on Image Segmentation: Region and Boundary Information Integration. In: Heyden, A., Sparr, G., Nielsen, M., Johansen, P. (eds.) ECCV 2002, Part III. LNCS, vol. 2352, pp. 408–422. Springer, Heidelberg (2002)
12. Rand, W.M.: Objective Criteria for the Evaluation of Clustering Methods. American Statistical Association Journal 66(336), 846–850 (1971)

One-Dimensional-Array Millimeter-Wave Imaging of Moving Targets for Security Purpose Based on Complex-Valued Self-Organizing Map (CSOM)

Shogo Onojima and Akira Hirose

Department of Electrical Engineering and Information Systems,
The University of Tokyo, 7-3-1 Hongo, Bunkyo-ku, Tokyo 113-8656, Japan
ahirose@ee.t.u-tokyo.ac.jp, onojima@eis.t.u-tokyo.ac.jp
http://www.eis.t.u-tokyo.ac.jp/

Abstract. We propose a millimeter-wave imaging system for moving targets consisting of one-dimensional array antenna, parallel front-end and complex-valued self-organizing map (CSOM) to deal with complex texture. Experiments demonstrate that the CSOM visualizes successfully a liquid-filled small plastic bottle even for measurement data in which we can see almost random phase and amplitude.

Keywords: Complex-valued neural network, millimeter-wave imaging, security imaging.

1 Introduction

Recent progress in electromagnetic-wave imaging using microwave, millimeter-wave, and X-ray has visualized many objects in various situations. Among them, millimeter-wave imaging is expected to be used more and more in human life including security scenes.

There are two ways in its physical realization. One is passive imaging. A passive system observes the black-body radiation of human body and other objects to show the temperature by taking into account the reflectance and absorptance of the object surface [1]. Passive images are obtained by using millimeter-wave antennas, amplifiers, and high-frequency detectors such as schottky diodes. The passive imaging process is similar to photography since it is incoherent, and suitable for capturing target shape directly [2]. However, the black-body power is so small and the contrast is low in most cases that the system is unsuitable for moving targets.

The other is active imaging where we illuminate the target area to receive scattered and/or reflected wave with much higher contrast [3]. It potentially realizes (i)quick image acquisition, (ii)use in combination with various types of modulation of illuminating wave, and (iii)utilization of coherence (phase information) [4]. In addition, the receiving power is so high that paralleled receiving circuits may be free from millimeter-wave amplifiers, resulting in large cost reduction. However, it suffers from so-called speckle noise originating from the high coherence [5] and, hence, not so good at capturing target shape in a straightforward manner. Then we employ synthetic aperture technique at a high calculation cost or neural networks for quick adaptive processing.

T. Huang et al. (Eds.): ICONIP 2012, Part V, LNCS 7667, pp. 229–236, 2012.

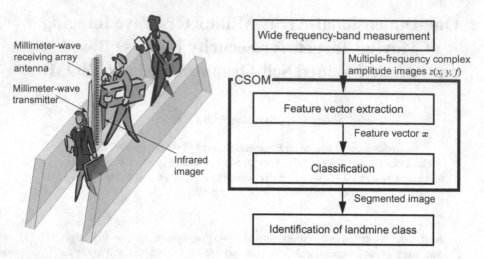

Millimeter-wave receiving array antenna

Millimeter-wave transmitter

Infrared imager

Wide frequency-band measurement

Multiple-frequency complex amplitude images $z(x, y, f)$

CSOM

Feature vector extraction

Feature vector x

Classification

Segmented image

Identification of landmine class

Fig. 1. Security imaging at Shinkansen railway ticket gate

Fig. 2. Flowchart of the total visualization processing

Passive imaging has been investigated by many research groups such as Mizuno's group [6]. Their system uses 35GHz wave, 100mm lens, and corrugated Fermi antennas. Active systems have also been developed by, for example, Sheen's group [7] to visualize weapons and liquid bottles concealed in clothes. Speckle reduction has also been investigated based on, e.g., physical Hadamard transform [8]. Presently there are three serious problems in active imaging, namely, long observation time for spatial scanning and aperture synthesis, high cost of millimeter-wave amplifiers and other electronics, and the privacy invasion in the synthetic aperture observation in which the body-line is visible.

This paper proposes a millimeter-wave active imaging system using one-dimensional array antenna and a complex-valued self-organizing map (CSOM) to visualize moving targets such as passengers walking through ticket gates of Shinkansen as shown in Fig.1. We assume 1,000 walking people per hour per gate. The use of an array antenna is one of the powerful solutions for moving-target visualization [9]. In this case, we have to develop a low-cost parallel front-end to mitigate the cost problem, which is one of the most serious problems in millimeter systems in general. Our new antenna, i.e., the bulk linearly-tapered slot antenna (bulk LTSA) [10] gives the solution in combination with envelope phase detection (EPD) technique.

To solve the privacy problem, we employ the CSOM so that we can visualize targets even with a low spatial resolution [11]. The CSOM has been effective in ground penetrating radar (GPR) systems to visualize landmines buried underground. In the GPR case, we pay attention to the variation in the complex-amplitude textures from stones, clods, metal fragments and landmines, though the reflectance values of them are very near to one another. The situation is similar to the present passenger case. A polyethylene terephthalate (PET) bottles filled with liquid explosives has a reflectance almost the same as our body. Then we again pay attention to the phase information, or to the complex-amplitude in total, in its texture. Previously we tried to observe targets with

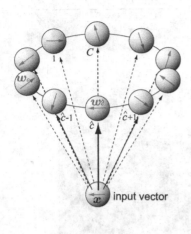

$$w_{\hat{c}}(t+1) = w_{\hat{c}}(t) + \alpha(t)(x - w_{\hat{c}}(t)) \qquad (1)$$

$$w_{\hat{c}\pm1}(t+1) = w_{\hat{c}\pm1}(t) + \beta(t)(x - w_{\hat{c}\pm1}(t)) \quad (2)$$

$$\alpha(t) = \alpha(0)\left(1 - \frac{t}{T}\right) \qquad (3)$$

$$\beta(t) = \beta(0)\left(1 - \frac{t}{T}\right) \qquad (4)$$

$w_{\hat{c}}(t)$: reference vector of the winner class $\hat{c}$
$w_{\hat{c}\pm1}(t)$: reference vector of the neighbor class $\hat{c} \pm 1$
x : input feature vector
t : iteration number in self-organization
T : maximum iteration number
$\alpha(t)$: self-organization coefficient for the winner
$\beta(t)$: self-organization coefficient for the neighbors
C : number of the neurons in the CSOM

Fig. 3. Construction of the ring-CSOM for textural feature vector classification [13] and the self-organization dynamics

a scanning pair of transmitter and receiver antennas [12] where we were successful in visualization. However, the target had to stay still.

In this paper we propose a millimeter-wave imaging system for moving targets consisting of one-dimensional array antenna, parallel front-end and CSOM processing unit. We present experimental results which are promising to develop future actually practical real-time visualization systems.

2 System Construction

2.1 Measurement and Processing in Total

Figure 2 shows the processing flowchart showing the total processing conducted in our previous or novel imaging system. First we acquire the scattering / reflection image in three dimension, i.e., (two-dimension in space) × (one-dimension in frequency), over a wide frequency band using coherent active imaging technique. Then we extract the textural features by calculating local correlations between pixel values in respective local areas in space and frequency domains [12]. We feed the obtained feature vectors to a CSOM to classify the features adaptively, and obtain a segmented space image by projecting backward the classification result to the space image.

2.2 Adaptive Classification of Three-Dimensional Complex-Valued Textures by Using the CSOM

A ring-CSOM [14] is used for the preprocessing to classify the texture. Here we explain the dynamics briefly. Figure 3 shows the ring structure of the ring-CSOM where we have 50 neurons for classification of features into 50 classes. In Fig. 3, $w_c \equiv [w_1, ..., w_N]_c^T$ denotes the N-dimensional weight vector representing the

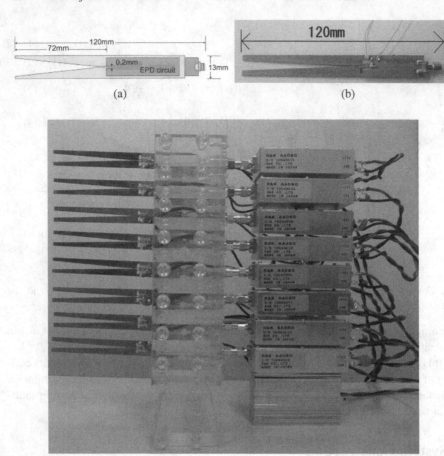

(a)

(b)

(c)

Fig. 4. (a)Design and (b)a photo of linearly tapered slot antennas (LTSAs) including directly-connected front-end circuit, and (c)one-dimensional millimeter-wave antenna array composed of eight LTSAs

textural feature of class c ($c = 1, .., C$), where $[\cdot]^\mathrm{T}$ stands for transpose, and C=50 in the present case. We may need the optimization of the class number C in the future investigation for practical development.

The dynamics of the self-organization in the CSOM is expressed as (1)–(4). Since the total neuron number is not so large (C=50), we define the neighbors of the winner class $\hat{c}$ as the nearest ones $\hat{c} \pm 1$. Unlike the K-means algorithm which we employed in our early works, the CSOM including the neighbors' self-organization improves the classification performance. At the same time, the dynamics places the reference vectors w_c in the order that reflects the similarity among the features, i.e., in such a manner that similar features belong to neighbor classes. The employment of the ring-CSOM is intended for the identification of the target based on example data in the next step.

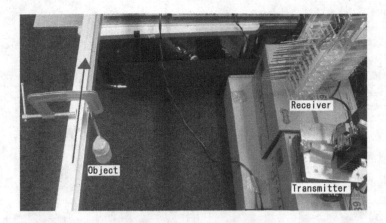

Fig. 5. Experimental setup

2.3 Array Antenna and Front-End

Figure 4 shows (a)the design and (b)a photo of the linearly tapered slot antennas (LT-SAs) including directly-connected detection and bias circuit [10]. Figure 4(c) shows the total array construction. The interval is 2cm, and the total height is 16cm for the eight-element array. Each antenna element receives 34GHz millimeter wave amplitude-modulated with lower frequency up to 900MHz for the envelope phase detection (EPD) to detect distance difference of several tens of centimeters [15]. It includes an envelope detector to realize direct detection without parallel millimeter amplifiers, resulting in low cost.

Figure 5 shows the top view of the experimental setup. The transmitter is a commercially available horn antenna connected with a single power amplifier. The target is a 8cm tall, 4cm in diameter small plastic bottle filled with water. It is attached to a bar moving at 3.0cm/s. The observation area is about 20cm×80cm with an observation time of 16s. There is nothing in the background in the present experiment.

2.4 Experimental Results

Figure 6 shows an example of captured raw data of normalized amplitude in dB and phase in rad. The horizontal scale corresponds to the movement of the target 3.0cm/s×16s, while the vertical scale equals to the height of the antenna array in total. The raw data does not present any significant footprint of the target.

Figure 7 shows photos and visualization images for two experimental setups. Figure 7(a) is the photo of the target passing in front of the array antenna located at position of 20cm horizontally. Figure 7(c) is the corresponding result, in which we can see the target at the expected position. Figure 7(b) and (d) are the photo and the result for position of 30cm. We can see the target again. The visualization results are very different from the raw data in which we could not find anything clearly.

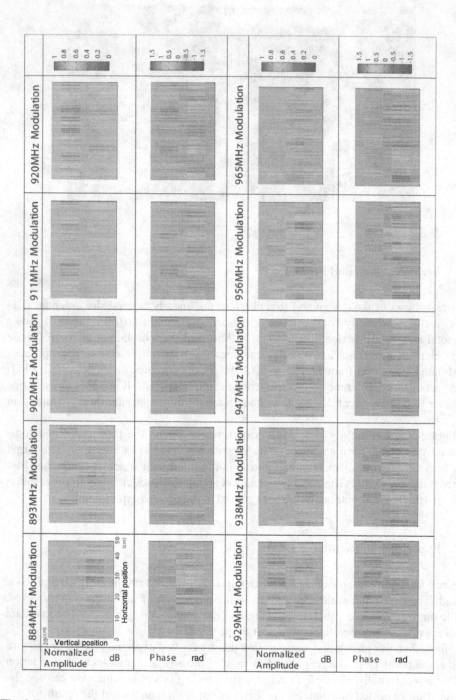

Fig. 6. Example of captured raw data showing normalized amplitude in dB and phase in rad for 10 modulation frequency points for 34GHz carrier wave

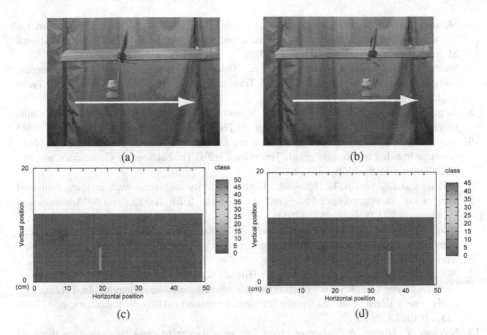

Fig. 7. Targets at positions of (a)x=20cm and (b)x=35cm, and the results of CSOM classification for (c)x=20cm and (d)x=35cm, respectively

3   Summary

We proposed a millimeter-wave imaging system for moving targets consisting of one-dimensional array antenna, parallel front-end and CSOM processing unit. Experiments demonstrated that the CSOM visualized successfully the target even for measurement data in which we cannot see almost random data. We will conduct further experiment with various background to elucidate strength and weakness of our system.

References

1. Sasaki, A., Nagatsuma, T.: Millimeter-wave imaging using an electrooptic detector as a harmonic mixer. IEEE Journal of Selected Topics in Quantum Electronics 6(5), 735–740 (2000)
2. Mizutani, A., Sakakibara, K., Kikuma, N., Hirayama, H.: Grating lobe suppression of narror-wall slotted hollow waveguide millimeter-wave planar antenna for arbitrarily linear polarization. IEEE Transactions on Antennas and Propagatoin 55(2), 313–320 (2007)
3. Tavakol, V., Feng, Q., Ocket, I., Schreurs, D., Nauwelaers, B.: System modeling for active MM-wave imaging systems using an enhanced calculation moethod. In: European Radar Conference (EuRAD), Amsterdam, pp. 56–59 (2008)
4. Qi, F., Ocket, I., Tavakol, V., Schreurs, D., Nauwelaers, B.: Millimeter wave imaging: System modeling and phenomena discussion. In: Internatinal Conference on Applied Electromagnetics and Communications (ICECom), Dubrovnik, pp. 1–4 (2007)
5. Koers, G., Ocket, I., Feng, Q., Tavakol, V., Jaeger, I., Nauwelaers, B., Siens, J.: Study of active millimeter-wave image speckle reduction by Hadamard phase pattern illumination. Journal of Optical Society of America, Part A 25(2), 312–317 (2008)

6. Mizuno, K., Matono, H., Wagatsuma, Y., Warashina, H., Sato, H., Miyanaga, S., Yamanaka, Y.: New applications of millimeter-wave incoherent imaging. In: IEEE MTT-S International Microwave Symposium, pp. 629–632 (2005)

7. Sheen, D.M., McMakin, D.L., Hall, T.E.: Three-dimensional millimeter-wave imaging for concealed weapon detection. IEEE Transactions on Microwave Theory and Techniques 49(9), 1581–1592 (2001)

8. Jaeger, I., Stiens, J., Koers, G., Poesen, G., Vounckx, R.: Hadamard speckle reduction for millimeter wave imaging. Microwave and Optical Technology Letters 48(9), 1722–1725 (2006)

9. Masuyama, S., Hirose, A.: Walled LTSA array for rapid, high spatial resolution, and phase sensitive imaging to visualize plastic landmines. IEEE Transactions on Geoscience and Remote Sensing 45(8), 2536–2543 (2007)

10. Radenamad, D., Aoyagi, T., Hirose, A.: High-sensitivity millimeter-wave imaging front-end using a low impedance linearly-tapered slot antenna. IEEE Transactions on Antennas and Propagation 59(12), 4868–4872 (2011)

11. Hara, T., Hirose, A.: Plastic mine detecting radar system using complex-valued self-organizing map that deals with multiple-frequency interferometric images. Neural Networks 17(8-9), 1201–1210 (2004)

12. Aoyagi, T., Radenamad, D., Nakano, Y., Hirose, A.: Complex-valued self-organizing map clustering using complex inner product in active mmillimeter-wave imaging. In: Proceedings of the International Joint Conference on Neural Networks (IJCNN), Barcelona, pp. 1346–1351. IEEE/INNS (2010)

13. Nakano, Y., Hirose, A.: Adaptive identification of landmine class by evaluating the total degree of conformity of ring-CSOM weights in a ground penetrating radar system. In: International Conference on Neural Information Processing (ICONIP), Sydney, Th10.30–Rm38 (2010)

14. Nakano, Y., Hirose, A.: Improvement of plastic landmine visualization performance by use of ring-csom and frequency-domain local correlation. IEICE Transactions on Electronics E92-C(1), 102–108 (2009)

15. Hirose, A., Hamada, S., Yamaki, R.: Envelope phase detection for millimeter-wave active imaging. Electronics Letters 45(6), 331–332 (2009)

Global Optimal Selection of Web Composite Services Based on UMDA

Shuping Cheng, Xiaoming Lu, and Xianzhong Zhou[*]

School of Management and Engineering, Nanjing University, Nanjing, China
chensp70@163.com, zhouxz@nju.edu.cn

Abstract. QoS model of composite services and Web services selection based on QoS are currently the hot issues in the web service composition area. Services selection based on QoS, which is a global optimal selection issue, has been proved a NP-HARD problem. Takes engine into account, this paper builds the QoS model of service selection in the Web composite services, uses the estimation of distribution algorithm to solve the NP-HARD problem of services selection, and presents a Web services selection method based on the UMDA. Example analysis and experimental analysis based on the UMDA method are performed; it's proved that the method is effective in solving the NP-HARD problem of Web services selection.

Keywords: Qos, Web Composite Services, Service Selection, UMDA.

1   Preface

Researches on Web composite services are attracting more and more attention. The QoS for the composite services and the Web services selection method based on QoS are the hot issues presently.

The present QoS for the composite services only takes the QoS of the member services and the structures of the composite services into account, while the operation of the composite services depends on the engine, which will affect the efficiency of the QoS. The paper, taking the engine into account, rebuilds the QoS for the composite services.

QoS for the Web services depends on QoS of individual service in the combination and the logical structures of the services composited. The current studies show that there are four basic composite structures among the composite services: sequential structure, loop structure, parallel structure and case structure. Based on these four structures, with the known QoS data of the member services, the QoS aggregation method [1] [2] has been put forward. This method supports several common QoS attributes mentioned in the literature [3], such as service price, response time, reliability, and reputation degree. It is common for the academic circle to transform the optimal selection method of the global QoS into a combinatorial optimization problem to find the solutions. There are several optimization methods, such as linear

---

[*] Corresponding author.

T. Huang et al. (Eds.): ICONIP 2012, Part V, LNCS 7667, pp. 237–246, 2012.

programming method[4][5], constrained optimization method[6], heuristic method[7][8] and genetic algorithm[9][10][11].

Although, to some extent, the current genetic algorithms have solved the service
selection problems, they haven't solved some special problems of the composite services. It is necessary to redesign the genetic algorithms of the Web services from the
perspective of QoS. The design of the QoS of the composite services will affect the
efficiency of the genetic algorithms deeply [12]. This paper builds the UMDA algorithm which supports the global optimal selection of Web composite services. Compared to the existing achievements, the method, based on the genetic algorithms, gives
a more complete and effective QoS scheme of the composite services, which has a
good adaptability.

2   QoS for the Composite Services

2.1   Service Price

Due to the introduction of execution engine, the price of the composite services
should be defined as the aggregate value of each member service's price adds the
price of the engine:

$$Pri(cws) = pri(s) + pri(e). \qquad (1)$$

$Pri(cws)$ means the price of the composite services(cws), $pri(s)$ means the
aggregation price of member services, $pri(e)$ means price of the engine.

According to the mainstream toll mode of the Web services, the toll mode of the
engine has two different modes: charging by time and charging by month. The charging by time mode happens when users use the engine, and the charging by month
mode means that users can use the engine a certain number of times in a certain period after charged.

2.2   Response Time

Besides the aggregate value of the time of the member services, the response time
should take the communication time and the treatment time of the engine into
account:

$$T(cws) = t(s) + tran(u,e) + pro(e). \qquad (2)$$

$T(cws)$ means the response time of the composite services, $t(s)$ means the aggregate value of the time of the member services, $tran(u,e)$ means the communication time between the users and the engine, $pro(e)$ means the treatment time of
the engine. The estimation process of response time is as follow:

The first step is estimating the $t(s)$. Engine server suppliers should provide the current network parameters, the users should provide the access amount of every member service, and the member service suppliers provide the environment data when the member services are running.

The second step is estimating the $tran(u,e)$. The $tran(u,e)$ depends on the network environment between the users and the engine, and this data can be calculated by the network performance and the transmission amount caused by the composite services used by the users.

The third step is estimating the $pro(e)$. The $pro(e)$ depends on the complexity of the control logic of the composite services and the performance of the engine.

2.3   Reliability

The reliability of the composite services is the probability of the right response of the services in a longest expected time. The present estimation methods of the Web services just take the aggregation of the reliabilities of the member services into account, while the engine controls the logic and the data flow of the composite services. If the engine fails, the whole operations of the composite services will fail, so the reliability of the engine should be more important than that of the member services. Here is the formula of the reliability of the composite services:

$$Rel(cws) = rel(s) \times rel(e).\tag{3}$$

$Rel(cws)$ means the reliability of the composite services, $rel(s)$ means the aggregate value of the reliabilities of the member services, and $rel(e)$ means the reliability of the engine. The reliability of the engine should be distributed by the engine suppliers after enough tests.

2.4   Reputation Degree

The reputation degree is the satisfaction of the composite services from users. In the problem of the member services selection, the whole reputation degree of the services after composited, as a constraint condition, should be taken into account in estimation of the reputation degree. The paper considers the reputation degree of the engine, besides the reputation degree of the member services. Here is the formula of the reputation degree:

$$Rep(cws) = (1-\alpha) \times rep(s) + \alpha \times rep(e)\tag{4}$$

$Rep(cws)$ means the reputation degree of services, $rep(s)$ means the composite of the reputation degree of member services, and $rep(e)$ means the reputation degree of the engine. α is a weight coefficient and $\alpha \in [0,1]$, and if α is bigger, the reputation degree of the engine is more important. Because the engine is very important and irreplaceable in the solution of the selection of the Web services, the range of α should be $\alpha \in (0.5,1]$.

3   Selection Method of Web Services Based on UMDA

3.1   Selection Method of Web Services Based on UMDA

The users consider the composite services as a whole and focus on satisfaction of the Qos of the whole under the global constraint condition, which is a global optimal selection problem.

Some assumptions are as follow: There are N tasks in the composite package, $task_i(i=1 \sim N)$ means the NO. i task, Set_i means the candidate service set for $task_i$, M_i means the service amount in Set_i, and $S_{ij}(j=1 \sim M_i)$ means it is NO. j service in the Set_i. The global optimal selection problem based on the QoS is a multi-objective optimization problem, and a simple solution for this problem is to change the problem to a single-objective optimization problem. The mathematical description for the problem is as follow:

$$MaxF(cws) = \sum\nolimits_{k=1}^{t} \omega_k \times Qf_k(cws) \qquad (5)$$

$$s.t.\ Qf_k(cws) \le B_k \ \text{Or} \ Qf_k(cws) \ge B_k$$

$$cws = (S_{1j}, S_{2j}, \cdots, S_{Nj})$$

$$S_{ij} \in Set_i, i=1 \sim N, j=1 \sim M_i$$

$$\sum\nolimits_{k=1}^{t} \omega_k = 1$$

cws means the composite of the services, $Qf_k(cws)$ means the QoS value of the No.k service in the composite services package, ω_k means weight coefficient, B_k means constraint condition of certain QoS.

In the Web service area, there are lots of researches on the traditional genetic algorithms and their improved versions, but there is little on the estimation of distribution algorithm. The paper discusses the solution of global optimal selection problem based on the QoS with the estimation of distribution algorithm.

3.2   Selection of the Web Services Based on UMDA

（1）Narrowing Solution Space

The users hope the global QoS of the composite services is the best and the QoS of every service corresponding to the task is the best too. If there are too many candidate services, the local optimal selection algorithm can be used to calculate the QoS of every candidate service, and all the services with a QoS value less than a certain threshold will be forsaken, to reduce the amount of the solutions. The threshold is given by the users. It is assumed that after the process above, the amount of the services in the candidate service set is still M_i.

（2） Coding

Binary code system is used in UMDA, and every variable is represented by a binary code. If the binary code system is used in the problem of Web service selection, every candidate service will be represented by a binary code, which will cause a long coding, and this is not suitable to the algorithm processing. In this case, the better solution is using the integer array for coding. Please see the structure in figure 1, and the details are as follow:

The integer array is used for representing the entities of the population, and the length of the array, which is the amount of the gens, is equal to the task amount N in the composite package. $G_i (i = 1 \sim N)$ means the NO. i gen, corresponding to the related $task_i$ and Set_i. G_i is the number j candidate service in Set_i, $j = \{1, 2, \cdots, M_i\}$; $G_i = j$ means the $task_i$ will be implemented by the NO. j service in the candidate service set.

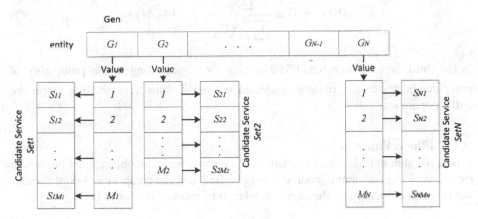

Fig. 1. Entity Structure

（3） Population Initialization

In the first step, the population size K should be determined. According to the research between the learning parameter and problem of the UMDA by Shapiro[14], only when the population size is huge enough, the UMDA algorithm can converge to the global optimal solution. The tests of the practical problems show that the population size should be the square root of the amount of the candidate solution. If there are total $\prod_{i=1}^{N} M_i$ candidate solutions of service selection, in order to make the algorithm converge to the global optimal solution, K should be approximately equal the square root of it.

In the second step, the initial probability of every possible value of every gen G_i should be determined. If users don't have any preference about certain service in the candidate Set_i, the chance of every service is the same. In that case, if the service in Set_i is chosen, the chance of every possible value of G_i is the same, and it will be showed as follow:

$$P(G_i = j) = \frac{1}{M_i}, j = 1 \sim M_i \tag{6}$$

If the users have certain preference, and could give the probability of every candidate, the initial probability of every possible value comes from users. But in the practice, although the users have the preference about some services, the users can't give the exact probability of every candidate service, especially when the amount of the candidate services is huge. The local optimal selection algorithm is used to calculate the QoS of every candidate service and the sum of QoS of every candidate service, and the probability of every candidate service will be calculated by certain formula.

$$P(G_i = j) = \frac{QS_{ij}}{\sum_{j=1}^{M_i} QS_{ij}}, j = 1 \sim M_i \tag{7}$$

In the third step, in every candidate service Set_i, according to the probability of every candidate service, roulette algorithm is used K times for sampling to get the initial population.

（4）Fitness Function

It is applicable that the fitness function is the same with the object function, which means when the constraint conditions are satisfied, and the composite QoS of composite services influenced by the engine is taken into account, the formula is:

$$f(cws) = \sum_{k=1}^{t} \omega_k \times Qf_k(cws) \tag{8}$$

Because the amount of the composite schemes is too huge to calculate all the QoS of the possible composite services, ω_k is given directly by the users, not by the objective data, which is different from the method of the local optimal selection algorithm.

（5）Selection Mechanism.

The selection mechanism is the foundation of the probability updating. Numbers of the distribution relations inherited from the old generation population by the new generation population are called selection pressure, which depends on selection mechanism. In the case that the proper selection mechanism is chosen, the selection pressure will keep population diversity and increase the possibility of the global optimal converging. It can also guarantee the high amount of the optimal entities during the evolution, and the algorithm converges fast. Similar with the genetic algorithm, the selection mechanisms of the EDA include sorting algorithm, roulette algorithm, tournament selection algorithm, and random traversal sampling algorithm.

The paper uses the sorting algorithm. First, the QoS of the composite services of every entity of the population is calculated, and those unsatisfied with the constraint condition will be discarded. Then, after calculating the fitness values of the left entities by the fitness function, the fitness values will be sorted from maximum to minimum, and the excellent ones with high fitness value will be selected according to certain rate. In order to keep the population diversity, the rate should be around 50%.

（6）**Probability Updating and Resample**
According to values of the gens of the selected optimal entities, the frequencies of the values of every gen are calculated. The probabilities of values of every gen are updated by the value calculated as the frequencies divided by total amount of the optimal entities. According to the new probabilities, every gen will be sampled K times by the roulette algorithm, and there will be K new entities then, which will form a new generation population.

（7）**Stopping Rule**
There are three common types of terminal conditions: the generation of evolution achieves the set value; the fitness value of the entity achieves the set value; and there is less likely a big change in the next generation. Normally, the generation of evolution of the population is large enough; the evolution should converge to certain optimal entity; or the evolution stops at the set value accepted by the users.

As the other evolutionary algorithms, the solution of the EDA depends on the related parameters and methods, and the solution may be not the optimal one. For the problem of the selection of the Web services, if there are too many composite schemes, the suboptimal solution calculated by the EDA is acceptable.

4   Simulation and Comparison

Some tests with the simulation data will be used in this part to verify that EDA is efficiency to solve the selection of the Web services based on QoS under the big solution space. Both the efficiency of the EDA and the efficiency of the genetic algorithm will be analyzed in this part.

There are 4 types of the QoS attributes, the values of the attributes are randomly generated, and the weight is $\omega = (0.4, 0.2, 0.2, 0.1, 0.1)$. The related data of the engine here is the same as that in the real example. There are N tasks in the composite services, the amount M of the candidate services of every task is the same, the generation size is

K, and the generation of the evolution is P. Different Ns and Ms are used in different experiments, and the algorithm will run 50 times in every experiment. If the convergence fails even the operation of the algorithm is over, the maximal fitness value of the entity in the last generation is used as the statistical data.

4.1   Simulation Result

The related data is as follow:

(1) N is 5, K is 100, the value of M is changing, and the stopping rule is that P is 200 or convergence finishes in advance.

（2）N is 5, K is 100, the value of M is changing, and the stopping rule is that the maximal fitness value achieve the set value or convergence finishes in advance.

（3）M is 10, K is 100, the value of N is changing, and the stopping rule is that P is 200 or convergence finishes in advance.

（4）M is 10, K is 100, the value of N is changing, and the stopping rule is that the maximal fitness value achieve the set value or convergence finishes in advance.

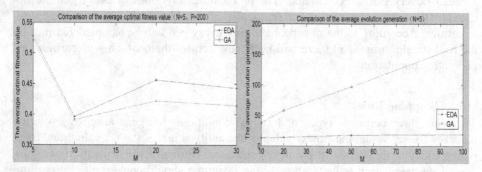

Fig. 2. Comparison of the average optimal fitness value

Fig. 3. Comparison of the average evolution generation

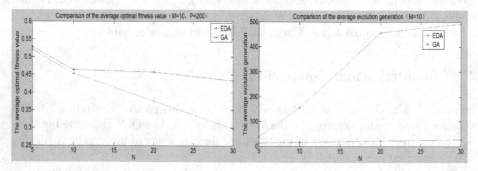

Fig. 4. Comparison of the average optimal fitness value

Fig. 5. Comparison of the average evolution generation

4.2   The Efficiency of EDA

From the comparison:

（1） When the stopping rule is the evolution generation, if the N is fixed, the fluc-tuation of the convergence evolution generation is small when M increases; if the M is fixed, the convergence evolution generation increases when N increases; and the EDA converges very fast to a good fitness value.

（2） If the stopping rule is a maximal fitness value, N is fixed, and the fluctuation of the evolution generation of EDA when the maximal fitness value achieves is small when M increases; M is fixed, and the evolution generation of EDA when the maxim-al fitness value achieves increases when N increases; and the maximal fitness value of EDA achieves fast under a small evolution generation, and it sometimes happens that the algorithm converges fast to the value less than the maximal fitness.

It shows that the satisfied suboptimal solution can be steadily achieved in a small evolution generation, and the better composite scheme can be found for the selection of the Web services with EDA.

4.3   Comparison

It shows that EDA is more efficiency than the traditional genetic algorithm in litera-ture[11] in the selection of the Web services:

The convergence speed of EDA is fast than that of GA. If the generation size and the stopping rules (the evolution generation) are the same, the optimal fitness value of EDA is better than that of GA. If the generation size and the stopping rules (the op-timal fitness value) are the same, the evolution generation of EDA is much less than that of GA.

Additionally, this experiment just proves that the EDA in this paper is better than the traditional genetic algorithm in literature[11] in the selection of the Web services, which doesn't mean the EDA is better than the traditional genetic algorithm. The operations of the two algorithms depend on the coding, evolution and the parameters.

5   Conclusion

The paper discusses the selection of the web services based on QoS, introduces the EDA which is a global optimal selection scheme to solve the NP-HARD problem. The paper introduces the solution of the selection of the Web services based on the UMDA, designs the algorithm, performs example analysis and experimental analysis, and verifies that the UMDA is efficient to solve the problem.

Acknowledgement. This research was supported by: (i) the National Planning Office of Philosophy and Social Science under Grant 11&ZD169; (ii) the NSFC under Grant 70971061 and 71171107.

References

1. Aalst, W.M.P.V.D.: Don't go with the flow: Web services composition standards exposed. IEEE Intelligent Systems 18(1), 72–76 (2003)
2. Yin, K.: Research on QoS-aware Services Composition in Internet Environment. Zhejiang University, Hangzhou (2010)
3. Michael, C.J., GeroMuhl, G.G.: QoS Aggregation for Web Service Composition using Workflow Patterns. In: EDOC, pp. 149–159 (2004)
4. Zeng, L.Z., Benatallah, B., et al.: QoS-Aware Middleware for Web Services Composition. Transactions on Software Engineering 30(5), 311–327 (2004)
5. Ardagna, D., Pernici, B.: Adaptive Service Composition in Flexible processes. IEEE Transactions on Software Engineering 33(6), 369–384 (2007)
6. Rosenberg, F., Celikovie, P., Michlmayr, A., Leitner, P., Dustdar, S.: An End-to-End Approach for QoS-Aware Service Composition. In: 2009 IEEE International Enterprise Distributed Object Computing Conference, pp. 151–160 (2009)
7. Tao, Y., Lin, K.: Service selection algorithms for Web services with end to end QoS constraints. Information Systems and e-Business Management 3(2), 103–126 (2005)
8. Berbner, R., Spahn, M., Repp, N., Heckmann, O.: Ralf Steinmetz. Heuristies for QoS-aware Web Service Composition. In: ICWS, pp. 72–82 (2006)
9. Liu, K., Wang, H., Xu, Z.: A Web Service Selection Mechanism Based on QoS Prediction. Computer Technology and Development 17(8), 103–109 (2007)
10. Xia, Y.: Research on Some Key Issues of Dynamic Service Composition. Beijing University of Posts & Telecommunications, Beijing (2009)
11. Wu, C.: Research on Dynamic Web Service Composition and Performance Analysis with QoS Assurances. Wuhan University, Wuhan (2007)
12. Ignacia, R., Jesu, S.G., Hector, P., et al.: Statistical analysis of the mainparameters involved in the design of a genetic algorithm. IEEE Transactions on Systems, Man, and Cybernetics-Part C: Applications and Reviews 32(1), 31–37 (2002)
13. Shapiro, J.L.: Drift and scaling in estimation of distribution algorithms. Evolutionary Computation 13(1), 99–123 (2005)
14. Al-Masri, E., Mahmoud, Q.H.: Discovering the best web service. In: Proc. of Int'l Conf. on World Wide Web (WWW), pp. 1257–1258 (2007)
15. Al-Masri, E., Mahmoud, Q.H.: QoS-based Discovery and Ranking of Web Services. In: Proc. of Int'l Conf. on Computer Communications and Networks (ICCCN), pp. 529–534 (2007)
16. Al-Masri, E., Mahmoud, Q.H.: Investigating Web Services on the World Wide Web. In: Proc. of Int'l Conf. on World Wide Web (WWW), Beijing, pp. 795–804 (2008)
17. Michael, C., Jaeger, G., Rojec-Goldmann, G.: QoS Aggregation in Web Service Composition using Workflow Patterns. EEE, 181–185 (2005)

Multistep Speaker Identification Using Gibbs-Distribution-Based Extended Bayesian Inference for Rejecting Unregistered Speaker

Yuta Mizobe, Shuichi Kurogi, Tomohiro Tsukazaki, and Takeshi Nishida

Kyushu Institute of Technology, Tobata, Kitakyushu, Fukuoka 804-8550, Japan
{mizobe@kurolab2.,kuro@,nishida@}cntl.kyutech.ac.jp
http://kurolab2.cntl.kyutech.ac.jp/

Abstract. This paper presents a method of multistep speaker identification using Gibbs-distribution-based extended Bayesian inference (GEBI) for rejecting unregistered speaker. The method is developed for our speaker recognition system which utilizes competitive associative nets (CAN2s) for learning piecewise linear approximation of nonlinear speech signal to extract feature vectors of pole distribution from piecewise linear coefficients reflecting nonlinear and time-varying vocal tract of the speaker. In this paper, we focus on the problem of Bayesian inference (BI) in multistep identification for rejecting unregistered speaker and introduce GEBI to solve the problem. The effectiveness of the present method is shown by means of experiments using real speech signals.

Keywords: Gibbs-distribution-based extended Bayesian inference, Multistep speaker identification, Competitive associative net.

1 Introduction

This paper presents a method of multistep speaker identification using Gibbs-distributi -on-based extended Bayesian inference (GEBI) for rejecting unregistered speaker. The method is developed for our speaker recognition system which utilizes competitive associative nets (CAN2s). Here, the CAN2 is an artificial neural net for learning efficient piecewise linear approximation of nonlinear function by means of using competitive and associative schemes [1,2,3]. Recently, we have shown that feature vectors of pole distribution extracted from piecewise linear predictive coefficients obtained by the bagging (bootstrap aggregating) version of the CAN2 reflect nonlinear and time-varying vocal tract of the speaker, and they are effective for speaker recognition [4,5]. Note that among the previous research studies of speaker recognition, the most common way to characterize the speech signal is short-time spectral analysis, such as Linear Prediction Coding (LPC) and Mel-Frequency Cepstrum Coefficients (MFCC), where spectral features of the speech are extracted from each of consecutive interval frames spanning 10-30ms [6,7,8,9]. Thus, a single feature vector of LPC and MFCC corresponds to the average of multiple piecewise linear predictive coefficients of the bagging CAN2. Namely, the bagging CAN2 has stored more precise information on the speech signal.

In this paper, we focus on Bayesian inference (BI) used in the multistep speaker identification [5], and analyze the problem for rejecting unregistered speaker. Namely,

T. Huang et al. (Eds.): ICONIP 2012, Part V, LNCS 7667, pp. 247–255, 2012.

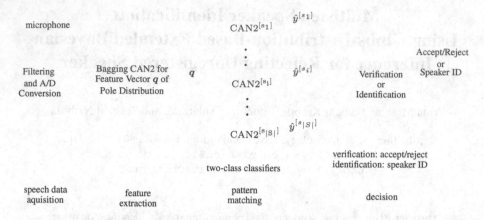

Fig. 1. Speaker recognition system using the CAN2s

from [10], we can say that this problem reflects the difficulty of open-set identification which is more complex than closed-set identification, where the latter determines the identification number (ID) from those of registered speakers and has no rejection scheme while the former has a rejection scheme. In this paper, we introduce GEBI to solve the problem. Incidentally, we can see that the multistep verification for rejecting unregistered speaker is achieved by using the result of the present identification method.

In the next section, we show an overview and a formulation of our singlestep speaker recognition system followed by Bayesian multistep speaker identification and its problem. And then we introduce GEBI and "void" speaker to solve the problem. In **3**, we present experimental results and examine the effectiveness of the present method.

2 Multistep Speaker Identification Using Gibbs-Distribution-Based Extended Bayesian Inference

2.1 Overview and Formulation of Singlestep Speaker Recognition

Fig. 1 shows our speaker recognition system using CAN2s. In the same way as general speaker recognition systems [6], it consists of four steps: speech data acquisition, feature extraction, pattern matching, and making a decision. The speaker recognition is classified into verification and identification, where the former is the process of accepting or rejecting the identity claim of a speaker, which is regarded as two-class classification. The latter, on the other hand, is the process of determining which registered speaker provides a given utterance, which is regarded as multi-class classification. In addition, speaker recognition has two modalities: text-dependent and text-independent. The former require the speaker to say key words or sentences with the same text for both training and recognition phases, whereas the latter do not rely on a specific text being spoken.

In this research study, we use a feature vector of pole distribution obtained from a speech signal (see [4] for details). Let $Q^{[s]}$ be a set of feature vectors $q = (q_1, q_2, \cdots, q_k)^T$ of a speech signal from a speaker $s \in S = \{s_i | i \in I_S\}$, where $I_S = \{1, 2, \cdots, |S|\}$.

We use a learning machine called CAN2, which learns to approximate the following target function:

$$f^{[s]}(q) = \begin{cases} 1, & \text{if } q \in Q^{[s]} , \\ -1, & \text{otherwise} . \end{cases} \tag{1}$$

Let CAN2$^{[s]}$ be the learning machine for the speaker s. Then, with a number of training data $(q, f^{[s]}(q))$ for $q \in Q^{[s]}$, we train CAN2$^{[s]}$ to approximate the above function by a continuous function as $\hat{y}^{[s]} = \hat{f}^{[s]}(q)$. After the training, for a feature vector q of a test speech signal, we execute singlestep identification by determining the identification number ID with the maximum detection as

$$\text{ID} = \underset{i \in I_S}{\text{argmax}} \{\hat{y}^{[s_i]} = \hat{f}^{[s_i]}(q)\} . \tag{2}$$

On the other hand, we execute singlestep verification or two class classification with the binarization of the output as

$$v^{[s]} = \begin{cases} 1, & \text{if } \hat{y}^{[s]} = \hat{f}^{[s]}(q) \geq y_\theta , \\ -1, & \text{otherwise} . \end{cases} \tag{3}$$

Namely, we accept the speaker s if $v^{[s]} = 1$, and reject otherwise. Here, the threshold y_θ is introduced for improving the performance of multistep identification shown below.

2.2 Bayesian Multistep Speaker Identification and Its Problem

Bayesian Inference (BI) for Multistep Speaker Identification. First, we obtain the probability $p(v^{[s_i]}|s)$ of the two class classifier CAN2$^{[s_i]}$ for a given training dataset as

$$p(v^{[s_i]} = 1|s) = \frac{n(v^{[s_i]} = 1|s)}{n(v^{[s_i]} = 1|s) + n(v^{[s_i]} = -1|s)} , \tag{4}$$

and $p(v^{[s_i]} = -1|s) = 1 - p(v^{[s_i]} = 1|s)$, where $n(v^{[s_i]} = 1|s)$ indicates the number of instances with the output of CAN2$^{[s_i]}$ being $v^{[s_i]} = 1$ for the speech of the speaker $s \in S$. Now, for speaker identification, let us suppose that the joint probability of the output vector $v^{[S]} = (v^{[s_1]}, \cdots, v^{[s_{|S|}]})$ of all classifiers $s_i \in S$ for an input speech of the speaker $s \in S$ is given by

$$p(v^{[S]}|s) = \prod_{s_i \in S} p(v^{[s_i]}|s) , \tag{5}$$

because any two probabilities $p(v^{[s_i]}|s)$ and $p(v^{[s_j]}|s)$ for $i \neq j$ are supposed to be independent. Let $v_{1:t}^{[S]} = v_1^{[S]}, \cdots, v_t^{[S]}$ be a sequence of $v^{[S]}$ obtained from a test speaker s, then we can execute Bayesian inference (BI) by

$$p_\text{I}^{[\text{Bys}]}(s|v_{1:t}^{[S]}) = \frac{p_\text{I}^{[\text{Bys}]}(s|v_{1:t-1}^{[S]}) \, p(v_t^{[S]}|s)}{\sum_{s_i \in S} p_\text{I}^{[\text{Bys}]}(s_i|v_{1:t-1}^{[S]}) \, p(v_t^{[S]}|s_i)} . \tag{6}$$

Here, we use naive BI or the conditional independence assumption $p(v_t^{[S]}|s, v_{1:t-1}^{[S]}) = p(v_t^{[S]}|s)$, which is shown effective in many real world applications of naive Bayes classifier [11].

Problem of BI for Multistep Speaker Identification. When the Bayesian probability $p_{\mathrm{I}}^{[\mathrm{Bys}]}(s|\boldsymbol{v}_{1:t}^{[S]})$ for the speaker $s = s_i$ becomes larger than a threshold, say $p_{\mathrm{I}\theta}$, for the increase of t, we would like to decide the identification number as ID $= i$. However, from the Bayesian inference for $t = 1, 2, \cdots$, we have

$$p_{\mathrm{I}}^{[\mathrm{Bys}]}(s|\boldsymbol{v}_{1:t}^{[S]}) = \frac{p_{\mathrm{I}}^{[\mathrm{Bys}]}(s|\boldsymbol{v}_{1:0}^{[S]})}{Z_t} \prod_{k=1}^{t} p(\boldsymbol{v}_k^{[S]}|s) = \frac{p_{\mathrm{I}}^{[\mathrm{Bys}]}(s|\boldsymbol{v}_{1:0}^{[S]})}{Z_t} \exp\left(-t\tilde{L}_{1:t}^{[s]}\right) , \quad (7)$$

where Z_t indicates the normalization constant for holding $\sum_{s \in S} p_{\mathrm{I}}^{[\mathrm{Bys}]}(s|\boldsymbol{v}_{1:t}^{[S]}) = 1$, and

$$\tilde{L}_{1:t}^{[s]} \triangleq -\frac{1}{t} \log L_{1:t}^{[s]} = -\frac{1}{t}\left(\sum_{k=1}^{t} \log p\left(\boldsymbol{v}_k^{[S]}|s\right)\right) \quad (8)$$

is the normalized negative log-likelihood. Here, $L_{1:t}^{[s]} \triangleq \prod_{k=1}^{t} L_k^{[s]}$ is the likelihood of s for the output sequence $\boldsymbol{v}_{1:t}^{[S]}$, and $L_k^{[s]} \triangleq p(\boldsymbol{v}_k^{[S]}|s)$ is the primitive likelihood for each $\boldsymbol{v}_k^{[S]}$. Since $\tilde{L}_{1:t}^{[s]}$ is the negative log of the geometric mean of the primitive likelihood $(\prod_{k=1}^{t} p(\boldsymbol{v}_k^{[S]}|s))^{1/t}$, let us suppose that $\tilde{L}_{1:t}^{[s]}$ converges for the increase of t. Then, the ratio of the Bayesian probability for $s_i \in S$ to $s_m = \underset{s_i \in S}{\operatorname{argmax}}\, p_{\mathrm{I}}^{[\mathrm{Bys}]}(s_i|\boldsymbol{v}_{1:t}^{[S]})$ becomes

$$r_i^{[\mathrm{Bys}]} \triangleq \frac{p_{\mathrm{I}}^{[\mathrm{Bys}]}(s_i|\boldsymbol{v}_{1:t}^{[S]})}{p_{\mathrm{I}}^{[\mathrm{Bys}]}(s_m|\boldsymbol{v}_{1:t}^{[S]})} = \frac{p_{\mathrm{I}}^{[\mathrm{Bys}]}(s_i|\boldsymbol{v}_{1:0}^{[S]})}{p_{\mathrm{I}}^{[\mathrm{Bys}]}(s_m|\boldsymbol{v}_{1:0}^{[S]})} \exp\left(-t(\tilde{L}_{1:t}^{[s_i]} - \tilde{L}_{1:t}^{[s_m]})\right) \rightarrow \begin{cases} 1, & s_i = s_m \\ 0, & s_i \neq s_m \end{cases}$$
$$(9)$$

because $\sum_{s \in S} p_{\mathrm{I}}^{[\mathrm{Bys}]}(s|\boldsymbol{v}_{1:t}^{[S]}) = 1$. This indicates that we will have a large $p_{\mathrm{I}}^{[\mathrm{Bys}]}(s_m|\boldsymbol{v}_{1:t}^{[S]})$ for a registered speaker s_m even when the current speech is of an unregistered speaker.

2.3 Gibbs-Distribution-Based Extended Bayesian Inference (GEBI)

Instead of (7), let us use the following probability given by Gibbs distribution:

$$p_{\mathrm{I}}^{[\mathrm{Gbs}]}\left(s|\boldsymbol{v}_{1:t}^{[S]}\right) \triangleq \frac{1}{Z_t} \exp\left(-\beta\left(\tilde{L}_{1:t}^{[s]} + \frac{1}{t} \log p_{\mathrm{I}}^{[\mathrm{Bys}]}(s|\boldsymbol{v}_{1:0}^{[S]})\right)\right) , \quad (10)$$

where β is a parameter called inverse temperature. Then, for the increase of t, the ratio of $p_{\mathrm{I}}^{[\mathrm{Gbs}]}(s|\boldsymbol{v}_{1:t}^{[S]})$ for $s = s_i$ to $s_m = \underset{s_i \in S}{\operatorname{argmax}}\, p_{\mathrm{I}}^{[\mathrm{Gbs}]}(s_i|\boldsymbol{v}_{1:t}^{[S]})$ converges to a constant value less than 1 as follows;

$$r_i^{[\mathrm{Gbs}]} \triangleq \frac{p_{\mathrm{I}}^{[\mathrm{Gbs}]}(s_i|\boldsymbol{v}_{1:t}^{[S]})}{p_{\mathrm{I}}^{[\mathrm{Gbs}]}(s_m|\boldsymbol{v}_{1:t}^{[S]})} \rightarrow \exp\left(-\beta(\tilde{L}_{1:t}^{[s_i]} - \tilde{L}_{1:t}^{[s_m]})\right) \rightarrow c_i^{\beta} < 1 . \quad (11)$$

Thus, we can avoid the problem of BI described above. Moreover, we can see that c_i is the converged value of the geometric mean of primitive likelihood ratios, or

$$c_i = \lim_{t \to \infty} \left(\prod_{k=1}^{t} \frac{L_k^{[s_i]}}{L_k^{[s_m]}} \right)^{1/t} = \lim_{t \to \infty} \left(\prod_{k=1}^{t} \frac{p(v_k^{[S]}|s_i)}{p(v_k^{[S]}|s_m)} \right)^{1/t} . \tag{12}$$

Therefore, the converged $r_{m_2}^{[\text{Gbs}]} = c_{m_2}^{\beta}$ for $s_{m_2} = \underset{s \in S \setminus \{s_m\}}{\arg\max} \; p_{\text{I}}^{[\text{Gbs}]}(s|v_{1:t}^{[S]})$ is considered to be smaller for the speech of a registered speaker than for unregistered speaker because it is expected that the likelihood of s_m for the former is bigger than for the latter while the likelihood of s_{m_2} is almost the same for both cases.

Here, from (10), we derive the following stepwise inference,

$$p_{\text{I}}^{[\text{Gbs}]}\left(s|v_{1:t}^{[S]}\right) = \frac{1}{Z_t} p_{\text{I}}^{[\text{Gbs}]}\left(s|v_{1:t-1}^{[S]}\right)^{\beta_t/\beta_{t-1}} p\left(v_t^{[S]}|s\right)^{\beta_t} , \tag{13}$$

where $\beta_t = \beta/t$ $(t \geq 1)$ and $\beta_0 = 1$. Note that the conventional BI is given by $\beta_t = 1$ $(t \geq 0)$, so that we name the above inference by Gibbs-distribution-based extended Bayesian inference (GEBI).

2.4 Introducing "Void" Speaker and Multistep Identification Procedure

The above GEBI as well as BI still suppose that the speech signal is spoken by a registered speaker because $\sum_{s \in S} p_{\text{I}}^{[\text{Gbs}]}(s|v_{1:t}^{[S]}) = 1$ and $\sum_{s \in S} p_{\text{I}}^{[\text{Bys}]}(s|v_{1:t}^{[S]}) = 1$. To deal with unregistered speaker, we introduce "void" speaker, s_{void}, and let $S_+ \triangleq S \cup \{s_{\text{void}}\}$. Furthermore, let us suppose the probability for the "void" speaker is given by the mean probability of registered speakers as follows;

$$p(v^{[s_i]} = 1|s) = \begin{cases} \left\langle p(v^{[\xi]} = 1|s) \right\rangle_{\xi \in S} , & \text{for } s_i = s_{\text{void}} \wedge s \neq s_{\text{void}} , \\ \left\langle p(v^{[s_i]} = 1|\xi) \right\rangle_{\xi \in S} , & \text{for } s_i \neq s_{\text{void}} \wedge s = s_{\text{void}} , \\ \left\langle p(v^{[\xi]} = 1|\xi) \right\rangle_{\xi \in S} , & \text{for } s_i = s = s_{\text{void}} , \end{cases} \tag{14}$$

where $\langle \cdot \rangle$ indicates the mean and the subscript indicates the range of the mean. Here, note that the probability of "void" speaker has been introduced differently in [5], but the above probability has provided better results in our experiments shown below. Furthermore, suppose the classifier to identify s_{void} has the output at t as

$$v_t^{[s_{\text{void}}]} = \begin{cases} 1, & \text{if } v_t^{[s_i]} = -1 \text{ for all } s_i \in S , \\ -1, & \text{otherwise} . \end{cases} \tag{15}$$

Here, we can see that in order to have $v_t^{[s_{\text{void}}]} = 1$, we had better reduce false positives (FPs), or $v_t^{[s_i]} = 1$, for all $s_i \in S$. To have this done, we tune y_θ in (3) by

$$y_\theta = \alpha_y \langle y_j^{[s]} \rangle_j + (1 - \alpha_y) \langle y_j^{[\bar{s}]} \rangle_j , \tag{16}$$

Table 1. Multistep speaker identification procedure for BI and GEBI. Here, t_B, t_E, $p_{I\theta}$ and $r_{m_2\theta}$ are constants, and $p_I(s|\boldsymbol{v}_{1:t}^{[S_+]})$ indicates either $p_I^{[Bys]}(s|\boldsymbol{v}_{1:t}^{[S_+]})$ or $p_I^{[Gbs]}(s|\boldsymbol{v}_{1:t}^{[S_+]})$.

---

step 1: Set $p_I(s|\boldsymbol{v}_{1:0}^{[S_+]}) := 1/|S_+|$ and $t := 1$.

step 2: Calculate $p_I(s|\boldsymbol{v}_{1:t}^{[S_+]})$.

step 3: If $t < t_B$, set $t := t + 1$ and go to **step 2**. Otherwise, if $p_I(s_{m_+}|\boldsymbol{v}_{1:t}^{[S_+]}) \geq p_{I\theta}$ for
$s_{m_+} = \underset{s \in S_+}{\mathrm{argmax}} \; p_I(s|\boldsymbol{v}_{1:t}^{[S_+]})$, decide the identification number as ID $= m_+$ and quit. If
$t < t_E$, set $t := t + 1$ and go to **step 2**. Otherwise, go to **step 4**.

step 4: If $r_{m_2} = \dfrac{p_I(s_{m_2}|\boldsymbol{v}_{1:t}^{[S_+]})}{p_I(s_m|\boldsymbol{v}_{1:t}^{[S_+]})} < r_{m_2\theta}$ for $s_m = \underset{s \in S}{\mathrm{argmax}} \; p_I(s|\boldsymbol{v}_{1:t}^{[S_+]})$ and
$s_{m_2} = \underset{s \in S\backslash\{s_m\}}{\mathrm{argmax}} \; p_I(s|\boldsymbol{v}_{1:t}^{[S_+]})$, decide ID $= m$. Otherwise, decide ID $=$ void.
Quit.

---

where α_y indicates the ratio of the mean output $\langle y_j^{[s]} \rangle_j$ of positive instances and $\langle y_j^{[\bar{s}]} \rangle_j$ of negative instances in the training dataset. We can expect that a large α_y reduces FPs and increases false negatives (FNs) for unknown test data (see **3.2** for details).

Now, Table 1 shows the procedure of multistep speaker identification for both BI and GEBI to identify registered speakers and reject unregistered speaker by means of the identification number ID $=$ void.

3 Numerical Experiments

3.1 Experimental Setting

We have used the same speech dataset examined in [5], where the speech data are sampled with 8kHz of sampling rate and 16 bits of resolution in a silent room of our laboratory. They are from five male speakers: $S =\{$SM, SS, TN, WK, YM$\}$. They involve five texts of Japanese words: $W = \{$/kyukodai/, /daigaku/, /kikai/, /fukuokaken/, /gakusei/$\}$ where each utterance duration of the words is about 1s. For each speaker and each text, we have ten samples of speech data, $L = \{1, 2, \cdots, 10\}$. So, we denote each speech data element as $x = x_{s,w,l}$ for $s \in S$, $w \in W$ and $l \in L$.

In order to evaluate the performance of the present method, we use the leave-one-out cross-validation (LOOCV). Precisely, for text-dependent tasks, we evaluate the performance with test dataset $X(S, w, l) = \{x_{s,w,l} \mid s \in S\}$ and training dataset $X(S, w, L_{\bar{l}}) = \{x_{s,w,i} \mid s \in S, i \in L\backslash\{l\}\}$ for each $w \in W$ and $l \in L$. On the other hand, for text-independent tasks, we use test dataset $X(S, w, l)$ and training dataset $X(S, W_{\overline{w}}, L) = \{x_{s,u,i} \mid s \in S, u \in W\backslash\{w\}, i \in L\}$ for each $w \in W$ and $l \in L$.

3.2 Experimental Results and Analysis

We have conducted experiments and obtained the error rates shown in Table 2. We can see that the error rate E_I of singlestep identification executed by (2) is not zero, but we have achieved the multistep identification error to be zero, or $E_I^{[Bys]} = E_I^{[Gbs]} = 0$,

Table 2. Error rates obtained in the experiments of speaker identification procedure shown in Table 1. E_{I} is the singlestep identification error rate. $E_{\mathrm{V[FN]}}$ and $E_{\mathrm{V[FP]}}$, respectively, are FN and FP rates of single step verification. $E_{\mathrm{I}}^{[\mathrm{Bys}]}$ and $E_{\mathrm{I}}^{[\mathrm{Gbs}]}$ indicate the multistep identification error rates for BI and GEBI, respectively, where the decision of identification is done at $t = t_{\mathrm{D}}$ in between $t_{\mathrm{B}} = 4$ and $t_{\mathrm{E}} = 10$. The column of n^{test} indicates the number of test data, and $\langle t_{\mathrm{D}}\rangle$ the mean t_{D}. The test of "speaker involving unregistered" is executed by LOOCV which leaves each classifier for $s_i \in S$ out as an unregistered speaker. The results of GEBI are obtained with the inverse temperature $\beta = 1$.

	/kyukodai/	/daigaku/	/kikai/	/fukuokaken/	/gakusei/	n^{test}	$\langle t_{\mathrm{D}}\rangle$
(a) text-dependent							
		singlestep identification and verification					
E_{I}	0.080	0.060	0.060	0.100	0.040	50	1
$E_{\mathrm{V[FP]}}$	0.005	0	0.010	0.015	0	200	1
$E_{\mathrm{V[FN]}}$	0.200	0.180	0.340	0.240	0.300	50	1
		speakers all registered					
$E_{\mathrm{I}}^{[\mathrm{Bys}]}$	0	0	0	0	0	5	5.04
$E_{\mathrm{I}}^{[\mathrm{Gbs}]}$	0	0	0	0	0	5	5.16
		speakers involving unregistered					
$E_{\mathrm{I}}^{[\mathrm{Bys}]}$	0	0	0	0	0	25	5.208
$E_{\mathrm{I}}^{[\mathrm{Gbs}]}$	0	0	0	0	0	25	5.792
(b) text-independent							
		single step identification and verification					
E_{I}	0.200	0.200	0.280	0.400	0.420	50	1
$E_{\mathrm{V[FP]}}$	0.010	0.010	0.020	0.060	0.050	200	1
$E_{\mathrm{V[FN]}}$	0.360	0.300	0.400	0.340	0.320	50	1
		speakers all registered					
$E_{\mathrm{I}}^{[\mathrm{Bys}]}$	0	0	0	0	0	5	6.84
$E_{\mathrm{I}}^{[\mathrm{Gbs}]}$	0	0	0	0	0	5	5.64
		speakers involving unregistered					
$E_{\mathrm{I}}^{[\mathrm{Bys}]}$	0	0	0	0	0	25	6.832
$E_{\mathrm{I}}^{[\mathrm{Gbs}]}$	0	0	0	0	0	25	6.112

for all cases. Here, note that our previous Bayesian multistep identification procedure shown in [5] could not achieve the error to be zero for some cases. As one of the reasons, we have tuned the FP rate $E_{\mathrm{V[FP]}}$ to be smaller than the FN rate $E_{\mathrm{V[FN]}}$ for the verification (two class classification) executed by (3) for both text-dependent and text-independent tests in the present experiments. Precisely, we have tuned α_y in (16) to 0.9 by means of increasing from 0.5 for reducing the identification error from the following points of view. First, since $p_{\mathrm{I}}^{[\mathrm{Bys}]}(s|\boldsymbol{v}_{1:t}^{[S]})$ and $p_{\mathrm{I}}^{[\mathrm{Gbs}]}(s|\boldsymbol{v}_{1:t}^{[S]})$ increase with the increase of positive responses consisting of FPs and TPs (true positives) for the sth classifier because the TP rate $p(v^{[s]} = 1|s)$ $(= 1 - (\text{FN rate}) > 0.5)$ for s increases $p(\boldsymbol{v}^{[S]}|s)$ in (5) even when it is FP. Thus, the increase of α_y contributes to reducing FPs and erroneous identifications, while reducing TPs to delay the decision. Actually, Table 2 shows that the mean decision time $\langle t_{\mathrm{D}}\rangle$ is bigger than the results shown in [5]. Next, the increase of α_y is considered to be necessary for "void" speaker as shown in **2.4**, and the above increased value has been effective to have the results for "speakers involving unregistered" test in Table 2.

Table 3. Probabilities $p_{I[t_0:t_1]}^{[Bys]}$ and $p_{I[t_0:t_1]}^{[Gbs]}$ in text-independent speaker identification for the speech data /fukuokaken/ of unregistered speaker SS. Here, $[t_0 : t_1]$ indicates the range of step time t. The values of y_i and v_i for SS are shown for analysis but unused for identification. We use $y_i(t+10) = y_i(t)$ for $t = 1, 2, \cdots, 10$ for analysis. The thresholds $p_{I\theta}^{[Bys]} = 0.9997$ and $p_{I\theta}^{[Gbs]} = 0.61$ are applied from $t = t_B = 4$ to $t_E = 10$, and $r_{m_2\theta}^{[Bys]} = 10^{-4}$ and $r_{m_2\theta}^{[Gbs]} = 0.41$ at $t = t_E$.

| | | y_i | | | | | v_i | | | | | | | | $P_{I[0:10]}^{[Bys]}$ | | | |
t	SM	SS	TN	WK	YM	SM	SS	TN	WK	YM	void		t	SM	TN	WK	YM	void
0	–	–	–	–	–	–	–	–	–	–	–		0	.200000	.200000	.200000	.200000	.200000
1	+.05	-.22	-.69	-.98	-.96	-1	+1	-1	-1	-1	+1		1	.105802	.109924	.109246	.039046	.635981
2	-.56	+.21	-.53	-.97	-.98	-1	+1	-1	-1	-1	+1		2	.025371	.027387	.027050	.003455	.916736
3	-.68	-.23	-.06	-.95	-.93	-1	+1	+1	-1	-1	-1		3	.003273	.464255	.095685	.000102	.436685
4	-.52	-.81	-.89	-.78		-1	-1	-1	-1	-1	+1		4	.001020	.150291	.030784	.000012	.817893
5	-.58	-.65	+.18	-.97	-1.00	-1	-1	+1	-1	-1	-1		5	.000043	.836318	.035746	.000000	.127893
6	-.56	-.29	-.07	-.96	-.95	-1	+1	+1	-1	-1	-1		6	.000000	.986956	.008803	.000000	.004241
7	-.46	-.43	-.08	-.94	-.95	-1	-1	+1	-1	-1	-1		7	.000000	.998022	.001858	.000000	.000121
8	-.75	-.15	-.11	-.86	-.93	-1	+1	+1	-1	-1	-1		8	.000000	**.999608**	.000388	.000000	.000003
9	-.42	-.31	-.61	-.77	-.83	-1	+1	-1	-1	-1	+1		9	.000000	.999595	.000386	.000000	.000020
10	-.37	-.48	-.75	-.65	-.75	-1	-1	-1	+1	-1	-1		10	.000000	.959772	.040178	.000000	.000049

$$r_{m_2}^{[Bys]} = \frac{.040178}{.959772} \simeq .042$$

| | $P_{I[0:10]}^{[Gbs]}$ | | | | | | $P_{I[10:20]}^{[Gbs]}$ | | | | | | | $P_{I[10:20]}^{[Bys]}$ | | | |
t	SM	TN	WK	YM	void	t	SM	TN	WK	YM	void	t	SM	TN	WK	YM	void
0	.200	.200	.200	.200	.200	10	.039	.452	.329	.011	.168	10	.000000	.959772	.040178	.000000	.000049
1	.106	.110	.109	.039	.636	11	.046	.423	.317	.013	.202	11	.000000	.959784	.039931	.000000	.000285
2	.106	.110	.109	.039	.636	12	.051	.396	.304	.015	.233	12	.000000	.958715	.039640	.000000	.001645
3	.068	.354	.209	.021	.347	13	.046	.436	.303	.013	.203	13	.000000	.991398	.008554	.000000	.000048
4	.080	.279	.188	.026	.427	14	.050	.412	.293	.015	.230	14	.000000	.991224	.008499	.000000	.000277
5	.058	.417	.222	.018	.286	15	.045	.446	.292	.013	.204	15	.000000	.998206	.001786	.000000	.000008
6	.043	.508	.231	.013	.205	16	.041	.475	.290	.012	.182	16	.000000	.999627	.000373	.000000	.000000
7	.034	.568	.231	.010	.157	17	.038	.501	.287	.010	.165	17	.000000	**.999922**	.000078	.000000	.000000
8	.028	**.609**	.228	.008	.126	18	.034	.523	.283	.009	.150	18	.000000	**.999984**	.000016	.000000	.000000
9	.036	.555	.232	.010	.166	19	.038	.502	.281	.011	.169	19	.000000	**.999984**	.000016	.000000	.000000
10	.039	.452	.329	.011	.168	20	.039	.452	.329	.011	.168	20	.000000	.998251	.001749	.000000	.000000

$$r_{m_2}^{[Gbs]} = \frac{.329}{.452} \simeq .73 \qquad r_{m_2}^{[Gbs]} = \frac{.329}{.452} \simeq .73 \qquad r_{m_2}^{[Bys]} = \frac{.001749}{.998251} \simeq .0018$$

Although we have achieved $E_I^{[Bys]} = 0$, we have had to set the threshold $p_{I\theta}^{[Bys]}$ to be very large as 0.9997 as described in **2.2**. To analyze more, we show Table 3, where the speech is of SS who is unregistered in this case. We can see that it is a difficult case for rejecting the classifier TN because v_i of TN shows five FPs, while the correct classifier SS shows seven TPs and is correctly identified in "speakers all registered" test as shown in Table 2. We see that $p_{I\theta}^{[Bys]} = 0.9997$ rejects $P_I^{[Bys]} = .999608$ for the classifier TN at $t = 8$ and "void" is identified at $t = t_E$ because $r_{m_2}^{[Bys]} \geq r_{m_2\theta}^{[Bys]}$ at $t = t_E$ (see **step 4** in Table 1). Here, we use $r_{m_2\theta}^{[Bys]} = 10^{-4}$ which is smaller than the minimum value of $r_{m_2}^{[Bys]} = 4.1 \times 10^{-4}$ for rejecting unregistered speaker and bigger than the maximum value $r_{m_2}^{[Bys]} = 5.0 \times 10^{-5}$ for identifying registered speakers among all text-independent tests. Although these threshold values $p_{I\theta}^{[Bys]} = 0.9997$ and $r_{m_2\theta}^{[Bys]} = 10^{-4}$ are effective for $t_E = 10$, we had to change them for $t_E = 20$, not only $p_{I\theta}^{[Bys]}$ as shown in Table 3 but also $r_{m_2\theta}^{[Bys]}$ for other cases.

On the other hand, we have achieved correct identification with $p_{I\theta}^{[Gbs]} = 0.61$ and $r_{m_2\theta}^{[Bys]} = 0.41$ for both $t_E = 10$ and 20. Here, note that Table 3 shows the result for the inverse temperature $\beta = 1$. For different β, the threshold $r_{m_2\theta}^{[Bys]}$ is derived by $r_{m_2\theta}^{[Bys]} = c_{m_2\theta}^\beta = 0.41^\beta$ (see (11) and (12)), while the threshold $p_{I\theta}^{[Gbs]}$ should be larger for large β because $p_I^{[Gbs]}(s|v_{1:t}^{[S]})$ for increasing t changes faster as is expected from (10). As a result, we can reduce the mean decision time $\langle t_D \rangle$, but the reduction is not so big. Actually, for $\beta = 2$ and 4, respectively, in the text-independent tests, we have $\langle t_D \rangle = 5.36$ and 5.32 ("speakers all registered") and $\langle t_D \rangle = 5.63$ and 5.46 ("speakers involving unregistered") with $r_{m_2\theta}^{[Bys]} = 0.17(\simeq 0.41^2)$ and $0.028(\simeq 0.41^4)$ and $p_{I\theta}^{[Gbs]} = 0.85$ and 0.98. As a result, we can say that GEBI has better properties than BI in multistep speaker identification for rejecting unregistered speaker.

4   Conclusion

We have presented a method of multistep speaker identification using GEBI for rejecting unregistered speaker. We have analyzed to show the problem of BI in multistep identification for rejecting unregistered speaker and introduced GEBI to solve the problem. We have examined the effectiveness of the present method using real speech signals. We would like to examine the performance of the present method with larger database and develop an online processing system in our future research studies.

Acknowledgements. This work was partially supported by the Grant-in Aid for Scientific Research (C) 24500276 of the Japanese Ministry of Education, Science, Sports and Culture.

References

1. Ahalt, A.C., Krishnamurthy, A.K., Chen, P., Melton, D.E.: Competitive learning algorithms for vector quantization. Neural Networks 3, 277–290 (1990)
2. Kohonen, T.: Associative Memory. Springer (1977)
3. Kurogi, S., Ueno, T., Sawa, M.: A batch learning method for competitive associative net and its application to function approximation. In: Proc. SCI 2004, vol. 5, pp. 24–28 (2004)
4. Kurogi, S., Mineishi, S., Sato, S.: An Analysis of Speaker Recognition Using Bagging CAN2 and Pole Distribution of Speech Signals. In: Wong, K.W., Mendis, B.S.U., Bouzerdoum, A. (eds.) ICONIP 2010, Part I. LNCS, vol. 6443, pp. 363–370. Springer, Heidelberg (2010)
5. Kurogi, S., Mineishi, S., Tsukazaki, T., Nishida, T.: Naive Bayesian Multistep Speaker Recognition Using Competitive Associative Nets. In: Lu, B.-L., Zhang, L., Kwok, J. (eds.) ICONIP 2011, Part I. LNCS, vol. 7062, pp. 70–78. Springer, Heidelberg (2011)
6. Campbell, J.P.: Speaker Recognition: A Tutorial. Proc. the IEEE 85(9), 1437–1462 (1997)
7. Furui, S.: Speaker Recognition. In: Cole, R., Mariani, J., et al. (eds.) Survey of the State of the Art in Human Language Technology, pp. 36–42. Cambridge University Press (1998)
8. Hasan, M.R., Jamil, M., Rabbani, M.G., Rahman, M.S.: Speaker identification using Mel frequency cepstral coefficients. In: Proc. ICECE 2004, pp. 565–568 (2004)
9. Bocklet, T., Shriberg, E.: Speaker recognition using syllable-based constraints for cepstral frame selection. In: Proc. ICASSP (2009)
10. Beigi, H.: Fundamentals of speaker recognition. Springer-Verlag New York Inc. (2011)
11. Zhang, H.: The optimality of naive Bayes. In: Proc. FLAIRS 2004 Conference (2004)

Chinese HowNet-Based Multi-factor Word Similarity Algorithm Integrated of Result Modification

Benbin Wu[1], Jing Yang[2,*], and Liang He[2]

Department of Computer Science and Technology, East China Normal University
Shanghai, 200241, P.R. China
bbwu@ica.stc.sh.cn {jyang,lhe}@cs.ecnu.edu.cn

Abstract. In this paper, we firstly describe a novel approach to calculate the Chinese sememe similarity based on the HowNet hierarchical sememe tree. When we calculate the sememe similarity, we not only take Semantic Distance, Node Depth and Semantic Coincidence Degree into consideration, but also propose two impact factors named Node Environment Dense (NED) and Node Layer Ratio (NLR) to optimize the calculation process. Secondly, quite a few words described by identical concept definition in HowNet should have a certain discrimination according to human perception, so we propose a hybrid modification algorithm integrated of TongYiCi CiLin (hereinafter, CiLin) to deal with this case. Experiment results of the HowNet-based multi-factor similarity hybrid algorithm shows that this approach improves the similarity of independent sememe words and the words having identical concept descriptions in HowNet, while no large bias influence on the similarity of other words.

Keywords: HowNet, TongYiCi CiLin, Density, Word Discrimination.

1 Introduction

Word similarity is an essential problem in the research fields of natural language process, information retrieval, text classification and so on[1]. In the field of Chinese word similarity study, it's popular to be carried out based on a semantic resource named HowNet. Liu and Li have proposed word similarity algorithm respectively [2, 3], and these algorithms are applied to further study or practice to a certain extent.

HowNet[4] is a large-scale detailed semantic knowledge dictionary created by Professor Dong. In HowNet, a word's concept definition(DEF), which is represented by Chinese or English words, is used to describe the relations between DEFs and the attributes of DEFs are revealed. A key term used in HowNet is sememe, which refer to some basic unit of sense and are organized on several hierarchical sememe trees. Liu proposed an algorithm to synthesize the similarity of words by that of their DEFs, and the similarity of DEFs is also synthesized by sememe similarity results [2]. Although this algorithm makes full use of the rich semantic information for each concept description and most of the similarity results are in line with human perception, there are still two shortcomings in Liu's approach.

---

*Corresponding author.

T. Huang et al. (Eds.): ICONIP 2012, Part V, LNCS 7667, pp. 256–266, 2012.

1). Owing to only Semantic Distance(SD) factor is considered sememe similarity calculation, similarity result of independent sememe words cannot receive a reasonable evaluation. The term Independent Sememe Word defines the words that also sememes in HowNet themselves. In Fig.1, sememe groups (animate, shape), (human, animal) and (crop, tree) will get the same similarity result because of their identical SD in the tree. There is a deviation between these results and hierarchical tree theory. The words tested here are just shown in English, while their corresponding Chinese words can be seen in Fig.1 or Fig.2.

2). Liu's approach simply computes similarity of the words with the same DEF as 1, which means that they are meaningfully identical. According to the theory, group (dad, brother) will get the similarity as 1.0 for these two words share the same DEF in HowNet, so there is no discrimination between them.

For shortcoming 1), lots of advanced methods have been proposed by scholars. Jiang Min [5] and Xia Tian [6] took Node Depth(ND) and Semantic Coincidence Degree(SCD) into consideration respectively, Shi Bin [7] proposed Node District Dense (NDD) factor and also proposed a sememe similarity calculation model based on the SD, SCD and NDD. But scholars did not fully consider the combined effect of various factors, the NDD factor proposed in [6] only make the son nodes of the given node in a tree play a role in similarity calculation, not take the effect of the brother nodes into consideration. So this paper proposes a factor named Node Environment Dense (NED) that contains effects of both node's son and brother nodes. As in Fig.1, NED(human, animal) is less than NED(vegetable, fruit), which means that similarity of the former less than that of the latter according to NED factor. Also, group (vegetable, fruit) gets the same SD as word group(vegetable, animate), but layer difference of the former is less than the latter, theoretically the similarity of (vegetable, fruit) should be greater than (vegetable, animate). So this paper proposes a factor named Node Layer Ratio (NLR) and eventually proposes a multi-factor sememe similarity calculation algorithm that contains SD, ND, SCD, NED and NLR factors.

For shortcoming 2), the words 'dad', 'grandpa', 'YoungerBrother', 'ElderBrother' and 'fellow' share the same DEF in HowNet as : DEF={human, family, male}.

According to the Liu's theory, these words will get the same similarity as 1.0 because of their identical DEF in HowNet, but they show no discrimination between. Other semantic resource should be introduced to deal with this problem, the resource chosen should not exert much influence on other words' similarity, while modifying the words with the same DEF in HowNet. Basing on this principle, the Chinese information resource CiLin is considered for its entry code of each word in CiLin is organized in the layered architecture and   the CiLin layer architecture is shown in Section 4. For example, the entry code of 'dad' and 'ElderBrother' is respectively 'Ah04A01=' and 'Ah09A01=' .Here we can see these two entry codes have the code difference begins at layer 3, which means certain discrimination. So with the introduction of CiLin some indistinguishable words in HowNet will receive discrimination to some extent, and the similarity will be in line with human perception. Simultaneously the similarity of some sememes, which in different HowNet hierarchical sememe trees, will be influenced by their identical CiLin entry codes, which means a degree of similarity, such as word group (thought, think).

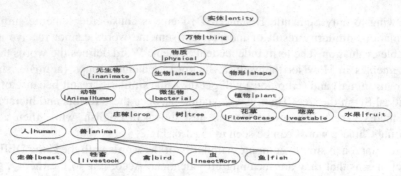

Fig. 1. Sememe tree with root 'entity'

The rest of this paper has the following structure. Section 2 introduces HowNet-based Chinese similarity algorithm adapted in this paper. Section 3 presents definition of factors used in our similarity algorithm and the proposed multi-factor similarity formula. Section 4 describes the hybrid modification algorithm integrated of CiLin, while Section 5 explains the conducted experiments and obtained results. Section 6 ends with some conclusions and future research lines. The words tested here will be just shown in English, their corresponding Chinese words can be checked in figures.

2   HowNet-Based Chinese Word Similarity Algorithm

Similarity means the similar degree between two words and can be represented as a real number in domain [0, 1]. Two words with identical DEF will get the similarity of 1.0. This paper follows the train of thought of Liu [2] that the similarity of two isolated words (based on no context) is their maximum value in all the concept similarity.

$$\text{Sim}(W_1, W_2) = \max Sim(C_{1i}, C_{2j}) \quad (i = 1 \cdots n, j = 1 \cdots m)$$

where C_{1i} is the ith DEF in W_1's n concepts, C_{2j} is the jth DEF in W_2's m concepts.

Li Feng's concept definition(DEF) similarity calculation method in [3] is adapted by this paper, and our multi-factor sememe similarity model to be propose in Section 3 will be used during the concept similarity process.

3   Multi-factor Sememe Similarity Calculation Model

In Section 3.1, we draws an initial formula with factors including SD, ND and SCD, meanwhile, these factors' definition are given in this section. Then in Section 3.2, we will propose two novel impact factors and the final similarity calculation model.

3.1   Initial Similarity Calculation Model

Definition 1. Semantic Distance(SD). SD means the number of edges to contact with two nodes (A, B) in a tree, and expressed as distance(A,B), e.g. distance(animal, plant)=3 in Fig.1. If two sememe nodes are not in the same tree, we define their similarity as a small value.

Definition 2. Node Depth(ND). ND means the layer of node A in a tree, and expressed as depth(A), e.g. Depth(plant)=5 in Fig.1. Here the root node depth is defined as 1. The deeper the node in hierarchical tree is, the more refined and specific concept definition gets and the greater their similarity is.

Jiang Min took ND into consideration while she calculated the sememe similarity based on HowNet [5], and her formula is:

$$Sim(C_1, C_2) = \frac{\alpha \times (h_1 + h_2)}{\alpha \times (h_1 + h_2) + distance(C_1, C_2) + |h_1 - h_2|}$$

where h_1 and h_2 are the ND of nodes C_1 and C_2 respectively. α is a parameter.

Definition 3. Semantic Coincidence Degree(SCD). SCD means the number of common ancestor nodes of nodes A and B in the tree, and expressed as $I_{share}(A, B)$, e.g. $I_{share}(animal, plant) = 4$ in Fig.1.

SCD shows the same degree of two nodes. Resnik[8] proposed a view that the maximum information of two nodes' common ancestor nodes can be used to measure their similarity. The more information two nodes share, the larger similarity they get.

According to the notions above, we get the preliminary sememe similarity calculation model between node C_1 and C_2 as below:

$$Sim'(C_1, C_2) = \frac{I_{share}(C_1, C_2) \times \min(depth(C_1), depth(C_2))}{I_{share}(C_1, C_2) \times \min(depth(C_1), depth(C_2)) + distance(C_1, C_2)} \tag{1}$$

3.2 Multi-factor Sememe Similarity Model

As a result of nodes' maldistribution in the hierarchical tree, the sparsity of tree may influence the similarity of two nodes. As the cases we have elaborated in Section 1, Owing to the factor Node District Dense proposed in [7] considers only the influence of one node's son nodes but ignores that of its brother nodes, here we define an improved dense factor and then propose our dense impact factor.

Definition 4. Node Environment Dense(NED). NED means sum of the number of node A's son nodes and that of its brother nodes, and expressed by dense(A). It's generally acknowledged that the larger the NED of two nodes' closest common ancestor node, the less their semantic distance is, and the larger their similarity will be. According to the idea mentioned above, we introduce a dense impact factor ρ, this factor can be used as a weight assigned to the Semantic Distance and define as:

$$\rho = \frac{\overline{dense}}{dense(c_r)} = \frac{\frac{1}{n}\sum_{i=1}^{n} dense(c_i)}{dense(c_r)} \tag{2}$$

where $\overline{dense}$ is the mean NED value of all the nodes in the tree. dense(c_r) is the NED of two nodes' closest common ancestor node c_r. In Fig.1, the closest common ancestor node of 'vegetable' and 'fruit' is 'plant', while that of 'human' and 'animal' is 'AnimalHuman'. According to formula (2),

$$\rho(\text{vegetable}, \text{fruit}) = 0.790, \quad \rho(\text{human}, \text{animal}) = 1.422.$$

As a weight, ρ will be assigned to the semantic distance, so:

$$\text{distance}(\text{vegetable}, \text{fruit}) = 0.790 \times 2 = 1.580$$
$$\text{distance}(\text{human}, \text{animal}) = 1.422 \times 2 = 2.844$$

The SD of the former is smaller than that of the latter, so the similarity of the former will be greater than the latter and this accords with the tree structure theory.

Also in Fig.1, word group (vegetable, fruit) gets the same semantic distance as word group (vegetable, animate), but layer difference of the former is less than the latter, theoretically the similarity of (vegetable, fruit) should greater than (vegetable, animate). Therefore this paper proposes a factor named node layer ratio (NLR) defined as blow:

Definition 5. Node Layer Ratio (NLR). NLR is equaled to the ratio of the mean layer and the maximum layer, and expressed as θ:

$$\theta = \frac{avg(depth(C_1), depth(C_2))}{max(depth(C_1), depth(C_2))}$$

Based on the preliminary sememe similarity calculation model proposed in Section 3.1 and the factors θ and ρ, we get the multi-factor similarity calculation model as blow:

$$\text{Sim}(C_1, C_2) = \theta \times \frac{I_{share}(C_1, C_2) \times min(depth(C_1), depth(C_2))}{I_{share}(C_1, C_2) \times min(depth(C_1), depth(C_2)) + \rho \times distance(C_1, C_2)} \tag{3}$$

4   Hybrid Modification Algorithm

For the second shortcoming of HowNet described in Section 1, this paper introduces TongYiCi CiLin to modify the HowNet-based similarity result. Here we make a brief introduction to CiLin, and then propose the hybrid modification algorithm.

4.1   Introduction to TongYiCi CiLin

CiLin is a semantic dictionary of synonyms and related terms, which organizes all collected terms together in accordance with five-layer tree-structured hierarchy as Fig.2. As the layer increases, the term semantic description is more detailed, the terms in Layer 5 are individually bag-of-words named atom term group or atom node.

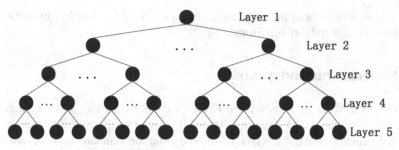

Fig. 2. Five-layer tree-structure of CiLin

CiLin supplies five layer coding pattern: Layer 1 and Layer 4 are indicated as a capital letter, Layer 2 is indicated as a lowercase letter, Layer 3 and Layer 4 are indicated as two digit decimal integer. Such as code 'Ae07C01=' stands for the term group {'fisherfolk', 'fisherman', 'fisherfolk', 'fisherman', 'fisherman'}, detailed coding layer segmentation is 'A/c/07/C/01='.

4.2 Methodology

If two words are actually synonyms, their similarity will be 1 not only in HowNet but also in CiLin. It shows discrimination when the similarity calculated based on either HowNet or CiLin is 1. To deal with this matter, a hybrid modification algorithm integrated HowNet and CiLin is proposed here, and the formula is:

$$\text{Sim}(A, D) = \varepsilon Sim_{hownet} + (1 - \varepsilon) Sim_{cilin} \tag{4}$$

where Sim_{hownet} and Sim_{cilin} are the similarity results based on HowNet or CiLin respectively. ε is a restrain factor and its value is discussed in Section 5.

The hybrid modification algorithm has two steps: Step 1, Calculating the similarity based on HowNet and CiLin by formula (3) separately. Step 2, According to formula (4) to form the final similarity result when one of the similarity in HowNet and CiLin is equivalent to 1, or the HowNet-based similarity is used as the final similarity.

The scenario that the similarity of words in HowNet is 1 and that in CiLin is not 1 means they should be discriminated. Words that 'dad', 'ElderBrother', 'grandpa' share the same DEF, which has showed in Section 1, results in the similarity between any two of them is 1 following Liu's algorithm in HowNet, but their different CiLin entry codes that 'dad=Ah04A01', 'ElderBrother =Ah09A01' , and 'grandpa=Ab02A04 or Ah02B01' perform their existing discrimination. So these words' HowNet-based similarity should be adjusted by formula (4).

Another scenario that the CiLin-based similarity is 1 but HowNet-based is not 1 means they are actually not synonyms but related terms. Hybrid modification algorithm can help tell the related terms from synonyms in CiLin atom term groups. Take 'thought' and 'think' for example, their identical CiLin entry codes mean their similarity or correlation, while their existing in different HowNet sememe hierarchical tree result in their small similarity in HowNet. It shows that 'thought' and 'think' should get certain correlation but not the very same concepts. So these words similarity result should also be adjusted through formula (4).

5   Experiments and Analysis

In consequence of using HowNet and CiLin, we only uses the words contained in both semantic resources. This paper follows Liu's word similarity calculation method based on no context, but there is lack of union evaluation standard, so we evaluate our experiments from different angles. And three experiments in this paper are :

1) Experiment aiming at the independent sememe words. We uses the contact of nodes in hierarchical tree to investigate the impact of factors NED and NLR, and compares some groups' similarity by our method with that by Liu' method.
2) Experiments aiming at the words with identical concept description in HowNet and those with identical CiLin entry codes.
3) Experiments aiming at the words not included in the above two circumstance. This paper using the word groups tested in Liu's algorithm to show the algorithm proposed here has no large impact on normal words.

5.1   Experiment One

In this experiment, formula (3) is used and the parameters used are the same as they defined   in [2, 3]. Here we show these parameters in Table 1.

Table 1. Parameters in experiment one

γ	0.0	Similarity of the word paired to null
δ	0.01	Similarity of sememes in different trees

This experiment is to investigate the impact of factors NED and NLR based on the tree showed in Fig.1 and Fig.3. There are 6 test groups, the former 3 groups are tested in the conditions of identical SD, ND, SCD and NLR factors, the latter 3 groups are tested in the conditions of identical SD, ND, SCD and NED factors.

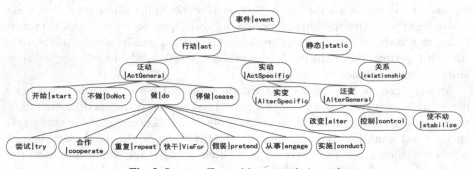

Fig. 3. Sememe Tree with root node 'event'

Table 2. Results of experiment one

Group	Word 1	Word 2	Liu	(3)
1	human	animal	0.348	0.813
	fruit	vegetable	0.444	0.950
2	start	relationship	0.186	0.222
	start	alter	0.242	0.379
3	alter	control	0.444	0.850
	cooperate	Repeat	0.444	0.948
4	human	crop	0.206	0.751
	human	inanimate	0.206	0.751
5	beast	inanimate	0.145	0.382
	beast	fruit	0.145	0.559
6	try	relationship	0.167	0.176
	try	alter	0.211	0.431

Take group 3 for example, the NED factor in group (alter, control) is smaller than that in group (cooperate, repeat), according to the NED theory, similarity of the former may smaller than that of the latter as Fig.1 shows. Compared with Liu's identical result 0.444, our method can reveal more about the dense impact on similarity.

In group 5, group (beast, inanimate) has the same SD as group (beast, fruit), but the NLR factor in the former is larger than that in the latter, so the similarity of (beast, inanimate) may have smaller similarity, which is also not shown in Liu's algorithm.

According to table 2, the similarity of words calculated using this paper's approach is more reasonable than that using Liu's method that considers single factor.

5.2   Experiment Two

In this experiment, hybrid modification algorithm proposed in Section 4 is used and the parameters used in this experiment are shown in Table 3.

Table 3. Parameters in experiment two

ε	**0.4**	Restrain Factor (if HowNet-based similarity is 1.0)
ε	**0.6**	Restrain Factor (if CiLin-based similarity is 1.0)

According to the idea discussed in Section 4, we assign the restrain factor with different value to have a test in similarity of groups shown in Fig.4.

It is shown that as the increasing value of ε makes the similarity get a trend to 1. When the HowNet-based similarity is 1, it should be assigned a small weight impact in formula (4). The value of restrain factor ε can be adjustable in different conditions. We assigns 0.4 to ε if HowNet-based similarity is 1, or it will be assigned 0.6.

Some special words with identical DEF in HowNet are enumerated in Table 4. These words' HowNet-based similarity is 1.0, after integrating with CiLin, the similarity shows the discriminations between each other. For example, group

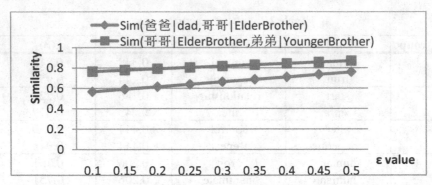

Fig. 4. Restrain factor test result

(ElderBrother, YoungerBrother) has a larger similarity than groups (dad, grandpa) and (dad, ElderBrother) by (4), this result is in line with people perception. The varieties of fish and fruit listed also get discrimination to an extent .

Table 4. Results of experiment two(1)

Word 1	Word 2	Same DEF	(4)	Word 1	Word 2	Same DEF	(4)
dad	grandpa	human, family, male	0.712	fish	hilsa	fish	0.812
Elder Brother	Younger Brother		0.843	sturgeon	crucian		0.808
dad	ElderBrother		0.710	butterfish	crucian		0.810
chestnut	fruit	fruit	0.808	apple	orange	fruit	0.816
chestnut	pineapple		0.811	apple	pear		0.807

Table 5. Results of experiment two(2)

Word 1	Word 2	Identical Entry Code	CiLin-based Similary	(4)
think	thought	Gb01B01=	1.0	0.406
straightforward	openly	Ee04A01=	1.0	0.932
surrender	compromise	Hi45B01=	1.0	0.902
stress	emphasize	Gb21A01=	1.0	0.487
PersonalLoyalty	sincere	Ee05D01=	1.0	0.733
fact	answer	Da21A02=	1.0	0.908
stress	value	Gb21A01=	1.0	0.952

Some words with identical CiLin coding are enumerated in Table 5. The similarity of words in CiLin is 1 and that in HowNet is not 1 means they are actually not synonyms but related terms. Hybrid modification algorithm can help tell the related terms from synonyms in CiLin atom term groups according to the results in Table 5. This result modification approach integrates both the roles of concept structure in HowNet and CiLin, which makes the similarity results tend to more reasonable.

5.3 Experiment Three

In table 6, we test some words that had been discussed in [2]. It shows that similarity generated by hybrid modification algorithm is approximate to Liu's result, most of the tested groups' similarity are greater than Liu's, it's resulted from sememes' five factors, which are taken into account in (3). It is also shown in table 6 that the hybrid algorithm doesn't change the actual synonyms' similarity, e.g. group(dad, father) still holds the similarity of 1.0. 'doctor' and 'patient' have a relation of doctor-patient, which only means associativity but not similarity. This case is also shown in the result of our algorithm and gets a smaller similarity than Liu's. This paper's algorithm not only has no large impact on overall words, but also outperforms Liu's algorithm to some extent, and the similarity is more reasonable.

Table 6. Results of experiment three

Word 1	Word 2	Liu	(4)	Word 1	Word 2	Liu	(4)
man	woman	0.833	0.976	invent	create	0.615	0.929
man	manager	0.657	0.704	dad	father	1.0	1.0
man	monk	0.833	0.916	teacher	worker	0.722	0.500
man	happy	0.013	0.166	doctor	patient	0.574	0.497

6 Conclusions

In this paper, we firstly propose the impact factors named Node Environment Dense and Node Layer Ratio, then introduce the multi-factor similarity model based on SD, ND, SCD, NED and NLR factors. Secondly, hybrid modification algorithm integrated with CiLin is proposed to make the similarity of words that share identical DEF in HowNet or identical CiLin entry codes more reasonable. Experiments show that our multi-factor similarity model and hybrid modification algorithm not only do well in the two shortcomings of Liu' algorithm, but also make other normal words' similarity closer to human sense. It's proved to be an effective improvement to Liu's algorithm.

The presented work is only the beginning of our exploration of how to make the Chinese word similarity more feasible. We plan to extend our analysis on the specialty of sememe structure in HowNet and investigate adaptation of Internet resources to obtain the discrimination between Chinese words.

Acknowledgments. This work is supported by the Shanghai Science and Technology commission Foundation (No. 10dz1500103, No. 11530700300, No. 11511504000).

References

1. Liu, Y.J., Xu, Y.: Automatic question answering system based on weighted semantic similarity model. Journal of Southeast University 34(5), 609–612 (2004)
2. Liu, Q., Li, S.J.: Word similarity computing based on How-net. Computational Linguistics and Chinese Language Processing 17(2), 59–74 (2002)

3. Li, F., Li, F.: An new approach measuring semantic similarity in HowNet 2000. Journal of Chinese Information Processing 21(3), 99–105 (2007)
4. Dong, Zh.D., Dong, Q.: HowNet (1999), http://www.keenage.com
5. Jiang, M., Xia, S.B., Wang, H.W.: An improved word similarity computing method based on HowNet. Journal of Chinese Information Processing 22(5), 84–88 (2008)
6. Xia, T.: Study on Chinese words semantic similarity computation. Computer Engineering 33(6), 191–194 (2007)
7. Shi, B., Yan, J.Z., Wang, P.: Ontology-based measure of semantic similarity between concepts. Computer Engineering 35(19), 83–85 (2009)
8. Resik, P.: Using information content to evaluate semantic similarity. In: Proceedings of the 14th International Joint Conference on Artificial Intelligence, pp. 448–453. IEEE Press, Montreal (1995)

Composite Data Mapping
for Spherical GUI Design: Clustering
of Must-Watch and No-Need TV Programs

Masaya Maejima, Ryota Yokote, and Yasuo Matsuyama

Waseda University, Department of Computer Science and Engineering,
Tokyo, 169-8555, Japan
{maejima_m,rrryokote,yasuo}@wiz.cs.waseda.ac.jp
http://www.wiz.cs.waseda.ac.jp

Abstract. Mapping tools applicable to big data of composite elements
are designed based on a machine learning approach. The central method
adopted is the multi-dimensional scaling (MDS). The data set is mapped
onto a continuous surface such as a sphere. For checking to see the effectiveness of this method, preliminary experiments on the local optimality
were conducted. Supported by those results, the main target for the application in this paper is the design for a spherical GUI (Graphical User
Interface) which presents "must-watch" and "no-need" program clusters
in TV big data. This GUI shows a certain genre of programs at around
the North Pole. Programs having an opposite genre placed at around the
South Pole. Since all-recording systems of TV programs are within the
realm of home appliances, this GUI can be expected to be one of necessary tools for a video culture.

Keywords: Clustering, multi-dimensional scaling, GUI design, big data,
TV programs.

1 Introduction

Rapid increase of the storage capacity with a reduced price has brought a tendency of "memorize everything in a device at hand." Because of such a trend,
ICT users have stepped to the world of big data. This indication can be found
also in consumer electronics. A typical example is a TV program recorder. Current TV program recorders which are sold at a price less than that of a TV set
can record all of

$$\{\text{all of 7 channels}\} \times \{\text{24 hours}\} \times \{\text{2 months}\} \gtrless \{\text{20,000 programs}\}$$

by the storage size of tera bytes. Such a system has become popular with the life
style change that less programs are watched real time so that program preferences
do not restrict one's daily life. But, this convenience needs to be supported by
intelligent user-aware interfaces so that the following problems can be coped with.

T. Huang et al. (Eds.): ICONIP 2012, Part V, LNCS 7667, pp. 267–274, 2012.

(a) There are so many programs stored or available. Search by keywords may not reflect user's preference. "Must-watch" programs are often forgotten away.

(b) Personal recorder's memory size is finite even though it is big. This means that too old programs are eventually erased disregarding user's implicit preference.

Note that problem (a) remains valid even if users have access rights to commercial on-demand program services. Thus, in the era of big data, user interface becomes a key existence for the real merit of the extensive information.

Usually, big data is a composite of heterogeneous objects. This is the main reason of its big size. Therefore, it becomes necessary to consider the following items.

(1) How do we measure the similarity between two records?
(2) Which graphical structure is user friendly?

In Section 2, we use the multi-dimensional scaling (MDS) on Item (1). On Item (2), a spherical GUI (Graphical User Interface) is presented. Following the discussions on Items (1) and (2), a set of preliminary experiments are presented. These experiments will show that the presented mapping method does not suffer from bad local optima. Section 3 shows results on the clustering of actual TV programs over all channels. Experiments will show that the presented mapping method and GUI are readily serviceable to recorded program users. In Section 4, discussions and concluding remarks towards the realization of more user friendliness are given.

2   Spherical MDS and Record Mapping

2.1   Graphical User Interface via MDS

MDS (Multi-Dimensional Scaling) [1], [2] is a class of multivariate analysis for the allocation of source data mapped onto a pre-selected geometric structure. Its required property is that the closeness of objects in a source space is preserved as the neighborhood in the mapped space. Source data set need not be a continuum if a similarity measure between two arbitrary records is provided. Such a case is this paper's target objects, recorded TV programs, whose mapping problem for GUI has not arisen until recently.

There are two types on MDS. One is the metric MDS which is based on a distance. The other is the non-metric MDS which uses a distortion measure. The distortion measure is free from the symmetry or the triangular inequality. But, it is required to generate a topological distortion space [3] where the neighborhood is preserved. The case of TV program database in Section 3 will use a distortion measure which need not maintain the triangular inequality.

On a GUI using MDS, we need to consider its serviceability. For this problem, there is a class of possible target spaces for GUI.

(a) Three dimensional Euclidian space
(b) Two dimensional plane
(c) Topological surfaces
(d) Combinations of the above

Space structures of Item (a) are inappropriate since records mapped to interior points are difficult to select by clicking. The plane structure of Item (b) suffers from the discontinuity of edges, which will not create an elegant interface. In this paper, therefore, we use Item (c) with the emphasis on the spherical surface structure. This is by considering the easiness of clicking an object by users.

2.2   Design of MDS onto a Spherical Surface

In this paper, the spherical surface onto which records are mapped is normalized to have a unit radius. The position of i-the record on this surface can be described by two angles, say θ_i and φ_i. Let n be the total number of records. Then, the positions of the total records on the unit sphere are expressed by an n by 2 matrix, say Ψ.

$$\Psi = \begin{bmatrix} \theta_1 & \varphi_1 \\ \vdots & \vdots \\ \theta_n & \varphi_n \end{bmatrix} \tag{1}$$

Its expression by the orthogonal axis is X whose i-th element $\mathbf{x}_i$ is

$$\mathbf{x}_i = \begin{bmatrix} x_i \\ y_i \\ z_i \end{bmatrix} = \begin{bmatrix} \sin\theta_i \cos\varphi_i \\ \sin\theta_i \sin\varphi_i \\ \cos\theta_i \end{bmatrix} . \tag{2}$$

Since there are many data points (records) mapped on the spherical surface, a distortion or dissimilarity is measured by the square of the Euclidian distance.

$$d_{ij}^2 = (x_i - x_j)^2 + (y_i - y_j)^2 + (z_i - z_j)^2 \tag{3}$$

This is equivalent to the following expression.

$$d_{ij}^2 = 2 - \{\cos(\theta_i + \theta_j) + \cos(\theta_i - \theta_j)\}$$
$$+ \frac{1}{2} \left\{ \begin{matrix} \cos(\theta_i + \varphi_i + \theta_j - \varphi_j) + \cos(\theta_i - \varphi_i + \theta_j + \varphi_j) \\ - \cos(\theta_i + \varphi_i - \theta_j - \varphi_j) - \cos(\theta_i - \varphi_i - \theta_j + \varphi_j) \end{matrix} \right\} \tag{4}$$

This means that a geodesic arc length is approximated by its chord length. Then, on the records i and j, the mapping distortion between the original dissimilarity o_{ij} and d_{ij} can be measured by the square of these distances. If we consider this to be an element of a matrix, say $\mathbf{A}$, we get to the following total distortion measure S.

$$S = \sum_{i=1}^{n-1} \sum_{j=i+1}^{n} (o_{ij} - d_{ij})^2 \tag{5}$$

Note that above summations reflect that the matrix $\mathbf{A}$ is symmetric and its diagonal elements are zero.

Given by the total distortion S, we need to find the optimal placement of the records on the unit sphere so that

$$\Psi^* = \arg\min_{\Psi \in R^{n \times 2}} S \tag{6}$$

is obtained. Since the total number n of records is allowed to be large, we use a gradient descent method.

$$\Psi^{\text{new}} = \Psi^{\text{old}} - \varepsilon \partial S / \partial \Psi \qquad (7)$$

Here, ε is a small positive number such as 0.01.

Most optimization problems require CPU time and may suffer from bad local optima. In the following two subsections, we will check to see if the method of Equations (5) and (7) is trapped by bad local minimums if we start from random initial states. On the computational feasibility large resources such as TV data, whole Section 3 is dedicated for its identification.

2.3   Spherical MDS for Inscribed Regular Tetrahedron

Since we use the chord distance as an approximation to the geodesic arc length, a test on the spherical MDS of Equations (5) and (7) is a necessary check point on the local optimality and approximations.

We prepare an inscribed regular tetrahedron for the unit sphere. Randomly placed four points got to the four vertexes of the tetrahedron successfully. Thus, the presented method could cope with bad local optimality well.

2.4   Color Placement by Spherical MDS

The second preliminary experiment is the mapping of hue elements. This experiment was conducted by the following steps.

Step 1: Select 100 hue points randomly spaced in $0 \le H < 360$ (mod 360).

Step 2: Place these hue points randomly on the unit sphere. Here, the distance of two hues H and H' is

$$d(H, H') = \{180 - |180 - |H - H'||\}/90 \qquad (8)$$

Step 3: Perform the optimization by repeating Equation (7).

The distance of Equation (8) is understood form the following properties.

(i) The difference of two hues is between 0 and 180.

(ii) Maximum of the chord length of the unit sphere is 2.

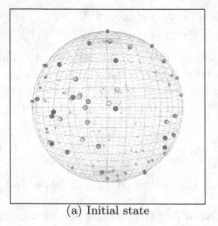
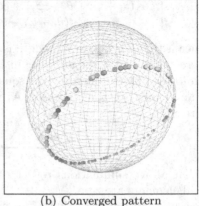

(a) Initial state   (b) Converged pattern

Fig. 1. Mapping of hues onto a sphere by MDS

Figure 1 (a) shows the initial placement. Note that this is a three dimensional color illustration: Readers are recommended to identify the color property through a pdf viewer. Therefore, the size of each element is used to identify the depth. Figure 1 (b) is the converged pattern of the MDS method of Equations (7) and (8). This is also a three dimensional color illustration mapped onto a two dimensional plane. The size of spherical dots reflects its depth position. The converged pattern generates a ring on the sphere. On this ring, continuity of hues can be observed reflecting the similarity. This is an important property that a GUI needs to possess.

It is also found that there is the indeterminacy on the orientation of the ring. But, this is not a problem as a GUI since users can rotate the sphere by a pointing device. Thus, the converged pattern is judged to be appropriate.

3   GUI Design for Recorded TV Programs

3.1   Definition of a Distortion Measure

There are tags and meta-data attached to recorded TV programs. On such a set of meta-information, we define dissimilarity between TV programs first. Then, the total distortion measures on the program set are applied to the spherical MDS. The training set contains 200 programs which are randomly selected from 2,081 programs broadcast for one week in all TV channels.

Adopted features are (i) broadcasting station, (ii) program genre, (iii) day of the week, (iv) time slot, and (v) program length. Such five categories give the following dissimilarity measures.

Broadcasting Stations: We use an example of Tokyo area. But, the figure allocation methods themselves are independent of districts.

There are two public stations. One is general and the other is educational. Besides, there are five commercial stations. Programs by two public stations look similar and they offer the same program sometimes. Commercial stations give programs considerably different from the public stations. Therefore, we defined the dissimilarity measure as is specified in Table 1.

Table 1. Dissimilarity between broadcasting stations

broadcasting stations	dissimilarity
same stations	0.0
two public stations	0.2
two commercial stations	0.8
public station and commercial station	1.0

Program Genre: There are two genres; the upper level and the lower level. The big genre is classified to 13 categories as is given in Table 2. Each category is further divided to around 10 small genres. The number of assigned genres is 0 to 3. Therefore, the dissimilarity is set to 1.0 if no genre is assigned in either side. If more than one genres are assigned to both programs, the dissimilarity measure of Table 3 becomes appropriate. This table is used for all combinations and finding their minimum. This minimum value is adopted to be the dissimilarity.

Table 2. Big genres

news/report	sports	information/wide show
drama	music	variety
movie	animations/SFX	documentary/culture
theater/performance	hobby/education	welfare
others		

Table 3. Dissimilarity between genres

two genres	dissimilarity
both big genres and small genres coincide	0.0
only big genres coincide	0.2
big genres differ	1

Day of the Week: Day of the week also gives a dissimilarity measure. Programs of week days and weekends considerably differ as is enumerated in Table 4.

Table 4. Dissimilarity on the day of the week

Days of the week of two programs	dissimilarity
same day of the week	0.0
different week days	0.6
Saturday and Sunday	0.4
weekdays and weekends	1.0

Broadcast Time Slot: Midnight programs end by 4 AM which is the start time of morning programs. Morning programs start from 4 AM. Therefore, one day is regarded as 24 hours starting from 4 AM. The dissimilarity between two programs with respect to the starting time is their difference divided by 24.

Length of Programs: Difference of program lengths gives a dissimilarity. We use the following measure:

$$length\_dissimilarity = (longer\_length - shorter\_length)/longer\_length \qquad (9)$$

Integration of All Dissimilarity for the Total Distortion Measure: Each of dissimilarity measures has considerably different distributions. This means that some of dissimilarity measures might be emphasized too much bringing about unexpected effects. Therefore, empirical dissimilarity sets are normalized to have a zero mean and a unit variance. Then, the genre dissimilarity is doubled because of its importance. On the other hand, the length dissimilarity is halved since it is less important than the rest.

For the convergence of the algorithm of Equation (7), total values are normalized so that the minimum value is 0 and the mean value is S_RandMean which is the mean value of the dissimilarity of randomly placed points on a unit sphere. Note that S_RandMean is around 1.32.

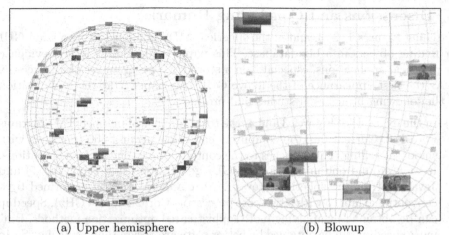

(a) Upper hemisphere   (b) Blowup

Fig. 2. Placement of programs on the northern hemisphere

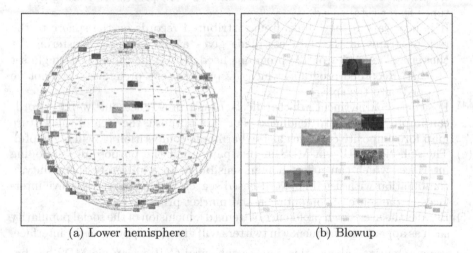

(a) Lower hemisphere   (b) Blowup

Fig. 3. Placement of programs on the southern hemisphere

3.2   Experiments on TV Program Set

Experiments of the spherical MDS were conducted on the TV program set with $\varepsilon = 0.01$ for Equation (7).

Figure 2 (a) is a view of the northern hemisphere. Programs are distributed well so that each one is visibly identifiable. Figure 2 (b) is a blowup of a square region of Figure 2 (a). Noon news programs are placed in this area.

Figure 3 (a) is a view of the opposite side, southern hemisphere. Figure 3 (b) is its blowup around the south pole (center). Wide show programs in the morning are placed in this area. These programs were broadcast in different days.

At first sight, one may consider that wide shows contain the contents of news programs. But, these two categories were place around the north and south poles respectively. This is mainly because of their genre difference.

4 Discussions and Concluding Remarks

Watching Figures 1, 2, 3, and 4, the spherical MDS is judged to be a good GUI for finding objects in a big database. This paper could show the first evidence for practical applications, especially on a stored TV program set of a large size.

As the first appearance of the above GUI, this paper is at the root position to the following branches of studies and products.

(a) In the example of Figures 3 and 4, the total number of mapped TV programs were 200. Current desktop computers can allow around 500 programs with smooth scrolling. This matches to the contemporary touch screen operations.

(b) It took around one minute for the GUI generation using Equation (7) and JavaScript for the spherical browser. All computation was performed by a conventional laptop with a two-core processor (Core2Duo, 1.4 GHz). Speedup can be obtained by applying more sophisticated optimization methods. But, more speedup will be obtained by better software design than using JavaScript which is just an interpreter. This is because we have preliminary experimental results that the usage of 4 core and 12 core processors show only 1.5 times faster total speed.

(c) It is necessary to have links from distributed records on the sphere to the rest programs in the database. This gives a new class of "clustering for clustering." A series of TV programs needs to be identified as a single set at least. Recorded programs which are out of user's preference need not be mapped onto the sphere.

(d) Position re-adjustment which reflects user's preference and watch records can improve the GUI friendliness. For this purpose, the idea of the feature map for composite cost such as the harmonic competition [4] will be helpful.

(e) The spherical GUI via MDS in this paper had the purpose of developing interfaces which can reflect human sensibility. In addition to this ability, a combination with usual keyword-based search will give less expensive interfaces in the sense of computation and market prices.

(f) Besides the user's own preference of Item (d), inclusion of the social popularity such as appearance frequency in twitters will give more power on the interface.

As was summarized above, this paper's spherical GUI design via MDS has become a prior evidence to developed versions having the properties of (a)~(f) appearing in the near future.

References

1. Cox, T.F., Cox, M.A.A.: Multidimensional scaling on a sphere. Communications in Statistics - Theory and Methods 20(9), 2943–2953 (1991)
2. Elad, A., Keller, Y., Kimmel, R.: Texture Mapping via Spherical Multi-dimensional Scaling. In: Kimmel, R., Sochen, N.A., Weickert, J. (eds.) Scale-Space 2005. LNCS, vol. 3459, pp. 443–455. Springer, Heidelberg (2005)
3. Matsuyama, Y.: Process Distortion Measures and Signal Processing. Ph.D. Dissertation, Stanford University, Stanford CA, pp. 18–21 (1978)
4. Matsuyama, Y.: Harmonic competition: A self-organizing multiple criteria optimization. IEEE Tnans. on Neural Networks 7, 652–668 (1996)

Analysis of Intrusion Detection in Control System Communication Based on Outlier Detection with One-Class Classifiers

Takashi Onoda and Mai Kiuchi

System Engineering System Laboratory,
Central Research Institute of Electric Power Industry,
2-11-1, Iwado Kita, Komae-shi, Tokyo 201-8511 Japan
{onoda,mai}@criepi.denken.or.jp

Abstract. In this paper, we introduce an analysis of outlier detection using SVM (Support Vector Machine) for intrusion detection in control system communication networks. SVMs have proved to be useful for classifying normal communication and intrusion attacks. In control systems, a large amount of normal communication data is available, but as there have been almost no cyber attacks, there is very little actual attack data. One class SVM and SVDD (Support Vector Data Description) are two methods used for one class classification where only information of one of the classes is available. We applied these two methods to intrusion detection in an experimental control system network, and compared the differences in the classification. To gain information of the kind of traffic that would be classified as an attack, the percentage of allowed outliers was changed interactively, adding human knowledge of the control system to the results. And our experiments clarified that sequence information in control system communication is very important for detecting some intrusion attacks.

Keywords: Intrusion Detection, Control System Communication, Support Vector Machine, Support Vector Data Description, Cyber Security.

1 Introduction

In the past, communication networks in control systems were built with proprietary protocols, and disconnected from other networks. This ensured the security of the system to a certain extent. Recently, control systems are beginning to use standard protocols, and connection to other communication systems for the sake of usability is more common. This has resulted in cyber security problems similar to that seen in business IT systems. Various security measures, such as firewalls, encryption and authentication, have been developed for IT systems, and use of similar technologies in control system communication networks has been considered[1]. An intrusion detection system, which monitors communication packets and system behavior to detect malicious behavior, can be one layer of cyber security. We have considered the use of intrusion detection systems in control systems[2]. Intrusion detection systems used in IT systems widely use signature based systems, which build a model based on available knowledge of attacks. This has proved to be effective in the IT system environment. In the case

T. Huang et al. (Eds.): ICONIP 2012, Part V, LNCS 7667, pp. 275–282, 2012.

of control systems, a large amount of normal communication data is available, but as there have been almost no cyber attacks, there is very little actual attack data. Therefore, anomaly detection using only the available normal data becomes effective. And we consider one class classification, using only normal communication data. One class SVM[6] and SVDD (Support Vector Data Description)[7] are one class classification methods based on SVM. In information systems, some research has been done on the use of one class SVM for intrusion detection[8]. The research focuses on anomalous data within legitimate normal communication. This paper considers the application of one class SVM and SVDD to intrusion detection in a control system communication network. The percentage of allowed outliers was changed interactively, utilizing human knowledge of the control system, and the differences in the classification are compared. Finally, this paper clarifies that sequence information in control system communication is very important for detecting some intrusion attacks.

2   Intrusion Detection in Control Systems

Many security measures used in the corporate business IT system, such as firewalls, access control, encryption, authentication and intrusion detection, are also applicable to control systems. However, special precautions must be taken when introducing the same solutions[1]. Issues particular to control systems include the demand for 100 percent uptime, and little tolerance for extra latency. Another issue is resource restrictions in controllers and other embedded devices commonly used in control systems. It is also necessary to analyze the network traffic flow to configure firewalls and intrusion detection systems accurately. Off the shelf security products such as firewalls, encryption and intrusion detection systems can be configured for a specific vendor application and protocol implementation to protect internal control system communication and detect attacks. However, it is not always possible to implement all the security measures, because of the issues in control systems previously mentioned. Intrusion detection systems, which monitor communication packets and system behavior to detect malicious behavior, can be one layer of cyber security. Furthermore, intrusion detection can be implemented in the system so that it does not add to the delays in the communication, making it an important choice in control system security. We have considered the use of intrusion detection systems in control systems[2]. Intrusion detection systems used in IT systems widely use signature based systems, which build a model based on available knowledge of attacks. This has proved to be effective in the IT system environment. In the case of control systems, a large amount of normal communication data is available, but as there have been almost no cyber attacks, there is very little actual attack data that can be used to build the model. Therefore, anomaly detection using only the available normal data becomes effective. Utilizing SVMs has been considered for intrusion detection, and has proved to be effective[3,4]. When SVMs are utilized for intrusion detection systems, it uses both the legitimate communication data and attack data for positive and negative examples. In the case of control systems, there have been almost no cyber attacks and very little attack data is available, so it is difficult to construct a binary class classification problem. Therefore, we consider one class classification, using only normal communication data. We considered one class SVM[6] and SVDD[7], both one class classification methods based on SVM.

3 One Class SVM and SVDD

In this section, we briefly introduce the two methods we used for intrusion detection; one class SVM and SVDD.

3.1 One Class SVM

One class SVM is a method that adapts SVM to a one class classification problem[6]. The available data is used as input data for the training, and assumed to belong to only one class. The input data is transformed by a kernel to a high dimension feature space. Then the data is separated from the origin with maximum margin, treating the origin as the second class. For input training data $\mathbf{x}$, the algorithm returns a function f which takes the value $f(\mathbf{x}) = +1$ in a small region capturing most of the training data points, and $f(\mathbf{x}) = -1$ elsewhere.

$$f(\mathbf{x}) = \operatorname{sgn}\left((\mathbf{w} \cdot \Phi(\mathbf{x})) - \rho\right) \tag{1}$$

Here ρ is the bias for the hyper-plane. For a new data point $\mathbf{x}$, the value $f(\mathbf{x})$ is determined by evaluating the side of the hyper-plane the data lies in feature space. Let the training data be $\mathbf{x}_1, \cdots, \mathbf{x}_\ell \in X$, where ℓ is the number of observations and the data belong to one class X. $\Phi : X \to H$ is the feature map which transforms the training data to a feature space. The dot product in the image of Φ can be computed by evaluating a simple kernel in the following form.

$$k(\mathbf{x}_i, \mathbf{x}_j) = (\Phi(\mathbf{x}_i) \cdot \Phi(\mathbf{x}_j)) \tag{2}$$

An example of such a kernel is the linear kernel, which we used in our experiments.

$$k(\mathbf{x}_i, \mathbf{x}_j) = \mathbf{x}_i \cdot \mathbf{x}_j \tag{3}$$

To separate the data set from the origin, the following quadratic problem needs to be solved.

$$\min_{\mathbf{w}, \boldsymbol{\xi}, \rho} \frac{1}{2}\|\mathbf{w}\|^2 + \frac{1}{\nu\ell} \sum_i \xi_i - \rho \quad \text{subject to} \quad (\mathbf{w} \cdot \Phi(\mathbf{x}_i)) \geq \rho - \xi_i, \quad \xi_i \geq 0 \tag{4}$$

Here $\mathbf{w}$ is the weight vector, ξ is a slack variable, and ν is the upper bound for the fraction of outliers in the data, and the lower bound for the fraction of support vectors. Since the nonzero slack variable ξ is penalized, if $\mathbf{w}$ and ρ solve the problem, we can expect that the decision function Eq. (1) will be positive for most data $\mathbf{x}$ in the training set, while the support vector type regularization term $\mathbf{w}$ will still be small. The trade off is controlled by ν. Using multipliers $\alpha_i, \beta_i \geq 0$, we introduce a Lagrangian L as follows.

$$L(\mathbf{w}, \boldsymbol{\xi}, \rho, \boldsymbol{\alpha}, \boldsymbol{\beta}) = \frac{1}{2}\|\mathbf{w}\|^2 + \frac{1}{\nu\ell} \sum_i \xi_i - \rho - \sum_i \alpha_i((\mathbf{w} \cdot \mathbf{x}_i) - \rho + \xi_i) - \sum_i \beta_i \xi_i \tag{5}$$

Setting the derivatives with respect to the primal variables $\mathbf{w}, \xi_i, \rho$ to 0, we have the following.

$$\mathbf{w} = \sum_i \alpha_i \Phi(\mathbf{x}_i), \quad \alpha_i = \frac{1}{\nu\ell} - \beta_i \leq \frac{1}{\nu\ell}, \quad \sum_i \alpha_i = 1 \tag{6}$$

In Eq. (6), all patterns satisfying $\{\mathbf{x}_i : i \in [\ell], \alpha_i > 0\}$ are called support vectors. Using Eq. (2), the support vector expansion transforms the decision function Eq. (1) into the following kernel expansion.

$$f(\mathbf{x}) = \mathrm{sgn}\left(\sum_i \alpha_i k(\mathbf{x}_i, \mathbf{x}) - \rho\right) \tag{7}$$

Substituting Eq. (6) into the Lagrangian Eq. (5), and using Eq. (2), we obtain the following dual problem.

$$\min_{\alpha} \frac{1}{2} \sum_{i,j} \alpha_i \alpha_j k(\mathbf{x}_i, \mathbf{x}_j) \quad \text{subject to } 0 \leq \alpha_i \leq \frac{1}{\nu\ell}, \quad \sum_i \alpha_i = 1 \tag{8}$$

3.2 SVDD

Another method used for one class classification is SVDD[7]. This method attempts to find a hyper-sphere with minimum volume, containing most of the input training data. Defining the hyper-sphere with center a and radius $r > 0$, we have the following problem.

$$\min_{r, \boldsymbol{\xi}, \mathbf{a}} r^2 + \frac{1}{\nu\ell} \sum_i \xi_i \quad \text{subject to } \|\Phi(\mathbf{x}_i) - \mathbf{a}\| 0 \leq r^2 + \xi_i, \quad \xi_i \geq 0 \tag{9}$$

The following Lagrangian can be constructed.

$$L(r, \mathbf{a}, \boldsymbol{\xi}, \boldsymbol{\alpha}, \boldsymbol{\beta}) = r^2 + \frac{1}{\nu\ell} \sum_i \xi_i \sum_i \alpha_i \{r^2 + \xi_i$$
$$- (\|\Phi(\mathbf{x}_i)\|^2 - 2\mathbf{a} \cdot \Phi(\mathbf{x}_i) + \|\mathbf{a}\|^2)\} - \sum_i \beta_i \xi_i \tag{10}$$

Setting the derivatives with respect to variables $r, \mathbf{a}, \boldsymbol{\xi}$ to 0, we have the following.

$$\mathbf{a} = \sum_i \alpha_i \Phi(\mathbf{x}_i), \quad \alpha_i = \frac{1}{\nu\ell} - \beta_i \leq \frac{1}{\nu\ell}, \quad \sum_i \alpha_i = 1. \tag{11}$$

Substituting Eq. (11) into Eq. (9) and using Eq. (2), we obtain the following dual problem.

$$\min_{\alpha} \sum_{i,j} \alpha_i \alpha_j k(\mathbf{x}_i, \mathbf{x}_j) - \sum_i \alpha_i k(\mathbf{x}_i, \mathbf{x}_i) \quad \text{subject to } 0 \leq \alpha_i \leq \frac{1}{\nu\ell}, \quad \sum_i \alpha_i = 1 \tag{12}$$

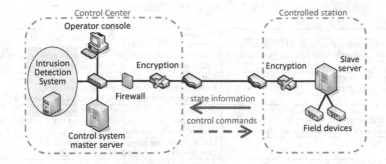

Fig. 1. An Overview of the Model System used for the experiment

This corresponds to the following decision function, where for any $\mathbf{x}_i$ with $0 \leq \alpha_i \leq \frac{1}{\nu\ell}$, the argument of the sgn is 0.

$$f(\mathbf{x}) = \text{sgn} \left(r^2 - \sum_{i,j} \alpha_i\alpha_j k(\mathbf{x}_i, \mathbf{x}_j) + 2\sum_i \alpha_i k(\mathbf{x}_i, \mathbf{x}_i) - k(\mathbf{x}, \mathbf{x}) \right) \qquad (13)$$

Comparing the problems Eq. (12) in SVDD and Eq. (8) in one class SVM, the difference is only the term $\sum_i \alpha_i k(\mathbf{x}_i, \mathbf{x}_i)$.

4   Experiment and Analysis

The model control system communication network used in the experiment is shown in Fig. 1 and was developed in our laboratory. It is assumed here that no connections exist between this control system network and the corporate business network. All the equipment, applications and security measures in the figure belong solely to the control system network, and the placement and settings mentioned are irrelevant to the corporate business network. The control system master server located in the control center is the direct control and data acquisition interface to the control system equipment, or field devices. To control the field devices located in the substation, the master server communicates with the server in the substation, which in turn sends the actual control signals to the field device. In this model system, the control system field devices are emulated in the same terminal as the substation server. The state information of the field devices and any responses to the control commands are sent through the substation server to the control system master server, and then sent to any communicating operator consoles, which in this case are located within the control center.

Using the model control system, we collected $10,000$ packets of normal communication data for training one class SVM and SVDD. The features used from the data are shown in Table 1. The data was calibrated based on prior knowledge of the control system, to have values closer to 0 for less probable data. This results in the less probable data lying closer to the origin when a linear kernel is used. This calibrated data was used as the training data $\mathbf{x}$ for one class SVM and SVDD, stated in the previous section. For

Table 1. Data used for Training

Data	Description
-Source IP address(IP address, MAC address) -Destination IP address(IP address, MAC address)	-Addresses belonging to devices in the system: 1 -Addresses belonging to the same network as the system: 0.5 -Addresses belonging to different network: 0
-Source port number -Destination port number	-Port numbers used in the system: 1 -Port numbers reserved for use in the system: 0.5 -Port numbers not expected to be used in the system: 0
-Protocol -Identification number of field device -Type of control command or state information	-Content specified to be used in the system: 1 -Content not clearly specified to be used in the system: 0.5 -Content clearly specified not to be used in the system: 0
-Data length	-Length within the range used in the system: 1 -Length out of the range used in the system: 0.5
-Interval between packets	-Time elapsed after previous packet
-Interval between use of field device	-Time elapsed after use of a certain field device

-Interval between control command or state
  information retrieval
	-Time elapsed after use of the same control command or
	  state information retrieval

-Frequency of use of field device	-Frequency of usage of a certain field device within the training dataset

Table 2. Data classified as Outliers

	Normal Data	1 Class SVM outliers	SVDD Outliers
Packet capture interval	over 0.07 sec.	under 0.05 sec.	under 0.05 sec.
Elapsed time after likewise control or data acquisition procedure	over 10 min.	under 6 min.	under 6 min.
Elapsed time after use of equipment	over 5 min (and. under 15 min. for SVDD)	under 3 min.	under 3 min. over 20 min.
Frequency of equipment use	under 12 times/h		over 18 times/h

the computations, we used libraries provided in MATLAB; LIBSVM[9], PRTools[10] and DDtools[11].

For ν, the upper bound for the fraction of outliers and the lower bound for the fraction of support vectors, we experimented within the range between 0.2% and 5%. Within this range, the results were checked with prior knowledge of control system communication, and primarily used the value 2% for detailed evaluations, where communication that should obviously be considered "normal" was classified as an intrusion. Further work is necessary for a method to automatically determine the value of ν.

Although there is no attack data in the control system, as a result of the learning, it is possible to gain insight into what kind of communication would be detected as an intrusion. This is valuable knowledge when implementing an intrusion detection system. We compared the difference of the results between one class SVM and SVDD. The features of the data classified as outliers are shown in Table 2. The ratio of data classified as outliers for one class SVM is 0.932 and the ratio of data classified as outliers for SVDD is 1.441.

For one class SVM, data with short intervals between packets, short interval between use of the field device, and short interval between control command or state information retrieval, are classified as outliers. The result supports the knowledge that control commands are not continuously sent in this control system. So the intrusion detection system is able to detect this kind of behavior.

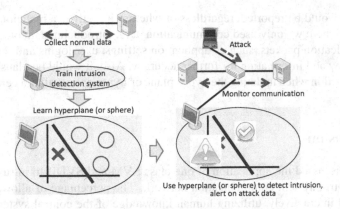

Fig. 2. Image of using machine learning in intrusion detection

For SVDD, in addition to the communication detected in one class SVM, data with very large intervals between use of the field device and frequent usage of the field device are also classified as outliers. As a result, SVDD has a higher ratio for data classified as outliers than one class SVM. This difference comes from the difference in the algorithms where one class SVM separates data close to the origin and low probability; while the data separated by SVDD does not necessary reside close to the origin. The results of SVDD indicate that the control system does not expect a field device to be used after a long period of being unused. Such behavior may result from legitimate but irregular use of a certain field device, or a cyber attack trying to control the device.

In general, SVDD extracts more data, which are rare data, than one class SVM. From Fig 2, the experimental results of one class SVM and SVDD show that sequence information in control system communication is very important for detecting some intrusion attacks, since Fig 2 consists of some sequence information in control system communication.

5   Image of Intrusion Detection Based on One Class SVM or SVDD

Fig. 2 shows an image of the actual usage of one class SVM and SVDD in an intrusion detection system.

First, normal communication data is collected from the communication network, and used as training data. One class SVM or SVDD are trained with the data, and the resulting hyper-plane or hyper-sphere is retained in the intrusion detection system. For the actual intrusion detection, the communication packets are monitored. With one class SVM, for a new data point x, the value $f(x)$ is determined by evaluating in feature space the side of the hyper-plane the data point is in. So if the data point is on the same side as the origin, then the data is classified as an intrusion. With SVDD, the same procedure is carried out with a hyper-sphere. If the data is classified as an intrusion, the intrusion detection system then sends an alert to the human machine interface.

In a control system, any possibility of disruption should be reported to a human operator. So when deploying an intrusion detection system, it is likely that all data classified as an outlier would be reported, regardless of whether it really is a malicious attack or not. Although here we only used communication data, introducing data other than just the communication packets, such as application settings, user logins and maintenance schedules may add information for further accuracy. Also, it would be plausible to add human interaction when updating the hyper-plane or hyper-sphere after a certain period of time.

6   Conclusion

This paper discussed the application of one class SVM and SVDD to intrusion detection in a control system communication network. The percentage of allowed outliers was changed interactively, utilizing human knowledge of the control system, and the differences in the classification were compared. Further work is necessary for a method to automatically determine the percentage.

In general, SVDD extracts more data, which are rare data, than one class SVM. From Fig 2, the experimental results of one class SVM and SVDD show that sequence information in control system communication is very important for detecting some intrusion attacks, since Fig 2 consists of some sequence information in control system communication.

References

1. Kiuchi, M., Serizawa, Y.: Security Technologies, Usage and Guidelines in SCADA System Networks. In: ICCAS-SICE 2009 (2009)
2. Kiuchi, M., Serizawa, Y.: Customizing Control System Intrusion Detection at the Application Layer. In: SCADA Security Scientific Symposium 2009. Digital Bond Press (2009)
3. Osareh, A., Shadgar, B.: Intrusion Detection in Computer Networks Based on Machine Learning Algorithms. International Journal of Computer Science and Network Security 8(11) (2008)
4. Wun-Hwa, C., Sheng-Hsun, H., Hwang-Pin, Sh.: Application of SVM and ANN for Intrusion Detection. Computers & Operations Research 32, 2617–2634 (2005)
5. Corinna, C., Vladimir, V.: Support-Vector Networks. Machine Learning 20, 273–295 (1995)
6. Schölkopf, B., Platt, J., Shawe-Taylor, J., Smola, A., Williamson, R.: Estimating the Support for a High-dimensional Distribution. Microsoft Research, One Microsoft Way Redmond WA 98052, Tech. Rep. MSRTR-99-87 (1999)
7. Tax, D., Duin, R.: Support Vector Data Description. Machine Learning 54, 45–66 (2004)
8. Zhang, R., Zhang, S., Muthuraman, S., Jiang, J.: One Class Support Vector Machine for Anomaly Detection in the Communication Network Performance Data. In: 5th WSEAS Int. Conference on Applied Electromagnetics, Wireless and Optical Communications (2007)
9. Chih-Chung, C., Chih-Jen, L.: LIBSVM: A Library for Vector Machines (2001), http://www.csie.ntu.edu.tw/~cjlin/libsvm
10. Duin, R.P.W., Juszczak, P., Paclik, P., Pekalska, E., de Ridder, D., Tax, D.M.J., Verzakov, S.: PRTools4.1, A Matlab Toolbox for Pattern Recognition, Delft University of Technology (2007)
11. Tax, D.M.J.: DDtools, the Data Description Toolbox for Matlab (2009), http://homepage.tudelft.nl/n9d04/dd_tools.html

Mobile Web Browsing Techniques

Zahiruddin Ahmad[1] and Jer Lang Hong[2]

[1,2] School of Computing and IT, Taylor's University
{zahiruddin.ahmad,jerlang.hong}@taylors.edu.my

Abstract. Society has changed over the years where people of all ages are now expected to own a mobile device. A mobile phone acts as a personal comfort zone for people to centralize their thoughts and commitments. From video, audio calls to SMS, everybody is somewhat interacting with one another which in turn generate emotional rewards. The personal conversation between close friends and family is an instant stimulus triggering mental satisfaction amongst them. Rapid development changes are happening in mobile technology advancement today. Smart phones lead the way by providing considerably large amounts of memory integrated within an Operating System to load web content across the mobile screen. From a desktop perspective, the screen is wide and a vast array of content can be displayed without much limitations. And here lies a problem where the small mobile screen intends to render every block of information from their associated desktop page. To date, there are serious compatibility concerns with web content failing to load successfully or even loading abnormally. Like many adoptions of standards, many are considered but most get dropped. Even for a specific platform say the BlackBerry, exist multiple screen-sizes which has to be accounted for. If the screen resolution fails to render the content accordingly, reliability of the phone will be compromised. As a whole, the entire writing discusses *why* Mobile Web Browsing is necessary to be considered and ultimately *how* we can further improve on the current standards. This paper proposes an advancement in mobile web browsing techniques which combines pattern analysis and visual separators. It will firstly cover a wide range of mobile browsing techniques which ought to be primary features in improving the user experience while critically analysing their implications.

Keywords: Mobile Web Browsing, Segmentation, Usability.

1   Introduction

Recently, there is a misconception plaguing us about the differences of the mobile web and a typical desktop web. People think the desktop web and the mobile web are two separate entities. However in essence, it is still the same Web. Think email. You don't use separate email accounts for both mobile and desktop. Some people might, but that is not a very typical practice. The telling differences is mobile devices are able to process significantly less resources, handle smaller bandwidth and screen can fit only minimal contents. Mobile web uses the same technology namely its network

T. Huang et al. (Eds.): ICONIP 2012, Part V, LNCS 7667, pp. 283–291, 2012.

protocols from the internet: HTTP, HTTPS, POP3, Wireless LAN, even TCP/IP. However it does not include entirely all. GSM, CDMA and UMTS are not protocols used in the desktop environment but this communication protocols are still operating at the lower layers.

Is the mobile market really worth our R&D? Here are some figures which justifies the countless studies.

"234 million Americans aged 13 and older used mobile devices for three month average period ending in December 2011."According to report published by comScore, Inc.

Adding to that, 47.5% of U.S. mobile subscribers are web browsing with their mobile devices, up to 4.6% from the preceding 3-month period.

Such as the statistics, mobile browsing is already embraced within the culture of the United States of America. Let us look at figures from a global perspective.

How many internet connections are there in the world?
2,267,233,742 (30% of the world's population) as of December 31st 2011
How many people have mobile devices?
4,851,038,607 (70% of the world's population)as of December 31st 2011

The vast amounts of users which will presumably keep increasing are enough reasons to undertake means to meet user's psychological and emotional expectations. As a direct comparison, the number of internet connections (30%) will increase closer to tally with the people with mobile devices figure of 70% should a suitable and reliable mobile interface get implemented. However, mobile browsing today has many limitations over convenience just by considering its physical design. The reason that people are not using their mobile browser frequently may be because of the web producers who are not offering them their actual needs (Design & Usability).

The fact of the matter is people are not using their mobile browsers enough. Access to the internet from any location indirectly promotes convenience in web browsing. However, a customer experience using a desktop page and a mobile page are worlds apart. Addressing users of all technical levels, most tend to expect the same exposure of a desktop page when using a mobile version. The obvious differences would be the small screen and input methods. A study by the Nielsen Norman Group concludes

"In user testing, Web site use on mobile devices got very low scores, especially when users accessed "full" sites that weren't designed for mobile."

This tells us that the general mobile browsing experience is not entirely user-friendly at the moment. To overcome this predicament, many web sites release corresponding mobile versions of their site which are designed to suit browsing from a mobile device. Official guidelines are written to facilitate the development of mobile optimized websites such as Mobile Web Best Practices, the Opera's Mobile Web Optimization Guid and etc.

Despite rising figures of mobile web pages being developed, the mass still prefer using the original desktop version to browse the Internet. A report published by Smashing Pumpkin indicates that, only 11% of Web sites have a mobile optimized version made. Measures have been debated on the topic of enhancing usability in regards to a desktop version converting in to a mobile web version.

This paper contains several sections. Section 2 describes the taxonomy of web browsing while Section 3 gives the techniques used in web browsing. Finally Section 5 summarizes our work.

2   Taxonomy on Web Browsing

2.1   Introduction

Mobile devices are currently undertaking a severe development phase. Abundant of ideas and solutions have risen suggesting ways to improve the usability of present mobile browsing practices. One point of view suggest to build a specific mobile optimized web site from the ground up while another major influence requires automation in adapting desktop pages to fit a mobile environment. Ultimately, there are two types of approaches which are manual and automatic ones. For automatic approaches, the proposed script intelligently realigns desktop Web pages to facilitate the specifications of a mobile device. Other approaches utilize navigation features for instance gesture based zooming used with a thumbnail overview.

In summary, mobile browsing approaches are classified into two points.

1. Manual efforts to support effective mobile browsing for a Web site individually.
2. Screenshot of a desktop Web page is used as an overview.

Based on the two points, our selection of mobile Web browsing techniques are defined in to three categories.

2.2   General Web Browsing Techniques

Manual Authoring
This includes mobile markup languages and interactive tools that support authoring a mobile Web site and its features. Examples of functions are the input methods and download speed. When mobile markup languages are implemented correctly, it optimizes mobile browsing for a better user experience. On the other hand, there are some downsides to adopting manual authoring. Most notably is the time and resource that have to be consumed to create a brand new mobile Web site from the beginning. Different development languages used also have high technical complexity which has to be addressed or it might bring about inconsistency issue that contributes to unnecessary maintenance cost.

Information Restructuring
Required information is extracted to be adjusted from a desktop Web page to a mobile environment. Information restructuring has diverted in to two steps: adaptive layout generation and page segmentation. For page segmentation, it operates by ingeniously identifying semantically related information and categorizing them. Transitioning to the second step which is (i.e. adaptive group generation) it renders an adaptive layout which adheres to a semantically grouped concept of relevant information placed in close-proximity with one another. This effectively minimizes the operations of a

mobile browser having scarce resources. As of today, a known challenge when it comes down to integrating many HTML standards is to design high-accuracy page segmentation algorithm that is applicable to various kinds of Web sites. A similar concern is the lacking of modifiable adaption specifically from the mobile users. A common presentation style practiced widely is the single column-presentation. This has become an almost universal standard but plenty of its limitations are still being addressed. Even if we could disregard universal preference of conventions, the one-style fits all still have problems integrating it-self on multiple devices which is spread amongst a variety of screen sizes and indefinite browsing situations.

Client Side Navigation

The basis for this approach is essentially to take an overview screenshot image of a desktop Web page as the precondition. With the image, a complete view of the entire web page will be rendered. If the users intend to do detailed reading on a certain portion of the page, they are required to manually adjust the focus by zooming in or out of the overview. The world-wide practice of the pinch and panning method using natural gestures on mobile phones are accommodated by the recent hardware breakthrough namely the capacitive multi-touch screen. Flaws with smaller screen browsing will still persist with users having to strain their eye-sight to recognize the significant contents apart from the less critical ones. People might be expert at typical desktop pages but when encountering an unfamiliar web page, the entire interface might be new to the users. To resolve the issue, many techniques have been developed to promote readability of the contents when users are browsing on their mobile devices. Page segmentation and the inclusion of text summarization are opted to improve visibility of sections within the overview.

3   Mobile Web Browsing Techniques

3.1   Manual Authoring

As previously mentioned, to create a mobile Web site consisting of language ruling and mobile mark-up languages is the essence of Manual Authoring. Common usages are in the form of XHTML (Extended Hyper Text Markup Protocol), CSS extensions(Cascading Style Sheet) and XML (Extended Markup Language) where desktop browsers widely embrace these practises. Such as the extreme variations of Web site presentational content, this technique is the tool to specifically cater for different interfaces to support mobile optimization.

A significant example is Unwired Planet (a company) who created the *Handheld Device Markup Language* (HDML) in 1996 which is fine-tuned to suit several factors such as: wireless network connection and drop time, control characteristics, transfer speeds and memory capability of the device. Specific mobile device manufacturers require different mobile mark-up language which is their own core proprietary standards such as the Nokia Tagged Text Mark-up Language (TTML) and Ericsson's proprietary mark-up language. However there was a suggestion from the WAP forum who proposed the Wireless Mark-up Language (WML) developed on HDML to represent the standard across the platform jungle. The upbeat reception WML received made it an international standard and has been adopted by the Wireless Application Protocol (WAP).

Further integrations were also proposed such as the *Rapid Serial Visual Presentation* (RSVP) Browser which is based on the existing WML technology. Its purpose was to support web browsing using a set of sub-pages (i.e.,cards), which purpose is to ease individuals in reading the contents. Besides that, different authoring tools [5, 12, 19] emerged claiming to adapt better to dynamically changing mobile browsing conditions. Constraint-based approaches [5, 12] enabled to dynamically accommodate multimedia presentations dependent on the change of media contents, display environment and user intentions. Zhang et al. [19] suggested a grammatical method to host adaptive layouts whereby allocating high level structural and spatial relations amongst multimedia objects via a graph grammar.

Without redesigning a Web site altogether, there exist conversion rules to transform a Web page from one presentational type to another. Sun et al. [16] created a system for mobile devices enabling online transactions. The process model is automata-based allowing a web developer to create a set of conversion rules which ensures only transaction-related information reaches the user. Data entry using speech is also supported. Nichols et al. [14] had a project (Highlight) which supports rapid prototyping and mobile application improvements. It consists of an embedded browser via a proxy server which transformed and clipped contents base on the conversion rules. The result of the conversion will then be sent to the user's mobile device. Dynamic contents such as Dynamic HTML and Ajax are supported with Highlight to certain extents.

3.2   Information Restructuring

Without the consideration of separate web sites, information restructuring aim to automatically adapt generic contents from a desktop version to a mobile size presentation. What content adaptation does is it fills the gap between device capabilities and the content formats [1]. The workings behind adapting to different contents are the grouping of semantically related information or meta-data. It can be a real daunting task because the conventional use of HTML is not entirely adopted with many different organization styles. Further challenges are the hardware limitations and types of browsing scenarios which exist to a mobile device user. Solutions were disputed such as the text summarization, image reduction, data extraction and block recognition to create a better viewing experience on mobile devices. Other approaches worth mentioning are documented in the sub-section below.

Previous mobile devices notably the PDAs had constraints of limited computing power accompanied with a miniature screen to project information. During those years, mobile phone screens could not even support colour images. It was necessary for information restructuring which addresses bandwidth usage to maximize utilization on a minimal screen scale. The Digestor system [3] became an influential project which emphasizes on heuristic planning algorithm and a set of heuristic rules (for instance reducing image size, discovering and highlighting significant headers, or sentence elision) to adapt the best presentation for a specific screen size. On the down-side this method is unable to support tables and applets. The Power Browser [7] eliminates white spaces and images to spare more screen space. The WEST browser

[4] removes JavaScript, Image maps, frames and at the same time adopts context and focus visualization with text reduction. Using multiple fixed sized blocks or "cards", this practise provides three display modes (namely i.e., thumbnail view, keyword view and link view). By focusing on one card, users can view the display in the central portion while the remaining cards are accessible along bordering areas.

Progressively, heuristic advances [8, 9, 11, 13] revises HTML structural tags to give a sense of structure to the Web page. Kaasinen et al. [11] enables automatic conversion of HTML arrangements to WML specifications. Buyukkokten et al. [8, 9] distinguished a number of semantic textual units based on HTML tags allowing the tag P to indicate a border line of two textual blocks. However, graphics were not supported in this text-focus initiative. SmartView[13] focuses on partitioning elements into table formats.Visual analysis was proposed for providing page segmentation in when considering the actual complexity of the HTML DOM structures. Yang et al. [18] did a paper on several matters namely, the HTML content visual similarities, detection of visual similarities pattern, and to derive a hierarchical organization of information within an HTML page. Chen et al. [10] categorized web pages by high level information blocks and later identified implicit and explicit separators residing in each block. CMo[6] has a different approach to segmenting Web pages where they rely on the geometrical alignment of frames. Hybrid analysis is been given more and more attention lately.

3.3   Client Side Navigation

Information loss due to analyzing and restructuring of a desktop Web page is a major concern for all parties involved. Addressing this matter, the client side navigation factor obtains the screenshot of a typical desktop Web page depicting a general overview. Focusing on the overview allows for a user-friendly interaction to scan through hefty information blocks on a small screen. Client side navigation technique can be further classified in to two sub-categories: Scaling and Zooming based navigation and Novel Interaction Techniques in mobile devices.

Scaling and Zooming Approaches
An alternative to restructuring desktop Web page layouts is to maintain the original structure and implement scaling and zooming facility to navigate through various sections of the overview. Assisting user to identify intended information is crucial when navigating on a small interface.

Thumbnail Based Zooming
Thumbnail based overview complements zooming-based interactions in client side navigating. Although a thumbnail consists of the structure and everything a Web page displays, the raw information rendered is not easily readable due to the lack of space. This brings about how essential it is to assist the mobile device users. Users should be guided to specific topic of interest while minimizing the frequency of zooming operations. WebThumb[17] enables basic panning and zooming capabilities on the thumbnail which originated from a desktop Web page. What this approach does is it introduces a "picking mode" to utilize information identification straight from the

overview. Whenever a specific point in the thumbnail is clicked, the tag which encloses content in that position will prompt a popup window displaying the text contained with readable attributes. This technique suggests a "text-mode" for better efficiency in reading text paragraph where the user: clicks on a block of text and the mode prompts out one word at a time. The advantage is that detailed reading on the text block can be done without heavy consumption of screen space. WebThumb is said to have achieve an optimal outcome when an end user had experience using a similar desktop Web page previously. In MiniMap[15], users are able to read detailed information and at the same time understand the overall page structure. This is made possible by overlaying a transparent page overview on top of the viewport. The viewport presentation has to be modified in order to host the maximum amount of information possible.

Slicing and Scaling
Basically, the main issue is to always minimize user navigation interactions. Some newer innovative methods aim to overcome this by combining thumbnail based zooming with page segmentation [13, 10, 2]. Result of page segmentation enables direct accessibility to generated information blocks. SmartView[13] divides an HTML Web page in to a number of logical partitions base on HTML table tags. Returning to thumbnail overviews, it highlights significant blocks of information to help users select the intended blocks for detailed reading. The steps taken in optimizing the information block is mainly to avoid horizontal scrolling. These steps are such as resizing graphics, rearrangement of content and re-flowing of text paragraph. Chen et al. [10] classified information in to two brands: a thumbnail overview as the first entity and fully descriptive information the second. The first renders the selected block in a new mobile Web page while the second loads the block in a central position of the screen.

4   Conclusions

Even though more people who are browsing the Web on mobile devices are increasing, fact is that desktop Web pages still dominate the Internet. Having a small screen, there are content and functional restrictions to browse desktop-originated Web pages on a mobile device. In order to enhance the overall user experience for mobile Web browsing, many specially configured systems are introduced in the fray.This paper proposes a classification for mobile browsing approaches according to the following two criteria: the degree of layout and structural changes. In the classification, three main categories are highlighted, i.e., manual authoring, information restructuring and client side navigation. To leverage manual efforts, interactive tools have been developed. Automatic restructuring approaches are developed based on a number of heuristic rules. These does not require manual efforts, however they are constrained by the complexity and diversity of HTML specifications. Client side navigation approach does not alter the page structure, and uses zooming based interaction to cater for the context plus focus visualization on a smaller screen.

References

1. Adzic, V., Kalva, H., Furht, B.: A Survey of Multimedia Content Adaptation for Mobile Devices. Multimedia Tools and Applications 51, 379–396 (2011)
2. Baluja, S.: Browsing on small screens: recasting web-page segmentation into an efficient machine learning framework. In: Proceedings of the 15th International Conference on World Wide Web (WWW 2006), pp. 33–42. ACM, New York (2006)
3. Bickmore, T.W., Schilit, B.N.: Digestor: device-independent access to the World Wide Web. In: Proceedings of International Conference on World Wide Web, Santa Clara, California, USA, pp. 1075–1082 (1997)
4. Björk, S., Holmquist, L.E., Redström, J., Bretan, I., Danielsson, R., Karlgren, J., Franzén, K.: WEST: a Web browser for small terminals. In: Proceedings of the 12th Annual ACM Symposium on User Interface Software and Technology (UIST 1999), pp. 187–196. ACM, New York (1999)
5. Borning, A., Lin, R.K., Marriott, K.: Constraint-based Document Layout for the Web. Multimedia Systems 8, 177–189 (2000)
6. Borodin, Y., Mahmud, J., Ramakrishnan, I.V.: Context browsing with mobiles – when less is more. In: Proceedings of the 5th International Conference on Mobile Systems, Applications and services (MobiSys 2007), pp. 3–15. ACM, New York (2007)
7. Buyukkokten, O., Garcia-Molina, H., Paepcke, A., Winograd, T.: Power browser: efficient Web browsing for PDAs. In: Proceedings of the SIGCHI Conference on Human Factors in Computing Systems (CHI 2000), pp. 430–437. ACM, New York (2000)
8. Buyukkokten, O., Garcia-Molina, H., Paepcke, A.: Accordion Summarization for End-Game Browsing on PDAs and Cellular Phones. In: Proceedings of ACM SIGCHI 2001, pp. 213–220 (2001)
9. Buyukkokten, O., Garcia-Molina, H., Paepcke, A.: Seeing the whole in parts: text summarization for web browsing on handheld devices. In: Proceedings of the 10th International Conference on World Wide Web (WWW 2001), pp. 652–662. ACM, New York (2001)
10. Chen, Y., Ma, W., Zhang, H.: Detecting web page structure for adaptive viewing on small form factor devices. In: Proceedings of the 12th International Conference on World Wide Web (WWW 2003), pp. 225–233. ACM, New York (2003)
11. Kaasinen, E., Aaltonene, M., Kolari, J., Melakoski, S., Laakko, T.: Two Approaches to Bringing Internet Services to WAP Devices. Computer Networks: The International Journal of Computer and Telecommunications Networking 33, 231–246 (2000)
12. Marriott, K., Meyer, B., Tardif, L.: Fast and Efficient Client-side Adaptability for SVG. In: Proceedings of WWW 2002, pp. 496–507 (2002)
13. Milic-Frayling, N., Sommerer, R.: SmartView: Flexible viewing of Web page contents. In: Proceedings of the 11th International World Wide Web Conference (WWW 2002), Honolulu, HI, USA (2002)
14. Nichols, J., Hua, Z., Barton, J.: Highlight: a system for creating and deploying mobile web applications. In: Proceedings of the 21st Annual ACM Symposium on User Interface Software and Technology (UIST 2008), pp. 249–258. ACM, New York (2008)
15. Roto, V., Popescu, A., Koivisto, A., Vartiainen, E.: Minimap: a web page visualization method for mobile phones. In: Proceedings of the SIGCHI Conference on Human Factors in Computing Systems (CHI 2006), pp. 35–44 (2006)
16. Sun, Z., Mahmud, J., Ramakrishnan, I.V., Mukherjee, S.: Model-directed Web transactions under constrained modalities. ACM Trans. Web 1(3), Article 12 (2007)

17. Wobbrock, J.O., Forlizzi, J., Hudson, S.E., Myers, B.A.: WebThumb: interaction techniques for small-screen browsers. In: Proceedings of the 15th Annual ACM Symposium on User Interface Software and Technology (UIST 2002), pp. 205–208. ACM, New York (2002)
18. Yang, Y.D., Zhang, H.J.: HTML Page Analysis Based on Visual Cues. In: Proceedings of 6th International Conference on Document Analysis and Recognition, pp. 859–864 (2001)
19. Zhang, K., Kong, J., Qiu, M., Song, G.: Multimedia Layout Adaptation Through Grammatical Specifications. ACM/Springer Multimedia Systems 10(3), 245–260 (2005)

Multiple Sections Extraction Using Visual Cue

Derren Wong and Jer Lang Hong

School of Computing and IT, Taylor's University
{derren.wong,jerlang.hong}@taylors.edu.my

Abstract. Current wrappers are unable to extract multiple sections data records from search engine results pages as sections usually have complicated layout and structure. Extracting data from search engine results pages is important for meta search engine applications and comparative shopping lists evaluation. In this paper, we present a novel data extraction technique which uses visual cue to check for the regularity of structure in multiple sections data records. Our findings show that though there are no regularity in structure for multiple sections data records, there is regularity in structure for multiple sections data records. Our technique is novel and can serve as a model for future multiple sections data extraction and it will be useful for meta search engine application, which needs an accurate tool to locate its source of information.

Keywords: Information Extraction, Automatic Wrapper, Search Engines.

1   Introduction

Current automatic wrappers are mostly designed to extract single section data records [5], [6], [7]. They use the principles that data records normally have similar data structures such as repetitive occurrence of similar nodes in the same tree level. However, automatic wrappers designed in this manner will not be able to extract multiple sections data records which are structured records containing important information and a large amount of such records can usually be found in a website.

Ordinary automatic wrappers are unable to detect and extract multiple sections data records because:

1.   Sections have different representation.
   Sections are usually presented in a number of ways in a web site. Furthermore, data records inside the section can also have different layout and format.
2.   There is no clear relationship between a section and a data record in the section.
   A reasonable assumption is that a HTML web page contains sections in its content and there can be a number of data records in the content of the sections. Thus, sections and data records are represented in a hierarchical form in Document Object Model (DOM) tree. In a DOM tree, HTML tag is the root tag. A sub-tree in a DOM tree will contains a HTML tag that represents section. Inside this HTML tag, it can further contain children tags which are data records. The examination of sample pages containing Multiple Section Data Records show that this example is not completely

T. Huang et al. (Eds.): ICONIP 2012, Part V, LNCS 7667, pp. 292–299, 2012.

true because the section and data records may not have a parent and child relationship as they can exist in other forms.

3.   It is difficult to partition sections from a web page.

Sections usually do not have a clear boundary in the HTML tags that can differentiate them. This is because a section can appear in a web page in many different ways, there are no general rules that can define the separation point between sections.

4.   Sections contain hidden template/schema.

Supervised and semi supervised wrappers require human labeling, this involves the correct identification of information to be extracted by the wrapper. Human labeling are usually done by marking the region of interest that containing data records in the HTML page so that the wrapper are able to generate extraction rules for data extraction. However, multiple section data records that contain sections with irregular structure and different search queries may cause the database server to generate different sections with different formats to be returned to user. Current wrappers (supervised and semi supervised) are not design to handle this kind of situation because these HTML pages need to be relabeled to cater for the new template/schema of these section every time a new section is generated. Hence, the labeling process is very tedious and time consuming. Automatic wrappers are design to extract data records without human labeling and support but they are not designed to handle the irregular structure in a multiple section data records.

Based on the above mentioned problems, we know that extracting multiple sections data records is a non trivial task. In this paper, we propose a novel and robust wrapper to extract relevant multiple sections data records from search engine results pages. Our findings show that multiple sections data records exhibit regularity in visual properties. Using visual cue obtained from the underlying browser rendering engine, we use visual properties such as width and height of the text node to determine the regularity of multiple sections data records. Our wrapper detects these data as regions and then filters out the relevant data as sections. Our wrapper is called Multiple SectionsWrapper (MultiSect).

This paper contains several sections. Section 2 describes the current work that is related to ours. Section 3 gives the implementation details of our wrapper. Finally Section 5 summarizes our work.

2   Related Work

Wrappers that detected data records using their repetitive sequence characteristics are unable to detect a section because a section can exist in different ways in a web page [1], [2], [3], [4], [10], [11]. Mining Data Region (MDR) [1] uses generalized nodes to detect data region. Generalized nodes are usually sibling nodes and located in the same level of a tree. Therefore, this wrapper is unable to identify the different formats of a section as a section might exist in a different level of a tree. Data Extraction based on Partial Tree Alignment (DEPTA) [12] uses data records boundaries in addition to generalized nodes to detect data records. Although visual information is incorporated in DEPTA, it is still unable to identify multiple sections data records as the visual information is only used to represent data records occur in same level only. Visual Perception based Extraction of Records (ViPER) [9] enhances the generalized nodes concept in MDR by using primitive tandem repeat. ViPER checks and determines repetitive patterns in a data records. Moreover, it is using X graph (a graph representing the horizontal position of every text elements in a HTML page) to determine separation between data records. However, the algorithm does not allow the partitioning of sections therefore it is unable to identify multiple sections data records.

Zhao improved and extended the work of Visual information aNd Tag (ViNT) [3] and developed Multiple Section Extraction (MSE) wrapper [4] to extract multiple sections data records. Although Zhao has successfully solved the problem found on extracting multiple sections data records, his approach has several limitations. Similar to ViNT, MSE uses content lines to determine the border of a particular text element. These content lines will form a block which contains data records. MSE is capable to determine the separation between data records and sections. The partitioning of sections could be carried out by using the separation distance (sections tend to be separated apart further than data records). The content lines used in ViNT and MSE are actually text/image nodes in a particular DOM tree. Therefore, MSE has not use much information of the DOM tree. Furthermore, Zhao also assumed the presence of Section Boundary Marker (SBM). SBM is a text that is located at the beginning or ending of a section (eg: "Search engine: LookSmart found 1 result. The query sent was history"). SBM usually exists in most of the sections but some sections do not have SBM as part of their contents.

In 2010, Hong presents a novel WMS DOM Tree and Visual Assisted based wrapper for multiple sections data records extraction [8]. WMS used a novel tree matching algorithm which calculates the frequencies of the tags in a particular tree to identify the similarity of data records. Unlike other existing works, WMS wrapper does not require the identification of Section Boundary Marker, and it can also works on single page record extraction. However, WMS wrapper suffers from several fundamental flaws. First, WMS is not able to identify dissimilar data records in different sections. Besides, WMS wrapper does not make full use of visual cue to match the similarity of data records located within the same section. Studies in [3] and [9] have shown that the use of visual cue in data extraction will help to improve the accuracy of a wrapper.

3   Proposed Solutions

3.1   Overview

We proposed a novel visual cue based wrapper to extract multiple sections data records from search engine results pages. Before extraction task could be carried out, we need HTML parser to parse through the HTML page and stored it in a DOM Tree for further processing. In this study, we use ICE Browser to parse through the HTML page and obtain the necessary DOM Tree with their associated visual cue information. Our proposed solutions are based on several intuitions:

1. Multiple Sections Data Records contain several patterns of interest in their data and layout. To detect and represent these data, we need visual cue from the HTML Tags in order to extract them.
2. Multiple Sections Data Records follows certain grouping rules that are they tend to group data records into section and group sections as an entity.
3. Multiple Sections Data Records exhibits similarity in their visual representation. For example, multiple sections data records contain headers and footers, and every data records in a section tend to exhibit the same visual information and layout (Fig. 1).

Fig. 1. Every data records in the section exhibit the same visual information and layout

3.2   Detection of Multiple Sections Data Records

Our wrapper uses the Adaptive Search extraction technique to determine and label potential tree nodes that represent data records (Fig. 2). Subtrees which store data records may be contained in potential tree nodes. The nodes in the same level of a tree are checked to determine their similarity (whether they have the same contents). If none of the nodes can satisfy this criterion, the search will go one level lower and perform the search again on all the lower level nodes. Our method involves the detection of repetitive nodes which may contain data records and the rearrangement of these nodes to form groups of potential records in a list in 2 steps:

In a particular tree level, if there are more than 2 nodes and a particular node occurs more than 2 times in this level, our wrapper will treat it as a potential data record irrespective of the distance between the nodes.

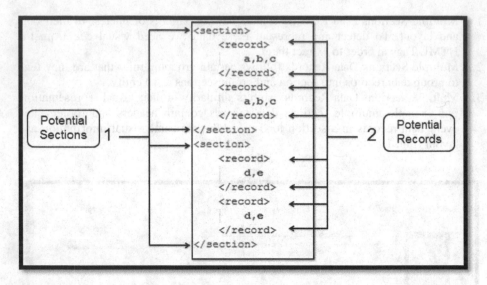

Fig. 2. Determine potential data record

These potential data records identified in this tree level are then grouped and stored in a list. The potential data records in this list are identified by the notation $[A_1, A_2, \ldots A_n]$ where A_1 denotes the position of a node in the potential data records where it first appears, A_2 is the position where the same node appears the second time and so on.

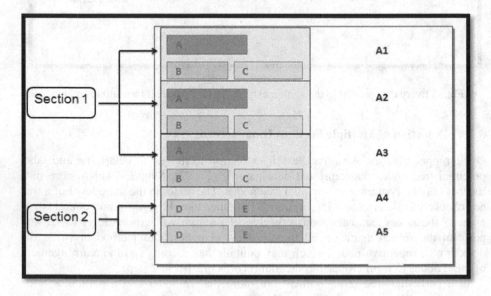

Fig. 3. Visual representation of data records in list

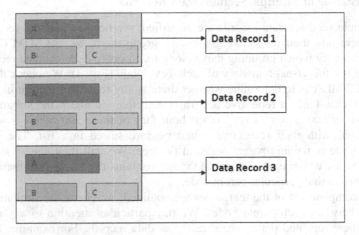

Fig. 4. Nodes appeared as a, b, c, a, b, c, a, b, c and the group {a, b, c} is chosen as data record

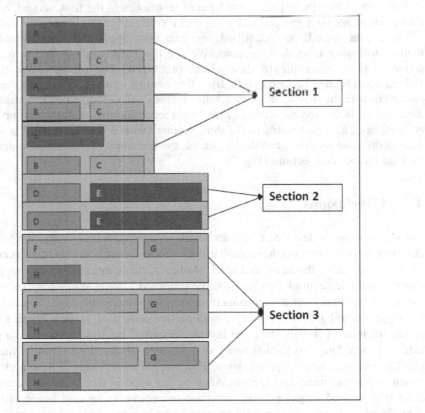

Fig. 5. SOM clustering method is used to identify the border of each section (represented in different colors). Sections with similar colors are treated as similar. After using SOM clustering method, the data records are grouped according to their specific section.

3.3   Clustering of Multiple Sections Data Records

Once multiple sections data records are identified, we need to sort the sections and grouped them into their respective groups. To group them, we use SOM Clustering method to identify which grouping they belong to. In order to detect data records in a section, we use the visual boundary of each Text Node (Fig. 3). We check the parent node of a HTML Tag to determine whether there is any repetitive pattern in the same level of the tree. Once a repetitive pattern is detected, we check the subtree of that parent node for text nodes. Every visual boundary of these text nodes is noted and their locations with their respective boundaries are stored in a list. The steps are repeated for the remaining parent nodes in the tree. We will then have n number of parent nodes list where each element in the list contains the visual information of all the text nodes in that particular parent node.

Once a complete list of the text nodes are obtained, we check the similarity of the text nodes between each parent nodes. We pay particular attention to any repeating patterns, where repeated patterns are treated as data records. For example, a section which has 3 data records may have pattern of text nodes occurring in the following orders: {a,b,c,a,b,c,a,b,c}.

For such a case, {a,b,c} are considered as text nodes in the first, second, and third data records. We then grouped {a,b,c} as data records (Fig. 4).

Once data records are identified, we can then apply the same formula and method to higher level data, sections. We use the same methods and principles to group sections according to their visual properties. For example, for the data {a,b,c,a,b,c,a,b,c,d,e,d,e,f,g,h,f,gh,f,g,h}, the group {a,b,c} is considered as data records in the first section (with 3 data records), the group {d,e} is considered as data records in the second section (with 2 data records), and finally the group {f,g,h} is considered as data records in the third section (with 3 data records). To partition data records and sections accordingly, we use the boundary of the groups to determine the data records and sections (Fig. 5).

4   Conclusions

Automatic wrapper designed to extract single sections data records use the principles that data records normally have similar data structures such as repetitive occurrence of similar nodes in the same tree level. However, this cannot apply to wrapper that extracts multiple sections data records. We proposed a novel visual based wrapper to extract multiple sections data records from search engine result page. This solution is based on several intuitions. Firstly, multiple sections data records contain several pattern of interest in their data and layout. Secondly, multiple sections data records tend to group data into section and group sections as an entity. Thirdly, they have similarity in their visual representation. The wrapper uses Adaptive Search extraction technique to determine data records. After we get a list of data records, we use SOM clustering method to group them into their respective group and hence extract the data. The techniques we proposed are novel and can be easily implemented in real world applications. This technique can also served as a model for future multiple sections data records extraction and it will be useful for meta search engine application, which needs an accurate tool to locate its source of information.

References

1. Liu, B., Grossman, R., Zhai, Y.: Mining data records in Web. In: ACM SIGKDD, pp. 601–606 (2003)
2. Miao, G., Tatemura, J., Hsiung, W.-P., Sawires, A., Moser, L.E.: Extracting Data Records from the Web Using Tag Path Clustering. In: ACM WWW, pp. 981–990 (2009)
3. Zhao, H., Meng, W., Wu, Z., Raghavan, V., Yu, C.: Fully automatic wrapper generation for search engines. In: ACM WWW, pp. 66–75 (2005)
4. Zhao, H., Meng, W., Yu, C.: Automatic extraction of dynamic record sections from deep web. In: ACM VLDB (2006)
5. Hong, J.L., Siew, E., Egerton, S.: Information Extraction for Search Engines using Fast Heuristic Techniques. DKE 69(2), 169–196 (2010)
6. Hong, J.L.: Deep Web Data Extraction. In: IEEE SMC (2010)
7. Hong, J.L.: Data Extraction for Deep Web using WordNet. In: IEEE TSMC (2011)
8. Hong, J.L., Siew, E., Egerton, S.: WMS- Extracting Multiple Sections Data Records from Search Engine Results Pages. In: ACM SAC (2010)
9. Simon, K., Lausen, G.: ViPER: augmenting automatic information extraction with visual perceptions. In: ACM CIKM, pp. 381–388 (2005)
10. Liu, W., Meng, X., Meng, W.: ViDE: A Vision-based Approach for Deep Web Data Extraction. IEEE TKDE 22(3), 447–460 (2009)
11. Su, W., Wang, J., Lochovsky, F.H.: ODE: Ontology-assisted Data Extraction. ACM TODS 34(12) (2009)
12. Zhai, Y., Liu, B.: Web data extraction based on partial tree alignment. In: ACM WWW, pp. 76–85 (2005)

Iterative Appearance Learning with Online Multiple Instance Learning

Bo Guo, Juan Liu, and Junpeng Chen

Wuhan University, School of Computer,
Wuhan 430072, China
guobo@whu.edu.cn,
liujuan@whu.edu.cn,
chenjp@whu.edu.cn

Abstract. A recent trend in object detection and tracking is using multiple-instance learning (MIL) to resolve the uncertainties in the training set. Though using online multiple instance learning instead of traditional instance-based learning can lead to a more robust appearance classier, but it also tends to drift or fail in case of wrong updates during the online self-learning process. In this work we propose a method to combine the benefit of online MIL learning and off-line/batch learning to get a robust appearance model which is able to effectively handle drifting problem. Our method not only copes with ambiguity with power of multiple instance learning, but also uses off-line learning with a sample weights descending in a iterative framework to suppress drifting in the result of online MIL. We demonstrate the effectiveness and robustness of our method on several challenging video clips and show performance improvement comparing to other state-of-art approaches especially to online MIL learning in a fully occlusion scene.

Keywords: Online learning, Adaptive Model, Multiple Instance Learning, Tracking By Detection.

1 Introduction

Machine learning techniques has been successfully used to detect and track many classes of object from videos such as cars, pedestrians, faces, and so on. Many methods have been proposed for this purpose. For examples, Wei *et al.* proposed an interactive off-line tracking system in an efficient global optimization framework[1]; Hasler *et al.* proposed a method called Flowboost [2] to learn the appearance model from a sparsely labeled training video (where at least one per 60 frames is manually labeled) using k-shortest paths optimization [3]; Li *et al.* proposed a hybrid boost model to learn the affinity in classification using future frames [4]. Traditionally these methods are off-line ones that require training sets to be provided in advance. In order to get a robust classifier, the training set should contain instances reflecting the appearance changes, illumination changes, view point changes and other possible variations of target object, which demand a lots of prohibitive effort of enormous manual annotation. More importantly, the classification model trained from the pre-determined and limited training set

T. Huang et al. (Eds.): ICONIP 2012, Part V, LNCS 7667, pp. 300–308, 2012.

generally is not adaptive enough and can not be applied to rapid and massive appearance changing scenarios.

This kind of difficulties can be partially handled with online learning methods which are able to incrementally update their representations. Online learning [5] training classifier online and incremental as new data incoming could overcome the drawback of traditional off-line learning's requirement of providing training set prior.

The online learning with appearance-based classifier is especially suitable for a tracking problem through recently dominant "Tracking by Detection" technique [6] because the entire video frames of clips are too difficult to be full labeled for traditional off-line learning. The online learning algorithm continually updates classifier as new frame coming and evolves an adaptive appearance model during the online learning process as appearance changing. Grabner et.al [7] proposed an online semi-supervised boosting algorithm which regularizes the online classifier [8] with first frame regulation, the semi-supervised learning are heavily rely on rst frame label regulation and tends to be unstable when rapid and massive appearance changing occurs; Stalder [9] proposes a method combining supervised classifier and semi-supervised classifier, where supervised classier is working as a detector for validation, but the supervised classier is often retrained in many challenging occasions and is inclined to lost target. Since the online classier performs self-learning and decides where to take positive and negative samples autonomously, the precision of each update of the classier is crucial for online learning.

Recently, multiple-instance learning (MIL) has been introduced to combine with online learning to increase the precision of classication by addressing the uncertainty of the training samples. Babenko et.al [10] proposed an online multiple instance learning method which can learn a robust adaptive model handling partial occlusion without large drifting in an online procedure. The online MIL in [10] conducts very robust results, but it still faces the drifting problem which online learning suffers from. More importantly, The online MIL can not handle scenarios where the object leaves the scene completely, it will start learning incorrect samples and finally lose the target object. Zeisl [11] proposed a model to combine online MIL and semi-supervised learning, but it still heavily relies on first frame regulation and is inclined to be unstable.

In this work we focus on learning an arbitrary object without prior knowledge, we propose an approach which combines online MIL learning and off-line learning to get a robust appearance model from a video sequence where only the first frame is annotated. Due to the integration of online and off-line learning framework, our method is expected to perform well in scenarios where the object leaves the scene completely. We first use online MIL learning to get positive and negative sample sets in a first frame labeled video sequences. Though the result sample sets retrieved by online MIL learning are robust due to the advantage of online MIL learning of processing ambiguity, it still inevitably contains noises induced by drifting. Therefore we then iteratively re-weight samples weight based on an importance weight descending strategy in an iterative off-line learning framework to finally get a appearance model. Due to the iterative processing, our method can avoid "forgetting factors" parameter tuning and make a reliable decision boundary when training set is provided appropriately.

We compare our method with several state-of-art approaches on some challenging video sequences, the experimental results show that our approach leads to more robust and accurate results. Out method can effectively handle drifting problem and leaving scene or fully occlusion problem where the comparing online learning methods unavoidable encounter.

2   Iterative Appearance Learning

2.1   Online Multiple Instance Boosting

Boosting is a framework for improving the accuracy of any learning algorithms. We use the boosting [12] as our main learning algorithm framework in online and off-line. Let D denotes a training set $\{(x_1, y_1), (x_2, y_2) \dots (x_n, y_n)\}$, $x_i \in \mathbb{R}^D$, $y_i \in \{0, 1\}$, weak classifier is a mapping $h : x \to y, h \in \mathcal{H}$, $\mathcal{H}$ is the set of all possible weak classifiers.

The character of boosting is combining many/T weak classifiers to produce a strong classifier H.

$$H(x) = \sum_{k=1}^{T} \alpha_k h_k(x) \tag{1}$$

where α is scalar weight of weak classifier h. The goal of H is to minimize the average exponential loss:

$$\mathcal{L}(H) = \frac{1}{n} \sum_{i=1}^{n} \exp\left(-y_i H(x_i)\right) \tag{2}$$

Every weak classifier is added in a greedy forward stage-wise fashion and all classifiers are trained sequentially using the following formula:

$$(\alpha_k, h_k) = \arg\min_{\alpha, h} \mathcal{L}(H_{k-1} + \alpha h) \tag{3}$$

In multiple instance learning [13], the data is presented as "bags" in the form of $\{(\mathcal{B}_1, y_1), (\mathcal{B}_2, y_2), \dots, (\mathcal{B}_n, y_n)\}$ rather than individual instances, where a bag $\mathcal{B}_i = \{x_{i1}, \dots, x_{im}\}$, y_i is bag label. All samples in a bag will share a label. A negative bag means all samples in the bag are negative samples for $y_i = 0$, while a positive bag is ambiguous because it only guarantees that there is at least one instance in the bag is positive for $y_i = 1$.

Multiple instance learning algorithm handles the ambiguity to decide which one is the most right instance in each positive bag. We follow the main MIL Boost framework proposed by Viola et.al [13] which use gradient boosting framework [14] to maximize the likelihood of bags:

$$\mathcal{L}_{bag} = \prod_{i} p_i^{y_i} (1 - p_i)^{(1 - y_i)} \tag{4}$$

where $p_i = p(y = 1 | \mathcal{B}_i)$ is the probability of the bag being positive, y_i is bag label. In [13,15], bag probability uses Noisy-OR(NOR) model. We adapt geometric mean function [11] to model the bag probability:

$$p(y = 1 | \mathcal{B}_i) = 1 - \left[\prod_{j} (1 - p(y_i | x_{ij}))\right]^{1/N_{bi}} \tag{5}$$

which is more suitable than NOR model because the posterior probability converges well, x_{ij} is the jth instance feature vector of bag $\mathcal{B}_i$, N_{bi} is instances number of bag $\mathcal{B}_i$.

The weak classifiers are selected sequentially in a greedy manner as in Eq.3, every weak classifier is added to maximize the log likelihood in Eq.4

$$
\begin{aligned}
likelihood &= \log \left(\prod_i p_i^{y_i} (1 - p_i)^{(1-y_i)} \right) \\
&= \sum_i y_i \log p_i + \sum_i (1 - y_i) \log (1 - p_i)
\end{aligned}
\tag{6}
$$

The log likelihood is defined over bags not instances, so we must define the instance probability to calculate the bag probability in Eq.5. we model the instance probability as a sigmoid function of H:

$$
p(y = 1 \mid x) = sigmoid(H(x)) = \frac{1}{1 + e^{-H(x)}}.
\tag{7}
$$

Just like [8,10], we use a weak stump classifier candidates pool $\mathcal{P}$, when new samples arrive through on-line learning, all weak classifier in the pool will be updated in parallel, then we choose K weak classifiers from M weak classifier $(M \gg K)$ candidates pool $\mathcal{P}$ sequentially by maximizing the log likelihood:

$$
h_k = \arg\max_{h \in \mathcal{P}} likelihood (H_{k-1} + h)
\tag{8}
$$

The online learning algorithm we implemented in this work is inspired by Babenko's work [10], and is summarized in Algorithm 1.

2.2   Iterative Appearance Learning

We propose our approach called Iterative Boost (IB) to take advantage of off-line learning to handle the drifting problem and leaving scene problem occurred in online MIL learning. We use Adaboost framework as our off-line framework. Off-line Adaboost could effectively handle the leaving-scene problem if the training samples are appropriately weighted. We model the drifting effect in the online MIL learning as a function in the form of sample weights. The intuition behind this is that drifting induced by online MIL learning could be viewed as a time accumulative effect. Drifting tends to happen in online MIL processing over time along with the appearance changing over time. At first, the on-line MIL learning can correctly identify the object location, but with changing of appearance, the model gradually can not capture the change of the object and begins to make the wrong decision, then drifting starts to be induced.

We assume the sample sets S^{pos} and S^{neg} which are retrieved by online MIL learning generally contain noise with drifting character, we use a sample weight descendent strategy to handle the sample noise on the behalf of drifting effect. As discussed before, drifting could be viewed as a timing accumulative effect of dramatic changing of appearance, if online detector starts drifting, it is more likely to drift more severe in later

---

Algorithm 1. Online MIL Boosting

Require: loc_t = Initial object location; video sequences V

1: $H = 0$, $S^{pos} = \emptyset$, $S^{neg} = \emptyset$
2: **while** next video frame I_i exists **do**
3:      Crop out a set of image patches $X^s = \{x : \lambda > \|loc(x) - loc_t\|\}$
4:      Computer every x haar feature vector, $x \in X^s$.
5:      **if** $H \neq 0$ **then**
6:          $p(y = 1|x) = sigmoid(H(x))$ for every $x \in X^s$.
7:          $x^{pos} = \arg\max_{x \in X^s} p(y = 1\,|x)$
8:          $loc_t = loc(x^{pos})$
9:      **else**
10:         $x^{pos} = x$ where $loc(x) = loc_t$
11:     **end if**
12:     $X^{pos} = \{x : \|loc(x) - loc_t\| < \rho\}$
13:     $X^{neg} = \{x : \rho < \|loc(x) - loc_t\| < \psi\}$
14:     $\mathcal{B} = X^{pos} \cup X^{neg}$
15:     Update all M weak classifiers in the candidates pool for every instance $\{x_{ij}, y_i\}$ in the bag set $\mathcal{B}$
16:     Initialize $H = 0$
17:     **for** $k = 1$ to K **do**
18:         **for** weak classifier h_m in M candidates pool **do**
19:             $p_{ij}^m = sigmoid(H + h_m(x_{ij}))$
20:             calculate bag probability p_i using Eq.5
21:             $L^m = \sum_i l(i)$ using Eq.6
22:         **end for**
23:         $m^* = \arg\max_m L^m$
24:         $H = H + h_{m^*}(x)$
25:     **end for**
26:     $S^{pos} = x^{pos} \cup S^{pos}$
27:     $x^{neg} = \{x_i^{neg} = random(X^{neg}), i = 1, ..., ni\}$
28:     $S^{neg} = x^{neg} \cup S^{neg}$
29: **end while**
30: **Output:** Positive sample set S^{pos} and Negative sample set S^{neg}

---

frames, so we assume samples which are got earlier have more important weights than samples which are got later in on-line MIL learning. That means The earlier samples are more important than the later samples, so we model the weights of samples as:

$$w_i = \exp\left(-\frac{i^2}{\left(2t\xi|S|\right)^2}\right), i = 1, ..., |S|, \tag{9}$$

where i is the ith sample, ξ is a const factor, t is the iteration number, $|S|$ is samples count. The sample index number i is ascending ordered according to the time stamp when it has been added in the sample set, i.e. $w_m > w_n$ if $m > n$ where m,n is the mth and nth sample respectively.

We propose our approach with pseudocode in Algorithm 2. We use a iterative process to handle the noise in the initial training set, and use Eq.9 as a sample weights descending model to initialize the weights of sample set. Once we initialize the sample weights, we then pass the weighted positive sample set S^{pos} and negative sample set S^{neg} to train a off-line classifier H^* which is combined of G weak classifiers. The strong classifier H^* is used to reclassify the video sequences V which is processed by online MIL learning before, then we collect the results as new positive sample set S^{pos} and negative sample set S^{neg}, one iteration ends here. Once the new training set is prepared, we start a new iteration to reinitialize the new training set and feed it to off-line learning process again. After all iteration, we finally get rened sample sets S^Pos and S^neg and a robust appearance model H^*.

Algorithm 2. Iterative Boosting

Require: Video sequences V with Initial object location
1: Use online MIL Learning to get positive and negative sets S^{pos}, S^{neg}
2: **for** $t = 1$ to N **do**
3:      Initialize weights of S^{pos} using Eq.9
4:      Initialize weights of S^{neg} using Eq.9
5:      Normalize positive and negative sample weights

$$w_{pi} = \frac{w_{pi}}{2\sum_{pos} w_{pi}}, w_{ni} = \frac{w_{ni}}{2\sum_{neg} w_{ni}}$$

6:      $S_t = S^{pos} \cup S^{neg}; W_t = \{w_{pi}\} \cup \{w_{ni}\}; H^* = 0$
7:      **for** $g = 1$ to G **do**
8:          Normalize sample weights for all $w_i \in W_t$

$$w_i = \frac{w_i}{\sum_i w_i}$$

9:          Choose weak classifier h^* with the lowest error $e = \sum_i w_i |h^*(x) - y_i|$ using weighted sample set S_t
10:         $\alpha = \frac{1}{2} log \frac{1-e}{e}$
11:         $w_i = w_i \log \frac{e}{1-e}$ for every sample classified correctly
12:         $H^* = H^* + \alpha h^*$
13:     **end for**
14:     Use H^* to classify video sequences V
15:     Crop out classified results into positive sample set S^{pos} and negative sample set S^{neg} respectively
16: **end for**
17: **Output:** Strong classifier H^*

---

3   Experiment

3.1   Experimental Setting

We apply our Iterative Boosting (IB) algorithm on several challenging video sequences and we use Haar features to allow for a comparison to other state-of-art algorithms:

Online-AdaBoost (OAB) [8], Beyond Semi-Supervised (BSER) [9], along with the original online-MIL model (MIL) [10] on the same context for comparison.

For OAB and BSER, we use all default setting suggested in [8][9]. For MIL, all parameters are tuned as author suggested in [10] In IB, const factor ξ is set to 0.1, we use $G = 100$ weak classifiers in IB to combine the strong classifier H^* and iterate $N = 10$ times to generate the final classifier.

We use average center location errors to indicate qualitative comparison on test video sequences. The center location error is defined in the form of Euclidean distance: $error = \|p - p^*\|_2^2$ where p is the object coordinate given by algorithms, p^* is ground truth.

Table 1. Average Center Location Errors (pixels)

Video Clip	OAB	BSER	SemiBoost*	Frag*	MIL	IB
coke11	30	14	**13**	63	57	<u>8</u>
david	45	73	39	46	**15**	<u>11</u>
dollar	15	61	67	59	**14**	<u>11</u>
faceocc	**17**	92	23	45	22	<u>12</u>
surfer	13	38	9	139	**8**	<u>6</u>
sylv	50	67	16	<u>11</u>	20	**13**
tiger1	51	92	42	39	**14**	<u>7</u>
tiger2	**12**	38	61	37	<u>8</u>	<u>8</u>

<u>underline</u> and **Bold** indicate best and second best performance respectively. SemiBoost* and Frag* result are directly used from [10].

We summary mean errors over all frames of every test clip in Table 1. Our method acquires 7 best. This clearly illustrates that our algorithm provides more accurate results compare to other state-of the art approaches. We also provide error location versus frame number plots for each test clip in Fig.1. In addition, Fig.2 show screen captures for some clips.

4   Discussion and Future Work

In this paper we present a method to learn an adaptive appearance model which combines the benet of online MIL learning and off-line learning in a rst frame labeled video sequences. Our method not only copes with ambiguity in training set with power of multiple instance learning, but also uses off-line learning with a sample weights descending in a iterative framework to suppress drifting in training set. Our method is applied to several challenging video clips, experiment results demonstrate that the performance promotion comparing to original online MIL boosting and other state-of-art models. Although our method focuses on single object detection, it is easily extended to multiple object appearance learning. We admit that integrating a off-line part into algorithm will lose real-time ability comparing to other pure online algorithms, but in many applications such as sport game strategy analysis, stable result is more important

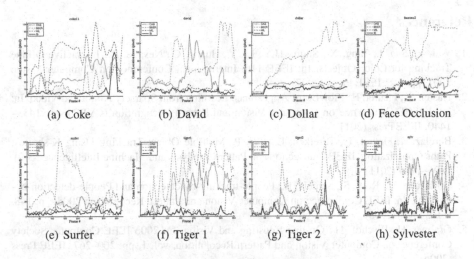

Fig. 1. Location error plots

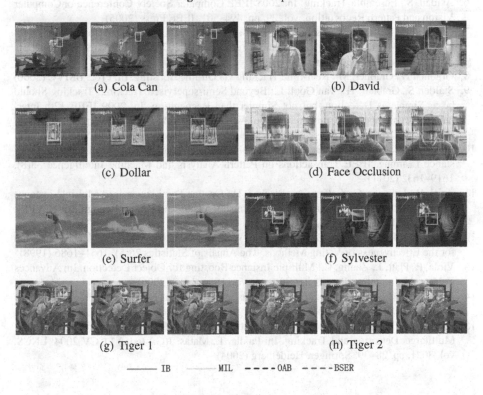

Fig. 2. Screen captures for test clips

than real-time ability. In the future, we will use a more complicate motion model such as particle lter to add a regulation to boost procedure. More information like color and edges will also be considered to use for feature representation.

References

1. Wei, Y., Sun, J., Tang, X., Shum, H.Y., Li, Y., Huang, C., Nevatia, R.: Interactive Offline Tracking for Color Objects. In: IEEE 11th International Conference on Computer Vision 2007, pp. 1–8. IEEE Press (2007)
2. Hasler, D., Fleuret, F.: FlowBoost Appearance Learning from Sparsely Annotated Video. In: 2011 IEEE Conference on Computer Vision and Pattern Recognition (CVPR), pp. 1433–1440. IEEE Press (2011)
3. Berclaz, J., Fleuret, F., Turetken, E., Fua, P.: Multiple Object Tracking Using K-shortest Paths Optimization. IEEE Transactions on Pattern Analysis and Machine Intelligence 33(9), 1806–1819 (2011)
4. Andriluka, M., Roth, S., Schiele, B.: People-tracking-by-detection and People-detection-by-tracking. In: IEEE Conference on Computer Vision and Pattern Recognition 2008, pp. 1–8. IEEE Press (2008)
5. Grabner, H., Bischof, H.: On-line Boosting and Vision. In: 2006 IEEE Computer Society Conference on Computer Vision and Pattern Recognition, vol. 1, pp. 260–267. IEEE Press (2006)
6. Avidan, S.: Ensemble Tracking. In: 2005 IEEE Computer Society Conference onComputer Vision and Pattern Recognition, vol. 29, pp. 494–501. IEEE Press (2005)
7. Grabner, H., Leistner, C., Bischof, H.: Semi-supervised On-Line Boosting for Robust Tracking. In: Forsyth, D., Torr, P., Zisserman, A. (eds.) ECCV 2008, Part I. LNCS, vol. 5302, pp. 234–247. Springer, Heidelberg (2008)
8. Grabner, H., Grabner, M.: Real-time Tracking via On-line Boosting. In: Proc. BMVC (2006)
9. Stalder, S., Grabner, H., Van Gool, L.: Beyond Semi-supervised Tracking: Tracking Should be as Simple as Detection, but not Simpler than Recognition. In: 2009 IEEE 12th International Conference on Computer Vision Workshops (ICCV Workshops), pp. 1409–1416. IEEE Press (2009)
10. Babenko, B., Yang, M.H., Belongie, S.: Robust Object Tracking with Online Multiple Instance Learning. IEEE Transactions on Pattern Analysis and Machine Intelligence 33(8), 1619–1632 (2011)
11. Zeisl, B., Leistner, C., Saffari, A., Bischof, H.: On-line Semi-supervised Multiple-instance Boosting. In: 2010 IEEE Conference on Computer Vision and Pattern Recognition (CVPR), p. 1879. IEEE Press (2010)
12. Schapire, R.E., Freund, Y., Bartlett, P., Lee, W.S.: Boosting the Margin: A New Explanation for the Effectiveness of Voting Methods. The Annals of Statistics 26(5), 1651–1686 (1998)
13. Viola, P., Platt, J., Zhang, C.: Multiple Instance Boosting for Object Detection. In: Advances in Neural Information Processing Systems, vol. 18, p. 1417 (2006)
14. Friedman, J.H.: Greedy Function Approximation: a Gradient Boosting Machine. Annals of Statistics, 1189–1232 (2001)
15. Okuma, K., Taleghani, A., de Freitas, N., Little, J.J., Lowe, D.G.: A Boosted Particle Filter: Multitarget Detection and Tracking. In: Pajdla, T., Matas, J(G.) (eds.) ECCV 2004. LNCS, vol. 3021, pp. 28–39. Springer, Heidelberg (2004)

Exploring Crude Oil Impacts to Oil Stocks through Graphical Computational Correlation Analysis

Anthony Lai, Lei Song, Yiming Peng, Peter Zhang,
Qili Wang, and Shaoning Pang

Department of Computing, Unitec Institute of Technology,
Private Bag 92025, New Zealand
alai@unitec.ac.nz

Abstract. This paper presented the relationship between the world price of crude oil and oil related stocks over 2011:4-2011:6 is analysed by using a graphical computational correlation analysis method. The Operation Neptune Spear happened in 2011:5 may change nature of the price connection between oil and stock, we evaluate and rank the impact of crude oil for the period before and after the event respectively. Over the statistical results, we find that graphic correlation is superior to typical point-to-point distance calculation for correlation analysis; and we discover stock market interesting knowledge on crude-oil to stock correlations.

Keywords: Crude oil to stock correlation, Graphical computational correlation analysis, Economical event impact analysis.

1 Introduction

The price of crude oil is often dynamic but it is considered one of the major factors for understanding fluctuations of related stock prices. In principle, the production of crude oil is subjected to "supply and demand" activities. The marketplace forces of supply and demand determine the price of crude oil. If demand grows or if a disruption in supply occurs, there will be an upward pressure on prices. On the other hand, if demand falls or there is an oversupply of crude oil in the market, there will be a downward pressure on prices.

Some researches show that the natural disaster may cause the crude oil to rise dramatically. Hurricane Katrina caused oil price to rise $3 a barrel, and gas price to reach $5 a gallon in 2005. Katrina affected 19% of the nation's oil production. It had destroyed 113 offshore oil and gas platforms, and 457 oil and gas pipelines were damaged [17]. Similarly, the Mississippi River flooding in May 2011 caused gas price growing to $3.98 a gallon [19]. Traders' concern lies at that, the flooding would damage oil refineries and the supply cannot accommodate the demand.

Also, economic events cause crude oil price change. For example, US's most recent military activities in the Middle East region have stirred up the political

T. Huang et al. (Eds.): ICONIP 2012, Part V, LNCS 7667, pp. 309–317, 2012.

unstable in the respective regions. These military actions will continue affecting crude oil price in short and long term, as the unstable political situation will cause a certain detrimental effect on the crude oil production from Middle-East countries. As seen in Figure 1, the WTI crude oil price fluctuates because Osama Bin Laden was killed in the Operation Neptune Spear on 02/05/2011.

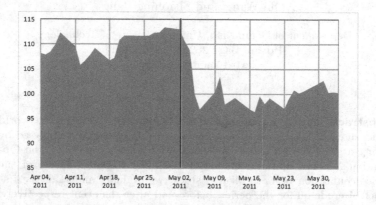

Fig. 1. The WTI crude oil price variation for the period of 03/04/2011 to 02/06/2011

There is a possible factor that the majority inventory oil producers and consumers build a storage capacity to store crude oil for immediate future needs. They also build some inventories to speculate on the price expectations and sale/arbitrage opportunities in case of any unexpected changes in supply chain. Any change in these inventory levels triggers volatilities in crude oil's prices which in turn creates ripples in the stock markets.

Another significant factor, which impacts on crude oil price, is the Organization of Petroleum Exporting Countries. A large part of the world's crude oil share is produced by OPEC (Organization of Petroleum Exporting Countries) nations. Apparently, they have the power in controlling the crude oil prices. Any decisions, made by OPEC countries to raise the prices or reduce production, would immediately lead to crude oil price shock in the global commodity market.

Importantly, the crude oil price is a major driver on its related stock market prices. Clearly, the crude oil is one of the most influencing evidences on macro-variables economic and commodity market, which implies that oil price variation is associated with most economic activities. This paper investigates in-depth the relationship between crude oil price and 6 oil-related stocks. A graphical computational correlation analysis method is introduced to discover those associations.

2   Related Work

2.1   Previous Oil Prices Impact Investigations

In literature, a number of previous researches have been conducted to investigate the effect of oil prices on stock market returns [3,5,4,1,7]. Jones research [5] indicated that oil price rises have a negative impact on stock market returns for all sectors except mining, oil and gas industries. Similarly supported by Robert[2] found that crude oil has significant positive correlation to the oil and gas and diversified resources industries. Nevertheless, he also found negative oil price correlation in the paper and packaging as well as transport industries. Kling[6] summarized that the increases in crude oil price may have been followed months later to see the impact of declining the stock prices. Isaac[4] used co-integrated vector error correction model to analyze the long run relationship between crude oil and international stock market which showed stock market respond negatively to increases in the oil price long run. However, Lutz[1] found the aggregate stock returns may differ greatly depending on whether the increase in the price of crude oil is driven by demand or supply shocks in the crude oil market. Huang[7] examined the lead-lag correlations between daily returns of oil futures contracts and stock returns. He found out there is no correlation between oil futures returns and the returns of various stock indexes. In the case of specific oil stocks, there is existing correlation on one day lead of oil futures returns. Both Perry and Huang used similar methods to evaluate the correlation.

2.2   Existing Correlation Analysis Methods

In recent years, correlation analysis methods have been with extensive and productive efforts. The research has been expanding from simple to more advanced techniques both in depth and in breadth. The computational correlation analysis is normally based on two variables (e.g., crude oil price and a stock price).

Point-to-point distance similarity is a straightforward correlation measurement. Given two variables X and Y, their correlation can be defined as their Euclidean mean distance,

$$d_s = \frac{\sum_{t=1}^{T} |y_t - x_t|}{T}, \tag{1}$$

where t is the dimensionality of variable X and Y.

The dot product is another simple approach to computational correlation analysis. It measures the angle between two vectors/variables that have the same initial point as,

$$cos(\theta) = \frac{X.Y}{\|X\|\,\|Y\|}, \tag{2}$$

where $cos(\theta)$ identifies the trend closeness of two vectors/variables.

Pearson's Correlation is the first formal correlation measure and it is still the most widely used measure of relationship[8]. It is a statistical measure of two variables movement relationship, which can be calculated as correlation coefficient $\rho_{X,Y}$,

$$\rho_{X,Y} = \frac{cov_{(X,Y)}}{\sigma_X \sigma_Y} = \frac{E((X - \mu_X)(Y - \mu_Y))}{\sigma_X \sigma_Y}, \tag{3}$$

where cov is the covariance; σ_X and σ_Y are standard deviations; μ_X and μ_Y are the expected value; and E is the expected value operator. Practically, except $\rho_{X,Y}$ returns a probability p-value. The advantage of using Pearson's correlation is that more accurate prediction can be made when a strong correlation exists amongst variables.

3   Graphical Computational Correlation Analysis

In this research, a graphical computational correlation analysis method is adopted which is called channel method [18] is used to model a concrete arc for graphically estimating the trend of the target stock price. Figure.2 shows 4 typical trend patterns: fast growing, slowly increasing, fast dropping and slowly decreasing. It is straightforward to describe each of the trend pattern by an arc which can be formulated to a sub-circle shown as in Figure 2. The 4 trend patterns can be obtained by the following equations respectively:

$$(x - x_0)^2 + (y - y_0)^2 = R^2 \left| \begin{array}{l} x_0 = 0, y_0 = R \\ x \in [0, sin\alpha \cdot R\sqrt{2(1 - cos2\alpha)}] \end{array} \right. \tag{4}$$

$$(x - x_0)^2 + (y - y_0)^2 = R^2 \left| \begin{array}{l} x_0 = R, y_0 = 0 \\ x \in [0, sin\alpha \cdot R\sqrt{2(1 - cos(\pi - 2\alpha))}] \end{array} \right. \tag{5}$$

$$(x - x_0)^2 + (y - y_0)^2 = R^2 \left| \begin{array}{l} x_0 = 0, y_0 = 0 \\ x \in [0, (1 - cos\alpha) \cdot R\sqrt{2(1 - cos(\pi - 2\alpha))}] \end{array} \right. \tag{6}$$

$$(x - x_0)^2 + (y - y_0)^2 = R^2 \left| \begin{array}{l} x_0 = R, y_0 = R \\ x \in [0, (1 - cos\alpha) \cdot R\sqrt{2(1 - cos2\alpha)}] \end{array} \right. \tag{7}$$

where $\alpha \in (0, \pi/4)$, $\angle\alpha$ is used to measure the speed of increasing and decreasing trend, and the radius R determines the length of the trend pattern corresponding to the time period of observation. In practice, a discrete arc can be obtained according to the length of time series for channel approximation.

Given the observation (i.e, a stock closing price) X within a time frame T, applying Eq.(4) - Eq.(7) to X, respectively, one of 4 types arc (i.e. functions) called 'channel pattern' is selected with its parameter α tuned to best suit the time series under observation,

$$p = \arg \min_{\alpha, i \in [1,4]} \frac{\sum_{t=1}^{T} \|p_t^i - x_t\|}{T}. \tag{8}$$

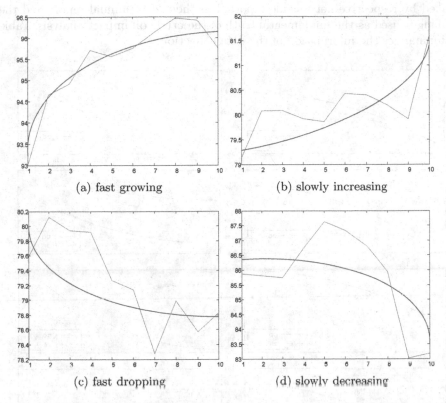

(a) fast growing   (b) slowly increasing

(c) fast dropping   (d) slowly decreasing

Fig. 2. Four trend patterns used for channel approximation

In addition, based on the extracted 'channel pattern' p, the channel distance of an observation X to the reference Y(i.e, WTI Crude Oil) can be defined as below:

$$d_c = \frac{\sum_{t=1}^{T-1} \left((p_{t+1}^y - p_t^y) + (p_{t+1}^x - p_t^x) \right)}{T-1}. \tag{9}$$

4   Computational Evaluations and Results

4.1   Data

In the research, the historical time frame 03/04/2011 to 02/06/2011 is selected as the study period. It is noticeable that within the period, the WTI crude oil price has a big fluctuation due to the Operation Neptune Spear event which led to the killing of Osama Bin Laden on 2 May 2011. WTI crude oil price data is collected from the oil-price.net. Within the same time frame, 6 crude oil related stocks price data is also collected from Google Finance. This includes ExxonMobil Cooperation (XO), Royal Dutch Shell Group (ZX), BP United Kingdom (BP), Total S.A. (TT), Chevron (CV) and Conoco Philips (CP). The 6 companies is ranked by respective net income reported in their 2011 annual report and the ranking is used as the fundamental truth of the crude oil impact analysis. Table 1 summarizes the information of the data collection.

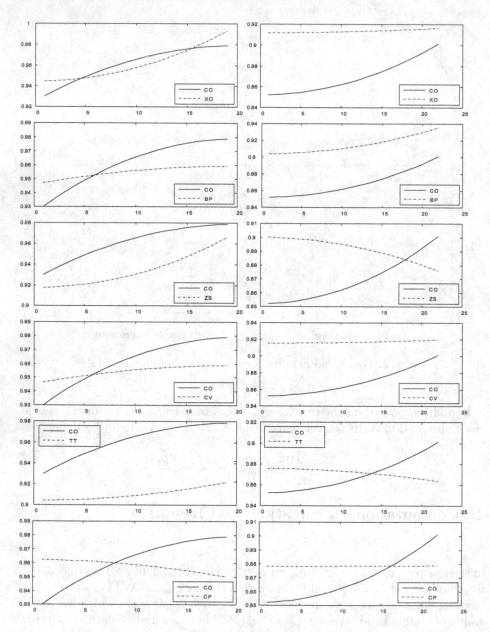

Fig. 3. Stock to crude oil correlation illustration by graphical channels. (left) before event, and (right) after event.

Table 1. The description of 6 crude oil related stocks and their ranks

Rank	Stock Name	Net Income ($ million)
1	Exxon Mobil Corporation (XO)	41,060 [14]
2	BP(BP)	38,463 [9]
3	Royal Dutch Shell(ZS)	31,185 [11]
4	Chervon(CV)	26,895 [10]
5	Total S.A.(TT)	15,902 [12]
6	Conoco Phillips (CP)	12,436 [13]

4.2 Results

In the crude oil impact analysis, the influence of the Operation Neptune Spear event is considered and the study period 03/04/2011-02/06/2011 are divided into two sub-periods, which are before event period 03/04/2011-01/05/2011, and after event period 02/05/2011-02/06/2011. As a preprocessing step, all prices are normalised into [0, 1] to ensure correlation calculation is comparable among different stocks. Then, the correlation of each stock is evaluated to the crude oil by the distance similarity (i.e., Eq.(1)) and channel similarity (i.e., Eq.(9)) calculation, respectively.

As a result, Figure 3 gives a stock to crude oil correlation illustration by the graphical channel method, where the *left* and *right* columns illustrate respectively the correlation for each stock before and after the Operation Neptune Spear event. Table 2 (a) and (b) present the results of crude oil to stock relevance evaluation and ranking by distance and channel similarity, respectively.

As seen from the tables, the overall rank by channel similarity is found consistent to the fundamental truth rank that were defined by the 2011 stock company net income comparison given in Table 1. In contrast, rank by distance similarity gives the result that could not be explained. This implies that the graphical channel approach is able to perceive the market better than the traditional point-to-point distance calculation. This can be on the other hand demonstrated by Figure 3, in which 4 graphical patterns are extracted as: 1) parallel pattern (e.g., BP *right*, XO *right*, and CV *right*); 2) double intersections, which includes actual crossing (e.g., XO *left*) or trend crossing (e.g., ZS *left*); 3) single intersection in same direction (e.g., CV *left*); and 4) single intersection in two reverse directions (e.g., CP *left*, TT *right*). This identifies 4 types of stock to crude oil correlation may occur at different period of the time. The strength of correlation decreases from pattern 1 (strongest positive correlation) to pattern 4 (strongest negative correlation), please find the statistical evidences in Table 2.

5   Impacts Discussion

As known from literature, oil price rises often cause a negative impact on all stock returns except mining, oil and gas industries [5,15,16]. This implies that crude oil to oil-related stocks is on a positive correlation (i.e., stock follows crude

Table 2. Stock to crude oil relevance evaluation and ranking by (a) distance similarity calculation, and (b)channel similarity calculation

Before event		After event		Overall	
Rank	S_d(Stock)	Rank	S_d(Stock)	Rank	S_d(Stock)
4	0.99870(CV)	5	0.99770(TT)	5	0.99530(TT)
1	0.99820(XO)	6	0.98790(CP)	3	0.96870(ZS)
5	0.99760(TT)	3	0.97780(ZS)	6	0.95070(CP)
2	0.99480(BP)	1	0.94630(XO)	1	0.94450(XO)
3	0.99090(ZS)	4	0.94410(CV)	4	0.94280(CV)
6	0.96280(CP)	2	0.91300(BP)	2	0.90780(BP)

(a)

Before event		After event		Overall	
Rank	S_c(Stock)	Rank	S_c(Stock)	Rank	S_c(Stock)
3	0.00542(ZS)	2	0.00380(BP)	1	0.00792(XO)
1	0.00542(XO)	1	0.00250(XO)	2	0.00719(BP)
5	0.00369(TT)	4	0.00250(CV)	3	0.00662(ZS)
2	0.00339(BP)	6	0.00230(CP)	4	0.00589(CV)
4	0.00339(CV)	5	0.00170(TT)	5	0.00539(TT)
6	0.00202(CP)	3	0.00120(ZS)	6	0.00432(CP)

(b)

oil on its price variation) As seen from the experimental results, for 5 of total 6 petroleum companies under discussion, the stock price variation matches well the above statement. The exception exits because the stock has been influenced by the Operation Neptune Spear event. It is worth to note that the event takes significant effect on the correlations of crude oil to oil-related stocks. For example, the correlation nature of the stock Shell has changed from positive to negative after the event occurs, meanwhile its correlation strength also drops from the top to the bottom. Among the 6 stocks, Exxon Mobil, BP and Shell are widely recognized as the top 3 largest petroleum companies. They are sensitive to crude oil shocks because they have a higher degree of internationalization, higher outputs of oil production and higher revenues per year as compared to Chevron, Total S.A. and ConocoPhilips. The strongest correlation to oil appears to be Exxon Mobil. The company dominate the market of 5 continents. In 2011, it produced 2.3 million barrels of oil per day and gained 486400 million dollars (US) revenue. On the other hand, ConocoPhilips shows the weakest correlation to crude oil. The company registers to the American market. Its oil production is only 0.7 million barrels per day and its revenue is just 12,436 million dollars (US).

6   Conclusions

The crude oil price has been widely sought has impact on stock market variation. It is practically very difficult for traditional technical and fundamental analysis approaches to discover the correlation of crude oil to an observed stock.

A graphical computational correlation analysis method is proposed for crude oil impact analysis. The comparison results demonstrate the significance of the proposed method against the traditional point-to-point distance calculation. More significantly, it helps discover and characterize the positive impacts to oil-related stocks.

References

1. Lutz, K., Cheolbeom, P.: The Impact of Oil price Shocks on the US Stock Market. International Economic Review 50(4), 1267–1287 (2009)
2. Faff, R.W., Brailsford, T.J.: Oil Price Risk and the Australian Stock Market. Journal of Energy Finance and Development 4, 69–87 (1999)
3. Sadorsky, P.: Oil Price Shocks and Stock Market Activity. Energy Economics, 449–469 (1999)
4. Miller, I.J., Ratti, R.A.: Crude Oil and Stock Markets: Stability, Instability and Bubbles. Energy Economics 31, 559–568 (2009)
5. Jones, C., Kaul, G.: Oil and the Stock Markets. Journal of Finance 51, 463–491 (1996)
6. Kling, J.L.: Oil Price Shocks and Stock Market Behaviour. Journal of Portfolio Management 12, 34–39 (1985)
7. Huang, R.D., Masulis, R.W., Stoll, H.R.: Energy Shocks and Financial Markets. Journal of Futures Markets 16, 1–27 (1996)
8. Pearson, K.: Mathematical Contributions to the Theory of Evolution. On a Form of Spurious Correlation Which Arise When Indices Are Used in the Measurement of Organs. Proceedings of the Royal Society of London 60, 489–498 (1896)
9. BP p.l.c., Summary Review 2011 (2012)
10. Chevron Corporation, 2011 Annual Report (2012)
11. Royal Dutch Shell p.l.c., Royal Dutch Shell plc Annual Review and Summary Financial Statements 2011 (2012)
12. TOTAL S.A. 2011 Annual Report (2012)
13. ConocoPhilips, 2011 Summary Annual Report (2012)
14. Exxon Mobil Corporation, 2011 Financial & Operating Review (2012)
15. Nandha, M., Faff, R.: Does Oil Move Equity Prices? A Global View. Energy Economics 30, 986–997 (2008)
16. Jones, D.W., Leiby, P.N., Paik, I.K.: Oil Price Shocks and the Macroeconomy: What Has Been Learned Since 1996. The Energy Journal 25, 1–3 (2004)
17. Vigdor, J.: The Economic Aftermath of Hurricane Katrina. Journal of Economic Perspectives 22(4), 135–154 (2008)
18. Pang, S., Song, L., Kasabov, N.: Correlation-aided Support Vector Regression for Forex Time Series Prediction. Neural Comput. Appl. 20(8), 1193–1203 (2011)
19. Master, J.: Mississippi River Floor of 2011 already a $2 Billion Disaster. Weather Underground (2011)

Learning Visual Saliency Based on Object's Relative Relationship

Senlin Wang[1], Qi Zhao[2], Mingli Song[1], Jiajun Bu[1],
Chun Chen[1], and Dacheng Tao[3]

[1] Zhejiang Provincial Key Laboratory of Service Robot, College of Computer Science,
Zhejiang University, Hangzhou 310027, China
{snail_wang,brooksong,bjj,chenc}@zju.edu.cn
[2] Department of Electrical and Computer Engineering, NUS
eleqiz@nus.edu.sg
[3] Centre for Quantum Computation and Information Systems, UTS
dacheng.tao@gmail.com

Abstract. As a challenging issue in both computer vision and psychological research, visual attention has arouse a wide range of discussions and studies in recent years. However, conventional computational models mainly focus on low-level information, while high-level information and their interrelationship are ignored. In this paper, we stress the issue of relative relationship between high-level information, and a saliency model based on low-level and high-level analysis is also proposed. Firstly, more than 50 categories of objects are selected from nearly 800 images in MIT data set[1], and concrete quantitative relationship is learned based on detail analysis and computation. Secondly, using the least square regression with constraints method, we demonstrate an optimal saliency model to produce saliency maps. Experimental results indicate that our model outperforms several state-of-art methods and produces better matching to human eye-tracking data.

Keywords: Visual attention, High-level Information, Low-level Information, Relative relationship.

1   Introduction

What is visual attention? Generally speaking, when humans watch a scene, which can be dynamical or statical, visual attention shows the areas we are interested in. There is far too much visual information during the interactions of humans in the world, but only a small part can be handled.

Previous works on producing saliency maps are mainly based on the bottom-up models[2][3]. Typical low-level visual stimulus such as color, intensity, orientation were firstly extracted from the image at different scales and orientations, then linear[2] or nonlinear[3] method was used to integrate them together to formulate the final saliency map. Some other low-level methods, e.g. self-information maximization[4], entropy maximization[5] were also proposed to solve this issue.

T. Huang et al. (Eds.): ICONIP 2012, Part V, LNCS 7667, pp. 318–327, 2012.

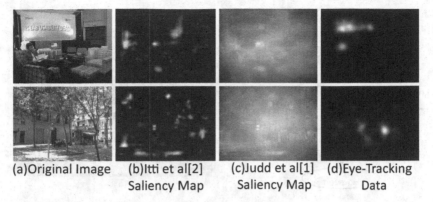

(a)Original Image   (b)Itti et al[2]   (c)Judd et al[1]   (d)Eye-Tracking
                     Saliency Map   Saliency Map   Data

Fig. 1. Different saliency maps

However, these methods neglected top-down cues in the image and could not get correct results in some complex scenes. Fig.1 shows that there is a big difference between Fig.1(b)(Itti's saliency maps) and Fig.1(d)(Eye-tracking data).

In addition, taking some high-level features into bottom-up saliency models could achieve better results too. Cerf et al.[6] and Judd et al.[1] added different high-level information like face, car information and outperformed original bottom-up models. But their models confused high-level and low-level information together by processing them in the same way. For example, the top right image in Fig.1 shows that face and text information are the saliency areas of top left image. In Judd's model, even though face was detected in Fig.1(c), the monitor which was not noticed by viewers in the eye-tracking data was the brightest area in their saliency map.

There are also some other methods by detecting salient objects in the images [7][8] to learn visual salience. These models concentrated on a single dominant object or one kind of objects. For many different kinds of objects, as their positions, quantities and sizes in the image are arbitrary and have a big influence on human's attention, it is still a great challenge to learn their visual attention. How do these objects influence our interesting areas, and is there any connection between them?

Two contributions are made in this paper. Firstly, we analyze the high-level information in the images, and concrete relationship between different objects are computed here. Secondly, an optimal linear saliency model is proposed by taking both low-level and high-level information into consideration. We test our model on the largest eye-tracking data set[1] and achieve better results than some state-of-art methods.

2   Motivation

The visual stimuli can be divided into two different types[9] based on the reaction time of visual neurons. One is independent of particular task and can be operated very rapidly for 25 to 50 ms per item. Image's colors and intensity

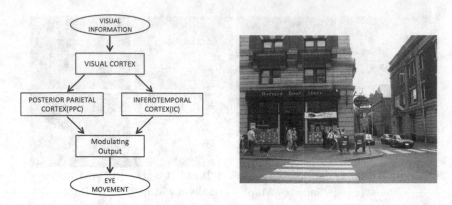

Fig. 2. The neuronal mechanism of processing visual information and a typical image

both belong to this stimulus which are also what bottom-up model concerned with. The other one is related to some cognitive factors such as knowledge, expectations or current goals, e.g.text or face information. While it takes 200ms or more for neurons to react. Fig.2 shows simply and briefly how two different visual information is processed by visual neurons[9]. The visual information enters the visual cortex and then is processed by two parallel neuronal cortex. The inferotemporal cortex(IC) and posterior parietal cortex(PPC) deal with the low-level and high-level information respectively. After that some other visual neurons(not shown) modulate them together to affect the final eye movement.

For example, the right image in Fig.2 is an ordinary street scene in our daily life. From the view of low-level saliency, the white banner in the middle, due to its intensity is different from the surroundings, will attract human's attention. For the same reason two telephone booths near the door can also be noticed. These deductions are in accordance with the experimental results from Itti's[2] saliency model. While on the basis of the analysis of high-level information, we can get different conclusions. People will tend to watch what is written on the wall, the signpost, persons or something like these which could provide more meaningful information for human.

Based on the analysis of visual stimuli and previous discussion, an effective saliency model should take both low-level and high-level information into consideration. Therefore, a unified model of both two information is proposed in this paper. And as there are already many saliency models[2][4][5]based on low-level information, here we mainly focus is on the analysis of high-level information and how to integrate them efficiently.

3   High-level Information Analysis

Current studies on how high level objects influence human's attention is insufficient and mainly focus on two aspects. One is to add some high level features into the bottom-up models[1][6], and the other one is to investigate a single or single kind of objects in the image[8][10]. The issue is still unclear and challenging that the concrete relationship between different kinds of objects during interactions of human in the real world.

In this paper we try to learn quantitative relationship between different objects from existing eye-tracking data set. Before our analysis, meaningful objects in the images are firstly annotated manually by a bounding box, and the area without any annotation is treated as "background object". These rectangular boxes could overlap each other, but that doesn't impact our analysis and computation.

By the union operations, the image I can be written as Eqn.1, n is the number of the image's objects.

$$I = I_1 \bigcup I_2 \bigcup ... \bigcup I_n \qquad (1)$$

Sal operator is used to calculate object's saliency areas, and the image's high-level saliency map Sal_H:

$$Sal_H(I) = Sal_H(I_1) \bigcup Sal_H(I_2) \bigcup ... \bigcup Sal_H(I_n) \qquad (2)$$

The object in the image can be treated as a "fixation choice" for human eyes. When there is only one choice, the object will pop out and attract human's more attention. However, if there are more than one choice, better ones will be made which means there exist some objects attracting more visual attention than other objects. Therefore, for each object, the probability of noticed is the product of conditional probability to the other objects. Eqn.3 is the corresponding mathematical expression of probability model.

$$Sal_H(I_i) = \prod_{j=1, j\neq i}^{n} P(I_i|I_j) \qquad (3)$$

Therefore, the overall image's high-level saliency map can be further written as Eqn.4. Our goal is to compute the conditional probability of arbitrary two objects in the image, which is also their relative relationship.

$$Sal_H(I) = \bigcup_{i=1}^{n} Sal_H(I_i) = \bigcup_{i=1}^{n} \prod_{j=1, j\neq i}^{n} P(I_i|I_j) \qquad (4)$$

It's quite difficult to analyze the relative relationship among variety kinds of objects. As their positions, quantities and sizes in the image are arbitrary and have a big influence on human's attention. Based on the analysis of mass eye-tracking data, some simple but efficient strategies are firstly used to remove these adverse influence.

3.1   Remove Position Effect

As center area has a natural advantage to attract human's attention, a reverse center model is firstly adopted before further processing. We use it to multiply the ground truth eye-tracking data to remove center position effect. As a normal bivariate gaussian model is used to simulate center impact[11], and we use Sal_C to present its center saliency map. The reverse model can be defined as $Sal_{RC}(I) = 1 - Sal_C(I)$ where Sal_C has been scaled between 0 and 1.

3.2   Remove Size and Quantity Effect

Besides position factor, objects with bigger sizes or larger quantities will get more opportunities to be noticed. Corresponding to a real saliency map, the brighter an area is, the more attention the area gets. As object's location coordinates can be obtained from the annotated rectangular box in the original color image, we sum up the pixel values of each object in the real eye-tracking data. Then the summation value will be divided by their size. For the objects appear more than once in the image, their average values will be calculated. The final computed value is called *Saliency Value(SV)* of this object. The bigger *SV* of an object, the more attention it attracts. After a series of computation, we could get accurate results for different objects. For example, the original image in the third row in Fig.5, face occupies a small size and quantity comparing with cars and humans, but it still has a bigger *SV* than the other two objects based on our computation, which is also consistent with the eye-tracking data.

3.3   Calculate Relative Relationship

For 1003 images in the MIT eye-tracking data set, 783 images are selected for our annotation. The quantity of images is sufficient enough to compute object's relationship as each eye-tracking data contains 15 people freely viewing for 3 seconds. More than 50 categories of objects are labeled and counted in our experiments, and finally four common objects are selected for two reasons: a)These objects appear most frequently. b)There exist effective recognition algorithms for these objects which make our model computable. The four objects are *text*, *face*, *human* and *car*. Their relative relationship is presented in Table 1.

The value in the table is the noticed probability of one object to another. For example, 0.35 in the second row, third column is the noticed probability of text when face also appears in the image, which is also denoted as $P(text|face) = 0.35$ as mentioned before. The computation method is not good enough, but this is the first time that quantifiable relative relationship is proposed. What's more, it also affords another thought to understand image's salience.

Table 1. Relative Relationship between Different Objects

	Text	Face	Person	Car
Text	\	0.35	0.81	0.85
Face	0.59	\	0.87	0.89
Person	0.11	0.09	\	0.64
Car	0.12	0.07	0.31	\

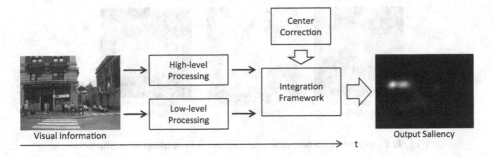

Fig. 3. The sketchy framework of the saliency model. From the input of visual information to the output saliency map.

4 Saliency Model

As aforementioned, an effective model should take both high-level and low-level information into consideration. And in Fig.2, we find that from input of visual information to output of the eye movement, the information is handled by two parallel channels. Therefore, in our proposed saliency model, two types of information are processed respectively at the same time. Moreover, for the statical scene, there is always an inherent advantage for center area as people naturally look at the middle of the image at first sight [1]. So center information is added into our saliency model to make it more reasonable and accurate. Furthermore, as general saliency maps are the accumulation of eye movement results in a short period of time(usually 3s), time is another factor that can not be ignored in the model either. Fig.3 gives a graphical overview of our saliency model. for an input image, high-level and low-level information processing will be implemented in parallel, after that, they are integrated together and center correction will be used for the output saliency map.

In order to simulate the process of visual information, Sal_X is used to represent image's saliency map while subscript X indicates the processing method used. There are three different methods H, L and C in this paper which are high-level, low-level and center information processing respectively.

$$Sal(I) = \lambda \left[\int_0^T \left(f_L * Sal_L(I) + f_H * Sal_H(I) \right) d_t \right] * Sal_C(I) \qquad (5)$$

Eqn.5 is our computational saliency model. On the left of equation is the final saliency map while I is the image. The saliency map is thus the integral over watching time T on the low-level and high-level saliency from the image, f is the corresponding time influence function for different visual stimulus. As these two information are processed in parallel by visual cortex, corresponding saliency maps are independent with each other. A plus operation is appropriate to describe their relationship. After integration, center correction is adopted to make the model more accurate.

Fig. 4. Contrast between different center information distribution

However, in general, a single statical image has no temporal information associated with it. It is convenient to replace the temporal integral by a weight coefficient for each map. Therefore, Eqn.5 can be further written as Eqn.6.

$$Sal(I) = \left[w_L * Sal_L(I) + w_H * Sal_H(I)\right] * Sal_C(I) \qquad (6)$$

We have discussed high-level information processing in the previous parts. And as low-level information in saliency models has been intensively studied[2][4] in recent years, we do not make detailed analysis on this issue and use existing model[11] to process low-level information. The center correction will be introduced for both high and low level information processing later, so the center bias prior is removed from Yang's model which can further speed up the computation.

After getting two information maps, the least square regression with constraints method is adopted to learn their weight coefficients. It's impossible to get the same optimal weight coefficients for each image. But by training half of the images, we could get their relationship roughly, the weights of high-level information are much larger than low-level's. For model's calculability, their average weights i.e. 0.63 for high-level and 0.27 for low-level are used in our experiments. Though average weights used here, the results still perform better than others without high-level information which further proves that our work to high-level information is quite necessary.

Though center area makes an effect on human's attention, it's still unclear that how it works. Judd et al.[1] use the square value of distance from center to simulate its impact. Fig.4(b) is the corresponding center saliency map in their model. And Fig.4(a) is the average center information map calculated from overall human fixations in MIT data set. On the one hand, it further proves that human tend to look at the center area in the image. On the other hand, it also shows that Judd's center model is not accurate enough to describe center effect. In this paper, a normal bivariate gaussian distribution is employed to simulate this process. Fig.4(c) is the simulated result which is very close to the real data.

5   Experimental Results

For objects recognition, face detector[12], person and car detector[13] and word detector[14] are used here. In the experiments, both qualitative and quantitative analysis are carried out to validate the effectiveness of our model.

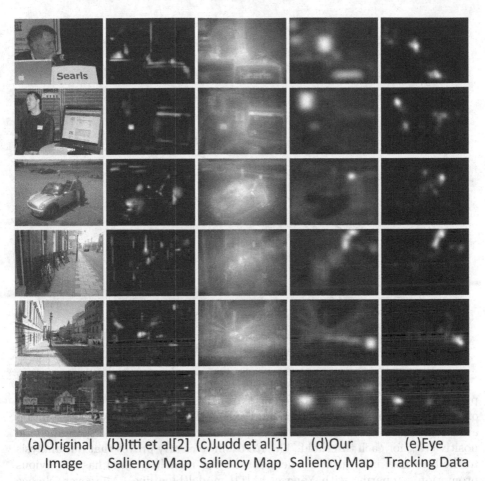

(a)Original (b)Itti et al[2] (c)Judd et al[1]   (d)Our   (e)Eye
Image   Saliency Map Saliency Map Saliency Map Tracking Data

Fig. 5. Qualitative comparison

5.1 Qualitative Analysis

Fig.5 affords qualitative comparisons between results of different models and
the eye-tracking data. Some typical methods are selected in this paper for comparison. Itti et al.[2] saliency model, which is the classical low-level analysis of
visual attention, is presented in Fig.5(b). Fig.5(c) shows Judd et al.[1] experimental results, which were also carried out on this data set. Our results and the
eye-tracking data are displayed in Fig.5(d) and Fig.5(e) separately. Whether in
the simple or complex scenes in Fig.5, our model achieves better results than
existing models and matches the eye-tracking data well.

5.2 Quantitative Analysis

Receiver operating characteristic curve(ROC) is adopted to evaluate eye fixation
prediction quantitatively. The curve could indicate how well the saliency map

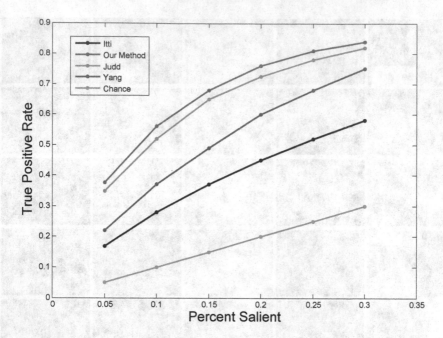

Fig. 6. Quantitatively comparison

predicting eye fixation with different thresholds. Fig.6 plots the ROC curves of different saliency models. Horizontal axis is the threshold of visual attention area from top 5% to 30% of saliency map. Vertical axis is the corresponding true positive rate for each threshold. The ROC curve clearly proves that our analysis to the high-level information is necessary as our model(red line) has an obvious promotion comparing with Yang et al.[11] model(blue line). Moreover, chance is also plotted for comparison. And we can easily find that the proposed model takes advantages over the conventional ones.

6   Conclusion

In this work, we stress the importance of high-level information in visual attention analysis. By annotating and analyzing nearly 800 images, four objects' concrete relative relationship are learned and used in our model. Besides, we also propose a saliency model based on both two types of information which achieves better experimental results than some other models.

Though good performance of our results, the number of studied objects here is still too small. And current object recognition algorithms are also incapable to distinguish all the objects in the image. In the future, better learning algorithm and more meaningful objects will be studied in our work.

Acknowledgments. This work is supported by National Natural Science Foundation of China(61170142), National Key Technology R&D Program (2011BAG05B04),the Zhejiang Province Key S&T Innovation Group Project (2009R50009)and the Fundamental Research Funds for the Central Universities(2012FZA5017).

References

1. Judd, T., Ehinger, K., Torralba, A.: Learning to predict where humans look. In: International Conference on Computre Vision, pp. 2106–2113 (2009)
2. Itti, L., Koch, C., Niebur, E.: A Model of Saliency-based Visual Attention for Rapid Scene Analysis. Pattern Analysis and Machine Intelligence 20, 1254–1259 (1998)
3. Zhao, Q., Koch, C.: Learning visual saliency. In: Conference on Information Sciences and Systems (CISS), pp. 1–6 (2011)
4. Bruce, N.D.B., Tsotsos, J.K.: Saliency, attention, and visual search: An information theoretic approach. Journal of Vision 9, 1–24 (2009)
5. Hou, X., Zhang, L.: Dynamic visual attention:Searching for coding length increments. In: Neural Information Processing Systems, vol. 21, pp. 681–688 (2008)
6. Cerf, M., Frady, E., Koch, C.: Faces and text attract gaze independent of the task: Experimental data and computer model. Journal of Vision 9, 1–15 (2009)
7. Goferman, S., Zelnik-Manor, L., Tal, A.: Contextaware saliency detection. In: Computer Vision and Pattern Recognition, pp. 2376–2383 (2010)
8. Liu, T., Yuan, Z., Sun, J., Wang, J., Zheng, N., Tang, X., Shum, H.: Learning to detect a salient object. Pattern Analysis and Machine Intelligence 33, 353 367 (2011)
9. Itti, L., Koch, C.: Computational modelling of visual attention. Nature Reviews. Neuroscience 2, 194–203 (2001)
10. Goferman, S., Zelnik-Manor, L., Tal, A.: Context-aware saliency detection. In: Computer Vision and Pattern Recognition, pp. 2376–2383 (2010)
11. Yang, Y., Song, M., Li, N., Bu, J., Chen, C.: What Is the Chance of Happening: A New Way to Predict Where People Look. In: Daniilidis, K., Maragos, P., Paragios, N. (eds.) ECCV 2010, Part V. LNCS, vol. 6315, pp. 631–643. Springer, Heidelberg (2010)
12. Viola, P., Jones, M.: Robust real-time face detection. International Journal of Computer Vision 57, 137–154 (2004)
13. Felzenszwalb, P., McAllester, D., Ramanan, D.: A discriminatively trained, multiscale,deformable part model. In: Computer Vision and Pattern Recognition, pp. 1–8 (2008)
14. Wang, K., Belongie, S.: Word Spotting in the Wild. In: Daniilidis, K., Maragos, P., Paragios, N. (eds.) ECCV 2010, Part I. LNCS, vol. 6311, pp. 591–604. Springer, Heidelberg (2010)

Template Matching Based Video Tracking System Using a Novel *N*-Step Search Algorithm and HOG Features

Tudor Barbu

Institute of Computer Science of the Romanian Academy, Iaşi, Romania
tudbar@iit.tuiasi.ro

Abstract. A novel video object tracking technique is proposed in this article. We consider a robust template-matching based video tracking technique that works satisfactory for both static-camera and moving-camera video sequences, being not influenced by the camera motions. In our approach, the first instance of the video object is selected interactively. Then, its successive instances in the video frames are detected using a novel and improved *N*-step search algorithm for motion estimation taking into account both the scaling and translation of the target. A HOG-based feature extraction approach is used by our algorithm.

Keywords: video tracking, human-computer interaction (HCI), *N*-Step Search algorithm, Histogram of Oriented Gradients, sliding-window, object matching.

1 Introduction

Video object tracking represents a very important and challenging computer vision domain. The tracking process has to locate video objects over time in a movie sequence, therefore its objective is to associate target objects in consecutive video frames [1]. Usually, video tracking is strongly correlated with video object detection [1]. While, the object detection process identifies image objects in the video frames, the object tracking procedure solves the temporal correspondence problem that is the task of matching the target object in successive frames.

Numerous video tracking techniques have been developed in recent years. Let us mention those based on Kalman filtering [1,2], Hidden Markov Models, optical flow [1], template matching, mean-shift tracking [1,3] and contour tracking [4]. Also, video object tracking has a wide variety of computer vision application areas. The most important of these fields are video compression, video surveillance, human-computer interaction, video indexing and retrieval, medical imaging, traffic control, augmented reality and robotics [1].

Video tracking can be a time consuming process due to the amount of data contained in movie streams. Also, video tracking is often a difficult process, due to some factors such as abrupt object motion, object occlusions and camera motion. These difficulties are usually faced by the approaches which track moving objects in static-camera video sequences. For this reason, we propose here a novel semiautomatic video tracking approach that is able to track successfully both the static and moving

T. Huang et al. (Eds.): ICONIP 2012, Part V, LNCS 7667, pp. 328–336, 2012.

objects, in both fixed camera and moving camera videos. Also, unlike many other methods which are concerned with tracking objects from a certain object-class (such as humans, animals or vehicles), our technique can track all kind of video objects.

We developed some automatic video object detection and tracking techniques in our previous works [5]. They detected all moving objects first, using approaches like temporal-differencing, then the correspondences between these objects being determined [5]. Obviously, their automatic character represented an important advantage, no interactivity being required in the tracking process, but the results were influenced by the video camera motion.

The video tracking technique proposed in this paper is not completely automatic, a HCI component being used to determine the first instance of the video object, but provides much better object tracking results than our previous methods [5] and many other automatic approaches, mainly because it is not dependable of the (undesired) video camera motions. Also, our technique contains a robust video object featuring approach, based on Histograms of Oriented Gradients [6]. The proposed HOG-based feature extraction is described in the next section.

Unlike other video tracking methods, our proposed video motion model takes into consideration not only the object position changes, but also the possible shape and scale transformations. In the third section one proposes an improved N - Step Search algorithm for motion estimation that successfully detects the next instance of the target, by treating both its scaling and translation. The search procedure developed here is computationally much less expensive than many other searching techniques, such as the full-search algorithms. The experiments performed using the described approach, are discussed in fourth section. The article ends with a conclusions section.

2   A HOG-Based Object Feature Extraction Technique

A proper feature extraction is very important in the tracking process. Each involved image object representing a video object instance has to be characterized by a robust feature vector. We model each instance as the grayscale conversion of the subimage corresponding to its bounding box. Such an object state is codified as a 4 –uple composed of coordinates of the upper-left and bottom-right corners:

$$Ob = [x_1(Ob), y_1(Ob), x_2(Ob), y_2(Ob)] \tag{1}$$

A video object may suffer changes in size from frame to frame, therefore the image objects representing its states could have various dimensions. So, the template-matching process of the video tracking technique must often compare different-sized image objects from different frames. There are various solutions to the same-size object matching task, most of them representing pixel differencing based techniques [7], such as *SAD* (sum of absolute differences), *MAD* (mean absolute distance), *MSD* (mean squared distance), and *NCC* (normalized cross-correlation) between sub-images representing objects. Unfortunately, the above mentioned approaches are not well-suited for our template matching process that should compare different-sized object instances. These pixel by pixel correspondence based methods require many object resizing operations, which may produce image information losses and increase the object matching error rate. A proper solution for this case would be the computation of some fixed-size feature vectors of these sub-images, like histogram-based vectors. Unfortunately, some histograms do not represent robust image content descriptors. For example, different objects may have the same color histogram. For this reason we consider *Histograms of Oriented Gradients* (HOG) for object featuring [6].

Histogram of Oriented Gradients represents a powerful image content descriptor used in computer vision for the purpose of object detection. The HOG features have been successfully used for human detection [6], numerous HOG-based person identification techniques being developed in recent years. While most of these techniques belong to image analysis domain, our approach applies HOG in the video detection/ tracking area. So, if the sub-image *Ob* represents an instance of a target video object, one computes a HOG-based feature vector for it. First, the image gradient values, representing directional changes in the intensity or color in the image, are computed, then, gradient vector is formed by combining the partial derivatives in x, y directions:

$$\nabla Ob = \left(\frac{\partial Ob}{\partial x}, \frac{\partial Ob}{\partial y} \right), \tag{2}$$

where the gradients in x and y directions can be computed by applying the $1D$ centered, point discrete derivative mask in the horizontal and vertical directions:

$$\frac{\partial Ob}{\partial x} = Ob * [-1 \ 0 \ 1], \ \frac{\partial Ob}{\partial y} = Ob * [-1 \ 0 \ 1]^T \tag{3}$$

The gradient orientations of the image are computed as $\theta = \arctan\left(\frac{\partial Ob}{\partial x}, \frac{\partial Ob}{\partial y} \right)$.

The image *Ob* is then divided into cells. For each cell, one computes a local $1D$ histogram of gradient directions (orientations) over the pixels of the cell. One considers 9 bins for the local histogram. The histogram channels are evenly spread over 0 to 180 degrees, so each histogram bin corresponds to a 20 degree orientation interval. The computed cell histograms have to be combined into a descriptor vector of the image. First, these cells should be locally contrast-normalized, due to the variability of illumination and shadowing in the image. This requires grouping the cells together into larger, spatially-connected blocks. Once the normalization is performed, all the histograms can be concatenated in a single image feature vector, representing the HOG descriptor. We use $[3 \times 3]$ cell blocks of $[6 \times 6]$ pixel cells with 9 histogram channels. The feature vector of each image *Ob* is computed as its HOG descriptor, with 81 coefficients being expressed as $V(Ob) = HOG(Ob)$.

The vector $V(Ob)$ represents a robust codification of the object content. All these $1D$ feature vectors have the same length, so the distances between the feature vectors of different video object states can be computed using the Euclidian metric.

3   A Novel *N*-step Search Algorithm for Object Tracking

In this section one proposes a novel search algorithm for tracking any target object in any video sequence. We formulate the following tracking problem: locate all the instances of a video object, no matter if it is static or moving with respect to camera, in the frames of a movie sequence, no matter if it is recorded with a static or moving camera, the location of the first instance of that video object being given. The first state of the target object is determined interactively. The proposed system contains a HCI component, allowing the selection of bbox of the target object in the first frame.

Then, one detects the location of the video object in the next frame of the sequence, using the information related to its first instance. The tracking process continues this way until the last frame is reached. A sliding-window based object detector could be used in this case, the sliding-windows representing a widely used tool for identifying and localizing image objects. The classic sliding-window approach, consisting of a full-search and using fixed-size sliding windows, is quite computationally expensive and does not treat the object scaling effect. Therefore, we propose an object localization approach based on a variable-sized sliding-window and an improved *N*-step search algorithm, that is less computationally expensive and time consuming.

3.1   Variable-Sized Sliding-Window Based Object Detection Approach

An image object could modify its position or its size from frame to frame in a video sequence. So, both translation and scale changes must be taken into account by the object tracking model. The object scaling aspect is treated using a variable-sized sliding-window while the object translation is approached using a template matching based on an *N*-step search algorithm [8]. We consider that any object in a video frame cannot differ much in shape and size from its state in the previous frame. The values of the width and the height of the bbox of that object are situated in some neighborhood intervals of the width and height of its previous state. We construct a novel algorithm that locates all the image objects determined by these intervals. It is the same process as applying a variable-sized sliding-window on a neighborhood region of that object. So, if the given image object Ob is described as in (1), one must determine all objects characterized by the variable-sized sliding-window based description:

$$Ob_{k,t,s,l} = [x_1(Ob) + k, y_1(Ob) + t, x_2(Ob) + l, y_2(Ob) + s], \qquad (4)$$

where $k,l \in [-M,M]$, $t,s \in [-N,N]$, and $M,N \geq 0$ represent quite small number of pixels. All image objects modeled by formula (4) contain the following sub-image:

$$Ob_{M,N} = [x_1(Ob) + M, y_1(Ob) + N, x_2(Ob) - M, y_2(Ob) - N] \qquad (5)$$

So, the set of needed objects is composed of $Ob_{M,N}$, the objects containing $Ob_{M,N}$ padded with an *upper zone*, those containing $Ob_{M,N}$ padded with a *bottom zone*, those containing $Ob_{M,N}$ padded with a *left zone* and those containing $Ob_{M,N}$ padded with a *right zone*. We propose a recursive object locating procedure in pseudo-code:

```
Procedure ObjectDetection ([x₁,y₁,x₂,y₂], S, [l₁,l₂,l₃,l₄])
{finds all objects containing object [x₁,y₁,x₂,y₂], with
coordinates constrained by l₁₋₄ and insert them as rows
in S}
```

$$S = Insert([x_1,y_1,x_2,y_2], S);$$

if $x_1 - 1 \geq l_1$ and $[x_1 - 1, y_1, x_2, y_2] \notin S$

$$S = Insert([x_1 - 1, y_1, x_2, y_2], S);$$

$$ObjectDetection ([x_1 - 1, y_1, x_2, y_2], S, [l_1,l_2,l_3,l_4]);$$

if $x_2 + 1 \leq l_2$ and $[x_1, y_1, x_2 + 1, y_2] \notin S$

$$S = Insert([x_1, y_1, x_2 + 1, y_2], S);$$

$$ObjectDetection ([x_1, y_1, x_2 + 1, y_2], S, [l_1,l_2,l_3,l_4]);$$

if $y_1 - 1 \geq l_4$ and $[x_1, y_1 - 1, x_2, y_2] \notin S$

$$S = Insert([x_1, y_1 - 1, x_2, y_2], S);$$

$$ObjectDetection ([x_1, y_1 - 1, x_2, y_2], S, [l_1,l_2,l_3,l_4]);$$

if $y_2 + 1 \leq l_1$ and $[x_1, y_1, x_2, y_2 + 1] \notin S$

$$S = Insert([x_1, y_1, x_2, y_2 + 1], S);$$

$$ObjectDetection ([x_1, y_1, x_2, y_2 + 1], S, [l_1,l_2,l_3,l_4]);$$

This algorithm is applied to object $Ob_{M,N}$ producing the set of all image objects $\{Ob_{k,t,s,l}\}$. If $l_1 = x_1(Ob) - M$, $l_2 = y_1(Ob) - N$, $l_3 = x_2(Ob) + M$, $l_4 = y_2(Ob) + N$, we get

$$S(Ob_{M,N}) = ObjectDetection(Ob_{M,N}, S, [l_1,l_2,l_3,l_4]) \qquad (6)$$

3.2   A Novel *N*-Step Object Matching Approach

As we have mentioned before, the tracking process starts with the first instance of the video object whose location and size is known, because it is interactively selected. The next instance must be identified in a search area from the next frame.

The best match of image object *Ob* could be detected by performing a full search of the corresponding objects contained by $S(Ob_{M,N})$, in that search window. The main disadvantage of the full-search approach is its high computational complexity that results in a high execution time. We propose a novel object searching technique running much faster than full search method. Various *N*-Step Search algorithms for motion estimation have been developed in recent years [8], such as the block

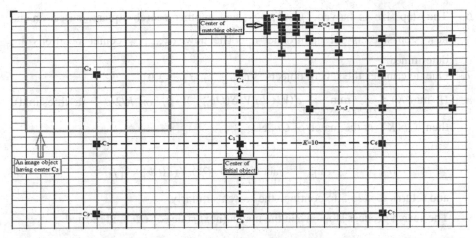

Fig. 1. Object matching example: N-step search with parameters 4 steps, initial $K=10$ and $T=1$

matching techniques 3-Step Search, 4-Step Search and their variants [8]. Our proposed N-step searching method derives from the 3-step search (*TSS*) approach, representing an extended version of it. The algorithm takes into account the translational motion, introducing the notation Ob^c for the object in next frame, having the same size as Ob and center $c = (x,y)$. Its corresponding objects resulted from padding operations are included as rows in $S(Ob^c_{M,N})$, each of them referred as $S(Ob^c_{M,N})[i], i \in \left[1, n(Ob^c_{M,N})\right]$. A N-step search example is described in Fig. 1. The steps of the proposed method are:

1. A square search area is established by selecting an initial step size K. It depends on the size of the motion, larger motions requiring higher K values.

2. One determines 9 points as pairs of coordinates c_i, as follows: c_1 is has the same coordinates as the center of Ob in the previous frame and is the center of the search square. The other 8 points are positioned in the corners and in the middle of the edges of the square, at a distance of step size, one from another: $d(c_i, c_{i+1}) = K, \forall i \in [1,8]$, d being the Euclidean metric.

3. One locates the image objects Ob^{c_i} and computes the sets $S(Ob^{c_i}_{M,N})$ using (6).

4. The best match, from all these objects is identified, by computing the distances between their HOG-based feature vectors and the feature vector of initial object:

$$\begin{cases} [i_K, j_K] = \arg \min_{i \in [1,9], j \in [1, n(Ob^{c_i}_{M,N})]} d(V(Ob), V(S(Ob^{c_i}_{M,N})[j])) \\ \min(K) = \min_{i \in [1,9], j \in [1, n(Ob^{c_i}_{M,N})]} d(V(Ob), V(S(Ob^{c_i}_{M,N})[j])) \end{cases} \qquad (7)$$

5. One computes the center of $Ob(K) = S(Ob^{c_{i_K}}_{M,N})[j_K]$ (it may differ from c_{I_K}) and assigns its value to c_1.

6. The step size is halved, $K \to \lfloor K/2 \rfloor$, and another search square, based on the new step size, is set up around c_1. The new $c_2, ..., c_9$ center positions are determined.

7. One determines the sets of objects $S(Ob_{M,N}^{c_i})$, $\forall i \in [2,9]$.

8. If $\min(K) < \min\limits_{i \in [2,9], j \in [1,n(Ob_{M,N}^{c_i})]} d(V(Ob), V(S(Ob_{M,N}^{c_i})[j]))$ or $K/2 < T$ (a threshold),

    then $Ob(K)$ is the match of Ob, else

$$\left\{ [i_{K/2}, j_{K/2}] = \arg \min\limits_{i \in [2,9], j \in [1,n(Ob_{M,N}^{c_i})]} d(V(Ob), V(S(Ob_{M,N}^{c_i})[j]) \right. \tag{8}$$

    and returns to step 5 to continue the search similarly.

9. The search process stops when the match (next state) of Ob is identified.

4   Experiments and Method Comparisons

We have performed numerous video tracking experiments using the proposed technique. It has been tested on various video data sets, containing several tens video sequences, satisfactory results being obtained. We have obtained good object tracking results for both static and moving camera videos. The approach described here produce a high video tracking rate, of approximately 90%. Choosing a high value for K and a low value for T increases the tracking rate, but also the computational complexity. We use for our tests parameter values $T = 3$, M, $N < 5$ and an initial K value depending on the target size: half of the diagonal of the interactively selected object.

Such a video tracking experiment, related to traffic monitoring [9], is described in Fig. 2. The moving car is interactively selected in the first frame, where is marked in red. The initial search square and K value are also displayed. In the second frame some results of our N-step search procedure are displayed. One can see that the orange rectangle, bounding the car, corresponds to the minimum distance value (0.84) between feature vectors. In the next two frames, the detected states of the target object are marked in orange and the corresponding distance values are displayed.

Fig. 2. Example of a target object tracking in a short video sequence converted to grayscale

The tracking technique described here was also compared with many other detection and tracking algorithms, the comparison results being encouraging. It provides better results than temporal-differencing and pixel differencing based methods [5,7]. Also, it performs much better than algorithms using other features than HOGs. Our approach runs much faster than exhaustive sliding window search (ESS), given its lower time complexity, while providing a better tracking than TTS and 4-Step Search.

5   Conclusions

A robust semiautomatic video object tracking system has been proposed in this paper. It uses a novel N-step search algorithm, taking into account both the object scaling and translation transforms, and a robust HOG-based object feature extraction.

Although the technique introduced here is not completely automatic, it has the important advantage of being insensitive to camera motion. For this reason it produces much better results than many automatic tracking methods. The proposed N-step object search algorithm represents the main contribution of this article. It constitutes an improved and extended version of the classic 3-Step Search block matching technique. Our method is used for image objects of various sizes, instead of small fixed-sized blocks of pixels [8]. The initial K value is not fixed as in the 3-step case and, as a result, the number of steps is not fixed at 3. Another important difference is that our algorithm does not necessary repeat the search process for one of the c_i centers. It uses the center of an object from neighborhood of the best match c_j, instead.

As resulting from our method comparisons, our technique performs slightly better than methods based on classic N-step searches, such as 3-Step Search (TTS), 4-Step Search and some of their variants, and runs much faster than tracking methods using Exhaustive Search. The video tracking approach proposed here can be successfully applied in some important domains, such as video surveillance [7], pedestrian detection [6], robotics and traffic monitoring [9].

References

1. Yilmaz, A.: Object Tracking: A Survey. ACM Computing Surveys 38, Article 13 (2006)
2. Peterfreund, N.: Robust Tracking of Position and Velocity with Kalman Snakes. IEEE Transactions on Pattern Analysis and Machine Intelligence 21 (1999)
3. Comaniciu, D., Meer, P.: Mean shift analysis and applications. In: Proc. of 7th IEEE Intl. Conf. on Computer Vision (ICCV 1999), Kerkyra, Greece, vol. 2, pp. 1197–1203 (September 1999)
4. Chen, Y., Rui, Y., Huang, T.: Multicue hmm-ukf for real-time contour tracking. IEEE Transactions on Pattern Analysis and Machine Intelligence 28, 1525–1529 (2006)
5. Barbu, T.: Multiple Object Detection and Tracking in Sonar Movies using an Improved Temporal Differencing Approach and Texture Analysis. U.P.B. Scientific Bulletin, Series A 74, 27–40 (2012)

6. Dalal, N., Triggs, N.: Histograms of Oriented Gradients for Human Detection. In: Proc. of the 2005 IEEE Computer Society Conf. on Computer Vision and Pattern Recognition (CVPR 2005), vol. 1, pp. 886–893 (2005)
7. Ching-Kai, H., Tsuhan, C.: Motion Activated Video Surveillance Using TI DSP. In: Proceedings of DSPS Fest 1999, Houston Texas (1999)
8. Gyaourova, A., Kamath, C., Cheung, S.-C.: Block matching for object tracking. Tech. Rep. UCRL-TR-200271, Lawrence Livermore Natl. Lab., Livermore, Calif, USA (2003)
9. Hsieh, J.W., Yu, S.H., Chen, Y.S., Hu, W.F.: Automatic traffic surveillance system for vehicle tracking and classification. IEEE Trans. Intell. Trans. Syst. 7, 175–187 (2006)

Local Structure Divergence Index
for Image Quality Assessment

Fei Gao[1], Dacheng Tao[2], Xuelong Li[3], Xinbo Gao[1], and Lihuo He[1]

[1] School of Electronic Engineering, Xidian University, Xi'an 710071, Shaanxi, China
{gaofeihifly,xbgao.xidian,lihuo.he}@gmail.com
[2] Centre for Quantum Computation and Intelligent Systems,
Faculty of Engineering and Information Technology,
University of Technology, Sydney, Broadway NSW 2007, Australia
Dacheng.Tao@uts.edu.au
[3] Center for OPTical IMagery Analysis and Learning (OPTIMAL),
State Key Laboratory of Transient Optics and Photonics,
Xi'an Institute of Optics and Precision Mechanics,
Chinese Academy of Sciences, Xi'an 710119, Shaanxi, China
xuelong_li@opt.ac.cn

Abstract. Image quality assessment (IQA) algorithms are important
for image-processing systems. And structure information plays a significant role in the development of IQA metrics. In contrast to existing
structure driven IQA algorithms that measure the structure information
using the normalized image or gradient amplitudes, we present a new
Local Structure Divergence (LSD) index based on the local structures
contained in an image. In particular, we exploit the steering kernels to
describe local structures. Afterward, we estimate the quality of a given
image by calculating the symmetric Kullback-Leibler divergence (SKLD)
between kernels of the reference image and the distorted image. Experimental results on the LIVE database II show that LSD performs
consistently with the human perception with a high confidence, and
outperforms representative structure driven IQA metrics across various
distortions.

Keywords: Contrast masking, full reference, local steering kernel,
symmetric Kullback-Leibler divergence.

1 Introduction

The past decades have witnessed a great development of image quality assessment (IQA) metrics. Until now, quantities of IQA algorithms have been conducted. And standard image quality evaluation metrics have been applied to
optimize parameter settings of image-processing algorithms and to benchmark
image-processing systems. Since most of these applications are in lab environments, the undistorted image is always in access, full reference (FR) image quality evaluation metrics play a dominant role. Thus we concentrate on FR-IQA
algorithms in this paper.

T. Huang et al. (Eds.): ICONIP 2012, Part V, LNCS 7667, pp. 337–344, 2012.
© Springer-Verlag Berlin Heidelberg 2012

Photographic images, on the whole, tend to contain a broad range of spatial structures. And structural information plays a significant role in the construction of IQA algorithms [1][2]. In the exiting literatures, there are broadly two methodologies to characterize structural information. One is by normalizing the image, such as the structural similarity index (SSIM) [3], multi-scale structural similarity index (MS-SSIM) [4] and so on. The other is using the gradient amplitudes of a given image, e.g. the gradient structure similarity (GSSIM) [5].

Although these structure driven IQA metrics obtained promising performances, they do not measure the structural information extensively. As claimed by Field [6], structures are mainly characterized in the form of textures and spatial patterns at a wide range of orientations. Edges and their orientations therefore play a significant role for the presentation of structures. In contrast, it is a little far-fetched to treat the normalized image as the structures. And the gradient based metrics do not count for the information of the edge orientation, which is an essential measure of structures.

To overcome the aforementioned problems, we present a new Local Structure Divergence (LSD) index based on the local structures contained in an image. In particular, we exploit the steering kernels to describe local structures. In addition, the steering kernel simulates the contrast masking effect of the human vision system (HVS) and the local dependency property of photographic images. Afterward, we estimate the quality of a given image by calculating the symmetric Kullback-Leibler divergence between kernels of the reference and the distorted image. Experimental results on the LIVE database II show the proposed IQA algorithm performs consistently with the human perception with a high confidence, and outperforms representative structure driven IQA metrics across various distortions.

The rest of the paper is organized as follows. Section 2 introduces the local steering kernel. In Section 3, we present the details of the proposed IQA algorithm. Experiments on the LIVE database II are presented in Section 4. Finally, Section 5 concludes this paper.

2   Local Steering Kernel

Local steering kernels have been demonstrated as an efficient technique to describe local structures in images and widely used in image-processing algorithms [7]. The key idea of local steering kernels is to robustly obtain the local structure of an image by analyzing the covariance matrix of gradients. The steering kernel is modeled as,

$$K_l(\mathbf{x}_l - \mathbf{x}_i) = exp\left\{\frac{(\mathbf{x}_l - \mathbf{x}_i)^T \mathbf{C}_l (\mathbf{x}_l - \mathbf{x}_i)}{-2h^2}\right\}, \qquad (1)$$

with

$$\mathbf{x}_i \in \mathbf{w}_l \text{ and } \mathbf{C}_l \in \mathbb{R}^{2\times 2}, \qquad (2)$$

where $\mathbf{x}_i$ is the location label contained in the window $\mathbf{w}_l$ around $\mathbf{x}_l$. $\mathbf{x}_l = [x_{1l}, x_{2l}]$, x_{1l} and x_{2l} are the coordinates along the x_1 and x_2 orientations, respectively. Let m denote the width of $\mathbf{w}_l$. The size of the kernel K_l is $m \times m$. $\mathbf{C}_l$ is the is the covariance matrix of the gradients in the local analysis window.

The covariance matrix $\mathbf{C}_l$ can be estimated by the gradients contained in the window as,

$$\hat{\mathbf{C}}_l = \mathbf{G}_l^T \mathbf{G}_l, \tag{3}$$

$$\mathbf{G}_l = \begin{bmatrix} \vdots & \vdots \\ z_1(\mathbf{x}_j) & z_2(\mathbf{x}_j) \\ \vdots & \vdots \end{bmatrix}, \mathbf{x}_j \in \mathbf{w}_l, \tag{4}$$

where, $z_1(\mathbf{x}_j)$ and $z_2(\mathbf{x}_j)$ are the first derivations along the two orientations, respectively. For the sake of robustness, in the calculation of the steering kernel, $\mathbf{C}_l$ is estimated using the singular value decomposition (SVD) of $\mathbf{G}_l$ [7]. $\mathbf{G}_l$ is given by

$$\mathbf{G}_l = \mathbf{U}_l \mathbf{S}_l \mathbf{V}_l^T = \mathbf{U}_l \begin{bmatrix} s_1 & 0 \\ 0 & s_2 \end{bmatrix} [\mathbf{v}_1, \mathbf{v}_2]_l^T \tag{5}$$

In this way, the covariance matrix can be expressed as,

$$\mathbf{C}_l = \gamma_l \mathbf{R}_{\theta_l} \Lambda_l \mathbf{R}_{\theta_l}^T$$

$$= \gamma_l \begin{bmatrix} \cos\theta_l & \sin\theta_l \\ -\sin\theta_l & \cos\theta_l \end{bmatrix} \begin{bmatrix} \sigma_l & 0 \\ 0 & \sigma_l^{-1} \end{bmatrix} \begin{bmatrix} \cos\theta_l & \sin\theta_l \\ -\sin\theta_l & \cos\theta_l \end{bmatrix}^T \tag{6}$$

with,

$$\sigma_l = \frac{s_1 + \lambda'}{s_2 + \lambda'}, \lambda' > 0, \tag{7}$$

$$\gamma_l = \frac{s_1 s_2 + \lambda''}{M}, \lambda'' > 0, \tag{8}$$

and

$$\theta_l = arctan(\frac{v_1}{v_2}), \tag{9}$$

where, λ' and λ'' are regularization parameters for kernel elongation. M is the number of pixels contained in the local analysis window. v_1 and v_2 are elements of $\mathbf{v}_2$, i.e.

$$\mathbf{v}_2 = [v_1, v_2]^T. \tag{10}$$

Here, θ_l indicates the main orientation of the local analysis window, σ_l denotes the energy of the dominant gradient direction. The scaling parameter γ_l modulates the size of the footprint. In addition, γ_l is large in the flat area, and small in the texture area. This is consistent with the contrast masking effect of the human visual system (HVS). In addition, the local steering kernel simulates the local dependency characteristic of photographic images. Thus, we can safely draw the conclusion that local steering kernels systematically characterize the local structures, evolving the contrast masking effect of HVS and the local dependency characteristic of photographic images.

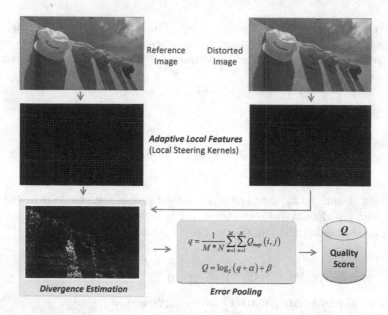

Fig. 1. Framework of the Local Structure Divergence (LSD) index

3   Local Structure Divergence Index

In this section, we present the framework of the proposed algorithm. The flowchart of the local structure divergence index is shown in Fig. 1. We first employ the technique of local steering kernels to extract adaptive local features of the reference and distorted images [8][9]. Afterward, we calculate the symmetric Kullback-Leibler divergence (SKLD) between kernels of the reference and distorted images. Finally, the quality of the distorted image is estimated by pooling all the structural divergences together. Details will be sequentially introduced in the following subsections.

3.1   Adaptive Local Features

As introduced in Section II, the steering kernel is a systematic measure of the local structure. Thus for both the reference and distorted images, a local steering kernel is estimated for each pixel to characterize the local structure. We use the term "kernel map" to denote the sequentially placed kernels of a given image. Figs. 2a and 2b show the kernel maps of a reference image and a distorted image derived from it. The values of the kernel are shown in the form of intensity. And a light point indicates a large value, verse vice. It is obvious that steering kernels reflect the changes of local structures contained in an image.

3.2   Divergence Estimation

Since each kernel is a two dimensional density function, we measure the difference between kernels of the reference and distorted images using the symmetric

Fig. 2. Associated steering kernels of the reference and distorted images at the same location. (a) Kernel map of the reference image; (b) kernel map of the distorted image.

Kullback-Leibler divergence (SKLD). And SKLD has been demonstrated as an effective way to estimate the distinction between possibility functions. SKLD is modeled as,

$$D_{SKL}\left(K_{i,j}^{ref}|K_{i,j}^{dis}\right)$$

$$= \frac{D_{KL}\left(K_{i,j}^{ref}\middle\| K_{i,j}^{dis}\right) + D_{KL}\left(K_{i,j}^{dis}\middle\| K_{i,j}^{ref}\right)}{2} \tag{11}$$

where, $D_{SKL}\left(K_{i,j}^{ref}|K_{i,j}^{dis}\right)$ denotes the SKLD between $K_{i,j}^{ref}$ and $K_{i,j}^{dis}$. $K_{i,j}^{ref}$ and $K_{i,j}^{dis}$ are the steering kernels at location (i,j) in teh reference and distorted images, respectively. $D_{KL}(\cdot)$ denotes the Kullback-Leibler divergence (KLD) between two distributions, i.e.

$$D_{KL}(p\| q) = \sum_k p(k) \log\left(\frac{p(k)}{q(k)}\right) \tag{12}$$

where p and q denote two possibility functions; and $p(k)$ and $q(k)$ are the k-th component of p and q, respectively.

We use $D_{SKL}\left(K_{i,j}^{ref}|K_{i,j}^{dis}\right)$ as a naive estimation of the quality at location (i,j), i.e.

$$Q_{map}(i,j) = D_{SKL}\left(K_{i,j}^{ref}\middle\| K_{i,j}^{dis}\right) \tag{13}$$

where Q_{map} denotes the quality map of the distorted image.

3.3   Error pooling

The quality of the distorted image is estimated using the mean of the quality map, which is given as,

$$q = \frac{1}{M*N} \sum_{i=1}^{M} \sum_{j=1}^{N} Q_{map}(i,j) \tag{14}$$

where, M and N are the size of the quality map, which is equal to that of the image. in addition, we modulate the quality score by using a logarithmic function, i.e.

$$Q = \log_2 (q + \alpha) + \beta \tag{15}$$

where α is the parameter to keep q from being zero, and β is used to keep the minimum Q being zero.

4   Exprimental Results

In this section, we compare the performances of LSD with a number of standard full reference (FR) IQA methods, e.g., peak signal to noise ratio (PSNR), SSIM [3], MS-SSIM [4], GSSIM [5], visual information fidelity measure (VIF) [10], and feature-similarity (FSIMc) [11]. For SSIM, MS-SSIM, VIF and FSIMc, we used the implementations provided by the original authors. For GSSIM, we implement it according to the related literature.

We use the LIVE database II [12] in the experiments. The LIVE database II is publicly accessible and includes 29 original images, from which 779 distorted images are generated. It contains five types of distortions, i.e., JPEG2000 compression (JP2k), JPEG compression (JPEG), white noise (WN), Gaussian blurring (Gblur), and fast fading (FF). It is noted that only the luminance component of an image is employed in the experiments. The differential mean opinion score (DMOS) of each image is available.

To evaluate the performances of the proposed metrics, four criteria are adopted: the Pearson's linear correlation coefficient (LCC), the Spearman's rank ordered correlation coefficient (ROCC), mean absolute error (MAE), and root mean square error (RMSE) between the estimated quality index, Q and the true DMOS. Before the computation of these indexes, the logistic function provided by the video quality experts group (VQEG) is employed.

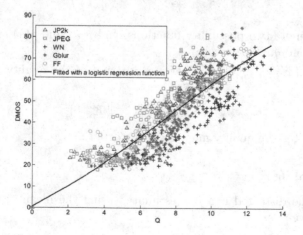

Fig. 3. Scatter plot of Q vs. DMOS on the LIVE database II

Fig. 3 shows the scatter plot of Q vs. DMOS on the LIVE database II. The predicted quality is in high consistency with the human perception for various distortions. The four criteria of each IQA algorithm are listed in Table I. For LCC and ROCC, the algorithms obtaining the best two performances are highlighted in boldface. It is clear that the proposed metric consistently performs well across all the five types of distortions and on the overall database. In addition, LSD performs better than SSIM and GSSIM, two classic structure driven metrics in the pixel domain. And its assessment accuracy is comparable with MS-SSIM, a structural similarity in wavelet domain, VIF and FSIMc.

Table 1. Performances of PSNR, SSIM, VIF, FSIMc and the proposed metric on LIVE Database II

	JP2k				JPEG			
	LCC	ROCC	RMSE	MAE	LCC	ROCC	RMSE	MAE
PSNR	0.8962	0.8898	7.1865	5.5283	0.8596	0.8409	8.1700	6.3797
SSIM [3]	0.9367	0.9317	5.6706	4.4332	0.9283	0.9028	5.9468	4.4846
GSSIM [5]	**0.9671**	0.9687	6.2039	4.8774	**0.9638**	0.9503	6.4666	4.7737
MS-SSIM [4]	0.9617	**0.9805**	5.6741	6.6871	0.9593	**0.9629**	5.4397	6.8473
VIF [10]	0.9615	0.9527	4.4493	3.4450	0.9430	0.9131	5.3212	3.8070
FSIMc [11]	0.9615	**0.9821**	6.7053	5.5924	0.9516	**0.9631**	7.4570	6.1185
LSD	**0.9658**	0.9652	6.3228	4.3739	**0.9598**	0.9474	6.8102	4.6239

	WN				Gblur			
	LCC	ROCC	RMSE	MAE	LCC	ROCC	RMSE	MAE
PSNR	**0.9858**	0.9853	2.6797	2.1639	0.7834	0.7816	9.7723	7.7425
SSIM [3]	0.9695	0.9629	3.9163	3.2566	0.8740	0.8942	7.6391	5.7595
GSSIM [5]	0.9533	0.9554	6.6394	5.6531	0.9461	0.9721	7.0439	5.8446
MS-SSIM [4]	0.9301	**0.9869**	6.9898	8.0733	0.9432	0.9739	6.1709	7.2267
VIF [10]	**0.9839**	**0.9857**	2.8514	2.3039	**0.9744**	0.9731	3.5334	2.8182
FSIMc [11]	0.9291	0.9799	8.1297	7.2865	0.9521	**0.9832**	6.6498	5.6070
LSD	0.9634	0.9498	5.8924	4.2799	**0.9746**	**0.9742**	4.8714	3.6171

	FF				All			
	LCC	ROCC	RMSE	MAE	LCC	ROCC	RMSE	MAE
PSNR	0.8895	0.8903	7.5158	5.8000	0.8240	0.8197	9.1236	7.3249
SSIM [3]	0.9428	0.9411	5.4846	4.2968	0.8634	0.8510	8.1262	6.2752
GSSIM [5]	**0.9682**	**0.9706**	5.5251	4.4063	0.9328	0.9306	23.116	19.647
MS-SSIM [4]	0.9207	0.9655	7.2063	8.6240	0.9376	**0.9519**	6.6842	8.0372
VIF [10]	0.9618	0.9649	4.5022	3.5469	**0.9501**	**0.9526**	5.0241	3.8866
FSIMc [11]	0.9213	**0.9708**	8.5897	7.4208	0.7314	0.9437	8.5921	6.4790
LSD	**0.9745**	0.9624	4.9581	3.5839	**0.9400**	0.9250	7.8869	5.6551

5   Conclusions

In this paper, we propose a local structure divergence (LSD) index for image quality assessment. Experimental results on the LIVE database II demonstrate that LSD is in high consistency with the human perception and is comparable with state-of-the-art FR-IQA metrics. Nevertheless, the computation complexity of LSD is high. Making it more efficient is a meaningful work. In addition, since the steering kernel is an effective and systematic presentation of local structures, how to apply it in the blind image quality prediction is another future work.

Acknowledgements. This work is supported by the National Basic Research Program of China(973 Program) (Grant No. 2012CB316400), by the National Natural Science Foundation of China (Grant Nos: 61125204, 61172146, 61125106, 91120302, and 61072093), and the Fundamental Research Funds for the Central Universities.

References

1. Tao, D.C., Li, X.L., Lu, W., Gao, X.B.: Reduced-reference iqa in contourlet domain. IEEE Trans. Systems, Man, and Cybernetics, Part B 39(6), 1623–1627 (2009)
2. Gao, X.B., Lu, W., Tao, D.C., Li, X.L.: Image quality assessment based on multi-scale geometric analysis. IEEE Trans. Image Processing 18(7), 1409–1423 (2009)
3. Wang, Z., Bovik, A., Sheikh, H., Simoncelli, E.: Image quality assessment: From error visibility to structural similarity. IEEE Transactions on Image Processing 13(4), 600–612 (2004)
4. Wang, Z., Simoncelli, E., Bovik, A.: Multiscale structural similarity for image quality assessment. In: Conference Record of the Thirty-Seventh Asilomar Conference on Signals, Systems and Computers, vol. 2, pp. 1398–1402 (2003)
5. Chen, G., Yang, C., Xie, S.: Gradient-based structural similarity for image quality assessment. In: 2006 IEEE International Conference on Image Processing, pp. 2929–2932 (2006)
6. Field, D.: Relations between the statistics of natural images and the response properties of cortical cells. Journal of the Optical Society of America 4(12), 2379–2394 (1987)
7. Takeda, H., Farsiu, S., Milanfar, P.: Kernel regression for image processing and reconstruction. IEEE Transactions on Image Processing 16(2), 349–366 (2007)
8. Tao, D.C., Li, X.L., Wu, X.D., Maybank, S.J.: Geometric mean for subspace selection. IEEE Trans. Pattern Anal. Mach. Intell. 31(2), 260–274 (2009)
9. Tao, D.C., Li, X.L., Wu, X.D., Maybank, S.J.: General tensor discriminant analysis and gabor features for gait recognition. IEEE Trans. Pattern Anal. Mach. Intell. 29(10), 1700–1715 (2007)
10. Sheikh, H., Bovik, A.: Image information and visual quality. IEEE Transactions on Image Processing 15(2), 430–444 (2006)
11. Zhang, L., Mou, X., Zhang, D.: Fsim: A feature similarity index for image quality assessment. IEEE Transactions on Image Processing 20(8), 2378–2386 (2011)
12. Sheikh, H., Wang, Z., Cormack, L., Bovik, A.: Live image quality assessment database release 2. Available (2005)

Feature and Signal Enhancement for Robust Speaker Identification of G.729 Decoded Speech

Kalpesh Raval[1], Ravi P. Ramachandran[1], Sachin S. Shetty[2],
and Brett Y. Smolenski[3]

[1] Rowan University, Glassboro, NJ, USA
kbr4957@gmail.com, ravi@rowan.edu
[2] Tennessee State University, Nashville, TN, USA
sshetty@tnstate.edu
[3] Assured Information Security, Rome, NY, USA
bsmolens@gmail.com

Abstract. For wireless remote access security, there is an emerging need for biometric speaker identification systems (SID) to be robust to speech coding distortion. This paper presents results on a Gaussian mixture model (GMM) based SID system that is trained on clean speech and tested on the decoded speech of the G.729 codec. To mitigate the performance loss due to mismatched training and testing conditions, five robust features, two enhancement approaches and three fusion strategies are used. The first enhancement method is feature compensation based on the affine transform. The second is the McCree signal enhancement approach based on the spectral envelope information in the G.729 bit stream. Ensemble systems using decision level, score fusion and Borda count are studied. The best performance is obtained by performing signal enhancement, feature compensation and decision level fusion. This results in an identification success rate (ISR) of 89.8%.

Keywords: Wireless security, robust speaker identification, G.729 coding distortion, ensemble systems.

1 Introduction

Biometric speaker recognition [1][2][3] finds many applications including remote access control, cybersecurity, border control, electronic commerce, forensics [4] and surveillance. The focus of this paper is on closed set, text-independent speaker identification for remote access applications. For comparison of a voice sample to a database of M speakers, the database is usually at a remote location from the person to be identified. The transmission of the speech can either be from a wired or wireless device (usually a mobile device) to a remote server via the internet. The remote server uses a biometric based speaker identification algorithm. The database acts as the central storage for all biometric data. The communication between the database and the server is secure and encrypted.

The biometric database containing the individual voiceprints of M speakers is assumed to be trained off-line on clean speech. The transmitted speech of

T. Huang et al. (Eds.): ICONIP 2012, Part V, LNCS 7667, pp. 345–352, 2012.

the person to be identified is subject to speech coding distortion for both cases of wired and wireless communication. The objective of this paper is to achieve a robust speaker identification system (a high performance) that is trained on clean speech and tested on speech subject to distortion introduced by the 8 kilobits/second ITU-T G.729 codec [6] (a standard in wireless and voice over IP applications).

In previous work, the emphasis has been on the mel frequency cepstrum (MFCC) and its first and second order derivatives as the features [7][8]. The features are either derived from the bitstream [7] or from the decoded speech [8] with the performance being better based on the latter approach. The approach in [8] also accomplishes waveform compensation based on the quantized linear prediction (LP) information to further augment performance.

In this paper, feature compensation using the affine transform and feature fusion using three techniques (decision level fusion, score fusion and Borda count) are used to accomplish robustness to mismatched training (clean speech) and testing (G.729 decoded speech) conditions. The overall system involves the four major steps of (1) feature extraction, (2) use of the affine transform, (3) the Gaussian mixture model (GMM) [2] classifier and decision logic and (4) feature fusion. Five features are considered, namely, the linear predictive cepstrum (CEP), adaptive component weighted cepstrum (ACW), the postfilter cepstrum (PFL), the mel frequency cepstrum (MFCC) and the line spectral frequencies (LSFs). The affine transform and feature fusion are critical in improving the performance over the case of using only a single feature. In fact, doing a sequential implementation of the waveform compensation approach of [8] and the affine transform followed by a decision level fusion strategy leads to the best performance.

2   Feature Extraction

Linear predictive (LP) analysis results in a stable all-pole model $1/A(z)$ from which the LSF and CEP features are calculated. The ACW cepstrum [9] is based on transforming $1/A(z)$ to a pole-zero form and then, obtaining the cepstrum. The postfilter (PFL) cepstrum is also obtained from a pole-zero transfer function and is equivalent to weighting the LP cepstrum by the factor $\alpha^n - \beta^n$ where $0 < \beta < \alpha \leq 1$ [9]. The MFCC is also used.

3   Affine Transform

The affine transform achieves feature compensation by mapping a feature vector $\mathbf{x}$ derived from G.729 speech to a feature vector $\mathbf{y}$ in the region of the p-dimensional vector space populated by the training vectors. This results in a better match between training and testing conditions and in effect, compensates for the coder distortion in the test speech. The mapping relating $\mathbf{x}$ and $\mathbf{y}$ is given by Eq. (1) as

$$\mathbf{y} = \mathbf{A}\mathbf{x} + \mathbf{b} \tag{1}$$

where $\mathbf{A}$ is a p by p matrix and $\mathbf{y}$, $\mathbf{x}$ and $\mathbf{b}$ are column vectors of dimension p.

The affine transform parameters $\mathbf{A}$ and $\mathbf{b}$ are determined from the training data only. Let $\mathbf{y}^{(i)}$ be the feature vector for the ith frame of the training speech utterance. Let $\mathbf{x}^{(i)}$ be the feature vector for the ith frame of the training speech utterance passed through the distortion encountered during testing (in this case, the clean utterance is compressed and decoded by the G.729 codec). By using a number of training speech utterances, N sets of vectors are collected, namely, $\mathbf{y}^{(i)}$ and $\mathbf{x}^{(i)}$ for $i = 1$ to N. A squared error function is formulated as

$$E(m) = \sum_{i=1}^{N} [y^{(i)}(m) - \mathbf{a}_m^T \mathbf{x}^{(i)} - b(m)]^2 \tag{2}$$

where $\mathbf{a}_m^T$ is the mth row of $\mathbf{A}$ and $y^{(i)}(m)$ and $b(m)$ are the mth components of $\mathbf{y}^{(i)}$ and $\mathbf{b}$, respectively. Minimization of $E(m)$ with respect to $\mathbf{a}_m^T$ and $b(m)$ results in the system of equations

$$\begin{bmatrix} \sum_{i=1}^{N} \mathbf{x}^{(i)} \mathbf{x}^{(i)T} & \sum_{i=1}^{N} \mathbf{x}^{(i)} \\ \sum_{i=1}^{N} \mathbf{x}^{(i)T} & N \end{bmatrix} \begin{bmatrix} \mathbf{a}_m \\ b(m) \end{bmatrix} = \begin{bmatrix} \sum_{i=1}^{N} y^{(i)}(m) \mathbf{x}^{(i)} \\ \sum_{i=1}^{N} y^{(i)}(m) \end{bmatrix}. \tag{3}$$

The function $E(m)$ is minimized for $m = 1$ to p. Hence, m different systems of equations of dimension $(p+1)$ as described by Eq. (3) are solved. The left hand square matrix of Eq. (3) is independent of m and hence, needs to be computed only once.

4 GMM Classifier and Decision Logic

A Gaussian Mixture model (GMM) λ is completely specified by a conditional probability density expressed as a linear combination of Gaussian densities as given by [2]

$$p(\mathbf{x}|\lambda) = \sum_{i=1}^{V} w_i c_i(\mathbf{x}) \tag{4}$$

where $\mathbf{x}$ is a p-dimensional feature vector, $c_i(\mathbf{x})$ is a p-variate Gaussian probability density function and w_i are the mixture weights ($\sum w_i = 1$) for $i = 1$ to V (V is the number of Gaussian mixtures). The Gaussian density $c_i(\mathbf{x})$ is specified by a vector of means μ_i and a covariance matrix Σ_i. The expectation maximization (EM) algorithm determines $\lambda = (w_i, \mu_i, \Sigma_i)$ for a particular speaker from the training set of feature vectors for that speaker only. No affine transform is applied to the feature vectors.

The GMM system for processing a test speech utterance is shown in Figure 1. A test utterance from one of the speakers is converted to a set of test feature vectors $X = (\mathbf{x}_1, \mathbf{x}_2, \cdots, \mathbf{x}_q)$. The affine transform is applied to each test feature vector to give $Y = (\mathbf{y}_1, \mathbf{y}_2, \cdots, \mathbf{y}_q)$. Given M speakers for which speaker i is

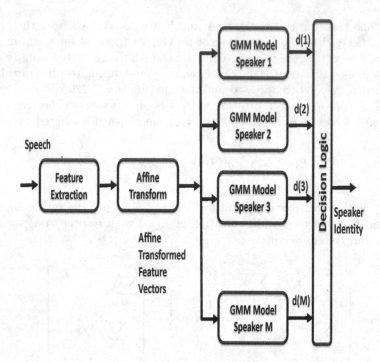

Fig. 1. Block diagram of speaker identification based on Gaussian mixture models

represented by GMM λ_i, the identified speaker $M^\star$ is chosen to maximize the posterior probability $\Pr(\lambda|Y)$, which is equivalent to [2]

$$M^\star = \arg \max_{1 \le j \le M} \sum_{i=1}^{q} \log p(\mathbf{y}_i|\lambda_j) = \arg \max_{1 \le j \le M} d(j) \tag{5}$$

where $p(\mathbf{y}_i|\lambda)$ is computed as given in Eq. (4) and $d(j)$ is the GMM score for speaker j. When many utterances are tested, the identification success rate (ISR) is the number of utterances for which the speaker is identified correctly divided by the total number of utterances tested.

5 Feature Fusion

Since five different features are used, each with a separate GMM classifier, an ensemble system [10] results, which naturally leads to the investigation of fusion. Decision level fusion is the simplest technique and involves taking a majority vote of the different features to get a final decision. For score fusion, the GMM scores $d(i)$ ($i = 1$ to M) of the five single feature classifiers depicted in Fig. 1 are each converted to normalized scores $r(i)$ that are in the range $[0,1]$ such that $\sum_i r(i) = 1$. For speaker i, the $r(i)$ generated by a subset of or all of the five different features are added to get a combined score. The maximum combined score identifies the speaker. The third fusion method is to use Borda count based on the original GMM scores $d(i)$.

6   Experimental Protocol

Ten sentences from each of 60 speakers from the TIMIT database are used for the experiments. The speech in this database is clean and first downsampled from 16 kHz to 8 kHz. The speech is preemphasized by using a nonrecursive filter $1 - 0.95z^{-1}$ and then divided into frames of 30 ms duration with a 20 ms overlap. For the LP analysis, the autocorrelation method is used to get a 12th order LP polynomial $A(z)$. The LP coefficients are converted into 12 dimensional LSF, CEP, ACW and PFL feature vectors. For the PFL feature, $\alpha = 1$ and $\beta = 0.9$. A 12 dimensional MFCC feature vector is computed in each frame. For each of the five features, a 12 dimensional delta feature [2] is computed in each frame using a frame span of 5 in order to derive first derivative information. Second derivative information is also obtained by calculating the delta-delta feature from the delta feature. Concatenation of the feature vector, the first derivative and the second derivative results in a 36 dimensional vector that is used for speaker identification. Energy thresholding is performed over all frames of an utterance to determine the relatively high energy speech frames. The 36 dimensional vectors are considered only in these high energy frames.

The GMM classifier is trained using the 36 dimensional vectors. A different classifier is used for each of the five features. The k-means algorithm is used to initialize the parameters of a 32 mixture GMM speaker model with a diagonal covariance matrix. Ten iterations of the EM algorithm results in the final GMM model.

For each speaker in the database, there are 10 sentences. The first eight are used for training the GMM classifier (clean speech). The remaining two sentences are individually used for testing thereby giving 120 test cases. The test speech is compressed by the 8 kilobits/second ITU-T G.729 codec [6] and the decoded speech is used for speaker identification.

The affine transform parameters are determined from 5 training utterances of each speaker. It is only used for the 12 dimensional feature vector and not for the first and second derivative information. The parameters, $\mathbf{A}$ and $\mathbf{b}$, are different for each feature. Consider a training utterance. The feature vectors of this clean utterance are first computed. The utterance is passed through the G.729 codec and the feature vectors of the decoded speech are computed. The vectors are matched up such that $\mathbf{y}^{(i)}$ and $\mathbf{x}^{(i)}$ correspond to the ith frame of the clean utterance and the G.729 utterance, respectively. Following the same procedure, the feature vectors of every clean and its corresponding G.729 utterance are matched up. Then, the entire set of N clean and G.729 vectors are used to calculate the affine transform parameters $\mathbf{A}$ and $\mathbf{b}$ using Eq. (3).

7   Results

The results are presented in terms of the identification success rate (ISR), which as mentioned earlier, is the number of utterances for which the speaker is identified correctly divided by the total number of utterances tested (expressed as a percent). In each experiment, the ISR is determined as an average over 5 trials.

The first experiment is to compare the performance of the five individual features with and without the affine transform. Also compared is the ISR obtained by McCree's method [8], which is based on enhancing the decoded speech by (1) computing the LP polynomial $A(z)$ and filtering the decoded speech on a frame-by-frame basis and (2) resynthesizing the speech by using the LP information in the G.729 bit stream. For McCree's method, the same GMM speaker identification system trained on 36 dimensional vectors derived from clean speech is used. However, no affine transform is applied to the test speech. Table 1 gives the identification success rate (ISR) results. The most pronounced differences in performance are for the LSF and MFCC features.

Table 1. Identification Success Rate (%) for Individual Features

Feature	No Affine Transform	With Affine Transform	McCree's Method
CEP	69.3	81.3	83.3
ACW	63.7	79.0	81.2
PFL	67.2	80.0	82.2
LSF	51.2	72.5	75.2
MFCC	55.8	74.8	70.8

Decision level fusion, score fusion and Borda count using all five features and all possible subsets of the features were attempted (31 feature combinations). The affine transform is applied. Table 2 shows the best cases for the different fusion methods. Although the best results are obtained with decision level fusion, the comparison of Tables 1 and 2 show that either feature enhancement (affine transform) or signal enhancement (McCree approach) gives more performance improvement than fusion. In fact, when fusion is performed without any enhancement (feature or signal), the best ISR is 71.7% using the CEP/ACW/PFL/MFCC combination. This motivates the question of whether fusion of the affine transform and the McCree approach can be accomplished.

Table 2. Identification Success Rate (%) for different fusion methods. The best feature combinations are shown for each method.

Number of Features	Decision Level Fusion	Score Fusion	Borda Count
All 5	82.3	71.5	77.3
4	CEP/ACW/LSF/MFCC 83.5	CEP/ACW/PFL/LSF 81.7	CEP/ACW/PFL/LSF 80.7
3	CEP/LSF/MFCC 84.0	CEP/PFL/LSF 82.3	CEP/ACW/PFL 81.3
2	CEP/PFL 80.7	CEP/PFL 82.0	CEP/PFL 81.7

Combining the affine transform and the McCree approach can be done in two ways.

- Method 1: Implement the affine transform for all the five features. In parallel, implement the McCree approach for the same five features. For a speech utterance, there are 10 opinions as to the speaker identity. Perform decision level fusion of the 10 opinions and all possible subsets. There are 1013 possible combinations.
- Method 2: Implement the McCree method to enhance the decoded speech. Then, implement the affine transform to make the features resemble more that of the clean training condition. This is a sequential implementation. In this case, the affine transform parameters $\mathbf{A}$ and $\mathbf{b}$ are caclulated using the clean speech used in training and the clean training utterances passed through the G.729 codec and enhanced by the McCree approach. Do this for all five features and perform a decision level fusion (31 possible combinations).

Table 3. Identification success rate for the two methods of combining the affine transform and the McCree approach. A combination of up to 5 features is possible for Method 2. NP: Not Possible

Number of Features	1	2	3	4	5	6	7	8	9	10
Method 1	83.3	83.5	85.7	87.3	87.8	87.7	87.7	86.8	86.2	85.2
Method 2	85.7	85.2	89.3	89.8	89.5	NP	NP	NP	NP	NP

Table 3 shows the maximum ISR achieved versus number of features used in decision level fusion for both Methods 1 and 2. For Method 1, the use of up to 10 features is possible, five from the affine transform and five from the McCree approach. The best performance (87.8%) is achieved by decision level fusion of the ACW, LSF and MFCC features resulting from the affine transform and the CEP and MFCC features resulting from the McCree approach. For Method 2, the use of up to 5 features is possible and the best performance (89.8%) is achieved by fusing the CEP, ACW, LSF and MFCC features. From Table 3, Method 2 is preferred.

A two sample statistical t-test with a 5% significance level and unequal variances is performed to determine if the best fusion combination of Method 2 (ISR of 89.8%) is significantly better than the other techniques. The test is based on the 5 trials that are performed for each experiment. Table 4 gives the results along with the p-values. For all comparisons, the p-value is less than the 5% significance level, thereby confirming the superiority of the fusion combination of Method 2.

8 Summary and Conclusions

Multiple techniques give the best performance for G.729 coding distortion. Signal and feature enhancement are equally significant in improving the ISR. However,

Table 4. Statistical t-test for comparison of best fusion combination of Method 2 to other techniques

Compared Technique	Mean (ISR)	Standard Deviation	p-value
Best fusion combination of Method 2 (best overall ISR)	89.8%	1.37%	-
Best fusion combination of Method 1	87.8%	0.95%	0.0154
CEP feature for Method 2 (no fusion)	85.7%	1.49%	8.9e-04
CEP feature for McCree method (no fusion)	83.3%	0.91%	2.48e-05
CEP feature for affine transform method (no fusion)	81.3%	2.28%	1.25e-04

performing both types of enhancement coupled with decision level fusion gives the best ISR of 89.8%. This is confirmed by a statistical t-test.

Acknowledgement. This work was supported by the U.S. Air Force Research Laboratory, Rome, NY, under contract FA8750-10-C-0249 and the National Science Foundation through Grants DUE-1122296 and DUE-1122344.

References

1. Jain, A.K., Ross, A., Nandakumar, K.: Introduction to Biometrics. Springer (2011)
2. Togneri, R., Pullella, D.: An overview of speaker identification: Accuracy and robustness issues. IEEE Circuits and Systems Magazine, 23–61 (2011)
3. Fazel, A., Chakrabartty, S.: An overview of statistical pattern recognition techniques for speaker verification. IEEE Circuits and Systems Magazine, 62–81 (2011)
4. Campbell, J.P., Shen, W., Campbell, W.M., Schwartz, R., Bonastre, J.-F., Matrouf, D.: Forensic speaker recognition. IEEE Signal Proc. Mag., 95–103 (2009)
5. Mammone, R.J., Zhang, X., Ramachandran, R.P.: Robust speaker recognition - A feature based approach. IEEE Signal Proc. Mag., 58–71 (1996)
6. ITU-T: Recommendation G.729 - coding of speech at 8 kbit/s using conjugate-structure algebraic-code-exited linear prediction, CS-ACELP (2007)
7. Moreno-Daniel, A., Juang, B.-H., Nolazco-Flores, J.A.: Robustness of bit-stream based features for speaker verification. In: IEEE Int. Conf. on Acoustics, Speech and Signal Proc., pp. I-749–I-752 (2005)
8. McCree, A.: Reducing Speech Coding Distortion for Speaker Identification. In: IEEE Int. Conf. on Spoken Language Proc. (2006)
9. Zilovic, M.S., Ramachandran, R.P., Mammone, R.J.: Speaker identification based on the use of robust cepstral features obtained from pole-zero transfer functions. IEEE Trans. on Speech and Audio Proc., 260–267 (1998)
10. Polikar, R.: Ensemble based systems in decision making. IEEE Circuits and Systems Magazine, 21–45 (2006)

Early-Vision-Inspired Method to Distinguish between Handwritten and Machine-Printed Character Images Using Hough Transform

Yuuya Konno[1] and Akira Hirose[2]

[1] Fuji Xerox Co., Ltd,
6–1 MinatoMirai, Nishi, Yokohama, Kanagawa 220-8668, Japan
[2] Department of Electrical Engineering and Information Systems,
The University of Tokyo, 7–3–1 Hongo, Bunkyo, Tokyo 113–8656, Japan
sayaka.yamauchi@inf.kyushu-u.ac.jp

Abstract. This paper proposes a method to distinguish handwritten character regions from machine-printed ones using Hough transform (HT). The Gabor filtering in the human early vision realizes a type of Fourier transform (FT). Previously we proposed a FT-based distinction method successfully. However, we noticed simultaneously that the HT, instead of FT, may extract more features when we deal with characters which are regarded as piles of line segments. Experiments show that HT-based method, in combination with real-space features, achieves higher accuracy than the FT-based method. At the same time, the total calculation cost is found lower.

Keywords: Handwritten character, machine-printed character, Hough transform, optical character reader (OCR).

1 Introduction

We have been pursuing methods to distinguish between handwritten and machine-printed character regions. We have two aims. One is to contribute to the technology of optical character recognition (OCR). It is true that some of OCR engines does not require any distinction between handwritten and machine-printed character regions. However, their performance is lower than those of handwritten / machine-printed specialized engines that are switched according to the distinction. In addition, by clarifying existing factors that leads to their difference, we will be able to contribute to the improvement of the recognition accuracy for the respective kinds of characters.

The other aim is applications to search and automatic classification of documents. Since electronic documents are widely available in recent years, document users are interested in wiser ways to use those documents for search, classification and analysis. For this purpose, it becomes more important to collect key strings efficiently. Besides higher-level processing such as the morphological analysis, we can also utilize the lower-level information such as whether characters are

T. Huang et al. (Eds.): ICONIP 2012, Part V, LNCS 7667, pp. 353–360, 2012.

handwritten or machine-printed. Characters written by hand often represent personal information such as name, address and telephone number in, for example, sales records. If we can determine handwritten character regions, we can enable the search and automatic classification based on such personal information.

This research realizes the distinction of handwritten and machine-printed character regions by image feature extraction and classification. We are inspired by the processing of early visual cortex. That is, we distinguish them by paying attention to the fluctuation of line segment directions. The method of Koyama et al. using Fourier transform is existing technology focused on the processing of early visual cortex. We expect that we can realize higher accuracy by employing the HT to extract more characteristics of the line segments.

In this paper, we propose HT-based method to distinguish between handwritten and machine-printed character regions. We assume to deal with low-quality office documents such as faxed sales records. Experiments demonstrate that our method achieves a distinction rate higher than the conventional FT-based method. In addition, the total calculation cost is lower because of the acceleration techniques specific to HT.

2   Conventional Methods

We can categorize the conventional methods to distinguish handwritten and machine-printed characters into the following three types based on the features used.

1. Methods using single-character features
   For example, method by comparing the horizontal and vertical run length of the character image [1], method by using matching character-image template [2].
2. Methods using relationship features among multiple characters
   Method using variances of baseline, median point, gap, aspect ratio of the bounding rectangles of characters [3].
3. Methods using the pixel features in arbitrary area
   Method using the angle fluctuations of line segments in the frequency domain by transforming the area image by the fast Fourier transform (FFT) [4] [5].

Methods 1 and 2 using real space (RS) features normally require hundreds of characters. Method 3 needs a high calculation cost for the FFT.

3   Proposal of the Hough-Transform-Based Method

To solve these problems, we propose a new method to distinguish between handwritten and machine-printed character regions using HT.

3.1   Feature Extraction Using Hough Transform

Koyama et al. have defined the fluctuation in handwritten characters as the nonuniformity of character size, unevenness in line spacing and misalignment of vertical or horizontal lines [4][5]. Information in spatial frequency area is used to extract the fluctuation. A type of information similar to that obtained by the method proposed by Koyama et al. is extracted using the line component extraction based on using HT.

3.2   Effective Utilization of Hough Transform

HT is a popular method to extract lines or circles from image[6]. The feature of HT is that it can detect

- Lines and circles from images with noise.
- Broken and/or crossing lines.
- Multiple lines at a time.

However, this procedure also has the following disadvantages:

- It uses large memory capacity for parameter space.
- It cannot be applied to all situations.
- It takes more time than real-space image processing.
- It does not include start and end point information of lines.

In the case of determining handwritten or machine-printed characters, this method is advantageous since we can overcome all the disadvantages except for the processing time.

3.3   Procedure to Extract the Features Using Hough Transform

The processing steps are shown in Fig.1 and explained as follows.

1. Apply thinning process to a character image.
2. Set the window to the specified size.
3. Apply HT and accumulate the Hough curves in the parameter space.
4. Remove the HT accumulation below threshold θ_A to cut noise to reduce the calculation cost in the next step.
5. Search peaks above another threshold θ_B, and compile the peak number in the angle domain.
6. Shift the window, and repeat steps 2 to 5. Continue the process until the window scans over all the image to make an angle-domain histogram.
7. Distinguish between handwritten or machine-printed characters based on the difference of the histogram.

The optimal parameters have been found as : Window size = 32×32 pixels, Threshold $\theta_A = 3$, Threshold $\theta_B = 9$, Shift interval for window scanning = 16 pixels.

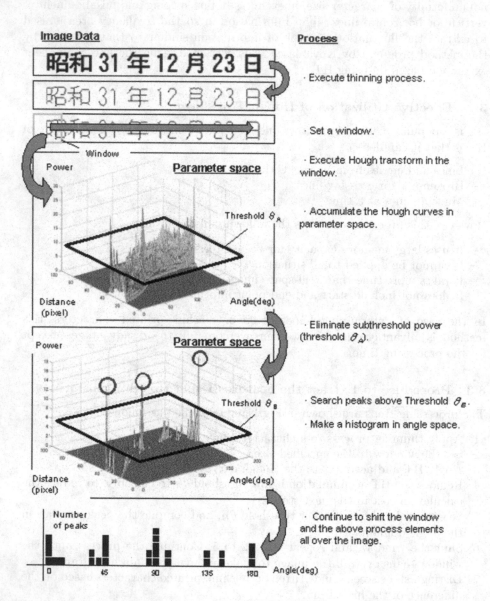

Fig. 1. Feature extraction procedure using Hough transform (HT)

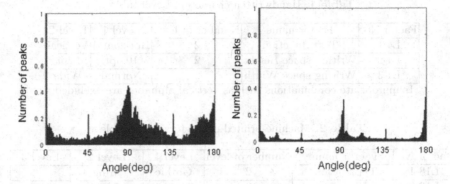

Fig. 2. Peaks histogram in angle space for (a) handwritten and (b) machine-printed characters

3.4   Feature Extraction Procedure Using Hough Transform

Fig. 2 shows a typical histogram obtained from handwritten and machine-printed images using HT. Fig. 2 indicates that the distribution is high at 0, 45, 90, 135, 180 degrees for both images. The variation of handwritten images is larger than that of machine-printed characters. This fact shows that it is possible to distinguish the characters based on the profile difference in the histogram made in the line-angle domain. We set observation regions at angles of 0, 45, 90, 135 and 180 degrees with ±7 degree width, in which we calculate local distribution variances. We feed the variances to an adaptive classifier (e.g., Support Vector Machine: SVM) to obtain the distinction decision.

4   Experiment

We implemented and compared the conventional methods and our HT-based method.

4.1   Experimental Condition

In order to evaluate performance, we prepared character image samples. It is desirable that a verification can cover various conditions. Then we defined the following condition for making the samples.

- 100 strings are selected. They are composed of digit, Kanji, Hiragana, Katakana, alphabet, mixture of number and Kanji, mixture of Kanji and Hiragana.
- A sample contains a single character, some characters or a few words.
- We make patterns of character descriptions according to the experimental design method, resulting in 18 patterns for machine-printed characters and 4 patterns for handwritten ones.

Table 1. Handwritten characters Level table

Factor No.	Factor name	Number of level	Level 1	Level 2
L4-1	Text direction	2	Horizontal	Vertical
L4-2	Writing space Height	2	10.5pt	22pt
L4-3	Writing space Width	2	Normal	Wide

Inappropriate combinations such as vertical alphabet are excluded.

Table 2. Machine-printed characters Level table

Factor No.	Factor name	Number of level	Level 1	Level 2	Level 3
L18-1	Type of font	2	Gothic	Ming	
L18-2	Font size	3	10.5pt	22pt	34pt
L18-3	Text decoration	3	Normal	Bold	Italic
L18-4	Proportional Font	3	Normal	Proportional	Normal
L18-5	Error	3	Error 1	Error 2	Error 3
L18-6	Text Direction	3	Horizontal	Vertical	Horizontal
L18-7	Character spacing	3	Narrow	Standard	Wide
L18-8	Error	3	Error 1	Error 2	Error 3

Inappropriate combinations such as vertical alphabet are excluded.

- Handwritten samples are written by 10 people so that the garphoanalytical aspects are taken into consideration.
- In total, we prepare (100 strings) × {(4 handwritten pattens) × (10 people) + (18 machine-printed pattens)} samples.
- We scan the samples to generate binary low-resolution (200dpi) images by assuming FAX quality.

Table 1 and 2 show the definition of the pattern descriptions . Inappropriate combinations such as vertical alphabet are excluded. Character image samples consist of 3,240 handwritten characters and 1,800 machine-printed characters. Fig. 3 shows a part of that set.

We compare accuracy and processing speed for the conventional methods and our HT-based method to distinguish the above character image samples as follows. We divide the sample images into ten groups. We use 9 groups as teachers for the learning, and test the performance for the last group. We change the groups so that all the groups experience one test, respectively. We define that the final result is the average of the correct distinction rates. We examine the following three methods, that is, RS-based method, FFT-based method which is the method proposed by Koyama et al. (conventional methods), HT-based method, and their combinations. We used a SVM as an adaptive classifier. The parameters of the SVM are optimized for respective methods.

4.2   Experimental Results

Table 3 shows the accuracy obtained in experiments. The accuracy of the HT-based method is lower than that of the FFT-based method. However, when it

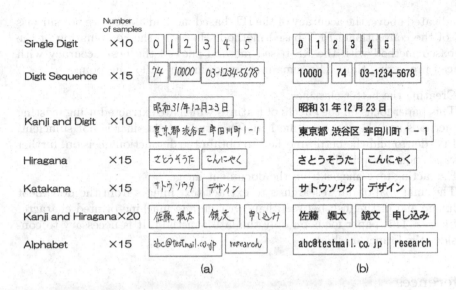

Fig. 3. (a) Handwritten and (b) machine-printed character test samples

Table 3. Comparison of accuracy and processing time

Used method	Accuracy()	Processing timeisecj
RS-based method	91.1	0.45
FFT-based method	93.9	7.72
HT-based method	90.4	0.52
FFTRS method	94.0	8.01
HTRS method	96.0	1.02

is combined with the RS-based method (HTRS method), the accuracy is higher than the FFT-based method. In the FFT-based case, the combination with the RS-based method (FFTRS method) does not change the accuracy.

Table 3 shows the processing time. The processing time of HTRS method is 7 times as fast as that of the FFT-based method.

5   Conclusions

In conclusion, the results suggest the following points.

- The accuracy of the HTRS method is higher than that of the FFT-based method.
- The HTRS method is quicker than the FFT-based method.
- The accuracy of FFTRS method is almost the same as that of the FFT-based method.

As indicated above, the accuracy of the HT-based method alone does not surpass that of the conventional FFT-based method. However, the combination of the HT-based method with the RS-based method showed the best accuracy with reduced processing time. The remaining issues are the following points.

- Creating the better classifier.
  This paper focus on the effects of features which are extracted using existing methods and the proposed method. The study of classifier is not sufficient. In order to find the better method to obtain the distinction decision, further validation is required.
- Extracting text images from the document image.
  The final goal of our research is to extract text images from the document image and distinguish between handwritten and machine-printed character images. This paper focus on only the latter method. It is necessary to consider the former method.

References

1. Ding, X., Chen, L., Wu, T.: Character independent font recognition on a single chinese character. In: IEEE Trans. Pattern Anal., pp. 195–204 (2007)
2. Kavallieratou, E., Stamatatos, S.: Discrimination of machine-printed from handwritten text using simple structural characteristics. In: International Conference on Pattern Recognition, pp. 437–440 (2004)
3. Wang, S.L., Fan, C.K., Tu, T.Y.: Classificationof machine-printed and handwritten texts using character block layout variance. Pattern Recognition, 1275–1284, 437–440 (1998)
4. Koyama, J., Kato, M., Hirose, A.: Distinction between handwritten and machine-printed characters without extracting characters or text lines. In: Proc. of World Congress on Computational Intelligence, International Joint Conference on Neural Networks (IJCNN), Hong Kong, pp. 4143–4150 (2008)
5. Koyama, J., Kato, M., Hirose, A.: Local-spectrum-based distinction between handwritten and machine-printed characters. In: Int'l Conf. on Image Processing, San Diego, pp. 12–15 (2008)
6. Hough, P.V.C.: U.S. Patent 3069654: method and means for recognizing complex patterns (1962)

Unitary Anomaly Detection for Ubiquitous Safety in Machine Health Monitoring

Muhammad Amar*, Iqbal Gondal, and Campbell Wilson

Faculty of Information Technology Monash University, Australia
{muhammad.amar,iqbal.gondal,campbell.wilson}@monash.edu

Abstract. Safety has always been of vital concern in both industrial and home applications. Ensuring safety often requires certain quantifications regarding the inclusive behavior of the system under observation in order to determine deviations from normal behavior. In machine health monitoring, the vibration signal is of great importance for such measurements because it includes abundant information from several machine parts and surroundings that can influence machine behavior. This paper proposes a unitary anomaly detection technique (UAD) that, upon observation of abnormal behavior in the vibration signal, can trigger an alarm with an adjustable threshold in order to meet different safety requirements. The normalized amplitude of spectral contents of the quasi stationary time vibration signal are divided into frequency bins, and the summed amplitudes frequencies over bin are used as features. From a training set consisting of normal vibration signals, Gaussian distribution models are obtained for each feature, which are then used for anomaly detection.

Keywords: machine health monitoring, anomaly detection, spectral content, ubiquitous safety.

1 Introduction

Machine Health Monitoring has gained considerable attention because of mandated high standards of safety. Manual observations of machinery are vulnerable to latency and misapprehension, which can lead to major damage. Automated and ubiquitous anomaly detection techniques for rotary machines are in demand to ensure improved performance, better maintenance and ultimately safety to avoid catastrophe. Several signals, such as current, voltage, speed, and vibration have been used for machine health monitoring tasks [1,2].The vibration signal has the advantage of having abundant and important information about the machine dynamics. For ubiquitous safety, the more plentiful the information, the greater the potential there is for better safety to be achieved. Fault sources in machines are numerous and the authors of [1] have classified these as internal and external fault sources, as shown in Fig. 1. In recent years, researchers have been looking, in particular, for fault signatures in the vibration signals which are attributed to only a specific element of the motor and thus developed

---

* Corresponding author.

T. Huang et al. (Eds.): ICONIP 2012, Part V, LNCS 7667, pp. 361–368, 2012.
© Springer-Verlag Berlin Heidelberg 2012

sophisticated signal analysis techniques under low SNR conditions [3]-[7]. But, the techniques deal only with fault-specific contents as signal and discard the ample portion of the vibration signal as noise. A range of diagnostic techniques have been applied to fault characterization, including: wavelet transforms, artificial intelligence, curve fitting, shock pulse method and many others [1]-[7].In order to characterize a system's overall behavior, many techniques are required to work together and the task is thus very exacting, because of the individual solutions required for each fault. The intuition of the proposed research is that, instead of looking for several solutions, one for each fault, a unitary anomaly detection technique should cover the inclusive behavior of the system using the abundance of contents of the vibration signal. The proposed technique is able to respond to assorted variations caused by sundry machine parts and environment, thus providing ubiquitous safety instead of orthodox categorical trends. The unitary anomaly detector is trained with the normal vibration signals of the motor and after training, gives an alarm signal and prompts for attention if it sees any abnormality in the vibration signal.

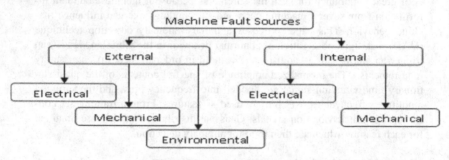

Fig. 1. Machine faults sources

Machine health monitoring requires certain features from input signals in order to make decisions. Normally, researchers use either time or frequency domain based features but the use of both time and frequency based features is also present in the literature [1], [2], [4] and [5]. For feature extraction, several transforms like Fourier, Wavelet and others are prevalent, whereby each one has its own advantages and disadvantages [2, 3]. The choice of transform method depends upon its complexity and its effect on end results. Vibration signals from motors are highly non-stationary in nature but the normal operations of the motors have been observed to be of quasi-stationary nature [7]. To better handle quasi-stationary vibration signals, the idea of using spectral bins as features has been adapted because it can mitigate the effect of small variations in spectral contents. Some of the machine health monitoring techniques require characteristic-fault-frequency patterns and others require system model in order to make these techniques work [6]. The proposed technique is independent of the constraints of a system's physical model or any feature characteristics of particular fault listed in [1, 2]. Rather, the system uses the machine's normal behavior as estimated by Gaussian models from a training dataset which is then used for anomaly detection. Keeping in mind the different safety level requirements at different locations and situations, the algorithm uses an adjustable safety threshold parameter.

As shown in Fig. 2,the proposed algorithm first segments the time domain vibration signal with a time window to form a training dataset and then, for each segment from the training dataset, calculates the normalized spectral contents and divides these contents into frequency bins. The sum of the amplitudes of constituent frequencies in each bin is used as feature input. Then by using this entire feature input dataset, one Gaussian model is obtained for each feature.

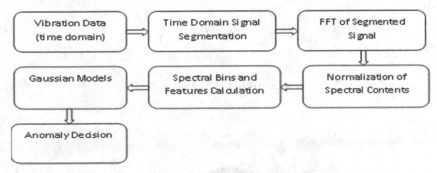

Fig. 2. Flow chart for ubiquitous anomaly detection

The rest of paper is organized as follows: section II of the paper describes segmentation of the time domain signal and bin formation from the spectral contents of each segment, section III presents the anomaly detection algorithm, section IV discusses the results and section V concludes the paper.

2   Time Domain Signal Segmentation and Spectral Contents Bins

The algorithm starts with the acquisition of the input time domain vibration signal from the machine. For further processing and feature extraction from the vibration signal, the time domain signal is divided into segments using a fixed rectangular window of 1024 samples. The size of window should be sufficient enough to capture at least one complete cycle of machine rotation. Each time domain segment is then transformed into frequency domain using the Fourier Transform to extract the spectral contents of time domain vibration. If the captured time domain signal is of the total length l and window size is w, then the number of training set examples or segments m is given by (1):

$$m = {}^{l}/_{w} \tag{1}$$

If x is the input signal with m time domain segments then training set is (2):

$$x = \{x^1, x^2, x^3, \dots, x^{m-1}, x^m\} \tag{2}$$

From time segments of the vibration signal spectral contents of down sampled time segments are calculated using FFT (3):

$$f^i = FFT(x^i) \quad \text{where } i = 1, 2, 3, \ldots, m \quad\quad (3)$$

Down sampling reduces the size of spectrum still retaining the shape of the spectrum. From the spectrum of input segments, only 512 frequencies are selected ranging from 0 to 511Hz because there is sufficient normal signal information in this spectral range for anomaly detection [7]. The spectral contents of each segment f^i are normalized to $\bar{f}^i$ using (4) to ensure the amplitudes range from 0 to 1 (5). The normalized spectral contents of healthy vibration signals of various segmented signals are shown in Fig. 3.

$$\bar{f}^i = f^i / \max(f^i) \quad\quad (4)$$

Thus

$$0 \leq Amp(\bar{f}^i) \leq 1 \quad\quad (5)$$

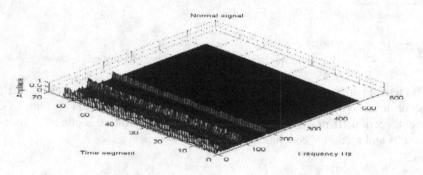

Fig. 3. Normal vibration signal frequency contents

The normalized spectrum is divided into n frequency bins using the frequency bin size b (6):

$$n = \frac{512}{b} \quad\quad (6)$$

Note that b is the number of adjacent frequencies to be combined in a bin and the frequency contents of $\bar{f}^i$ over entire training set m having n bins with bin of size b are given by (7):

$$\bar{f}_j^i(k) = \bar{f}^i\big(((j-1) \times b) + k\big) \quad\quad (7)$$

$$i = 1, 2, 3, \ldots, m; \quad j = 1, 2, 3, \ldots, n \quad \text{and} \quad k = 1, 2, 3, \ldots, b;$$

Now, we will use Eq. (8) to sum the amplitudes of constituent frequencies for total of n bins given by (6):

$$a_j^i = \sum_{k=1}^{b} A\left(\bar{f}_j^i(k)\right) \quad\quad (8)$$

The function $A\left(\bar{f}_j^i(k)\right)$ gives the amplitude of the kth frequency and a_j^i is the sum of frequencies amplitudes of the jth bin of ith training example to serve as a feature.

3 Anomaly Detection Algorithm

In this section we will develop n Gaussian models, where the Gaussian model for each feature input is developed by using the jth feature input from the entire training set of size m. For the jth Gaussian model, we need mean $(ì_j)$ and variance $(ó_j^2)$ which are calculated using Eq. (9) and (10) respectively:

$$ì_j = \frac{1}{m}\sum_{i=1}^{m} a_j^i \tag{9}$$

$$ó_j^2 = \frac{\sum_{i=1}^{m}(a_j^i - ì_j)}{m} \tag{10}$$

For n features we have $ì = \{u_1, u_2, u_3, \dots u_n\}$ and $ó^2 = \{ó_1^2, ó_2^2, ó_3^2, \dots, ó_n^2\}$.

The larger the training set size m, the better is the training, i.e. the estimation of the parameters $(ì, ó^2)$ .The trained Gaussian models for n features are (11):

$$G(ì, ó^2) = \{G(ì_1, ó_1^2), G(ì_2, ó_2^2), G(ì_3, ó_3^2), \dots, G(ì_n, ó_n^2)\} \tag{11}$$

In Eq. (11), j^{th} Gaussian model parameterized by $(ì_j, ó_j^2)$, describes the probability distribution of jth feature over the entire training set of size m.For, illustration purpose, $n = 4$, the trained Gaussian models are shown in Fig. 4.

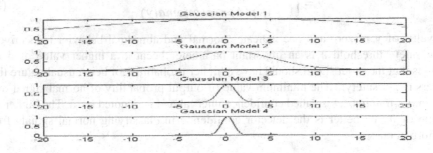

Fig. 4. Gaussian models for $n = 4$

From these models, we can find the output probability for a new feature input value $á_j$ as given by Eq. (12)

$$p(á_j) = G(á_j; ì_j; ó_j^2) \tag{12}$$

If the output probability value $p(á_j)$ of the jth Gaussian model for a new feature input $á_j$ is large it means that the jth bin's contents are similar to that of usual behavior and smaller output values indicates the abnormal nature of the bin contents.

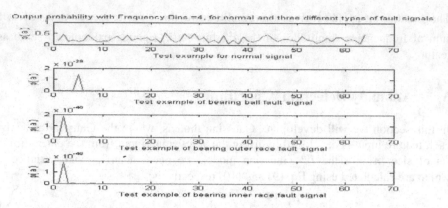

Fig. 5. $p(\acute{a})$ for normal vibration signal and three different abnormal signals

For overall output probability $p(\acute{a})$ Eq. (13) is used by passing each feature of test signal through the relevant Gaussian model

$$p(\acute{a}) = \prod_{j=1}^{n} G(\acute{a}_j; \grave{1}_j; \acute{o}_j^2)$$
(13)

The larger $p(\acute{a})$ value indicates the probability of the input being of a usual nature is high. Fig. 5 compares $p(\acute{a})$ of normal and several abnormal vibration signals. To trigger an alarm, either the input signal belongs to a normal or abnormal class; a decision boundary value, denoted by safety threshold å, is compared with $p(\acute{a})$: (14)

$$y = \begin{cases} 0 & \text{if } p(\acute{a}) \geq \text{å } (normal) \\ 1 & \text{if } p(\acute{a}) < \text{å}(anomaly) \end{cases}$$
(14)

A value of y of zero and one indicates normal and abnormal behavior respectively. The safety threshold å is an important parameter because a higher value of å demands that the input signal should have a higher probability of being usual nature thus offers better safety. The minimum value of output probability of normal signal over the entire training set is called confidence interval and is denoted by ã. The higher the value of ã, the better is the detector confidence in classifying normal signals from abnormal ones.

4   Results and Discussion

Unitary anomaly detection technique for ubiquitous safety has shown promising results over various test signals. The effect of various parameters on the performance of UAD will be studied using bearing faults which constitute 40% of total machine fault [1], [2]. We have used publically available datasets [8] to compare the test results of the UAD. Three different bearing faults: inner race, outer race and ball; have been used to test the performance of UAD in distinguishing these faults from normal behavior. It is evident that UAD can classify these signals as abnormal with high accuracy and confidence Fig. 5. One of the performance parameters, confidence interval of UAD, can be investigated by varying the number of bins. Fig. 6 shows the variations in the $p(\acute{a})$ of a normal signal for different numbers of bins depicting a noticeable trend of decrease in confidence interval with corresponding increase in the number of bins. This is because as we increase the number of bins or features, the size of each frequency bin decreases (6), increasing the chance of the spectral contents of the one bin to spill into the adjacent ones due to nonstationarity, thus ultimately decreasing the confidence level of the anomaly detector. For larger bin size spectral contents are more likely to be confined in the same frequency bin, thus increasing the confidence interval but for larger bin size it is possible for two faults having different frequency signature to reside in the same frequency bin causing difficulties in anomaly detection. From the above discussion, it is clear that there exists a compromise between the capability of the number of faults to be addressed for anomaly detection and the confidence interval.

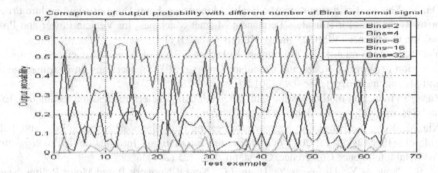

Fig. 6. Effect of frequency bins on confidence interval

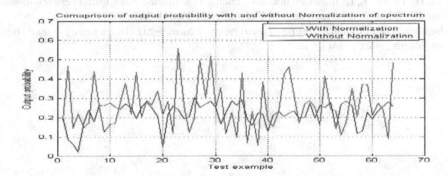

Fig. 7. Effect of spectral amplitude normalization on confidence interval

Spectral amplitude normalization has considerable effect on the output probability $p(\acute{a})$ and thus on confidence interval. Fig. 7 shows the $p(\acute{a})$ comparison of normalized and non-normalized features, used for training, and it is evident that normalization improves the confidence.

5   Conclusions

In this paper, a unitary anomaly detection algorithm for machine health monitoring has been presented. Spectral features of time segmented signal have been used to develop Gaussian models which are then used for normal or abnormal behavior classification of test input. Using an adjustable safety threshold, to meet desirable safety levels, the algorithm has shown that the successful classification of normal vibration signals against numerous abnormal signals can be made with great confidence, thus providing ubiquitous safety.

References

1. Singh, G.K., Ahmed Saleh Al Kazzaz, S.: Induction machine drive condition monitoring and diagnostic research—a survey. Electric Power Systems Research 64(2), 145–158 (2003)
2. Da, Y., Shi, X., Krishnamurthy, M.: Health Monitoring, Fault Diagnosis and Failure Prognosis Techniques for Brushless Permanent Magnet Machines. In: Vehicle Power and Propulsion Conference (VPPC), Chicago, IL, pp. 1–7 (2011)
3. Yaqub, M.F., Gondal, I., Kamruzzaman, J.: Inchoate Fault Detection Framework: Adaptive Selection of Wavelet Nodes and Cumulant Orders. IEEE Transactions on instrumentation and Measurement 61(3), 685–695 (2012)
4. Zarei, J.: Induction motors bearing fault detection using pattern recognition techniques. Expert Systems with Applications 39(1), 68–73 (2012)
5. Xia, M., Kong, F., Hu, F.: An approach for bearing fault diagnosis based on PCA and multiple classifier fusion. In: 2011 6th IEEE Joint International Information Technology and Artificial Intelligence Conference, China, pp. 321–325 (2011)
6. Li, B., Chow, M.Y., Tipsuwan, Y., Hung, J.C.: Neural-Network-Based Motor Rolling bearing fault diagnosis. IEEE Transactions on Industrial Electronics 47(5), 1060–1069 (2000)
7. Su, H., Chong, K.T.: Induction machine condition monitoring using neural network modeling. IEEE Transactions on Industrial Electronics 54(1), 241–249 (2007)
8. Bearing Data Center, http://www.eecs.case.edu/laboratory/bearing/welcome_overview.htm

FusGP: Bayesian Co-learning
of Gene Regulatory Networks
and Protein Interaction Networks

Nizamul Morshed, Madhu Chetty, and Nguyen Xuan Vinh

Monash University,
Australia
{nizamul.morshed,madhu.chetty,vinh.nguyen}@monash.edu

Abstract. Understanding gene interactions is a fundamental question
in uncovering the underlying biological relations that enable successful
functioning of living organisms. The modeling of gene regulations is usually done using DNA microarray data. However, presence of noise and
the scarcity of microarray data affect the reconstruction of gene regulatory networks. In this paper, we propose a novel co-learning based fusion
algorithm using the dynamic Bayesian netowrk (DBN) formalism for reconstruction of gene regulatory networks which incorporates knowledge
obtained from protein-protein interaction networks to improve network
accuracy. The proposed approach is efficient and naturally amenable to
parallel computation. We apply the algorithm on the well-known *Saccharomyces cerevisiae* gene expression data that shows the effectiveness
of our approach.

Keywords: Bayesian network, Gene regulatory network, Binary
Markov network, Protein interaction network.

1 Introduction

Gene regulatory networks (GRN) depict the regulatory interactions among the
genes in a living cell. One of the main obstacles in deciphering the regulatory relationships is that microarray data is inherently noisy, and moreover, the number
of samples from microarray is very low. Due to these problems, it would be advantageous to the modeling process to use alternate sources of knowledge rather
than using microarray data alone. Attempts have been made to use prior knowledge from location binding data [1] and protein-protein interaction networks
(PPIN) [2,3] for better reconstructing GRNs. However, due to the fact that the
data for PPI may themselves be erroneous, it is appropriate to use diverse knowledge sources for the reconstruction in conjunction with using protein-protein
interaction (PPI) data[1].

---

[1] Meaning in addition to PPI data, other information sources (e.g., genomic data)
should be used during PPIN reconstruction.

T. Huang et al. (Eds.): ICONIP 2012, Part V, LNCS 7667, pp. 369–377, 2012.
© Springer-Verlag Berlin Heidelberg 2012

A number of techniques have been proposed in the literature with a view to using both microarray and PPI data for reconstructing GRNs and PPINs. Nariai et al. [2] propose a static BN based framework using the concept of formation of protein complexes, based on results from principal component analysis. However, it is difficult to interpret the results to know about whether the estimated causal relationships show gene regulations or protein-protein interactions [3]. The improvement of the method [3] considers it as a three component model, and by combining these three components into a single statistical model, the authors maximise the joint posterior probability with a view to getting the optimal GRN and PPIN. However, both these methods work on static Bayesian networks. As a result, they cannot model feedback loops which are inherent in biological networks. Also, the improved model [3] uses parameters for controlling the balance between microarray and PPI data, which need to be set up based on conducting simulation experiments, and there is no theoretical means of determining the optimal value of the parameter. Chaturvedi et al. [4] model time delayed gene interactions using a skip-chain based dynamic Bayesian network model, which finds missing edges between non-consecutive time points based on knowledge from PPIN using Viterbi approximation. However, the method does not work with multiple sources of prior knowledge (e.g., both PPI data and TF binding location data). Further, knowledge sources for GRNs and PPINs might contain noisy information and thus unlike [4], it is better to consider the information probabilistically.

In this paper, we propose a probabilistic framework for jointly constructing a GRN and a PPIN. We use information from multiple sources of prior knowledge (PPI data, functional category data, essentiality phenotype information etc) probabilistically. Because our method marginalizes over the parameters, we do not need to include the balance parameter during PPI network's posterior probability calculation, which is done in Nariai et al. [3]. Rather than using *conjunctive* approaches like [3], we use a *disjunctive* approach, noting that given the PPIN, the GRN depends only on the microarray, and does not depend on the PPI data (and vice versa). This essentially facilitates us to work in parallel for the PPIN and GRN construction, at the same time maintaining coherent and flexible fusion of information among the parallel threads. This also has the advantage that effectively we have to deal with roughly half of the structure space (considering PPI networks are non directional) compared to approaches where both networks are considered simultaneously. We show the effectiveness of our approach by using two different networks from yeast.

2   Background

In this section, we briefly discuss the formalizations related to the modeling concepts used in this paper: dynamic Bayesian Networks (DBN) and binary Markov Networks.

2.1 Dynamic Bayesian Network (DBN)

Considering X to be a set of attributes changing in a temporal process of T time slices, a DBN represents the joint probability distribution over the variables $X[0] \bigcup X[1] \bigcup \cdots \bigcup X[T-1]$, where random variable $X_i[t]$ denotes the value of node X_i at time slice t, and $X[t]$ denotes the set of variables $\{X_i[t] | 1 \leq i \leq n\}$, for $0 \leq t \leq T-1$ [5]. In this paper, we work with first-order Markov DBN, which is based on two assumptions: (i) *First Order Markov Property*, and (ii) *Stationarity* (see [5] for details).

2.2 Binary Markov Network

Given a set of n nodes, let Y denote any random graph on those nodes and y denote a particular graph on those nodes. A general form for binary Markov networks can then be defined as follows [6]:

$$P(Y = y) = \frac{1}{Z(\theta)} e^{\sum_t \theta_t s_t(y)} \tag{1}$$

where θ_t is the unknown parameter related to $s_t(y)$, and $s_t(y)$ is a known vector of graph statistic (of type t) on y. $Z(\theta)$ is the normalizing constant. Albeit difficult to calculate, the $Z(\theta)$ quantity is essentially constant for all possible networks, and thus it can be safely ignored when we do comparison based network searching.

3 FusGP: Fusion of Gene Regulatory Networks and Protein-Protein Interaction Networks

Consider Figure 1, where the symbols D_r, D_p, G_r and G_p denote GRN data, PPI information, GRN and PPIN, respectively. Dashed arrows among gene regulatory networks (GRN) and protein interaction networks (PPIN) denote transfer of structural information between the corresponding structures.

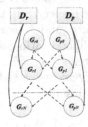

Fig. 1. Schematic of the fusion based co-learning approach

From the figure, noting that G_r depends only on D_r and G_p, and similar for G_p, the posterior probability of the gene regulatory networks and protein interaction networks can be defined by the following formula:

$$P(G_p|\{G_r, D_p\}) \propto \frac{P(G_p)\,P(D_p|G_p)P(G_r|G_p)}{P(G_r)} \tag{2}$$

and similarly for G_r,

$$P(G_r|\{G_p, D_r\}) \propto \frac{P(G_r)\,P(D_r|G_r)P(G_p|G_r)}{P(G_p)} \tag{3}$$

Using above relationships, we can optimize the posterior probability iteratively. In this paper, we propose an evolutionary computation based iterative fusion/co-learning algorithm that achieves this task. We will describe how we calculate different quantities in these equations, and then provide algorithms that can optimize these quantities.

The likelihood of the gene expressionns data can be calculated in a straightforward manner using multinomial conditionals and Dirichlet priors, i.e., using the Bayesian Dirichlet (BD) based scores, assuming parameter independence. Regarding the prior probability of G_p under a given G_r, it can be defined as follows:

$$P(G_p|G_r) \propto e^{-\sum\limits_{e(i,j)\in G_p}\xi c_{ij}} \tag{4}$$

where

$$c_{ij} = \begin{cases} 1 \ if \ e(i,j) \in G_r \\ 2 \ if \ e(i,j) \notin G_r \end{cases} \tag{5}$$

and $e(i,j)$ denote edges between genes X_i and X_j in the GRN. The inverse $(P(G_r|G_p))$ can be defined in a similar manner [3].

Next we consider the calculation of $P(D_p|G_p)$. We need to define a few quantities that are required to calculate this. First, we define likelihood ratio [7] based on the assumption that each genomic features is conditionally independent:

$$L(i,j) = \frac{P(y_{ij}(1)|pos)}{P(y_{ij}(1)|neg)} \cdots \frac{P(y_{ij}(N)|pos)}{P(y_{ij}(N)|neg)} \tag{6}$$

where $y_{ij}(k)$ is an element of D_p that shows a genomic feature of protein pair, X_i and X_j, and 'pos' and 'neg' are respectively the positive and negative sets of protein pairs constructed in advance, and N is the number of genomic features that we consider. Noting that likelihood ratio provides noisy evidence regarding the existence of edges among proteins, we define the $l-$value of an edge $e\{i,j\}$ on the interval $[0, 1]$, which is inversely related to the probability of an edge being present in the true PPIN. Formally,

$$l(i,j) = \frac{\min\limits_{i,j} L(i,j)}{L(i,j)} \tag{7}$$

Based on this definition, we define:

$$\beta(i,j) = \frac{\lambda e^{-\lambda l(i,j)}}{\lambda e^{-\lambda l(i,j)} + 1 - e^{-\lambda}} \tag{8}$$

where λ is the parameter controlling the scale of the truncated exponential distribution. Rather than using (8) in the raw format, we use marginalization [1] over the parameter λ, to get:

$$\beta(i,j) = \frac{1}{\lambda_H - \lambda_L} \int_{\lambda_H}^{\lambda_L} \frac{\lambda e^{-\lambda l(i,j)}}{\lambda e^{-\lambda l(i,j)} + 1 - e^{-\lambda}} d\lambda \tag{9}$$

which is numerically tractable.

Since we have a finite set of $L(i,j)$ values, we can pre-compute these integrals for each $L(i,j)$ value and store the results for later use. The computational overhead associated with marginalizing over λ is thus constant. The net effect of marginalization is an edge probability distribution that is a smoother function of the reported $\beta(i,j)$ values than without marginalization. This results in a much heavier tailed distribution [1], which is advantageous.

Based on these definitions, we can now define the probability $P(D_p|G_p)$:

$$P(D_p|G_p) \propto \prod_{e\{i,j\}\in G_p} \beta(i,j) \tag{10}$$

Since the $\beta(i,j)$ values are marginalized probabilities, we do not need to use parametrized values of the likelihood ratio $(L(i,j)^\alpha)$, which is unlike the method described in [3]. This novel approach reduces the number of parameters that we need to consider during computation, and thus saves computation time.

Finally, the prior probability of G_p is defined to encourage sparsity, using the following equation:

$$P(G_p) \propto e^{\sum_{e(i,j)} \xi_p} \tag{11}$$

Based on the above definitions, we propose an iterative Bayesian co-learning algorithm, *FusGP*. The algorithm first generates initial estimates for the GRN and PPIN. The estimates are then fed to a routine called *FGP*, which performs the task of co-learning based fusion. In the next sub-section, we describe the working procedure of the algorithm.

3.1   The Search Strategy for Initial Network Generation

A genetic algorithm (GA), applied to explore this structure space, begins with a sample population of randomly selected network structures and their fitness calculated. Iteratively, crossovers and mutations of networks within a population are performed and the best fitting individuals of the population are kept for future generations. During crossover, two random edges are chosen and swapped.

Mutation is applied on a randomly chosen individual edge of the network. For our study, we incorporate the following two types of mutations: i) Flipping existence of a random edge in the network, and ii) Changing direction of a randomly selected edge.

Keeping a parallel execution in mind, the genetic algorithm can be called with either of two parameters: R and P. If it is called with a R parameter, this means it is supposed to build a GRN, given the PPIN. However, the P parameter denotes constructing a PPIN. In this case, the operations that the GA can perform becomes restricted (e.g., Markov networks are non directional; during mutation, this needs to be taken into consideration).

3.2   The Algorithm for Co-learning, *FGP*

After the initial networks have been generated, we start the co-learning of the two networks in two parallel threads based on (2) and (3). Since a PPIN cannot give any direction information for use in a GRN, there are only two possible operations in a particular stage of the algorithm: adding an edge if it was not there, and vice versa (flip operation). In a similar manner, since the direction information of a GRN is not useful while constructing a PPIN, only flip of edge existence operations are permitted while constructing PPINs. The overall algorithm is shown in Table 1.

Table 1. Algorithm *FGP*

1. Evaluate the network from the previous iteration. If input is R, evaluate score of $G_{r(m-1)}$ based on $G_{p(m-1)}$, using (3) for the calculation. If input is P, score $G_{p(m-1)}$ based on $G_{r(m-1)}$, using (2).
2. Generate new network by applying flip operation on each possible edge. Store the score of the changed network ($P(G_{pm}|\{G_{r(m-1)}, D_p\})$) using (2) if input is P, or $P(G_{rm}|\{G_{p(m-1)}, D_r\})$) using (3) if input is R).
3. Find the changed network with the maximum score. Keep this as the new "best solution" if score increases compared to the "best solution" from the previous iteration. Otherwise, set the "best solution" from the previous iteration as the new "best solution" and send an "end flag" to the counterpart thread.
4. Repeat steps 1) - 3) until the stopping criteria (new "best solution" is same as previous "best solution", or an "end flag" from the counterpart thread) is reached.   ◁

4   Simulation and Results

We apply our algorithm to the well known *Saccharomyces cerevisiae* cell cycle data obtained from Spellman et al. [8]. We use two different networks, a KEGG pathway of yeast consisting of 11 genes, and a genetically modified network of yeast called IRMA, for the investigation. Both these networks have been used previously for assessment of reconstruction techniques. For all the experiments, the λ_H and λ_L parameters were set to 1 and 1000 respectively, to avoid problems near terminal values [1]. The ζ_p values were set to the log of the cut-off parameter (calculated to be 600) obtained from the PPIN datasets of Jansen et al. [7]. Finally, to set the ζ_1 and ζ_2 parameters, we made the practical assumption that physical protein-protein interactions should be considered as part of PPIN instead of GRN, making them mutually exclusive [3], and as a result ζ_1 can be set to 0 and ζ_2 to ∞. For the initial network generation, we used a basic GA as outlined in Section 3.1 with an initial population size of 100 and a stopping criteria of the completion of 400 generations. To compare our algorithm, we consider two other methods, namely, BANJO [9] and BNFinder [10] (with both BDe and MDL).

4.1 Yeast KEGG Pathway Reconstruction

In order to test the proposed method's performance on yeast *S. cerevisiae* cell cycle, we selected a eleven gene network of the G1-phase, using data obtained from the *cdc28* experiment of Spellman et al. [8]. In Figure 2(B)-(E), we report network graphs reconstructed by our proposed approach, BNFinder(BDe and MDL) and BANJO. We also report the KEGG pathway [11] of the cell-cycle in yeast in 2(A). Since the exact ground truth for this network is not known, instead of applying performance measures as a means of determining network accuracy, we refer to the available correct interactions obtained from the KEGG pathway [11]. We observe from the results that overall none of the methods perform particularly well on this network. However, the number of correct predictions by our method (7) is higher than the other methods.

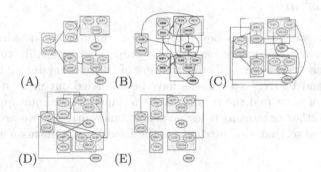

Fig. 2. Reconstruction of Yeast KEGG Pathway [11]. (A) Target Network. (B) Network Inferred by FusGP. (C) Network Inferred by BANJO. (D) Network Inferred by BNFinder+BDe. (E) Network Inferred by BNFinder+MDL.

4.2 Real-Life Biological Data of Yeast, IRMA

To validate our method with another real-life biological gene regulatory network, we investigate a recent network reported in [12]. In that significant work, the authors built a network, called IRMA, of the yeast *Saccharomyces cerevisiae* [12]. A 'simplified' network, ignoring some protein level interactions, is also reported in [12]. We use four well known performance measures to assess the algorithms, namely Sensitivity (*Se*) , Specificity (*Sp*), Precision (*Pr*) and F-Score (*F*) for performance comparison.

IRMA ON Dataset. The performance comparison amongst various method based on the ON dataset is shown in Table 2. We observe that our method reconstructs the gene network with high precision. Specificity is also high, implying that the inference of false positives is low. Overall, FusGP outperforms the other methods in terms of most of the performance measures.

IRMA OFF Dataset. The OFF dataset lacks the presence of 'stimulus' (applied during the experiments); however, it contains more samples compared to the ON dataset (21 versus 16). The comparison of the performance of the algorithms using the OFF dataset is shown in Table 3. We observe from the results that in terms of all the four performance measures, FusGP has the best performance.

Table 2. Results-IRMA ON dataset

	Original Network				Simplified Network			
	Se	Sp	Pr	F	Se	Sp	Pr	F
FusGP	0.25	0.94	0.67	0.37	0.33	1.00	1.0	0.50
BNFinder+BDe	0.13	0.82	0.25	0.17	0.33	0.80	0.50	0.40
BNFinder+MDL	0.13	0.82	0.25	0.17	0.33	0.80	0.50	0.40
BANJO	0.38	0.88	0.60	0.46	0.33	0.90	0.67	0.44

Table 3. Results-IRMA OFF dataset

	Original Network				Simplified Network			
	Se	Sp	Pr	F	Se	Sp	Pr	F
FusGP	0.38	0.82	0.50	0.43	0.50	0.95	0.75	0.60
BNFinder+BDe	0.13	0.82	0.25	0.17	0.17	0.80	0.33	0.22
BNFinder+MDL	0.13	0.82	0.25	0.17	0.17	0.80	0.33	0.22
BANJO	0.25	0.76	0.33	0.27	0.50	0.70	0.50	0.50

5 Conclusion

In this paper, we propose a novel approach to fusing the knowledge of PPI networks to GRNs and vice versa. We also propose an algorithm that works in parallel threads to achieve a coherent transfer of information during the building of the GRN and PPIN. Experiments have been carried out using different real life networks of yeast and the results show the superiority of our approach. The PPI data for other organisms is less available, and currently we are building a knowledge base so that this method can be used for organisms which are less studied.

References

1. Bernard, A., Hartemink, A., et al.: Informative structure priors: joint learning of dynamic regulatory networks from multiple types of data. In: Pac. Symp. Biocomput., vol. 10, pp. 459–470 (2005)
2. Nariai, N., Kim, S., Imoto, S., Miyano, S., et al.: Using protein-protein interactions for refining gene networks estimated from microarray data by bayesian networks. In: Pacific Symposium on Biocomputing, vol. 9, pp. 336–347 (2004)
3. Nariai, N., Tamada, Y., Imoto, S., Miyano, S.: Estimating gene regulatory networks and protein–protein interactions of saccharomyces cerevisiae from multiple genome-wide data. Bioinformatics 21(suppl. 2), ii206–ii212 (2005)
4. Chaturvedi, I., Rajapakse, J.: Fusion of gene regulatory and protein interaction networks using skip-chain models. Pattern Recognition in Bioinformatics, 214–224 (2008)

5. Morshed, N., Chetty, M.: Combining Instantaneous and Time-Delayed Interactions between Genes - A Two Phase Algorithm Based on Information Theory. In: Wang, D., Reynolds, M. (eds.) AI 2011. LNCS, vol. 7106, pp. 102–111. Springer, Heidelberg (2011)
6. Bulashevska, S., Bulashevska, A., Eils, R.: Bayesian statistical modelling of human protein interaction network incorporating protein disorder information. BMC Bioinformatics 11(1), 46 (2010)
7. Jansen, R., Yu, H., et al.: A bayesian networks approach for predicting protein-protein interactions from genomic data. Science 302(5644), 449–453 (2003)
8. Spellman, P., Sherlock, G., et al.: Comprehensive identification of cell cycle–regulated genes of the yeast saccharomyces cerevisiae by microarray hybridization. Molecular Biology of the Cell 9(12), 3273–3297 (1998)
9. Yu, J., Smith, V., Wang, P., Hartemink, A., Jarvis, E.: Advances to Bayesian network inference for generating causal networks from observational biological data. Bioinformatics 20(18), 3594 (2004)
10. Wilczyński, B., Dojer, N.: BNFinder: exact and efficient method for learning Bayesian networks. Bioinformatics 25(2), 286 (2009)
11. Kanehisa, M., Goto, S., Kawashima, S., Nakaya, A.: The kegg databases at genomenet. Nucleic acids research 30(1), 42–46 (2002)
12. Cantone, I., Marucci, L., et al.: A yeast synthetic network for in vivo assessment of reverse-engineering and modeling approaches. Cell 137(1), 172–181 (2009)

An Image Representation Method Based on Retina Mechanism for the Promotion of SIFT and Segmentation

Hui Wei[*], Bo Lang, and Qing-Song Zuo

Department of Computer Science and Technology, University of Fudan, Shanghai, China
{weihui;09110240025;09210240055}@fudan.edu.cn

Abstract. In this paper, a bio-inspired neural network was constructed. It could represent images effectively and provide a processing method for image understanding. Our model adopted the retinal ganglion cells (GCs) and their non-classical receptive field (nCRF) can dynamic self-adjusts according to the characteristics of the image. Extensive experimental evaluations to demonstrate that this kind of representation method was able to make for SIFT detector for focus on foreground object more correctly, and promote the result of image segmentation significantly.

Keywords: image representation, non-classical receptive field, ganglion cell, SIFT, segmentation.

1   Introduction

An image is not just a random collection of pixels for human vision system; it is a meaningful arrangement of regions and objects [1]. Efficient representation of visual information lies at the heart of many image processing task, including compression, denoising, feature extraction, and inverse problems. Efficiency of a representation refers to the ability to capture significant information about an object of interest from scene using a description with minimum cost. From this perspective, image representation may be designed to rely on the specific constrains and knowledge available for the restricted task at hand.   An appropriate representation scheme must be Generality, Reliability, Precision, Concision and Explicitness [1]. It is widely assumed that a general-purpose visual system must entail multiple types of representation which share data through a variety of computation.

In biological visual systems, the classical receptive field (CRF) of GC is sensitive to brightness and contrast. It demonstrates the attributes that has spatial summation properties enabling the boundaries to be detected. The expanded area around CRF, it is referred to as the non-classical receptive field (nCRF), can compensate for the loss of low-frequency information caused by the antagonistic center-surround mechanism of CRF [2]. The receptive field (RF), including CRF and nCRF, is the basic structural and functional unit of visual information processing. Any GC merges all stimuli occurring in its RF, and report a composite signal upwards for further processing [3,4].

---

[*] Corresponding author.

T. Huang et al. (Eds.): ICONIP 2012, Part V, LNCS 7667, pp. 378–385, 2012.

By means of these dense and regular RFs, a GC-array can produce a general representation of any external stimulus.

Many studies have examined the representation of natural images. Previously proposed methods of image representation range from color histograms to feature statistics, from spatial frequency-based to region-based and from color-based to topology-based approaches for an extensive review of image representation techniques, see [5,6]. Some studies have attempted to construct models simulate GCs. Li et al [2]. constructed a 3-Gaussian model using it to simulate spatial responses of GCs. Qiu et al. [7] reported a new model of the mechanisms underlying mutual inhibition within disinhibitory nCRFs. Ghosh et al. [8]modeled the nCRF as a combination of 3-Gaussian at three different scales and seeking to explain certain brightness-contrast illusions. In particular, none of these models specifies the relationship between the disinhibitory nCRF and neural circuits in the retina and not take into account of the dynamic adjust nature of the RF of GCs. They are also not proposed mechanism s for joining pixels or organizing them fragmentally for clustering an explicit and tentatively assembled representation.

In this paper, we describe both microcircuits and macro neural networks based on the structural and functional properties of the nCRF, which can achieve self-adaptive resizing of RFs and image representation, respectively. In our computational network, for the purpose of integration, top-down control and the neighborhood properties of input stimuli are taken into account. Our experimental results revealed that our representation method facilitates SIFT detector and hierarchical segmentation significantly.

2 The Design of Computational Model Based on the nCRF

Biophysical studies [9,10] have suggested a brief mathematical form with three concentric convolutions to model the positional effect of receptive field (Fig. 1). A conventional mathematical model for the CRF is the difference of two Gaussian functions (DOG). Due to the existence of the nCRF, the response of the GC can simulate using three-Gaussian model.

Fig. 1. The structure of receptive field of GC

For a numerical calculation consideration, we also adopt a DOG (Difference of Gaussian) –like formula. The output of a GC can be expressed as follows:

$$GC(x, y) = \sum_{x, y \in S_1} W_{center} \cdot I(x, y) - \sum_{x, y \in S_2} W_{surround} \cdot I(x, y) + \sum_{x, y \in S_3} W_{extend} \cdot I(x, y) \quad (1)$$

Where the response of GC is denoted by $GC(x, y)$, and $I(x, y)$ represents the input image.

$$W_{center} = \frac{A_1}{\sqrt{2\pi}\sigma_1} e^{\frac{(x-x_0)^2 + (y-y_0)^2}{2\sigma_1^2}} \qquad W_{surround} = \frac{A_2}{\sqrt{2\pi}\sigma_2} e^{\frac{(x-x_0)^2 + (y-y_0)^2}{2\sigma_2^2}}$$

$$W_{extend} = \frac{A_3}{\sqrt{2\pi}\sigma_3} e^{\frac{(x-x_0)^2 + (y-y_0)^2}{2\sigma_3^2}} \qquad\qquad (2)$$

Where, x and y are the coordinates of the photoreceptor cells (RCs) position; x_0 and y_0 are the center coordinates of RF; X and Y are the position coordinates of retinal GCs; $GC(x, y)$ is the response of a GC at position (x,y); $I(x, y)$ is the brightness projected onto the RCs within each RF; S_1, S_2 and S_3 are the CRF center, the CRF surround and the nCRF, respectively; $W_{center}, W_{surround}$ and W_{extend} are the weighting functions of the RCs within S_1, S_2 and S_3, respectively; A_1, A_2 and A_3 are the peak sensitivities of S_1, S_2 and S_3, respectively; σ_1, σ_2 and σ_3 are the standard deviations of the three weighting functions, respectively. Based on the physiological data reported in previous studies [9,10], we set $\sigma_3 = 4\sigma_2, \sigma_2 = 5\sigma_1, A_1 = 1, A_2 = 0.18$ and $A_3 = 0.05$.

From the neurophysiological perspective, Receptor cells (RC) compose the center area of RF of each bipolar cell (BC). Horizontal cells (HC) integrate the response of RFs, transfer them to BC, inhibit the CRF center of BC, and form the surround zone of CRF. In the same way, BCs transfer their response to a GC, and GC's CRF with antagonistic center-surround structure is constructed. Amacrine cells (AC) integrate the response of BC and transfer them to GC and inhibit the CRF surround of the GC, thus forming the GC's nCRF (extra-surround). Interplexiform cell (IC) exerts feedback control over HC and BC, such that the cessation of the IC's activity enhances the activity of HC and BC. The number of HC and BC was increased, that form the CRF center and surround of GC, in other words, increasing the size of GC's RF. conversely, an increase in IC's activity inhibits the activity of HC and BC, such that it decreases the number of HC and BC that form the CRF center and surround of GC, reducing the size of RF of GC. Based on the neurophysiological mechanisms of the GC's RF, a simplified neural circuit shown in Fig.2 was designed to enable the RF to be adjusted automatically and dynamically.

Through touring, sampling and comparing at several close positions, GCs can dynamically adjust the perceptual scope of its RF, finally maintaining an appropriate

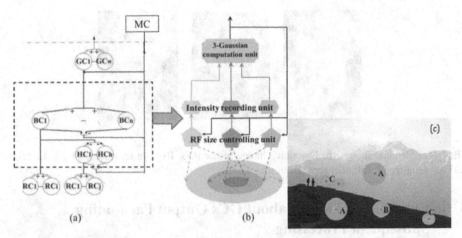

Fig. 2. The computational model based on GC. (a) is a micro neural circuit for RF constitution, where the solid lines represent forward transfer and feedback control, the hollow circle and solid circle represents synapse of facilitation and inhibition, respectively. The simplified computation model based on (a) is shown in (b). While (c) is the mimic diagram of several RFs cast into a real image. A, B and C represents the maximum size RF, the medium size RF and the minimum size RF, respectively. It can be seen that if stimuli are uniform, the RF tends to expand its size to cover as large an area as possible. In contrast, if stimuli are sufficiently diverse, the RF tends to shrink in size, to obtain a finer resolution that can accurately identify detail.

size, enabling the formation of a more distinctive representation. The general principle of adjustment dynamic is presented in the following according to Fig.2 (a) and (b).

- At first stage, the RGB values of the image pixels were converted to the wave length by the photoreceptor cells (RCs). The CRF center area, the CRF surround area and the nCRF have their initial sizes.
- All summed RGB information is transmitted upwards, through different channels of RFs size controlling units.
- All information was integrated in GC layer (3-Gaussian computation unit). According to the output of GC, the upper layer send feedback signal to RF size controlling units and the RF can dynamically adjust their size.

To test whether the RFs in our model exhibited self-adaptation, we conducted an experiment with a natural image (see Fig. 3(a)). At the beginning, some RFs with of random size were located evenly on the image (Fig. 3(b)), and they became stabilized after self-adaption. Fig. 3(c) is the visualized result of GCs. From this figure, it can be seen that if stimuli in a local area are uniform, the RF tends to expand its size to cover as large an area as possible. In contrast, if stimuli are sufficiently diverse, the RF tends to shrink in size, to obtain a finer resolution that can accurately identify detail. This type of self-adjustment ability provides an ideal sampling mechanism for representing images.

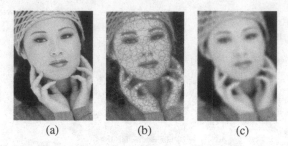

(a)   (b)   (c)

Fig. 3. (a) Original image. (b) Dynamic change of the RFs. RFs are represented by red circles. (c)The visualized result after representation of GCs.

3   Experimental Result about GCs Output Facilitating Subsequent Processing

3.1   Promoting Concentration to Target Object

SIFT [11] is a widely used feature detector and descriptor. We tested whether our new representation method could facilitate SIFT. Table 1 shows the results, in which the red clock-like circles denote SIFT detectors and their orientations. The results revealed that the GC-array led SIFT detectors to distribute more on the object and less on the background. This indicates that our method can help to locate an object and focus representation on it. (In our experiment, all testing images are come from BSD of UC Berkeley).

This facilitation effect was likely related to the dynamic sampling of the RF mechanism, by which not all stimuli are equally represented. As such, tiny details are

Table 1. GC-array facilitating SIFT

Image	SIFT on image	SIFT on GC-array

less prominently represented. If the uneven distribution of GCs in the retina is taken into consideration, then an object highlighting effect will be even more obvious, because peripheral RFs are large, making fine detail more ambiguous.

3.2   Promoting Hierarchical Image Segmentation Algorithm

In [12], nCRF mechanism had been proved to be an effective way to facilitate segmentation. And [13] proposed a new segmentation algorithm (gPb-OWT-UCM) and its performance is excellent in comparison with other segmentation method. We tested the effects of running this algorithm both on original image and on a GC-array output.

Table 2. Facilitation of the hierarchical segmentation by the GC processing

Image		Segmentation result
	Result on original image	
	Our result: *The building and hillside is segmented*	
	Result on original image	
	Our result: *The deer is segmented from background*	
	Result on original image	
	Our result: *The boundary of meadow is segmented*	

Table 2 shows comparison of the result. Those rows headed by "our result" show that the results based on the new representation were either improved or of the same with the original one. Thus, a good GC-based representation can improve the efficiency of segmentation without sacrificing performance.

3.3   Performance Evaluation

In order to evaluate the segmentation results against the human ground truth of Berkeley Segmentation Dataset [14], we use the precision-recall framework, a standard
evaluation technique in the information retrieval community. In order to captures the
tradeoff as the weighted harmonic mean of P and R, the F-measure is defined as
$F = PR/(\alpha R+(1-\alpha)P)$. A particular application will define a relative cost
α between these quantities, which focus attention at a specific point on the precision-
recall curve (In this experiment, $\alpha =0.5$). To provide a basis of comparison for the
segmentation result, we make use of the gPb [15] as detector, and OWT-UCM [13] as
segmentation algorithm operated on original image[13] and our representation method, respectively, meanwhile, some segmentation algorithm, such as Ncuts [16] and
UCM [17] are compared too. Fig. 4 summarizes our main results. It demonstrates that
the performance of our segmentation result is optimal.

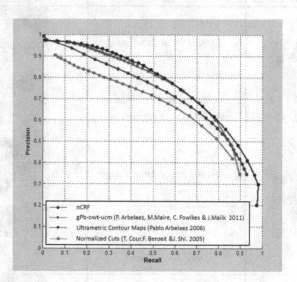

Fig. 4. Evaluation of segmentation algorithms on the BSD Benchmark [14]. gPb [15] contour
detector as input, our representation method based on nCRF whose boundaries match ground-
truth than those produced by other methods [13,16,17].

4   Conclusion

The current paper examined the possibility of an image representation system based
on the physiological mechanisms of GCs and their RFs. Although some biological
details of RFs and retina remain unclear, this does not prevent us from designing realistic simulations using appropriate algorithms. Experiments on segmentation, and the
application of the SIFT to GC-array images revealed that this novel representation
schema is help of their performance significantly. This representation method can

substantially facilitate upcoming processing. Future research should consider more biological mechanisms in the design of similar models of representation.

Reference

1. Elder, J.H.: Are edges incomplete? International Journal of Computer Vision 34(2), 97–122 (1999)
2. Shou, T., Wang, W., Yu, H.: Orientation biased extended surround of the receptive field of cat retinal ganglion cells. Neuroscience 98, 207–212 (2000)
3. Hoiem, D., Efros, A.A., Hebert, M.: Geometric context from a single image. In: IEEE International Conference on Computer Vision, vol. 651, pp. 654–661. IEEE Press (2005)
4. Saxena, A., Chung, S.H., Ng, A.Y.: 3-d depth reconstruction from a single still image. International Journal of Computer Vision 76(1), 53–69 (2008)
5. Fauqueur, J., Boujemaa, N.: Region-based image retrieval: Fast coarse segmentation and fine color description. Journal of Visual Languages & Computing 15(1), 69–95 (2004)
6. Deng, Y., Manjunath, B., Kenney, C., Moore, M.S., Shin, H.: An efficient color representation for image retrieval. IEEE Transactions on Image Processing 10(1), 140–147 (2001)
7. Qiu, F.T., Chao-Yi, L.: Mathematic simulation of disinhibitory properties of concentric receptive field. Acta Biophysica Sinica 11, 214–220 (1995)
8. Ghosh, K., Sarkar, S., Bhaumik, K.: A possible mechanism of zero-crossing detection using the concept of the extended classical receptive field of retinal ganglion cells. Biological Cybernetics 93, 1–5 (2005)
9. Chao-Yi, L., Wu, L.: Extensive integration field beyond the classical receptive field of cat's striate cortical neurons–classification and tuning properties. Vision Research 34(18), 2337–2355 (1994)
10. Chao-Yi, L.: Integration field beyond the classical receptive field: organization and functional properties. News Physiol. Sci. 11, 181–186 (1996)
11. Lowe, D.G.: Object recognition from local scale-invariant features. In: IEEE International Conference on Computer Vision, vol. 1152, pp. 1150–1157. IEEE Press (1999)
12. Fernandes, B.J.T., Cavalcanti, G.D.C., Ren, T.I.: Nonclassical receptive field inhibitonapplied to image segmentation. Neural Network World 19, 21 (2010)
13. Arbelaez, P., Maire, M., Fowlkes, C., Malik, J.: Contour detection and hierarchical image segmentation. IEEE Transactions on Pattern Analysis and Machine Intelligence 33, 1 (2011)
14. Martin, D., Fowlkes, C., Tal, D., Malik, J.: A database of human segmented natural images and its application to evaluating segmentation algorithms and measuring ecological statistics. In: Proceeding Eighth IEEE International Conference Computer Vision, vol. 412, pp. 416–423. IEEE Press (2001)
15. Maire, M., Arbelácz, P., Fowlkes, C., Malik, J.: Using contours to detect and localize junctions in natural images. In: IEEE Conference Computer Vision and Pattern Recognition, pp. 1–8. IEEE Press (2008)
16. Shi, J., Malik, J.: Normalized cuts and image segmentation. IEEE Transactions on Pattern Analysis and Machine Intelligence 22(8), 888–905 (2000)
17. Arbelaez, P.: Boundary extraction in natural images using ultrametric contour maps. In: IEEE Vision and Pattern Recogniton Workshop, pp. 182–182. IEEE Press (2006)

Attach Topic Sense to Social Tags

Junpeng Chen, Juan Liu, and Bo Guo

School of Computer, Wuhan University, P.R. China
{chenjp,liujuan,guobo}@whu.edu.cn

Abstract. Social tagging system, also noted as folksonomies, is an important way for users to describe resources on the Web. Because of the continually changing and informal definition, the semantics of these social tags are ambiguous and hard to adopt for web applications. In this paper, we propose a method to attach semantic topic sense to tags. The non-negative matrix factorization (NMF) is performed to find the hidden topics in the folksonomy. A novel automatic evaluation method is also proposed to measure our approach. Our evaluation shows that the topic sense induction in a folksonomy allows for precise and complete search, which is one of the key functionalities in social tagging systems.

Keywords: Topic Sense, Latent Semantic Space, Social Tags.

1 Introduction

Collaborative or social tagging systems, such as Delicious, Flickr, or Last.fm, allow Internet users to annotate online resources. Many annotations take together results in a complex network of user-tag-resource triplets commonly referred to as a folksonomy. As a valuable source of knowledge, these folksonomies are now widely studied in many fields, such as information retrieval [1,2,3], ontology learning [4,5]. For the nature of free-form and lack of explicit semantic in folksonomy, one tag may have multiple senses, or multiple tags may have the same sense. Therefore folksonomies are difficultly adopted by users and web applications to leverage the knowledge.

To tackle this issue, some approaches use Word Sense Disambiguation (WSD) me-thods to assigning a sense label to a tag in folksonomies. However, these WSD me-thods rely on a fixed sense inventory to link terms to concepts, which may be not applicable to folksonomies. In folksonomies, new concepts and terms appear so quickly that a pre-defined sense inventory is hard to cover them effectively. Additionally, different folksonomies may belong to different domains, and it is hard for a fixed sense inventory to cover all such domains. As shown in [6], WordNet, a most widely used sense inventory in WSD, covers less than half of the terms used by the users of the folksonomy. Compared with WSD, Word Sense Induction (WSI), a task of automatically identifying the senses of words in context without the need for manually building a sense inventory, is more applicable to attach semantic to folksonomies. However, most of the present WSI methods mainly focus on texts, cannot be directly adapted in folksonomies, the network of user-tag-resource triplets.

T. Huang et al. (Eds.): ICONIP 2012, Part V, LNCS 7667, pp. 386–393, 2012.

By applying an automatic WSI procedure, we are able to only extract the senses that are objectively present in a particular folksonomy, and the sense inventory auto-matically built may be straightforwardly adapted to a new domain. The intuition is that there are different hidden topics represented in those tags. We regard these se-mantic meaning contained in these tags as topic sense that can describe different topic semantics belong to the same or various domains. The topic senses may be regarded as the different word senses to a tag, or just the different profile of the same thing. These topic senses may be hard to be detected in dictionaries for two reasons. Firstly, their lifetime may be short compared to the traditional word senses. The topic senses completely rely on the topics that users are interested in. These interested topics emerge and disappear everyday and nobody can tell how long users will be interested in them. Secondly, since the interested topics will change, the corresponding topic senses of the same tag will change, too. Hence, traditional methods may be failed to find these topic senses. At the same time, the topic senses may be the users interests and will give help to web search, recommendation and other applications, for their rapid catching of the hot spot in the folksonomy.

In this paper, we present a unified model to automatically induce the topic senses of tags from folksonomies in a fully unsupervised way. Our contributions are mainly on two fields: To begin with, we adopt non-negative matrix factorization (NMF) to detect the topic senses of tags in folksonomy; Secondly, we proposed an automatic evaluation method to measure the performance of the topic sense induction in folk-sonomy, which can avoid the laborious and erroneous human evaluations.

2   Related Works

There are several approaches for disambiguate the word sense of tags in folksonomy and recommendation. Andrews et al.[6] and Sofia Angeletou et al. [7] used word sense disambiguation methods to resolve tag ambiguity based on WordNet. Lee et al. [8] use Wikipedia as the tag vocabulary to disambiguate the tags in folksonomies. In [9], Garca-Silva et al. use DBpedia, which was a complied version of Wikipedia, to label tag senses. These formal or pre-defined semantic word senses can cover a little among tags and reach a pool recall.

In addition, Weinberger et al. [10] introduce a probabilistic framework to find the ambiguous tags that appear in different contexts and ask their colleagues to evaluate their method. In [11], Zorn and Gurevych use hierarchical agglomerative clustering to induce the tag senses from folksonomies, but they lack an evaluation to measure the quality of the induced senses. Andrews et al. [12] adopt DBScan algorithm to detect new sense of tags in folksonomies, and use a human disambiguation dataset [6] for evaluation. But the dataset is relative small to cover the domains of a folksonomy.

In our methods, tags are matched to different topics which can represent the interest and common sense of the user to some resources. The topic sense is more flexible and nimble to catch the interest of users compare to traditional word sense. And our automatic evaluation similar to tag expansion is scalable and objective compared to human evaluations.

3   Topic Sense Induction

In this paper, we propose a method that automatically induces the topic senses of tags. Our method assumes that the tags in a folksonomy are comprised of k clusters and each of which corresponds to a coherent topic, or topic sense. Each tag in the folksonomy either completely belongs to a particular sense, or is more or less related to several senses. To accurately cluster the tags in a folksonomy, it is ideal to project the tags into a k-dimensional semantic space in which each axis corresponds to a particular sense. In such a semantic space, each tag can be represented as a linear combination of the k senses. Because it is more natural to consider each tag as an additive rather than subtractive mixture of the underlying senses, the linear combination coefficients should all take non-negative values. In addition, it is also quite common that the senses comprising the tags in a folksonomy are not completely independently of each other, and there are some overlaps among them. In such a case, the axes of the semantic space that capture each of the senses are not necessarily orthogonal. Based on the above suggestions, we perform non-negative matrix factorization (NMF) [13] to find the latent semantic structure for the tags in a folksonomy, and identify topic senses in the derived latent semantic space.

3.1   Preprocession

As formalized in [14], a folksonomy F is a four-tuple. There exists a set of users, $U = \{u_1, \ldots, u_q\}$; a set of resources, $R = \{r_1, \ldots, r_c\}$; a set of tags, $T = \{t_1, \ldots, t_m\}$; and a set of annotations, A. Then F can be defined as $F =< U, R, T, A >$. The annotations A are represented as a set of triples containing a user, resource and tag defined as $A \in \{< u, r, t >: u \in U, r \in R, t \in T\}$. In particular, an element $a \in A$ is a triple $< u, r, t >$, indicating that user u labeled resource r with tag t.

In our method, we adopt the vector space model to work with folksonomies. Each resource r is a vector over the set of tags. While each weight $v(t_i)$ in each dimension corresponds to the importance of a particular tag t_i. Here, we use the modified term frequency (tf) and inverse document frequency (idf) to calculate the vector weights. The tag frequency $tf(t, r)$ is the number of times the tag t labels the resource r. And the $idf(t, r)$ is measured by the total number of resources N, and the number of resources to which the tag was applied, n_t. We define $tf \times idf$ as:

$$tf \times idf(t, r) = tf(t, r) \times \log(\frac{N}{n_t}) \tag{1}$$

Then, using $r_i \in R$ as the $i'th$ column, we construct the $m \times n$ tag-resource matrix B, whose cell is the value of $tf \times idf(t, r)$. We use the matrix B to implement the non-negative factorization and obtain the tag cluster result.

3.2 Topic Sense Induction Based on NMF

Here, we use NMF to factorize $B_{m \times n}$ into $W_{m \times k}$ and $H_{k \times n}$:

$$B_{m \times n} \approx W_{m \times k} H_{k \times n} \tag{2}$$

In particular, k is the number of tag clusters of a folksonomy. So each row i of W denotes the relative weights of one tag t_i within the k clusters. Then t_i can be represented as a k dimensions vector, $vector(t_i) = (w_{i1}, w_{i2}, \ldots, w_{ik})$. In addition, k is much smaller than m and n so that both instance and feature are expressed in terms of a few components.

The non-negative matrix factorization is carried out by minimizing an objective function, where Euclidean distance and Kullback-Leibler divergence are often used. In our method, Kullback-Leibler divergence is adopted for its better performance on natural language. Hence, the problem is to find the W and H to make the Kullback-Leibler divergence between B and WH to be the smallest. Lee[15] has proved that the objective function using Kullback-Leibler divergence is non-increasing under the iterative updating rule (3) and (4), whose convergence is guaranteed.

$$H_{\alpha \mu} \leftarrow H_{\alpha \mu} \frac{\sum_i W_{i\alpha} \frac{B_{i\mu}}{(WH)_{i\mu}}}{\sum_k W_{k\alpha}} \tag{3}$$

$$W_{i\alpha} \leftarrow W_{i\alpha} \frac{\sum_\mu H_{\alpha \mu} \frac{B_{i\mu}}{(WH)_{i\mu}}}{\sum_v H_{\alpha v}} \tag{4}$$

We summarize our topic sense induction algorithm in following steps:

Firstly, for a given folksonomy system, we first build the tag-resource matrix B, whose cell is the $tf \times idf$ value of (t, r), where $t \in T$, and $r \in R$.

Secondly, perform the NMF on B to obtain the two non-negative matrices of W and H using rule (3) and (4). In each iteration, each vector is adequately normalized. So all dimensions values sum to 1.

Thirdly, From W, we induce a set of the topic sense $S = \{s_1, s_2, \ldots, s_k\}$, a topic sense $s_j \in S$ represent as a set of the tags: $s_j = \{t_i \in T, (1 \le i \le m)$, and $w_{ij} = \max_{1 \le p \le k} w_{ip}\}$.

At last, set a threshold $\varepsilon > 0$, then $sense(t_i) = \{s_j \in S, and w_{ij} \ge \varepsilon\}$.

The underlying idea is that for each tag t_i, it can have k number latent topic senses, and we can choose the most important topic senses $sense(t_i)$ by measuring the value of $w_{ij}(1 \le j \le k)$.

After perform our algorithm, we can get the topic senses of the tags. For example, when the tag job takes the word sense the principal activity in your life that you do to earn money in WordNet, it can have three topic senses after performing topic sense induction:

1. career, jobsearch, employ, recruit, linkedin, trabajo
2. resume, cv, curriculum, apply
3. freelance, outsource, consult, contract, selfemploy, contractor.

For the topic sense 1, it focuses on the job hunting, the famous job-hunting web site linkedin is listed out. For the topic sense 2, these tags are more about the cv or resume. For the topic sense 3, the different forms of employment are mentioned, such as freelance, selfemploy. Another instance, the three topic senses for pop can be:

1. culture, popculture, opinion, cultura, slate, kulture, digitalculture, cyberculture
2. photoshop, adobe, tutorial, popart, photoshoptutorial, firework,ps, psd
3. music, mp3blog, mp3, bootleg, musicblog, indie.

The first sense focuses on pop culture or opinion. The second means Point of Pur-chase for point of purchase is usually made by image processing softwares such as photoshop, firework, etc. The last is mainly about music. Compared to the topic senses of job, the topic senses of pop describes different things and can be re-garded as different word senses of pop. Since the topic senses is totally relying on the topics the users interested, it may not be equal to the word sense of the tag.

Hence, adopting our approach, not only the different word senses can be discrimi-nated, but the similar semantic senses belong to different topics can also be detected. Compared with the word senses defined in dictionary, the topic senses induced by our method may be more interesting to the web users and can be more flexible and helpful in resource search and recommendation.

4   Evaluation and Results

We evaluated our word sense induction method based on del.icio.us, a collaborative tagging systems in which the resources are web pages. The dataset is from the PINTS experimental dataset [16] containing a systematic crawl of delicious in 2006 and 2007. The dataset contains 532,924 users, 2,481,698 tags, and 17,262,480 resources. There are 140,126,586 annotations with one user, resource, and tag per annotation.

There is a hypothesize that out-of-the norm peak at fifteen tags per bookmark made by one user is too strong to be coincidental and may be created by spam bots [6]. Hence, we remove these users from our dataset. Then we choose the top 50,000 most frequent tagged resources to build the dataset. It contains 50,000 resource, 382,778 tags, and 35,594,962 annotations. As a general experience in natural language processing, lemmatization can reduce data sparsity. So we made lemmatization and removed stop words and meaningless tags on 382,778 tags. Finally 268,564 tokens are extracted.

Then the token-resource matrix is constructed for training. It is a 66118702 matrix that contains 18,976,322 annotations. The matrix cell is the $tf \times idf$ value of the corresponding token-resource. Let $tSet = \{t_1, t_2, \ldots, t\}, 1 \le \pi \le m$, be the set of user-selected tags to label a resource. A bookmark can be represented as a tuple $(u, r, tSet)$. We randomly extracted a test dataset compose of 66,000 bookmarks from the remained dataset after got rid of the training dataset from the whole dataset. The test set composed of 160,000 annotations.

4.1   Evaluation

We constructed an automatic evaluation by expanding the test set with similar tag from training set. A comparison between the query ranks on the resource of the dif-ferent versions is performed. This automatic evaluation can avoid laborious and erroneous human evaluation.

We get the original $tSets$ $F = \{f_1, f_2, \ldots, f_\mu\}$ from all bookmarks of test dataset. Each $f_\sigma \in F$ denotes the set of tags originally assigned by the user u_σ to label the resource r_σ. For σth $tSet$, $f_\sigma = (t_{\sigma 1}, t_{\sigma 2}, \ldots, t_{\sigma \pi})$, there is :
$vector (f_\sigma) = (w_{\sigma 1,1} \times w_{\sigma 2,1} \times \ldots \times w_{\sigma \pi,1}, w_{\sigma 1,2} \times w_{\sigma 2,2} \times \ldots \times w_{\sigma \pi,2}, \ldots, w_{\sigma 1,k} \times w_{\sigma 2,k} \times \ldots \times w_{\sigma \pi,k}), |f_\sigma| = \pi$.

As mentioned in Section 3.2, we set $s_h = \max sense (vector (f_\sigma))$, $s_h \in S$. We select top e tags from s_h, here $e = 5$. For each t_i in the top e tags of s_h, we caculate the similarity between t_i and f_σ using KullbackLeibler divergence:

$$simi (t_i, f_\sigma) = \frac{(kl (vector(t_i)||vector (f_\sigma)) + kl(vector (f_\sigma) ||vector(t_i)))}{2} \quad (5)$$

Here, the function $kl (vector(t_i)||vector (f_\sigma))$ denotes the KL divergence from t_i to f_σ , while $kl(vector (f_\sigma) ||vector(t_i))$ denotes the KL divergence from f_o to t_i. Finally, we enrich f_σ with the tag t_j to get $t_j = \max_{1 \leq i \leq e} simi (t_i, f_\sigma)$. Hence, $f_\sigma^+ = f_\sigma + \{t_j\}$.

The intuition in the evaluation is that a user labeled a specific resource r_σ with $tSet$ f_σ. f_σ would very likely be employed by user to query folksonomy to get r_σ. So in this evaluation, we use both $f_\sigma \in F$ and $f_\sigma^+ \in F^+$ to query the whole dataset and get the rank of the target resource r_σ. To determine the relevance of a resource to f_σ, the $tf \times idf$ coefficient assigned to such resource for each tag in f_σ is computed and summed up.

The resource rank based on f_σ^+ is also reorderd. If the user had annotated the resource r_σ with the tags in f_σ, it is known that the user is interested in r_σ. Hence, the resource r_σ is regarded as relevant to the user. Being enriched from the induced topic senses, f_σ^+ reflects the quality of the induced topic sense. So good induced topic senses should move the target resource higher in the ranking. Querying with f_σ^+ enriched from training data should improve the position of the resource r_σ in the whole returned set. As adopted in query answer task [17], imp is to measure the difference in the inverse of the two ranks can be used to judge the improvement provided by the method:

$$imp_\sigma = \frac{1}{r_{f_\sigma^+}} - \frac{1}{r_{f_\sigma}} \quad (6)$$

If our approach moves the resource higher in the ranking, the improvement will be positive. And the imp measure shows how good f_σ^+ is at moving relevant resource to the front of the list. As to the whole $tSets$ F or F^+, the total improvent $Imp = \sum_{1 \leq \sigma \leq \mu} imp_\sigma$.

4.2   The Impact of Topic Sense Number

We report our result in Figure 1, $k = (100, 200, \ldots, 1000)$ is the number of topic sense and Imp is the improvement when query 66,000 bookmark of the whole dataset.

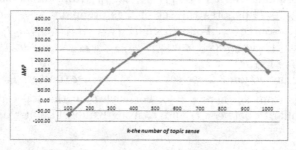

Fig. 1. The Impact of Topic Sense Number Selection on Query Improvement

From Figure 1, we can see when $k = 100$, the Imp is negative and that implies there are no improvement but push the relevant resource away from user. When $k200$, the Imp is positive and our method give a rise to the rank of the relevant resource. While $k = 600$, Imp mostly achieves the best result. The explanation to this observation is probably because that we get 6,611 tags for training. If the value of k is about 10% of the number of training tags, the classification of topic sense would properly represent the real topic assignment in the folksonomy. On one hand, if the value of k is too small, the granularity of classification is too big that there will introduce too much noise. On the other hand, if the value of k is too large, the edge of the classification will be too vague to be clearly discriminated so that it may fail to give effective topic sense to the tags. These results show that our method can obviously improve the performance of resource recommendation in folksonomy. Hence, the topic sense induction in our method is effective and can give positive effect on web applications.

5   Conclusion

In this paper, we present a model based on latent semantic space to perform topic sense induction in folksonomy. This model is able to detect polysemous and to homograph tags and assign different topic senses to them. By adopting the topic sense we detected, the highly related tags can be automatically enriched or recommended when users select certain tags to query. Since there are lacks of standard evaluation methodology in the state of the art, we also propose a new methodology for evaluating topic sense induction. As our evaluations showed, the topic sense induction in a folksonomy allows for more precise and complete search, which is one of the key functionalities in social tagging systems.

References

1. Bao, S.H., Xue, G.R., Wu, X.Y., Yu, Y., Fei, B., Su, Z.: Optimizing web search using social annotations. In: Proceedings of the 16th International Conference on World Wide Web, WWW 2007, New York, NY, USA, pp. 501–510 (2007)
2. Heymann, P., Koutrika, G., Hector, G.M.: Can social bookmarking improve web search? In: Proceedings of the International Conference on Web Search and Web Data Mining, WSDM 2008, New York, NY, USA, pp. 195–206 (2008)
3. Zhou, D., Bian, J., Zheng, S.Y., Zha, H.Y., Lee, C.: Exploring social annotations for information retrieval. In: Proceedings of the 17th International Conference on World Wide Web, WWW 2008, New York, NY, USA, pp. 715–724 (2008)
4. Mika, P.: Ontologies are us: A unified model of social networks and semantics. Web Semant. 5, 5–15 (2007)
5. Schmitz, P.: Inducing Ontology from Flickr Tags. In: Proceedings of the 15th International Conference on World Wide Web (WWW), Edinburgh, UK (2006)
6. Pierre, A.J., Zaihrayeu, I., Pane, J.: Semantic disambiguation in folksonomy: a case study (2011)
7. Marta, S.: Semantically enriching folksonomies with FLOR (2008)
8. Lee, K., Kim, H., Shin, H., Kim, H.J.: Tag Sense Disambiguation for Clarifying the Vocabulary of Social Tags (2009)
9. Silva, A.G., Szomszor, M., Alani, H., Corcho, G.: Preliminary Results in Tag Disambiguation using DBpedia (2009)
10. Quirin, W.K., Malcolm, S., Van, Z.R.: Resolving tag ambiguity (2008)
11. Zorn, H.P., Gurevych, I.: A Study of Sense-Disambiguated Networks Induced from Folksonomies. In: Proceedings of the 25th Pacific Asia Conference on Language, Information and Computation, Singapore, pp. 323–332 (2011)
12. Juan, P.A., Zaihrayeu, I., Pane, J.: Sense Induction in Folksonomies (2011)
13. Lee, H.: Learning the parts of objects by non-negative matrix factorization. Nature 401, 788–791 (1999)
14. Giovanni, Q., Licia, C., Pasquale, M., Emilio, F., Domenico, U.: Effective retrieval of resources in folksonomies using a new tag similarity measure (2011)
15. Lee, H.: Algorithms for non-negative matrix factorization. In: Neural Information Processing Systems. MIT Press (2001)
16. Görlitz, O., Sizov, O., Staab, S.: Infrastructure for Tagging Systems (2008)
17. Voorhees, E.: The TREC-8 Question Answering Track Report. In: TREC, pp. 77–82 (1999)

Emotion Recognition Using the Emotiv EPOC Device

Trung Duy Pham and Dat Tran

Faculty of Information Sciences and Engineering,
University of Canberra, ACT 2601, Australia
dat.tran@canberra.edu.au

Abstract. Emotion plays an important role in the interaction between humans as emotion is fundamental to human experience, influencing cognition, perception, learning communication, and even rational decision-making. Therefore, studying emotion is indispensable. This paper aims at finding the relationships between EEG signals and human emotions based on emotion recognition experiments that are conducted using the commercial Emotiv EPOC headset to record EEG signals while participants are watching emotional movies. Alpha, beta, delta and theta bands filtered from the recorded EEG signals are used to train and evaluate classifiers with different learning techniques including Support Vector Machine, k-Nearest Neighbour, Naïve Bayes and AdaBoost.M1. Our experimental results show that we can use the Emotiv headset for emotion recognition and that the AdaBoost.M1 technique and the theta band provide the highest recognition rates.

Keywords: EEG, Emotion recognition, Emotiv EPOC headset, AdaBoost.M1.

1 Introduction

Emotions play an essential role in many aspects of our daily lives, including decision marking, perception, learning, rational thinking and actions. Therefore, study of emotion recognition is indispensable.

The first approach to emotion recognition is based on text, speech, facial expression and gesture that were studied in the past few decades [1]. However, these methods are not reliable to detect emotion, especially when people want to conceal their feelings. Some emotions can occur without corresponding facial emotional expressions, emotional voice changes or body movements, especially when the emotion density is not very high. On the contrary, such displays could be faked easily. In order to overcome that disadvantage, the multi-modality approach has been introduced. A peripheral neurons system including heart rate variations, skin conductivity and respiration is a typical system for this approach [2, 3]. The advantage is that those modalities can hardly be deceived by voluntary control and are available all the time, without needing any further action of the users. Anttonen and Surakka found that heart rate decelerated in response to emotional stimulation, especially in response to negative stimuli compared to responses to positive and neutral stimuli [2]. Leng et al. verified that amusement produces a larger average and standard deviation of the heart rate

T. Huang et al. (Eds.): ICONIP 2012, Part V, LNCS 7667, pp. 394–399, 2012.

than fear [3]. However, the user has to wear measurement devices which are hard to use and very expensive, and the emotion classification results are not high.

In recent years, researchers have attempted to develop hardware and software systems that can capture emotions automatically. This approach is called affective computing. One of those systems employs electroencephalography (EEG) signals recorded when users perform some brain activities. The advantages of this approach include 1) Brain activities have direct information about emotion, 2) EEG signals can be measured at any moment and are not dependent on other activities of the user such as speaking or generating a facial expression, and 3) Different recognition techniques can be used.

It has been found in the literature that the Naïve Bayes technique provides recognition accuracy of 70% for two classes [4]. Petrantonakis and Hadjileontiadis [5] studied changes in the EEG signal of subjects when presented with images of faces expressing six basic emotions. They showed that a recognition accuracy of 83% could be achieved using features based on higher-order crossings and support vector machine (SVM). Lin et al [6] extracted power spectrum density of different EEG sub-bands as features during different emotions induced during listening to music and a classification accuracy of 82% for four emotions was achieved. Using k-Nearest Neighbor (kNN) technique for two different sets of EEG channels (62 channels and 24 channels), Murugappan obtained an accuracy of 82.87% on 62 channels and 78.57% on 24 channels, respectively for five emotions [7].

Finding the frequency bands that most related to emotions is one of the main goals of emotion recognition. Li and Lu found that gamma band plays an important role in emotion recognition [9]. Dan Nie et al. suggested that higher frequency bands contributed to human emotional response rather than lower frequency bands [10].

In general there is no method based on EEG signals has been identified as the best [8]. Furthermore, it should be noted that once emotion recognition systems are more widely used in practice, new properties will have to be taken into consideration, such as the availability of large data sets or long term variability of the EEG signal. One difficulty encountered in such a study concerns the lack of published objective comparisons between classifiers. Ideally, classifiers should be tested within the same context, i.e., with the same users, using the same feature extraction method and the same protocol. Currently, this is a crucial problem for emotion recognition research [8].

On the other hand, providing reliable recommendation is the main task of recommender systems in e-commerce, entertainment and society research. For subjective and complexes products such as movies, music, news, user emotion plays surprising critical roles in the decision process. The use of EEG data to recognise user emotion will contribute to building reliable recommender systems.

In this paper, we present an automatic EEG-based emotion recognition system that can record the EEG signals from users and measure their emotions when they are watching movies. We propose to use the commercial Emotiv EPOC headset since it is significantly less expensive than other EEG devices. The EEG data are then filtered to get separate frequency bands to train emotion classifiers with the four well-known classification techniques that are SVMs, Naïve Bayes, kNN and AdaBoost.M1.

2   The Proposed EEG-Based Emotion Recognition System

The proposed emotion recognition system records EEG data using the Emotiv EPOC wireless headset [9]. This Emotiv headset has 14 electrodes locating at AF3, F7, F3, FC5, T7, P7, O1, O2, P8, T8, FC6, F4, F8 and AF4 following the American EEG Society Standard. The Emotiv EPOC headset does not require a moistened cap to improve conduction. The sampling rate is 128Hz, the bandwidth is 0.2-45Hz, and the digital notch filters are at 50Hz and 60Hz.

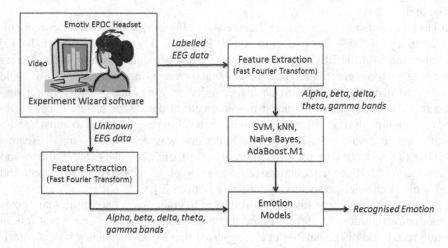

Fig. 1. The proposed EEG-based emotion recognition system

We used the Experiment Wizard software tool [10] for acquiring raw EEG data from the Emotiv headset while a participant is watching videos. This software is useful to design an experiment, to prepare and configure multimedia, and to collect the EEG data in a structured and systematic way.

The raw EEG data is found in the frequencies below 30Hz. There is not much brain activity with very low frequencies and artefact occurs at a lower frequency below 2Hz, therefore a 2-30 Hz band-pass filter was applied to separate the EEG data in to 4 bands that are 2-4Hz (delta), 4-7Hz (theta), 8-12Hz (alpha) and 12-30Hz (beta). Fast Fourier Transform (FFT) with 5s non-overlapping window was used to compute signal power in each of four frequency bands for each channel and the mean of each channel. The average EEG signal and the 4 frequency bands (alpha, beta, delta and theta) obtained from the 14 electrodes will provide up to 70 features (5 × 14) for the recorded EEG data.

In the training phase, the labelled EEG data, i.e. the data from a known emotion, are recorded and sent to the four well-known algorithms that are SVM, kNN, Naïve Bayes and AdaBoost.M1 to build a model for that emotion.

In the recognition phase, an unknown EEG data will be compared with the built emotion models and the recognised emotion is the label of the best matching model.

3   Data Acquisition

3.1   Subjects

The EEG data in our experiments were recorded from subjects aged around 30, who were healthy and right-handed. All the subjects were informed about the purpose of this experiment. None of them suffered at the time of experiment from a chronic disease, mental disorder, drugs or alcohol abuse, depression or anxiety, hearing defects or neurological disorder and none of them were on medication.

3.2   Stimuli

There are many different methods to induce emotion like films, music, pictures or imagination. In order to obtain affective EEG data, experiments were carried out with different kinds of stimuli as audio, visual, and combined ones to induce emotions. As stimuli, video clips are extracted from films (video and audio) taken from the Film-Stim database [11]. The emotional impact of the stimuli from the database has been scientifically assessed. The reason why we chose movies as stimuli is that audio-visual stimuli are highly benefit for arousing human emotion [12]. In a meta-analysis Westermann et al. found that film/story is the most effective method to induce positive and negative emotional states [13]. In our experiments, we were mainly concerned about sadness and happy emotions. Thus each of the movie clips was classified into these two kinds of emotions. The advantages of this approach are that there is no need for a professional actor and that responses should be closer to the ones observed in real life.

3.3   Procedure

A set of movie clips classified in to amusement and fear emotions was used. About 8 movie clips (4 video clips for each of the emotions) were randomly selected in each experiment and presented to a participant. The participant was alerted and was asked to focus on a movie clip during its presentation. There is a rest period of 2 minutes between two consecutive movie clips. We also recorded the EEG data when the participant was watching non-emotional movie clips and labelled this data set as neutral. In total, there were 3 classes which were amusement, fear and neutral for recognition.

3.4   Data Recordings

The Emotiv EPOC wireless headset was used in our experiments to record the EEG data. We used the Experiment Wizard software tool for acquiring the raw EEG data. The default EEG sampling rate from the Emotiv headset is 128Hz, which provides enough samples for the frequency ranges of the 4 frequency bands. The EEG data were also filtered to remove noise and artefacts. All of the EEG data from the 14 channels (electrodes) were used for band pass filtering and fast Fourier transform to extract up to 70 features as described in the previous section.

4   Experimental Results and Discussion

The four techniques SVM, kNN, Naïve Bayes and AdaBoost.M1 to build emotion models were obtained from WEKA [14]. The parameter k was set to 3 and Euclidean distance was used in the kNN technique. The linear kernel was selected for the SVM technique. For the AdaBoost.M1 technique, the kernel J48 was used. We chose those settings to obtain the highest performance for those techniques. In our experiments, the 5-fold cross validation was used to have recognition rates for the four techniques and the results are presented in Table 1. The highest recognition rate was achieved by the AdaBoost.M1 technique. The SVM technique performed better than the kNN and Naïve Bayes techniques.

Table 1. Emotion recognition rates for SVM, kNN, Naïve Bayes and AdaBoost.M1

	SVM	kNN	Naïve Bayes	AdaBoost.M1
Emotion recognition rate	89.25%	83.35%	66%	**92.8%**

We also investigated emotion recognition rates for each of the delta, theta, alpha and beta bands. Table 2 shows the emotion recognition rates for each of the 4 bands. The average recognition result for those bands is also presented. The last columns in Table 2 present the recognition rates for the non-filtered EEG data which include the data for the 4 above-mentioned bands and the data for gamma band.

The system performances for the alpha and beta bands are obviously better than those for the delta and theta bands. This result suggests that human emotional response is mainly related to high frequency bands rather than low bands. This finding is consistent with the studies of other researchers with a note that the frequency band that provides the highest performance is subject-dependent [12].

Table 2. Emotion recognition rates for the 4 frequency bands with AdaBoost.M1

	Delta	Theta	Alpha	Beta	All
Emotion recognition rate	69.95%	68.4%	75.5%	89.7%	92.8%

5   Conclusion

We have presented our emotion recognition system using the EEG data recorded when the participants were watching emotional movies to build emotion models with the four techniques Naïve Bayes, KNN SVM, and AdaBoost.M1. The recognition results have shown that the low-cost Emotiv EPOC headset is good for implementing emotion recognition applications for recommender systems in e-commerce, entertainment and society.

References

1. Anderson, K., McOwan, W.P.: A real-time automated system for the recognition of human facial expressions. IEEE Trans. System, Man, and Cybernetics, Part B: Cybernetics 36(1), 96–105 (2006)
2. Anttonen, J., Surakka, V.: Emotions and Heart Rate while Sitting on a Chair. In: CHI 2005: Proceedings of the SIGCHI Conference on Human Factors in Computing Systems, pp. 491–499 (2005)
3. Leng, H., Lin, Y., Zanzi, L.A.: An Experimental Study on Physiological Parameters Toward Driver Emotion Recognition. In: Dainoff, M.J. (ed.) HCII 2007 and EHAWC 2007. LNCS, vol. 4566, pp. 237–246. Springer, Heidelberg (2007)
4. EEG-based emotion recognition (Online),
   http://hmi.ewi.utwente.nl/verslagen/capita-selecta/CS-Oude_Bos-Danny.pdf
5. Petrantonakis, C.P., Hadjileontiadis, J.L.: Emotion Recognition From EEG Using Higher Order Crossings. IEEE Transactions on Information Technology in Biomedicine 14(2), 186–197 (2010)
6. Lin, Y.: EEG-Based emotion recognition in music listening. IEEE Transactions 57(7), 1798–1806 (2010)
7. Murugappan, M.: Human emotion classification using wavelet transform and KNN. In: International Conference on Pattern Analysis and Intelligent Robotics (ICPAIR), vol. 1, pp. 148–153 (2011)
8. Lotte, F., Congedo, M., Lécuyer, A.: F. Lamarche and B. Arnaldi, A review of classification algorithms for EEG-based brain–computer interfaces. Journal of Neural Engineering (2007)
9. Emotiv EPOC headset, http://www.emotiv.com/
10. Experiment Wizard software tool, http://code.google.com/p/experiment-wizard/
11. FilmStim database, http://www.ipsp.ucl.ac.be/recherche/FilmStim/
12. Li, M., Lu, B.L.: Emotion classification based on gamma-band EEG. In: IEEE International Conference Engineering in Medicine and Biology Society, Minneapolis, pp. 1223–1226 (2009)
13. Nie, D., Xiao, W.W., Li, C.S., Baoliang, L.: EEG-based emotion recognition during watching movies. In: 5th International IEEE/EMBS Conference on Neural Engineering (NER), pp. 667–670 (2011)
14. WEKA, http://sourceforge.net/projects/weka/
15. Zeng, Z., Pantic, M., Roisman, I.G., Huang, S.T.: A survey of affect recognition methods: audio, visual, and spontaneous expressions. IEEE Trans. Pattern Analysis and Machine Intelligence 31(1), 39–58 (2009)
16. Westermann, R., Spies, K., Stahl, G., Hesse, F.W.: Relative effectiveness and validity of mood induction procedures: A meta-analysis. European Journal of Social Psychology 26, 557–580 (1996)

Botnet Detection Based on Non-negative Matrix Factorization and the MDL Principle

Sayaka Yamauchi, Masanori Kawakita, and Jun'ichi Takeuchi

[1] Institute of Systems, Information Technologies and Nanotechnologies (ISIT)
[2] Graduate School of ISEE, Kyushu University
sayaka.yamauchi@inf.kyushu-u.ac.jp

Abstract. We propose a method for botnet detection from darknet data by non-negative matrix factorization (NMF), which can decompose the vector valued time series data into several components. In addition, we propose a new method to estimate the number of components in the data, by the minimum description length (MDL) principle. Our method for botnet detection consists of change point detection and analysis based on variance of the decomposed data.

Keywords: botnet, darknet, NMF, the MDL principle.

1 Introduction

We propose a method of botnet detection by analyzing traffic data of the Internet. Our method is based on non-negative matrix factorization (NMF) [6] and the minimum description length (MDL) principle [8]. NMF is a method to decompose a non-negative vector valued time series data into linearly independent non-negative components, and has been used for text mining, spectral data analysis, scalable Internet distance prediction, speech denoising, etc. We use it to decompose the Internet traffic data. When using NMF, we usually have to set the number of components in advance. In this paper, we employ the MDL principle to determine the number of components and derive a form of the MDL criterion specialized for NMF.

Damage caused by botnets is increasing recently. A botnet is a network comprised of a large number of computers infected by malwares called "bot," which spread worldwide. Bots do not behave unless they receive an order from malicious attacker called "bot-master." However, once bots receive commands from a bot-master, they perform malicious activities simultaneously. Therefore, a botnet can cause significant damage. Moreover, there are serious problems that criminals use botnets to benefit by them [11]. For example, they buy personally Identifiable Information stolen by botnets. However, since there are a lot of subspecies of bots, detection using the signature matching method does not work for most of the bots. Also, since the number of packets released by each bot is small, it is difficult to detect anomaly from the Internet traffic.

To overcome these problems, it is effective to use the data of darknets. A darknet is an accessible and unused IP address space. Since the packets which

T. Huang et al. (Eds.): ICONIP 2012, Part V, LNCS 7667, pp. 400–409, 2012.

reach a darknet are due to infection activities by malwares or misconfiguration of networks, darknets are more useful to detect the behavior of botnets than real networks. However, even though we use the darknet, it is not always easy to detect botnet activities from traffic data because darknet data is also mixed with various components besides botnets.

For this problem, Kitagawa et al. [5] proposed a method to decompose the traffic data into independent components using NMF [6] and analyze each component. We expect that it is possible to detect botnet activities with high precision by applying a change point detection method such as ChangeFinder [10] [4] to the obtained components. In the previous works [6] [5], the parameter of NMF corresponding to the number of independent components is set by the user beforehand. However, the number is unknown and we have to set the parameter suitably. We regard this problem as the statistical model selection and employ MDL as an information criterion [8], i.e. we determine the number of components by minimizing MDL. Moreover, the proposed technique in the previous work [5] only decomposes traffic data into some patterns by NMF, while we propose a method to detect botnet activities in those patterns. We demonstrate that the proposed method successfully detected botnets activities via experiments with real data sets.

2   Features of Botnet

Akiyama et al [1] pointed out that botnets have the following properties:

- Relationship: A botnet has a one-to-many relationship between the botmaster and bots.
- Response: After bots receive commands from their master, they perform with a constant response time.
- Synchronization: All bots may be synchronized with each other.

We focus on synchronization out of the three and discuss based on the assumption that all bots perform same activity simultaneously.

3   Proposed Method

3.1   Problem Setting

For each μ $(\mu = 1, \cdots, m)$, let $v^{\mu 1}, v^{\mu 2}, \cdots, v^{\mu n}$ a time series of the number of packets sent by the μth source host in a unit time interval and $\mathbf{v}^\mu$ be a vector $(v^{\mu 1}, v^{\mu 2}, \cdots, v^{\mu n})^T$. Let $V \in \Re^{n \times m}$ be a matrix such that $\mathbf{v}^\mu$ is the μth column vector. Given a data matrix V, we are to find

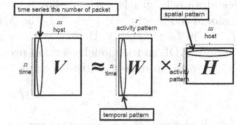

Fig. 1. The role of each matrix of NMF in our work

two non-negative matrices $W \in \Re^{n \times r}$ and $H \in \Re^{r \times m}$ so that $V \approx WH$ using NMF. Here, r corresponds to the number of activity patterns which are contained in the data matrix V.

The role of each matrix is shown in Fig. 1. As we observe the column vectors of W, we can see the temporal behavior of the number of packets of each potential pattern. We refer to it as a "temporal pattern". Similarly, as we observe the row vectors of H, we can see the distribution of the number of packets released by each host in the potential pattern. We refer to it as a "spatial pattern".

3.2   Non-negative Matrix Factorization

Non-negative Matrix Factorization (NMF) [6] is a method to approximately decompose a non-negative data matrix into a product of two low-rank non-negative matrices. The problem setting of NMF is as follows:

[Problem Setting]
Given a data matrix $V \in \Re^{n \times m}$, solve the following optimization problem, where r is a positive integer less than $\min\{n, m\}$, $W \in \Re^{n \times r}$, and $H \in \Re^{r \times m}$.

$$\min_{W,H} ||V - WH||^2 \text{ (Frobenius norm)}, \tag{1}$$

$$\text{subject to } W \geq 0, H \geq 0 .$$

For this problem, the algorithm proposed by Lee et al. [6] gives a good solution and is widely used. Their algorithm is an iterative type one based on the supplementary function method with initial values set randomly. However for initialization, usage of Singular Value Decomposition (SVD) of V was proposed and its effectivity was reported [3]. Therefore, we employ SVD for the initialization.

3.3   Model Selection Problem

We propose a method to find the optimal value of r in terms of model selection problem.

The minimum description length (MDL) principle is an information criterion introduced by Rissanen [8]. The MDL principle says that, when the data is compressed with the help from a statistical model, the model with shortest description length is optimal. For a data set $x^N = x_1 x_2 \ldots x_N$ and a k-dimensional model $q(x^N|\theta)$ ($\theta \in \Re^k$), the MDL criterion is generally known as $MDL = -\ln f + (k/2) \ln N$. Here, $f = \max_\theta q(x^N|\theta)$ is the maximum likelihood.

To apply MDL to the model selection problem finding the optimal parameter r in NMF, we introduce a statistical model corresponding to NMF as

$$V = WH + \epsilon . \tag{2}$$

Here, each element of ϵ is a normal random variable independently generated according to a Gaussian distribution with mean 0. Let $p(V|\theta)$ ($\theta = (W, H)$)

denote the density function of the model (2). In this model, the formula of MDL can be given as

$$MDL_{NMF} = -\ln f + \frac{nr + rm - r^2}{2} \ln \frac{nm}{2\pi} , \tag{3}$$

where the maximum logarithmic likelihood $\ln f$ is given as

$$\ln f = -\frac{nm}{2} \ln 2\pi\sigma^2 - \frac{1}{2\sigma^2} \sum_{i=1}^{n} \sum_{\mu=1}^{m} (V_{i\mu} - (\hat{W}\hat{H})_{i\mu})^2 , \tag{4}$$

where $(\hat{W}, \hat{H})$ is an output of NMF algorithm, $\sigma^2 = (nm)^{-1} \sum_{i=1}^{n} \sum_{\mu=1}^{m} (V_{i\mu} - (\hat{W}\hat{H})_{i\mu})^2$, and the factor $nr + rm - r^2$ in (3) is the dimension of the model defined by (2). Note that the dimension is less than the sum of the numbers of components of W and H, since they contains r^2 redundant components. Let $\hat{r}$ denote the r which minimizes (3). We refer to $\hat{r}$ as the MDL estimate of r. We employ it in this work.

To show (3), we make use of an expression of stochastic complexity (SC) [9], which is a strict value of MDL. For a regular model, SC is evaluated as

$$SC = -\log q(x^N|\hat{\theta}) + \frac{k}{2} \log \frac{N}{2\pi} + \log \int \sqrt{|J(\theta)|}d\theta + o(1) , \tag{5}$$

where $\hat{\theta}$ is the maximum likelihood estimate given x^N and $J(\theta)$ is the Fisher information matrix. Though the model (2) is not a regular model, we assume that this formula approximately holds. Regard the data matrix V as a single nm-dimensional vector valued datum. Then, (5) can be transformed to

$$SC_{NMF} = -\log f + \frac{nr + rm - r^2}{2} \log \frac{1}{2\pi} + \log \int \sqrt{|J(\theta)|}d\theta . \tag{6}$$

The third term can be evaluated by noting that the following holds for each element of $J(\theta)$.

$$
\begin{aligned}
J_{ij}(\theta) &= -E_\theta \left[\frac{\partial^2}{\partial\theta_i\partial\theta_j} \left(\frac{1}{2}(x^N - \mu)\Sigma^{-1}(x^N - \mu)^T + \frac{1}{2}\ln|\Sigma| \right) \right] \\
&= -E_\theta \left[\frac{\partial^2}{\partial\theta_i\partial\theta_j} \left(\frac{1}{2}(x^N - \mu)\Sigma^{-1}(x^N - \mu)^T \right) \right] \\
&= -E_\theta \left[\frac{\partial^2}{\partial\theta_i\partial\theta_j} \frac{1}{2}\mathrm{Tr}(\Sigma^{-1}(x^N - \mu)(x^N - \mu)^T) \right] \\
&= nmA_{ij}(\theta) + O(1) .
\end{aligned}
\tag{7}
$$

Here, $A_{ij}(\theta)$ is a quantity independent of nm. The factor nm in (7) appears because the dimension of the datum V is nm. Therefore, we have

$$\int \sqrt{|J(\theta)|}d\theta = (nm(1 + o(1)))^{\frac{nr+rm-r^2}{2}} C , \tag{8}$$

where C is a certain constant. Neglecting a term of $o(\frac{nr+rm-r^2}{2})$, we obtain (3).

We demonstrate by comparing MDL with AIC that MDL is reasonable to solve the model selection problem. The formula of AIC is given by

$$AIC_{NMF} = -\ln f + (nr + rm - r^2),$$ (9)

since the number of free parameters is $nr + rm - r^2$. Therefore, AIC tends to select a larger model than true one because the penalty term $nr + rm - r^2$ in AIC is smaller than that of MDL. This is demonstrated by simulation in Sec. 4.1.

3.4 Detection Method for Botnet Activity Pattern

We describe our method to detect botnet patterns using the spatial patterns and the temporal patterns obtained by NMF. We employ the assumption that the infected hosts which belong to an identical botnet act simultaneously on receiving a command emitted by a bot-master as explained in Sec. 2. Hence we expect that the temporal pattern which corresponds to a botnet has a 'change point' and that the spatial pattern looks like uniformly distributed. Here, a change point means a point in a time series where its property (typically mean or variance) suddenly changes.

To find a change point, we employ ChangeFinder (CF) [10], which is one of data analysis technique to detect change points. CF learns the time series in online manner with the AR model and calculates the change point score based on it. See [10] for its detailed property.

For the uniformity in the spatial pattern, we examine the following quantity u_a^2 (variance).

$$u_a^2 = \frac{1}{|M| - 1} \sum_{\mu \in M} (\bar{H}_a - H_{a\mu})^2.$$ (10)

Here, $\bar{H}_a = \sum_\mu H_{a\mu}$ and M denotes the set of index μ for which $H_{a\mu}$ is not less than 0.5. The reason why we ignore $H_{a\mu}$ less than 0.5 is that it should be an integer originally.

It should be noticed that the values of W and H are not fixed uniquely, since it can be rewritten with a diagonal matrix $\Lambda = (\lambda_a \delta_{ab}) \in \Re^{r \times r}$ as $WH = W\Lambda\Lambda^{-1}H$. Hence we normalize the rows of H such that the sum of elements of each row equals 1, when applying CF to W. By this manipulation, the value of an element of H has a meaning as the number of packets. Similarly when analyzing the variance in H, we normalize the columns of W such that the sum of elements of each column equals 1.

Finally, we release an alert for botnet activity when the following two conditions are satisfied for a pair of temporal and spatial patterns $(\{W_{ia}\}_i, \{H_{a\mu}\}_\mu)$ at the same time: 1) CF outputs change point scores larger than the pre-determined threshold and 2) the variance u_a^2 is less than the pre-determined threshold. We set the threshold of variance to 1000, because it worked well in preliminary experiments.

4   Experiments

We evaluate our method using synthetic and real data.

[Simulation data]
We prepare the synthetic data as follows, where $\mathbf{w}_i \in \Re^n$ and $\mathbf{h}_i \in \Re^m$ denote the ith column vector of W and ith row vector of H, respectively.

1. Initialize $\mathbf{w}_i$ and $\mathbf{h}_i$ $(i = 1, \cdots, r)$ manually.
2. Let a data matrix V be the product of W and H.

[Real data]
As for the real data, we use several data sets observed in years 2008 and 2009 by the darknet managed by National Institute of Information and Communications Technology (NICT) [4]. For those data sets, NICT provides the labels which indicate whether each packet is sent by bots or not. Such the labels were obtained owing to detailed analysis by a researcher. Using these data sets, we can verify if the botnet activity can be detected by the proposed method.

We show how to make a data matrix V below.

1. Split the whole data into every 60 minutes.
2. For each split data, select 2000 source hosts randomly from the hosts which emit more than 2 packets [2].
3. Count the packets of each selected host every minute and make matrix $V \in \Re^{60 \times 2000}$ whose rows correspond to time and column corresponds to hosts.

[Procedure of NMF]
We carry out the NMF algorithm with parameter $r = 2, \cdots, 10$, calculate MDL, and select the output of NMF with r selected by the MDL criterion. Here, the maximum times of iteration is $T_{\mathrm{Max}} = 10000$ and the termination condition is that the difference between approximation error in the Tth iteration and that in the $(T - 1)$th is less than 0.00001.

For the initial value, we use Singular Value Decomposition (SVD) [3] as follows. Decompose the data matrix V with SVD as $V = S\Sigma U^T$, and let S^r and U^r denote the first r columns of S and U respectively. Then, set $W_{ia} = |S^r{}_{ia}|$ and $H_{a\mu} = |U^r{}_{\mu a}|$.

[Procedure of ChangeFinder]
We apply CF to each column of W. We set each parameter as follows.
Discounting parameter: $r_{\mathrm{CF}} = 0.03$
Width of window: $T = 5$ (the first stage), $T = 5$ (the second stage)
Order of AR model: $k = 2$ (the first stage), $k = 3$ (the second stage)
For the meaning of those parameters, see [10].

4.1   Results of Simulation

We conduct an experiment with the synthetic data to evaluate the precision of $\hat{r}$. With changing the true (W, H) we did 4 trials respectively for $r = 4$, $r = 6$, and $r = 8$ cases. Also we conduct the experiment with AIC to compare with MDL.

Table 1 shows the results of model selection by MDL or AIC, where the numbers in each column denote the averaged values of $|\hat{r} - r|$. We can see that the results of MDL are better than that of AIC.

We also perform experiments to verify whether we can split into each activity pattern by NMF. Fig. 3 and 5 show the results of NMF with MDL. Here, $r = 4$ and $\hat{r} = 4$. It can be seen that the output of NMF and the original data have the similar behaviors. From these results, it was found that we can estimate the number of activity patterns by MDL and we can split the input data into each activity pattern by NMF if we use the result of MDL.

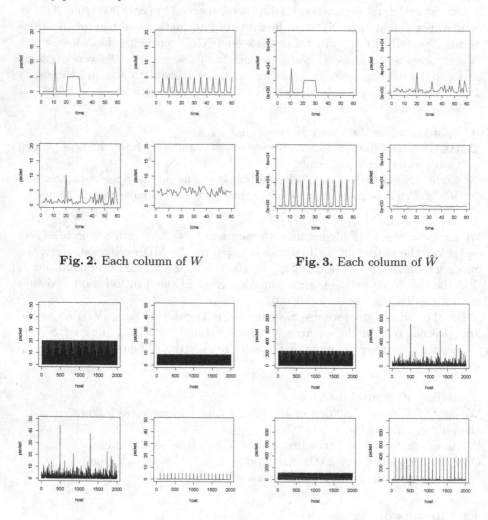

Fig. 2. Each column of W **Fig. 3.** Each column of $\hat{W}$

Fig. 4. Each column of H **Fig. 5.** Each column of $\hat{H}$

Table 1. Precision of $\hat{r}$ by MDL or AIC

		\multicolumn{6}{c}{matrix size}					
		30*500	30*1000	30*2000	60*500	60*1000	60*2000
$r = 4$	MDL	0	0	0	0	0	0
	AIC	0	0	0	0.75	0.75	0.75
$r = 6$	MDL	0	0.5	0.75	0.25	0.25	0
	AIC	1	0.75	0.75	1	1	0.75
$r = 8$	MDL	1.5	1.25	0.75	1.25	0	1
	AIC	2.75	2.75	3.75	3	5	3.25

4.2   Results for Real Data

Fig. 6 and 7 show the results of NMF with the selected $\hat{r}$ by MDL. For this data set, $\hat{r} = 9$ was selected. We show 4 patterns among them, which have the largest numbers of packets. In Fig. 6, the solid lines indicate each column of W, and the dotted lines indicate the actual botnet activity. In Fig. 7, each graph indicates each row of H and the actual botnet activity. In both figures, pattern 8 corresponds to the actual botnet pattern.

The computation time for an NMF execution was within 5 minutes for all setting of parameters, with using a PC of Intel(R) Core(TM) i5 3.2GHz with 8GB RAM.

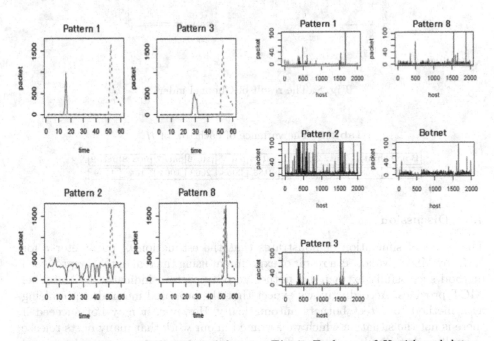

Fig. 6. Each column of W with real data   **Fig. 7.** Each row of H with real data

Fig. 8 shows the outputs of ChangeFinder. In Fig. 8, the X-axis represents time, the left Y-axis represents the number of packets, the right Y-axis represents the change point score, the solid lines indicate each column of W, and the dotted lines indicate the change point score. In applying CF to columns of W, change points are detected in the 8th column. Table 2 shows the variance of each row of H. The variance of only the pattern 8 is within the threshold. Therefore, we can judge that the pattern 8 corresponds to the botnet activity.

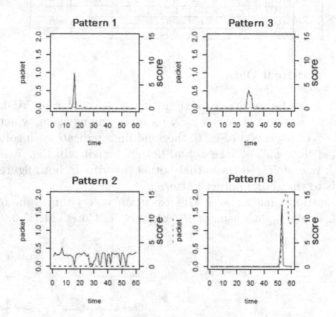

Fig. 8. The result of ChangeFinder

Table 2. The variance of each row of H

Botnet	pat. 1	pat. 2	pat. 3	pat. 4	pat. 5	pat. 6	pat. 7	pat. 8	pat. 9
19	7634	4216	5469	2183	2053	2069	1964	164	1739

4.3   Discussion

The result of simulation demonstrates that the estimation of parameter r for NMF by MDL is valid. As a result of experiment using the real data, the proposed method successfully extracted a botnet activity pattern indicated by the label NICT provides. Accordingly, we expect that the proposed method is promising as a method to detect botnets automatically. However, it may not succeed if there is not the situation which we assumed in our work that many hosts release a small packet simultaneously. We have to do more experiments and improve the method.

5   Conclusion

This paper describes a method to detect botnet activity from darknet traffic data. We use NMF to split the data into each activity pattern. Additionally, we use ChangeFinder for temporal patterns and variance of spatial patterns to detect botnet activity pattern from the result of NMF. Moreover, we propose a method to estimate the parameter of NMF corresponding to the number of activity patterns by MDL. By some experiments with synthetic and real data, we have shown that the proposed method works well.

Acknowledgment. We thank everyone of Cybersecurity Laboratory, NICT who provides the darknet data. We also thank them and Dr. Shinichi Nakajima for their helpful comments. Our work is partially supported by "Research and development of predictive technology for cyber attack by international cooperation" administered by the Ministry of Internal Affairs and Communications.

References

1. Akiyama, M., Kawamoto, T., Shimamura, M., Yokoyama, T., Kadobayashi, Y., Yamaguchi, S.: A Proposal of Metrics for Botnet Detection Based on Its Cooperative Behavior. In: SAINT (2007)
2. Hamasaki, H., Kawakita, M., Takeuchi, J., Yoshioka, K., Inoue, D., Etoh, M., Nakao, K.: Proposal of Botnet Detection Based on Structure Learning and Its Application to Darknet Data. In: SCIS (2011)
3. Hotta, S., Miyahara, S.: An Initialization Method for non-negative matrix factorization and Its Applications. Technical report, IEICE (2003)
4. Inoue, D., Yoshioka, K., Eto, M., Yamagata, M., Nishino, E., Takeuchi, J., Ohkouchi, K., Nakao, K.: An Incident Analysis System NICTER and Its Analysis Engines Based on Data Mining Techniques. In: Köppen, M., Kasabov, N., Coghill, G. (eds.) ICONIP 2008, Part I. LNCS, vol. 5506, pp. 579–586. Springer, Heidelberg (2009)
5. Kitagawa, J., Kawakita, M., Takeuchi, J., Yoshioka, K., Inoue, D., Etoh, M., Nakao, K.: Extraction of Botnet Communication Based on Non-negative Matrix Factorization. In: SCIS (2010)
6. Lee, D.D., Seung, H.S.: Algorithms for Non-negative Matrix Factorization. Neural Inf. Process. Syst. 13, 556–562 (2001)
7. Nikolaus, R.: Learning the Parts of Objects Using Non-negative Matrix Factorization.Term Paper (2007)
8. Rissanen, J.: Modeling by Shortest Data Description. Automatica 14, 465–471 (1978)
9. Rissanen, J.: Fisher Information and Stochastic Complexity. IEEE Transactions on Information Theory 42, 40–47 (1996)
10. Takeuchi, J., Yamanishi, K.: Unifying Framework for Detection Outliers and Change Points from Time Series. IEEE Transactions on Knowledge and Data Engineering 18, 676–681 (2006)
11. Cyber Clean Center, https://www.ccc.go.jp

Fusion of Multiple Texture Representations for Palmprint Recognition Using Neural Networks

Galal M. BinMakhashen and El-Sayed M. El-Alfy

College of Computer Sciences and Engineering,
King Fahd University of Petroleum and Minerals, Dhahran 31261, Saudi Arabia
binmakhashen@gmail.com
alfy@kfupm.edu.sa

Abstract. During the last decade, palmprint recognition has received an increasing attention due to the abundant features that can be extracted from the captured palmprint image. However, a single palmprint texture representation may not be sufficient for reliable recognition. Therefore, in this paper we propose a computational model for palmprint recognition/identification by fusing different categories of feature-level representations using a multilayer perceptron (MLP) neural network. Features are extracted using Gabor filters and Principle Component Analysis (PCA) is used to reduce the dimensionality of the feature space by selecting the most relevant features for recognition. The proposed model has shown promising results in comparison with naïve Bayes, rule based and $K*$ algorithm.

Keywords: Machine Learning, Pattern Recognition, Biometrics, Palmprint, Multilayer Perceptron, Principle Component Analysis, Gabor Filter.

1 Introduction

Personal recognition based on biometric technology is an emerging field of research in security systems. Many researchers are attracted to this technology due to two main advantages as compared to traditional methods; it provides a good resistance to spoof attacks and it is hard to be lost or stolen. Typically, this technology uses human physiological or behavioral characteristics such as face, iris, palmprint, fingerprint, voice, gait, etc. Palmprint recognition is relatively recent in comparison to other biometric technologies. It has various distinctive features (e.g. principal lines, wrinkles, minutia, and textures). Previous work on palmprint can be divided into two main categories based on the acquired images and the target application. The first category is high-resolution based techniques which typically extract and compare minutiae features. Such features were basically used for fingerprint analysis. However, extracting these features is very difficult [1] since the high-resolution images can be easily affected by any little noise. The other category is based on low-resolution images, and typically uses structural features such as principal lines, wrinkles and textures [2], [3]. This category is generally employed in commercial and civil systems. Fig. 1 illustrates two

T. Huang et al. (Eds.): ICONIP 2012, Part V, LNCS 7667, pp. 410–417, 2012.
© Springer-Verlag Berlin Heidelberg 2012

palmprint images of different resolutions, where the high-resolution image is adopted from [2].

Whether the palmprint images are high or low resolution, recognition consists of four phases: acquisition, pre-processing, feature extraction and decision making. For image acquisition, high or normal resolution scanners can be used to capture hand images with CCD cameras such as in [4] or [5]. Images are then pre-processed to determine the region of interest (ROI) within a high-dimensional image space. Zhang et al. [5] have established a coordinate system using the finger gaps to determine the center of the palm. Then, they extract a square shape of the palmprint. Kumar et al. [6] have used a free hand acquisition with similar palmprint extraction algorithm as in [5]. As soon as the ROI is determined, the feature extraction phase has to be started to represent each palmprint image by a distinctive smaller amount of data called feature vector. Finally, the feature vectors are used for classification or recognition purposes.

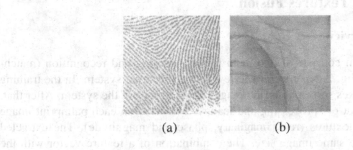

(a)   (b)

Fig. 1. Different resolution palmprint images: (a) high, (b) low

There are several research works on fusion for palmprint recognition and can be divided into score-level and feature-level fusion; as shown in Table 1. Kumar and Zhang [3] conducted a study on combining different palmprint techniques at the score level. Wu et al. [1] have done similar work by integrating two palmprint representations at the score level. On the other hand, feature-level fusion using multiple elliptical Gabor filters for palmprint domain has been suggested to enhance performance [7]. In [8], another method has been proposed to fuse different feature types (textures, palm lines and appearance) using wavelets. Moreover, a log multiplication fusion rule at feature level has been adopted in [9] where appearance and texture features are combined by Dual Tree Complex Wavelet Transform (DT-CWT).

Unlike earlier work, we are proposing a personal recognition system based on multilayer perceptron neural network. Moreover, we are consolidating different palmprint textures extracted by Gabor filters. The features are also reduced by applying Principal Component Analysis (PCA) directly on the Gabor domain. The proposed system is evaluated using a publicly available dataset of palmprint images.

The rest of the paper is organized as follows. Section 2 presents an overview of the palmprint biometric recognition system. In Section 3, we describe the experimental work and discuss the results. Finally, Section 4 concludes the work.

Table 1. Pervious work on multi-representation fusion using palmprint and distance measure

Feature Extraction Technique	Matching/ Classification	Fusion Level	Year	Ref
FusionCode	Verification	Feature	2004	[10]
Multiple elliptical Gabor filters	Identification & Verification	Feature	2005	[7]
PalmCode , Palm lines	Verification	Score	2005	[1]
Gabor, palm lines, PCA	Verification	Score	2005	[3]
OrientationCode, diffCode	Verification	Feature	2006, 2007	[11], [12]
Tensor Locality Preserving Projection (TLLP), Competitive Code	Identification	Score	2008	[13]
PCA, Dual Tree-Complex Wavelet transform.	SVM, Random Forest	Feature	2011	[9]

2 Palmprint Textures Fusion

2.1 System Overview

The proposed system consists of two main phases: training and recognition (matching/classification). Fig. 2 depicts an overview of the proposed system. In the training phase, the system takes some palmprint images for each user of the system. After that, the system carries out pre-processing and feature extraction on each palmprint image to extract multiple textures (real, imaginary, phase and magnitude). The extracted feature sets have the same image size. The combination of a feature vector with the ID for each person is known as his/her template. We also selected from 40 to 80 principle components using PCA for each texture features. These features are integrated in different combinations to solidify the palmprint image representation. Finally, a multilayer perceptron (MLP) is trained over these feature sets to generate a computational model with better generalization.

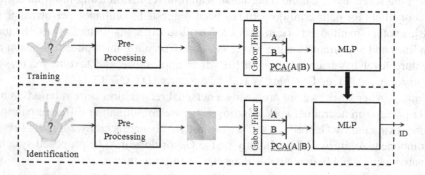

Fig. 2. An overview of the proposed system

2.2   Gabor Filters and Feature Extraction

Gabor filters are one of the most powerful texture analyses that have been studied in the image processing field [14]. They are named after Dennis Gabor. In this technique, images are transformed from spatial domain to frequency domain. Moreover, the filter's parameters can be adjusted to achieve multi-resolution analysis. Usually a complex 2D Gabor filter is defined as the product of a complex sinusoid (known as carrier) $s(x, y)$ and a Gaussian-shaped kernel (known as Gaussian envelope or window) $w(x, y)$:

$$g(x, y) = s(x, y)w(x, y) \qquad (1)$$

The complex sinusoid is defined as a two separate functions as follows:

$$Re(s(x, y)) = \cos(2\pi (ux + vy) + P) \qquad (2)$$

$$Im(s(x, y)) = \sin(2\pi (ux + vy) + P) \qquad (3)$$

where $Re(.)$ and $Im(.)$ represent the real and imaginary parts, respectively; and (u, v) and P are the spatial frequency and phase of the sinusoid. The spatial frequency can be expressed in polar coordinates as magnitude and phase as follows:

$$F = \sqrt{u^2 + v^2}, \quad \omega = \tan^{-1}(v / u) \qquad (4)$$

These equations can be rewritten as:

$$u = F \cos \omega, \quad v = F \sin \omega \qquad (5)$$

Thus, the complex sinusoid can be rewritten as:

$$s(x, y) = \exp\left(j\left(2\pi F(x\cos \omega + y\sin \omega)\right) + P \right) \qquad (6)$$

where $j = \sqrt{-1}$. On the other hand, the Gaussian envelope is as follows:

$$w(x, y) = K \exp\left(-\pi\left(a^2(x - x_o)_r^2 + b^2(y - y_o)_r^2\right)\right) \qquad (7)$$

where (x_o, y_o) is the peak of the function; a and b are scaling parameters of the Gaussian envelope; and r stands for the θ-rotation, i.e., in matrix form,

$$\begin{bmatrix} (x - x_0)_r \\ (y - y_0)_r \end{bmatrix} = \begin{bmatrix} \cos\theta & \sin\theta \\ -\sin\theta & \cos\theta \end{bmatrix} \begin{bmatrix} x - x_0 \\ y - y_0 \end{bmatrix} \qquad (8)$$

The Gabor filter is tuned to reduce the effect of brightness by adjusting the DC (direct current) component to zero by subtracting the output of a low pass filter. To extract discriminant palmprint features, the Gabor filter is convolved with the palmprint images. The Gabor parameters can be found by trial and error or by a tuning process to optimize their selection [5], [7]. In our work, we have adjusted these parameters to: $b = a = 0.3$, $P = \theta = 0$, $(x_0, y_0) = (0, 0)$, $u = v = 0.2592$ and $K = a^2$. Fig. 3 illustrates the process of feature extraction for one palmprint image.

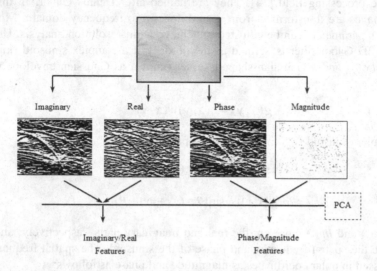

Fig. 3. An example of palmprint texture representations using Gabor filters

2.3   Feature-Level Fusion

Using a Gabor filter, four different features can be represented (imaginary, real, angle and magnitude) of the palmprint image. Fusion at the feature level can be performed in different forms depending on the nature of the extracted features. For example, when features are extracted from multiple samples (but in the same domain), fusion can be done using weighted sum or averaging to these multiple sample features to produce one single feature vector. Another feature-level fusion technique that have been recommended when the domain of the two feature vectors are different is by concatenation [15]. Moreover, applying simple concatenation will require a feature selection to reduce dimensionality. We applied concatenation and used PCA to produce a single fused feature vector for each image.

2.4   MLP-Based Recognition

Multilayer perceptron (MLP) is a feed-forward neural network capable of modeling the nonlinearity problem into a linear one with high generalization [16]. It consists of one or more hidden layers between its input and output layers. Fig. 4 shows a typical structure of MLP neural network. Each node is a processing element that computes its output as a nonlinear function of its inputs. The number of nodes in the input layer is determined from the number of elements in the feature vector. The number of nodes in the output layer is set equal to the number of different palmprint IDs in the system. The number of hidden layers and the number of nodes in each layer can vary.

For example, one can use one hidden layer with number of nodes equal to sum of the number of nodes in input layer and number of nodes in the output layer. Different types of activation functions can be used but in our identification problem, all nodes use sigmoid functions. The network is trained using supervised learning from a set of pre-classified instances. During training, the network parameters are adjusted to minimize the recognition error using back propagation algorithm. The amount of adjustment for the network parameters depends on a decaying learning rate. Training terminates after a certain number of training epochs or until the error observed on a validation set is getting worse. Once a computational model is constructed, it can be used to classify unseen data (generalization).

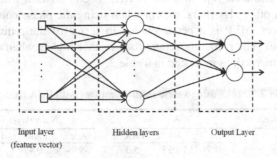

Input layer   Hidden layers   Output Layer
(feature vector)

Fig. 4. A typical structure of a multilayer perceptron

3   Experiments

In the following section, we describe the palmprint database. We also discuss the conducted experiments to evaluate and compare the performance of the proposed approach.

3.1   Database Description

The adopted palmprint images are publicly available at [17]. The dataset contains 1000 palmprint images of 100 people. Each person provides 10 left hand images. The images have been captured using a CCD digital camera. The hand images are preprocessed and ROIs are extracted and stored as text files. The text files are converted to palmprint images in different image scales (445×442 and 404×400). We have adjusted the scales of all palmprint images to (128×128) using MATLAB.

3.2   Experiments and Performance Analysis

A feature-level fusion is carried out by combining multiple texture representations of the palmprints. Several experiments have been conducted to evaluate the performance of a MLP neural network with and without PCA. The MLP has a number of neurons in the input layer equal to the dimensionality of the input feature vector.

In addition, the number of neurons in the hidden layer depends on the number of palmprint IDs and length of the feature vector. Each neuron at (input and hidden layers) applies a nonlinear processing function (sigmoid). The training process has 0.2 as a learning rate and it is executed for 500 epochs. Also, we have compared the performance of the proposed system with three other classifiers: rule based (JRip), naïve Bayes and K^* algorithm [16], [18]. The first set of experiments was conducted on full feature representation of palmprints (imaginary, real, phase and magnitude). We have executed it for 12 hours without recording results, but after reducing the feature vector to only 40-80 features, experiments results can be obtained in a couple of minutes. Table 2 summarizes all the cases in terms of recognition accuracy (Acc), false acceptance rate (FAR) and false rejection rate (FRR). It can be seen from the results that MLP outperforms other classifiers, except when using the phase components with K^* algorithm. However, MLP is a better choice in terms of the required storage (not shown in the table). The recognition rate goes further higher when different textures of palmprints are merged; as indicated in Table 3.

Table 2. Results using single texture features with PCA reduction

Feature	K^*			NaiveBayes		
	Acc	FAR	FRR	Acc	FAR	FRR
Imag	94.7	0.053	5.3	95.4	0.45	4.6
Real	99.6	0.004	0.4	98.4	0.016	1.6
Phase	**98.6**	0.014	1.4	88.9	0.111	11.1
Mag	94.8	0.052	5.2	94	0.068	6
	JRip			MLP		
	Acc	FAR	FRR	Acc	FAR	FRR
Imag	58.9	0.411	49.1	**98.50**	0.015	2.5
Real	65.8	0.342	34.2	**99.8**	0.002	0.2
Phase	45.7	0.543	54.3	90.4	0.096	9.6
Mag	48.3	0.517	51.7	**96.7**	0.033	3.3

Table 3. Feature-level fusion of multiple palmprint textures

Feature PCA	K^*			NaiveBayes		
	Acc	FAR	FRR	Acc	FAR	FRR
Imag + Real	98.7	0.013	1.3	98.5	0.015	1.5
Phase + Mag	**99**	0.02	1	94.5	0.055	5.5
	JRip			MLP		
	Acc	FAR	FRR	Acc	FAR	FRR
Imag + Real	65.2	0.348	34.8	**100**	0	0
Phase + Mag	55.1	0.449	44.9	97	0.03	3

4   Conclusion

In this paper, we have investigated a neural network approach for feature-level fusion of different palmprint texture representations for human recognition. Multilayer perceptron neural network was evaluated and compared with three different classifiers:

rule based (JRip), naïve Bayes and K^* algorithm. We found that fusion with multilayer perceptron can be a good candidate to improve the recognition rate. For future work, we are exploring other means of extracting distinguishing palmprint features and fusion techniques.

Acknowledgements. The authors would like to thank King Fahd University of Petroleum & Minerals (KFUPM), Saudi Arabia, under Grant no. RG1106-1&2, and the Hadhramout Establishment for Human Development, Yemen, for funding and support during this work.

References

1. Wu, X., Zhang, F., Wang, K., Zhang, D.: Fusion of the textural feature and palm-lines for palmprint authentication. In: Advances in Intelligent Computing, pp. 1075–1084 (2005)
2. Kong, A., Zhang, D., Kamel, M.: A Survey of palmprint recognition. Pattern Recognition 42, 1408–1418 (2009)
3. Kumar, A., Zhang, D.: Personal authentication using multiple palmprint representation. Pattern Recognition 38, 1695–1704 (2005)
4. Han, C.C., Cheng, H.L., Fan, K.C., Lin, C.L.: Personal authentication using palmprint features. Pattern Recognition 36, 371–381 (2003)
5. Zhang, D., Kong, W.K., You, J., Wong, M.: Online palmprint identification. IEEE Trans. on Pattern Analysis and Machine Intelligence 25, 1041–1050 (2003)
6. Kumar, A., Wong, D., Shen, H., Jain, A.K.: Personal verification using palmprint and hand geometry biometric. In: Audio- and Video-Based Biometric Person Authentication, pp. 1060–1060 (2003)
7. Kong, A., Zhang, D., Kamel, M.: Palmprint identification using feature-level fusion. Pattern Recognition 39, 478–487 (2006)
8. Krishneswari, K., Arumugam, S.: Intra-modal feature fusion using wavelet for palmprint authentication. Int. J. of Eng. Science and Tech. (IJEST), 1597–1605 (2011)
9. Shashikala, K.P., Ashwin, P., Raja, K.B.: Palmprint identification based on fusion of PCA and DT-CWT features. In: Proc. Int. Conf. Image Inf. Proc (ICIIP), pp. 1–6 (2011)
10. Kong, A.W.-K., Zhang, D.: Feature-Level Fusion for Effective Palmprint Authentication. In: Zhang, D., Jain, A.K. (eds.) ICBA 2004. LNCS, vol. 3072, pp. 761–767. Springer, Heidelberg (2004)
11. Wu, X.Q., Wang, K.Q., Zhang, D.: Fusion of multiple features for palmprint authentication. In: Proc. Int. Conf. Machine Learning and Cybernetics, pp. 3260–3265 (2006)
12. Pan, X., Ruan, Q., Wang, Y.: Palmprint recognition using fusion of local and global features. In: Int. Symp. Intell. Signal Proc. and Commun. Systems, pp. 642–645 (2007)
13. Jia, W., Huang, D.S., Tao, D., Zhang, D.: Palmprint identification based on directional representation. In: IEEE Int. Conf. Systems, Man and Cybernetics, pp. 1562–1567 (2008)
14. Movellan, J.: Tutorial on Gabor filters. Technical Report, MPLab Tutorials, Univ. of California, San Diego (2005)
15. Ross, A., Jain, A., Nandakumar, K.: Handbook of Multibiometrics. Springer (2006)
16. Witten, I.H., Frank, E., Hall, M.A.: Data Mining: Practical Machine Learning Tools and Techniques, 3rd edn. Morgan Kaufmann, Burlington (2011)
17. http://www.cse.ust.hk/~helens
18. Cleary, J.G., Trigg, L.E.: K*: An instance-based learner using an entropic distance measure. In: Proc. 12th International Conference on Machine Learning, pp. 2–12 (1995)

Adaptive Neural Networks Control
on Ship's Linear-Path Following[*]

Wei Li[**], Jun Ning, Zhengjiang Liu, and Tieshan Li

Navigation College, Dalian Maritime University, Dalian, P.R. China
muziyuri@tom.com, jun-ning@139.com, liuzjj@online.ln.cn,
tieshanli@126.com

Abstract. In this paper, we investigate the problem of linear tracking control for an underactuated surface ship with rudder actuator dynamics. By using Radial Basis Function (RBF) Neural Networks (NN) to approximate the uncertainties of the systems, the problem of singularity is avoided and the trouble caused by "explosion of complexity" in traditional backstepping methods is removed by taking advantage of dynamic surface control (DSC) technique. Also, it is proved that all the signals of the closed-loop system are uniformly ultimately bounded(UUB), and the tracking error converges to the neighborhood of zero. The simulation results on an ocean-going training ship 'YULONG' are shown to validate the proposed algorithm.

Keywords: RBF Neural Networks, DSC, Ship's Linear-Tracking Control, Backstepping.

1   Introduction

Nowadays, traditional automatic heading helm can't meet the requirements of the ships for the reasons that it can't control the tracking error directly[1], so ship linear-tracking control become an important practice and has caused considerable attention[2], and many remarkable results have shown the rapid development of systematic design methods for ship tracking control,such as[3]-[7]. At the same time, intelligent control which has the ability of adaptive, self learning, self optimization, self-adjusting becomes a hot spot in the research of ship's tracking control and gained good effect [8]- [12] .

In this paper, combining the DSC and minimum learning parameters(MLP) techniques in[13], a simple robust adaptive neural tracking control algorithm is proposed for nonlinear ship motion system with the rudder actuator dynamics. The scheme can force the ship to follow a desired path and ensure the uniformly ultimate boundedness of the closed-loop system. Finally, simulation results on an ocean-going training ship are given to demonstrate the performance of the proposed scheme.

---

[*] This work was supported in part by the National Natural Science Foundation of China(No.51179019), the Natural Science Foundation of Liaoning Province (No. 20102012) and the Program for Liaoning Excellent Talents in University(LNET).

[**] Corresponding author.

T. Huang et al. (Eds.): ICONIP 2012, Part V, LNCS 7667, pp. 418–427, 2012.
© Springer-Verlag Berlin Heidelberg 2012

2   RBF Neural Network

It has been proved that RBF neural networks can be used to approximate an arbitrary smooth function $F(x)$: $R^q \to R$. The approximator $H_{nn}(x)$ can be expressed as

$$H_{xx}(x) = \varpi^T S(x) \tag{1}$$

where $x \in \Omega_x \subset R^q$, weight vector $\varpi = [\varpi_1 \ldots \ldots \varpi_l]^T \in R^l$, $S(x) = [S_1(x), \ldots \ldots S_l(x)]^T$, and the NNs node number $l > 1$. We usually choose $S_i(x)$ as the Gaussian function.

$$S_i(x) = \frac{1}{\sqrt{2\pi}\,\sigma_i} \exp\left[\frac{-(x - u_i)^T (x - u_i)}{\sigma_i^2}\right] \tag{2}$$

Any continuous function with $\forall x \in \Omega_x \subset R^q$ can be approximated as

$$F(x) = \varpi^{*T} S(x) + \varepsilon, \forall x \in \Omega_x \tag{3}$$

where ϖ^* is the ideal weight vector, ε is the approximation error with an assumption of $|\varepsilon| \le \varepsilon^*$, and the unknown constant $\varepsilon^* > 0$ for all $Z \in \Omega_Z$. With the estimate value ϖ minimizing $|\varepsilon|$ for all $x \in \Omega_x$, the ideal weight vector ϖ^* can be typically defined as

$$\varpi^* = \arg \min_{\varpi \in R^l} \left\{ \sup_{Z \in \Omega_Z} \left| F(x) - \varpi^T S(x) \right| \right\} \tag{4}$$

Now, we bring in a very important lemma.

Lemma 1. [14] : For any given function $f(x)$ with $f(0) = 0$, if the continuous function separation technique [14] and the RBF NN approximation technique are used, then $f(x)$ can be denoted as

$$f(x) = \bar{S}(x)Ax \tag{5}$$

where $\bar{S}(x) = [1, S(x)] = [1, s_1(x), s_2(x), \cdots s_l(x)]$, $A^T = [\varepsilon, \varpi^T]$, $\varepsilon^T = [\varepsilon_1, \varepsilon_2, \cdots \varepsilon_n]$ is a vector of the approximation error, and ϖ is a weight matrix as follows.

$$\varpi = \begin{bmatrix} \varpi^*_{11} & \varpi^*_{12} & \cdots & \varpi^*_{1n} \\ \varpi^*_{21} & \varpi^*_{22} & \cdots & \varpi^*_{2n} \\ \vdots & \vdots & \cdots & \vdots \\ \varpi^*_{l1} & \varpi^*_{l2} & \cdots & \varpi^*_{ln} \end{bmatrix}$$

3   Problem Formulation

In this passage, according to the references[7], we introduce the nonlinear ship straight-line motion equation plus the rudder actuator dynamics in the following form:

$$\begin{cases} \dot{y} = U\sin(\psi) \\ \dot{\psi} = r \\ \dot{r} = \eta_2(r) + g_2\delta + d \\ \dot{\delta} = -\dfrac{1}{T_E}\delta + (\dfrac{k_E}{T_E})\delta_E \end{cases} \tag{6}$$

where y denote the sway displacement (cross-track error), ψ denote the heading angle, r denote the yaw rate, U denote the cruise speed respectively. δ denotes the control rudder angle, d denotes uncertain external perturbations which is bounded, $\eta_2(r)$ is an unknown nonlinear function for r. g_2 is the control gain. T_E and k_E are the time delay constant and the control gain of the rudder actuator. δ_E is the order angle of rudder. In order to make the design work much easier, the coordinate transformation are defined as follows [3], [7].

$$x_1 = \psi + \arcsin\left(\frac{ky}{\sqrt{1+(ky)^2}}\right) \tag{7}$$

Due to the equation transformation, we can obtain a class of nonlinear uncertain system as follows:

$$\begin{cases} \dot{x}_1 = \eta_1 + g_1 x_2 + w_1 \\ \dot{x}_2 = \eta_2 + g_2 x_3 + w_2 \\ \dot{x}_3 = \eta_3 + g_3 u + w_3 \\ y = x_1 \end{cases} \tag{8}$$

$x_1 = \psi + \arcsin\left(\dfrac{ky}{\sqrt{1+(ky)^2}}\right)$, $\eta_1 = \dfrac{k}{1+(ky)^2}U\sin\psi$, $\eta_3 = -\dfrac{1}{T_E}x_3$, $x_2 = r$,

$x_3 = \delta$, $g_1 = 1$,

$g_3 = \dfrac{K_E}{T_E}$, $\bar{x}_i = [x_1, x_2, x_3]^T$, $d_1 = d_3 = 0$, $d_2 = w$, $u = \delta_E$ and

$x = [x_1, x_2, x_3]^T \in R^q$ is the system state vector. $u, y \in R$ are the system's input and the output. Now we quote some assumptions.

Assumption 1: The unknown virtual control-gain function g_i are strictly either positive or negative, its absolute value is positive, for the purpose of analysis easier, without loss of generality, we make further assumption that

$$0 < g_{\min} \leq g_i \leq g_{\max} \tag{9}$$

while $g_{\min}$ and $g_{\max}$ are the lower and upper bound of $|g_i|$, respectively.

Assumption 2: $y_r(t)$ is a sufficiently smooth function of t, and y_r, $\dot{y}_r$, and $\ddot{y}_r$ are bounded, that means, there exists a positive constant Y_0 such that:

$$\Pi_0 = \left\{ (y_r, \dot{y}_r, \ddot{y}_r) : (y_r)^2 + (\dot{y}_r)^2 + (\ddot{y}_r)^2 < Y_0 \right\}.$$

Assumption 3: $|w_i|$ is bounded, that is, there exists a positive unknown constant γ_i, satisfying $|w_i| < \gamma_i, i = 1, \cdots, n$.

4   Control Design and Stability Analysis

4.1   Control Design

Step1: Consider the first equation $\dot{x}_1 = \eta_1 + g_1 x_2 + w_1$, and define the tracking error variable $s_1 = x_1 - y_r$. According to Lemma 1 given above, we have:

$$\dot{s}_1 = g_1 x_2 + \eta_1(x_1) + w_1 - \dot{y}_r \tag{10}$$

$$\eta_1(x_1) = S_1(x_1) A_1 x_1 + \varepsilon_1 = S(x_1) A_1 s_1 + S(x_1) A_1 y_r + \varepsilon_1 \tag{11}$$

where ε_1 denote the approximation error. Let $b_1 = \|A_1\|$, the normalized term, $A_1^m = \dfrac{A_1}{\|A_1\|} = \dfrac{A_1}{b_1}$ and $\theta_1 = A_1^m s_1$. Substituting (12) into (10), then we get:

$$\eta_1(x_1) = b_1 S_1(x_1) \theta_1 + S_1(x_1) A_1 y_r + \varepsilon_1 \tag{12}$$

$$\dot{s}_1 = g_1 x_2 + b_1 S_1 \theta_1 + \Delta_1 - \dot{y}_r \tag{13}$$

where $\Delta_1 = S_1(x_1) A_1 y_r + \varepsilon_1 + w_1$. According to the Assumption above mentioned, we have:

$$\|\Delta_1\| \leq \|S_1(x_1) A_1 y_r + \varepsilon_1^* + w_1\| \leq g_{\min} \xi_1 \varphi_1(x_1) \tag{14}$$

where, $\xi_1 = g_{\min}^{-1} \max(\|A_1 y_r\|, \|\varepsilon_1^* + w_1\|)$ and $\varphi(x_1) = 1 + \|S_1\|$. It is obvious that $\|\Delta\|$ is bounded, because the bound of y_r, ε_1^* and w_1. Then choose a virtual controller α_2 for x_2.

$$\alpha_2 = -k_1 s_1 + \dot{y}_r - \frac{\hat{\lambda}_1}{4\gamma_1^2} S_1(x_1) S_1^T(x_1) s_1 - \hat{\vartheta}_1 \varphi_1(x_1) \tanh(\frac{\hat{\vartheta}_1 \varphi_1(x_1) s_1}{\delta_1}) \tag{15}$$

For the sake of avoiding calculation explosion caused by repeated derivations, the DSC technique is bring in, and virtual control law is replaced by its estimation using the following first-order filter with the time constant of τ_2.

$$\tau_2 \dot{z}_2 + z_2 = \alpha_2, z_2(0) = \alpha_2(0) \tag{16}$$

By defining the output error of this filter as $y_2 = z_2 - \alpha_2$, it yields:

$$\dot{y}_2 = \dot{z}_2 - \dot{\alpha}_2 = -\frac{y_2}{\tau_2} + (-\frac{\partial \alpha_2}{\partial s_1} \dot{s}_1 - \frac{\partial \alpha_2}{\partial x_1} \dot{x}_1 - \frac{\partial \alpha_2}{\partial \hat{\lambda}} \dot{\hat{\lambda}} - \frac{\partial \alpha_2}{\partial \hat{\vartheta}} \dot{\hat{\vartheta}} + \ddot{y}_d) = -\frac{y_2}{\tau_2} + B_2(s_1, s_2, y_2, \hat{\lambda}, \hat{\vartheta}, y_r, \dot{y}_r, \ddot{y}_r) \tag{17}$$

Where $B_2(\cdot)$ is a smooth bounded function and has a maximum value M_2 [15].

Step2: Consider the second equation $\dot{x}_2 = \eta_2 + g_2 x_3 + w_2$, and define the error variable $s_2 = x_2 - z_2$, we have:

$$\dot{s}_2 = g_2 x_3 + \eta_2(x_2) + w_2 - \dot{z}_2 \tag{18}$$

$$\eta_2(x_2) = S_2(\overline{x}_2) A_2 \overline{x}_2^T + \varepsilon_2 = S_2 A_2 \begin{bmatrix} s_1 + y_r \\ s_2 + z_2 \end{bmatrix}^T + \varepsilon_2 = b_2 S_2 \theta_2 + \Delta_2' \tag{19}$$

where $b_2 = \|A_2\|$, the normalized term $A_2^m = \frac{A_2}{\|A_2\|}$, $\theta_2 = A_2^m s_2$ with $\overline{s}_i = [s_1, s_2]^T$, and $\Delta_2' = S_2(\overline{x}_2) A_2 y_r + S_2(\overline{x}_2) A_2 z_2 + \varepsilon_2$, then we have:

$$\dot{s}_2 = g_2(\overline{x}_2) x_3 + b_2 S_2(\overline{x}_2) \theta_2 + \Delta_2 - \dot{z}_2 \tag{20}$$

$$\|\Delta_2\| \le \|S_2(\overline{x}_2) A_2 y_r + S_2(\overline{x}_2) A_2 z_2 + \varepsilon_2 + w_2\| \le g_{\min} \xi_2 \varphi_2(\overline{x}_2) \tag{21}$$

where $\Delta_2 = \Delta_2' + w_2$, $\xi_2 = g_{\min}^{-1} \max(\|A_2 y_r\|, \|A_2 z_2\|, \|\varepsilon_2^* + w_2\|)$ and $\varphi_2(\overline{x}_2) = 1 + \|S_2(\overline{x}_2)\|$, also $\|\Delta_2\|$ is bounded. Then we choose virtual controller α_3 for x_3:

$$\alpha_3 = -k_2 s_2 + \dot{z}_2 - \frac{\hat{\lambda}_2}{4\gamma_2^2} S_2(\overline{x}_2) S_2^T(\overline{x}_2) s_2 - \hat{\vartheta}_2 \varphi_2(\overline{x}_2) \tanh(\frac{\hat{\vartheta}_2 \varphi_2(\overline{x}_2) s_2}{\delta_2}) \tag{22}$$

Similarly to step 1, let α_3 passing through a first-order filter with time constant τ_3

$$\tau_3 \dot{z}_3 + z_3 = \alpha_3, z_3(0) = \alpha_3(0) \tag{23}$$

$$\dot{y}_3 = \dot{z}_3 - \dot{\alpha}_3 = \frac{y_3}{\tau_3} + (\frac{\partial \alpha_3}{\partial x_2}\dot{s}_2 - \frac{\partial \alpha_3}{\partial x_2}\dot{x}_2 - \frac{\partial \alpha_3}{\partial \hat{\lambda}_2}\dot{\hat{\lambda}}_2 - \frac{\partial \alpha_3}{\partial \hat{\vartheta}_2}\dot{\hat{\vartheta}}_2 + \ddot{y}_r) = \frac{y_3}{\tau_3} + B_3(s_1, s_2, s_3, y_2, y_3, \hat{\lambda}_1, \hat{\vartheta}_1, \hat{\lambda}_2, \hat{\vartheta}_2, y_r, \dot{y}_r, \ddot{y}_r) \quad (24)$$

Step3: Consider the third equation $\dot{x}_3 = \eta_3 + g_3 u + w_3$, and define the error variable $s_3 = x_3 - z_3$, we have:

$$\dot{s}_3 = g_3 u + \eta_3 (x_3) + w_3 - \dot{z}_3 \quad (25)$$

$$\eta_3(x_3) = S_3(\overline{x}_3)A_3\overline{x}_3^{\mathrm{T}} + \varepsilon_3 = S_3 A_3 \begin{bmatrix} s_1 + y_r \\ s_2 + z_2 \\ s_3 + z_3 \end{bmatrix}^{\mathrm{T}} + \varepsilon_3 = b_3 S_3 \theta_3 + \Delta'_3 \quad (26)$$

where $b_3 = \|A_3\|$, the normalized term $A_3^m = \dfrac{A_3}{\|A_3\|}$, $\theta_3 = A_3^m s_3$ with

$\overline{s}_i = [s_1, s_2, s_3]^{\mathrm{T}}$, and $\Delta'_3 = S_3(\overline{x}_3)A_3 y_r + S_3(\overline{x}_3)A_3 z_2 + S_3(\overline{x}_3)A_3 z_3 + \varepsilon_3$, then we have:

$$\dot{s}_3 = g_3(\overline{x}_3)\mu + b_3 S_3(\overline{x}_3)\theta_3 + \Delta_3 - \dot{z}_3 \quad (27)$$

$$\|\Delta_3\| \le \|S_3(\overline{x}_3)A_3 y_r + S_3(\overline{x}_3)A_3 z_2 + S_3(\overline{x}_3)A_3 z_3 + \varepsilon_3 + w_3\| \le g_{\min}\xi_3\varphi_3(\overline{x}_3) \quad (28)$$

Where $\Delta_3 = \Delta'_3 + w_3$, $\xi_3 = g_{\min}^{-1} \max(\|A_3 y_r\|, \|A_3 z_2\|, \|A_3 z_3\|, \|\varepsilon_3^* + w_3\|)$ and

$\varphi_3(\overline{x}_3) = 1 + \|S_3(x_3)\|$, also $\|\Delta_3\|$ is bounded. Then we choose virtual control law μ as:

$$u = -k_3 s_3 + \dot{z}_3 - \frac{\hat{\lambda}_3}{4\gamma_3^2}S_3(\overline{x}_3)S_3^{\mathrm{T}}(\overline{x}_3)s_3 - \hat{\vartheta}_3\varphi_3(\overline{x}_3)\tanh(\frac{\hat{\vartheta}_3\varphi_3(\overline{x}_3)s_3}{\delta_3}) \quad (29)$$

Now the update laws of $\hat{\lambda}_i$ and $\hat{\vartheta}_i$ are given as follows:

$$\dot{\hat{\lambda}}_i = \Gamma_{i1}\left[\frac{1}{4\gamma_i^2}S_i(\overline{x}_i)S_i^{\mathrm{T}}(\overline{x}_i)s_i^2 - \sigma_{i1}(\hat{\lambda}_i - \lambda_i^0)\right], \dot{\hat{\vartheta}}_i = \Gamma_{i2}\left[\varphi_i(\overline{x}_i)\|s_i\| - \sigma_{i2}(\hat{\vartheta}_i - \vartheta_i^0)\right] \quad (30)$$

where λ_i^0 , ϑ_i^0, σ_{i1} and σ_{i2} are design parameters, k_i , γ_i , δ_i , Γ_{i1} , Γ_{i2} , σ_{i1} , and σ_{i2} are positive design constants, $\hat{\lambda}_i$ and $\hat{\vartheta}_i$ are the estimates of $\lambda_i = g_{\min}^{-1}b_i^2$ and ϑ_i, and λ_i^0 , ϑ_i^0 are the initial values of $\hat{\lambda}_i$ and $\hat{\vartheta}_i$.

4.2   Stability Analysis

[Theorem 1]: Formula (13), (20) and (27), the virtual controllers (15) and (22), the controller (29), and the updated laws (30). Given any positive number $p_i, i = 1, 2, 3$ for

all   initial   conditions
satisfying $\Pi_i=\left\{\sum_{j=1}^{i}(s_j^2+\tilde{\vartheta}_j^{\mathrm{T}}g_{\min}\tau_{ji}^{-1}\hat{\vartheta}_j+\tilde{\lambda}_j^{\mathrm{T}}g_{\min}\Gamma_{j2}^{-1}\tilde{\lambda}_j)+\sum_{j=2}^{i}y_j^2<2p_i\right\}, i=1,2,3$ there

exist $k_i, r_i, \delta_i, \tau_i, \sigma_{i1}, \sigma_{i2}, \Gamma_{i1}$, and Γ_{i2} such that all the signals in the closed-loop control system are uniformly ultimately bounded. Furthermore, given any $\mu > 0$, we can make the output error s_1 satisfies $\lim_{t\to\infty} |s_1(t)| \le \mu$

Proof: Choose Lyapunov function candidate as:

$$V = \frac{1}{2}\sum_{i=1}^{3}\left(s_i^2 + \tilde{\vartheta}_i^{\mathrm{T}}g_{\min}\Gamma_{i1}^{-1}\tilde{\vartheta}_i + \tilde{\lambda}_i^{\mathrm{T}}g_{\min}\Gamma_{i2}^{-1}\tilde{\lambda}_i\right) + \frac{1}{2}\sum_{i=1}^{2}y_{i+1}^2, i=1,2,3 \qquad (31)$$

Where $\tilde{\vartheta}_i = \vartheta_i - \hat{\vartheta}_i$, and $\tilde{\lambda}_i = \lambda_i - \hat{\lambda}_i$. By mentioning $x_{i+1} = s_{i+1} + z_{i+1}$ and $z_{i+1} = y_{i+1} + \alpha_{i+1}$, the time derivative of V along the system trajectories is:

$$\dot{V} = \sum_{i=1}^{n}\left(s_i\dot{s}_i - \vartheta_i^{\mathrm{T}}g_{\min}\Gamma_{i2}^{-1}\hat{\vartheta}_i - \lambda_i^{\mathrm{T}}g_{\min}\Gamma_{i1}^{-1}\hat{\lambda}\right) + \sum_{i=1}^{2}y_{i+1}\dot{y}_{i+1}$$

$$\le \sum_{i=1}^{n-1}\left(-g_{\min}k_is_i^2 + b_iS_i(\overline{x}_i)\theta_is_i + \Delta_is_i - g_{\min}\frac{\tilde{\lambda}_i}{4\gamma_i^2}S_iS_i^{\mathrm{T}}s_i^2 - \tilde{\lambda}_i^{\mathrm{T}}g_{\min}\Gamma_{i2}^{-1}\hat{\lambda}_i + g_iy_{i+1}s_i + g_is_{i+1}s_i - \tilde{\vartheta}_i^{\mathrm{T}}g_{\min}\Gamma_{i1}^{-1}\hat{\vartheta}_i\right.$$

$$\left. -g_i\hat{\vartheta}_i\varphi_is_i\tanh\left(\frac{\hat{\vartheta}_i\varphi_is_i}{\delta_i}\right)\right) - g_{\min}k_ns_n^2 + \sum_{i=2}^{n}(g_i\dot{z}_is_i - \dot{z}_is_i) + (g_i\dot{y}_ds_1 - \dot{y}_ds_1) + b_nS_n(x)\theta_ns_n + \Delta_ns_n \qquad (32)$$

$$-g_{\min}\frac{\tilde{\lambda}_n}{4\gamma_n^2}S_nS_n^{\mathrm{T}}s_n^2 - \tilde{\lambda}_n^{\mathrm{T}}g_{\min}\Gamma_{n2}^{-1}\hat{\lambda}_n - g_n\hat{\vartheta}_n\varphi_ns_n\tanh\left(\frac{\hat{\vartheta}_n\varphi_ns_n}{\delta_n}\right) - \tilde{\vartheta}_n^{\mathrm{T}}b_{\min}\Gamma_{n1}^{-1}\hat{\vartheta}_n + \sum_{i=1}^{n-1}\left(-\frac{y_{i+1}^2}{\tau_{i+1}} + |y_{i+1}B_{i+1}|\right)$$

It is worth noticing that:

$$b_iS_i\theta_is_i = b_iS_i\theta_is_i - \gamma_i^2\theta_i^{\mathrm{T}}\theta_i + \gamma_i^2\theta_i^{\mathrm{T}}\theta_i \le g_{\min}\frac{\hat{\lambda}_i}{4\gamma_i^2}S_iS_i^{\mathrm{T}}s_i^2 + g_{\min}\frac{\tilde{\lambda}_i}{4\gamma_i^2}S_iS_i^{\mathrm{T}}s_i^2 + \gamma_i^2\theta_i^{\mathrm{T}}\theta_i \qquad (33)$$

$$\Delta_is_i \le g_i\hat{\vartheta}_i\varphi_i(\overline{x}_i)\|s_i\| + g_{\min}\tilde{\vartheta}_i\varphi_i(\overline{x}_i)\|s_i\| \qquad (34)$$

$$g_i\dot{z}_is_i - \dot{z}_is_i \le \frac{1+g_{\max}}{\tau_i}s_i^2 + \frac{g_{\max}+1}{4\tau_i}y_i^2 \qquad (35)$$

$$g_1\dot{y}_rs_1 - \dot{y}_rs_1 \le \frac{b_{\max}+1}{4}s_1^2 + (g_{\max}+1)B_0^2 \qquad (36)$$

Then we have:

$$\tilde{\vartheta}_i^{\mathrm{T}}(\hat{\vartheta}_i - \vartheta_i^0) \ge \frac{1}{2}|\tilde{\vartheta}_i|^2 - \frac{1}{2}|\vartheta_i^* - \vartheta_i^0|^2 \qquad (37)$$

$$\dot{V} \le \sum_{i=2}^{n-1}(-(g_{min}k_i-2\frac{1+g_{max}}{\tau_i})s_i^2+\frac{g_{max}}{4}s_{i+1}^2)-(g_{min}k_1-2\frac{g_{max}+1}{4})s_1^2-(g_{min}k_n-\frac{1+g_{max}}{\tau_i})s_n^2+\sum_{i=2}^{n}(\gamma_i^2\theta_i^T\theta_i+\delta_i)$$

$$-\sum_{i=1}^{n}\frac{\sigma_{i2}}{2\lambda_{max}(g_{min}\Gamma_{i1}^{-1})}\tilde{\vartheta}_i^T\Gamma_{i1}^{-1}\tilde{\vartheta}_i)-\sum_{i=1}^{n}\frac{\sigma_{i1}}{2\lambda_{max}(g_{min}\Gamma_{i2}^{-1})}\tilde{\lambda}_i^T\Gamma_{i2}^{-1}\tilde{\lambda}_i)+\sum_{i=1}^{n-1}(\frac{g_{max}}{4}y_{i+1}^2-\frac{(3-g_{max})}{4\tau_{i+1}}y_{i+1}^2+|y_{i+1}B_{i+1}|) \quad (38)$$

where $\quad \delta_i=(g_{max}+1)B_0^2+g_{max}\delta_i+(\sigma_{i1}/2)|\tilde{\lambda}_i-\lambda_i^*|^2+(\sigma_{i2}/2)|\tilde{\vartheta}_i-\vartheta_i^*|^2 \quad$ Since   the

sets $\Pi_0 \in R^3$ and $\Pi_i \in R^{\sum_{j=1}^{i}(N_j+2i-1)}$, where N_j is the dimension of $\tilde{\vartheta}_j$, are

compact. $\Pi_0 \times \Pi_i \in R^{\sum_{j=1}^{i}(N_j+2i+2)}$ is also compact. Therefore $|B_{i+1}|$ has a

maximum $\quad M_{i+1} \quad$ on $\quad \Pi_0 \times \Pi_i \quad$. $\quad$ Then $\quad$ let

$1/\tau_{i+1}=(3-b_{max}/4)^{-1}((b_{max}/4)+(M_{i+1}^2/2\alpha)+\alpha_0)$, $\quad$ and $\quad$ note $\quad$ that

$|B_{i+1}y_{i+1}|\le(y_{i+1}^2B_{i+1}^2/2\alpha)+(\alpha/2)$, where α_0 and α are positive constants. Then, we arrive at:

$$\frac{g_{max}}{4}y_{i+1}^2-\frac{(3-g_{max})}{4\tau_{i+1}}y_{i+1}^2+|B_{i+1}y_{i+1}|\le-(\frac{g_{max}}{4}+\frac{M_{i+1}^2}{2\alpha}+\alpha_0)y_{i+1}^2+\frac{g_{max}}{4}y_{i+1}^2+\frac{M_{i+1}^2y_{i+1}^2B_{i+1}^2}{2\alpha M_{i+1}^2}+\frac{\alpha}{2}\le-\alpha_0y_{i+1}^2+\frac{\alpha}{2} \quad (39)$$

Setting $\qquad \sigma_{i1}/2\lambda_{max}(g_{min}\Gamma_{i1}^{-1})=\sigma_{i2}/2\lambda_{max}(g_{min}\Gamma_{i2}^{-1})=\alpha_0 \qquad$ and

$k_1=g_{min}^{-1}(2+(g_{max}+1/4)+\alpha_0)$,

$k_i=g_{min}^{-1}(2+(1+g_{max}/\tau_i)+(g_{max}/4)+\alpha_0)(i=2,...,n-i)$ and

$k_n=g_{min}^{-1}((g_{max}+1/\tau_i)+\alpha_0)$ then (37) can be expressed as:

$$\dot{V}\le-\alpha_0\sum_{i=1}^{n}s_i^2-\alpha_0\sum_{i1}^{n}(\tilde{\vartheta}_i^Tg_{min}\Gamma_{i1}^{-1}\tilde{\vartheta}_i+\tilde{\lambda}_i^Tg_{min}\Gamma_{i2}^{-1}\tilde{\lambda}_i)-\alpha_0\sum_{i=1}^{n-1}y_{i+1}^2+\sum_{i=1}^{n}(\gamma_i^2\theta_i^T\theta_i)+\rho\le-2\alpha_0V+\gamma^2\|\theta\|^2+\rho \quad (40)$$

Note that $\|\theta_i\|\le\|A\|\|S\|=\gamma'\|S\|$, and $\gamma\gamma'<1$, then

$$\dot{V}\le-2a_0+\gamma'^2\gamma^2\|s\|^2+\rho\le-2a_0+\|s\|^2+\rho\le-c_1V+\rho \quad (41)$$

Where $c_1=(2a_0-1)$ and we have $V(t)\le\frac{\rho}{c_1}+(v(t)-\frac{\rho}{c_1})e^{-(t-t_0)}$. It

follows that, for any $\mu_1>\left(\rho/c_1\right)^{1/2}$, there exists a constant $T>0$ such that

$\|s_1(t)\|\le\mu_1$ for all $t\ge t_0+T$, and the tracking error can be made small

since $\left(\rho/c_1\right)^{1/2}$ can arbitrarily be made small if the design parameters

$k_1,\gamma_1,\delta_1,\tau_2,\sigma_{11},\sigma_{12},\Gamma_{11}$, and Γ_{12} are appropriately chosen.

5   Application Examples

In this section, oceangoing training vessel YULONG is used as the simulation example. The initial conditions for x_0, y_0, ψ_0 and r_0 are $[0\text{m}, 200\text{m}, -0.2, 0]$. The desired reference signal $y_r = 0$, and design parameters are chosen as $k = 0.002$, $k_1 = 0.3$, $k_2 = 5$, $k_2 = 5$. The external disturbance signal is chosen as $w = 0.001\,(1 + \sin(\,0.1t))$. $\Gamma_{11} = \Gamma_{21} = \Gamma_{31} = 0.2$, $\sigma_{11} = \sigma_{12} = \sigma_{13} = 0.005$, $\tau_2 = \tau_3 = 0.5$. The initial values of the weights vectors $\hat{\theta}_i^0$, $\hat{\lambda}_i^0$, $i = 1, 2, 3$ are zero. The simulation results with Figs. 1~2 illustrate the effectiveness of the proposed scheme.

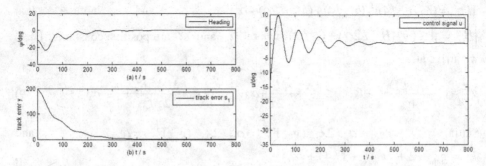

Fig. 1. (a) heading ψ, (b) cross-track error y   **Fig.2.** rudder signal u

6   Conclusion

In this paper, ship linear-tracking control has been considered. By using RBF neural networks and dynamic surface control, a scheme of adaptive backstepping control has been developed for ship's nonlinear motion system. The proposed scheme can force the ship to follow a desired path and ensure the uniformly ultimate boundedness of the closed-loop system. At last, simulation results on ocean-going vessel "YULONG" validate the effectiveness of the proposed scheme.

References

1. Luo, L.W., Zou, J.Z., Li, S.T.: Robust tracking control of nonlinear ship steering. Control Theory and Applications 26, 894–895 (2009)
2. Breivik, M., Fossen, I.T.: Path Following of Straight Lines and Circles for Marine surface Vessels. In: Proc. of the IFAC CAMS 2004, Ancona, Italy, pp. 65–70 (2004)
3. Do, D.K., Jiang, P.Z., Pan, J.: Robust global stabilization of understand ships on a linear course: state and output feedback. Int. J. Control 76, 1–17 (2003)

4. Pettersen, Y.K., Lefeber, E.: Way-point tracking control of ships. In: Proc. of 40th IEEE CDC, Orlando, USA, pp. 940–945 (2001)
5. Li, S.T., Yang, S.Y., Zheng, F.Y.: Input-output linearization design for straight-line tracking control of underactuated ships. Systems Engineering and Electronics, 945–948 (2004)
6. Bu, R.X., Liu, Z.J., Li, T.S.: Increment feedback control algorithm of ship track based on nonlinear sliding mode. Journal of Traffic and Transportation Engineering, 75–79 (2006)
7. Ghommam, J., Mnif, F., Benali, A., Derbel, N.: Nonsingular Serret-Frenet based path following control for an underactuated surface vessel. Journal of Dynamic Systems, Measurement and Control 131, 1–8 (2009)
8. Yang, L.S.T., Hong, S.Y., Robust, B.G.: adaptive fuzzy design for ships track-keeping control. J. Control Theory and Applications, 445–448 (2007)
9. Xu, H.J., Liu, J.Y.: the application of Hybrid intelligence system for ship-tracking control. Master Dissertation of Shanghai Maritime University, 53–61 (2006)
10. Velagic, J., Vukic, Z., Omerdic, E.: Adaptive fuzzy ship autopilot for track-keeping. Control Engineering Practice, 433–443 (2003)
11. Wang, S.J., Liu, S.: Study of Intelligent Control System of Ship-track. Master Dissertation of Harbin Engineering University, 39–57 (2008)
12. Li, T.S., Yu, B., Hong, B.G.: A Novel Adaptive Fuzzy Design for Path Following for Underactuated Ships with Actuator Dynamics. In: ICIEA, pp. 2796–2800 (2009)
13. Yang, Y., Li, T., Wang, X.-F.: Robust Adaptive Neural Network Control for Strict-Feedback Nonlinear Systems Via Small-Gain Approaches. In: Wang, J., Yi, Z., Żurada, J.M., Lu, B.-L., Yin, H. (eds.) ISNN 2006. LNCS, vol. 3972, pp. 888–897. Springer, Heidelberg (2006)
14. Lin, W., Qian, C.: Adaptive control of nonlinear parameterized systems: The smooth feedback cade. IEEE Trans. Autom. Control, 1249–1266 (2002)
15. Wang, D., Hang, J.: Neural network-based adaptive dynamic surface control for a class of uncertain nonlinear systems in strict-feedback form. IEEE Trans. Neural None, 195–202 (2005)

Direct Robust Adaptive NN Tracking Control
for Double Inverted Pendulums

Wenlian Yang[1], Ye Tao[1], and Tieshan Li[2]

[1] Educational Technology & Computing Center, Dalian Ocean University, Dalian, P.R. China
lotusyangwl@163.com,
taoye@dlou.edu.cn
[2] Navigation College, Dalian Maritime University, Dalian, P.R. China
tieshanli@126.com

Abstract. In this paper, based on Lyapunov stability theory and Backstepping technique, a novel direct robust adaptive neural network (NN) controller is proposed for double inverted pendulums (DIPs). By incorporating dynamic surface control (DSC) technique into a neural network based adaptive control design framework, the control design is achieved. The problem of "explosion of complexity" inherent in the conventional backstepping method is avoided, and the controller singularity problem is removed completely by utilizing a special property of the affine term. In addition, it is proved that all the signals in the closed-loop system are bounded and the tracking error converges to a small neighborhood of the origin. Finally, simulation results for the trajectory tracking of the DIPs are given to demonstrate the effectiveness of the proposed scheme.

Keywords: Double inverted pendulums, adaptive control, neural networks, dynamic surface control.

1   Introduction

The design of non-linear controllers is important for their application because most existing physical systems are non-linear. The double inverted pendulums system is the most typical representation. Due to its unique characteristics, it is used to verify new control design methods in many of control laboratories, especially for nonlinear interconnected systems. The studies of inverted pendulum (s) has been studied greatly for investigating effectiveness of various kinds of control schemes and demonstrating ideas emerging in the area of non-linear control. The control objective consists of swinging up and balancing it (them) about the vertical [1-5], and the references therein. But they are not applicable when the system model behaves with complicated uncertainty or completely unknown structures. Recently, based on NNs and DSC [6], a novel decentralized indirect adaptive neural control scheme was proposed for DIPs in [7], and both problems of "dimension curse" and "explosion of complexity" were avoided.

Motivated by the above observations, a direct robust adaptive NN controller is developed for DIPs based on DSC technique. The simulation experimental results demonstrate the effectiveness of the proposed scheme.

T. Huang et al. (Eds.): ICONIP 2012, Part V, LNCS 7667, pp. 428–436, 2012.
© Springer-Verlag Berlin Heidelberg 2012

2   Problem Formulation

In this paper, we will consider the trajectory tracking problem of DIPs connected by a spring, described by the following TITO systems [8]:

$$\dot{x}_{1,1} = x_{1,2},$$

$$\dot{x}_{1,2} = \frac{m_1 g r}{J_1}\sin(x_{1,1}) - \frac{k}{J_1}x_{1,1} + \frac{u_1}{J_1} + \frac{k}{J_1}x_{2,1} + \frac{v_1}{J_1}$$

$$\dot{x}_{2,1} = x_{2,2},$$

$$\dot{x}_{2,2} = \frac{m_2 g r}{J_2}\sin(x_{2,1}) - \frac{k}{J_2}x_{2,1} + \frac{u_2}{J_2} + \frac{k}{J_2}x_{1,1} + \frac{v_2}{J_2}$$

(1)

where, each pendulum may be positioned by a torque input u_1, $i = 1,2$, applied by a servomotor at its base. v_i, $i = 1,2$, are the torque disturbances. It is assumed that both θ_i and $\dot{\theta}_i$ (angular position and rate) are available to the i th controller for $i = 1,2$. $x_{i,1} = \theta_i$ being the angular displacement of the i th pendulums from the vertical reference, and m_i, J_i being the end masses of the pendulums and the moment of inertia, respectively. k is the constant of the connecting torsional spring. r denotes the pendulum length. $g = 9.81$ is the gravitational acceleration.

The objective is to drive the angular position of each pendulum to track a reference signal $y_{i,d}$, $i = 1,2$ under the torque disturbances [7].

Before the beginning of controller design, we convert the system (1) into a more general TITO form as follows

$$\dot{x}_{i,1} = f_{i,1}(x_{i,1}) + g_{i,1}(x_{i,1})x_{i,2} + \Delta_{i,1}(t,x)$$

$$\dot{x}_{i,2} = f_{i,2}(\overline{x}_{i,2}) + g_{i,2}(\overline{x}_{i,2})u_i + \Delta_{i,2}(t,x)$$

$$y_{i,1} = x_{i,1}, i = 1,2$$

(2)

where $\overline{x}_{i,2} = [x_{1,1}, x_{1,2}, x_{2,1}]^T$, $x = [x_{1,1}, x_{1,2}, x_{2,1}, x_{2,2}]^T$ are vectors of the system states. $y_{i,1}$ is the output of the system. This general system (2) will be used as a design model later. And we assumed that $f_{i,j}$ and $g_{i,j}$ with $i = 1,2$, $j = 1,2$ represent unknown non-linear smooth functions, respectively, where $g_{i,j}$ is referred to as virtual control gain function.

In this paper, for the development of control laws, the following assumptions are imposed on the system (2).

Assumption 1. The reference signal $y_{i,d}(t)$ is a sufficiently smooth function of t, $y_{i,d}(t)$, $\dot{y}_{i,d}(t)$ and $\ddot{y}_{i,d}(t)$ are bounded, that is, there exists a known positive constant $B_{i,0}$, such that $\Pi_{i,0} := \{(y_{i,d}, \dot{y}_{i,d}, \ddot{y}_{i,d}) : y_{i,d}^2 + \dot{y}_{i,d}^2 + \ddot{y}_{i,d}^2 \leq B_{i,0}\}$ with $i = 1,2$.

Assumption 2. The disturbances $\Delta_{i,j}$ with $i = 1, 2$, $j = 1, 2$ are bounded by some unknown constants $d_{i,j}$, respectively, that is, $\left| \Delta_{i,j} \right| \le d_{i,j}$.

Assumption 3. $g_{i,j}$, $i = 1, 2$ is an unknown virtual control gain function. The signs of $g_{l,j}$ are known, and there exist constants $g_{i,j1} \ge g_{i,j0} > 0$ such that $g_{i,j1} \ge \left| g_{i,j} \right| \ge g_{i,j0}$. There exist constants $g_{i,jd} > 0$ such that $\left| \dot{g}_{i,j} \left(\overline{x}_{i,j} \right) \right| \le g_{i,jd}$, $\forall \overline{x}_{i,j} \in \Omega \subset R^n$.

The above assumption implies that smooth functions $g_{i,j}$ are strictly either positive or negative. Without losing generality, we assume $g_{i,j1} \ge g_{i,j} \left(\overline{x}_{i,j} \right) \ge g_{i,j0} > 0$.

3   NN-Based Controller Design and Stability Analysis

3.1   Controller Design

For simplicity, the procedure of the controller design for $i = 1$ pendulum subsystem in (2) is developed as follows. Similar procedure for $i = 2$ is omitted.

Step 1: Define error variable $z_{1,1} = x_{1,1} - y_{1,d}$, then, the time derivative of $z_{1,1}$ is

$$\dot{z}_{1,1} = f_{1,1} \left(x_{1,1} \right) + g_{1,1} \left(x_{1,1} \right) x_{1,2} + \Delta_{1,1} - \dot{y}_{1,d} \tag{3}$$

Here, we define an unknown function $h_{1,1}(\cdot) = \dfrac{1}{g_{1,1}} \left(f_{1,1}(x_{1,1}) - \dot{y}_{1,d} \right)$. Based on the well known universal approximation of an RBF NN to any continuous function over a compact set, construct an RBF NN to approximate it in the compact set $\Omega_{x_{1,1}} \in R^1$, as follows:

$$h_{1,1}(Z_{1,1}) = \frac{1}{g_{1,1}} \left(f_{1,1}(x_{1,1}) - \dot{y}_{1,d} \right) = \theta_{1,1}^{*T} \xi_{1,1} \left(Z_{1,1} \right) + \delta_{1,1}^{*} \tag{4}$$

where $Z_{1,1} \overset{\Delta}{=} \left[x_{1,1}, \dot{y}_{1,d} \right]^T \subset R^2$. $\theta_{1,1}^{*}$ is the weight vector of RBF NN, $\xi_{1,1} \left(Z_{1,1} \right)$ is the Gaussian functions vector, and $\delta_{1,1}^{*}$ represents the network reconstruction error with an assumption of $\left| \delta^{*} \right| \le \delta_m$, where the unknown constant $\delta_m > 0$.

Now, choose the intermediate stabilizing function $\alpha_{1,2}$ of $x_{1,2}$ for the first subsystem as follows

$$\alpha_{1,2} = -c_{1,1} z_{1,1} - \hat{\theta}_{1,1}^T \xi_{1,1}(Z_{1,1}) - D_{1,1} \tanh(\frac{D_{1,1} z_{1,1}}{s_{1,1}}) \tag{5}$$

where design parameters $c_{1,1}$, $s_{1,1} > 0$. $D_{1,1} = d_{1,1} / g_{1,10}$, $d_{1,1}$ is the known upper limit of $\Delta_{1,1}$. $\hat{\theta}_{1,1}$ is the estimation of $\theta_{1,1}^*$ and is updated as follows:

$$\dot{\hat{\theta}}_{1,1} = \Gamma_{1,1}\left[\xi_{1,1}(Z_{1,1})z_{1,1} - \sigma_{1,1}\hat{\theta}_{1,1}\right] \tag{6}$$

with constant matrix $\Gamma_{1,1} = \Gamma_{1,1}^T > 0$, $\sigma_{1,1} > 0$ is a small constant.

To avoid repeatedly differentiating $\alpha_{1,2}$, which leads to the so called "explosion of complexity" in the sequel steps, the DSC technique first proposed in [6] is employed here. Introduce a first-order filter $\beta_{1,2}$, and let $\alpha_{1,2}$ pass through it with time constant $\tau_{1,2}$, i.e.,

$$\tau_{1,2}\dot{\beta}_{1,2} + \beta_{1,2} = \alpha_{1,2} \quad \beta_{1,2}(0) = \alpha_{1,2}(0) \tag{7}$$

Define $z_{1,2} = x_{1,2} - \beta_{1,2}$, then

$$\dot{z}_{1,1} = f_{1,1}(x_{1,1}) + g_{1,1}(x_{1,1})(z_{1,2} + \beta_{1,2}) + \Delta_{1,1} - \dot{y}_{1,d} \tag{8}$$

and define

$$\eta_{1,2} = \beta_{1,2} - \alpha_{1,2} = \hat{\theta}_{1,1}^T\xi_{1,1}(Z_{1,1}) + c_{1,1}z_{1,1} + D_{1,1}\tanh(\frac{D_{1,1}z_{1,1}}{s_{1,1}}) + \beta_{1,2} \tag{9}$$

Substituting (5) and (9) into (8), one has

$$\dot{z}_{1,1} = g_{1,1}(x_{1,1})(z_{1,2} - \tilde{\theta}_{1,1}^T\xi_{1,1}(Z_{1,1}) - c_{1,1}z_{1,1} - D_{1,1}\tanh(\frac{D_{1,1}z_{1,1}}{s_{1,1}}) + \delta_{1,1}^* + \eta_{1,2}) + \Delta_{1,1} \tag{10}$$

Also, from the definition of $\eta_{1,2} = \beta_{1,2} - \alpha_{1,2}$, one has

$$\begin{aligned}
\dot{\eta}_{1,2} &= \dot{\beta}_{1,2} - \dot{\alpha}_{1,2} \\
&= -\frac{\eta_{1,2}}{\tau_{1,2}} + \left(-\frac{\partial\alpha_{1,2}}{\partial x_{1,1}}\dot{x}_{1,1} - \frac{\partial\alpha_{1,2}}{\partial z_{1,1}}\dot{z}_{1,1} - \frac{\partial\alpha_{1,2}}{\partial\hat{\theta}_{1,1}}\dot{\hat{\theta}}_{1,1} - \frac{\partial\alpha_{1,2}}{\partial y_{1,d}}\dot{y}_{1,d} - \frac{\partial\alpha_{1,2}}{\partial\dot{y}_{1,d}}\ddot{y}_{1,d}\right) \\
&= -\frac{\eta_{1,2}}{\tau_{1,2}} + B_{1,2}(z_{1,1},z_{1,2},\eta_{1,2},\hat{\theta}_{1,1},y_{1,d},\dot{y}_{1,d},\ddot{y}_{1,d})
\end{aligned} \tag{11}$$

where $B_{1,2}(\cdot)$ is a continuous function and has a maximum value $M_{1,2}$ (please refer to [7] for details).

Remark 1. $\dot{\beta}_{1,2}$ with a simple and concise form, derived from the DSC approach, will appear in the following step instead of $\dot{\alpha}_{1,2}$, which avoids repeated differentiations of $\alpha_{1,2}$. Thus, the substantial problem of "explosion of complexity" within the conventional backstepping technique is circumvented [7].

Step 2: Define $z_{1,2} = x_{1,2} - \beta_{1,2}$, then

$$\dot{z}_{1,2} = f_{1,2}\left(\overline{x}_{1,2}\right) + g_{1,2}\left(\overline{x}_{1,2}\right)u_1 + \Delta_{1,2} - \dot{\beta}_{1,2} \tag{12}$$

Similarly, we construct an RBF NN to approximate the unknown function as follows:

$$h_{1,2}(Z_{1,2}) = \frac{1}{g_{1,2}}\left(f_{1,2}(x_{1,2}) - \dot{\beta}_{1,2}\right) = \theta_{1,2}^{*T}\xi_{1,2}\left(Z_{1,2}\right) + \delta_{1,2}^{*} \tag{13}$$

where $Z_{1,2} \overset{\Delta}{=} \left[x_{1,1}, x_{1,2}, x_{2,1}, \dot{\beta}_{1,2}\right]^{T} \subset R^{4}$.

Choose the actual controller as follows

$$u_1 = -c_{1,2}z_{1,2} - \hat{\theta}_{1,2}^{T}\xi_{1,2}(Z_{1,2}) - D_{1,2}\tanh(\frac{D_{1,2}z_{1,2}}{s_{1,2}}). \tag{14}$$

where design parameters $c_{1,2}$, $s_{1,2} > 0$. $D_{1,2} = d_{1,2} / g_{1,20}$, $d_{1,2}$ is the known upper limit of $\Delta_{1,2}$. $\hat{\theta}_{1,2}$ is the estimation of $\theta_{1,2}^{*}$ and is updated as follows:

$$\dot{\hat{\theta}}_{1,2} = \Gamma_{1,2}\left[\xi_{1,2}(Z_{1,2})z_{1,2} - \sigma_{1,2}\hat{\theta}_{1,2}\right] \tag{15}$$

with constant matrix $\Gamma_{1,2} = \Gamma_{1,2}^{T} > 0, \sigma_{1,2} > 0$.

Substituting (13)-(15) into (12), yields

$$\dot{z}_{1,2} = g_{1,2}\left(\overline{x}_{1,2}\right)(-\tilde{\theta}_{1,2}^{T}\xi_{1,2}(Z_{1,2}) - c_{1,2}z_{1,2} - D_{1,2}\tanh(\frac{D_{1,2}z_{1,2}}{s_{1,2}}) + \delta_{1,2}^{*}) + \Delta_{1,2} \tag{16}$$

3.2   Stability Analysis

Now, we are in a position to state our main result in this paper.

Theorem 1. Consider the closed-loop system composed of (10), (16), the virtual controllers (5), the final controller (14), and the updated laws (6), (15), given δ_m, let $\theta_{i,j}^{*} \in R^{N_j}, i = 1, 2, j = 1, 2$, be such that (4) and (13) holds in the compact set $\Omega_{xi,j} \in R^{j}$ with $\left|\delta_{i,j}^{*}\right| \le \delta_m$. Given any positive number p_i, for all initial conditions satisfying

$\Pi_i := \left\{ \sum_{j=1}^{i} (z_{i,j}^2 / g_{i,j}(\overline{x}_{i,j}) + \tilde{\theta}_{i,j}^T \Gamma_{i,j}^{-1} \tilde{\theta}_{i,j}) + \eta_{i,2}^2 \right\} \le 2p_i$, $i = 1, 2, j = 1, 2$, there exist $c_{i,j}, \tau_{i,2}, \lambda_{\max} \left\{ \Gamma_{i,j} \right\}, s_{i,j}$ and $\sigma_{i,j}$ such that all the signals in the closed-loop system are bounded. Furthermore, given any $\mu_{i,1} > 0$, we can tune our controller parameters such that the output error $z_{i,1} = y_{i,1}(t) - y_{i,d}(t)$ satisfies $\lim_{t \to \infty} |z_{i,1}(t)| = \mu_{i,1}$.

Proof. Choose the Lyapunov function candidate as

$$V = \frac{1}{2} \sum_{j=1}^{2} z_{i,j}^2 / g_{i,j}(\overline{x}_{i,j}) + \frac{1}{2} \sum_{j=1}^{2} \tilde{\theta}_{i,j}^T \Gamma_{i,j}^{-1} \tilde{\theta}_{i,j} + \frac{1}{2} \eta_{i,2}^2 \tag{17}$$

Then, the time derivative of V along the system trajectories is

$$\dot{V} = \sum_{j=1}^{2} \left(z_{i,j} \dot{z}_{i,j} / g_{i,j}(\overline{x}_{i,j}) - \dot{g}_{i,j}(\overline{x}_{i,j}) z_{i,j}^2 / 2 g_{i,j}^2(\overline{x}_{i,j}) \right) + \sum_{j=1}^{2} \tilde{\theta}_{i,j}^T \Gamma_{i,j}^{-1} \dot{\hat{\theta}}_{i,j} + \eta_{i,2} \dot{\eta}_{i,2}$$

$$\le -c_{i,1} z_{i,1}^2 + z_{i,1} z_{i,2} + z_{i,1} \eta_{i,2} + z_{i,1} \delta_{i,1}^* - \dot{g}_{i,1}(\overline{x}_{i,1}) z_{i,1}^2 / 2 g_{i,1}^2(\overline{x}_{i,1})$$

$$+ z_{i,2} \delta_{i,2}^* - c_{i,2} z_{i,2}^2 - \dot{g}_{i,2}(\overline{x}_{i,2}) z_{i,2}^2 / 2 g_{i,2}^2(\overline{x}_{i,2}) + \sum_{j=1}^{2} \left(-z_{i,j} D_{i,j} \tanh(\frac{D_{i,j} z_{i,j}}{s_{i,j}}) \right) \tag{18}$$

$$+ \sum_{j=1}^{2} \left((z_{i,j} d_{i,j}) / g_{i,j0}(\overline{x}_{i,j}) \right) + \left(-\frac{\eta_{i,2}^2}{\tau_{i,2}} + |\eta_{i,2} B_{i,2}| \right)$$

$$+ \sum_{j=1}^{2} \left(-\tilde{\theta}_{i,j}^T \left(\xi_{i,j}(Z_{i,j}) z_{i,j} - \Gamma_{i,j}^{-1} \dot{\hat{\theta}}_{i,j} \right) \right)$$

According to the fact $0 \le |\phi| - \phi \tanh(\frac{\phi}{\varepsilon}) \le 0.2785\varepsilon$ with $\varepsilon > 0$ and $\phi \in R$, and after a series of manipulations, one has

$$\dot{V} \le -c_{i,11} z_{i,1}^2 + z_{i,1} z_{i,2} + z_{i,1} \eta_{i,2} + z_{i,1} \delta_{i,1}^* - \left(c_{i,10} + z_{i,1}^2 \dot{g}_{i,1}(\overline{x}_{i,1}) z_{i,1}^2 / 2 g_{i,1}^2(\overline{x}_{i,1}) \right)$$

$$+ z_{i,2} \delta_{i,2}^* - c_{i,21} z_{i,2}^2 - \left(c_{i,21} z_{i,2}^2 + \dot{g}_{i,2}(\overline{x}_{i,2}) z_{i,2}^2 / 2 g_{i,2}^2(\overline{x}_{i,2}) \right) \tag{19}$$

$$+ \sum_{j=1}^{2} (0.2785 s_{i,j}) + \left(-\frac{\eta_{i,2}^2}{\tau_{i,2}} + |\eta_{i,2} B_{i,2}| \right) + \sum_{j=1}^{2} \left(-\sigma_{i,j} \tilde{\theta}_{i,j}^T \hat{\theta}_{i,j} \right)$$

where $c_{i,11} = 3 + \alpha_0 - c_{i,10}^*$, $c_{i,21} = 1 \frac{1}{4} + \alpha_0 - c_{i,20}^*$ with α_0 is a positive constant, $c_{i,j0}^* \stackrel{\Delta}{=} \left(c_{i,j0} - g_{i,jd}(\overline{x}_{i,j}) / 2 g_{i,j0}^2(\overline{x}_{i,j}) \right) > 0$.

After a further manipulation, one can obtain

$$\dot{V} \le \sum_{j=1}^{2}\left(-\alpha_0 z_{i,j}^2 - \frac{\sigma_{i,j}}{2}\left(\left\|\tilde{\theta}_{i,j}\right\|^2 - \left\|\hat{\theta}_{i,j}^*\right\|^2\right) + \frac{1}{4}\delta_{i,j}^{*2}\right)$$

$$+ \sum_{j=1}^{2}\left(0.2785 s_{i,j}\right) + \left(\frac{1}{4}\eta_{i,2}^2 - \frac{\eta_{i,2}^2}{\tau_{i,2}} + \left|\eta_{i,2}B_{i,2}\right|\right)$$

$$\le \sum_{j=1}^{2}\left(-\alpha_0 z_{i,j}^2 - \frac{\sigma_{i,j}}{2\lambda_{\max}\left(\Gamma_{i,j}^{-1}\right)}\tilde{\theta}_{i,j}^{\mathrm{T}}\Gamma_{i,j}^{-1}\tilde{\theta}_{i,j} + \frac{1}{4}\delta_{i,j}^{*2} + \frac{\sigma_{i,j}}{2}\left\|\theta_{i,j}^*\right\|^2\right)$$

$$+ \sum_{j=1}^{2}\left(0.2785 s_{i,j}\right) + \left(\frac{1}{4}\eta_{i,2}^2 - \frac{\eta_{i,2}^2}{\tau_{i,2}} + \left|\eta_{i,2}B_{i,2}\right|\right)$$

(20)

Let $(1/4)\delta_{i,j}^{*2} + \left(\sigma_{i,j}/2\right)\left\|\theta_{i,j}^*\right\|^2 = e_{i,j}$, $1/\tau_{i,2} = (1/4) + \left(M_{i,2}^2/2\kappa\right) + \alpha_0 g_{i,j0}$. Noting that $\left|\delta_{i,j}^*\right| \le \delta_m$ and $\left\|\theta_{i,j}^*\right\| \le \theta_M$ gives $e_{i,j} \le (1/4)\delta_m^2 + \left(\sigma_{i,j}/2\right)\theta_M^2 = e_M$. For any positive number κ, $\left(\eta_{i,2}^2 B_{i,2}^2/2\kappa\right) + \left(\kappa/2\right) \ge \left|\eta_{i,2}B_{i,2}\right|$. Since on the level set defined by $\left|B_{i,2}\right| < M_{i,2}$, $V\left(z_{i,1}, z_{i,2}, \eta_{i,2}, \tilde{\theta}_{i,1}, \tilde{\theta}_{i,2}\right) = p_i$ (please refer to [7] for details). Then

$$\dot{V} \le \sum_{j=1}^{2}\left(-\alpha_0 z_{i,,j}^2\right) + \sum_{j=1}^{2}\left[-\frac{\sigma_{i,j}}{2\lambda_{\max}\left(\Gamma_{i,j}^{-1}\right)}\tilde{\theta}_{i,j}^{\mathrm{T}}\Gamma_{i,j}^{-1}\tilde{\theta}_{i,j}\right] + 2e_M$$

$$+ \sum_{j=1}^{2}\left(0.2785 s_{i,j}\right) + \left(\frac{1}{4}\eta_{i,2}^2 - \frac{\eta_{i,2}^2}{\tau_{i,2}} + \left|\eta_{i,2}B_{i,2}\right|\right)$$

$$\le \sum_{j=1}^{2}\left(-\alpha_0 z_{i,,j}^2\right) + \sum_{j=1}^{2}\left[-\frac{\sigma_{i,j}}{2\lambda_{\max}\left(\Gamma_{i,j}^{-1}\right)}\tilde{\theta}_{i,j}^{\mathrm{T}}\Gamma_{i,j}^{-1}\tilde{\theta}_{i,j}\right] + 2e_M$$

$$+ \sum_{j=1}^{2}\left(0.2785 s_{i,j}\right) + \frac{\kappa}{2} - \alpha_0 g_{i,j0}\eta_{i,2}^2$$

(21)

If we choose $\alpha_0 \ge C/\left(2g_{i,j0}\right)$, where C is a positive constant, and choose $\sigma_{i,j}$ and $\Gamma_{i,j}$ such that $\sigma_{i,j} \ge C\lambda_{\max}\left(\Gamma_{i,j}^{-1}\right) i = 1,2$. Let $D \overset{\Delta}{=} 2e_M + \frac{\kappa}{2} + \sum_{j=1}^{2}\left(0.2785 s_{i,j}\right)$. Then from (21) we have the following inequality:

$$\dot{V} \le -\sum_{j=1}^{2}\frac{C}{2g_{i,j0}}z_{i,j}^2 - \sum_{j=1}^{2}\frac{C\tilde{\theta}_{i,j}^{\mathrm{T}}\Gamma_{i,j}^{-1}\tilde{\theta}_{i,j}}{2} - \frac{C}{2}\eta_{i,2}^2 + D$$

$$\le -\left(\sum_{j=1}^{2}\frac{C}{2g_{i,j}}z_{i,j}^2 + \sum_{j=1}^{2}\frac{C\tilde{\theta}_{i,j}^{\mathrm{T}}\Gamma_{i,j}^{-1}\tilde{\theta}_{i,j}}{2} + \frac{C}{2}\eta_{i,2}^2\right) + D \le -CV + D$$

(22)

Actually, the Equation (22) means that $V\left(t\right)$ is bounded (please refer to [9] for details). Thus, all signals of the closed-loop system. Moreover, we can appropriately choose the design parameters $c_{i,j}, \lambda_{max}(\Gamma_{i,j}^{-1}), \tau_{i,2}, \sigma_{i,j}, s_{i,j}$ etc. make the tracking error arbitrarily small. This concludes the proof.

4   Simulation Results

In the simulation, the system (1) is uses for simulation model, $v_1(t) = v_2(t) = 5\sin(2\pi t)$ with $i = 1, 2$ in (2). The parameters of the pendulums are $m_1 = 2$ kg, $m_2 = 2.5$ kg, $J_1 = 2$ kg, $J_2 = 2.5$ kg $k = 2$ Nm/rad, and $r = 1$ m. The reference signals $y_{1,d} = y_{2,d} = \sin(t)$ [8].

In addition, the two RBF neural networks contain 25 and 135 nodes, respectively.

The Gaussian function is chosen as $\xi_i(Z) = \exp\left[\dfrac{-(Z - \mu_i)^T(Z - \mu_i)}{\eta_i^2}\right]$, $i = 1, 2, ..., l$.

The initial conditions $\left[x_{1,1}(0), x_{2,1}(0)\right]^T = \left[30°, -30°\right]^T$, $\left[x_{1,2}(0), x_{2,2}(0)\right]^T = \left[0, 0\right]^T$.

The controllers parameters are chosen as $\Gamma_{i,j} = \{0.01,\ 0.01,\ 0.01,\ 0.01\}$, $c_{i,j} = \{10,\ 80,\ 10,\ 80\}$, $\sigma_{i,j} = \{15,\ 15,\ 15,\ 15\}$, $s_{i,j} = \{0.5,\ 0.5,\ 0.5,\ 0.5\}$ with $i = 1, 2$, $j = 1, 2$, and $\tau_{1,2} = \tau_{2,2} = 0.5$. The effectiveness and good performance of the proposed algorithm are illustrated in Fig. 1.

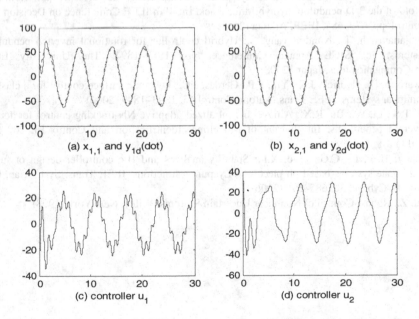

Fig. 1. Simulation results for DIPs: angular positions, desired signals and controllers

5   Conclusion

The trajectory tracking control problem has been studied for DIP in this paper. A direct adaptive NN-based DSC control scheme has been proposed by using adaptive back-stepping, DSC techniques, and the approximation of RBF NNs. In the proposed algorithm, both the "explosion of complexity" problem and the controller singularity problem are removed. And it is shown that the stability of the closed-loop system are guaranteed. Finally, simulation results validate the effectiveness and the good transient performance of the proposed scheme.

Acknowledgements. This work was supported in part by the National Natural Science Foundation of China (Nos.60874056 & 51179019).

References

1. Angeli, D.: Almost global stabilization of the inverted pendulum via continuous state feedback. Automatica 37, 1103–1108 (2001)
2. Matsuda, N., Izutsu, M., Furuta, K.: Simultaneous swinging-up and stabilization of double Furuta pendulums. In: Proc. of the SICE Annual Conference, pp. 110–115 (2007)
3. Xin, X., Kaneda, M.: Analysis of the energy-based control for swinging up two pendulums. IEEE Trans. on Automatic and Control 50(5), 679–684 (2005)
4. Mayhew, C.G., Teel, A.R.: Global asymptotic stabilization of the inverted equilibrium manifold of the 3-D pendulum by hybrid feedback. In: 49th IEEE Conference on Decision and Control, pp. 679–684. IEEE, Atlanta (2010)
5. Benjanarasuth, T., Nundrakwang, S.: Hybrid controller for rotational inverted pendulum systems. In: SICE Annual Conference, pp. 1889–1894. The University Electro-Communications, Japan (2008)
6. Swaroop, D., Hedrick, J.K., Yip, P.P., Gerdes, J.C.: Dynamic surface control for a class of nonlinear systems. IEEE Trans. Autom. Control 45, 1893–1899 (2000)
7. Li, T.S., Li, W., Bu, R.X.: A novel decentralized adaptive NN tracking control for double inverted pendulums. Int. Journal of Modelling, Identification and Control 13, 269–277 (2011)
8. Zhang, H.B., Li, C.G., Liao, X.F.: Stability analysis and H∞ controller design of fuzzy large-scale systems based on piecewise Lyapunov functions. IEEE Trans. Syst., Man, Cybern. B, Cybern. 36, 685–698 (2006)
9. Qu, Z.: Robust Control of Nonlinear Uncertain Systems. Wiley, New York (1998)

Clustering with Uncertainties:
An Affinity Propagation-Based Approach

Wenye Li

Macao Polytechnic Institute,
Rua de Luís Gonzaga, Macao SAR, China
wyli@ipm.edu.mo
http://staff.ipm.edu.mo/~wyli

Abstract. Clustering is a classical unsupervised learning technique which has wide applications. One popular clustering model seeks a set of centers and organizes the data into different groups, with an objective to maximize the net similarities within each cluster. In this paper, we first formulate a generalized form of the clustering model, where the similarity measure has uncertainties or changes in different states. Then we propose an affinity propagation-based algorithm, which gives an efficient and accurate solution to the generalized model. Finally we evaluate the model and the algorithm by experiments. The results have justified the usefulness of the model and demonstrate the improvements of the algorithm over other possible solutions.

Keywords: Multi-state Clustering, Belief Propagation, Affinity Propagation.

1 Introduction

Clustering [1, 2] is a classical unsupervised learning technique and has many applications in scientific analysis and engineering systems. The principle of clustering is to partition a set of objects into different groups, so that the data in each cluster share some common traits.

To provide the intra-cluster proximity, a widely-used approach is to organize the data by learning a set of centers such that the net similarity between each point and its closest center is large. One classical example is k-means, where the center of a cluster is the mean of all the data points in that cluster, and this center does not have to be any data examples. However, in many real-world problems, we want the cluster centers to be selected from actual data points, where the centers are also called *exemplars*.

An exact solution to this exemplar-based clustering problem is computationally expensive, which is only feasible for small problems. For large problems, people generally use heuristics. Recently, based on the advances in probabilistic inference and graphical models [3–6], Frey and Dueck [7] proposed a method called *affinity propagation* and gave a novel solution to the problem. On many benchmark tests, this approach has reported improved results over other heuristics.

T. Huang et al. (Eds.): ICONIP 2012, Part V, LNCS 7667, pp. 437–446, 2012.

In this paper a generalized exemplar-based model is studied, where we are interested in the changing similarities between data points across different periods of time. The problem is to determine the centers such that they are able to serve different cluster members at different states. This model has many applications, such as location problems and transportation problems in operations research and management science [8–10].

When uncertainties are taken into consideration, the computation becomes even more demanding. To make the problem tractable, this paper proposes a natural extension of the affinity propagation algorithm and gives an efficient solution to the problem.

2   Background

2.1   Exemplar-Based Clustering

Recently, the work of [7] triggers some research in exemplar-based clustering models [11, 12]. Starting from a similarity matrix between data points, people try to detect a set of actual points as exemplars, and connect every point to the exemplar that best represents it. As noted in [13], detecting exemplars in fact goes beyond simple clustering. The exemplars themselves have stored compressed information, and the extraction of this information is meaningful. It is conceivable that a broad range of potential applications will benefit from this technique.

A definition of the exemplar-based clustering can be given as follows. Given a set of data points $X = \{x_1, \cdots, x_N\}$ and a similarity measure $s(i,j)$[1] that indicates how well x_i is represented by x_j (for example, the negative Euclidean distance $- \|x_i - x_j\|^2$), the exemplar-based clustering model seeks an optimal set of centers such that the sum of similarities of each point to its center is maximized. Or, we seek $C \subseteq X$ to maximize the objective:

$$\sum_{i=1}^{N} \max_{x_c \in C} s(i,c) \ \ s.t. \ \ |C| = K \tag{1}$$

where $|C|$ denotes the cardinality of C and K $(K \leq N)$ is a given number. Then we assign each point in X to its closest center in C and get the K clusters.

A related model studies the maximization of the objective:

$$\sum_{i=1}^{N} \max_{x_c \in C} s(i,c) - \gamma |C| . \tag{2}$$

Rather than requiring the number of centers to be given a priori, this model automatically determines the number of clusters by taking into consideration of the center cost γ (> 0), and thus provides a mechanism in automatic model

---

[1] Without loss of generality, we assume $s(i,j) \leq 0$ and $s(i,i) = 0$ $(1 \leq i, j \leq N)$ here.

selection. Sometimes $-\gamma$ is also referred as a *preference* value, which quantifies a kind of willingness or likelihood for a data point to be chosen as a center.

Optimal solutions to the two problems in Equ. (1) and Equ. (2) are well-known to be computationally expensive. For small problems ($N \leq 500$), we can use integer linear programming and relaxation to find exact solutions. However, for larger problems, we have to resort to heuristics due to the inherent *NP*-hardness.

Among the heuristics, a popular *k-centers clustering* algorithm begins with an initial set of randomly selected exemplars and iteratively refines the set so as to increase the total similarities. This technique is quite sensitive to the initial selection of centers, so it is usually re-run many times with different initializations in an attempt to find a good solution.

A more complicated heuristic is the *vertex substitute heuristic* [14], which often achieves better accuracy than *k*-centers clustering. It arbitrarily selects centers from the data set, and then relocates centers, each time substituting in the points so as to yield the greatest improvement in the objective function, until no more improvement can be made.

2.2 Affinity Propagation

A different approach to the problem is proposed in [7], called *affinity propagation* (AP). It is in fact an application of the *belief propagation* method, which has been used to achieve state of the art performances in many different disciplines [15].

The method does not fix the number of centers. Instead, all data points are considered as potential centers and each point is assigned a number that represents a priori knowledge of how good the point is as a center. In most cases all points are equally suitable, so all numbers take the same value, which provides a control parameter: the larger the value, the more centers one is likely to find.

Then the algorithm operates by viewing each point as a node in a network and exchanging messages between the nodes until a good set of centers emerges. Two types of real-valued messages (*responsibility* and *availability*) are transmitted along edges of the network and each undertakes a different kind of of competition. The messages are updated by simple rules. The magnitude of each message reflects the current *affinity* that one point has for choosing another point as its center. At any time, the centers can be identified by combining the messages flowing in and out.

The model and algorithm mentioned above are very effective in dealing with static data objects and affinities. However in practice, many applications involve clustering when data objects and affinities change dynamically with time, and the conventional technique loses its advantage. Therefore, we propose to study a generalized exemplar-based model and derive a extended AP algorithm to achieve that.

3   Clustering with Uncertainties

To solve the clustering problems with uncertainties, we can consider the similarity measure between any pair of points to be a discrete random variable, yielding a finite number of states, where one state differs from another if at least one similarity differs. The problem is then formulated as follows. Given a set of data points $X = \{x_1, \cdots, x_N\}$. Suppose there are Q states. Let p_q $(1 \leq q \leq Q)$ denote the probability of being in state q and a function $s_q(j, j')$ measure the similarity between x_j and $x_{j'}$ $(1 \leq j, j' \leq N)$ in state q. We are seeking a set of points $C \subseteq X$ to maximize

$$L(C) = \sum_{q=1}^{Q} \sum_{i=1}^{N} p_q \max_{x_c \in C} s_q(i, c) - \gamma |C|. \tag{3}$$

The objective in this generalized clustering problem is to locate a number of centers from the dataset so that the weighted similarity between each point and its closest centers in different states is maximized, subject to the cost constraint on the number of centers. In each state, a point is assigned to a unique center. However, a point may be assigned to different centers at different states. In this paper we call this generalized model *multi-state clustering* while calling the conventional model in Equ. (1) and Equ. (2) *single-state clustering*.

When all states have the same probability, i.e. $p_1 = \cdots = p_Q = \frac{1}{Q}$, the objective becomes the maximization of $\frac{1}{Q} \sum_{q=1}^{Q} \sum_{i=1}^{N} \max_{x_c \in C} s_q(i, c) - \gamma |C|$, or equivalently, the maximization of

$$\sum_{q=1}^{Q} \sum_{i=1}^{N} \max_{x_c \in C} s_q(i, c) - \gamma Q |C|. \tag{4}$$

In the sequel, we assume that all states have the same probability for simplicity. However the algorithm introduced in Section 4 can be easily extended to different probabilities.

4   Solution

4.1   A Factor Graph Representation

It is obvious that seeking a set C is equivalent to seeking the cluster labels $\mathbf{c} = (c_{11}, \cdots, c_{1Q}, \cdots, c_{NQ})$ with each c_{iq} given by $\arg_c \max_{x_c \in C} s_q(i, c)$, and vice versa. Maximizing Equ. (4) is equivalent to maximizing

$$\sum_{q=1}^{Q} \sum_{i=1}^{N} s_q(i, c_{iq}) - \gamma Q |C| \tag{5}$$

satisfying $c_{iq} = \arg_c \max_{x_c \in C} s_q(i, c)$. To maximize Equ. (5), we consider the following function:

$$g(\mathbf{c}) = \prod_{q=1}^{Q} \prod_{i=1}^{N} e^{s'_q(i, c_{iq})} \prod_{k=1}^{N} \delta_k(\mathbf{c}) \tag{6}$$

where

$$s'_q(i, c_{iq}) = \begin{cases} s_q(i, c_{iq}) & if\ c_{iq} \neq i \\ -\gamma & if\ c_{iq} = i \end{cases}$$

and

$$\delta_k(\mathbf{c}) = \begin{cases} 0\ if\ \exists 1 \leq j \leq N, 1 \leq q_1, q_2 \leq Q : c_{jq_1} = k\ and\ c_{kq_2} \neq k \\ 1 \qquad\qquad\qquad\qquad otherwise \end{cases}.$$

Taking logarithm of Equ. (6), we have

$$\log g(\mathbf{c}) = \sum_{q=1}^{Q} \sum_{i=1}^{N} s'_q(i, c_{iq}) + \sum_{k=1}^{N} \log \delta_k(\mathbf{c}). \tag{7}$$

Each $\log \delta_k(\mathbf{c})$ is a penalty term and forces a constraint on the label assignments. The constraint can be understood via follows. Suppose a point x_k has been chosen as a center by some point x_i in a state, then x_k must decide to be its own center in all states. These constraints together guarantee a valid configuration of the assignments.

When the objective is maximized, each $\log \delta_i(\mathbf{c})$ is forced to be 0 and the penalty items are eliminated. The result becomes $\sum_{q=1}^{Q} \sum_{i=1}^{N} s'_q(i, c_{iq})$, which can be easily verified to be equivalent to the maximum of Equ. (5).

With the equivalence between the maximizers of Equ. (5) and Equ. (6), we represent Equ. (6) by a factor graph. In Fig. 1, each term in $g(\mathbf{c})$ is represented by a function node and each label c_{iq} by a variable node. Edges exist only between function nodes and variable nodes. A variable node is connected to a function node if and only if the function's term depends on the variable. So the term $s'_q(i, c_{iq})$ has a corresponding function connected to the single variable c_{iq}. The term $\delta_k(\mathbf{c})$ has a corresponding function connected to variables $c_{11}, \cdots, c_{1Q}, \cdots, c_{NQ}$. The global function, $g(\mathbf{c})$, is given by the product of all the functions represented by function nodes.

4.2   An Affinity Propagation-Based Solution

To solve the problem, we use the max-product procedure [4], which becomes the max-sum rule after taking logarithm, to search over valid label configurations in the graph to maximize $g(\mathbf{c})$. Messages are sent iteratively from each c_{iq} to $\delta_k(\mathbf{c})$, and vice versa. In Fig. 2, the message sent from c_{iq} to $\delta_k(\mathbf{c})$ consists of N real numbers, one for each possible value j of c_{iq} and is given by:

$$m_{\delta_k \leftarrow c_{iq}}(j) = \sum_{k' \neq k} m_{\delta_{k'} \rightarrow c_{iq}}(j) + s'_q(i, j). \tag{8}$$

The message sent from $\delta_k(\mathbf{c})$ to c_{iq} is given by

$$m_{\delta_k \rightarrow c_{iq}}(j) = \max_{j_{11}, \cdots, j_{i,q-1}, j_{i,q+1}, \cdots, j_{NQ}} \left[\begin{matrix} \log \delta_k(j_{11}, \cdots, j_{i,q-1}, j, j_{i,q+1}, \cdots, j_{NQ}) \\ + \sum_{i', q' : i'q' \neq iq} m_{\delta_k \leftarrow x_{i'q'}}(j_{i'q'}) \end{matrix} \right]. \tag{9}$$

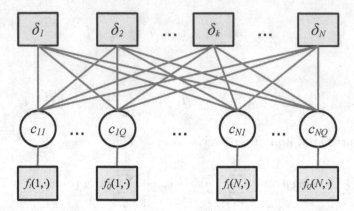

Fig. 1. A factor graph representation of the similarity maximization problem. In the graph, $f_q(i,j)$ denotes a function $e^{s'_q(i,j)}$.

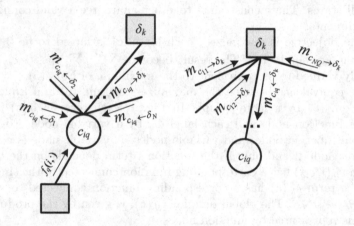

Fig. 2. The messages flowing between variable nodes and function nodes

The message update of N numbers can be reduced to a single number. This simplified message involves a *responsibility* message $r_q(i,k)$, which is sent from c_{iq} to δ_k, reflects the accumulated confidence for point i in choosing point k as its center in state q, combining the opinions from other points that point k should be a center, and an *availability* message $a_q(i,k)$, which is sent from δ_k to c_{iq}, collects the evidences from data points in deciding whether point k would be an appropriate center.

The algorithm with simplified message passing is given below:

1. Set $t = 0$. Initialize messages $a_q^t(i,k) = 0$ $(1 \le i, k \le N, 1 \le q \le Q)$.
2. For each i, q, k

$$r_q^{t+1}(i,k) = s'_q(i,k) - \max_{k' \ne k} \left[s'_q(i,k') + a_q^t(i,k') \right]$$

and

$$a_q^{t+1}(i,k)$$

$$= \begin{cases} \sum_{q':q'\neq q} r_{q'}^{t+1}(i,k) + \sum_{i',q':i'\neq i} \max\left(0, r_{q'}^{t+1}(i',k)\right) & k = i \\ \min\left\{0, \sum_{q'} r_{q'}^{t+1}(k,k) + \sum_{i',q':i'\neq k \& i'q'\neq iq} \max\left(0, r_{q'}^{t+1}(i',k)\right)\right\} & k \neq i \end{cases}$$

3. Set $t = t + 1$. For each point i, estimate the value of $\hat{c}_{iq}^t$ by

$$\hat{c}_{iq}^t = \arg_j \max\left[a_q^t(i,j) + s_q'(i,j)\right].$$

4. Iterate step 2 until some convergence criterion is satisfied and return all $\hat{c}_{iq}^t$'s.

This algorithm can be regarded as an extended AP algorithm. When $Q = 1$, this is exactly the standard AP algorithm [7].

The message updates in step 2 involve only simple computations. A naïve implementation would have $O\left(QN^3\right)$ operations for all updates per iteration. However, since certain operations can be re-used, the actual computational requirement is $O\left(QN^2\right)$. For sparse problems where a subset of M ($\ll QN^2$) similarities are given, the computational requirement can be further reduced to $O\left(M\right)$, which is appealing for many real applications.

Similar to the standard AP algorithm, a damping factor λ is often used when updating the messages to avoid numerical difficulties in certain circumstances. Thus each message updating rule in step 2 is followed by $r_q^{t+1}(i,k) = \lambda r_q^t(i,k) + (1-\lambda) r_q^{t+1}(i,k)$ and $a_q^{t+1}(i,k) = \lambda a_q'(i,k) + (1-\lambda) a_q^{t+1}(i,k)$. A valid damping factor takes value from $[0.5, 1)$.

After each iteration, the value of a variable c_{iq} can be estimated in step 3 by summing up all messages flowing into node c_{iq} and taking the value that maximizes the summation.

In our implementations, we used a default damping factor 0.9. The message flow is terminated after a maximum of 2000 iterations, or the identified centers remain unchanged for 50 iterations.

5   Experiments

To evaluate the model and the algorithm, we have carried out a series of experiments. Four datasets (RandomS, Routing, ECB and FaceVideo) with 400 to 1965 points, from affinity propagation homepage[2], were used.

We have tested $Q = 2$ and $Q = 5$ respectively. The first state similarity $s_1(i,j)$ $(1 \leq i, j \leq N)$ is read directly or computed (as negative Euclidean norm of a pair of points) from a dataset. In the following states q $(2 \leq q \leq Q)$, each similarity $s_q(i,j)$ is calculated as the product of $s_1(i,j)$ and a non-negative random number $\left|r_{ij}^q\right|$, where r_{ij}^q is from a normal distribution with mean 1.0 and variance 0.2.

---

[2] http://www.psi.toronto.edu/affinitypropagation/

Table 1. Comparison of relative errors by multi-state clustering and single-state clustering. The error is measured relative to the minimum negative net similarity.

Data Set	N	K	Q	Error(%) Multi:Single	Time Multi:Single
RandomS	400	32	2	0 : 0.72	2.20 : 1
	400	34	5	0 : 1.05	4.75 : 1
Routing	456	38	2	0 : 0.64	2.33 : 1
	456	39	5	0 : 0.72	4.55 : 1
ECB	1272	101	2	0 : 1.05	3.02 : 1
	1272	102	5	0 : 1.28	6.29 : 1
Face Video	1965	103	2	0 : 4.04	1.78 : 1
	1965	192	5	0 : 19.9	5.21 : 1

5.1   Clustering with/without Uncertainties

Let us start from a classical application of locating ambulance stations in a city. We are to select a number of spots as cluster centers (to establish ambulance stations). The objective is to balance the establishment costs and the averaged travel times from the closest station to each spot. However, the travel times differ significantly over the time of day. This is modeled as the multi-state optimization of net similarities, which are defined as the negative travel times. Based on the travel time, we have a number of states such as quiet hours, rush hours, and normal hours. For simplicity, we assume each state is equally likely. Generally, the optimal solution to this multi-state clustering problem often has a smaller average response time than the optimal solution if we model the problem as single-state clustering, where the average time required to travel is used to develop a single similarity matrix.

We simulated this experiment on the datasets, and compared the net similarities by multi-state and single-state modeling. Because N was too large to compute the exact solution, we report the relative error. For single-state clustering, the standard AP algorithm was applied. As shown in Table 1, the multi-state modeling achieves better solutions in all scenarios.

This comparison also reflects the computational requirements on multi-state and single-state clustering problems by AP. Roughly speaking, in these experiments a clustering problem with Q states would require a computation around Q times as much as the problem with one state.

5.2   Comparison with Other Possible Solutions

We also compared the performance of our AP-based solution with the k-centers clustering (KCC) and the vertex substitute heuristic (VSH). To compare with KCC, we applied AP to each dataset for a fixed cost parameter γ, and then applied KCC to produce the same number of clusters. As shown in Table 2, the relative errors between the result by one run of AP and the best result by 100 runs of KCC indicates a considerable advantage of AP based multi-state

Table 2. Comparison of performances using AP, KCC and VSH

Data Set	N	K	γ	Q	Error(%) AP:KCC	Error(%) AP:VSH	Time AP:VSH
RandomS	400	32	0.8	2	0 : 39.4	0 : 0.35	1 : 2.68
	400	34	0.8	5	0 : 44.3	0.29 : 0	1 : 2.86
Routing	456	38	225	2	0 : 49.0	0 : 0.2	1 : 7.34
	456	39	225	5	0 : 99.7	0 : 0.2	1 : 7.79
ECB	1272	101	200000	2	0 : 101	0 : 0.15	1 : 14.6
	1272	102	200000	5	0 : 164	0 : 0.07	1 : 15.2
Face Video	1965	189	20000	2	0 : 132	0 : 0.13	1 : 24.4
	1965	357	20000	5	0 : 53.4	0 : 0.00004	1 : 41.4

clustering over KCC. The improvement by AP in multi-state clustering is even more significant than the result reported in [7] in single-state clustering.

To compare with VSH, Table 2 shows the relative error between the result by one run of AP and the best result by 20 runs of VSH. Both AP and VSH have reported substantial improvement over KCC algorithm. Although the accuracies of the two algorithms are comparable, AP has an evident advantage over VSH in terms of scalability. As the number of points and the number of centers get larger, AP becomes much faster than VSH. A speedup of over 40 is achieved for a problem with 1965 nodes and 357 centers.

6   Conclusion

This paper studies a generalization of the exemplar-based clustering model, where multi-state similarities are considered. To solve the problem, we proposed an algorithm based on the idea of affinity propagation. The experiment results show that this solution has improved performances over related search strategies in accuracy and scalability, and thus provides a reasonable solution.

The explicit treatment of multi-state similarities can be traced back to the age-old planning problems in operations research [16, 17]. Very recently this idea has also been applied in sensor networks [18]. In this paper, we investigate this idea on clusterings and propose a generalized clustering model which models uncertainties directly. We anticipate a number of recent machine learning problems such as in computer visions may benefit from this model. The study of these applications deserves our further study.

Acknowledgments. The work is partially supported by The Science and Technology Development Fund, Macao SAR.

References

1. MacQueen, J.: Some methods for classification and analysis of multivariate observations. In: Le Cam, L.M., Neyman, J. (eds.) Proceedings of the Fifth Berkeley Symposium on Mathematical Statistics and Probability, vol. 1, pp. 281–297. University of California Press, Berkeley (1967)
2. Jain, A.K., Murty, M.N., Flynn, P.J.: Data clustering: A review. ACM Computing Surveys 31(3), 264–323 (1999)
3. Pearl, J.: Probabilistic Reasoning in Intelligent Systems: Networks of Plausible Inference. Morgan Kaufmann Publishers, inc, San mateo (1988)
4. Kschischang, F.R., Frey, B.J., Loeliger, H.A.: Factor graphs and the sum-product algorithm. IEEE Transactions on Information Theory 47(2), 498–519 (2001)
5. Aji, S.M., McEliece, R.J.: The generalized distributive law. IEEE Transactions on Information Theory 46(2), 325–343 (2000)
6. Yedidia, J.S., Freeman, W.T., Weiss, Y.: Constructing free-energy approximations and generalized belief propagation algorithms. IEEE Transactions on Information Theory 51(7), 2282–2312 (2005)
7. Frey, B.J., Dueck, D.: Clustering by passing messages between data points. Science 315 (2007)
8. Teitz, M.B., Bart, P.: Heuristic methods for estimating the generalized vertex median of a weighted graph. Operations Research 16(5), 955–961 (1968)
9. Tansel, B.C., Francis, R.L., Lowe, T.J.: Location on networks: A survey; part I: The p-center and p-median problems. Management Science 29(4), 482–497 (1983)
10. Hansen, P., Mladenović, N.: Variable neighborhood search for the p-median. Location Science 5(4), 207–226 (1997)
11. Lashkari, D., Golland, P.: Convex clustering with exemplar-based models. In: Platt, J., Koller, D., Singer, Y., Roweis, S. (eds.) Advances in Neural Information Processing Systems 20, pp. 825–832. MIT Press, Cambridge (2008)
12. Leone, M., Weigt, M.: Clustering by soft-constraint affinity propagation: Applications to gene-expression data. Bioinformatics 23, 2708 (2007)
13. Mézard, M.: Where are the exemplars? Science 315, 949–951 (2007)
14. Lin, S., Kernighan, B.W.: An effective heuristic algorithm for the traveling salesman problem. Operations Research 21, 498–516 (1973)
15. Mézard, M.: Passing messages between disciplines. Science 301(5640), 1685–1686 (2003)
16. Handler, G.Y., Mirchandani, P.B.: Location on Networks: Theory and Algorithms. MIT Press, Cambridge (1979)
17. Li, W., Xu, L., Schuurmans, D.: Facility locations revisited: An efficient belief propagation approach. In: Proceedings of the 2010 IEEE International Conference on Automation and Logistics (2010)
18. Li, W.: Probabilistic inference over sensor networks for clusters: Extension to multiple states. In: Proceedings of the 2012 IEEE International Conference on Information and Automation (2012)

Online Vigilance Analysis Combining Video and Electrooculography Features

Ruo-Fei Du[1], Ren-Jie Liu[1], Tian-Xiang Wu[1] and Bao-Liang Lu[1,2,3,4,*]

[1] Center for Brain-like Computing and Machine Intelligence
Department of Computer Science and Engineering
[2] MOE-Microsoft Key Lab. for Intelligent Computing and Intelligent Systems
[3] Shanghai Key Laboratory of Scalable Computing and Systems
[4] MOE Key Laboratory of Systems Biomedicine
Shanghai Jiao Tong University
800 Dongchuan Road, Shanghai 200240, China
bllu@sjtu.edu.cn

Abstract. In this paper, we propose a novel system to analyze vigilance level combining both video and Electrooculography (EOG) features. For one thing, the video features extracted from an infrared camera include percentage of closure (PERCLOS) and eye blinks, slow eye movement (SEM), rapid eye movement (REM) are also extracted from EOG signals. For another, other features like yawn frequency, body posture and face orientation are extracted from the video by using Active Shape Model (ASM). The results of our experiments indicate that our approach outperforms the existing approaches based on either video or EOG merely. In addition, the prediction offered by our model is in close proximity to the actual error rate of the subject. We firmly believe that this method can be widely applied to prevent accidents like fatigued driving in the future.

Keywords: Vigilance Analysis, Fatigue Detection, Active Shape Model, Electrooculography, Support Vector Machine.

1 Introduction

There is no doubt that vigilance plays a crucial role in our daily life. It is essential to ensure the vigilance level of the drivers since a large number of accidents are resulted from fatigued driving, which has been investigated by [1] and [2]. Meanwhile, vigilance is an indispensable characteristic for other occupations such as policemen, soldiers and operators who have to deal with hazardous equipments. Thus, an efficient and accurate system is badly in need in order to prevent accidents by warning the users in advance.

In the last two decades, extensive researches have been conducted regarding vigilance analysis [3]. These studies can be broadly classified into 3 categories based on the techniques which are adopted during the procedure of feature extraction: (infrared) video, electrooculography (EOG) and electroencephalography (EEG).

Firstly, video is the most convenient approach. Compared with EOG and EEG based systems, in which the subjects have to interact directly with the equipments, cameras

---

* corresponding author.

T. Huang et al. (Eds.): ICONIP 2012, Part V, LNCS 7667, pp. 447–454, 2012.

are much less intrusive. Moreover, not only eye movement can be measured through video, but yawn state and facial orientation can be estimated [4][5][6]. Unfortunately, there are 3 apparent drawbacks for video approach: the accuracy would decrease due to various luminance; the range of the video is limited to the horizon of the cameras; the recognition usually fails on account of the appearance of the subjects such as wearing sunglasses.

Secondly, EOG signals is a moderate method since it is irrelevant to the environment and easier to analyze. In order to estimate the vigilance level, features like eye blinks, blink duration, slow eye movement (SEM), rapid eye movement (REM) are utilized, which has been proved to be accurate according to [7].

Finally, vigilance analysis based on EEG signals gains a good accuracy. It has been shown that both delta waves (0-4 Hz), which are relevant to slow wave sleep (SWS), and theta waves (4-7 Hz), which are relevant to drowsiness of older children and adults play significant roles in vigilance analysis [3]. Nevertheless, it is ineffective as a result of its poor user experience of the interaction.

From our perspective, in order to utilize comprehensive information like yawn state and improve the accuracy of vigilance analysis, we construct an online vigilance analysis system combining both video and EOG features. On one hand, we extract the displacements of feature points around eyes and mouth based on the well-known Active Shape Model (ASM) and the location of the eyelids based on a binarization approach. Afterwards, we calculated the average height, area and PERCLOS of eyes and mouth. Finally we extract features of blinks and eye movements for vigilance analysis from the information we have. On the other hand, we preprocess the EOG signals using a low-pass filter with the frequency of 10Hz and a normalization procedure. After that, features of blinks and eye movements are extracted according to the algorithm proposed in [8]. In our experiments, based on the actual error rate signals, we employ SVM regression to offer the prediction of the vigilance level. Compared with the actual error rate, our algorithm is proved to work rather well for the online vigilance estimation. In addition, although the accuracy provided by the video features is not so good as that of EOG features, the combination of both video and EOG features outperform both models running alone.

In Section 2, our online vigilance analysis system is introduced in details. In section 3, the preprocess in our system is illustrated. We describe the approaches of features extraction in Section 4. Experiments and analysis are conducted in section 5, followed by the conclusion and further discussions in section 6.

2   Vigilance Analysis System

2.1   Equipments

In our vigilance analysis system, the video signals are extracted from infrared cameras in the front. Meanwhile, the EOG signals are extracted by 8 electrodes and recorded by the NeuroScan system [1].

---

[1] Neuroscan Inc, Herndon, VA, USA.

There are 2 vertical and 2 horizontal electrodes on the forehead. Different traffic signs in four colors (red, green, blue and yellow) are displayed on the screen every 6 seconds and each sign lasts for only 0.5 seconds. Meanwhile, the subject is ought to press the button with the same color as the signs flashed on the screen. Therefore, not only the actions of the subjects could be recorded but the actual error rate could be calculated for further analysis.

2.2 Overall Process

The overall process of our vigilance analysis system is illustrated in Fig. 1. In summary, four procedures including preprocessing, feature extraction, model training and prediction of vigilance are followed for both video and EOG signals.

3 Preprocess

The video signals are captured by infrared cameras which is in front of the subject. The recorded video has the resolution of 640×480 pixels with 30 frames per second in RGB color. In order to extract features based on video, we need to preprocess the images in each frame. For a single test of 67 minutes, the raw data is approximately 70 GB. For each frame, we apply the Active Shape Model [9]. After acquiring landmarks of eyelids, eyeballs and mouth, we employ an extraction algorithm to get blink, eye movement and yawn information.

3.1 Face Recognition

Firstly, we employ a cascade Adaboost classifier based on Haar-like features, which is proposed by [10], for the face recognition procedure. If multiple faces are successfully recognized in the frame, the one with the largest area is regarded as the subject's face.

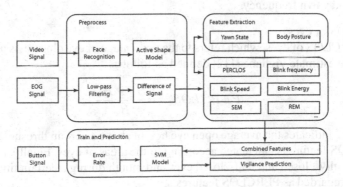

Fig. 1. The proposed vigilance analysis system by combining video and EOG features

3.2 Active Shape Model

After getting the approximate location of the subject in the image, we adopt the Active Shape Model (ASM), which is a statistical model to recognize the shape of deformable object, to get the displacements of both eyes and mouths [9]. Generally speaking, ASM is used to locates landmarks on the image by fitting the parameters of the shape using optimization techniques like gradient descent. The shape S provided by ASM is consisted of n landmarks, explicitly $S = [x_1, y_1, x_2, y_2, ..., x_n, y_n]$, where x_i, y_i are the coordinates of the i_{th} landmark. In our vigilance system, there are totally 68 landmarks for the ASM Model. The distance $D(i, j)$ between landmarks i and j is defined as $D(i, j) = \sqrt{(x_i - x_j)^2 + (y_i - y_j)^2}$. Furthermore, we calculate the current average height of both eyes and that of the mouth according to: $H_e = \frac{D(35,33) - D(34,32)}{2}$ and $H_m = D(64, 57)$. Similarly, the approximate areas A_e, A_m are calculated according to the polynomial shape of the eyes and mouth.

3.3 Preprocess of EOG Signals

Firstly, in order to extract blink features from the EOG signals, we filter the vertical EOG signal by a low-pass filter with a frequency of 10Hz using EEGLab [11]. Afterwards, a multiplier is used to adjust the amplitude of the signals. In order to obtain the variance ratio from the EOG signals, we compute the difference of signals for blink feature extraction. Denote D as the difference signal, V as the signal and R as the sampling rate, we have $D(i) = (V(i + 1) - V(i)) \times R$.

4 Feature Extraction

In this section, we will briefly introduce how we extract the features based on video and EOG signals. Both kinds of features are extracted based on a time window of 8 seconds from a 67-minutes test.

4.1 Video Features

The main features we extracted from the video include PERCLOS, blink frequency, eye movement and yawn frequency.

– PERCLOS of eyes
  The PERCLOS of eyes, which refers to the percentage of eye closure, is an efficient feature to estimate vigilance [12] defined as follows:

$$PERCLOS_e = \frac{\overline{H}_e - H_e}{\overline{H}_e}$$

where $\overline{H}_e$ indicates the average open eye height above a certain threshold. Another PERCLOS feature is calculated according to the areas of eyes. Finally, the proportions of fully-closed and half-closed eyes and mouth during a certain time window are also regarded as PERCLOS features.

- Blink frequency

The frequency of blinks also has a strong relationship to vigilance. We setup four thresholds H_{c1}, H_{c2}, H_{o1}, H_{o2}, , which indicate the relative height when eyes are about to close, already closed, about to open and fully open. This procedure suggests a complete blink. The times of eye blinks during a time window is calculated as vigilance features.

- Eye Movement

The relative position of the pupil can be recognized by ASM. Thus the moving frequency of eyes are recorded as an important feature. During each time window, we calculate the movement of eye pupils and its the amplitude. The speed of the eye movement is also calculated as a feature.

- Yawn frequency

Since an action of yawn suggest significantly that the subject has already been fatigued. The window size w for yawn state should be large enough such as 16 seconds $\times$ 30 frames. Denote the average of the least k heights of mouth as H_m^k

$$Y_i = \frac{\sum_{j=i-w}^{i}(H_j/H_m^k) > C}{w}$$

Here C is a threshold and indicates the ratio between open mouth height and normal mouth height when the subject is about to yawn.

- Body Posture

We estimate the posture of body by locating the relative position of eyes, nose and mouth on the face. Denote orientation of the face as α, θ and β, corresponding to different reference of eyes, nose and mouth. This degree can be calculated as follows:

$$\alpha = \frac{D(67,2)}{D(67,12)}; \theta = \frac{D(31,0)}{D(36,14)}; \beta = \frac{D(66,3)}{D(66,11)}$$

where points $67, 66, 31, 36$ denote the center of the nose, mouth, left and right pupil separately while the others indicate the left and right side of the face horizontally and correspondingly.

4.2 EOG Features

- Blink Features

After the preprocess of EOG signals, every blink is marked at four time points $c1, c2, o1$ and $o2$, which indicate the time when the eye is to close, closed, to open and opened. Denote V as the signal, D as the difference of signal, we have the following features:

$$T_{blink} = T_{o2} - T_{c1}; \quad T_{close} = T_{c2} - T_{c1}$$

$$T_{open} = T_{o2} - T_{o1}; \quad T_{closed} = T_{o2} - T_{c2}$$

$$S_{close} = \frac{\sum_{i=T_{c1}}^{T_{c2}} D_i}{T_{close}}; \quad S_{open} = \frac{\sum_{i=T_{o1}}^{T_{o2}} D_i}{T_{open}}; \quad E_{blink} = \sum_{i=T_{c1}}^{T_{o2}} V_i^2$$

where T indicates the time during a window size, S indicates the speed, and E indicates the energy of blinks.

- Eye Movements
  Two kinds of eye movements, Slow Eye Movement (SEM) and Rapid Eye Movement (REM) are extracted, according to different kinds of time threshold in [13]. In order to get these features more accurately, two methods of Fourier transformation and wavelet transformation are used. In the Fourier transformation method, we use a band-pass filter with frequency 0.5Hz and 2Hz to process the horizontal EOG signal. The sampling rate is 125Hz and the period is 8 seconds.

4.3   Linear Dynamic System

Considering the fact that both video and EOG features introduce much noise, we adopt the linear dynamical system, which is proposed in [14] to process them. As an unsupervised learning method, LDS can increase the main component of the features and reduce the noise, leading to a higher correlation with vigilance.

Finally, all the features are normalized between 0 and 1. Afterwards, the features are eventually used for training and prediction.

5   Experiments

There are totally five healthy subjects for our experiments, including four men and 1 woman, all of whom are around 23 years old. Particularly, we ensure that none of the subjects is color-blind. Each subject is asked to have sufficient sleep at night and get up early in the morning. Each experiment is conducted after lunch and lasts for 67 minutes so that the subject behaves sober at first and sleepy after a period of about half an hour in the experiment. The room is in silence and the light is soft.

As is stated in Section 2, the subject is asked to press the button with the same color as the traffic sign displayed on the screen. With the help of NeuroScan Stim Software, we can collect error rate data and EOG signals, as well as video images from the infrared camera.

We use LibSVM [15] to train and test data and use the square correlation coefficient and mean squared error to evaluate the model. The parameters of the SVM Model are selected as

$$s = 3(\epsilon - \text{SVR}), \quad t = 2(\text{RBF kernel}), \quad c = 8, \quad g = 1/64, \quad p = 1/1024$$

Data for each subject is 400 points long. It is divided into 2 parts with the same length. After the data is divided, the first part is used as the testing set while the other one is used as the training set. Fig. 2 indicates the example of prediction result for the second part of the subject 1.

The correlation and squared error are displayed in Table 1.

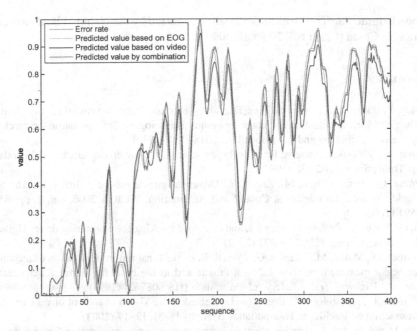

Fig. 2. Example of comparison between error rate and different prediction methods

Table 1. Squared correlation coefficient and Mean squared error of regression result

Subject	Video-based	EOG-based	Combinination
1	0.731/0.0256	0.843/0.0136	**0.852/0.0117**
2	0.778/0.0129	0.892/0.0064	**0.919/0.0170**
3	0.750/0.0151	0.866/0.0148	**0.882/0.0111**
4	0.750/0.0175	0.929/0.0091	**0.937/0.0045**
5	0.756/0.0170	0.809/0.0051	**0.921/0.0072**
Average	0.752/0.0882	0.88/0.0098	**0.898/0.0089**

6 Conclusions and Future Work

In this paper, we have proposed a novel system for vigilance analysis based on both video and EOG features. From the experimental results, we can arrive at the conclusion that our new system offers a good prediction of the actual vigilance level. Moreover, this method outperforms the existing approaches using either video features or EOG features alone, since our proposed method utilizes both the accuracy of EOG signals but the yawn state and body postures provided by video as well. In the future, we plan to utilize comprehensive features including depth information and grip power to get a better performance. Besides, more experiments will be conducted and the stability and robustness of the algorithms are expected to be improved.

Acknowledgments. This work was partially supported by the National Basic Research Program of China (Grant No. 2009CB320901).

References

1. May, J., Baldwin, C.: Driver fatigue: The importance of identifying causal factors of fatigue when considering detection and countermeasure technologies. Transportation Research Part F: Traffic Psychology and Behaviour 12(3), 218–224 (2009)
2. Stutts, J., Wilkins, J., Vaughn, B.: Why do people have drowsy driving crashes. A Foundation for Traffic Safety, 202-638 (1999)
3. Wang, Q., Yang, J., Ren, M., Zheng, Y.: Driver fatigue detection: a survey. In: The Sixth World Congress on Intelligent Control and Automation, WCICA 2006, vol. 2, pp. 8587–8591 (2006)
4. Ji, Q., Yang, X.: Real-time eye, gaze, and face pose tracking for monitoring driver vigilance. Real-Time Imaging 8(5), 357–377 (2002)
5. Dinges, D., Mallis, M., Maislin, G., Powell, I., et al.: Final report: Evaluation of techniques for ocular measurement as an index of fatigue and as the basis for alertness management. National Highway Traffic Safety Administration (HS 808762) (1998)
6. Fletcher, L., Apostoloff, N., Petersson, L., Zelinsky, A.: Vision in and out of vehicles. IEEE Transaction on Intelligent Transportation Systems 18(3), 12–17 (2003)
7. Ma, J., Shi, L., Lu, B.: Vigilance estimation by using electrooculographic features. In: Proceedings of 32nd International Conference of the IEEE Engineering in Medicine and Biology Society, pp. 6591–6594 (2010)
8. Wei, Z., Lu, B.: Online vigilance anaysis based on electrooculography. In: The 2012 International Joint Conference on Neural Networks, pp. 1–7 (2012)
9. Cootes, T., Taylor, C., Cooper, D., Graham, J., et al.: Active shape models - their training and application. Computer Vision and Image Understanding 61(1), 38–59 (1995)
10. Viola, P., Jones, M.: Rapid object detection using a boosted cascade of simple features. In: Proceedings of Computer Vision and Pattern Recognition (2001)
11. Delorme, A., Makeig, S.: Eeglab: an open source toolbox for analysis of single-trial eeg dynamics including independent component analysis. Journal of Neuroscience Methods 134(1), 9–21 (2004)
12. Dinges, D., Grace, R.: Perclos: A valid psychophysiological measure of alertness as assessed by psychomotor vigilance. Federal Highway Administration. Office of Motor Carriers, Tech. Rep. MCRT-98-006 (1998)
13. Bulling, A., Ward, J., Gellersen, H., Troster, G.: Eye movement analysis for activity recognition using electrooculography. IEEE Transactions on Pattern Analysis and Machine Intelligence (99), 1–1 (2011)
14. Shi, L., Lu, B.: Off-line and on-line vigilance estimation based on linear dynamical system and manifold learning. In: Proceedings of 32nd International Conference of the IEEE Engineering in Medicine and Biology Society, pp. 6587–6590 (2010)
15. Chang, C., Lin, C.: Libsvm: a library for support vector machines. ACM Transactions on Intelligent Systems and Technology 2 (2011)

Sensorless Speed Control of Hystersis Motor Based on Model Reference Adaptive System and Luenberger Observer Techniques

Abolfazl Halvaei Niasar[1], Hassan Moghbelli[2], and Mojtaba Yavari[3]

[1,3] Dept. of Electrical Engineering, University of Kashan, Kashan, P. Box: 87317-51167, Iran
halvaei@kashanu.ac.ir
[2] Dept. of Mathematics, Texas A&M University, TX, USA
hamoghbeli@yahoo.com

Abstract. There is no available sensorless technique for hysteresis motors. This paper presents a comparative study of two different speed observer techniques for hysteresis motor drive. Luenberger observer (LO) and model reference adaptive system (MRAS) techniques are considered. The speed estimation algorithm based on both techniques are derived and implemented in Simulink. The robustness of both speed estimation algorithms against load torque changes is investigated. Hysteresis motors are super high speed motors that are used in different type of applications such as micro gas turbines, high-speed centrifuges, and machine tool spindle drives. They are basically synchronous motors; however they can also work in asynchronous mode. For some special applications, hysteresis motor may work at asynchronous mode, and knowing the motor's speed characteristics is necessary. Moreover, in these cases, using speed sensors is impossible due to the load and difficulty of installation; therefore, using sensorless methods is the proper solution. Simulation results are presented and discussed.

Keywords: Control, Estimation, Hysteresis Motor, Sensorless Control, Luenberger Observer, MRAS.

1 Introduction

Hysteresis motors have a comparatively small output for their mechanical dimensions compared with other types of motor. But they offer the advantages of extremely low vibration and noise levels, and so are widely used as driving motors in acoustic equipment and uranium gas centrifuges. The construction of the hysteresis motor is of stator and rotor parts. The stator has conventional stator windings while the rotor comprises a solid rotor hysteresis ring of permanent-magnet material with no teeth or polar projection. The starting of the motor is due to the hysteresis losses induced in the rotor of the motor. The starting current is at most 200% of its rated load current, and it can pull into synchronism any load inertia coupled to its shaft. So, hysteresis motor under full load condition doesn't need to any speed sensor or closed-loop control systems [1-2].

T. Huang et al. (Eds.): ICONIP 2012, Part V, LNCS 7667, pp. 455–464, 2012.

In some high speed applications, due to load considerations, speed rising and falling of the motor to synchronous speed takes some hours. The failure of the power supply or frequency inverter causes the motor works at asynchronous mode. If the lack of the power is lasting for prolong time, slip frequency crosses a certain level, motor cannot reach to synchronous frequency and falls down. In this condition, the inverter frequency should decrease around the speed motor. However, installation of the speed sensor is difficult or impossible or increases the cost and complexity of the drive system and reduces the robustness of the overall system. Therefore, using suitable speed estimation techniques is effective solution [3-4].

Speed sensorless estimation has been used extensively over the past few decades in motor drives. Various speed estimation algorithms and speed sensorless control methods have been reported in literature. Most of the reported methods have been developed for PMS motors, IM, BLDC motors. The main techniques can be grouped into the following categories: using observers, MRAS techniques, using back-emf methods, using measurable variables including voltage and current, intelligent methods [5-6].

Unfortunately, no methods have been reported for hysteresis motors yet. Due to similarity of hysteresis motor to PM and IM motors, it seems wisely to employ effective sensorless methods of PM and IM motors for hysteresis motor such as MRAS and Luenberger observers. Model reference adaptive system (MRAS) is one of the famous speed observers usually used for estimation motor speed. MRAS speed observer defines two models: reference model and adjustable model. That yield and compares two similar output signals like flux, back-emf or stator current. One of the models does not involve the rotor speed while the other model needs the estimated rotor speed. The error between two models is driven to zero by speed adaptive law. A Luenberger observer is another method proposed for motor speed estimation. In this method, first dynamic model of system and system state equations derived, and Observable state and unobservable state identified, then gain of the observer obtained. Observer gain usually determined by pole placement methods [7-8].

In this paper, the dynamic model of hysteresis motor is developed firstly. Then, an adaptive model reference system (MRAS) speed estimator is developed to estimate of the hysteresis motor's speed. Afterwards, speed estimator based on Luenberger observer is developed. Finally the capability of both methods is confirmed via simulation in Matlab/ Simulink toolbox.

2   Dynamic Modeling of Hysteresis Motor

Figure 1 shows a general equivalent circuit of hysteresis motor. The difference between the model of hysteresis and general synchronous motors comes from the modeling of rotor material that for the hysteresis motor is so different. In this circuit, L_g is the value of air gap inductance, L_o is unsaturated incremental inductance and L_p is saturated incremental inductance are calculated from:

$$L_g = \frac{3\pi}{8} \frac{N_s^2 \mu_o r_g l}{g_e} \tag{1}$$

$$L_o = \frac{3\pi}{8} \frac{N_s^2 \mu_o \mu_{ro} t_r l}{r_h} \tag{2}$$

$$L_p = \frac{3\pi}{8} \frac{N_s^2 \mu_o \mu_p t_r l}{r_h}$$ (3)

The lag angle is independent of the frequency of rotor magnetization but it depends on the area of the hysteresis loop [2,9]. At synchronous speed, the fundamental eddy current torque is zero and the operation of the motor is accomplished exclusively by the hysteresis torque that is developed from hysteresis power of B-H loop. The hysteresis power can be represented as hysteresis resistance R_h that may be approximated from the power loss approach as:

$$R_h = \frac{E_g^2}{P_h} = \frac{mE_g^2}{4B_r H_c f V_r}$$ (4)

However, at any speed except to synchronous speed, the motor torque is due to both the hysteresis and eddy current effects as shown in Fig 3. The representation of the rotor eddy current is carried out by equivalent resistance R_e as:

$$R_e = \frac{12\rho l_h}{10^4 A_h}$$ (5)

Figure 2 shows the dynamic model of hysteresis motor in d-q reference frame. This model is the general d-q model of a synchronous motor. The difference between the model of hysteresis and general synchronous motors comes from the modeling of rotor material that for the hysteresis motor is so different. For this purpose the steady-state model of the rotor in hysteresis motor from Figure 1 is used. The hysteresis motor voltage and linkage flux equations in the synchronously d-q rotating reference frame is represented by [10-11]:

$$\begin{cases} v_{qs} = p\lambda_{qs} + \omega_r \lambda_{ds} + r_s i_{qs} \\ v_{ds} = p\lambda_{ds} - \omega_r \lambda_{qs} + r_s i_{ds} \\ v_{os} = p\lambda_{os} + r_s i_{os} \end{cases}$$ (6)

$$\begin{bmatrix} \lambda_{qs} \\ \lambda_{ds} \\ \lambda_{os} \\ \lambda_{qr}' \\ \lambda_{dr}' \\ \lambda_{or}' \end{bmatrix} = \begin{bmatrix} L_{ls}+L_m & 0 & 0 & L_m & 0 & 0 \\ 0 & L_{ls}+L_m & 0 & 0 & L_m & 0 \\ 0 & 0 & L_{ls} & 0 & 0 & L_m \\ L_m & 0 & 0 & L_{lr}'+L_m & 0 & 0 \\ 0 & L_m & 0 & 0 & L_{lr}'+L_m & 0 \\ 0 & 0 & L_m & 0 & 0 & L_{lr}' \end{bmatrix} \begin{bmatrix} i_{qs} \\ i_{ds} \\ i_{os} \\ i_{qr}' \\ i_{dr}' \\ i_{or}' \end{bmatrix}$$ (7)

which the magnetizing and the rotor parameters, using figure 2 are derived from the following:

$$x_m = (\omega L_g) \| (\omega L_o) \tag{8}$$

$$x'_{lr} = \omega (L_p + L_h) \tag{9}$$

$$\omega L_h = \frac{R_h}{\tan \beta} \tag{10}$$

$$r'_r = R_h \| \frac{R_e}{s} \tag{11}$$

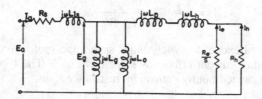

Fig. 1. Equivalent circuit of hysteresis motor

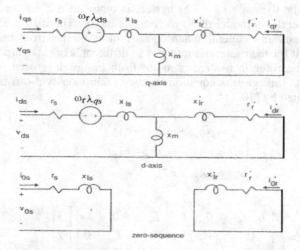

Fig. 2. Hysteresis motor model in d-q reference frame

where $\omega L_h = X_h$ is the equivalent reactance of hysteresis ring in rotor and x'_{lr}, r'_r are the leakage reactance and equivalent resistance of the rotor. The definitions and corresponding values of the parameters for employed hysteresis motor are listed in Table 1. The electromagnetic torque and the rotor speed are obtained from:

$$T_{em} = \frac{3}{2}\frac{P}{2}(\lambda_{ds}i_{qs} - \lambda_{qs}i_{ds})$$ (12)

$$T_{em} - T_{mech} = \frac{2J}{p}\frac{d\omega_r(t)}{dt}$$ (13)

Table 1. Parameters of Hysteresis Motor

Number of poles	p	2	
Rated output power	P	60	[W]
Rated voltage	V_s	400	[V]
Rated frequency	f_s	100	[Hz]
Stator leakage reactance	X_{ls}	152	[Ω/ph]
Stator resistance	r_s	36	[Ω/ph]
Equivalent resistance due to eddy current	R_e	3288	[Ω/ph]
Equivalent resistance due to hysteresis ring	R_h	127	[Ω/ph]
Equivalent reactance due to hysteresis ring	X_h	163.7	[Ω/ph]
Unsaturated incremental reactance	X_o	451	[Ω/ph]
Saturated incremental reactance	X_ν	13	[Ω/ph]
Air gap equivalent reactance	X_g	1217	[Ω/ph]

3   MRAS Based Speed Estimator

MRAS is based on the comparison between the outputs of two observers. The observers are used to calculate the instantaneous reactive power. Figure 3 illustrates the structure of MRAS for speed estimation.

The input data required for this model is the stator voltage and stator current. In the d-q reference frame, two sets of equation are developed to compare reactive power of the hysteresis motor in the reference model and adaptive model. The reference model doesn't involve the rotor speed, while the adaptive model needs the estimated rotor speed to adjust the computed reactive power to that computed from the reference model.

3.1   Reference Model

The reactive power in the reference model is computed from cross product of the stator current and voltage as follows:

$$Q_{ref} = 1.5 \times (v_{qs}i_{ds} - v_{ds}i_{qs})$$ (14)

3.2   Adaptive Model

The active power in the adaptive model is computed by substituting (6) and (7) in (14) as follows:

$$Q_{est} = 1.5 \times [\, i_{ds} \times (l_s p i_{qs} + \omega_r l_s i_{ds}) - i_{qs} \times (l_s p i_{ds} - \omega_r l_s i_{qs}) + r_s i_{ds}\,] \tag{15}$$

The error between two models is used to drive a suitable proportional integrator PI controller which an estimation for rotor speed ω_r.

4   Speed Estimator Based on Luenberger Observer

As respect of the dynamic model of hysteresis model rotor flux and stator current are calculated as follows:

$$\frac{di_s}{dt} = \frac{1}{\sigma}\left[\begin{array}{l} V_s - \left(r_s - \dfrac{l_m^2}{l_r^2} r_r\right) i_s + \dfrac{\hat{\omega}_r l_m}{l_r}\lambda_r \\[2mm] + \hat{\omega}_r \sigma i_s + \dfrac{l_m r_r}{l_r^2}\lambda_r \end{array} \right] \tag{16}$$

$$\frac{d\lambda_r}{dt} = -\frac{r_r}{l_r^2}\left(\lambda_r - l_m i_s\right) \tag{17}$$

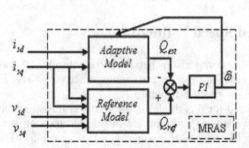

Fig. 3. Block diagram of MRAS estimator

At first, the state observer is designed for flux and current estimation. Then with an adaptive algorithm speed is estimated. Since these equations are nonlinear it is not possible to use traditional linear Luenberger observer for speed estimation. So an adaptive Luenberger is used as shown in figure 4. This adaptive observer can be expressed as:

$$\frac{di_s}{dt} = \frac{1}{\sigma}\left[\begin{array}{l} V_s - \left(r_s - \dfrac{l_m^2}{l_r^2} r_r\right) i_s + \dfrac{\hat{\omega}_r l_m}{l_r}\hat{\lambda}_r \\[2mm] + \hat{\omega}_r \sigma \hat{i}_s + \dfrac{l_m r_r}{l_r^2}\hat{\lambda}_r + G_1(i_s - \hat{i}_s) \end{array} \right] \tag{18}$$

$$\frac{d\hat{\lambda}_r}{dt} = -\frac{r_r}{l_r^2}\left(\hat{\lambda}_r - l_m \hat{i}_s\right) + G_1(i_s - \hat{i}_s) \tag{19}$$

$$\sigma = l_s - \frac{l_m^2}{l_r} \tag{20}$$

Where $\hat{i}_s$ and $\hat{\lambda}_r$ are the estimated stator current and rotor flux respectively. G_1 and G_2 are the observer gain. The gain selection of the observer is based on pole placement method proposed in [7].

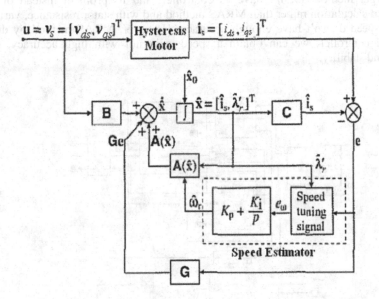

Fig. 4. Block Diagram of adaptive Luenberger observer

They can be computed as follows:

$$g_{1r} = 2a \tag{21}$$

$$g_{1i} = 0 \tag{22}$$

$$g_{1r} = a\sigma l_s l_r / l_m \tag{23}$$

$$g_{2i} = 0 \tag{24}$$

with a being the coefficient of proportionality. Using (6) and (7), stator linkage flux can be computed. Adaptive mechanism for speed estimation is obtained as following:

$$\hat{\omega}_r = \frac{\hat{\lambda}_{ds}\left(v_{qs} - R_s i_{qs}\right) - \hat{\lambda}_{qs}\left(v_{ds} - R_s i_{ds}\right)}{\hat{\lambda}^2{}_{ds} + \hat{\lambda}^2{}_{qs}} \qquad (25)$$

5   Simulation Results

The block diagram of the hysteresis motor and speed estimator methods is shown in figure 5. The system parameters for simulation are listed in table I. To show the rapid response and stability of an estimator, a variable load torque is applied to motor shaft, which it's not happened in practical (practical load varied slowly). The real and estimated speed is shown in figure 6 to 9.

Luenberger method doesn't have PI controller, and its problem instead of it is complicated calculation rather than MRAS method and with stator resistance, variation estimated speed doesn't have enough accuracy. However, MRAS method by design suitable PI controller, we can obtain a speed estimator with high accuracy, rapid response and stability.

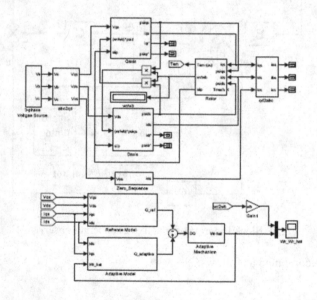

Fig. 5. Speed estimation diagram of hysteresis motor in Simulink

6   Conclusion

Comparison among the speed estimation of hysteresis motor in asynchronous conditions using Luenberger observer (LO) and model reference adaptive system (MRAS) methods was presented. These methods are beneficial, as they are able to increase the performance of the sensorless systems in terms of low speed behavior

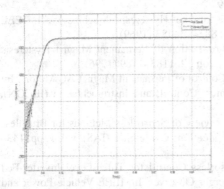
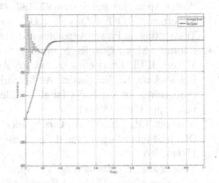

Fig. 6. Real speed (blue) and estimated speed with MRAS (red) at no load

Fig. 7. Real speed (blue) and estimated speed with Luenberger observer (red) at no load

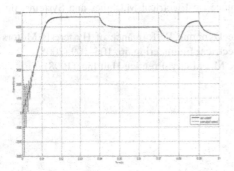
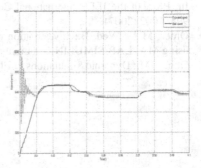

Fig. 8. Real speed (blue) and estimated speed with MRAS (red) at Variable load

Fig. 9. Real speed (blue) and estimated speed with Luenberger observer (red) at Variable load

and under different load conditions. The speed of hysteresis motor at asynchronous condition was estimated under no load and under variable load conditions. Estimated speed can be used for closed loop vector /scalar control of hysteresis motor. Developed Luenberger observer is an adaptive estimator. Comparison of the results shows that (MRAS) method has better performance than Luenberger observer (LO).

References

1. Copeland, M.A., Slemon, G.R.: An Analysis of the Hysteresis Motor: part-II-The Circumferential-flux Machine. IEEE Transaction on Power Apparatus and Systems 83, 619–625 (1964)

2. Rahman, M.A.: Analytical Models for Polyphase Hysteresis Motor. IEEE Transaction on Power Apparatus and System 92, 137–242 (1973)
3. Robertson, S.D.T., Zaky, S.Z.G.: Analysis of the Hysteresis Machine-part-I. IEEE Transaction on Power Apparatus and Systems 88, 474–483 (1969)
4. Rahman, M.A., Qin, R.: Starting and Synchronization of Permanent Magnet Hysteresis Motors. IEEE Transaction on Industry Application 32, 1183–1189 (1996)
5. Rajashekara, K., Kawamura, A.: Sensorless Control of Permanent Magnet AC Motors. In: Proceedings of the IEEE Industrial Electronics, Control, and Instrumentation (IECON), vol. 3, pp. 1589–1594 (1994)
6. Johnson, J.P., Ehsani, M., Guzelgunler, Y.: Review of Sensorless Methods for Brushless DC. In: Proceedings of the IEEE Industry Applications Conference (IAS), vol. 1, pp. 143–150 (1999)
7. Yongchang, Z., Zhengming, Z.: Speed Sensorless Control for Three-Level Inverter-Fed Induction Motors Using an Extended Luenberger Observer. In: IEEE Vehicle Power and Propulsion Conference (VPPC), pp. 3–5 (2008)
8. Sai Kumar, P., Siva Kumar, J.S.V.: Model Reference Adaptive Controlled Application to the Vector Controlled Permanent Magnet Synchronous Motor Drive. International Journal of Power System Operation and Energy Management (IJPSOEM) 1, 35–41 (2011)
9. Darabi, A., Lesani, H.: Modeling and Optimum Design of Disk-Type Hysteresis Motors. In: Proc. of the IEEE International Conference on Electrical Machines and Systems, pp. 998–1002 (2007)
10. Omer, M.A., Wed, B.: Investigation of the Dynamic Performance of Hysteresis Motors using MATLAB/SIMULINK. Journal of Electrical Engineering 56, 106–109 (2005)
11. Ong, C.M.: Dynamic Simulations of Electric Machinery. Prentice-Hall Inc. (1998)

Using Hybrid Neural Networks for Identifying the Brain Abnormalities from MRI Structural Images

Lavneet Singh, Girija Chetty, and Dharmendra Sharma

Faculty of Information Sciences and Engineering
University of Canberra, Australia
{Lavneet.singh,Girija.chetty,Dharmendra.sharma}@canberra.edu.au

Abstract. In this study, we present the investigations being pursued in our research laboratory on magnetic resonance images (MRI) of various states of brain by extracting the most significant features, and to classify them into normal and abnormal brain images. We propose a novel method based on deep and extreme machine learning on wavelet transform to initially decompose the images, and then use various features selection and search algorithms to extract the most significant features of brain from the MRI images. By using a comparative study with different classifiers to detect the abnormality of brain images from publicly available neuro-imaging dataset, we found that a principled approach involving wavelet based feature extraction, followed by selection of most significant features using PCA technique, and the classification using deep and extreme machine learning based classifiers results in a significant improvement in accuracy and faster training and testing time as compared to previously reported studies.

Keywords: Deep Machine Learning, Extreme Machine Learning, MRI, PCA.

1   Introduction

Magnetic Resonance Images (MRI) is an advance technique used for medical imaging and clinical medicine and an effective tool to study the various states of human brain. MRI images provide the rich information of various states of brain which can be used to study, diagnose and carry out unparalleled clinical analysis of brain to find out if the brain is normal or abnormal. However, the data extracted from the images is very large and it is hard to make a conclusive diagnosis based on such raw data. In such cases, we need to use various image analysis tools to analyze the MRI images and to extract conclusive information to classify into normal or abnormalities of brain. The level of detail in MRI images is increasing rapidly with availability of 2-D and 3-D images of various organs inside the body.

Fully automatic normal and diseased human brain classification from magnetic resonance images (MRI) is of great importance for research and clinical studies. Recent work [2, 5] has shown that classification of human brain in magnetic resonance (MR) images is possible via machine learning and classification techniques such as artificial neural networks and support vector machine (SVM) [2], and

T. Huang et al. (Eds.): ICONIP 2012, Part V, LNCS 7667, pp. 465–472, 2012.

unsupervised techniques such as self-organization maps (SOM) [2] and fuzzy c-means combined with appropriate feature extraction techniques [5], [16], 17]. Other supervised classification techniques, such as k-nearest neighbors (k-NN), which group pixels based on their similarities in each feature image [1, 6, 7, 8] can be used to classify the normal/pathological T2-weighted MRI images. Inspired by new segmentation algorithms in computer vision and machine learning, we propose an efficient semi-automatic and deep learning algorithm for white matter (WM) lesion segmentation around Region of Interest (ROI) based on extreme and deep machine learning. Further, we compare this novel approach with some of the other supervised machine learning techniques reported previously.

2   Materials and Methods

2.1   Datasets

The input dataset consists of axial, T2-weighted, 256 X 256 pixel MR brain images (Fig. 1). These images were downloaded from the (Harvard Medical School website (http:// med.harvard.edu/AANLIB/) [9]. Only those sections of the brain in which lateral ventricles are clearly seen are considered in our study. The number of MR brain images in the input dataset is 60 of which 6 are of normal brain and 54 are of abnormal brain. The abnormal brain image set consists of images of brain affected by Alzheimer's and other diseases. The remarkable feature of a normal human brain is the symmetry that it exhibits in the axial and coronal images. Asymmetry in an axial MR brain image strongly indicates abnormality.

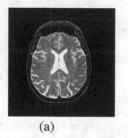 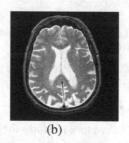 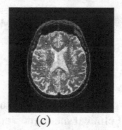

(a)   (b)   (c)

Fig. 1. (a) T2, weighted an axial MRI Brain Image; (b) T2, weighted an axial MR brain image as abnormal brain; (c) T2, weighted an axial MR brain image as normal brain after Wavelets Decomposition and denoising

A normal and an abnormal T2-weighted MRI brain image are shown in Fig. 1(a) and 1(b), respectively. Indeed, for multilayer learning models like deep and extreme machine learning algorithms needed big datasets for training, however due to lack of availability of proper datasets in MRI imaging, we used this dataset for examining the performance of proposed approaches for this paper, but acquiring other suitable datasets for future studies. At each decomposition level, the length of the decomposed signals is half the length of the signal in the previous stage. Hence the size of the approximation component obtained from the first level decomposition of an NXN image is N/2 X N/2, second level is N/4 X N/4 and so on. As the level of decomposition is increased, compact but coarser approximation of the image is obtained. Thus, wavelets provide a simple hierarchical framework for interpreting the image information.

2.2   Deep Belief Nets

DBNs [10] are multilayer, stochastic generative models that are created by learning a stack of Restricted Boltzmann Machines (RBMs), each of which is trained by using the hidden activities of the previous RBM as its training data. Each time a new RBM is added to the stack, the new DBN has a better variation lower bound on the log probability of the data than the previous DBN, provided the new RBM is learned in the appropriate way [11].

A Restricted Boltzmann Machine (RBMs) is a complete bipartite undirected probabilistic graphical model. The nodes in the two partitions are referred as hidden and visible units. An RBM is defined as

$$p(v, h) = \frac{e^{-E(v,h)}}{\sum_u \sum_g e^{-E(u,g)}} \tag{1}$$

Where $v \in V$ are the visible nodes and $h \in H$ are the latent random variables. The energy function E (v,h,W) is described as

$$E = -\sum_{i=1}^{D} \sum_{j=1}^{K} v_i W_{ij} h_j \tag{2}$$

Where $W \in R^{DXK}$ are the weights on the connections, and where we assume that the visible and hidden units both contain a node with value of 1 that acts to introduce bias. The conditional distribution for the binary visible and hidden units are defined as

$$p(v_i = 1/h, W) = \sigma(\sum_{j=1}^{K} W_{ij} h_j) \tag{3}$$

$$p(h_j = 1/v, W) = \sigma(\sum_{i=1}^{D} W_{ij} v_i) \tag{4}$$

Where σ is the sigmoid function. Using above equations, it easy to go back and forth between the layers of RBM. While training, it consists of some input to the RBM on the visible layer, and updating the weights and the biases such that p(v) is high. In generalized way, in as set of C training cases $\{v^c \mid c \in \{1,....,C\}\}$, the objective is to maximize the average log probability defined as

$$\sum_{c=1}^{C} logp(v^c) = \sum_{c=1}^{C} log \frac{\sum_g e^{-E(v^c,g)}}{\sum_u \sum_g e^{-E(u,g)}} \tag{5}$$

The whole training process involves updating the weights with several numbers of epochs and the data is split in 20 batches which we take it randomly and the weights are update at the end of every batch. We use the binary representation of hidden units activation pattern for classification and visualization. The autoencoder with N_h hidden nodes is trained and fine-tuned using back-propagation to minimize squared reconstruction error, with a term encouraging low average activation of the units.

2.3   Extreme Machine Learning

The Extreme Learning Machine [12, 13, 14] [15] [18] is a Single hidden Layer Feed forward Neural Network (SLFN) architecture. Unlike traditional approaches such as Back Propagation (BP) algorithms which may face difficulties in manual tuning control parameters and local minima, the results obtained after ELM computation are extremely fast, have good accuracy and has a solution of a system of linear equations. For a given network architecture, ELM does not have any control parameters like stopping criteria, learning rate, learning epochs etc., and thus, the implementation of this network is very simple. Given a series of training samples (x_i, y_i) $_{i=1, 2 \ldots N}$ and $\hat{N}$ the number of hidden neurons where $x_i = (x_{i1},\ldots x_{in})$ ϵR^n and $y_i = (y_{i1},\ldots y_{in})$ ϵR^m , the actual outputs of the single-hidden-layer feed forward neural network (SLFN) with activation function $g(x)$ for these N training data is mathematically modeled as

$$\sum_{k=1}^{\hat{N}} \beta_k g\big((w_k, x_i) + b_k\big) = 0_i , \forall = i = 1, \ldots, N \tag{6}$$

Where $w_k = (w_{k1},\ldots, w_{kn})$ is a weight vector connecting the kth hidden neuron, $\beta_k = (\beta_{k1},\ldots \beta_{km})$ is the output weight vector connecting the kth hidden node and output nodes. The weight vectors w_k are randomly chosen. The term (w_k, x_i) denotes the inner product of the vectors w_k and x_i and g is the activation function. The above N equations can be written as $H\beta = O$ and in practical applications $\hat{N}$ is usually much less than the number N of training samples and $H\beta \neq Y$, where

$$H = \begin{bmatrix} g\big((w_1, x_1) + b_1\big) & \cdots & g\big((w_{\hat{N}}, x_1) + b_{\hat{N}}\big) \\ \vdots & \ddots & \vdots \\ g\big((w_1, x_{1N}) + b_1\big) & \cdots & g\big((w_{\hat{N}}, x_N) + b_{\hat{N}}\big) \end{bmatrix}_{N \times \hat{N}} \tag{7}$$

The matrix H is called the hidden layer output matrix. For fixed input weights $w_k = (w_{k1},\ldots, w_{kn})$ and hidden layer biases b_k, we get the least-squares solution $\hat{\beta}$ of the linear system of equation $H\beta = Y$ with minimum norm of output weights β, which gives a good generalization performance. The resulting $\hat{\beta}$ is given by $\hat{\beta} = H+$ Ywhere matrix H^+ is the Moore-Penrose generalized inverse of matrix H [14].

2.4   Trained Classifiers and Feature Selection Evaluators

In this study, apart from deep learning based on Restricted Boltzmann machines and extreme machine learning based on Single hidden Layer Feed forward Neural Network (SLFN) architecture as classifiers, several other classifiers are also examined in terms of accuracy and performance, including K-nearest neighbor, SVM , Naive Bayes, MultiboostAB, Rotation Forest, VFI, J48 and Random Forest.

To reduce the dimensionality of the large set of features of dataset, in our study, we propose the use of three optimal attribute selection algorithms: correlation based feature selection (CFS) method, which evaluates the worth of a subset of attributes by considering the individual predictive ability of each feature along with the degree of redundancy between them, secondly an approach based on wrappers which evaluates attribute sets by using a learning scheme. Also in this study, three search methods are also examined: the Best First, Greedy Stepwise and Scatter Search algorithms. These search algorithms are used with attribute selector's evaluators to process the greedy forward, backward and evolutionary search among attributes of significant and diverse subsets. In total, these feature selection algorithms were tested to select nearly 10 optimal and significant features out of 1024 features. The whole proposed method is implemented using Weka 3.6 platform.

3   Experiments and Results

3.1   Level of Wavelet Decomposition

We obtained wavelet coefficients of 60 brain MR images, each of whose size is 256 X 256. Level-1 HAR wavelet decomposition of a brain MR image produces 16384 wavelet approximation coefficients; while level-2 and level-3 produce 4096 and 1024 coefficients, respectively. The preliminary experimental analysis of the wavelet coefficients through simulation in Matlab 7.10., we showed that level-2 features are the best suitable for different classifiers, whereas level-1 and level-3 features results in lower classification accuracy. We also use the DAUB-4 (Daubachies) as mother wavelets to get decomposition coefficients of MRI images at Level 2 for comparative evaluation of two wavelets decomposition methods in terms of classification accuracy.

3.2   Attribute Selection and Classification

The second step after Wavelet decomposition of MRI images is to select significant features among whole set of coefficients. Table 1 shows the accuracy of classification (percentage of correctly classified samples), True Positive Rate (TP), False Positive Rate (FP) and Average Accuracy (ACC) over all pair-wise combination with different feature evaluators and search algorithms with respect to multi-class classification.

Table 1 shows the performance of several learning classifiers, including K-nearest neighbor, SVM, Naive Bayes, MultiboostAB, Rotation Forest, VFI, J48 and Random Forest. Among the pair-wise classification, the lowest accuracy is observed for the classification VFI classifiers of 74.16% and the highest accuracy for the classification by Rotational forest of 97.06%. Moreover, the combination of CFS feature evaluator with the Best First search algorithm gives the highest classification accuracy compared to other feature evaluators and search algorithms. While Table 1 shows the performance of indivual classifiers, Table 2 compares the proposed method against a popular dimensionality reduction method, known as Principal Component Analysis (PCA). PCA applies an orthogonal linear transformation that transforms data to a new coordinate system of uncorrelated variables called principal components. We have applied PCA to reduce the number of attributes or feature to 18 attributes and plotted the ROC curves using several above mentioned learning classifiers in terms of True Positive and False Positive Rate, as seen in figure 2. As can be seen in figure 2, ROC curves for all the trained learning classifiers examined in this study, the curves lie above the diagonal line describing the better classification rather than any other random classifiers. The optimal points of various trained classifiers are indicated by bold solid points as False Positive rate (FP) and True Positive rate (TP). These optimal points in ROC curves show the maximum optimal value (FP, TP) of all trained classifiers.

Table 1. Various Classifiers comparision with respect Average Classification Accuracy(%) and other parameters

Classifiers	TP Rate	FP Rate	Precision	Recall	F-Measure	(ACC %)
KNN	0.935	0.917	0.826	0.853	0.839	91.04
SVM	0.912	0.912	0.831	0.912	0.87	91.17
Naive Bayes	0.868	0.916	0.828	0.868	0.847	86.76
MultiboostAB	0.91	0.91	0.829	0.91	0.868	91.04
Rotation Forest	0.971	0.285	0.971	0.971	0.968	97.06
VFI	0.742	0.049	0.93	0.742	0.796	74.16
J48	0.96	0.314	0.958	0.96	0.957	95.98
Random Forest	0.97	0.271	0.97	0.97	0.968	97.01

Table 2. Comparison using PCA and other feature attribute evaluators in terms of ACC (%)

Classifier	PCA (%)	CFS-Best First (%)	Wrapper-Best First (%)
KNN	91.38	91.04	89.32
SVM	96.24	91.17	90.65
Naive Bayes	85.63	86.76	85.44
MultiboostAB	94.52	91.04	89.39
Rotation Forest	97.06	97.06	93.78
VFI	77.12	74.16	72.22
J48	95.34	95.98	95.98
Random Forest	97.34	97.01	96.25

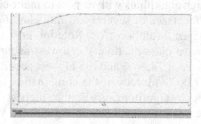

Fig. 2. Shows the ROC curve of the above mentioned trained classifiers

Table 3 describes the classification results using Extreme Machine Learning and Deep Machine Learning. In table 3, we compared the training time, testing time and classification error using extreme and deep machine Learning. As we can see in the table both learning algorithms are processed to many hidden layers and their evaluations is done in terms of various factors. As depicted in Table 3, it clearly shows that deep machine learning plays a major role in reducing the classification error. As Deep and extreme machine learning are designed to work on large datasets for it is difficult to compare the performance. However, they result in acceptable accuracy levels, and we are currently examining several other publicly available large MRI datasets for enhancing the performance of these two novel approaches (Deep learning and Extreme machine learning approaches).

Table 3. Classification results using Extreme Machine Learning and Deep Machine Learning

	Training Time(s)			Testing Time(s)			Classification Error		
Hidden Layers	**10**	**15**	**20**	**10**	**15**	**20**	**10**	**15**	**20**
Deep Learning	0.56	0.47	0.72	0.51	0.34	0.64	0.083	0.065	0.071
Extreme Learning	0.31	0.31	0.61	0.41	0.31	0.56	0.042	0.042	0.061

However, the deep learning networks do not need any particular feature reduction algorithms because of the inherent capability for feature reduction in terms of deep learning (learning through multiple layers). In case of extreme machine learning, the learning proceeds through random assignment of weights and hidden nodes (unlike gradient descendent based techniques). Due to this, there is a significant improvement in training and testing time as depicted in Table 3.

4   Conclusions

In this study, we have presented a principled approach for investigating brain abnormalities based on wavelet based feature extraction, PCA based feature selection and deep and extreme machine learning based classification comparative to various others classifiers. Experiments on a publicly available brain image dataset show that the proposed principled approach performs significantly better than other competing methods reported in the literature and in the experiments conducted in the study. The classification accuracy of more than 93% in case of deep machine learning and 94% in case of extreme machine learning demonstrates the utility of the proposed method. In this paper, we have applied this method only to axial T2-weighted images at a particular depth inside the brain. The same method can be employed for T1-weighted, proton density and other types of MR images. With the help of above approaches, one can develop software for a diagnostic system for the detection of brain disorders like Alzheimer's, Huntington's, Parkinson's diseases etc. Further, the proposed approach uses reduced data by incorporating feature selection algorithms in the processing loop and still provides an improved recognition and accuracy. The training and testing time for the whole study used by deep and extreme machine learning is much less as compared to SVM and other traditional classifiers reported in the literature. Further work will be pursued to classify different type of abnormalities, and to extract new features from the MRI brain images on various parameters as age, emotional states and their feedback.

References

1. Fletcher, H.L.M., Hall, L.O., Goldgof, D.B., Murtagh, F.R.: Automatic segmentation of non-enhancing brain tumors in magnetic resonance images. Artificial Intelligence in Medicine 21, 43–63 (2011)
2. Sandeep, C., Patnaik, L.M., Jagannathan, N.R.: Classification of magnetic resonance brain images using wavelets as input to support vector machine and neural network. Biomedical Signal Processing and Control 1, 86–92 (2006)

3. Gorunescu, F.: Data Mining Techniques in Computer-Aided Diagnosis: Non-Invasive Cancer Detection. PWASET 25, 427–430 (2007)
4. Kara, S., Dirgenali, F.: A system to diagnose atherosclerosis via wavelet transforms, principal component analysis and artificial neural networks. Expert Systems with Applications 32, 632–640 (2007)
5. Maitra, M., Chatterjee, A.: Hybrid multi-resolution Slantlet transform and fuzzy c-means clustering approach for normal-pathological brain MR image segregation. Med. Eng. Phys. (2007), doi:10.1016/j.medengphy.06,009
6. Abdolmaleki, P., Futoshi, M., Kouji, M.: Lawrence Danso Buadu.: Neural networks analysis of astrocyticgliomas from MRI appearances. Cancer Letters 118, 69–78 (1997)
7. Rosenbaum, T., Volkher, E., Wilfried, K., Ferdinand, A.D., Hoehn-Berlagec, M., Lenard, H.G.: MRI abnormalities in neuro-bromatosis type 1 (NF1): a study of men and mice. Brain & Development 21, 268–273 (1999)
8. Cocosco, C., Alex, Z.P., Evans, A.C.: A fully automatic and robust brain MRI tissue classification method. Medical Image Analysis 7, 513–527 (2003),
9. Database taken, http://med.harvard.edu/AANLIB/
10. Hinton, G.E., Salakhutdinov, R.R.: Reducing the dimensionality of data with neural networks. Science 313, 504–507 (2006)
11. Hinton, G.E., Osindero, S.: A fast learning algorithm for deep belief nets. Neural Computation 18, 1527–1554 (2006)
12. Lin, M.B., Huang, G.B., Saratchandran, P., Sudararajan, N.: Fully complex extreme learning machine. Neurocomputing 68, 306–314 (2005)
13. Huang, G.B., Zhu, Q.Y., Siew, C.K.: Extreme Learning Machine: Theory and Applications. Neurocomputing 70, 489–501 (2006)
14. Serre, D.: Matrices: Theory and Applications. Springer Verlag, New York Inc. (2002)
15. Anurag, M., Lavneet, S., Girija, C.: A Novel Image Water Marking Scheme Using Extreme Learning Machine. In: Proceedings of IEEE World Congress on Computational Intelligence (WCCI 2012). IEEE Explore, Brisbane (2012)
16. Lavneet, S., Girija, C.: Hybrid Approach in Protein Folding Recognition using Support Vector Machines. In: Proceedings of International Conference on Machine Learning and Data Mining (MLDM 2012), Berlin, Germany. LNCS. Springer (2012)
17. Lavneet, S., Girija, C.: Review of Classification of Brain Abnormalities in Magnetic Resonance Images Using Pattern Recognition and Machine Learning. In: Proceedings of International Conference of Neuro Computing and Evolving Intelligence, NCEI, Auckland, New-Zealand, LNCS Bioinformatics. Springer (2012)
18. Lavneet, S., Girija, C.: A Novel Approach for protein Structure prediction Using Pattern Recognition and Extreme Machine Learning. In: Proceedings of International Conference of Neuro Computing and Evolving Intelligence, NCEI, Auckland, New-Zealand. LNCS Bioinformatics. Springer (2012)

ANN for Multi-lingual Regional Web Communication

Kolla Bhanu Prakash[1], M.A. Dorai Rangaswamy[2], and Arun Raja Raman[3]

[1] Sathyabama University, Chennai, India
[2] AVIT, Chennai, India
[3] IIT Madras, Chennai 600 036, India
bhanu_prakash231@rediff.com

Abstract. In India web development and communication are very much on the rise with cheaper mobile communication being the catalyst. Use of mobile phones has transformed the culture of communication with even villagers using sophisticated computer-related words like SMS. But the major complexity arises when web documents in regional languages are displayed. Understanding the content of the document and later communication through oral or text means, becomes difficult and this is the area the current paper addresses and in the process tries a model for how the knowledge is created in the minds of illiterate user. The paper first presents how letters and words which form the basis of text-based communication can be used for content and later content-related words are chosen as bases for training in ANN. A comparison with statistical – termed algorithmic approach, here is made to bring out how ANN could be more effective.

Keywords: Media Mining, Multi-Lingual, Web Communication.

1   Introduction

In India use of cell phones and consequently communication in different levels is on an unprecedented rise and this will definitely pave way to use of internet and e-commerce in a big way. But the major hurdle is the variations in language, customs, tradition and established practices which result in sticking to a particular language-mother tongue-making web communication more complex. So, web pages are developed in different languages with varying contents to suit different regions and dialect. For example the word 'chai' which means tea is used in almost all Indian languages with local script. The focus of the current study is to extract content of such web documents without going through translation or conventional data mining approaches as most of the web documents have media-related data like images or handwritten texts. This led to the basic premise on communication where the mind translates a text or media and later interacts or just interacts with content based understanding. The latter approach is taken and studies related to a particular discipline are presented to focus attention on how neural processing and later training will make content extraction faster.

T. Huang et al. (Eds.): ICONIP 2012, Part V, LNCS 7667, pp. 473–478, 2012.

2   Feature of Indian Regional Languages

Indian languages are very much different from European –German or Russian-or other Asian languages- like Japanese or Persian- in that regional customs and practices bring in certain commonalities like the scripts of Tamizh or Telugu or Kannada have similarities of different kinds as compared to the north-ern Hindi or Punjabi scripts. But English being the link language both in com-munication and forms the basis in higher education, some complexities in migrating from English to regional language or vice versa exist like the ones shown in Fig.1. Here a letter 'a' which can be used as a word in English, if written in three languages Hindi, Tamizh and Telugu as shown in Fig 1(a), has no meaning and if one wants to replicate 'a' as it sounds, different sets of cha-racters are needed. Similarly letters in English like 'x' need one character while in the other three languages several as shown in Fig.1(b). Similarly character in regional language if written in others poses different context level.

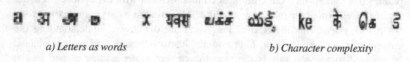

a) Letters as words   *b) Character complexity*

Fig. 1. Complexities in Indian languages with English

Web documents developed in Indian regions vary profusely in content as well in form and Fig.2 gives three different examples. The first one in Fig. 2(a) shows the

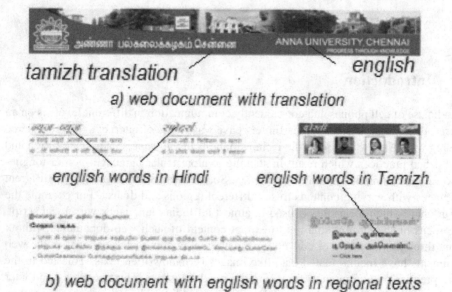

Fig. 2. Web document variations in form and text-local level

English text 'ANNA UNIVERSITY, CHENNAI' on the right and translated Tamizh version on the left. Whereas in Fig.2(b) we have two language texts-Hindi and Tamizh which use English words 'news and views' and 'on-line' written in Hindi and Tamizh. Hence a person looking at such web sites has to go by content rather than translation. Similar problem exists with news in the web as shown in Fig.3. Here CNN news on the same day is shown in three different levels world, Asia and USA.

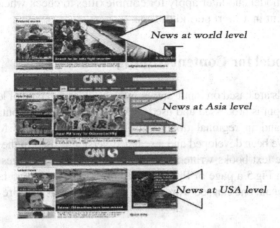

Fig. 3. Web documents -same language, content different

It may be seen clearly the content is different and similar complexity with regional news exists in India and Fig.4 shows this with national, Hindi and Telugu regions. So if one wants to continue surfing and later interact content of the web is the only way to go about [1].

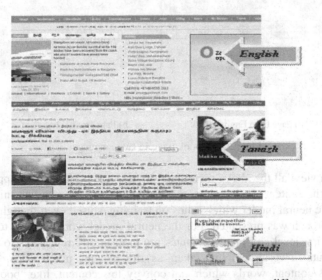

Fig. 4. Web documents in regional India –different languages, different contents

It may be seen clearly that regional web documents pose different problems in terms of comprehension, understanding and interaction in another language. Hence it is preferable to assess the content even before looking at the document fully. Images and figures do help as seen in Fig.2 (a) but many times texts and sketches with words pose problems as they reflect local dialect and flavor. So it is necessary to assess the content irrespective of the language and the way text is produced. Hence the objective is to develop a neural model and later apply for complexities to check whether it is possible to assess the content in a short period of time.

3   ANN Model for Content Mining

Since neural models are based on input-output, weights, bias and type of learning rules, it is preferable if the input is processed and used. A web document may contain texts, images, audio/video files and in regional documents hand-generated ones. Many a time the document may have been developed and a scanned version is used in the web. This is true in education where text books written by authors in regional languages are used in web pages. Typically in Fig.5 a page of Physics text book is shown in two languages English and Telugu. Here again one can see words in English are used as they are in Telugu scripts

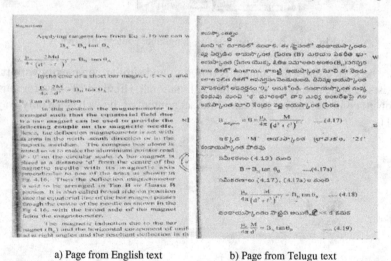

a) Page from English text   b) Page from Telugu text

Fig. 5. Text book page in two languages

Keeping in view all the above mentioned issues, it is preferable an approach based on extracting the content of the document rather than translation or data-mining seemed better and here neural model which relies on training is used even though elsewhere a statistical approach was presented. For the neural model words related to content are chosen I the first level and pixel attributes are used in different ways to get different kinds of inputs. The targets are taken as the content pixel maps. Considering 'education' as content, words which can convey this content, like 'book', 'school', 'college', 'teacher', 'student' are taken as inputs and 'education' as the target. All words in four languages are shown in Fig. 7 and the content word 'education' is also taken in four languages. Now neural network model needs to be developed using selective words from web pages so that whether they can match the content word 'education'. Since no translation is used, the only way is to get the pixel-map attributes and develop a pattern recognition approach. As an initial step the density of pixels is taken as inputs like number of pixels describing the word and normalizing it with the word size as shown in Fig.6. where four similar words 'school' in four languages are used as inputs to train a net for getting the output of 'education' in 4 languages. As the inputs are not sufficient more details are taken and results are shown for more words and more pixel attributes which were taken in three groups as described in an earlier paper [2]. Results are shown as bars in Fig.8

Fig. 6. Four words used as input/outputs

The words chosen relate to one particular word in input-'school' and output 'education'. But input words can vary and some of them may not relate to content, which needs to be brought out. Six words are taken as shown in Fig.7, which definitely reflect the content education and these were used as inputs and compared with earlier ANN model. Further if the details of input are sub divided into regions-three each to reflect upper and lower extensions in letters, better models could be obtained. The pixel-map of each is divided into three segments to account for extensions at top and bottom and central portion accounting for 50% giving full text information. For example letters like 'g' have extensions at top and bottom while letters like 'h' have only top extensions. This approach was used for statistical interpretation earlier [3.].

Tamizh	Telugu	Hindi	English
ஆசிரியர்	ఉపాధ్యాయుడు విద్యార్థి	शिक्षक	teacher
மாணவர்		छात्र	student
புத்தகம்	పుస్తకం	किताब	book
பள்ளி	పాఠశాల తరగతి	स्कूल	school
வார்க்கம்	కళాశాల	वर्ग	class
கல்லூரி		कॉलेज	college

Fig. 7. Six words belonging to one content

4 Results and Discussion

Fig.8 gives an idea of the performance of the base neural model with four input-output words and variations in terms of less number of words and sub-divided pixel-maps. The values are around 1 showing clearly that matching is good and can be improved with more words and more nodes in the network. Values falling outside .9 to 1.1 are clearly indicative that they do not belong to the content.

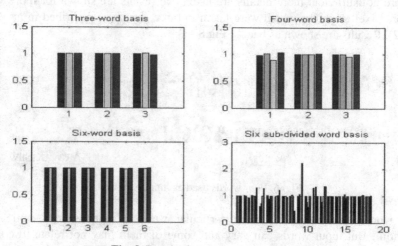

Fig. 8. Results for content extraction

References

1. Li, Y., Kuo, C.C.J., Wan, X.: Introduction to content-based image retrieval — Over view of key techniques. In: Castelli, V., Bergman, L.D. (eds.) Image Databases: Search and Retrieval of Digital Imagery, pp. 261–284. John Wiley, New York (2002)
2. Prakash, K.B., Dorai Ranga Swamy, M.A., Raja Raman, A.: Mining Approach for Documents Containing Multilingual Indian Texts. In: NCRTAC 2009. Bharath University, Chennai (2009)
3. Prakash, K.B., Dorai Ranga Swamy, M.A., Raja Raman, A.: Two-Input Neuron Model for Documents Containing Multingual Indian Texts. In: EPPCSIT 2009, Guru Nanak Dev Engineering College, Ludhiana (2009)

Integration of Face Detection and User Identification with Visual Speech Recognition

Alaa Sagheer[1,2,*] and Saleh Aly[1,3]

[1] Center for Artificial Intelligence and Robotics (CAIRO)
[2] Department of Mathematics, Faculty of Science, Aswan University, Aswan, Egypt
alaa@cairo-svu.edu.eg
[3] Department of Electrical Engineering, Faculty of Engineering,
Aswan University, Egypt
saleh@cairo-svu.edu.eg

Abstract. The use of visual features to help acoustic speech recognition (ASR) is an appropriate tool to enhance ASR. In this paper, we propose a novel system integrates face detection, user identification and visual speech recognition. Here we use the self organizing map to achieve visual features extraction. Then, the extracted features are recognized using K-nearest neighbor classifier. Experimental results, using a database includes Arabic digits, show that the proposed system is promising and effectively comparable with other reported systems.

Keywords: Face detection, user identification, mouth detection, visual speech recognition, self organizing map, K-nearest neighbor.

1 Introduction

Today's trend is to make the interaction between humans and their artificial assistants easier and closer to the natural means of human communication. Automatic speech recognition (ASR) technology has reached a maximum of performance and good recipes for building speech recognizers have been written [1]. However, the two major problems of background noise and reverberations are still insurmountable [2]. Therefore, inspecting other sources for complementary information which could diminish these problems is a thinkable solution. The known source capable to compensate these two problems is visual speech recognition.

Visual speech recognition (VSR) [3-4], *sometimes denoted as speechreading* [5] *or lip reading* [6], can be seen both as a complementary process to ASR and as a stand-alone application. VSR is an essential stage in many multimedia systems such as audio visual speech recognition AVSR [3-4], recovery of speech from deteriorated or mute movie clips [7], multimedia or mobile phone for hearing impaired people [8], sign language recognition systems, person identification [9], security by "video surveillance", home health care systems for elderly and handicapped [10-11], driver assistant systems [12], etc.

It is widely demonstrated that, traditional VSR systems are concerned only with the mouth region and the movements of lips [13-14]. In other words, VSR system

---

* Corresponding author.

T. Huang et al. (Eds.): ICONIP 2012, Part V, LNCS 7667, pp. 479–487, 2012.

neglects important visual elements such as face movement, face recognition and so on, which shouldn't be avoided in real life [15].

In this paper, we introduce a novel system achieves lip reading and integrates face detection, mouth detection, user identification with the visual word recognition. In other words, the proposed system adapts to the user facial movements that cannot be avoided in real life. Here we apply the system in order to recognize Arabic digits; however, it can be used for other languages as well. If we combine an acoustic channel with the proposed system, it will yield audio-visual speech recognition system.

The paper is organized as follows: Section 2 presents outline of the system. The three main elements of our system, face detection, user identification and visual speech recognition are described in sections 3, 4 and 5 respectively. Section 6 shows the database used in this paper. Experimental results and comparisons with other reported systems are provided in section 7. Finally, section 8 concludes the paper.

2   System Outline

It is known that, traditional VSR systems are concerned only with the mouth regions and neglect the user ID as well as other information from the user's face [13-14]. Additionally, they force the user to fix himself/herself in front of the computer and avoid any possible movements. All these aspects represent real obstacles in the way of developing recent technology based on VSR, such as audio VSR, or AVSR.

In this paper, we are motivated to release these restrictions by assuming unconstrained environment where the user feels free to move his/her face as long as he/she still in the scope of the camera. The proposed system contains three consecutive elements: Face and mouth detection, user identification and visual speech recognition.

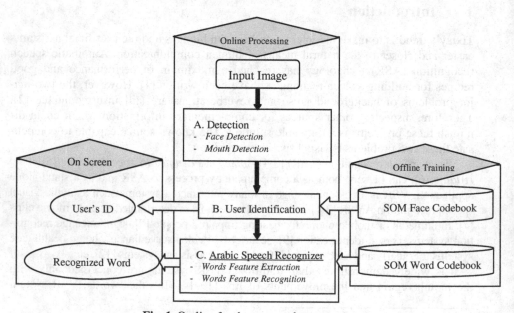

Fig. 1. Outline for the proposed system

3   Face and Mouth Detection

Face detection represents the first step in our system and we perform it using the face detection module provided in [16]. This detection module is much faster than any of its contemporaries. Its performance can be attributed to the use of an attentional cascade, using low feature number of detectors based on a natural extension of Haar wavelets. Each detector in this cascade fits objects to simple rectangular masks. In order to reduce the number of computations, while moving through their cascade, the authors of [16] introduced a new image representation called integral image.

For each pixel in the original image, there is exactly one pixel in the integral image, whose value is the sum of the original image values above and to the left. The integral image can be computed quickly which drastically improves the computation costs of the rectangular feature models. At the highest levels of the attentional cascade, where most of the comparisons are made, the rectangular features are very large.

The attentional cascade classifiers are trained on a training set as the authors have explained in [16]. As the computation progresses down the cascade, the features can get smaller and smaller, but fewer locations are tested for faces until detection is performed. Same procedure is applied for mouth detection, except that the object is different and we search about mouth only on the lower half of the image.

In detection part of our experiments, we do not ask the user to stick himself/herself in a specific place. The user is free to move as long as he/she in the scope of the camera. Also, there are no image restrictions such as background, where the video is captured in office environment. In addition, the rectangle around the mouth is flexible, which means that it can extend up or down and right or left in order to fit with the movement of lips when the user starts to speak.

Furthermore, the proposed method here for extracting the mouth region is simple and has the advantage of providing a reliable mouth region without any geometric model assumptions or using certain values derived by mathematical calculations needed to be decided for every user separately, as the system given in [8].

4   User Identification

After the user's face has been detected, the system should identify the user, i.e. show the user's name. In fact, user identification element distinguishes our system than other systems given in [8] and [17-18] which neglect the user identification element. The proposed system shows the ID of the user once his/her face has detected. This is done automatically, where the system saves one frame for the detected face and then two tasks are performed on this frame: feature extraction and feature recognition.

4.1   Feature Extraction

Feature extraction maps the salient aspects of the data into a new space includes less number of features than original space. These few features should include the main information of the original data structure. In this paper, we use self organizing map (SOM) to achieve feature extraction. The SOM is one of the most widely used artificial neural networks applies an unsupervised competitive learning approach [19].

SOM has a two dimensional map whose units, usually called neurons, become tuned to different input vectors I. A weight vector w_u is associated with each neuron u. In each learning step, one sample input vector I from the input data is considered and a similarity measure, usually taken as the Euclidian distance, is calculated between the input and all the weight vectors of the neurons of the map. The best matching neuron (BMN) c, usually denoted as *winner*, is the one whose weight vector w_c has the greatest similarity (or least distance) with the input sample I; i.e. which satisfies:

$$\|I - w_c\| = \min_u(\|I - w_u\|) \tag{1}$$

Then, the weight vectors of the SOM map are updated according to the rule:

$$w_u(t+1) = w_u(t) + h_{cu}(t)[I(t) - w_u(t)] \tag{2}$$

where
$$h_{cu}(t) = \alpha(t).\exp\left(\frac{\|r_c - r_u\|}{2\sigma^2(t)}\right) \tag{3}$$

$h_{cu}(t)$ is the neighborhood kernel around the *winner* c at time t, $\alpha(t)$ is the learning rate and is decreased gradually toward zero and $\sigma^2(t)$ is a factor used to control the neighborhood kernel. The term $\|r_c - r_u\|$ represents the difference between the locations of both the winner neuron c and the neuron u. After training phase, the neuron sheet is automatically organized into a meaningful two-dimensional order map denoted as a feature map (or codebook). The SOM codebook has the merit of preserving the topographical order of the training data. In other words, similar features in the input space are mapped into nearby positions in the feature map.

For our experiments here, we used a two dimensional SOM includes 7x8 neurons. Each neuron has the size of 48x48 pixels (same size of the input image). Here, once the system detects the user's face, one face frame is taken and saved. This frame is applied as input to the SOM. Then, the SOM starts to extract features from the face included in the applied frame. Using the SOM's feature map, which we got during training phase before the testing experiments, SOM seeks about the winner neuron to the given input using Eqs. (1)-(3). Then, SOM keeps the weight vector of the winner neuron, which will be used later in feature recognition task.

4.2 Feature Recognition

In this paper, we achieve feature recognition using K-nearest neighbor (K-NN) classifier [20]. The K-NN has a simple structure and exhibits effective classification performance, especially, when variance is clearly large in the training data. It has a labeled reference pattern set (RPS) for each class being determined during training phase. For the user identification task, the observed features, i.e. the weight vector of the winner neuron of SOM, is compared with each reference feature (the SOM's feature map) using the Euclidian distance. Then, we choose the K-nearest neighbors and determine the class of the input feature using a majority voting procedure. Here, to construct the best K-NN classifier, it is not practical to use all training sample data as the reference pattern set. Instead, we construct the K-NN using Hart's condensing algorithm [21] which effectively reduces the number of the reference pattern set.

5   Visual Speech Recognition (VSR)

We proceed now to the VSR element. Once the system identifies the user, the user starts to utter a specific word. The system captures the *visual* phon*emes* (*visemes*) included in this word across a number of frames of the mouth movements. Then, by the same way described in section 4, two tasks are performed on these frames: viseme features extraction and, then, viseme features recognition.

In the first task, we re-use the SOM by the same way given in section 4.1, except that the object here (mouth region) is different. In the second task, also we re-use the K-NN classifier by the same way described in section 4.2. We believe that the K-NN is so suitable for the speech classification task here since the visual speech samples are, usually, sparse and their sample variance is large.

6   Database

The database utilized in this paper includes the samples of 20 subjects (13 female and 7 male). Each subject uttered the Arabic digits, from one to nine, such that each digit takes 20 frames. Then, the total number of images has been used is: 20 subjects x 9 words x 20 frame/word = 3600 images. Each frame has a 48x48 pixel as a resolution. To build the SOM map (or codebook) for the user identification task, we just used ten samples for each subject to be in total 200 images. However, to build the map for VSR we used the samples of 10 subjects, whereas the samples of the other 10 subjects are devoted for testing.

The samples of the database are captured in office environment through different sessions, day and night and, sometimes, different illumination conditions. There were no restrictions on the user movement as long as his/her face in the scope of the camera. Fig. 2 shows samples of the used database.

Regarding the Arabic words, each Arabic word includes between two or four visemes. As a total, the database has 32 visemes. These visemes are represented in an SOM feature map includes 7 x 8 = 56 neurons, which means that each viseme is

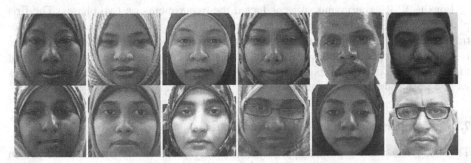

Fig. 2. Database samples

Table 1. A list of the Arabic digits with their English meanings

Arabic word	Pronunciation	English meaning
واحد	/wa-he-d/	One
اتنين	/et-nee-n/	Two
ثلاثة	/ta-la-ta/	Three
أربعة	/ar-ba-aa/	Four
خمسة	/kha-m-sa/	Five
ستة	/se-taa/	Six
سبعة	/sa-ba-aa/	Seven
ثمانية	/ta-ma-nya/	Eight
تسعة	/te-se-aa/	Nine

represented by at least one neuron. For the K-NN, the number of nearest neighbors K is chosen to be 3. Table 1 shows a list of the Arabic digits, where each word is described by three ways: the Arabic word, its pronunciation and its English translation.

7   Experimental Results

7.1   Overview of the Experimental Results

The experiments of this paper are conducted in office environment with normal lightening conditions using a webcam with a resolution 1.3 MP. The webcam is installed into a laptop with a CPU Intel(R) core i3 CPU with 2.4 GHz and a RAM with 4GB. The source code is built using Microsoft Visual C++ version 2008. For each subject/word we used videos as input to the system. In each video, the system detects the user face and mouth region, where two rectangles are drawn around the user's face and mouth. Then, the system identifies the user by writing his/her name on the program console (see the left panel in Fig. 3).

Every subject tries the system 5 times in different sessions. Then the total number of trials is: 20 (subjects) x 5 (trials) = 100 trials. The system achieved user identification successfully in 97 trials out of the total 100 trials, i.e. the identification average rate 97%.

For the VSR experiments, we conducted the experiments on two phases. The first phase is person dependent experiments where we asked the ten subjects who used to train the SOM, to test the system in the final (testing) experiments. The second phase is a person independent phase where the subjects used in testing phase are different than those used in training phase. The recognition rates in the first and second phases were 82.3% and 55.7%, respectively.

7.2   Comparison with other Reported Systems

There are two other reported approaches are very close to the system presented in this paper. The first is designed for Korean language given in [17] and the other is designed for Japanese language given in [18]. Table 2 below shows the comparison among the three systems. Here, we should highlight that the proposed system is the only system which concerns the user identification. Therefore there is no comparison can be held with the other systems in this element.

Fig. 3. Video experiment (Right) Face and mouth detection (Left) User identification

Regarding to the person dependent experiment of VSR element, both systems in [17] and [18] are outperform our system. However, we can notice easily that our system outperforms the other systems in person independent experiments. Needless to say that, the person independent experiment is more important and general than person dependent experiment since it measures the generalization of the approach.

Table 2. Performance of user identification and VSR of the proposed system and the systems given in [17] and [18]

User Identification Rate (%)		
SOM+KNN	System in [19]	System in [20]
97	N/A	N/A
Visual Speech Recognition (VSR) Rate (%)		
Type of Experiment	Proposed	System in [19]	System in [20]
Person-dependent	82.3	**92.7**	88.5
Person-independent	**55.7**	46.5	N/A

8   Conclusion and Future Works

In this paper, we proposed a novel system integrates face detection and user identification elements with the visual speech recognition element. Here we used the self organizing map to achieve visual features extraction and K-nearest neighbor for recognition. This combination is used in both elements: user identification and visual speech recognition. Experimental results show that the proposed system is promising and effectively comparable with other reported systems, especially, in the case of person independent phase. In future, we are planning to increase the number of words to be 25 words. Also, we will combine an audio channel with the current visual channel to yield an overall Audio-Visual Speech Recognition (AVSR) system.

Acknowledgment. The presented work is funded by Science and Technology Development Fund (STDF) project- Ministry of Higher Education, Egypt via the research grant no. 1055.

References

1. Potamianos, G., Neti, C., Gravier, G., Garg, A., Senior, A.: Recent advances in the automatic recognition of audiovisual speech. Proc. of the IEEE 91(9), 1306–1326 (2003)
2. Potamianos, G.: Audio-Visual Speech Processing: Progress and Challenges. In: Goecke, R., Robles-Kelly, A., Caelli, T. (eds.) Proceeding of the HCSNet Workshop (VisHCI 2006), Conferences in Research and Practice in Information Technology (CRPIT), vol. 56 (2006)
3. Matthews, I., Cootes, T., Bangham, A., Cox, S., Harvey, R.: Extraction of Visual Features for Lip-Reading. IEEE Transaction on PAMI 24(2), 198–213 (2002)
4. Dupont, S., Luettin, J.: Audio-Visual Speech Modeling for Continuous Speech Recognition. IEEE Transaction on Multimedia 2(3), 141–151 (2000)
5. Luettin, J., Thacker, N.: Speechreading using Probabilistic Models. Computer Vision and Image Understanding 65(2), 163–178 (1997)
6. Hassanat, A.: Automatic Lip-Reading System. Lap Lambert Academic Publishing (2011)
7. Guitarte, J.F., Frange, A.F., Solano, E.L., Lukas, K.: Lip Reading for Robust Speech Recognition on Embedded Devices. In: Proceedings of the 30th IEEE International Conference on Acoustics, Speech, and Signal Processing, ICASSP 2005, vol. 1, pp. 473–476 (2005)
8. Puviarasan, N., Palanivel, S.: Lip reading of hearing impaired persons using HMM. Expert Systems with Applications 38, 4477–4481 (2011)
9. Sanderson, C., Paliwal, K.K.: Identity verification using speech and face information. Digital Signal Processing 14, 449–480 (2004)
10. Takahashi, S., Morimoto, T., Maeda, S., Tsuruta, N.: Dialogue Experiment for Elderly People in Home Health Care System. In: Matoušek, V., Mautner, P. (eds.) TSD 2003. LNCS (LNAI), vol. 2807, pp. 418–423. Springer, Heidelberg (2003)
11. Vergados, D.: Service personalization for assistive living in a mobile ambient healthcare-networked environment. Personal and Ubiquitous Computing 14, 575–590 (2010)
12. Sun, Z.: On-road vehicle detection: a review. The IEEE Transaction on PAMI 28(5), 694–711 (2006)
13. Sagheer, A., Tsuruta, N., Taniguchi, R., Maeda, S.: Appearance Features Extraction vs. Image Transform for Visual Speech Recognition. International Journal of Computational Intelligence and Applications, IJCIA 6(1), 101–122 (2006)
14. Hazen, T.: Visual Model Structures and Synchrony Constraints for Audio-Visual Speech Recognition. IEEE Transaction on Speech and Audio Processing 14(3), 1082–1089 (2006)

15. Çetingül, H.E., Erzin, E., Yemez, Y., Tekalp, A.M.: Multimodal speaker/speech recognition using lip motion, lip texture and audio. Signal Processing 86(12), 3549–3558 (2006)
16. Viola, P., Jones, M.: Robust real-time object detection. The IEEE Transactions on Computer Vision 57(2), 137–154 (2004)
17. Shin, J., Lee, J., Kim, D.: Real-Time Lip Reading System for Isolated Korean Word Recognition. Pattern Recognition 44, 559–571 (2011)
18. Saitoh, T., Konishi, R.: Real-Time Word Lip Reading System Based on Trajectory Feature. IEEJ Transactions on Electrical and Electronic Engineering 6, 289–291 (2011)
19. Kohonen, T.: Self-Organizing Maps, 3rd edn. Springer (2001)
20. Xindong, W.: Top 10 algorithms in data mining. Know. Info. Systems 14, 1–37 (2008)
21. Hart, P.: The condensed nearest neighbor rule. The IEEE Transaction on Information Theory 14, 515–516 (1968)

Smart Phone Based Machine Condition Monitoring System

Iqbal Gondal, Muhammad Farrukh Yaqub[*], and XueliangHua

Faculty of Information Technology Monash University, Australia
{iqbal.gondal,farrukh.yaqub,xueliang.hua}@monash.edu

Abstract. Machine condition monitoring has gained momentum over the years and becoming an essential component in the today's industrial units. A cost effective machine condition monitoring system is need of the hour for predictive maintenance. In this paper, we have developed a machine condition monitoring system using smart phone, thanks to the rapidly growing smart-phone market both in scalability and computational power. In spite of certain hardware limitations, this paper proposes a machine condition monitoring system which has the tendency to acquire data, build the fault diagnostic model and determine the type of the fault in the case of unknown fault signatures. Results for the fault detection accuracy are presented which validate the prospects of the proposed framework in future condition monitoring services.

Keywords: machine condition monitoring, smart phone, fault diagnosis, wavelet transform.

1   Introduction

Machine condition monitoring (MCM) is crucial in all industrial processes to achieve high reliability, reduced man power and scheduled maintenance. MCM specifically deals with abnormality detection and diagnosis. To diagnose the abnormality, it is important to record certain physical parameters which vary according to the variation in the operation of the machine and 'vibration' is one of such parameters. Among different physical parameters such as vibrations, temperature, thermal imaging etc, vibration signatures are the most suitable for fault diagnosis, particularly under incipient fault conditions [1]. Machine fault diagnosis is well-explored by the researchers' community and many works have already been published [2-7], and also proposed by the authors [1, 8-11].

The new advances in the field of condition based maintenance has reduced the trend towards reactive maintenance in which the maintenance cost is five times higher than well planned and schedule maintenance. Though the condition monitoring devices are becoming more and more popular and cost is decreasing accordingly, still it is one of the vital considerations for the industrial investors and the solutions providers

---

[*] Corresponding author.

T. Huang et al. (Eds.): ICONIP 2012, Part V, LNCS 7667, pp. 488–497, 2012.
© Springer-Verlag Berlin Heidelberg 2012

to reduce the cost of the monitoring equipment. This paper proposes a novel condition monitoring framework based on 'smart phone'. The proposed technique has the potential to give a new direction to the research in the area of condition monitoring. Knowingly that the smart phone devices are not assembled to bear the industrial operating conditions, still the comprehensive sensor bank, enhanced computational power and reconfigurablilty of the hardware resources of these devices make them suitable choices in future industrial application.

Overall machine condition monitoring system could be divided into following four steps: data acquisition, signal processing, feature extraction/selection and classification as in Fig. 1. Vibration signal is non-stationary in nature, i.e., its spectral contents vary with respect to time. Particularly, wavelet packet transform (WPT) [12-21] can decompose the signal into multiple frequency nodes and provide multi-resolution analysis. Once the signal is decomposed using wavelet transform into multiple subbands, the next phase is to extract the feature. Root Means Square (RMS) is one of the most extensively used feature value in the literature for fault diagnosis [10-11, 15-16]. Not all the feature in the vibration frequency spectrum contain use information about fault diagnosis, the application develops an adaptive criterion to optimally select the wavelet nodes which contain relatively dominant fault related information [10]. Lastly, a classifier is developed based on K-nearest neighbor approach and type of the fault is determined [22].

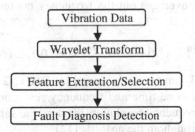

Fig. 1. Framework for machine health monitoring

In this paper, all the stages in Fig. 1 are implemented on smart-phone in the android operating system. The availability of accelerometer in the smart phone and computation capability make them suitable choice for developing the whole condition system with sufficient level of reconfigurabilty and agility. The reconfiguarbility lies in the ease of manipulating the functionality of the whole system by bringing the change in the software. The built-in accelerometer is used to capture the vibration data. All the stages are implemented using the phone resources, both the computations and memory. The performance of the proposed application is determined using the vibrations from three different laptops, though the vibrations are very nominal and hard to distinguish. The proposed condition monitoring system classifies the vibration pattern with a certain level of accuracy.

2   Framework

Figure 1 presents the overall system for fault diagnosis. In this section, we discuss different components to build the fault diagnosis system. Digitized vibration data acquired using the accelerometer of the smart phone, is first decomposed using WPT. Wavelet decomposes the signal into different frequency bands or nodes. Root mean square (RMS) value is computed for feature extraction which has been extensively used as feature extraction metric in the literature. From all the extracted features, the dominant features are selected based on the dominancy level. A classifier is built using the extracted features and which is used for fault diagnosis and prognosis of unknown fault instances. All of these modules are implemented on android architecture, and the developed application has the tendency to implement the fault diagnosis modules according to the baseline data, configure the prominent feature vector and perform test for future instances.

2.1   Time/Frequency Domain Analysis for Vibration Signal

Among the time-frequency domain signal processing techniques, e.g., DFT, STFT and WT, WT can be used for comprehensive analysis of non-stationary vibration signal to reliably extract time and frequency domain contents [12]. DFT of a non-stationary signal $x[n]$ (1) does not exploit the variation in frequency contents with respect to time. Rather it averages out the frequency content over the whole signal range [12].

$$X(k) = \sum_{n=0}^{N-1} x[n]\, e^{-j\left(\frac{2\pi}{N}\right)kn}, \quad k = 0,1, \dots (N-1). \tag{1}$$

The shortcomings of DFT can be compensated by STFT (2) but STFT suffers from the problem, that it gives same time and frequency resolution for low and high frequencies. The time and frequency resolution remains the same because window size $w[n]$ remains constant throughout the analysis [12].

$$X(m,k) = \sum_{n=0}^{N-1} x[n]w[m-n]e^{-j\left(\frac{2\pi}{N}\right)kn}, k = 0,1, \dots (N-1). \tag{2}$$

In order to overcome the drawback of fixed time-frequency resolution in STFT, wavelet transformation can be used which has the tendency to perform multi-resolution analysis, wavelet transformation can be used which has the tendency to perform multi-resolution analysis. Wavelet packet transform provides multi-resolution analysis, i.e., time and frequency resolution can be adjusted [12-21]. Figure 2 gives the decomposition tree for WPT. Digitized vibration data are passed through high pass $h[n]$ and low pass $g[n]$ Quadrature Mirror Filters (QMFs) (3)-(4) and then down sampled. QMFs are finite impulse response (FIR) filters or infinite impulse response (IIR) filters. Filter selection is a very crucial part in case of analysis using WT. In the proposed scheme, Daubechies (Db5) filter [1,10-11] is used which is an FIR filter. Figure 2 also shows that the total number of nodes at any decomposition level is given by (5):

$$y_{approx}[n] = x[n] * g[n],$$

$$\text{or} \quad y_{approx}[n] = \sum_{k=-\infty}^{k=\infty} x[k] \times g[n-k]. \tag{3}$$

$$y_{detailed}[n] = x[n] * h[n],$$

$$\text{or} \quad y_{detailed}[n] = \sum_{k=-\infty}^{k=\infty} x[k] \times h[n-k]. \tag{4}$$

$$N_j = 2^{j_{decomp}}. \tag{5}$$

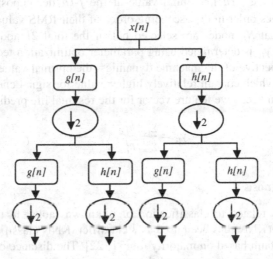

Fig. 2. Wavelet packet transform (WPT)

2.2 Feature Extraction

RMS value of wavelet decomposition nodes is computed as in (6) which is extensively used as feature extraction metric in the literature [10-11, 15-16].

$$R_i^j = \sqrt[2]{\frac{1}{N}\sum_{i=1}^{N} x_i^2}. \tag{6}$$

Where R_i^j represents the RMS value of the i-th node at the j-th decomposition level. Vibration data are divided till decomposition level-4 [11] and RMS value of the resultant '16' features (5) is computed.

2.3 Feature Selection

It has been investigated that only a small portion of the overall vibration spectrum contains dominant information regarding the fault-induced-vibrations, i.e., resonant frequency band and the rest is noise [11, 23]. In the proposed framework, a criterion is proposed to select the nodes containing relatively larger signal energy. To study the analytical relations and expressions for the proposed scheme, let us assume that R^j is the vector with RMS values for the nodes at j-th decomposition level as represented in (7).

$$R^j = \{ R_1^j, R_2^j, R_3^j, \dots , R_{2^j}^j \}. \tag{7}$$

In order to achieve high energy frequency bands for the fault signature-signal, nodes are reordered in the descending order of their energy values at their respective level. Let $\hat{R}_i^j$ represents the i-th maximum value at the j-th decomposition level. If $\hat{R}_i^j$ represents the nodes order in the ascending order of their RMS values in (7), ratio γ_n is defined such that N_D nodes are selected out of the total 2^j nodes as in (8). The optimal value for γ_n is determined using parameter optimization techniques given in Section 2.5. Irrespective of the machine dynamics, the optimal value for γ_n gives the number of nodes which contain relatively higher value for signal energy, and the corresponding node indices give feature vector for the residual life prediction model.

$$\gamma_n = \frac{\sum_{i=1}^{N_D} \hat{R}_i^j}{\sum_{i=1}^{2^j} R_i^j}. \tag{8}$$

2.4   Fault Diagnosis

Fault diagnosis is related to classification of unknown faults; in this paper $K-th$ Nearest Neighbor (KNN) has been used as a classifier. KNN classifier determines the type of unknown fault based on majority rule [1, 22]. The distance of the test point is computed from K nearest neighbors of the training data points. The fault type of unknown test data point is classified as the class which contains maximum number of neighbors out of K nearest neighbors to the test data point [1, 22]. The dominant feature vector obtained from subsection 2.3 is used as input for the KNN classifier for fault diagnosis. The classification accuracy of the proposed fault diagnostic model is computed as given in (9):

$$H_r = \frac{Accurate\ Detection}{Total\ Trials} \times 100\ \%. \tag{9}$$

where H_r represents the 'hit ratio' and it measures the percentage of the accurately detected unknown events to the total number of unknown events. Better classification accuracy ensures robustness and reliability of the fault diagnostic model. H_r is used as a quantification metric in parameters optimization as in subsection 2.5.

2.5   Implementation on the Smart Phone

All the stages in the proposed framework are implemented on the smart phone architecture using JAVA. Vibration data is acquired using the built in accelerometer. The application is configured for three types of fault as shown in Fig. 3. Baseline_data1-3 represent the historical data which are used to build the fault diagnostic model. The proposed framework determines the dominant features according to the criterion presented in subsection 2.3. Dominant features are determined by 5-fold cross validation [24] in which the base line data are split in training data and validation data. For any equipment to be monitored, the developed application requires the historical baseline data to build the fault diagnostic model, it determines the dominant feature vector automatically using 5-fold cross validation and maximizing H_r in (9).

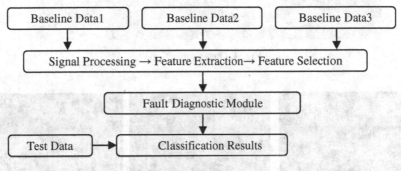

Fig. 3. Implementation on Smart Phone

3   Experimental Results

This section presents results for the proposed condition monitoring system, it presents the experimental setup for data acquisition (subsection 3.1) and fault detection accuracy (subsection 3.2). Fault detection accuracy results are presented by varying different parameters and there impact is characterized.

3.1   Data Acquisition

In order to test the working of the proposed system, the datasets is achieved by 3 different laptops as in Fig. 4: 1) a perfectly running brand-new laptop, 2) a laptop with loose keypad, 3) a laptop with mechanical problem in the cooling fan. Vibration data are captured using the application with the built-in accelerometer.

3.2   Fault Detection Accuracy

Figure 5 gives the snapshots for different operations on smart-phone, i.e., data acquisition, signal processing, feature extraction/selection and fault diagnosis.

In order to determine the performance of the proposed system, fault detection accuracy is measured using the test data belonging to different types of the faults, each test is conducted 100 times. Fault diagnostic model is configured using the baseline training data. Overall vibration datasets are split into windows containing 1024 vibration

Fig. 4. Data acquisition

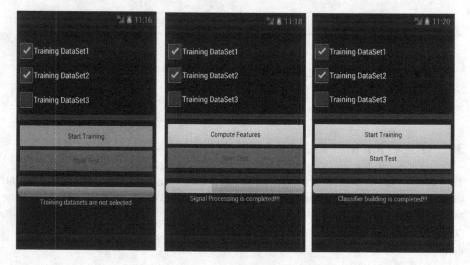

Fig. 5. Screen shots under different operations

samples and feature vectors are computed by wavelet decomposition and computing RMS value. Table 1 lists the fault detection accuracy for each of the test-data individually. Results in Table 1 are obtained by using the optimal number of features. The fault detection accuracy for the data with fan fault is relatively better, because vibration signatures are more prominent. Overall fault detection accuracy is poor, because the vibrations generated by the laptops are very nominal.

In order to validate the significance of features selection, Table 2 lists the average fault detection accuracy in case of three tests scenario by varying the number of features. Since the vibration data are decomposed till level '4', it results into '16' features according to (5). Table 2 shows that the fault detection accuracy reaches at maximum at optimal number of features which decreases if the number of features is increased or decreased. Increasing the number of features beyond the optimal level does not only result into degradation in the fault detection accuracy because of

Table 1. Performance Evaluation

Test Data	Accuracy (%)
Brand-new	59.00
Keypad looseness	58.00
Fan fault	71.00

Table 2. Fault Detection Accuracy Vs. Features

Number of Features	Average Accuracy (%)
2	48.00
4	53.00
6	51.00
8	55.00
10	61.00
12	63.00
14	59.00
16	60.00

redundancy but also enhances the computational complexity. Whereas, decreasing the number features beyond the optimal threshold results into poor fault detection accuracy because some of the features carrying important information about fault diagnosis are discarded.

4 Conclusion

This paper proposes a smart-phone based condition monitoring system. The advancement in the smart-phone technology in terms of computational complexity and the availability of the sensor-bank makes it easier to realize such a system. In this paper, vibration data are captured using built-in accelerometer and classification results are presented. Knowingly that the architecture of these phones is not robust enough to bear the industrial environment, it is expected that the robustness will achieved in future technologies. The future works include the usage of smart-phone not only for localized fault diagnosis and prognosis but also extending it towards data-logging using wireless communication. Results obtained from the proposed system validate the prospects of extensive usage of smart-phones in the future condition monitoring services, and other related industrial application.

References

1. Yaqub, M.F., Gondal, I., Kamruzzaman, J.: Inchoate Fault Detection Framework: Adaptive Selection of Wavelet Nodes and Cumulant Orders. IEEE Trans. Instrum. Meas. 61, 685–695 (2011)
2. Li, W., Mechefske, C.K.: Detection of Induction Motor Faults: A Comparison of Stator Current, Vibration and Acoustic Methods. Journal of Vibration and Control 12, 165–188 (2006)
3. Chen, Z., Mechefske, C.K.: Diagnosis of Machinery Fault Status using Transient Vibration Signal Parameters. Journal of Vibration and Control 8, 321–335 (2002)
4. Marichal, G.N., Artés, M., García-Prada, J.C.: An intelligent system for faulty-bearing detection based on vibration spectra. Journal of Vibration and Control 17, 931–942 (2011)

5. Xi, F., Sun, Q., Krishnappa, G.: Bearing Diagnostics Based on Pattern Recognition of Statistical Parameters. Journal of Vibration and Control 6, 375–392 (2000)
6. Hu, Q., He, Z., Zhang, Z., Zi, Y.: Fault diagnosis of rotating machinery based on improved wavelet package transform and SVMs ensemble. Mechanical Systems and Signal Processing 21, 688–705 (2007)
7. Teotrakool, K., Devaney, M.J., Eren, L.: Adjustable-Speed Drive Bearing-Fault Detection Via Wavelet Packet Decomposition. IEEE Trans. Inst. Meas. 58, 2747–2754 (2009)
8. Yaqub, M.F., Gondal, I., Kamruzzaman, J.: Severity Invariant Feature Selection for Machine Health Monitoring. International Review of Electrical Egnineering 6, 238–248 (2011)
9. Yaqub, M.F., Gondal, I., Kamruzzaman, J.: Machine Health Monitoring Based on Stationary Wavelet Transform and 4th Order Cumulants. Australian Journal of Electrical & Electronics Engineering 9 (2011)
10. Yaqub, M.F., Gondal, I., Kamruzzaman, J.: An Adaptive Self-Configuration Scheme for Severity Invariant Machine Fault Daignosis. IEEE Trans. Rel. (accepted, under press)
11. Yaqub, M.F., Gondal, I., Kamruzzaman, J.: Multi-Step SVR and Optimally Parameterized WPT for Machine Residual Life Prediction. Journal of Vibration and Control (2012) (published online, 2012)
12. Peng, Z.K., Chu, F.L.: Application of the wavelet transform in machine condition monitoring and fault diagnostics: a review with bibliography. Mechanical Systems and Signal Processing 18, 199–221 (2004)
13. Eren, L., Devaney, M.J.: Bearing damage detection via wavelet packet decomposition of the stator current. IEEE Trans. Inst. Meas. 53, 431–436 (2004)
14. Teotrakool, K., Devaney, M.J., Eren, L.: Adjustable-Speed Drive Bearing-Fault Detection Via Wavelet Packet Decomposition. IEEE Trans. Inst. Meas. 58, 2747–2754 (2009)
15. Lau, E.C.C., Ngan, H.W.: Detection of Motor Bearing Outer Raceway Defect by Wavelet Packet Transformed Motor Current Signature Analysis. IEEE Trans. Inst. Meas. 59, 2683–2690 (2010)
16. Yen, G.Y., Kuo-Chung, L.: Wavelet packet feature extraction for vibration monitoring. In: IEEE Internation Conference on Control Applications, vol. 2, pp. 1573–1578 (1999)
17. Li, F., Meng, G., Ye, L., Chen, P.: Wavelet Transform-based Higher-order Statistics for Fault Diagnosis in Rolling Element Bearings. Journal of Vibration and Control 14, 1691–1709 (2008)
18. Zhao, F., Chen, J., Xu, W.: Condition prediction based on wavelet packet transform and least squares support vector machine methods. Part E: Journal of Process Mechanical Engineering 223, 71–79 (2009)
19. Eren, L., Cekic, Y., Devaney, M.J.: Enhanced feature selection from wavelet packet coefficients in fault diagnosis of induction motors with artificial neural networks. In: IEEE Conference on Instrumentation and Measurement Technology Conference (I2MTC), pp. 960–963 (2010)
20. Jianhua, Z., Zhixin, Y., Wong, S.F.: Machine condition monitoring and fault diagnosis based on support vector machine. In: IEEE International Conference on Industrial Engineering and Engineering Management (IEEM), pp. 2228–2233 (2010)
21. Dongfeng, S., Gindy, N.N.: Industrial Applications of Online Machining Process Monitoring System. IEEE/ASME Trans. Mechatronics 12, 561–564 (2007)
22. Umamaheswari, K., Sumathi, S., Sivanandam, S.N., Anburajan, K.K.N.: Neuro Genetic-Nearest Neighbor Based Data Mining Techniques for Fingerprint Classification and Recognition System. Journal of ICGST-GVIP 7, 1–8 (2007)

23. Yaqub, M.F., Gondal, I., Kamruzzaman, J.: Resonant Frequency Band Estimation using Adaptive Wavelet Decomposition Level Selection. In: IEEE International Conference on Mechatronics and Automation (2011)
24. Hsu, C.W., Chang, C.C., Lin, C.J.: A practical guide to Support Vector Classification: Technical report. Department of Computer Science and Information Engineering. National Taiwan University (2003)

Spatio-temporal LTSA and Its Application to Motion Decomposition

Hongyu Li[1,2], Junyu Niu[3,*], Lin Zhang[2], and Bo Hu[1]

[1] Electronic Engineering Department, Fudan University, Shanghai, China
[2] School of Software Engineering, Tongji University, Shanghai, China
[3] School of Computer Science, Fudan University, Shanghai, China

Abstract. This paper describes a STLTSA-based framework to analyze and decompose human motion for synthesis. In this work, we mainly intend to extend a manifold learning method, local tangent space alignment, to a spatio–temporal version for manifold analysis and offer an effective method of estimating the intrinsic dimensionality of motion data. Based on an assumption that a long sequence of motion is composed of a number of short motion units, we can decompose a motion into several basic motion units in a low-dimensional manifold space and extract motion cycles from the cyclic unit. The generation of new complex movement using obtained motion units is feasible and promising.

Keywords: Manifold Learning, Spatio-Temporal Neighborhood, Motion Learning, Motion Decomposition.

1 Introduction

When data lies on a low-dimensional manifold, its structure may be highly non-linear, hence linear dimensionality reduction methods such as principal component analysis (PCA) [1] and metric multi-dimensional scaling (MDS) [2] often fail in finding the nonlinear embeddings. This has motivated extensive efforts toward developing nonlinear dimensionality reduction methods which is known as manifold learning. Manifold learning methods can be categorized into two main groups: global and local techniques. Global techniques attempt to preserve global properties of the data lying on manifolds [3]. Local techniques attempt to retain global properties of the data by preserving local properties obtained from neighborhoods around data points [4,5].

In general, motion data is of high dimensionality and difficult to understand and analyze, its intrinsic DOFs, however, are essentially supposed to be quite few and easy to visualize. Linear methods have been widely used in [6,7] to reduce the dimensionality of human motion for motion analysis. Recently, Wang et al [8] introduce Gaussian process dynamical models to learn nonlinear models of human motion from high-dimensional motion capture data. Li et al [9] propose a method to learn a nonlinear low-dimensional manifold for high-dimensional time series and model the dynamical process in the manifold space.

---

* Corresponding author.

T. Huang et al. (Eds.): ICONIP 2012, Part V, LNCS 7667, pp. 498–505, 2012.

In this paper, we aim to extend locality-based manifold learning techniques to a spatio-temporal version for motion capture data. The proposed method could significantly reduce the time cost of constructing the similarity graph, which facilitates to handle the large scale data sets. According to the work [10], the local tangent space alignment (LTSA) [5] method generally performs best among the popular manifold learning techniques, it is therefore chosen to test the spatio-temporal similarity graph. In addition, an effective method is offered to estimate the intrinsic dimensionality of motion data in this study.

2   Spatio-temporal LTSA

2.1   Spatio-temporal Neighborhood Construction

To estimate local tangent spaces of a manifold, the original LTSA method requires to first construct a similarity neighborhood through selecting nearest neighbors of each point. The simplest way to construct a similarity neighborhood is to identify a fixed number k of nearest neighbors per data point according to spatial distance. Although the k-nearest neighborhood is good at describing local structure of data, such neighborhood construction has its own drawbacks to handle large scale data sets. The construction cost for k-nearest neighborhood is $O(kn^2)$, which is expensive in large scale situations.

However, for time-dependent data such as motion data, the variation of data in two continuous frames is fairly low. Therefore, the temporal neighborhood implicitly contains much cue about spatial neighbors. If the temporal distances is also taken into consideration, the construction cost for k-nearest neighborhood will be dramatically reduced. Our neighborhood construction strategy is to first select $2k$ sequential frames as initial neighbors backward and forward from the current frame, and then find the k-nearest neighbors to construct a similarity graph using the spatial distance. For example, taking $k = 5$, the nearest neighbors of the i-th frame will be found from frames between $i-5$ and $i+5$ according to their spatial distance.

Given n time-dependent data points, the time complexity is $O(k^2n)$ if the k-nearest neighbors of each point are selected in terms of both the spatial and temporal distance. Since $k \ll n$ in general, using the spatio-temporal distance will greatly improve the construction efficiency of the neighborhood in comparison with solely using the spatial distance. Meanwhile, as two continuous frames vary little in motion data, our construction strategy can faithfully describe the local geometrical structure in data.

2.2   Summary of the Algorithm

Next we briefly describe in Table 1 how to extract low-dimensional coordinates Y from a set of high-dimensional motion data X with STLTSA.

It is worth that there are two free parameters, k and d, as input in the proposed method. It has been discussed in [5] that if the parameter k is too small, the

Table 1. The STLTSA algorithm

Input: the dataset $X = \{x_i\}_{i=1}^n$ where $x_i \in \mathbb{R}^m$, the number k of nearest neighbors, and the dimensionality d of the embedded manifold.

---

1. Construct the spatio-temporal neighborhood represented in the form of a $m \times k$ matrix, $X_i = (x_i^j)$, for each point x_i. Column vector x_i^j is the j-th nearest neighbor of x_i.
2. Calculate the d largest eigenvectors $g_1, \ldots, g_d$ of the correlation matrix $(X_i - \bar{x}_i e^T)^T (X_i - \bar{x}_i e^T)$. e is an all-one column vector, and $\bar{x}_i$ represents the average of the neighborhood of x_i: $\bar{x}_i = \frac{1}{k} \sum_j x_i^j$.
3. Extract the local geometry G_i by setting $G_i = [e/\sqrt{k}, g_1, \ldots, g_d]$.
4. Construct the $n \times n$ alignment matrix B by locally summing as follows: $B(I_i, I_i) \leftarrow B(I_i, I_i) + I - G_i G_i^T, i = 1, \ldots, n$ with initial $B = 0$. I is a $k \times k$ identity matrix, I_i denotes the set of indices for the k-nearest neighbors of x_i.
5. Compute the $d+1$ smallest eigenvectors of B and pick up the eigenvector matrix $[u_2, \ldots, u_{d+1}]$ corresponding to the $2nd$ to $d+1st$ smallest eigenvalues.

Output: the global coordinates $Y = [y_1, \ldots, y_n] = [u_2, \ldots, u_{d+1}]^T$.

---

mapping will not reflect any global properties of data; if k is too large, the mapping will lose its nonlinear character and behave like traditional PCA as the entire data set is seen as the local neighborhood. However, the algorithm is essentially stable over a wide range of values of k. How to determine the intrinsic dimensionality d is introduced in the following section.

2.3 Intrinsic Dimensionality

It is well known that PCA [1] exploits the number of large singular values of the covariance matrix of input data to estimate intrinsic dimensionality. Furthermore, a similar estimate of LLE was also proposed by Polito and Perona [11], where $d+1$ should be less than or equal to the number of eigenvalues of a kernel matrix that are close to zero.

Likewise, one could estimate the dimensionality d with the eigengap trick in spatio-temporal LTSA. The number of eigenvalues of B that are close to zero gives us an answer that d should be no more than this number. Precisely speaking, the minimal d is considered as the intrinsic dimensionality if it satisfies that the eigengap, $|\lambda_d - \lambda_{d+1}|$, between the eigenvalues λ_d and λ_{d+1} of matrix B is more than the threshold τ,

$$|\lambda_d - \lambda_{d+1}| > \tau.$$

Due to the fact that DOFs of human motion are quite few, the intrinsic dimensionality of motion data is clearly very low. Using the walking motion data, we find that the dimensionality d is supposed to be equal to 3. This agrees with the fact that there exist exactly three DOFs that describe the rotation of joints in the motion capture data we use.

2.4   Dynamic Mapping

In this work, we provide a way of dynamic mapping, which can project new data points between the low-dimensional manifold space and the original high-dimensional space. The basic assumption of this mapping is that there exists a locally linear mapping between the original space and the manifold space, which is consistent with the derivation of LTSA [5]. Therefore, once x_j (y_j) and x_{n+1} (y_{n+1}) lie close enough to each other, the transformation matrix of x_j (y_j) is naturally applicable to x_{n+1} (y_{n+1}). Note that this assumption requires that the input data must be dense enough to sufficiently cover the whole surface of the embedded manifold, otherwise the generalization can not perform well.

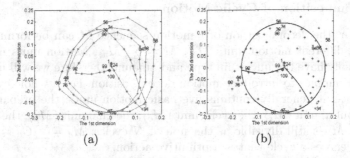

| (a) | (b) |

Fig. 1. The 2D manifold description of the walking motion with STLTSA. Red stars denote the mapping results of the input motion data into the 2D manifold space, the number is the frame index. (a): the normal walking path connected along the frame order, (b): the simplification of the walking path as a circle plus two curves.

2.5   Analysis

Using the walking motion containing 109 frames of 54-dimensional motion capture data with three walking cycles as an example, we applied the STLTSA method to such data and projected them into a 2D manifold space for visual analysis. Fig. 1 presents the 2D description of the walking motion in this manifold space, where a red star corresponds to an action in the walking motion sequence and the number is the frame index. The manifold curve in Fig. 1(a) depicts a transition path of actions connected along the time order. The transition path can be considered as the concatenation of three sub-paths: first transferring from the beginning (No. 1) to the initial action of walking (No. 39); then walking three cycles(from No. 40 through No. 88) and finally returning to the end (No. 109) close to the beginning. Although the walking trajectory is not completely identical in each cycle due to the liberty of human motion, the cyclic characteristic of such motion is still clearly expressed in the manifold space. Thus this walking path can be simplified and approximately sketched as a circle plus two curves, as shown in Fig. 1(b).

3   Motion Decomposition

According to the geometrical analysis of the reduced motion data, it is easy to deconstruct human motion in the manifold space. Assuming that a complex human behavior is always composed of several small motion units each of which represents a simpler behavior, one can decompose a long motion sequence into some small motion units. Sometimes the low-dimensional manifold curve may not be continuous and smooth due to the liberty and instability of human motion. So slight modification of the manifold coordinates is often required for denoising. In this section, we will put a special emphasis on how effectively decomposing cyclic motion and how removing noise from the manifold coordinates.

3.1   Decomposition of Cyclic Motion

In particular, the decomposition of a motion sequence M can be formulated as the sum of different motion units M_{u_i}, $M = \sum_i M_{u_i}$. Motion units in acyclic motion usually have the uncertain type and amount, therefore we will use cyclic motion as examples to introduce motion decomposition. For a whole cyclic motion sequence, it generally contains five basic motion units: the preparing unit (M_p), the initial unit (M_i), the cyclic unit (M_c), the final unit (M_f), the wind-up unit (M_w). M_c is still divisible as the sum of N cycles: $M_c = \sum C_u = N * C_u$, where C_u denotes a cycle that is a primitive action.

If the manifold curve computed with STLTSA is relatively smooth and continuous, motion decomposition can be easily completed with the derivative of this curve. In particular, the preparing unit starts from the beginning through the position where the first derivative of this curve changes sharply; the initial unit ends at the first position where the second derivative is zero; the cyclic unit continues until the last position with the second derivative being zero; the final unit is over if the manifold coordinates do not change obviously; the remainder is the wind-up unit. In the real applications, we use the first and second order difference of manifold coordinates with respect to time instead of the derivative of the manifold curve,

$$f_y'(t) \approx f_y(t+1) - f_y(t),$$

$$f_y''(t) \approx f_y(t+1) - 2f_y(t) + f_y(t-1),$$

where t is the frame index ranging in $[1, n]$, and $f_y(t)$ denotes manifold coordinates $Y = \{y_1, \cdots, y_n\}$ computed with STLTSA. In this means, the extraction of a cycle contained in the cyclic unit would be fairly simple. With the first order difference $f_y'(t)$ being close 0, we can find the valley and peak in the manifold curve. Between each pair of peak and valley alternately appearing in the time order, a point where the second order difference $f_y''(t)$ is closest to 0 will be picked out as a candidate of the margin of a cycle. Generally, the first and third candidates construct a motion cycle in the cyclic unit. As shown in Fig. 2(b), a pirouette cycle is automatically extracted in this means.

3.2 Denoising

Each motion cycle in real applications is just approximately consistent rather than completely identical due to the randomness and liberty of human action. Therefore, the cyclic unit may not take on such perfect periodicity in the manifold space. To recover the intrinsic shape of the manifold curve, one has to remove noise from manifold coordinates through smooth curve fitting. The task of curve fitting can be completed in terms of the least square method. The noise in manifold coordinates can be removed through translating noisy points toward the fitted curve. Noisy manifold coordinates $N_p(t)$ are defined as those points that deviate the fitted curve $f_c(t)$: $N_p(t) = \{f_y(t) \mid \|f_c(t) - f_y(t)\| > \gamma)\}$, where γ is a predefined threshold close to 0. The translation distance d_t of noisy manifold coordinates can be simply computed through the average deviation distance of the noise points from the fitted curve, After translation, the noisy manifold coordinates are changed into

$$\tilde{f}_y(t_i) = \begin{cases} f_y(t_i) - d_t, \; if \; f_y(t_i) \geq f_c(t_i); \\ f_y(t_i) + d_t, \; if \; f_y(t_i) < f_c(t_i). \end{cases} \tag{1}$$

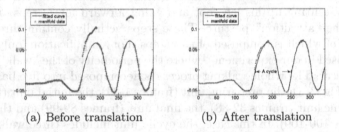

(a) Before translation   (b) After translation

Fig. 2. An example of translation about the pirouette data. The reduced manifold coordinates are marked with red stars and blue curve represents the fitted curve. (a): before translation, (b): after translation.

Fig. 2 gives an example of translation for denoising. The pirouette data is tested with STLTSA and the 1D manifold coordinates are displayed with red stars in this figure. The blue curve represents the fitted curve and the x-axis the time order. As in Fig. 2(a), some points in these two pirouette cycles obviously deviate from the fitted manifold curve, arising from the surrounding noise. These disjointed points are subsequently translated to the proximity of the fitted curve according to the formula (1). The results after translation are shown in Fig. 2(b), which reveals better periodic regularity.

4 Experimental Results

Here two examples regarding walking and running are offered to illustrate motion decomposition in Fig. 3. The 1D manifold coordinates (y-axis) are extracted

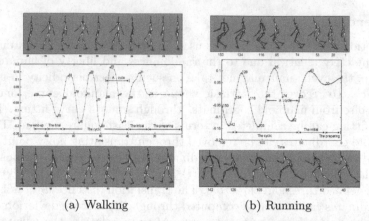

(a) Walking   (b) Running

Fig. 3. The periodicity of human walking and running. The middle shows the variation of 1D manifold coordinates obtained by STLTSA with time (from right to left for visual consistency). Some postures corresponding to numbered points in the middle are shown in the top and bottom.

using with STLTSA and are shown with the variation of time (x-axis). The walking data describes a series of human actions, beginning with the "attention" posture, then lifting up his left foot and going forward for 7 steps, and finally returning the "attention" posture. These steps actually contain three walking cycles each of which is composed of two steps. For simplification, only 109 data points are used in our experiments, where the periodicity of the "walk" motion is still quite clear. The whole walking process is decomposed into five basic motion units as in Fig. 3(a): the preparing unit (frames 1-24), the initial unit (frames 25-31), the cyclic unit (frames 32-88), the final unit (frames 89-99) and the wind-up unit (frames 100-109). In this case, the cyclic unit includes three walking cycles each of which needs two steps respectively with the left and right legs. A walking cycle C_u starts from frames 39 to 56.

The used running data, composed of 150 frames, are cut off from a long sequence. Although the input running motion is not a complete process, i.e., it does not include all five motion units, the periodic regularity of its cyclic unit is still clear, as shown in Fig. 3(b). The running sequence can be divided into four basic motion units: the preparing, initial, cyclic, and final units.

If some motion is fully understandable and can be decomposed into several different motion units, conversely, directly connecting these motion units can also restore the motion, or even create a new complex motion. For self-connection, a cycle is repeated directly and arbitrarily during the synthesis. For the connection of different motion units, it requires a common posture between two units. The results of motion synthesis are recorded in our supporting video.

5   Conclusion

In this work, we propose the spatio-temporal LTSA (STLTSA) method to analyze and decompose human motion. This method extends the original LTSA to handling temporal sequences, where the nearest neighborhood is constructed through the temporal consistency. In essence, the STLTSA method is also applicable in behavior classification [12] or motion analysis [13] with video data. Combining video data with motion capture data could effectively improve the performance of human motion tracking and recognition.

Acknowledgement. This work was partially supported by Natural Science Foundation of China Grant 60903120, 863 Project 2009AA01Z429, Shanghai Natural Science Foundation Grant 09ZR1434400, and Innovation Program of Shanghai Municipal Education Commission.

References

1. Jolliffe, I.T.: Principal Components Analysis. Springer (1986)
2. Cox, T.F., Cox, M.A.A.: Multidimensional Scaling, 2nd edn. Chapman and Hall/CRC, Boca Raton (2001)
3. Tenenbaum, J., Silva, V.D., Langford, J.: A global geometric framework for non-linear dimension reduction. Science 290, 2319–2323 (2000)
4. Roweis, S., Saul, L.: Nonlinear dimension reduction by locally linear embedding. Science 290, 2323–2326 (2000)
5. Zhang, Z., Zha, H.: Principal manifolds and nonlinear dimension reduction via local tangent space alignment. SIAM Journal of Scientific Computing 26, 313–338 (2004)
6. Park, M.J., Shin, S.Y.: Example-based motion cloning: Research articles. Comput. Animat. Virtual Worlds 15, 245–257 (2004)
7. Shin, H.J., Lee, J.: Motion synthesis and editing in low-dimensional spaces. Computer Animation and Virtual Worlds 17, 219–227 (2006)
8. Wang, J.M., Fleet, D.J., Hertzmann, A.: Gaussian process dynamical models for human motion. IEEE Trans. Pattern Anal. Mach. Intell. 30, 283–298 (2008)
9. Li, R., Tian, T.P., Sclaroff, S.: Divide, conquer and coordinate: Globally coordinated switching linear dynamical system. IEEE Trans. Pattern Anal. Mach. Intell. 34, 654–669 (2012)
10. van der Maaten, L., Postma, E.O., van den Herik, H.J.: Dimensionality reduction: A comparative review (2008)
11. Polito, M., Perona, P.: Grouping and dimensionality reduction by locally linear embedding. In: Neural Information Processing Systems, pp. 1255–1262 (2001)
12. Jenkins, O.C., Mataric, M.J.: Deriving action and behavior primitives from human motion data. In: IEEE/RSJ International Conference on Intelligent Robots and Systems (IROS), pp. 2551–2556 (2002)
13. Laptev, I., Belongie, S.J., Perez, P., Wills, J.: Periodic motion detection and segmentation via approximate sequence alignment. In: Proceedings of ICCV 2005 (2005)

Cost-Effective Single-Camera Multi-Car Parking Monitoring and Vacancy Detection towards Real-World Parking Statistics and Real-Time Reporting

Katy Blumer, Hala R. Halaseh, Mian Umair Ahsan,
Haiwei Dong, and Nikolaos Mavridis[**]

Department of Computer Engineering, New York University Abu Dhabi, Abu Dhabi, UAE
{kb1379,hrh243,ma2795,haiwei.dong,nikolaos.mavridis}@nyu.edu

Abstract. Parking is a huge problem in densely populated areas and drivers spend a significant amount of time finding a suitable place to park their cars. A system that could show drivers the nearest available space would result in enormous savings of time, fuel, and street space. In order to achieve that, real-world periodic statistical analysis of car parking areas could help increase efficiency. Ideally, real-time information could also be used to create personalized suggestions to drivers, thus enabling satisficing of a wide range of possible criteria of optimality. We propose a system that uses a single camera for a wide-area external parking, followed by a combination of two kinds of algorithms: static image analysis of parking lot spaces using a combination of histogram classification and edge detection, and dynamic image analysis using blob analysis. Our system thus achieves monitoring of parking spaces and reports statistics as well as empty slots in real-time. Our results indicate that almost 90% of empty spots are reported correctly, resulting in significant savings through a highly cost-effective single-camera system which can monitor more than 100 spaces.

Keywords: Parking Vacancy Static Analysis, Car Movement Dynamic Analysis, Image Processing.

1 Introduction

In crowded cities, the problem of finding an empty parking space can be so dominant that it often enters everyday conversations of citizens. A driver often has to search for several minutes in order to find an empty parking spot close to his intended destination and sometimes after spending enormous amount of time searching in parking lot, the driver realizes that there is actually no space available. In many cases, the projected future does not seem to be more attractive: with rapid increase in population size in many cities, parking spaces are getting filled up fast so that the above situation can only deteriorate further, if no ingenious solutions are found and applied. Furthermore it is estimated that the part of city traffic generated by the vehicles looking for a

---

[*] Corresponding author.

T. Huang et al. (Eds.): ICONIP 2012, Part V, LNCS 7667, pp. 506–515, 2012.

space could represent from 5 to 10% of global traffic – translating to huge losses. Thus, suppression of searching time is an important goal to be pursued [1].

Even more so, this is an important goal, in light of the fact that the current unfortunate state of affairs, has multiple side effects: First of all, driving in circles around the parking lot to find any empty space is time consuming with – and millions of man-hours are spent globally. Second, gasoline and diesel fuel is consumed from the act of examining all the spaces in a parking lot; it is a waste of natural resources and also adds to air pollution and other forms of environmental degradation. And third, this situation causes traffic accidents and frustration for the driver. As a further consequence of these problems, during peak hours, a driver might become tempted to park in authorized areas – which in turn can intensify traffic problems even further.

On the positive side, there do exist many ways this problem can start to be tackled: for example, assigning fixed parking numbers, toll parking, valet parking etc. However, in practice, and even more so when routes are not highly regular, such methods are very inflexible and can be highly inefficient. In reality, when viewed from a higher level, the real overarching goal is optimizing a combination of important parameters: wasted driving time, wasted energy, psychological effects, etc. while providing the right framework for technology adoption, in view of organizational (government institutions and controls, parking space ownership and regulation), as well as financial, cultural, and behavioral considerations. No matter what stance one takes on this highly important matter, towards effective solutions, there is one central point of agreement: towards any solution, long-term *real-world statistics* as well as *real-time information* is vital.

Concentrating on the more technical aspects of the problems involved, several projects in the past have attempted to provide solutions. Focusing on exterior parking spaces, and one-sensor-per-many-car solutions, there is a number of existing vision-based approaches. For example, there exist systems based on a complex algorithmic process using a stereo camera system that aim to create a 3D model of the lot. Also, it is worth noticing that the changing lighting conditions of the exterior spaces make the camera data more difficult to analyze as compared to the interior spaces with much more stable lighting.

Generally, the algorithms used fall into three basic categories [2]. The first category of algorithms centers on car detection, meaning recognizing objects in the image that look like cars, without taking into account temporal dynamics. This approach can be problematic because cars with unusual shapes and sizes may be ignored. The second category utilizes motion and blob detection [3-5]. Usually, methods such as background subtraction, or even simpler ideas such as usage of reference images taken when there are no cars in the parking lot and compared to each frame are used. Another common problem is vehicular occlusion [6-7], making it hard to detect cars because they are partially blocked from the camera. Yet, there is also the possibility of combining aspects of both categories towards achieving better results.

In this paper, a streamlined and computationally efficient static/dynamic combination method using histogram classification, edge counting, and blob analysis is proposed. The method is robust to varying lighting conditions, and thus suitable for outdoors parking. Furthermore, it only requires a single inexpensive camera, and low processing power, and can provide real-time statistics and data for parking lots of more than one hundred cars, upon suitable placement of the camera.

Our system, in comparison to many existing projects, has several advantages. The vehicular occlusion problem, for example, which is important in other approaches, was almost totally alleviated through appropriate camera placement at high floors of surrounding buildings, and as our results indicate, even under the partial occlusions that take place in such a setting, our system still provides good results. Although edge detection and blob tracking have been used before, our overall processing pipeline as well as illumination compensation and combination methods are different as compared to existing approaches. Furthermore, our overall system is highly cost robust, easy to install and calibrate, and last but not least, cost effective.

2   Methods

2.1   Experimental Setup

Our real-world experiment was carried out in a crowded outdoors car park in Down Town Abu Dhabi, namely the parking space behind the multi-floor Sama Tower. In order to obtain a video of the parking lot from a suitable viewpoint, the camera was placed at 12th floor of the Tower. A data set of 45 minutes of video was captured, during 26th April 2012, around 2PM. Then, our algorithms were coded in MATLAB R2011a, and were used to analyze the initial 45 minutes video dataset.

The parking lot consists of 4 lanes with 14 parking spots in each (56 spots total); each spot of each parking lane that was not occluded by other cars was marked by coordinates (Fig.1). In order to enhance the processing speed of code, the video was processed only at the 29 instances when there was any activity or change in the parking lot. The parking states of all 56 spots for each of the 29 frames were initially labeled manually, in order to provide ground truth to be checked against the states observed by the code.

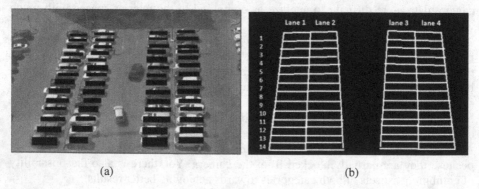

(a)   (b)

Fig. 1. Parking space slots

Each frame was then processed, and the final output was a list of spots and their associated vacancy status. For greater accuracy, three different methods of image analysis were combined. These methods can be characterized as two for static and one for dynamic analysis. We utilized edge counting and histogram classification as our static analysis methods, which rely on information available in a single frame for each decision, whereas for our dynamic across-frames method we utilized a specially crafted algorithm for blob tracking, which also served as a corrective mechanism for our static analysis-driven results. These methods are presented in detail in the next section. As a first stage, after the installation of he camera, the parking spots are hand-marked in terms of coordinates of vertices of bounding quadliterals (Fig.1).

2.2   Static Analysis

Edge Counting

The Canny edge detection algorithm was used to count the percentage of edges in each of the 52 parking spots (Fig.2). This algorithm looks for local maxima of the gradient of the grayscale image of the frame from a video. The algorithm is applied to each of the 29 frames of interest. The total area occupied by the edge for each marked parking spot is calculated and the percentage of the edge area with respect to the total area of spot is found. Spots with a higher percentage of edges are more likely to be occupied because cars have more sharp edges than empty pavement.

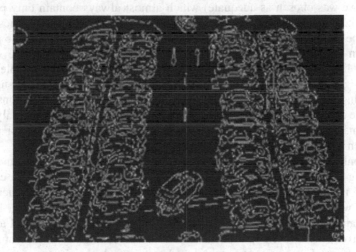

Fig. 2. Edge counting of parking spots

Histogram Classification

This method compares the pixels of parking spots with those of an empty parking pavement with road texture. If there is a high degree of similarity in both then it means that the parking spot is probably empty whereas if there is a low degree of similarity then it means that there is a car parked in the spot (Fig.3).

Fig. 3. Pavement pixels subtracted

Each of the 29 frames was converted to the YCbCr color space, and then several regions (five was chosen as adequate) which almost always contain only pavements were marked out. There is a very small probability that the majority of these regions will ever be covered by cars. These five regions act as a reference for the pavement color given the current illumination conditions. Thus, in order to decide whether a pixel most likely belongs to the pavement or to another object (for example, car), we can use the color of these regions as a reference for how the pavement is supposed to be appearing given the current lighting conditions. By having five regions, we can discard one or two of them as outliers – and in our video samples it was only the case that maximally one of these happened not to act as a good reference for pavement color, given passage of a human or other object over it.

Using only these regions, we made histograms of each channel of pavement region by dividing them into 64 bins, and then selected the highest peak from each. This allowed us to find the most prevalent bin value for the pavements. For each pixel of the frame, we calculated the color space distance from the prevalent bin. Pixels below a certain distance threshold were marked as pavement (Fig.3). Total area occupied by pavement for each spot is calculated and percentage of pavement area with respect to the total area of spot is found to determine whether it was empty. This allowed minimizing the error due to different light conditions in different times of the day in determining whether a parking spot looks more like empty pavement.

Weka Combination

The above 'Edge counting' method provides us with the percentage of the parking spot occupied by edges due to a car, whereas the above 'histogram classification' method gives us the percentage of the spot whose color resembles the current pavement color (and is thus conjectured to be empty). The two static image processing methods thus provide us with two feature values for each parking bin (features in the sense that the word is used in pattern recognition literature). These two features were combined to give an estimate of whether a parking spot is empty or not, using machine learning algorithms, using Weka [8] (Fig.4).

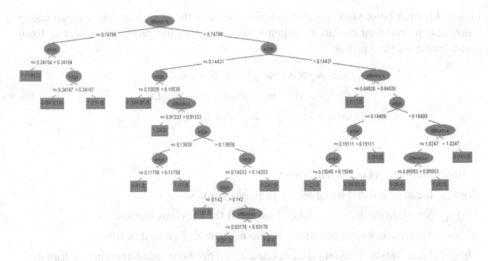

Fig. 4. Empty/Full decision tree based on Weka (1 indicate full and 0 indicates empty)

This decision tree algorithm that we created takes input from both features and decides the parking state of a spot based on the ranges in which both the percentages lie. The tree diagram in Fig.4 illustrates the mechanism of algorithm. For example, if percentage from histogram classification is less than 74%, then left branch is chosen from top box, and if percentage from edge counting is less than 24% for a spot, then next branch is chosen which marks the spot as occupied by giving it the value of 1. Based on this algorithm the parking state of each spot is recorded in a matrix.

2.3   Dynamic Analysis

Blob Analysis
In order to detect motion in parking spot where cars were parking or unparking, background/foreground estimation was used. This method starts looking from a few

(a)   (b)

Fig. 5. Tracking of moving blobs

hundred frames before our frame of interest and uses the frames to find a mean frame value and its standard deviation. When a new frame is analyzed, its difference from mean frame is calculated as

$$Mean_t = \gamma \times Frame_t + (1 - \gamma) \times Mean_{t-1}$$
$$Diff_t = \gamma \times |Frame_t - Mean_t| + (1 - \gamma) \times Diff_{t-1}$$
$$Change_t = Diff_t - (Mean_t - Frame_t)$$

where

$Frame_t$ pixel matrix of current frame

$Mean_{t-1}$ mean calculated using all the previous frames

$Diff_{t-1}$ Standard deviation calculated using all the previous frames

$Mean_t$ New mean calculated using previous mean and current frame

$Diff_t$ New standard deviation calculated using previous mean and current frame

γ Weightage of current frame

$Change_t$ Change in the value of current frame from mean value of previous frames

Pixels whose values differ beyond a threshold from the mean frame value calculated using previous frames, indicate that some motion took place in those pixels. These pixels are extracted and converted into white blobs whereas the rest of the pixels are discarded (being assumed to be not moving, i.e. background). Then by applying twice dilution and once erosion to the blobs with a circular structuring element, we effectively unify small blobs that happened to be disconnected and remove noise. We took this measure as it was found that often a moving car would appear as two disconnected blobs moving with similar velocity – but with a very small gap between them, and thus the dilation/erosion morphological operation was helpful in unifying them. Blobs below a certain minimum size were neglected as pedestrians and only large blobs were considered to be cars. Blobs whose centers were moving into or out of a parking were determined to be cars moving into or out of a parking space (Fig.5). Hence if a blob's center was moving into a parking spot, the spot was marked full, and if the center was moving out of a parking spot, the spot was marked empty.

2.4   Composite Analysis

Static analysis is not very useful in telling whether a parking spot is empty if there is a car moving in or out of the parking spot since it uses static analysis. Therefore dynamic analysis is needed that can estimate from the motion of the car, whether it is parking in the spot or unparking, hence whether the spot is empty or full (Fig.6). At the beginning of the dynamic analysis, each parking space has already been determined by the static analysis; some of which could be wrong. Output from dynamic analysis updates the state of those parking spots where some motion was occurring during the time of estimation.

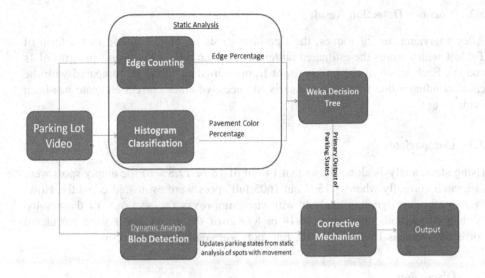

Fig. 6. Block diagram of algorithm mechanism

3   Results

3.1   Car Parking Status Measurements

During the 45min video of car parking, 29 instances were investigated when there was a change in the parking spots. For each of these 29 frames, the empty or full state of each of the 52 spots was recorded manually in matrix in the form of 1(full) or 0 (empty). This matrix was imported in MATLAB program to be compared against the output matrix of observed states estimated by the algorithms. The matrix of actual results resembles the format of output as Table 1.

Table 1. Format for storing parking states

Frame	Lane	Spot 1	Spot 2	Spot 3	Spot 4	......	Spot 13	Spot 14
	1	1	1	1	1	......	1	1
	2	1	1	1	1	......	1	1
6600	3	1	1	1	1	......	1	1
	4	1	1	1	1	......	1	1
	.	.	.	.	.	.	.	.
	.	.	.	.	.	.	.	.
	.	.	.	.	.	.	.	.
	1	0	1	1	1	......	1	1
	2	1	1	1	1	......	1	1
156720	3	1	1	1	1	......	1	1
	4	1	1	1	1	......	1	1

3.2 Vacancy Detection Result

After analyzing the 29 frames, the algorithm gives an output matrix in the form of Table 1 which stores the estimated full/empty states of all parking spots in terms of 1s and 0s. Each value of the parking spot from estimated matrix is compared with the corresponding value from actual matrix to check whether correct estimate has been made or not.

3.3 Comparison

Using static analysis alone shows that 14 out of 18 or 77.8% of the empty spots were estimated correctly whereas 1597 out 1605 full spots were estimated correctly. However combining dynamic analysis with static improves the accuracy of the results. With both methods combined, 16/18 or 88.8% of the empty spots were calculated correctly, whereas 1596/1605 of the full spots were calculated correctly.

3.4 Discussion

We are thus achieving almost 90% detection of empty spots. Inaccuracies in the estimation can be attributed to many reasons. First, the video sample had few empty slots test because the parking slot was heavily occupied at all times. Second, the video was recorded from the side view of cars, which meant that blocked view of the car parked behind it. This led to many errors because if there was any empty slot behind a car, it was covered by small part of image of a car that was not parked inside it and hence led to presence of false edges and color in the empty slot. The results can be tremendously improved if the video is recorded from an angle perpendicular to the parking lanes. Most importantly, at the moment we are experimenting with much larger data sets that we have collected and which we are labeling, spanning over several days, and fine-tuning our algorithms, as well as combining them with larger-scale trajectory tracking for cars, enabling us to get quantitative estimates of searching times, waiting times, search strategies and other such interesting features. Furthermore, we are developing a real-time sms notification system for drivers entering the park that are subscribing to the service, which instantly assigns the nearest unassigned free slot to them, if one is available, and informs them of its position.

4   Conclusion

After having discussed about the multi-faceted importance of parking statistics and real-time management techniques in densely populated areas, in this paper we presented a cost-effective camera-based system that can supervise open air parking lots, can provide statistics, and that can show drivers the nearest available space. In our system, a single camera is placed on a tall building adjacent to the lots, and its output is fed to a combination of two kinds of algorithms: static image analysis of parking lot spaces using two types of features (derived from histogram classification and edge detection), and dynamic image analysis using blob analysis (based on background subtraction). Our system achieves monitoring of parking spaces, and reports statistics as well as empty slots in real-time. Our results indicate that almost 90% of empty spots are reported correctly, and acts as a highly cost-effective single-camera unit which can monitor more than 100 spaces. The widespread application of the currently evolving versions of our system could well potentially result in enormous savings of time, fuel, and street space, and thus help citizens save time and fuel, preserve our environment, and enjoy a much more pleasant everyday driving experience.

Acknowledgement. The authors would like to thank George Chaidos for his great help in coding, and to thank Leonard Helmrich and Mo Ogrodnik for their huge support.

References

1. Gantelet, E., Lefauconnier, A.: The Time Looking for A Parking Space: Strategies, Associated Nuisances and Stakes of Parking Management in France. In: European Transport Conference, pp. 6 (2006)
2. Ichihashi, H., Notsu, A., Honda, K., Katada, T., Fujiyoshi, M.: Vacant Parking Space Detector for Outdoor Parking Lot by Using Surveillance Camera and FCM Classifier. In: IEEE International Conference on Fuzzy Systems, pp. 127–134 (2009)
3. Salem, M., Watson, D.: Person Tracking for Parking Space Vacancy Prediction. Technical Report, University of California (2007)
4. Wah, C.: Catherine. Parking Space Vacancy Monitoring. Technical Report, University of California (2009)
5. Liu, S.: Robust Vehicle Detection from Parking Lot Images. Technical Report, Boston University (2005)
6. Mejia-Inigo, R., Barilla-Perez, M.E., Montes-Venegas, H.: A Color-based Texture Image Segmentation for Vehicle Detection. In: Image Segmentation, pp. 273–290 (2011)
7. True, N.: Vacant Parking Space Detection in Static Images. Technical Report, University of California (2007)
8. Hall, M., Frank, E., Holmes, G., Pfahringer, B., Reutemann, P., Witten, I.: The WEKA Data Mining Software: An Update. SIGKIDD Explorations 11, 10–18 (2009)

Effect of Facial Feature Points Selection on 3D Face Shape Reconstruction Using Regularization

Ashraf Y.A. Maghari, Iman Yi Liao, and Bahari Belaton

School of Computer Sciences, Universiti Sains Malaysia, Pulau Pinang, Malaysia
myashraf2@gmail.com

Abstract. This paper aims to test the regularized 3D face shape reconstruction algorithm to find out how the feature points selection affect the accuracy of the 3D face reconstruction based on the PCA-model. A case study on USF Human ID 3D database has been used to study these effect. We found that, if the test face is from the training set, then any set of any number greater than or equal to the number of training faces can reconstruct exact 3D face. If the test face does not belong to the training set, it will hardly reconstruct the exact 3D face using 3D PCA-based models. However, it could reconstruct an approximate face shape depending on the number of feature points and the weighting factor. Furthermore, the accuracy of reconstruction by a large number of feature points (>150) is relatively the same in all cases even with different locations of points on the face. The regularized algorithm has also been tested to reconstruct 3D face shapes from a number of feature points selected manually from real 2D face images. Some 2D images from CMU-PIE database have been used to visualize the resulted 3D face shapes.

Keywords: Representational Power, statistical face modeling, PCA, Regularization, feature points.

1 Introduction

The use of 3D data in face image processing applications has received great attention among scholars and researchers during the last few years [14]. The need for 3D face reconstruction has grown in applications like virtual reality simulations, plastic surgery simulations, and face recognition [6,7]. 3D facial reconstruction systems are to recover the three dimensional shape of individuals from their 2D pictures or video sequences. Until now, in most popular commercially available tools, the 3-D facial models are obtained not directly from images but by laser-scanning of the people's faces [16]. These scanners are usually expensive and working in some restricted circumstances, otherwise it would fail in the cases; for example, in plastic surgery a person's face has been damaged during an accident and only his/her pictures taken prior to the accident are available. There are many approaches for reconstruction 3D faces from images such as Shape-from-Shading (SFS) [15,13], shape from silhouettes and shape from motion [1].

T. Huang et al. (Eds.): ICONIP 2012, Part V, LNCS 7667, pp. 516–524, 2012.

There are also non statistical learning-based methods, such as neural network [11] and statistical learning-based methods, such as analysis by synthesis using 3D Morphable Model (3D MM) [5]. Analysis by synthesis is an approach in which the parameters of the 3D statistical model are adjusted to increase the accuracy between the reconstructed face and the 2D face image [14]. The presence of 3D scanning technology lead to create a more accurate 3D face model examples [10]. Examples based modeling allows more realistically face reconstruction than other methods [14,9]. In the simplest form, example-based 3D face reconstruction methods have two main stages: The model building stage and the model fitting stage. In this paper, PCA-based 3D face model is used for model building and the regularized algorithm is used for model fitting.

Although in some statistical modeling methods both shape and texture are modeled separately using PCA (e. g. 3DMM), it is suggested that shapes are more amenable to PCA based modeling than texture, as textures varies dramatically than the shape. In many situations, the 2D texture can be warped to the 3D geometry to generate the face texture [8]. Therefore, in this paper we focus on shape modeling. The regularized algorithm that uses the 3D Morphable Model to reconstruct 3D face shape from facial 2D points has been presented in [4] and [3]. However, they did not conduct any statistical analysis on how the feature points selection affect the reconstruction.

This study address the problem of how the feature points affect the 3D face shape reconstruction using regularized algorithm. USF Human ID 3D database has been used to train and test PCA-model. 80 and 20 face scans have been used for training and testing respectively.

We used the 20 testing face shapes to test and compare the influence of feature points selection on the accuracy of 3D reconstruction. For each tested 3D face shape, the RP was calculated for the PCA-model. We defined the representational power (RP) of the model as the Euclidian Distance between the reconstructed shape surface vector and the true surface shape vector divided by the number of points in the shape vector. For 3D face shape reconstruction from real 2D images, some 2D images from CMU-PIE database [12] have been used to visualize the resulted 3D face shapes.

This paper is organized as follows: Reconstruction based on regularization is introduced in Section 2, whereas in Section 3 the Experiments and statistical analysis are explained. Discussion and conclusion are reported in section 4.

2   Reconstruction Based on Regularization

The regularized algorithm was categorized as one of the main existing four methods for 3D facial reconstruction [9]. The algorithm was used by Dalong [8] to reconstruct 3D face for face recognition purposes. A number of selected facial points are used to compute the 3D face shape coefficients of the eigenvectors. Then, the coefficients are used to reconstruct the 3D face shape.

2.1 Modeling Shape Using PCA

In order to build shape PCA-based 3D model, PCA decomposition is applied to model shape variability. Each training sample is represented by the 3D coordinates of all vertices in the triangulated mesh. Based on this representation, the mean 3D face among the training set is estimated and the deviation of each training sample from the mean is calculated. PCA is then applied on the covariance matrix. As a result of the analysis, new 3D shape s can be generated using:

$$s = s_0 + \sum_{i=1}^{m} \alpha_i e_i \,, \tag{1}$$

where s_0 is the mean 3D shape, e_i represent the $i - th$ eigenvectors of the covariance matrix, α_i is the coefficient of the shape eigenvector e_i and m is the number of significant eigenvectors.

2.2 Regularized Algorithm

Let t be the number of points that can be selected from the input 2D face image, $s_f = (p_1, p_2, ..., p_t) \in R^{2t}$ be the set of selected points on the 2D face image, whereas every point p_i has 2 axis x and y, $S_{f0} \in R^t$ is the t corresponding points on s_0 (the average 3D face shape) and $E_f \in R^{t \times m}$ is the t corresponding columns on $E \in R^{3n \times m}$ (the matrix of row eigenvectors). Then the coefficient α of a new 3D face shape can be derived as

$$\alpha = (E_f^T E_f + \lambda \Lambda^{-1})^{-1} E_f^T (S_f - S_{f0})) \,, \tag{2}$$

where Λ is a diagonal $m \times m$ matrix with diagonal elements being the eigenvalues and λ is the weighting factor [8]. Then we apply α to equation (1) to obtain the whole 3D face shape.

2.3 Reconstruction of 3D Face Shapes from 2D Images

The following subsections will describe the reconstruction procedure of 3D face shapes from a single 2D face images.

Feature Extraction. Human faces share a common geometric facial landmarks such as eye corners, eyebrow, mouth corners, nose and face contour. 78 landmarks are selected manually from the input 2D face to be used for face alignment and 3D shape reconstruction.

Feature Points Alignment. In our work, the input 2D images are in frontal pose and neutral expression. 78 feature points, which have been manually selected, can be aligned using Procrustes Analysis. Procrustes determines a linear transformation (translation, reflection, orthogonal rotation, and scaling) of the points in shape A to best conform them to the points in Shape B. The mathematical theory of the Procrustes Analysis is similar to Iterative Closest Point (ICP) [2]. Fig. 1 demonstrate the alignment procedure.

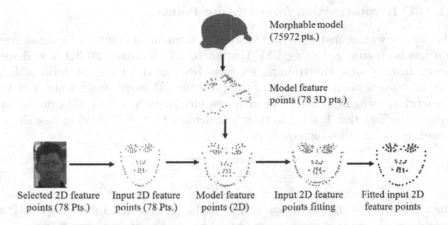

Fig. 1. input 2D feature points fitting process

3D Face Shape Reconstruction. The aligned feature points have been used to compute the 3D shape coefficients of the eigenvectors using Equation 2. Then, the coefficients were used to reconstruct the 3D face shape using Equation 1.

2.4 Reconstruction of Testing and Training Face Shapes

In this study, the 3D faces in the database are already aligned with each other as explained by [5]. So the selected points can be chosen randomly from each face inside and outside the training dataset. The points can directly used to calculate the 3D shape coefficients of the eigenvectors using Equation 2. Then, the coefficients are used to reconstruct the 3D face shape using Equation 1. The resulted face can be compared with original face by calculating the RP. It is worth noting that any selected point does not represent a vertex on the 3D face shape as it matches a mere number in the shape vector.

3   Experiments and Statistical Analysis

This study aims at exploring the influence of feature points on the accuracy of 3D face shape reconstruction using regularized algorithm. We evaluated the accuracy of reconstruction by calculating the RP for the resulted 3D face compared with that of original 3D face shape. The USF Raw 3D Face Data Set is used in the evaluation since it contains 3D faces that can be numerically compared with the represented 3D face shapes. The USF database includes shape and texture information of 100 3D faces obtained by using Cyberware head and face 3D color scanner. They are aligned with each other as explained by [5]. This study uses only the shape vectors for training and testing the models. To visually evaluate the accuracy of 3D reconstruction from single 2D images, CMU-PIE database is used. The regularized algorithm is used to reconstruct 3D face shapes from 2D facial points.

3.1   3D Reconstruction from Feature Points

The accuracy of reconstructed 3D face shape from facial points using regularization has been analyzed using USF Human ID 3D database. 20 3D face shapes chosen from outside the training set have been used for testing process. The selected feature points are used to compute the 3D shape coefficients α of the eigenvectors using Equation 2. Then, a new face shape s_r can be obtained by applying α to Equation 1. RP can then be calculated for all 20 testing face shapes according to the following equation

$$RP = \frac{\sum_{i=1}^{3*75972} \sqrt{(S_i - S_{ri})^2}}{3*75972} \,, \tag{3}$$

where s is the testing face shape and s_r is the reconstructed face shape. RP of reconstructed new face shape represents the accuracy of reconstruction.

Number of Feature Pints. The PCA-model has been trained with 80 face shapes. The selected feature points are between 10 and 300 and randomly selected from face shapes inside and outside the training set (the selected point represents a mere number in the shape vector). If the test face is from the training set, any number of selected feature points greater than or equal the number of training face shapes can reconstruct exact 3D face. If the test face is not from the training data set, it hardly reconstructs the exact 3D face.

Fig. 2 shows the relation between the number of feature points and the accuracy of the reconstructed face shapes using PCA-model trained with 80 faces. The algorithm was analyzed for different values of weighting factor $\alpha = 0.005$, 0.05, 0.1, 1, 10, 20, 50, 100, and 200 as shown in Fig. 2. The 3D shape error

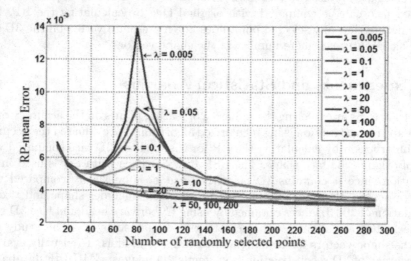

Fig. 2. The relationship between the number of feature points and the accuracy of the reconstructed face shapes with different values of the weighting factor λ

(RP) is determined 30 times for each shape by a constant number of feature points f selected randomly for every time. Then, RP-mean is the mean of the 30 resulted RPs. RP-mean Error, the average of RP-means for every number of feature points, represents the accuracy of reconstruction by the number of feature points.

3D face shapes reconstructed from different number of feature points $f =$ 10, 50, 80, 120 and 200 are shown in Fig. 3. The reconstructed faces are compared with the original and the mean face shapes.

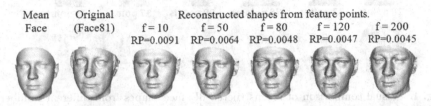

Fig. 3. Reconstructed 3D faces among different number of feature points ($\lambda = 200$)

Location of Feature Points. For each of the 20 face shapes chosen from outside the training set, a constant number of points were randomly selected 50 times to reconstruct face shapes. The 50 selections of feature points were repeated with each of the weighting factors 0.05, 1, 10, and 100. The reconstruction results of all repeated 50-time selections showed the similarity of RP.

Table 1. ANOVA results with $\alpha = 1$

Source of Variation	Between Groups	Within Groups	Total
SS	5.82E-05	0.000968	0.001027
Df	49	0.000968	999
MS	1.19E-06	1.02E-06	
F	1.165887		
P-value	0.206254		
F crit	1.367567		

Single factor ANOVA was run using MS Excel to compare the 50 alternatives for each of the selected 30, 40, 50, 60, 80 and 100 feature points. In all cases, the ANOVA results show that there is no significant difference among the 50 feature selections. Table 1 shows sample results using 60 feature points which have been chosen randomly 50 times with $\alpha = 1$.

3.2   3D Face Shape Reconstruction from Real 2D Images

From the CMU-PIE database, we selected 2 frontal images with neutral expression. 78 2D feature points were manually selected for reconstruction. The comparisons between reconstructed 3D face shapes using different number of feature points are illustrated in Fig. 4. Three 3D shapes are reconstructed for every input image with numbers of 20, 37, and 78 feature points. From the results in

Fig. 4, some differences are noted in the reconstructed 3D face shapes depending on the number of feature points. In case of real 2D images, it must be noted that the accuracy of reconstruction depends not only on the number of feature points, but it also depends on the accurate selection of feature points from the 2D image and the correspondence points on the reference 3D face shape.

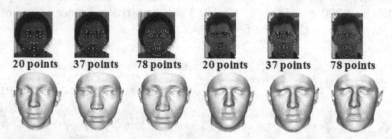

Fig. 4. Visual comparison of reconstructed 3D face shapes from different number of facial points selected manually from real 2D images

4   Discussion and Conclusion

The accuracy of 3D face shape reconstruction from feature points was evaluated by analyzing the 3D face database obtained from University South Florida (USF) with a series of experiments and statistical analysis. The current USF Human ID 3D database has 100 faces. 80 faces used for training and 20 for testing purposes. The regularized algorithm was analyzed to find the relationship between the number of feature points and the accuracy of reconstructed 3D face shapes. The extensive experimental results showed that if the test face is from the training set, then any set of any number greater than or equal to the number of training faces can reconstruct exact 3D face. If the test face does not belong to the training set, it will hardly reconstruct the exact 3D face using 3D PCA-based models. However, it could reconstruct an approximate face depending on the number of feature points and the weighting factor. Based on this relationship, we found that, regardless of the value of the weighting factor λ, the accuracy of reconstruction by a large number of feature points (greater than 150 points) is relatively the same in all cases associated with slight improvement related to the larger number of selected points (see Fig. 2) even with different locations of points on the face. The visualized results in Fig. 3 are consistent with the findings on the feature points. Accordingly, the greater the points used to reconstruct a face shape the closer the reconstructed face shape is to the original face shape. Meanwhile, whenever the number of feature points is fewer, the reconstructed face is closer to the mean face. This is because the regularized algorithm being used causes the reconstructed face to be closer to the mean face whenever the number of feature points approaches zero.

For 3D face shape reconstruction from real 2D face images, some 2D images from CMU-PI database have been used to visualize the reconstructed 3D face shapes. The visual results in Fig. 4 show better recovering of the 3D shape by larger number of feature points. With limited information about the face shape, the regularized algorithm cannot capture details of the face shape. However, the overall 3D face shape can be recovered well related to large number of accurately selected facial points.

Acknowledgments. The work presented in this paper is sponsored by RU grant 1001/PKCOMP/817055, Universiti Sains Malaysia.

References

1. Amin, S., Gillies, D.: Analysis of 3d face reconstruction. In: 14th International Conference on Image Analysis and Processing, ICIAP 2007, pp. 413–418 (September 2007)
2. Besl, P., McKay, H.: A method for registration of 3-d shapes. IEEE Transactions on Pattern Analysis and Machine Intelligence 14(2), 239–256 (1992)
3. Blanz, V., Mehl, A., Vetter, T., Seidel, H.: A statistical method for robust 3d surface reconstruction from sparse data. In: Proceedings of 2nd International Symposium on 3D Data Processing, Visualization and Transmission, 3DPVT 2004, pp. 293–300. IEEE (2004)
4. Blanz, V., Vetter, T.: Reconstructing the complete 3d shape of faces from partial information (rekonstruktion der dreidimensionalen form von gesichtern aus partieller information). it-Information Technology (2002)
5. Blanz, V., Vetter, T.: A morphable model for the synthesis of 3d faces. In: Proceedings of the 26th Annual Conference on Computer Graphics and Interactive Techniques, New York, NY, USA, pp. 187–194 (1999)
6. Elyan, E., Ugail, H.: Reconstruction of 3d human facial images using partial differential equations. In: JCP, pp. 1–8 (2007)
7. Fanany, M.I., Ohno, M., Kumazawa, I.: Face Reconstruction from Shading Using Smooth Projected Polygon Representation NN. In: Proceedings of the 15th International Conference on Vision Interface, Calgary, Canada, pp. 308–313 (2002)
8. Jiang, D., Hu, Y., Yan, S., Zhang, L., Zhang, H., Gao, W.: Efficient 3d reconstruction for face recognition. Pattern Recogn. 38, 787–798 (2005)
9. Levine, M.D., Yu, Y(Chris): State-of-the-art of 3d facial reconstruction methods for face recognition based on a single 2d training image per person. Pattern Recogn. Lett. 30, 908–913 (2009),
   http://dl.acm.org/citation.cfm?id=1552570.1552692
10. Luximon, Y., Ball, R., Justice, L.: The 3d chinese head and face modeling. Computer-Aided Design 44(1), 40–47 (2012)
11. Nandy, D., Ben-Arie, J.: Shape from recognition: a novel approach for 3-d face shape recovery. IEEE Transactions on Image Processing 10(2), 206–217 (2001)
12. Sim, T., Baker, S., Bsat, M.: The cmu pose, illumination, and expression database. IEEE Transactions on Pattern Analysis and Machine Intelligence 25(12), 1615–1618 (2003)
13. Smith, W., Hancock, E.: Recovering facial shape using a statistical model of surface normal direction. IEEE Transactions on Pattern Analysis and Machine Intelligence 28(12), 1914–1930 (2006)

14. Widanagamaachchi, W., Dharmaratne, A.: 3d face reconstruction from 2d images. In: Digital Image on Computing: Techniques and Applications, DICTA 2008, pp. 365–371 (December 2008)
15. Zhang, R., Tsai, P.S., Cryer, J., Shah, M.: Shape-from-shading: a survey. IEEE Transactions on Pattern Analysis and Machine Intelligence 21(8), 690–706 (1999)
16. Zhang, Z., Hu, Y., Yu, T., Huang, T.: Minimum variance estimation of 3d face shape from multi-view. In: 7th International Conference on Automatic Face and Gesture Recognition, pp. 547–552 (April 2006)

Human Posture Recognition with the Stochastic Cognitive RAM Network

Weng Kin Lai[1], Imran M. Khan[2], and George G. Coghill[3]

[1] School of Technology, TARC, 50932 Kuala Lumpur, Malaysia
laiwk@mail.tarc.edu.my
[2] Dept. of Electrical and Computer Engineering, IIUM, 53100 Kuala Lumpur, Malaysia
electronicluddite@gmail.com
[3] Department of Computer and Electrical Engineering, University of Auckland, New Zealand
g.coghill@auckland.ac.nz

Abstract. This paper examines a weightless neural network (WNN) for human posture recognition. Like all earlier weightless neural network models, the Cognitive RAM Network (CogRAM) learns in one pass through the data and due to its simplicity it can be fabricated in hardware. While it has shown good performance in earlier studies, it still suffers from the common problem of network saturation especially when it comes to high dimensional and poorly separated data in the feature space. Hence, we proposed the Stochastic CogRAM which has shown significant improvements when tested on the challenging human postures recognition problem. We also present some comparisons of the experimental results obtained from the popular *K-Means* clustering algorithm. Future research is outlined at the end of the paper.

Keywords: Pattern recognition, weightless neural networks, computer vision, video analytics.

1 Introduction

Human ability to predict the intentions of individuals from their posture has important uses in the security industry. However, posture recognition is a difficult and challenging problem. When performed by human guards, it can be quite a mundane and boring task, especially as the human attention span is known to be notoriously short. However, an automated human posture recognition system can have many potentially useful applications. An example of such an application is in automated video surveillance (AVS) which can be deployed to help strengthen security, especially in public spaces. This has been motivated by the need for better public safety, availability of powerful computing hardware, as well as the relatively lower cost of good cameras in the market. Moreover, AVS systems can play a very effective role in assisting their human operators by combating operator fatigue especially in circumstances where the ratio of the monitors being used for surveillance versus the number of human operators is large. Ideally, we want to recognise postures from images that were obtained from one camera and in real time. These constraints placed on the system can be rationalised because in the majority of video surveillance systems, only one camera is

T. Huang et al. (Eds.): ICONIP 2012, Part V, LNCS 7667, pp. 525–532, 2012.

used to observe the scene and any analysis of the captured images is done in real time 1. This means that the processing speed of the system is also a crucial aspect to be considered. However for our investigations, the first priority is achieving high accuracy of correct posture recognition. Only after a satisfactory level of accuracy has been achieved, the system will be optimised to improve its speed – and this is something in which weightless neural networks can play a major role.

The remainder of this paper is organised as follows. The next section discusses the problem of human posture recognition and some of the related work. In section 3 we examine the concept of weightless neural networks (WNN's) for classification, with a brief description of related work. It also introduces the Cognitive RAM network (CogRAM) and details its architecture and learning rules. Following this, Section 4 discusses the major design issues as well as the important aspects of the experiments and the results obtained. Finally, in section 5 we present some conclusions and potential areas for further work.

2   Human Posture Recognition

Human posture refers to the arrangement of the body and its limbs 2. There are several agreed types of human postures such as standing, sitting, squatting, lying down, and kneeling. However, for the purpose of our investigation we focused on recognizing five postures, viz. lying down, jumping, fighting (pushing/punching), climbing and pointing. A sample of these various postures is shown in Fig. 1.

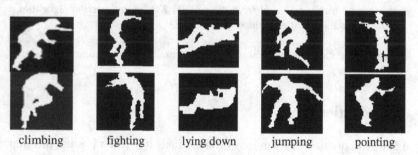

climbing   fighting   lying down   jumping   pointing

Fig. 1. 5 Human postures used in the experiments

One of the earlier human motion analysis systems was based on a template matching approach. Bobick and Davis had adopted a view-based approach to investigate the representation and recognition of human actions using temporal templates 3. Other techniques include the efforts by Spagnolo et. al. 4 who proposed a fast and reliable approach to estimate body postures in outdoor visual surveillance systems. The sequences of images coming from a static camera is trained and tested for recognition. The system uses a clustering algorithm which needed manual labelling of the resulting clusters after training. The features extracted are the horizontal and vertical histograms of binary shapes associated with humans. After training, the Manhattan distance is used to develop the clusters and for recognition. The main strengths of their method are high classification performance and relatively low computational time which allows the system to perform well in real time. On the other hand, Buccolieri et. al. 5 explored the combination of active contours and a radial basis function neural network to recognise postures. The primary advantage of their approach is that it has low sensitivity to noise, fast processing speed, and the ability to handle some degree of occlusion. However, their system can only recognise three postures, namely standing, bending and squatting postures. Our recent work involves the use of particle swarm optimization (PSO) to recognise five different postures 6 7.

3   Weightless Neural Nets

Weightless Neural Networks (WNN's) which only require a one pass learning have been around in various configurations since the 1960's. Ludermir et al 8 give an interesting review of the topic. The WNN used here, shown in Fig. 2(a) is derived from the work of Igor Aleksander 9.

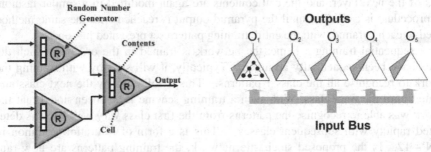

(a)   The Probabilistic Logic Neuron (PLN)   (b) Typical pyramid structure of the PLN

Fig. 2. Weightless neural network

In Fig. 2, we chose a very simple example for the purpose of easy explanation. In reality there may be many input layer cells, with 4, or even 8-bit input address lines. The network will always be triangular shaped (and pyramid for two-dimensional input data) as the network moves towards the output layer – Fig. 2(b). If we require several output classes, we have to construct the appropriate number of pyramids, each with its own desired output value, but with a common input vector.

During training, the input vector, which (in this case) is a 4 bit binary number, acts as an address generator. Initially, all of the cells' contents in all three cells are set to an undefined state. When an undefined location is addressed, a random number generator is called. This produces an output from the cell of either a 1 or a 0 with equal probability. Further developments saw Filho et. al 10 proposing a different approach which does not use random number generators. Instead, when an undefined location is selected by the input address to a cell, the output generates an undefined value and this is propagated forward to the next layer. Their Goal Seeking Neural Network then provides rules to deal with this.

The idea of having an addressable set and a set of simple rules was developed further to produce the Deterministic Adaptive RAM Network (DARN) 10. Although the generalisation performance improved, the DARN's capabilities were still poorer than those of the MLP. The Cognitive RAM network (CogRAM), which is an enhancement to the DARN has produced very promising results 12 13. In the CogRAM, each addressed content is a signed register where all of the registers in the network's cells are initially set to zero. The zero is interpreted as an undefined value. During learning the counter register may become positive, or negative.

3.1   Learning

The desired output of every cell in the pyramid is the same as the desired output of the pyramid. If the desired output of the neurons at the input layer is 1, the addressed location is incremented by one. If the desired output is 0, the location is decremented by one. The method then involves calculating the address vectors for the next layer, moving towards the output. The input address vector to each cell in the next layer is constructed by invoking the Recall function (described below), on the previous layers from each connecting line in that next layer. The address vector is then applied to the cell inputs of the next layer, and the cell contents are again modified in a similar fashion. This procedure is continued until the pyramid output is reached with the same method applied to each pyramid, with the entire training pattern set presented just once.

In a sequential training scheme, the network is trained in the order in which the classes have been sequentially arranged. Typically, it will start by with training the network to recognise all the class 1 patterns. This is followed by the next class and repeated until the final class. With such a training scheme it has been noted that the network was able to recognise the patterns from the first class very well but this deteriorated rapidly with subsequent classes. This is a form of saturation common in WNN's 12. In the proposed stochastic network, the training patterns are now randomly selected from each class, without any one class monopolising the available memory locations. More importantly, the order in which these patterns are presented to the network is no longer in the same sequence of ascending order.

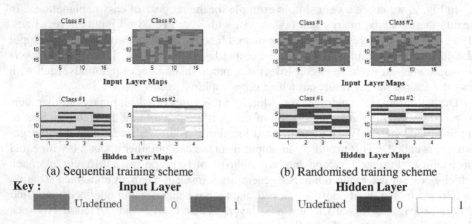

(a) Sequential training scheme   (b) Randomised training scheme

Fig. 3. Memory maps for the trained CogRAM (2 classes)

This may be seen in Fig. 3 where the *Stochastic CogRAM* has been trained to recognise two different types of human postures. Each column represents the memory locations for one 4-bit neuron (resulting in sixteen locations), and there are sixteen of such neurons in the input layer and 4 for the hidden layer.

3.2   Recall

In the recall phase, again, the content of each addressed location, starting from the input layer, is interpreted as U (undefined), 0, or 1. In the event that the output of each cell propagates an undefined(U) output forward, the Goal Seeking Network approach of propagating the addressable set forwards to the next layer will be adopted. Even though several locations may be selected at the same time, the following rules, which are essentially the same as for the GSN are adopted.

- If at least one *zero* is addressed and no *ones* are addressed, the output is *zero*.
- If at least one *one* is addressed and no *zeros* are addressed, the output is *one*.
- Otherwise the output is *undefined*.

Recall is propagated forward through the pyramid until the output is reached

4   Experimental Setup and Results

For our study, we have selected six features from the human body to be used in representing a single sample of silhouette image. These include the locations of the head (H), left arm (AL), right arm (AR), left leg (LL), right leg (LR) and torso (T). The points of the features, $S = ([H_x,H_y], [AL_x , AL_y], [AR_x ,AR_y], [LL_x ,LL_y], [LR_x ,LR_y], [T_x, T_y])$ represent the feature points as illustrated in Fig. 4. x and y are the coordinates of the major reference points where $0 <$ x, y < 1.

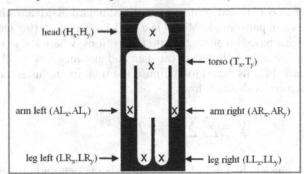

Fig. 4. Reference points of key features for human posture

When collecting data for the various postures, we have ensured that no occlusion occurs and the entire body of the person is visible. There are 20 images per posture class. Furthermore it is essential that the distance of the person from the camera is kept constant too. These limitations will ensure that the system is more orientated towards posture recognition and does not result in long computation time spent just on preprocessing the data.

For convenience, a 64 cell network was used. With 12 features, we chose a *4-4-8-8-8-8-8-4-4-4-4-4* representation scheme for each input data. This means that for a 4-input line cell, there would be 16 cells at the input layer. Furthermore, the data was partitioned into a mix of training and test data set with 85 and 15 for each set respectively. The

Stochastic Cognitive RAM network was then trained and tested for each run. Following this, the average as well as the best classification accuracy for each of the 1,000 runs was obtained. In addition, we also computed the confusion matrix for all the various postures. This is illustrated in Fig. 5, where a lighter colour represents a higher correlation for that particular posture pair. It can be seen that the conventional CogRAM was not able to clearly differentiate between the 'jumping' and 'fighting' postures as well as the 'pointing' and 'fighting' postures. In contrast, the Stochastic CogRAM recorded the highest recognition for the 'lying down' posture. This is shown in Fig. 5(b).

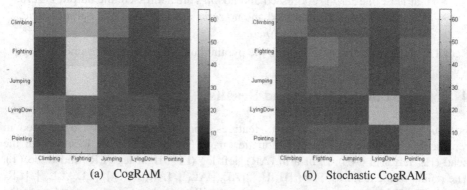

(a) CogRAM   (b) Stochastic CogRAM

Fig. 5. Confusion matrix for CogRAM

Moreover, we also obtained results from the standard K-Means clustering algorithm to provide a comparison. K-Means is a well-known clustering technique which will cluster n objects based on attributes into k partitions, where k < n. Its main aim is to find the centres of natural clusters in the data and assumes that the object attributes form a vector space 14. Its aim is to minimise the total intra-cluster variance, or, the squared error function as shown below,

$$V = \sum_{i=1}^{k} \sum_{x_j \in S_i} (x_j - \mu_i)^2 \tag{1}$$

where there are k clusters S_i, i = 1, 2, ..., k, and μ_i is the centroid or mean point of all the points $x_j \in S_i$. As the K-Means is an unsupervised clustering technique, it will result in clusters where the individual members of each cluster may not directly map back to the desired clusters. To map the actual postures identified by K-Means to the true postures, we compute the centroid of each cluster to the nearest desired clusters. Nevertheless, some clusters may end up equal-distance between two or more centroids of the desired clusters, indicating a poor separation of the various classes in the problem space. In such cases, we chose to ignore the results. The results obtained from these three approaches are summarised and illustrated in the radar plot of Fig 6.

It may be seen that the Stochastic CogRAM (SCogRAM) was able to return significantly better average and highest recognition accuracies as compared with the conventional CogRAM. Even though the average recognition accuracy of the Stochastic CogRAM may be lower than that obtained from K-Means, nevertheless, the maximum recognition accuracy was on par with that obtained by K-Means.

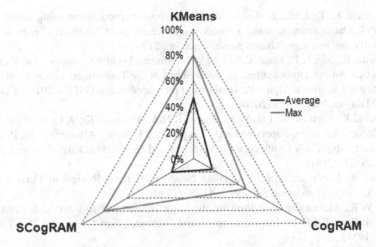

Fig. 6. Radar plot for the best and average recognition for each technique

5   Conclusions

The CogRAM has shown good performance for a variety of different problems. Nevertheless, it still suffers from network saturation especially when it comes to complex or poorly separable data. This provided the motivation for the proposed Stochastic CogRAM to study on a set of challenging human postures which are representative of typical applications found in video analytics. A comparison was also made between the two CoGRAM networks, as well as with the popular K-Means. Even though the average recognition accuracy for the K-Means may be better than both the CogRAM models, nevertheless, the Stochastic CoGRAM was able to match the best solutions obtained by K-Means. In the future, a priority area is to investigate a better selection of the feature set to represent the postures. In addition, we would also like to explore building a hardware implementation of this proposed new model, especially for human posture recognition.

References

1. Boulay, B., Bremond, F., Thonnat, M.: Human Posture Recognition in Video Sequence. In: Joint IEEE International Workshop on Visual Surveillance, Performance Evaluation of Tracking & Surveillance (VS-PETS), pp. 23–29 (2003)
2. Girondel, V., Bonnaud, L., Caplier, A., Romaut, M.: Static Human Body Postures Recognition in Video Sequences Using the Belief Theory. In: International Conference on Image Processing (ICIP 2005), Genoa, Italy (2005)
3. Bobick, A., Davis, J.: Real-time recognition of activity using temporal templates. In: Proceedings of IEEE CS, Workshop on Applications of Computer Vision, pp. 39–42 (1996)
4. Spagnolo, P., Leo, M., Leone, A., Attolico, G., Distante, A.: Posture es-timation in visual surveillance of archaeological sites. In: Proceedings of the IEEE Conference on Advanced Video and Signal Based Surveillance, pp. 277–283 (2003)

5. Buccolieri, F., Distante, F., Leone, A.: Human posture recognition using active contours and radial basis function neural network. In: Proceedings of the IEEE Conference on Advanced Video and Signal Based Surveillance, pp. 213–218 (2005)
6. Maleeha, K., Sin, L.T., Chan, C.S., Lai, W.K.: Human Posture Classi-fication Using Hybrid Particle Swarm Optimization. In: Proceedings of the Tenth International Conference on Information Sciences, Signal Processing and their application (ISSPA 2010), Kuala Lumpur, Malaysia, May 10-13 (2010)
7. Maleeha, K., Chan, C.S., Lai, W.K., Kyaw, K.H.A., Othman, K.: A Comparison Of Posture Recognition using Supervised and Unsupervised Learning Algorithms. In: Proceedings of 24th EUROPEAN Conference on Modeling and Simulation, Kuala Lumpur, June 1-4, pp. 226–232 (2010)
8. Teresa, B., Ludermir, et al.: Weightless Neural Models: A Review of Current and Past Works. Neural Computing Surveys 2, 41–60 (1999)
9. Kan, W.K., Aleksander, I.: A Probabilistic Logic Neuron Network for As-sociative Learning. In: Proc. IEEE 1st Int. Conf. on Neural Networks, San Diego, vol. II, pp. 541–548 (June 1987)
10. Filho, E.C.D.B.C., Fairhurst, M.C., Bisset, D.L.: Adaptive Pattern Recognition using Goal Seeking Neurons. Pattern Recognition Letters 12, 131–138 (1991)
11. Bowmaker, R.G., Coghill, G.G.: Improved Recognition Capabilities for Goal Seeking Neuron. Electronics Letters 28(3), 220–221 (1992)
12. Lai, W.K., Coghill, G.G.: Data Classification With An Improved Weight-less Neural Network. Journal of Information Technology in Asia 1(1), 17–34 (2005)
13. Yong, S., Lai, W.-K., Goghill, G.: Weightless Neural Networks for Typing Biometrics Authentication. In: Negoita, M.G., Howlett, R.J., Jain, L.C. (eds.) KES 2004. LNCS (LNAI), vol. 3214, pp. 284–293. Springer, Heidelberg (2004)
14. Costa, L., Cesar, R.M.: Shape Analysis and Classification: Theory and Practice. CRC Press (2001)

Entropy Based Image Semantic Cycle for Image Classification

Hongyu Li[1,2], Junyu Niu[3,*], and Lin Zhang[2]

[1] Electronic Engineering Department, Fudan University, Shanghai, China
[2] School of Software Engineering, Tongji University, Shanghai, China
[3] School of Computer Science, Fudan University, Shanghai, China
{hongyuli,jyniu}@fudan.edu.cn, cslinzhang@tongji.edu.cn

Abstract. This paper proposes a novel framework for image classification with an entropy based image semantic cycle. Entropy minimization leads to an optimal image semantic cycle where images are connected in the semantic order. For classification, the training step is to find an optimal image semantic cycle in an image database. In the test step, the suitable position of an unknown image in this cycle is first found. Then, the class membership is determined through recognizing the nearest neighbors at this position. Experimental results demonstrate that the proposed framework achieves higher classification accuracy.

Keywords: Entropy, Image Semantic Cycle, Image Classification.

1 Introduction

As an integral part of image management and retrieval, image classification aims to classify an unknown image by the object category, has been the subject of many recent papers [1,2,3,4,5]. Most state-of-the-art image classification work focuses on the description of images. One of the popular approaches to image classification is to model images with bag-of-features (BOF) [6,7] and classify them with general classifiers, such as the Support Vector Machine (SVM) method [2]. Although it has been reported that SVM classifiers have a good performance in image classification, the classification accuracy still needs to improve for practical applications.

Motivated by [8], this paper proposes a novel framework for image classification. The core of this framework is to design an entropy based image semantic cycle, i.e., a classifier in essence, for an image database. The training procedure in the proposed framework employs the prior information of image class membership to find the optimal image semantic cycle, different from the idea in [8] where image retrieval is an unsupervised learning problem and does not require any priori.

For a test image, the strategy of classification here is to find its best position in the trained optimal image semantic cycle, where the "best" means that the entropy increases the least when inserting the image into this position.

---

* Corresponding author.

T. Huang et al. (Eds.): ICONIP 2012, Part V, LNCS 7667, pp. 533–540, 2012.
© Springer-Verlag Berlin Heidelberg 2012

2   Entropy Based Cycle

Statistical mechanics views entropy as the amount of uncertainty, or mixedupness in the phrase of Gibbs. In the field of computer vision, the concept of entropy is also widely used, e.g., the literature [9] and [10].

2.1   Entropy Definition

Given a set of vertices $V = \{v_i | i = 1, \cdots, n\}$, there exists a *cycle* graph C, a closed path without self-intersections, among them in which each vertex has degree 2. Suppose the cycle starts from vertex v_1 and ends at vertex v_n, we can write the connection order of vertices in this cycle as

$$O = \{o_1, o_2, \ldots, o_n, o_1\},$$

where the entry corresponds to the index of vertices. The path along the connection order O implies the relationship between vertices and embodies the degree to which the cycle is disordered and mixed up. If the path is smooth and desirable, the cycle should be clean and in order; otherwise, the cycle is disordered.

Like in statistical mechanics, to measure the degree to which the cycle is disordered and mixed up, this paper proposes a novel definition of entropy descriptor. The entropy definition is quite general, and is expressed in terms of the spatial position and local discrete curvature of data points.

In particular, for a set of vertices V, if the spatial coordinate x_i of each vertex v_i and the connection order O are known, the entropy of a cycle C is represented as the average of the entropy on every point in the cycle as,

$$E(V, O) = \frac{1}{n} \sum_{i=1}^{n} e(V, O, i). \tag{1}$$

$e(V, O, i)$ is defined as follows,

$$e(V, O, i) = 1 - \exp(-\frac{G(V, O, i)}{\sigma}), \tag{2}$$

where the symbol $\exp(\cdot)$ denotes an exponential function and σ is an empirical constant representing the variance. $G(V, O, i)$ is composed of two parts: the spatial distance $D(V, O, i)$ and local discrete curvature $L(V, O, i)$ as,

$$G(V, O, i) = D^2(V, O, i) + \alpha L^2(V, O, i), \tag{3}$$

weight coefficient α is used to adjust the contribution of L to the entropy. The definition of these two parts is same as that in [8].

The entropy E stated above is essentially a quantity of measuring the uncertain state of the cycle C and contains the smoothness and sharpness of the path with the connection order O. Therefore, for a set of vertices V, the implicit *cycle* can be formally defined as a triplet in this paper,

$$C = (V, O, E).$$

In addition, from the viewpoint of pattern recognition, the entropy is also a metric of distinction and similarity of the data. That is, the smaller the entropy, the more ordered the path and the cycle, the more similar the continuous points along the order.

2.2 Finding an Optimal Cycle Using Tabu Search

From the definition of entropy, if the cycle connecting vertices are ordered enough, the entropy is supposed to be quite small. Other possible paths incline towards the more disorder and the higher entropy. Therefore, to find an optimal cycle, we need to search for the cleanest order O, which can be achieved through minimizing the entropy defined above,

$$\min_{O} E(V, O). \tag{4}$$

That is, the order O^* for the optimal cycle C^* is obtained as follows,

$$O^* = \arg\min E(V, O). \tag{5}$$

Finding a globally optimal solution to equation (5) is an NP-hard problem and completely impossible in practice as there exist $\frac{(n-1)!}{2}$ possible combinations of order for $n(n \geq 4)$ vertices. In this study, we approximate the global minimum of the entropy through a simplified tabu search method [8].

3 Framework for Image Classification

In the proposed framework for image classification, we view an image as a point in the image feature space, and train the images in the way of searching for the optimal cycle through entropy minimization. In this context, the connection order O represents the relevance of images in the entropy based cycle. If local or global features of images are extracted and used to represent images, the entropy based cycle is essentially *an image semantic cycle* containing the content of images. This paper adopts the Pyramid of Histograms Orientation Gradients (PHOG) [11] as the content of images for PHOG pays more attention to such high-level features as edge or contour. In the following experiments, the PHOG descriptor turns an image into a 252-dimensional vector.

3.1 Training Algorithm

The fact that continuous points along the optimal cycle usually belong to the same category except for those boundary points enlightens us that the nearest neighbors of an image along the optimal image semantic cycle should be more similar and relevant except for those boundary images between different categories. Therefore, to do image classification, the key step in our framework is to search for the optimal image semantic cycle, which is actually our training procedure.

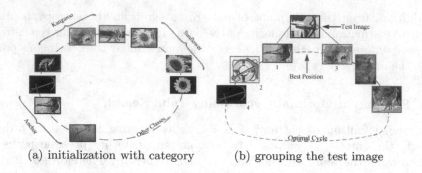

(a) initialization with category   (b) grouping the test image

Fig. 1. An example of image classification

Since classification is a supervised learning problem, the categories of training images are already known and can be made full use of during the training. Therefore, the initialization of the image semantic cycle in the training does not have to be completely random. The idea of "locally random" initialization is brought up. That is, images in the same class are randomly linked together and hold a separate section of the initial cycle, different classes (sections) are randomly placed in the initial cycle. The idea of initialization with category information is illustrated in Fig.1(a). Images from category *kangaroo* and category *sunflower* are put together respectively and different categories, *kangaroo, sunflower, anchor etc.* are randomly placed in the cycle. This initialization strategy considers the known class membership during training, the speed of searching for the optimal image semantic cycle is thus obviously improved.

Given an image database with category information, the training procedure is briefly described as in Table 1.

Table 1. Training Algorithm

Step 1 Initialize the image semantic cycle with a *locally random* order O.
Step 2 Construct the entropy descriptor with Eq.(1) in the image feature space.
Step 3 Minimize the entropy with tabu search, and search for the optimal image semantic cycle with the order O^*.

3.2   Test Algorithm

The classification strategy here is to find the best position of a test image Q in the optimal image semantic cycle and determine the class membership with its nearest neighbors along this optimal cycle. By "best", we mean that the entropy increase of the optimal image semantic cycle will be the least if inserting the test image into this position.

The principle of grouping the test image is borrowed from in the idea of k-NN. More specifically, k nearest neighbors of image Q are first picked from the optimal cycle; then image Q is considered as a member of the class to which most neighbors belong. The measure of determining the nearest neighbors can be the Euclidean distance for simplicity. For example, the 4 nearest neighbors of the test image in Fig.1(b) come from two different categories respectively, three from *anchor* and one from *kangroo*. As a consequence, the test image is grouped into category *anchor*.

Given a test image Q, the test procedure in the proposed framework is briefly described as in Table 2.

Table 2. Test Algorithm

Step 1 Calculate the entropy variation ΔE when inserting the unknown image Q at each position in the optimal image semantic cycle O^*.

Step 2 Find the best position for image Q from the optimal cycle O^*. The measure of determining the best position is that ΔE must be the least if inserting Q at this position.

Step 3 Select k nearest neighbors of Q along the optimal cycle according to Euclidean distance.

Step 4 Assign image Q to the class from which most neighbors come.

3.3 Dynamic Update

An obvious advantage of the proposed framework is that the dynamic update of the trained model, the image semantic cycle, is quite simple and easy. Suppose that we already had an optimal image semantic cycle, $C = (V, O, E)$, and now a new image Q comes to this cycle. To update this cycle, the image Q is first added to V. As a result, the new vertex set V' is equal to the union of V and Q, i.e., $V' = union(V, Q)$. Secondly, find the best position of Q in the current order O. Since the "best" means that the entropy variation ΔE is the least at this position, the new entropy E' remains the least if inserting Q at this best position, $E' = E + \Delta E$. In other words, the new cycle C' after inserting will keep the optimal property. At this time, the connection order O of images is also changed to O'. Therefore, the new image semantic cycle C' is reformulated as, $C' = (V', O', E')$.

In fact, the training procedure can also be completed in the way of dynamic update. A subset of images are first trained to get the optimal image semantic cycle, which is very quick due to the small number of images. The other images gradually come and dynamically update the image semantic cycle until all images join in the training procedure. This dynamic update for training is essentially a procedure of incremental learning, which thus has the potential application to video tracking.

4   Experimental Results

Since the calculation of entropy of each element is independent, parallel computing is a good way to speed up such an optimization problem. Like the work [8], this study also makes full use of the ability of graphic processing unit (GPU) in parallel computing and implements the training procedure with the CUDA

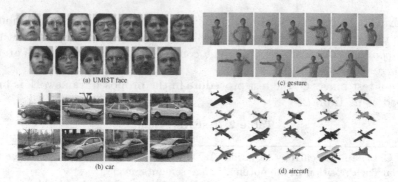

Fig. 2. Example images of four datasets used in our experiments

programming technology. The convergence speed of entropy minimization can become faster 100 times with GPU than with CPU. In the parallel way, the test procedure is as quick as realtime and is hardly affected by the data scale.

This section presents the results of experiments conducted to evaluate and analyze the performance of the proposed framework for image classification. The datasets used are first presented, followed by the comparison of the proposed framework with the SVM based framework.

4.1   Datasets

Five datasets were tested respectively for image classification. The first dataset is the UMIST face database [12]. To track the smooth change of shooting angles, 360 face images belonging to 13 persons were manually picked in our experiments from the UMIST face. The car and gesture datasets were acquired by ourselves. The former contains 14 different types of cars, each of which has about 30 photos. The latter was generated in the laboratory environment, composed of 11 different human gestures and 20 images for each gesture. Another dataset concerns 700 images of 7 aircraft models. The images were produced through projecting three-dimensional models into different planes, which is equivalent to changing the viewpoint through rotating the models. Some example images of the above four datasets are shown in Fig.2. The last dataset includes 3286 images of 32 categories with significant variance in shape, picked out from the Caltech-101 Object Categories dataset[13].

4.2   Classification Accuracy

To evaluate the performance of the proposed approach to image classification, 10-fold cross-validation experiments were here conducted on five datasets mentioned above. To compare with the state of the art, the SVM based classifier [14] was also tested on same datasets due to its advantages in speed and excellent performance in image classification. All test results are presented in Table 3, where the second and third columns respectively represent the number of classes

Table 3. Classification accuracy on different datasets

datasets	class	image	Ours(%)	SVM(%)
UMIST	13	360	98.89(0.77)	85.28(3.13)
car	14	400	99.25(2.08)	74.00(2.79)
gesture	11	220	99.60(1.11)	78.64(5.85)
aircraft	7	700	99.00(1.35)	74.57(4.84)
Caltech-101	32	3286	88.88(2.94)	78.77(2.98)

and images in each dataset. The last two columns are the average classification accuracy respectively obtained using our approach and SVM. The value in parentheses denotes the confidence interval. Although it is well known that SVM has the strong ability of generalization in image classification, the proposed approach obviously has the better performance than SVM, higher accuracy and smaller confidence interval.

The accuracy for the first four datasets is very high and almost reaches 100%. Such good classification performance on the UMIST face and gesture datasets demonstrates that the proposed framework is quite promising in the application of face and gesture analysis. The success in classifying the projection images of aircraft models makes possible the future image-based model retrieval.

In spite of the existence of background clutter in the Caltech-101 (or car) images, our method can still work well with the average accuracy 88.88% (99.25%) and confidence interval 2.94 (2.08), much better than SVM with the average accuracy 78.77% (74.00%) and confidence interval 2.98 (2.79).

To sum up, no matter how different the viewpoint of images is or how cluttered the image background is, the performance of our method is always better than the SVM based classifier, which proves the feasibility and robustness of the proposed framework for image classification.

5 Conclusion

In this paper, we propose an entropy based image semantic cycle for image classification. The training problem in classification is treated as searching for an ordered image semantic cycle for an image database. The optimal cycle is found by minimizing the entropy through tabu search. For classification, the test image is assigned to the class to which most neighbors belong. The use of GPU in our framework yields a very considerable speedup. Experiments demonstrate that the proposed method is feasible and robust to the cluttered background and the viewpoint variation of images.

Acknowledgement. This work was partially supported by Natural Science Foundation of China Grant 60903120, 863 Project 2009AA01Z429, Shanghai Natural Science Foundation Grant 09ZR1434400, and Innovation Program of Shanghai Municipal Education Commission.

References

1. Bosch, A., Zisserman, A., Munoz, X.: Image classification using random forests and ferns. In: ICCV, pp. 1–8 (2007)
2. Goh, K.S., Chang, E., Cheng, K.T.: SVM binary classifier ensembles for image classification. In: CIKM 2001, pp. 395–402 (2001)
3. Wang, J., Yang, J., Yu, K., Lv, F., Huang, T.S., Gong, Y.: Locality-constrained linear coding for image classification. In: CVPR, pp. 3360–3367 (2010)
4. Gao, S., Tsang, I.W.H., Chia, L.T., Zhao, P.: Local features are not lonely - laplacian sparse coding for image classification. In: CVPR, pp. 3555–3561 (2010)
5. Cai, H., Yan, F., Mikolajczyk, K.: Learning weights for codebook in image classification and retrieval. In: CVPR, pp. 2320–2327 (2010)
6. Sivic, J., Zisserman, A.: Video google: A text retrieval approach to object matching in videos. In: ICCV, pp. 1470–1477 (2003)
7. Csurka, G., Dance, C.R., Fan, L., Willamowski, J., Bray, C.: Visual categorization with bags of keypoints. In: Workshop on Statistical Learning in Computer Vision, ECCV, pp. 1–22 (2004)
8. Zhang, C., Li, H., Guo, Q., Jia, J., Shen, I.F.: Fast active tabu search and its application to image retrieval. In: IJCAI, pp. 1333–1338 (2009)
9. Wang, W., Wang, Y., Huang, Q., Gao, W.: Measuring visual saliency by site entropy rate. In: CVPR, pp. 2368–2375 (2010)
10. Lin, R.S., Ross, D.A., Yagnik, J.: Spec hashing: Similarity preserving algorithm for entropy-based coding. In: CVPR, pp. 848–854 (2010)
11. Bosch, A., Zisserman, A., Munoz, X.: Representing shape with a spatial pyramid kernel. In: CIVR 2007, pp. 401–408 (2007)
12. Graham, D.B., Allinson, N.M.: Characterizing virtual eigensignatures for general purpose face recognition. Face Recognition: From Theory to Applications 163, 446–456 (1998)
13. Fei-Fei, L., Fergus, R., Perona, P.: Learning generative visual models from few training examples: An incremental bayesian approach tested on 101 object categories. Comput. Vis. Image Underst. 106, 59–70 (2007)
14. Chang, C., Lin, C.: LIBSVM: a library for support vector machines (2001)

Fuzzy Particle Swarm Optimization
for Intrusion Detection

Dalila Boughaci, Mohamed Djamel Eddine Kadi, and Meriem Kada

USTHB- FEI- Department of Computer Science,
Laboratory of Artificial Intelligence LRIA,
BP 32 EL-Alia, Beb-Ezzouar, Alger, 16111
dboughaci@usthb.dz, dalila_info@yahoo.fr

Abstract. This paper tries to propose a fuzzy particle optimization algorithm (FPSO) for intrusion detection. The proposed FPSO classifier works on a knowledge base modelled as a fuzzy rule *if-then* and improved by a PSO algorithm. The objective is to obtain good quality solutions by optimizing the fuzzy rules generation. The method is tested on the benchmark KDD'99 intrusion dataset and compared with the fuzzy genetic algorithm and with other existing techniques available in the literature. The obtained results show the efficiency of the proposed approach.

Keywords: Particle swarm optimization, genetic algorithm, fuzzy logic, classification, intrusion detection, DARPA dataset.

1 Introduction

The intrusion detection systems (IDS) are becoming indispensable for effective protection against attacks that are constantly changing in magnitude and complexity. An intrusion detection system is software or hardware or both of them designed to monitor computer system or network activities for malicious activities or policy violations. Intrusion detections methods can be divided into two main models: anomaly detection and misuse detection approaches.

The anomaly detection model is based on user's profiles. It describes the usual behaviour of a user to detect this user's anomalous or unaccustomed action. Among these methods, we find: the statistical methods [13], the expert systems [22] and the neural networks [8].

The misuse detection model defines some anomalous behaviour to analyze data susceptible to be attacks. The approach often uses known attacks called signatures. Among these methods, we cite: the expert systems [18], the genetic algorithm [19] and the pattern matching method that provides signatures of attacks. Various algorithms are used to localize these signatures in the audit trail [16]. In the last few years, various techniques have been applied extensively for intrusion detection such as agents-based detection intrusion [7, 9], the Data mining approaches [17], the clustering techniques [20], the naive Bayesian classifier [14, 4], and the fuzzy evolutionary algorithms [1, 6].

T. Huang et al. (Eds.): ICONIP 2012, Part V, LNCS 7667, pp. 541–548, 2012.
© Springer-Verlag Berlin Heidelberg 2012

In this work, we focused on fuzzy particle swarm optimization for intrusion detection. The method works on fuzzy rules which are improved by using the particle swarm optimisation method [15].

The method has been implemented and evaluated on the KDD99 Benchmarks (http://kdd.ics.uci.edu/databases/kddcup99/kddcup99.html).

The KDD99 dataset contains 22 different attack types which could be classified into four main categories namely Denial of Service (DOS), Remote to User (R2L), User to Root (U2R) and Probing.

The paper is organized as follows. The second section presents the fuzzy particle swarm optimization algorithm for intrusion detection. The implementation and some numerical results are given in the third section. Finally, the fourth section concludes the work.

2   The Proposed Approach

The Fuzzy logic [23] is an intelligent method that has been successfully employed for many IDSs [11, 10, 2]. In this work, we proposed a fuzzy particle swarm optimization for intrusion detection. The proposed approach consists of two main steps. The first one is the data normalization and the second step is the fuzzy PSO algorithm.

2.1   Data Pre-processing and Normalization

The data pre-processing is launched before calling the FPSO process. The goal is to normalize the dataset to be analyzed. Each line of the KDD'99 dataset called *connection* includes a set of 41 features and a label which specifies the status of connection as either normal or specific attack type. The features of a connection include the duration of the connection, the type of the protocol (TCP, UDP, etc), the network service (http, telnet, etc), the number of failed login attempts, and the service and so on. These features had all forms of continuous, discrete, and symbolic, with significantly varying ranges. After having analyzed the KDD99 dataset, among the 41 attributes of the connection, we consider only sixteen attributes which are: $A8$, $A9$, $A10$, $A11$, $A13$, $A16$, $A17$, $A18$, $A19$, $A23$, $A24$, $A32$, $A33$, $A1$, $A5$ and $A6$. These attributes are the important ones that can help in classifying a connection correctly. These attributes are normalized. The normalization formula given in (1) is applied in order to set attribute numerical values in the range [0.0, 1.0].

$$X = \frac{(X - MIN)}{(MAX - MIN)} \tag{1}$$

Where X: is the numerical attribute value, MIN is the minimum value that the attribute X can get and MAX is the maximum one.

For the numerical attributes $A1$, $A5$ and $A6$, we have observed a big value of MAX hence the need to modify the normalization formula given in (1). The logarithmic scaling (with base 10) is applied to these features to reduce the range. We used all the sixteen features as the inputs of The FPSO classifier which is detailed in the next section.

2.2   The Particle Swarm Optimization Algorithm

The Fuzzy particle swarm optimization algorithm starts with a swarm (a population) of candidate solutions (called particles) generated randomly. Each particle is an *if-then* fuzzy rule. Then, for each particle, in the swarm, a simple formula is applied on it to obtain a new particle for the next iteration. The movements of the particle are guided by its best known position in the search-space (called *pbest*) and by the global best solution (called *gbest*). The process is repeated for a certain number of iterations fixed empirically.

The components of the FPSO for intrusion detection are defined in following:

2.3   The Fuzzy Rule Encoding

A fuzzy rule *if-then* is encoded as a string. We have used a vector of 16 bits where each bit corresponds to an attribute. Fig. 1-(a) draws the Membership functions of the five possible linguistic values used for each attribute which are: S: Small, MS: Medium Small, M: Medium, ML: Medium Large and L: Large.

2.4   The Membership Function $\mu(x)$

The membership function for each attribute X noted $\mu(X)$ is calculated by a projection on the graph of the fuzzy set as depicted in Fig.1-(b). Formula (2) shows how we can calculate the $\mu(X)$ value.

$$\mu(x) = Max\{0, 1 - |X - X0|/b\}. \tag{2}$$

where: b is the base of the triangle, $b = 0.5$. $X0 = 0, 0.25, 0.5, 0.75, 1$ corresponding to S, MS, M, ML, L. X: is the attribute value after normalization.

2.5   The Evaluation of a Fuzzy Rule and the Fitness Function

To evaluate an *if-then* rule R_j and classify a connection X_p with a certain confidence degree, we have used the method introduced in [12]. To evaluate a fuzzy rule Rj, we give the following steps:

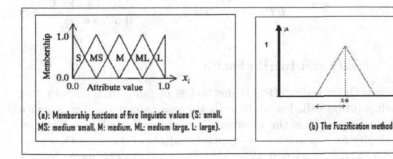

Fig. 1. Membership functions and the Fuzzification method

1. Calculate the compatibility of connections with the rule R_j: Let us consider the fuzzy *if-then* rule R_j denoted "$A_{j1}A_{j2}\ldots A_{jn}$", we calculate the compatibility of each connection X_p of the dataset with the rule R_j by using the Formula (3).

$$\mu_{Rj}(X_P) = \mu_{Aj1}(X_{P1}) \times \mu_{Aj2}(X_{P2}) \times .. \times \mu_{Ajn}(X_{Pm}). \quad p = 1,2\ldots,m \text{(3)}$$

where μ is the membership function. m: is the total number of connections. X_i: are the attributes. X_p is the current connection and n is the number of attributes which equals to 16.

2. Calculate the sum of the compatibilities for each class of the five categories: for each class h belonging to the five classes *DoS, R2L, U2R, Probing* and *Normal*, we calculate the sum of compatibilities as given in Formula (4).

$$\beta_{CLASS_h}(Rj) = \sum_{Xp\in CLASSh} \mu_{Rj}(X_P) \qquad h = 1,2..C \qquad C = 5. \quad \text{(4)}$$

After having calculated the sum of compatibilities of a rule R_j for each class h, we selected the class having the maximum value (C_j) as given in Formula (5). This class C_j is considered the suitable class for the rule R_j. If two classes had the same maximum value then the class is not specified $(C_j = $ null) and $CF_j = 0$. The confidence degree CF_j of the class C_j for the rule R_j is computed as shown in the Formula (6).

$$\beta_{CLASS_{C_j}}(Rj) = max\{\beta_{CLASS_1}(Rj)...\beta_{CLASS_C}(Rj)\} \qquad \text{(5)}$$

$$CF_j = \frac{\beta_{CLASSCj}(Rj) - \bar{\beta}}{\sum_{h=1}^{C}\beta_{CLASSh}(Rj)} \quad Where \quad \bar{\beta} = \frac{\sum_{h\neq Cj}\beta_{CLASSh}(Rj)}{C - 1} \quad \text{(6)}$$

The formula (7) shows how the fitness of a fuzzy rule is obtained. The PPF represents the Positive Power Rule. The fitness value of a rule is the sum of the *PPF* for all considered classes.

$$fitness(R_j) = \sum_{P\in CLASS_{Cj}} PPF_P^{R_j} \qquad PPF_P^{R_j} = \{ \begin{matrix} 1 \text{ if } \mu_{Rj}(X_P) > 0 \\ 0 \text{ otherwise} \end{matrix} \quad \text{(7)}$$

2.6   Position, Velocity and Inertia Factor

The Particle Swarm Optimization (PSO) method is inspired from social interactions [15]. Each solution called a particle in the swarm can move in a space of dimension n, where $n = 16$ is the number of considered attributes. The position and the velocity for a particle i are represented respectively by these two vectors $X_i = (X_{i1}, X_{i2}, \ldots X_{in})$ and $V_i = (V_{i1}, V_{i2}, \ldots V_{in})$. The move from a position to another position can be given as: $X_i = X_i + V_i$. To avoid the particles do not move too quickly in the search space, possibly passing near the optimum, it may be necessary to fix the maximum and the minimum velocities values $[Vmax, Vmin]$ which can improve the convergence of the algorithm. The PSO method used also another parameter called: the inertia factor γ which permits to define the capacity of each particle to explore the search space. A small value of γ (<1) imply a local exploration. When γ (>1), the space is better searched which imply a global exploration. The velocity is computed by using the following formula.

$$V_i(t) = \gamma.V_i(t-1) + \varphi_1(X_{pbest} - X_i(t)) + \varphi_2(X_{gbest} - X_i(t)) \qquad (8)$$

Where X_{pbest} : is the best local position visited by the particles. X_{gbest} : is the global best position, γ is the inertia factor. φ_1 and φ_2 are the coefficients of confidence that permit to weigh the trends of the particle, $\varphi_1 = r_1.c_1$, $\varphi_2 = r_2.c_2$ where r_1 et r_2 follow an uniform law on $[0..1]$, c_1 and c_2 are two positive constants valued fixed empirically with $c_1 + c_2 \leq 4$.

2.7   The FPSO Algorithm

The fuzzy particle swarm optimization algorithm is sketched in Algorithm 1.

Algorithm 1. FPSO for intrusion detection.

1: Start from a random population of particles (a random set of *if-then* rules)
2: S $\Leftarrow$ The first particle.
   Repeat
3: **for** each particle **do**
4:    Evaluate its quality by using the evaluation function
5:    **if** the current particle is the first one **then**
6:       pbest=gbest $\Leftarrow$ quality (S); S$\Leftarrow$ Xgbest
7:    **else**
8:       **if** pbest < quality (S) **then**
9:          pbest $\Leftarrow$ Quality (S) ;
10:      **end if**
11:      **if** gbest < quality (S) **then**
12:         gbest $\Leftarrow$ Quality (S) ;S$\Leftarrow$ Xgbest
13:      **end if**
14:   **end if**
15:   Update the particle's velocity:
        $V_i(t) = \gamma.V_i(t-1) + \varphi_1(X_{pbest} - X_i(t)) + \varphi_2(X_{gbest} - X_i(t))$
16:   Update the particle's position: $X_i = X_i + V_i$
17: **end for**
    Until a maximum number of iterations return the best rule with the highest fitness;

2.8   Numerical Results

In this section, we give some experimental results of the FPSO applied on the KDD dataset. The implementation was done on NetBeans IDE 7.0. The source codes are written in Java. All experiments were performed on a PC AMD dual core 2.1 GHz with 3 GB of Ram under Windows 7. The number of particles is fixed to 10, $Vmin = -4$, $Vmax = 4$. First, we have created five matrices: the matrix containing the U2R-events, the matrix containing R2L-events, the matrix containing the Probing events, the matrix containing the DOS events and the matrix containing the normal connections. Then, the normalization phase is launched where the various attributes of connections of all matrices are normalized. We have obtained five normalized matrices U2R, R2L, Probing, Normal and DOS. The next step is the generation of fuzzy rules. To do this, we used the *rand* function (random number to generate random numbers that must be among the five values (0, 1, 2, 3, 4) which correspond to (Small, Medium Small, Medium, Medium Large and Large).

To evaluate the performance of the proposed approach, we used the following measures:

- The presence rate (TP) is the proportion of false alarms reported by the approach.
- The Absence rate (TA) is the proportion of unreported cases of attacks by the approach.
- The Success rate (SR) is the percentage of the number of connections correctly classified by the approach.

Table 1. The results found by the FPSO

CLASS	TP %	TA %	SR%
Normal	8.33	8.33	91.66
DOS	2.77	2.77	97.22
U2R	30.55	0.0	69.44
R2L	2.77	2.77	97.22
Probing	22.22	0.0	77.77

Table 1 summarizes the measures found by the FPSO on the five classes. According to the results, we can see a good performance in favour of the proposed approach.

2.9 Comparative Study

In order to situate our contribution, we compared our results with some well-know methods for intrusion detection such as: FGA [6], FLS [5], Hybrid EFS [2], C4.5 [20], 5-NN [3], EFRID (Evolving Fuzzy Rules for Intrusion. Detection) proposed in [11], NB [14] and Naive Bayesian classifier [4]. We note that, the fuzzy genetic algorithm we used (FGA) performs for each generation:

- A random one-point crossover on two randomly selected individuals.
- A random mutation of all genes of an individual randomly selected.
- A selection of individuals having a fitness value >0. So all individuals having a fitness value equals to zero are discarded and eliminated from the population. We consider only individuals with a fitness value superior to zero.

Table 2 presents the results obtained for the five classes.

Table 2. Comparison of some algorithms for intrusion detection

Algorithm	Normal %	U2R%	R2L %	DOS%	Probing%
FPSO	**91.66**	**69.44**	**97.22**	**97.22**	**77.77**
FGA	92,5	92,5	92,5	99,99	92,5
FLS	10	95	85	80	80
Hybrid EFS	98.5	76.3	89	98.5	82.5
C4.5	95.9	21.1	30.2	97.1	76.3
5-NN	96.3	25.4	3.8	96.7	87.5
EFRID	92.78	88.13	7.41	98.91	50.35
NB	94.2	25	5.4	79.4	90.4
Naive Bayesian	97.68	11.84	8.66	96.65	88.33

Table 2 shows the effectiveness of the proposed FPSO for intrusion detection. For all the five classes U2R, R2L, DOS, Probing and Normal, the FPSO and FGA are comparable. They find good results compared to FLS, Hybrid EFS, C4.5, 5-NN, EFRID, NB and Naive Bayesian for intrusion detection.

3   Conclusion

In this paper, we proposed a fuzzy particle swarm optimization for intrusion detection. The proposed method has been implemented and tested on KDD99 Benchmarks. The results are encouraging and demonstrate the benefit of the proposed technique of classification in the field of intrusion detection. We plan to study in future work a parallelization approach in the hope to minimize the computation time.

References

1. Abadeha, S., Habibia, J., Lucasb, C.: Intrusion detection using a fuzzy genetics-based learning algorithm. Journal of Network and Computer Applications 30, 414–428 (2007)
2. Abadeha, S., Habibia, J., Soroush, E.: Induction of Fuzzy Classification systems via evolutionary ACO-based algorithms. IJSSST 9(3) (September 2008)
3. Aha, D., Kibler, D.: Instance-based learning algorithms. Machine Learning 6, 37–66 (1991)
4. Ben-Amor, N., Benferhat, S., Elouedi, Z.: Naive Bayes vs Decision Trees in Intrusion Detection Systems. In: Proceedings of the ACM Symposium on Applied Computing, pp. 420–424. ACM Press (2004)
5. Boughaci, D., Bouhali, S., Ordeche, S.: A Fuzzy Local Search for intrusion detection. In: Proceedding of the ACIT (2011)
6. Boughaci, D., Herkat, M. L., Lazzazi, M.A.: A specific fuzzy genetic Algorithm for intrusion detection. In: Proceedings of ICCIT (2012)
7. Boughaci, D., Drias, H., Bendib, A., Bouznit, Y., Benhamou, B.: Distributed Instrusion Detection Framework Based on Mobile Agents. In: Proceedings of the International Conference on Dependability of Computer Systems, pp. 248–255. IEEE Press (2006)
8. Debar, H., Becker, M., Siboni, D.: A neural network component for an intrusion detection system. In: Proceedings of the IEEE Symposium of Research in Computer Security and Privacy, pp. 240–250 (May 1992)

9. Nath, K.G.B., Ramamohanarao, K.: Layered Approach Using Conditional Random Fields for Intrusion Detection. IEEE Trans. Dependable Sec. Comput. 7(1), 35–49 (2010)
10. Gao, M., Zhou, M. C.: Fuzzy intrusion detection based on fuzzy reasoning Petri Nets. In: Proceeding of the 2003 IEEE International Conference on Systems, Man and Cybernetics, 5-8, Washington D.C., pp. 1272-1277 (October 2003)
11. Gomez, J., Dasgupta, D.: Evolving Fuzzy Classifies for Intrusion Detection. In: Proceedings of the 2002 IEEE Information Assurance Workshop (2002)
12. Ishibuchi, H., Murata, T.: Techniques and applications of genetic algorithms-based methods for designing compact fuzzy classification systems. Fuzzy Theory Systems Techniques and Applications 3(40), 1081–1109 (1999)
13. Javitz, H.S., Valdes, A., Lunt, T.F., Tamaru, A., Tyson, M., Lowrance, J.: Next generation intrusion detection expert system (NIDES). Technical Report A016-Rationales, SRI (1993)
14. John, G.H., Langley, P.: Estimating Continuous Distributions in Bayesian Classifiers. In: Proceedings of the Eleventh Conference on Uncertainty in Artificial Intelligence, pp. 338–345. Morgan Kaufmann, San Mateo (1995)
15. Kennedyand, J., Eberhart, R.: Particle Swarm Optimization. In: Proceedings of IEEE International Conference on Neural Networks IV, pp. 1942–1948 (1995)
16. Kumar, S., Spafford, E.H.: A pattern-matching model for misuse intrusion detection. In: Proceedings of the International Computer Security Conference, pp. 11–21 (1994)
17. Lee, W., Stolfo, S., Mok, K.: Mining Audit Data to build Intrusion Detection Models. In: Proceedings of the 4th International Conference on Knowledge Discovery and Data Mining, pp. 66–72. AAAI Press (1998)
18. Lunt, T.F., Jagannathan, R.: A prototype real-time intrusion-detection expert system. In: Proceedings of the IEEE Symposium on Security and Privacy, pp. 59–66 (1988)
19. Ludovic, M.: GASSATA, A genetic algorithmas an alternative tool for security audit trails analysis. In: First International Workshop on the Recent Advances in Intrusion Detection (1998)
20. Quinlan, R.: C4.5: Programs for Machine Learning. Morgan Kaufmann Publishers, San Mateo (1993)
21. Shah, H., Undercoffer, J., Joshi, A.: Fuzzy Clustering for Intrusion Detection. In: Proceedings of the 12th IEEE International Conference on Fuzzy Systems, vol. 2, pp. 1274–1278. IEEE Press (2003)
22. Vaccaro, H.S., Liepins, G.E.: Detection of anomalous computer session activity. In: Proceedings of the IEEE Symposium on Security and Privacy (May 1989)
23. Zadeh, L.A.: Fuzzy sets. Information and Control 8, 338–353 (1965)

Dual-Feature Bayesian MAP Classification: Exploiting Temporal Information for Video-Based Face Recognition

John See[1], Chikkannan Eswaran[1], and Mohammad Faizal Ahmad Fauzi[2]

[1] Faculty of Computing and Informatics, Multimedia University,
Persiaran Multimedia, 63100 Cyberjaya, Selangor, Malaysia
[2] Faculty of Engineering, Multimedia University,
Persiaran Multimedia, 63100 Cyberjaya, Selangor, Malaysia
{johnsee,eswaran,faizal1}@mmu.edu.my

Abstract. Machine recognition of faces in video is an emerging problem. Following recent advances, conventional exemplar-based schemes and image set approaches inadequately exploit temporal information in video sequences for the classification task. In this work, we propose a new dual-feature Bayesian *maximum-a-posteriori* (MAP) classification method for face recognition in video sequences. Both cluster and exemplar features are extracted and unified under a compact probabilistic framework. To realize a non-parametric solution, a joint probability function is modeled using relevant similarity measures for matching these features. Extensive experiments on two public face video datasets demonstrate the good performance of our proposed method.

Keywords: Bayesian MAP classification, feature fusion, similarity measures, video-based face recognition.

1 Introduction

In the past few decades, automatic machine recognition of faces in still images have vastly matured, with numerous commercial applications and a variety of state-of-art techniques reported in a notable survey [16]. Under complex face appearance variations and adverse environments particularly in videos, these methods encounter severe limitations and tend to perform rather poorly. The emergence of video media coupled with these new challenges have presented a rapidly growing research area in video-based face recognition (VFR).

Many works [1,3,5,6,17] have been motivated to improve face recognition in video by exploiting temporal information, an intrinsic property only available in videos. Some of these methods represent video as a complex face manifold to extract a variety of features such as exemplars (or images that summarizes a video) and image sets represented as local models or subspaces. Due to a wide taxonomy of approaches in literature, focus will be given to the discussion of exemplar-based and image set-based approaches, both which, form the barebones of our approach. For generality, we will also discuss recent methods that employ a Bayesian probabilistic classification framework for recognition.

T. Huang et al. (Eds.): ICONIP 2012, Part V, LNCS 7667, pp. 549–556, 2012.

2   Related Work

Exemplar-based representations deal with the abundance of face images in training videos by selecting a small number of appearance-specific representative face images or exemplars, to summarize each subject class. Conventionally, clusters or patches are extracted from the video face manifold and the sample means are chosen as exemplars. Then, an aggregation of matching scores between the training exemplars and test video frames is performed. Exemplar-based approaches for VFR are common in literature and have garnered much attention recently. Various approaches have been proposed to perform exemplar selection, notably using radial basis function network [17], k-means clustering [1,3], Hierarchical Agglomerative Clustering (HAC) [2,9] and Maximal Linear Patch (MLP) [12]. More recently, Spatio-Temporal Hierarchical Agglomerative Clustering (STHAC) [11] was proposed, utilizing both spatial and temporal information to good effect. While reported results have shown good promise, these approaches are highly dependent on the accuracy of clustering and the number of exemplars chosen.

Representation by image sets involved the extraction of images to form temporally ordered or unordered groups. Image sets from both training and test videos are usually represented by subspaces or manifolds that are learned from the original vectors. Classification or matching between the learned subspaces or manifolds are accomplished using subspace distances or similarity measures. The use of principal angles have seen tremendous appeal due to its simplicity of representation. Yamaguchi et al.[15] was first to propose a Mutual Subspace Method (MSM) that utilizes the cosine of the smallest principal angle (or largest canonical correlation) for recognizing faces in image sets. Later approaches using kernel-based [13] and discrimination-based [4] extensions demonstrated marked improvements in recognition accuracy. These approaches can capture variations within and between image sets very well but they have holistically poor representation at the face appearance level, thus causing possible deterioration of performance in sequences that are longer, or contain a larger variety of samples.

In attempt to combine both image sets and exemplars for recognition, Wang et al. [12] proposed a weighted distance measure called Manifold-Manifold Distance (MMD) that considers both subspace and exemplar distances within the face manifold. However, this simplistic fusion is subjected to a tunable weight parameter while its recognition framework does not harness temporal continuity.

Bayesian classification frameworks offer an alternative method of encoding temporal dependencies between video frames, a core motivation of our work. Fan et al. [1] fitted a Bayesian inference on the recognition task by transforming maximum likelihood estimation to non-parametric distance measures. Liu et al. [6] formulated a spatio-temporal embedding based on Bayesian keyframe learning and statistical classification. In our earlier work [10], we presented an exemplar-driven Bayesian *maximum-a-posteriori* (MAP) classifier capable of encoding within-class influence of extracted exemplars.

In this paper, we propose a dual-feature approach to Bayesian MAP classification by unifying both exemplar and image set features into a temporally-driven framework for VFR. Applying state-of-art methods, appearance-specific clusters are first constructed followed by feature extraction. The subjects in video sequences are recognized by maximizing the posterior probability, where a joint probability function is modeled using similarity measures that match these features. Evaluation was performed extensively on two public face video datasets – Honda/UCSD and NICTA ChokePoint. Experimental results demonstrate the superiority of the proposed approach in comparison with related approaches.

3 Video-Based Face Recognition

Video-based face recognition (VFR) can be achieved through various methodologies [7]. Exemplar-based approaches accomplish a complete *video-to-video* recognition by simplifying it to an *image-to-image* recognition task, where each test video frame is matched with a set of class-specific image exemplars. Meanwhile, image set-based approaches perform direct matching between test and training image sets to identify subject in video. Intuitively, our fusion approach utilizes similarity measures in a frame-by-frame basis (similar to that of exemplar-based setting), which can then be aggregated using a Bayesian framework to achieve full *video-to-video* recognition.

3.1 Problem Setting

For general notation, we define a sequence of face images from a video as

$$\mathbf{X}_c = \{\mathbf{x}_{c,1}, \mathbf{x}_{c,2}, \ldots, \mathbf{x}_{c,N_c}\}, \tag{1}$$

where N_c is the number of face images in the video, with the subject label of a C-class problem, $c \in \{1, 2, \ldots, C\}$ and each video is assumed to contain faces of the same person.

Assuming one training video per subject, M number of clusters are extracted from each training video to form the training cluster set

$$\mathbf{Z}_c = \{\mathbf{z}_{c,1}, \mathbf{z}_{c,2}, \ldots, \mathbf{z}_{c,M}\}, \tag{2}$$

where $\mathbf{z}_{c,m} = \{\mathbf{x}_{c,1}, \mathbf{x}_{c,2}, \ldots, \mathbf{x}_{c,N_m}\}$ is the m-th cluster of the c-th class containing N_m images. From each cluster, an exemplar image $\mathbf{e}_{c,m}$ is then selected, resulting in an exemplar set,

$$\mathbf{E}_c = \{\mathbf{e}_{c,1}, \mathbf{e}_{c,2}, \ldots, \mathbf{e}_{c,M}\} \tag{3}$$

for each class, where $\mathbf{e}_{c,m} \subseteq \mathbf{z}_{c,m}$.

For test videos, each extracted sequence $\mathbf{X}'$ consists of an array of test face images $\mathbf{x}'$, as similarly denoted in (1)). Meanwhile, the partitioning of test clusters is done by dividing the sequence into k temporally continuous segments

$$\mathbf{z}'_k = \{\mathbf{x}'_{k,1}, \mathbf{x}'_{k,2}, \ldots, \mathbf{x}'_{k,L}\}, \tag{4}$$

each of a fixed length L. The test cluster set for the c-th class is denoted by

$$\Theta_c = \{\boldsymbol{\theta}_{c,k,1}, \boldsymbol{\theta}_{c,k,2}, \ldots, \boldsymbol{\theta}_{c,k,N_c}\}, \tag{5}$$

where the test cluster vector corresponding to the i-th frame is assigned the k-th cluster segment $\boldsymbol{\theta}_{c,k,i} = \mathbf{z}'_k$ based on its associated sequence image $\mathbf{x}'_i \in \mathbf{z}'_k$.

(a) Honda/UCSD   (b) NICTA ChokePoint

Fig. 1. Extracted exemplars of two subjects from the evaluated datasets

3.2   Feature Extraction

In feature extraction, the state-of-art Locally Linear Embedding (LLE) [8] is first applied to learn a meaningful low-dimensional embedding from the original data space of each training video. LLE is capable of uncovering the underlying nonlinear manifold structure posed by the large appearance variations in a training video. The projected faces in LLE-space are then partitioned into M clusters using the recently proposed spatio-temporal hierarchical agglomerative clustering (STHAC) algorithm [11]. STHAC is a spatio-temporal extension of the HAC algorithm where both spatial and temporal distances are utilized to produce more relevant clusters. The global scheme of STHAC is used in this work. For each cluster, the face image nearest to the cluster mean is chosen as the face exemplar. Some sample extracted exemplars are shown in Fig. 1.

To reduce dimensional space, suitable subspace representations for exemplar and cluster features are employed. Features of the training exemplar set are learned using a nonlinear dimensionality reduction method called Neighborhood Discriminative Manifold Projection (NDMP) [9] while test sequence images can be projected to the feature space by simple linear transformation. For training and test cluster sets, features can be described using a subspace representation spanned by the images where distance between subspaces or similarity measures can be derived. Common methods include the MSM [15], Kernel Principal Angles (KPA) [13], and Discriminative Canonical Correlations (DCC) [4].

4   Dual-Feature Bayesian MAP Classification

4.1   MAP Classifier

Generally, a recognition task can be modeled using a Bayesian inference model, by estimating the *maximum-a-posteriori* (MAP) decision rule,

$$c^* = arg \max_C P(c|\mathbf{X}) \qquad (6)$$

to determine the subject identity of a test video $\mathbf{X}$. Assuming conditional independence between all observations, i.e. $\mathbf{x}_i \perp\!\!\!\perp \mathbf{x}_j | c$ where $i \neq j$, the posterior distribution over the class hypotheses at video time frame N (which is also the length of the test sequence) can be expressed as

$$P(c|\mathbf{X}) \equiv P(c|\mathbf{x}_1, \ldots, \mathbf{x}_N) \propto P(c) \prod_{i=1}^{N} P(\mathbf{x}_i|c) \qquad (7)$$

Motivated by our previous work [10], conditional dependencies between image and cluster features are modeled through a new joint probability function

$$P(c, \mathbf{E}, \mathbf{Z}, \mathbf{\Theta}, \mathbf{X}) = P(\mathbf{X}|\mathbf{E}, \mathbf{\Theta})P(\mathbf{\Theta}|\mathbf{Z})P(\mathbf{Z}|c)P(\mathbf{E}|c)P(c) \ . \tag{8}$$

By marginalizing the feature variables, we define the cluster-level MAP classifier that maximizes the posterior probability

$$P(c|\mathbf{E}, \mathbf{Z}, \mathbf{\Theta}, \mathbf{X}) \propto P(c) \prod_{i=1}^{N} P(\mathbf{x}_i|\boldsymbol{\theta}_i) \sum_{j=1}^{M} P(\boldsymbol{\theta}_i|\mathbf{z}_{c,j})P(\mathbf{x}_i|\mathbf{e}_{c,j})P(\mathbf{e}_{c,j}|c)P(\mathbf{z}_{c,j}|c)$$

$$\propto P(c) \prod_{i=1}^{N} \sum_{j=1}^{M} P(\boldsymbol{\theta}_i|\mathbf{z}_{c,j})P(\mathbf{x}_i|\mathbf{e}_{c,j})P(\mathbf{e}_{c,j}|c), \tag{9}$$

assuming conditional independence between observations in $\mathbf{X}$ and law of total probability. $P(\mathbf{x}_i|\boldsymbol{\theta}_i)$ and $P(\mathbf{z}_{c,j}|c)$ are non-informative terms as they are intuitively insignificant to the decision rule. The non-informative class priors $P(c)$ are assumed to be uniformly distributed at the start of sequence.

4.2   Feature Similarity Measures

Estimating sample distributions is a non-optimal task that can easily result in overfitting or underfitting of data especially with limited samples. Thus, we construct relevant similarity measures between the various extracted features.

The exemplar likelihood, or likelihood of the observed face image $\mathbf{x}_i$ given training exemplar $\mathbf{e}_{c,j}$ can be formulated as

$$P(\mathbf{x}_i|\mathbf{e}_{c,j}) = \frac{S_i^{em}(\mathbf{x}_i, \mathbf{e}_{c,j})}{\sum_{k=1}^{C} \sum_{j=1}^{M} S_i^{em}(\mathbf{x}_i, \mathbf{e}_{c,j})} \ , \tag{10}$$

where the exemplar similarity score $S_i^{em}(\mathbf{x}_i, \mathbf{e}_{c,j}) = 1/(\mathbf{x}_i - \mathbf{e}_{c,j})\mathbf{\Sigma}^{-1}(\mathbf{x}_i - \mathbf{e}_{c,j})^T$ is the inverse-squared Mahalanobis distance in NDMP feature space. Similar to [10], the weight of influence for each exemplar within its own class can be determined via an exemplar prominence measure, characterized by

$$P(\mathbf{e}_{c,j}|c) = \frac{S_{c,j}^{pr}(\mathbf{e}_{c,j}, \mathbf{E}_c)}{\sum_{j=1}^{M} S_{c,j}^{pr}(\mathbf{e}_{c,j}, \mathbf{E}_c)} \ , \tag{11}$$

where the prominence similarity measure, $S_{c,j}^{pr}(\mathbf{e}_{c,j}, \mathbf{E}_c) = 1/(\min_{\mathbf{e}' \in \mathbf{E}_c} \|\mathbf{e}_{c,j} - \mathbf{e}'\|)$ is the inverse normalized ℓ^2-Hausdorff distance in exemplar subspace. This term can be pre-computed offline since it is independent of observation the sample $\mathbf{X}$.

Finally, the cluster likelihood, or likelihood of the observed cluster subspace $\boldsymbol{\theta}_i$ at frame i given training cluster subspace $\mathbf{z}_{c,j}$ is denoted as

$$P(\boldsymbol{\theta}_i|\mathbf{z}_{c,j}) = (1 - \alpha)S_i^{cl}(\boldsymbol{\theta}_i, \mathbf{z}_{c,j}) + \alpha \ , \tag{12}$$

where the parameter α is the normalization lower bound and the cluster similarity measure is computed by averaging the first r canonical correlations σ (or cosines of principal angles), $S_i^{cl}(\boldsymbol{\theta}_i, \mathbf{z}_{c,j}) = \sum_{l=1}^{r} \sigma_l/r$, where r is the minimum dimension of the two subspaces.

Table 1. Average recognition rates (%) of various evaluated methods (with parameters used for building augmented test set)

Methods	Dataset (W, T, C)		
	Honda/UCSD (5,10,20)	ChokePoint-LL (9,9,25)	ChokePoint-EL (9,9,25)
Majority Voting-NDMP	62.3	73.2	78.6
MSM	62.8	67.9	67.7
DCC	73.9	85.5	83.5
MMD	76.2	87.7	80.1
Exemplar Bayes MAP	75.9	92.8	78.8
Dual-Feature Bayes MAP	84.3	93.4	84.7

5   Experiments

The proposed classification method was evaluated on two public face video datasets – Honda/UCSD [5] and NICTA ChokePoint [14]. The first dataset, Honda/UCSD was specifically collected for face recognition in video, and it consists of challenging pose and expression variations with significant out-of-plane head rotations. We consider their first subset, which has 59 sequences of 20 people (each person has at least 2 videos), with video lengths about 300-600 frames. The second dataset is the recently created NICTA ChokePoint dataset, collected for person identification in real-world conditions. We use their first subset (Portal 1) that has 600 sequences of 25 different people. Each person has 24 sequences, from a combination of 4 sequence shots, 3 camera angles, and 2 movement modes (entering and leaving portal). Faces from this dataset contain a variety of lighting conditions, pose, video quality, and also the presence of slight occlusions. For this dataset, we construct the training-test samples in two different settings: leaving-leaving (LL) and entering-leaving (EL). The second setting in particular, contains two entirely different sets of environment and camera view angles. For both datasets, faces were extracted and resized to 32×32 pixel grayscale images, followed by histogram equalization to normalize lighting effects.

By convention, one sequence per subject is used for training and the remaining sequences are used for testing. To ensure extensive evaluation, we build an augmented test set by randomly sampling W subsequences of T different lengths for all C subjects, from the test videos. This can accommodate different starting frame positions and also reduce sequence length bias. The number of clusters per subject were determined empirically (details can be found in [11]) *i.e.* $M = 7$ for Honda/UCSD and $M = 6$ for ChokePoint. Temporal segment length and cluster similarity parameter are set to $L = 20$ and $\alpha = 0.75$ by experiments. Cluster subspaces are represented using DCC with the PCA subspaces learned for each cluster set to 10 (or $\approx 98\%$ of data energy). Parameters of other compared approaches were set using good values suggested by their respective authors.

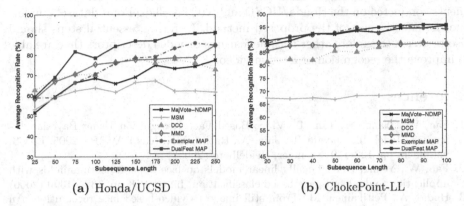

(a) Honda/UCSD   (b) ChokePoint-LL

Fig. 2. Performance of various methods with different test subsequence lengths

5.1   Results and Discussion

Overall, the Bayesian MAP classifiers yielded relatively better results, with the proposed dual-feature method achieving best recognition performance across all datasets, as shown in Table 1. Unlike crude voting strategies or rigid subspace methods, temporal dependencies between frames are well-exploited in our framework. Furthermore, the dual-feature utilization of clusters and exemplars ensures that coarser set variations and finer appearance cues both contribute towards the classification decision. This is evident from its ability to address the shortcomings of its exemplar-based counterpart.

In closer analysis, the proposed method gradually outperformed the other approaches as the subsequence length increases, as observed in Fig. 2. This is characteristic of our framework where temporal accumulation of frames can lead to better convergence of an identified subject. It is also worth noting that while the performance of image set-based approaches seemed uncorrelated to sequence lengths, they fared poorly against Bayesian methods. As expected, the ChokePoint dataset on EL setting is more challenging than on LL setting (see Table 1) as it struggles to match faces in adversely different conditions with a variety of new poses previously not encountered. This appears to be a viable research challenge in future.

6   Conclusion

In this paper, we present a new dual-feature Bayesian MAP classification method for video-based face recognition. The main contributions of this work is twofold. Firstly, the usage of both exemplar (image) and cluster (image set) features inherently ensures that both variational and appearance information of faces in video are well-captured. Secondly, similarity measures for matching these features are compactly fused in a probabilistic framework for classification, which capably exploit temporal information in video sequences. Comprehensive experiments conducted on the Honda/UCSD and NICTA ChokePoint datasets underline the effectiveness of the proposed method. In future, essential steps should also be taken to explore other feasible features or video descriptors that are able to improve the recognition of faces in video.

References

1. Fan, W., Wang, Y., Tan, T.: Video-Based Face Recognition Using Bayesian Inference Model. In: Kanade, T., Jain, A., Ratha, N.K. (eds.) AVBPA 2005. LNCS, vol. 3546, pp. 122–130. Springer, Heidelberg (2005)
2. Fan, W., Yeung, D.Y.: Locally linear models on face appearance manifolds with application to dual-subspace based classification. In: CVPR, pp. 1384–1390 (2006)
3. Hadid, A., Peitikäinen, M.: From still image to video-based face recognition: An experimental analysis. In: Proceedings of IEEE FGR, pp. 813–818 (2004)
4. Kim, T., Kittler, J., Cipolla, R.: Discriminative learning and recognition of image set classes using canonical correlations. IEEE Trans. PAMI 29(6), 1005–1018 (2007)
5. Lee, K., Ho, J., Yang, M., Kriegman, D.: Visual tracking and recognition using probabilistic appearance manifolds. CVIU 99(3), 303–331 (2005)
6. Liu, W., Li, Z., Tang, X.: Spatio-temporal Embedding for Statistical Face Recognition from Video. In: Leonardis, A., Bischof, H., Pinz, A. (eds.) ECCV 2006. LNCS, vol. 3952, pp. 374–388. Springer, Heidelberg (2006)
7. Poh, N., Chan, C.H., Kittler, J., Marcel, S., Mc Cool, C., Argones Rua, E., Alba Castro, J.L., Villegas, M., Paredes, R., Struc, V., Pavesic, N., Salah, A.A., Fang, H., Costen, N.: An evaluation of video-to-video face verification. IEEE Transactions on Information Forensics and Security 5(4), 781–801 (2010)
8. Roweis, S., Saul, L.: Nonlinear dimensionality reduction by locally linear embedding. Science 290, 2323–2326 (2000)
9. See, J., Ahmad Fauzi, M.F.: Learning Neighborhood Discriminative Manifolds for Video-Based Face Recognition. In: Maino, G., Foresti, G.L. (eds.) ICIAP 2011, Part I. LNCS, vol. 6978, pp. 247–256. Springer, Heidelberg (2011)
10. See, J., Ahmad Fauzi, M.F., Eswaran, C.: Video-based face recognition using exemplar-driven bayesian network classifier. In: Proc. of ICSIPA (2011)
11. See, J., Eswaran, C.: Exemplar extraction using spatio-temporal hierarchical agglomerative clustering for face recognition in video. In: Proc. of ICCV, pp. 1481–1486 (2011)
12. Wang, R., Shan, S., Chen, X., Gao, W.: Manifold-manifold distance with application to face recognition based on image set. In: Proc. IEEE CVPR (2008)
13. Wolf, L., Shashua, A.: Learning over sets using kernel principal angles. Journal of Machine Learning Research 4, 913–931 (2003)
14. Wong, Y., Chen, S., Mau, S., Sanderson, C., Lovell, B.C.: Patch-based probabilistic image quality assessment for face selection and improved video-based face recognition. In: CVPR Workshops (2011)
15. Yamaguchi, O., Fukui, K., Maeda, K.: Face recognition using temporal image sequence. In: Proc. of IEEE FGR, pp. 318–323 (1998)
16. Zhao, W., Chellappa, R., Phillips, P., Rosenfeld, A.: Face recognition: A literature survey. ACM Computing Surveys 35(4), 399–485 (2003)
17. Zhou, S., Krüeger, V., Chellappa, R.: Probabilistic recognition of human faces from video. Computer Vision and Image Understanding 91(1-2), 214–245 (2003)

Robot Dancing:
Adapting Robot Dance to Human Preferences

Qinggang Meng, Ibrahim Tholley, and Paul W.H. Chung

Department of Computer Science, Loughborough University, UK
{Q.Meng,P.W.H.Chung}@lboro.ac.uk

Abstract. In this paper, we investigate an approach for robots to extract the preferences of human observers, and combine them to generate new moves in order to improve robot dancing. Human preferences can be extracted even when a reward is given a few steps after a dance movement. With the feedback the robots perform more of what was preferred and less of what was not preferred. Human observers watch the robot generated dance movements and provide feedback in real time; then the robot learns the observers' preferences and creates new dance movements based on varying percentage of their preferences; and finally the observers rate the new robot's dancing. Experimental results show that the robot learns, using Interactive Reinforcement Learning, the expressed preferences of human observers and dance routines based on preferences of multiple observers are rated more highly.

Keywords: Robot dancing, robot interaction with humans, robot adaptation.

1 Introduction

In the robot entertainment industry, there is a growing interest in the idea of robots expressing rhythmic behaviours to musical signals. This has been expressed in many different ways [1,2,3,4]. Some robots contain pre-programmed dance motions, which are either selected at random, or choreographed to follow a particular music signal [2]. Others imitate dance using motion capture [3] or predict their human partner's dance directions based on the force applied by the human partners [4]. Whilst these different approaches to robot dance have shown to be successful in achieving dancing robots, little work is done to show how a robot can allow human partners to specify the dance parts they like and dislike in the robot's dance. Adaptive behaviour in dancing robots is still in its infancy. Work done in this area has largely focused on the audio or visual changes in the robots environment, such as changes in the dynamics of the music [5] or in the rhythmic motions detected from humans partners [6]. There is very little work done regarding the adaptation of dance based on receiving human feedback. This is an important area to explore because it would provide observers the ability to influence the way in which a robot improves its dancing without having to program the robot each time or have knowledge of dance or robots. This paper proposes a methodology to achieve this goal.

T. Huang et al. (Eds.): ICONIP 2012, Part V, LNCS 7667, pp. 557–565, 2012.
© Springer-Verlag Berlin Heidelberg 2012

2   Main Technologies for Adapting Robot Dance towards Human Preferences

There are two main technologies that have been used to achieve adaptation in dancing robots based on human input: Interactive Evolutionary Computation (IEC) and Interactive Reinforcement Learning (IRL).

IEC is a biologically inspired interactive learning approach based on evolutionary computations and works by using genetic algorithms or neural networks and the evaluation is determined by human feedback. During the process, different variations of generated behaviour are tested to see if a solution closer to the observer's preferences has been found. Randomness is introduced into the process by allowing a certain percent of generations to occur so that no part of the search space is excluded for exploration. This approach has proved satisfactory in different areas including robot dancing [7].

For many researchers who have explored the idea of IEC on human feedback to robotic dance systems, the common approach is to limit the feedback of dance partners to one dance partner at a time, and limiting human feedback to only positive rewards. For example, in the work of [8], the robot interacted with one observer for a limited time (4 seconds). [7] implemented an interactive system, in which human agents observed the dance of seven humanoid robots, each with varying dance moves. The work in both papers showed successful results and suggested that the robots could adapt their movements to a human observer. However, the results were all based on the feedback of a single observer and participants were limited to give one type of feedback, to express one type of preference. Little was shown to determine the result of the robot responding to more than one observer or to preferred and non-preferred dance behaviours. This however, has been demonstrated using Interactive Reinforcement Learning.

IRL [9] is a psychologically inspired interactive learning approach, based on traditional reinforcement learning algorithms and like IEC, the reward signals are replaced by a human agent as opposed to a pre-programmed model. The human agent can interact with the learning agent (e.g. a robot) at anytime and vary the rewards as they wish, during the robots learning process.

Like IEC, IRL has shown to be successful in areas that require adaptive interactive learning, but little has been used in robot dance. Of particular interest is the work by [10]. The authors implemented a reinforcement learning approach allowing humans to interact with a real robot and a computer game. They used the idea of positive and negative rewards in their work to guide learning. To satisfy the possibility that reinforcements might be received late during the execution of each action, the robot system had a "small delay to allow for human reward". [10] limited human input to scalar values in the scale of -1 to +1 for non-preferred and preferred actions respectively. Their results showed that participants gave more positive rewards than negative rewards. This same approach of positive and negative rewards for IRL had been explored by others such as [8]. The use of positive and negative rewards is a simple way for a robot to gain knowledge of the preferred and unwanted dance steps of a dance. However, this has a number of problems. In particular, if the rewards are not consistent then contradictions can occur in the observer's feedback [9]. For example, a "good" dance move in one part of the robot's dance maybe a "bad" dance move in another, which could cause contradiction in the robots understanding of the true reward. Therefore, an effective way of processing such rewards would have to be adopted. In this paper, the solution was to sum the reinforcement values so as to reduce this problem of conflicting reward values and have different actions rewarded differently.

3   Processing of Human Preferences

The two technologies described above have two main issues. The first is how best to capture human preferences as in real time feedback may be received after several dance steps of the preferred move. The second issue is determining how good actions are in comparison to other actions. In this paper, buffering and pattern matching using IRL technology are used to deal with the two issues.

A. Buffering and Pattern Matching

Buffering refers to the process of keeping a record of all the dance motions that were performed prior to the most recent feedback given, and after the previous feedback, and reducing them to a predefined maximum length. The buffer was a cache of n number of dance motions which constituted to a sequence. Each time the robot received feedback, the last n dance motions were saved in the buffer keeping the sequences at maximum lengths.

The robot associated the entire feedback sequence with the reinforcement given. The number of dance motions of a feedback sequence that could be stored in the buffer could be any value equal to 1 or greater. After reducing the feedback sequence to the length of the predefined buffer size, these would then be stored in a preference database along with the reinforcement given for each feedback sequence. The buffered sequences are then compared to obtain any common patterns among them. This is what's known as pattern matching. Found matches (common patterns) were then extracted and stored in the action database for the reinforcement learning algorithm to select and update the database, and generate a new dance sequence. On the other hand, if the rewarded dance motions had only received feedback once, i.e. no match was found, then the feedback would be ignored and the movements of that feedback would not be promoted for selection for the next dance.

The dance sequence would then be sent to the robot alongside with the music so that the dance could be observed and rewarded/ punished again by the observer. Observers could then have another opportunity to further fine-tune the robots dance in the next generated dance, indicating their dislike for unwanted moves and encouraging preferred dance steps.

B. Learning

The Sarsa algorithm [11] from traditional reinforcement learning was used as the underlining learning algorithm. For action-selection, the Softmax algorithm [11] was used. The actions were the matched patterns that were extracted from participants and only one state was recorded for the system to be in, i.e. the beat detected from the music signal.

Observers could provide feedback on dance moves they preferred and did not prefer in real time. These preferences had scalar values of positive (+1) and negative (-1) reinforcements. This was so that the observers were not overloaded with multiple feedback inputs. The approach taken was to weigh the preferences by summing all subsequent rewards given by the observers for some movement patterns. This was so that these movement patterns contained different accumulated rewards and, more highly rewarded dance actions in the action database encouraged the robot to do the dance movements more often than those with low reinforcements. All the generated movements were given a chance to be performed, but in the case where observers preferred not to see particular movements at all, this was only achieved if the total reward value for matched patterns was negative. The complete algorithm used for this system is illustrated below in Figure 1. Here, $m(sF_t)$ and $m(sF_{t+1})$ denote the number of dance motions at time t and $t + 1$, $Q(s_t, a_t)$ is the Q-value, α is the learning rate, and γ is the discount factor.

4   Experimental Procedure

Three experiments were conducted to explore the effects of human feedback on the robot's dance based on the proposed approach, each consisting of participants observing a number of robot dances containing varying percentage of their preferences. The dancing is by a virtual Sony AIBO robotic dog using the Webots simulation application [12] and displayed on a computer screen.

Experiment 1: The robot produced one random dance, which was shown to all participants who then observed the dance and provided their preferences on it. Each participant was then required to observe the newly generated dances which incorporated varying percentages of their own preferences, combined with the varying percentages of another participant who had observed the same dance.

Experiment 2: The is was the same as experiment 1, except that participants each observed a different initial random dance instead of the same dance.

For both experiments 1 and 2, the aim was to determine if combining a participant's preferences with another participant's preferences in a dance made a difference in the robot's performance in terms of the perception of newly preferred generated dance behaviours, after observing the same dance and different initial dances. Five different dances were generated which incorporated varying percentages of preferences from the participant taking part in the experiment and the preferences saved from another participant. Participant preferences were divided beginning with a dance whose dance motions were entirely selected from one participant's own preferences. The remaining other four generated dances included a gradual reduction of one participants own preferences and an increase of another participants preferences. These were compared to the initial random dance and observed by the participant to determine if perceptually, participants

```
1.    Initialise parameters: Q(s_t,a_t)=0, state (s_t) = beat; buffer size; time step (k)
2.    Play music
3.    While music is playing
4.        Read in predefined dance sequence
5.        If observer input (r_t) then
6.            Cache number of dance actions (n) as a feedback sequence (F_t)
7.            Save the sequence (F_t) to preference database
8.    End While
9.    Clear action database
10.   Repeat (for each feedback sequence (F_t))
11.       Initialise parameters:
              Number of dance motions in dance sequence (m(sF_t)=1) at time t
              Number of dance motions in dance sequence (m(sF_{t+1})=1) at time t+1
12.       While m(sF_t) < buffer size + 1
13.           Action a_t ← sF_t
14.           Repeat (for each sub-sequence (sF_{t+1}) in next feedback
                                       sequence (F_{t+1}))
15.               m(sF_{t+1}) ← m(sF_t)
16.               While m(sF_{t+1}) < buffer size + 1
17.                   Action a_{t+1} ← sF_{t+2}
18.                   If a_t and a_{t+1} are the same
19.                       If a_t in action database
20.                           R(a_t) ← R(a_t)+R(a_{t+1})
21.                       Else
22.                           Store a_t in action database
23.                           R(a_t) ← R(a_t)+R(a_{t+1})
24.                   Else
25.                       If m(sF_{t+1}) <= buffer size
26.                           Observe next sub-sequence (F_{t+2})
27.                           sF_{t+1} ← sF_{t+2}
28.               End Repeat
29.               Increment m(sF_t)
30.       End While
31.       Choose action (a_t) from action database (using Softmax)
32.       Repeat (until k is reached)
33.
34.           Take action (a_t) from action database, observe r
35.           Add a_t to new dance sequence
36.           Choose next action (a_{t+1}) (using Softmax)
37.           Q(s_t,a_t) ← Q(s_t,a_t)+α[r+γQ(s_{t+1},a_{t+1})-Q(s_t,a_t)]
                  s_t ← s_{t+1}; a_t ← a_{t+1}
38.       End Repeat
39.   End Repeat
```

Fig. 1. Complete system algorithm

could observe new dance combinations that they preferred. These combinations were done to determine if any perceived improvements were arbitrary or actually attributed to the preferences of the participant. All participants gave a minimum of 15 preferences, and only the first 15 preferences were selected for dance combinations so that the preferences were not biased because different individuals gave different amounts of feedback.

Experiment 3: In this experiment, the robot produced one random dance for each participant (as in experiment 2). For each participant who took part in the experiment, after observing the robot's initial random dance, three other dances were shown to them. The first was a newly generated random dance. The second dance contained the preferences of another participant, and the third dance contained 50% of two other participants. In all the three dances, none of them contained the participant's own preferences. These were compared to the initial random dance observed by the participant to determine how satisfied the participant was with the dances based on the number of newly preferred combinations they observed.

Each participant was asked to view each of the generated dances in all three experiments and provide their judgement on how satisfied they were with each dance. The participants were unaware of which of the robot's dances were as a result of only their own preferences or if the robot had undergone any training at all. This was to determine how well the robot responded to the preferences and improved in its dancing. After observing the dances, participants were then required to rate the robot's dance, by answering an online questionnaire. For experiments 1 and 2, participants were specifically asked two questions for each of the dances they observed: 1) How well did each dance incorporate your preferences? 2) How many newly generated combinations did you prefer? For question 1, responses were based on a five point ordinal scale with the labels "very bad", "bad", "neither bad nor good", "good" and "very good". The qualitative ratings were then converted into a numerical scale with 1 corresponding to "very bad", and 5 corresponding "very good" for the purpose of analysing the results. Question 2 was asked in experiments 1, 2 and 3. A count of the total number of newly preferred dance action combinations that participants observed (compared to the initial random dance) for each dance except in the initial random dance was required. It would have been very difficult for participants to keep count of preferred moves whilst observing the dances, so this was achieved with the help of the experiment controller, who kept a count as each participant, indicated their preference. Experiment 3 also required that participants kept a count of the total number of newly generated dance actions that were not preferred. 31 participants took part in all three experiments. The participants were from different professions and aged 17 to 40. None of them had any experience of interacting with robots. The results of the experiments are summarised below.

5   Results Analysis

Recall, the aim of experiments 1 and 2 were to determine if combining a participant's preferences with another participant's preferences in a dance made a difference in the observation of their preferences, and for all experiments, the aim was to determine if it was better for the robot to learn from other people than through random exploration.

Experiment 1 and 2 were first compared to determine how well the robot followed participant's preferences. An average was taken of the ratings from each participant. Figure 2 below shows the results.

The y-axis here is a rated measure of how well the participants felt that their preferences were performed in each of the observed dances, the x-axis is the percentage of their own preference in the combination. In both experiments 1 and 2, the decrease in rating was directly related to the decrease in the participant's own preferences. This makes sense since there would be more be diversity in

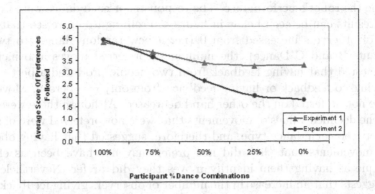

Fig. 2. Score of how well participants felt their preferences were followed

dance behaviours in experiment 2 than in experiment 1. A count of the number of perceived newly performed dance steps was also made for each subsequent dance after observing the initial random dance. Figure 3 below shows the average number of newly preferred dance combinations for each observed dance.

In experiment 1 there was a clear preference for dances that included 100% of the participant's preferences. As the percentage of that participant's preferences decreased the perception of preferred newly generated dance combinations decreased. Experiment 2 on the other hand, had a much more higher perception of newly performed dance combinations. This, like figure 2, supports the hypothesis that there would be more diversity in dance behaviours for experiment 2 than in experiment 1. The comparison between experiment 1 and experiment 2, demonstrated that an increase in the diversity of dance moves from others who have observed different dances, does increase the number of newly preferred generated dance steps. The possibility that the perceived improvements in dance were due to random movements is clearly ruled out. This has built a strong case suggesting that the quality of the dance had improved as a result of incorporating human preferences in the robot's dance.

In experiment 3, participants were required to provide the number of newly generated moves they preferred and did not prefer for each dance as a measure to

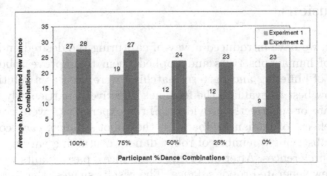

Fig. 3. Average score of the number of newly generated preferred combination

determine the robot's performance. The responses for each dance were averaged and the results can be seen below in figure 4. From figure 4, we see that as the number of observers increased from 0 (i.e. a new random dance) to two (i.e. Participants B and C Dance), the number of preferred moves also increased. This suggested that having feedback from two people produced a better dance than having no feedback or having feedback from only one person. Movements that were not preferred on the other hand decreased. Although these were initial results, the declined values of movements that were not preferred were of similar values between each dance type, and therefore, suggested that having observers identify movements that they did not prefer may not have been as effective a technique as having them identify moves they did prefer. Nevertheless, the result suggests that an increase in the number of observers giving feedback would increase the quality of new dances generated. More experiments and results can be found in [13]. The results suggest that training with feedback from more observers produced better learning than no training at all. This also suggests that learning occurred faster if the feedbacks from at most two observers were combined, although it would be interesting to know whether by increasing the number of people giving feedback learning would continue to increase or not.

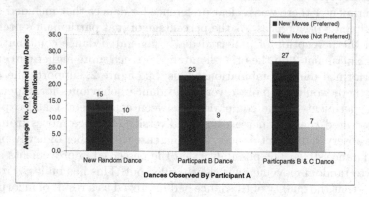

Fig. 4. Score of average number of newly preferred combinations in experiment 3

6   Conclusions

In this paper, we have introduced a way of capturing the desired and undesired preferences of human observers and applied them to improve robot dancing. The concept of buffering and pattern matching were used as a solution to help determine how best to evaluate the feedback effectively, particularly if feedback is received late or incorrectly provided. Three experiments were conducted to explore the effects of human feedback on the robot's dance, each consisting of participants observing a number of robot dances containing varying percentage of their own preferences. After observing the dances, participants were required to rate the new generated robot's dance. The results suggest that a robot dance that incorporates the preferences of others will improve the quality of the dance.

References

1. Aucouturier, J.J.: Dancing robots and AI's future. IEEE Intelligent Systems 23(2), 74–84 (2008)
2. Santiago, C., Oliveira, J., Reis, L., Sousa, A.: Autonomous robot dancing synchronized to musical rhythmic stimuli. In: 2011 6th Iberian Conference on Information Systems and Technologies (CISTI), pp. 1–6 (2011)
3. Shiratori, T., Ikeuchi, K.: Synthesis of dance performance based on analyses of human motion and music. IPSJ Transactions on Computer Vision and Image Media 1(1), 34–47 (2008)
4. Jens, H., Peer, A., Buss, M.: Synthesis of an interactive haptic dancing partner. Control, 527–532 (2010)
5. Solis, J., Chida, K., Suefuji, K., Takanishi, A.: Improvements of the sound perception processing of the anthropomorphic flutist robot (WF-4R) to effectively interact with humans. In: IEEE International Workshop on Robot and Human Interactive Communication, pp. 450–455 (2005)
6. Tanaka, F., Movellan, J.R., Fortenberry, B., Aisaka, K.: Daily HRI evaluation at a classroom environment: Reports from dance interaction experiments. In: Proceedings of the 1st ACM SIGCHI/SIGART Conference on Human-Robot Interaction, pp. 3–9 (2006)
7. Vircikova, M., Sincak, P.: Dance choreography design of humanoid robots using interactive evolutionary computation. In: 3rd Workshop for Young Researchers: Human Friendly Robotics for Young Researchers (2010)
8. Dozier, G.: Evolving robot behavior via interactive evolutionary computation: From real-world to simulation. In: Proceedings of the 2001 ACM Symposium on Applied Computing, pp. 340–344 (2001)
9. Thomaz, A.L., Hoffman, G., Breazeal, C.: Real-time interactive reinforcement learning for robots. In: AAAI 2005 Workshop on Human Comprehensible Machine Learning (2005)
10. Thomaz, A., Breazeal, C.: Asymmetric interpretations of positive and negative human feedback for a social learning agent. In: The 16th IEEE International Symposium on Robot and Human Interactive Communication, pp. 720–725 (2007)
11. Sutton, R., Barto, A.: Reinforcement Learning: An Introduction. MIT Press, Cambridge (1998)
12. Cyberbotics: Webots 6 for fast prototyping and simulation of mobile robots (2011), http://www.cyberbotics.com
13. Tholley, I.: Towards A Framework To Make Robots Learn To Dance. PhD thesis, Loughborough University, UK (2012)

Secure Distributed Storage for Bulk Data

Tadashi Minowa and Takeshi Takahashi

National Institute of Information and Communications Technology, Tokyo Japan
{minowa,takeshi_takahashi}@nict.go.jp

Abstract. Distributed types of data storage techniques are important especially for the cases where data centers are compromised by big natural disasters or malicious users, or where data centers consist of nodes with low security and reliability. Techniques using secured distribution and Reed-Solomon coding have been proposed to cope with the above issue, but they are not efficient enough for dealing with big data in cloud computing in terms of return-on-investment. This paper proposes a secure distributed storage system architecture. It maintains high security level by using packaging techniques that need not require key management inherent in AES encryption. Moreover, it scales out so that it is capable of storing a large amount of data safely and securely. The performance of the architecture is also dealt with in terms of storage efficiency and security evaluation.

Keywords: distributed storage, cloud computing, secret sharing, network coding, erasure codes.

1 Introduction

Recently, online storage services such as Dropbox, web services such as Google, and Social Networking Services such as Facebook became more and more available from various users' terminals. At the same time, many companies are getting started storing their data, which include not only emails and document files but also a large amount of multimedia or scientific data at data centers through the web cloud computing [4]. It is a challenging technique that provides scalable resources to users; instead of scale-up of server capabilities. Scalability may increase the number of servers to scale out the overall system performance in order to cope with temporal surge of server burdens and applications that uses big data. The cloud services provided by Google, Amazon, and Salesforce became successful in reducing initial and operational costs of users.

Cloud computing does not necessarily need to use Relational Database Management Systems that handle only structured data. It expects to have architecture that provides scalability (in terms of storage size), that maintains reliability by avoiding Single Point of Failure, and that does not leak information even if some data may be leaked from some of the storages, in order to cope with an increasing number of unstructured big data [3]. Various schemes, such as the ones using secret sharing and Reed-Solomon coding, have been proposed by now, and

T. Huang et al. (Eds.): ICONIP 2012, Part V, LNCS 7667, pp. 566–575, 2012.

they distribute data among storages to maintain high reliability and security. However, they do not scale out and do not provide efficient storage [17,13,7,15].

This paper is organized as follows: Section 2 reviews the existing data distribution schemes in application for secure and reliable storage systems. Section 3 describes the details of proposed scheme that uses a new class of erasure codes. Section 4 analyzes the performance of our scheme in comparison to the existing schemes in terms of security, storage efficiency, and complexity. Section 5 concludes the paper.

2   Dispersal Algorithms

A (k, n) threshold scheme are at the heart of all dispersal algorithms. The scheme encodes data into n separate symbols to be distributed among storages in such a way that knowledge of at least any k of these n symbols can be used to reconstruct the data.

We assume no error due to malice or accident exists in the k pieces by detecting and eliminating them using error detection based on a cryptographic hash function (or checksum algorithm). Furthermore, a message authentication code can be used if an attacker can change the message as well as its hash value [12]. However, no error correction is considered in this paper.

In the case of the error corruption of messages, it is possible that a (n, k) scheme could be designed to possess the properties of error correcting codes. The scheme could detect any of up to t erroneous pieces, and correct up to $\lfloor t/2 \rfloor$ pieces. Then by applying an errors-and-erasures decoding algorithm [11], it can recover data provided that $n - 2t \geq k$.

Let $\mathbb{F}_q$ be a finite field of q elements. Let α be primitive in $\mathbb{F}_q$. The distinct nonzero elements are $(q-1)$ consecutive powers of α, that is, $\{1, \alpha, \alpha^2, \ldots, \alpha^{q-2}\}$. The number of elements in $\mathbb{F}_q$ is of the form p^m, where p is a prime integer and m is a positive integer. In order that symbols may fit into computer words of eight-bit bytes, it is convenient that $p = 2$ and $m = 8$ are chosen.

Let $F = (d_1, d_2, \ldots, d_N)$ be a file that is a string of N symbols. The symbols d_i is considered as integers taken from elements in $\mathbb{F}_q$. The file F is segmented into messages of k symbols. Thus,

$$F = (d_1, \ldots, d_k), (d_{k+1}, \ldots, d_{2k}), \cdots, (d_{N-k+1}, \ldots, d_N). \tag{1}$$

Let $\mathbf{d}_i = (d_{(i-1)k+1}, \ldots, d_{ik})$ be the ith message of k message symbols. Given the message $\mathbf{d}_i$, a dispersal algorithm can produce the codeword $\mathbf{c}_i$ consisting of n codeword symbols, that is $\mathbf{c_i} = (c_{(i-1)n+1}, \ldots, c_{in})$. Thus the file F is encoded into the encoded file F' consisting of codewords of n symbols. Thus,

$$F' = (c_1, \ldots, c_n), (c_{n+1}, \ldots, c_{2n}), \cdots, (c_{N-n+1}, \ldots, c_N). \tag{2}$$

Without loss of generality, we can focus on encoding the first message $\mathbf{d}_1$ into the first codeword $\mathbf{c}_1$.

Choose n vectors $\mathbf{a}_i = (a_{i1}, \ldots, a_{ik}) \in \mathbb{F}_q^k$, $1 \le i \le n$ such that every subset of k different vectors are linearly independent. Note that n cannot be larger than $q - 1$. Then, the (k, n) threshold scheme encodes a message $\mathbf{d}_1 = (d_1, \ldots, d_k)$, $d_i \in \mathbb{F}_q$, into the codeword $\mathbf{c}_1 = (c_1, \ldots, c_n)$, where

$$c_i = \mathbf{a}_i \cdot \mathbf{d}_1 = a_{11}d_1 + \cdots + a_{1k}d_k. \tag{3}$$

Each c_i of the codeword $\mathbf{c}_1$ is stored on a different storage node.

If any k pieces of $\mathbf{c}_1$, say, $c_1, \ldots, c_k$ are given, we can reconstruct $\mathbf{d}_1$. Let $A = (a_{ij})_{1 \le i,j \le k}$ be the $k \times k$ matrix whose ith row is a_i. It is seen that

$$A \cdot \begin{bmatrix} d_1 \\ \vdots \\ d_k \end{bmatrix} = \begin{bmatrix} c_1 \\ \vdots \\ d_n \end{bmatrix}. \tag{4}$$

Since any subset of k different $\mathbf{a}_i$ are linearly independent, A is invertible because the determinant of A can be reduced to that of a *Vandermonde* matrix that is nonsingular. Thus,

$$\begin{bmatrix} d_1 \\ \vdots \\ d_k \end{bmatrix} = A^{-1} \cdot \begin{bmatrix} c_1 \\ \vdots \\ c_n \end{bmatrix}. \tag{5}$$

2.1 Shamir's Scheme

A (k, n) threshold dispersal algorithm in Shamir's scheme [17] encodes a secret symbol d into n codeword symbols $c_1, \cdots, c_n$, each of the same size as d. The secret d can be reconstructed from any set of the k codeword symbols.

Let s_0 in $\mathbb{F}_q$ be the secret symbol d to be implicitly shared among n storages. Thus $s_0 = d$. Generate the first $k - 1$ shares $s_1, \ldots, s_{k-1}$ in $\mathbb{F}_q$ at random. Shamir's scheme constructs a polynomial $p(x)$ as

$$p(x) = s_0 + s_1 x + \cdots + s_{k-1} x^{k-1} \tag{6}$$

and evaluates $p(x)$ at n distinct nonzero elements $\alpha^1, \ldots, \alpha^n$ of $\mathbb{F}_q$. Notice that we do not evaluate $p(x)$ at the zero element because s_0 appears as it is without secrecy. The n codeword symbols $c_1, \ldots, c_n$ that are distributed to n different storages are just these polynomial values $p(\alpha^1), \ldots, p(\alpha^n)$. Thus, for $1 \le i \le n$,

$$c_i = p(\alpha^i) = s_0 + s_1 \alpha^i + \cdots + s_{k-1} \alpha^{i(k-1)}. \tag{7}$$

Let $\mathbf{s} = (s_0, s_1, \ldots, s_{k-1})$ and $\mathbf{a}_i = (1, \alpha^i, \ldots, \alpha^{i(k-1)})$. Then, we can express codeword symbols as $c_i = \mathbf{a}_i \cdot \mathbf{s}$.

Given k of n codeword symbols, we can form a $k \times k$ matrix A whose rows consist of distinct $\mathbf{a}_i$. The matrix A is invertible because every subset of k different vectors a_i are linearly independent. Then, the secret $\mathbf{s}$ (and hence s_0) can be recovered by

$$\begin{bmatrix} s_0 \\ \vdots \\ s_{k-1} \end{bmatrix} = A^{-1} \cdot \begin{bmatrix} c_1 \\ \vdots \\ c_n \end{bmatrix}. \tag{8}$$

2.2 Rabin's Scheme

Rabin's scheme weakens the security of Shamir's, but improves storage efficiency by a factor of k.

A (k, n) threshold dispersal algorithm in Rabin's scheme [13] encodes a message $\mathbf{d}=(d_1, \ldots, d_k)$, $d_i \in \mathbb{F}_q$, into the codeword $\mathbf{c}=(c_1, \ldots, c_n)$, each symbol of the same size. The message $\mathbf{d}$ can be recovered from any set of k codeword symbols.

With the message $\mathbf{d}$, Rabin's scheme constructs a polynomial $p(x)$ as

$$p(x) = d_1 + d_2 x + \cdots + d_k x^{k-1}. \tag{9}$$

It can be seen that $p(x)$ is filled up with k message symbols instead of only s_0 being a message symbol in Shamir's scheme. It can be seen that Shamir's scheme is a special case of Rabin's scheme, and both schemes belong to the same form as Reed-Solomon codes.

Rabin's scheme forms the codeword $\mathbf{c}$ by evaluating $p(x)$ at n distinct elements $\alpha^1, \ldots, \alpha^n$ in $\mathbb{F}_q$. Notice that the n elements can include the zero element. Thus, for $1 \le i \le n$,

$$c_i = p(\alpha^i) = d_1 + d_2 \alpha^i + \cdots + d_k \alpha^{i(k-1)}. \tag{10}$$

Let $\mathbf{a}_i = (1, \alpha^i, \ldots, \alpha^{i(k-1)})$. Then, we have codeword symbols as $c_i = \mathbf{a}_i \cdot \mathbf{d}$.

Given k of n codeword symbols, we can form a invertible $k \times k$ matrix A whose rows consist of distinct $\mathbf{a}_i$ as with Shamir's. Then, $\mathbf{d}$ can be recovered by

$$\begin{bmatrix} d_1 \\ \vdots \\ d_k \end{bmatrix} = A^{-1} \cdot \begin{bmatrix} c_1 \\ \vdots \\ c_n \end{bmatrix}. \tag{11}$$

2.3 Krawczyk's Scheme

Krawczyk's scheme [7] is a blending of Rabin and Shamir. It encrypts data with a key-based encryption algorithm, and then disperses the encryption key with a secret sharing algorithm.

Let K be a random encryption key. Let E_K be the encryption algorithm such as AES under the key K. Encrypt a message $\mathbf{d}=(d_1, \ldots, d_k)$, $d_i \in \mathbb{F}_q$, into the encrypted message $\mathbf{e}=(e_1, \ldots, e_k)$, $e_i \in \mathbb{F}_q$, that is, $\mathbf{e} = E_K(\mathbf{d})$.

A (k, n) threshold dispersal algorithm in Krawczyk's scheme [7] encodes the encrypted message $\mathbf{e}=(e_1, \ldots, e_k)$ into the codeword $\mathbf{c}=(c_1, \ldots, c_n)$ using Rabin's scheme, and encodes the key K into the codeword $\mathbf{s}=(s_1,\ldots,s_n)$ using Shamir's scheme. The pair (c_i, s_i) of each codeword symbol c_i and s_i is distributed into a distinct storage.

Notice that the size of s_i is much larger than that of an element in $\mathbb{F}_q$ because the size of K is sufficiently larger than that of d_i. Thus in practice, with the same key, the subsequent messages $\mathbf{d}_i$ are encoded into the encrypted messages $\mathbf{e}_i$.

The recovery of the message $\mathbf{d}$ can be made as follows. Collect any set of k codeword symbols $c_1,\ldots,c_n$ and any set of k codeword symbols $s_1,\ldots,s_k$. Using Rabin's scheme reconstruct $\mathbf{e}$ out of the collected symbols $c_1,\ldots,c_k$. Using Shamir's scheme recover the key K out of the collected symbols $s_0,\ldots,s_k$. Finally, decrypt $\mathbf{e}$ using K to recover the original message $\mathbf{d}$.

2.4 Resch's Scheme

Compared to existing approaches to data distribution, Resch's scheme [15] has attractive properties. Its storage and computational efficiency is much less than Shamir's secret sharing, while maintaining that compromise is computationally infeasible. Compared to Rabin's scheme, Resch' scheme achieves a far greater degree of security.

Let $F = (\mathbf{d}_1, \mathbf{d}_2, \ldots, \mathbf{d}_N)$ be a file that is a string of N messages, each message in $\mathbb{F}_{2^{256}}$. Let $\mathbf{d}_i = (d_{(i-1)w+1}, \ldots, d_{iw})$ be the ith message of w symbols, each symbol in $\mathbb{F}_{2^8}$. It follows that $w = 32$, and the size of each block $\mathbf{d}_i$ is 32 bytes, i.e., 256 bits.

Then the file F can be split into Nw message symbols of $\mathbb{F}_{2^8}$ as

$$F = (d_1,\ldots,d_w),(d_{w+1},\ldots,d_{2w}),\cdots,(d_{(N-1)w+1},\ldots,d_{Nw}). \qquad (12)$$

Let $F' = (\mathbf{e}_1, \mathbf{e}_2, \ldots, \mathbf{e}_{N+1})$ be the encrypted file of $N+1$ encrypted messages, each message in $\mathbb{F}_{2^{256}}$. Let $\mathbf{e}_i = (e_{(i-1)w+1}, \ldots, e_{iw})$ be the ith encrypted message of w symbols, each symbol in $\mathbb{F}_{2^8}$. Then the encrypted file F' can also be split into $(N+1)w$ encrypted message symbols of $\mathbb{F}_{2^8}$ as

$$F' = (e_1,\ldots,e_w),(e_{w+1},\ldots,e_{2w}),\cdots,(e_{Nw+1},\ldots,e_{(N+1)w}). \qquad (13)$$

The file F' can also be split into $(N+1)w$ encrypted message symbols, each symbol in $\mathbb{F}_{2^8}$. Thus

$$F' = (e_1,\ldots,e_k),(e_{k+1},\ldots,e_{2k}),\cdots,(e_{(N+1)w-k+1},\ldots,e_{(N+1)w}). \qquad (14)$$

Let $F' = (\mathbf{e'}_1, \mathbf{e'}_2, \ldots, \mathbf{e'}_{(N+1)w/k})$ be the encrypted file of $(N+1)w/k$ encrypted messages. Let $\mathbf{e}'_i = (e'_{(i-1)k+1}, \ldots, e'_{ik})$ be the ith encrypted message of k symbols, each symbol in $\mathbb{F}_{2^8}$.

Let $\mathbf{c}_i = (c_{(i-1)n+1}, \ldots, c_{in})$ be the ith codeword of n symbols, each symbol in $\mathbb{F}_{2^8}$. Then a (k, n) threshold dispersal algorithm in Resch's scheme [15] encrypts the file F into the encrypted file F' using Rivest's All-Or-Nothing Transform (AONT) [16], and for $1 \le i \le (N+1)w/k$, encodes each encrypted message $\mathbf{e}'_i$ of k symbols into the codeword $\mathbf{c}_i$ of n symbols using Rabin's scheme.

Let K be a randomly chosen key chosen for the AONT block cipher. For the messages $\mathbf{d}_i$, $1 \le i \le N$, the AONT calculates the encrypted messages $\mathbf{e}_i$ as

$$\mathbf{e}_i = \mathbf{d}_i \oplus E_K(I + i), \tag{15}$$

where I is the initialization vector and $E_K(\cdot)$ is a key-based encryption algorithm such as AES [5]. The AONT calculates the final encrypted message $\mathbf{e}_{N+1}$ as the exclusive-or function of the key K and a hash of all previous encrypted messages $\mathbf{e}_1, \dots, \mathbf{e}_N$. Thus

$$\mathbf{e}_{N+1} = K \oplus H(\mathbf{e}_1, \dots, \mathbf{e}_N), \tag{16}$$

where $H(\cdot)$ is a cryptographic hash algorithm such as SHA-256 [1]. Using Rabin's scheme, for $1 \le i \le (N+1)w/k$, the (k, n) threshold dispersal algorithm encodes the input messages $\mathbf{e}'_i$ of k symbols into the output codewords $\mathbf{c}_i$ of n symbols.

It is easy to recover the original file $F = (\mathbf{d}_1, \mathbf{d}_2, \dots, \mathbf{d}_N)$ by reversing Rabin's scheme and the AONT, respectively. The first step is to get F' by decoding the message $\mathbf{e}'_i$, $1 \le i \le (N + 1)w/k$, from any set of k codeword symbols. The second step is to compute a hash $H(\mathbf{e}_1, \dots, \mathbf{e}_N)$ by using the first N messages of $F' = (\mathbf{e}_1, \mathbf{e}_2, \dots, \mathbf{e}_{N+1})$. The third step is to calculate the key K by using the final message $\mathbf{e}_{N+1}$ of F' and the hash, such that

$$K = \mathbf{e}_{N+1} \oplus H(\mathbf{e}_1, \dots, \mathbf{e}_N), \tag{17}$$

The final step is to compute the original messages $\mathbf{d}_i$, $1 \le i \le N$, by using the encrypted messages $\mathbf{e}_i$ and the decryption with the key K such that

$$\mathbf{d}_i = \mathbf{e}_i \oplus E_K(I + i). \tag{18}$$

2.5 Scalability Issues in Existing Erasure Codes

Traditional erasure codes typically input k symbols to generate $n - k$ redundant symbols for a total of n encoding symbols with a fixed rate k/n. Any k of the n encoding symbols is sufficient to recovering the original k input symbols. As erasure codes, Reed-Solomon codes [14] and Low-Density Parity-Check (LDPC) codes [6] that are well-known as error-correcting codes can be used, but both or either of k and n of the codes need to be fixed before the encoding process begins. Either feedback or retransmission of lost encoding symbols would be necessary for a reliable distribution over erasure channel. Furthermore, Reed-Solomon codes in practice are only efficient for relative small settings of k and n, because typical algorithms need $O(k^2)$ symbol operations in decoding and $O(kn)$ symbol operations in encoding [10]. It is clear that the use of Reed-Solomon erasure codes over large files for bulk data distribution leads to prohibitive encoding and decoding overhead. Thus all existing dispersal algorithms, which use Shamir's or Rabin's or ones such combination of both or either of the schemes as Krawczyk's or Resch's, are prohibitively expensive for large files.

The concept of digital fountain codes are first introduced by Byers *et al.* [2] in 1998 for information distribution. Luby Transform (LT) codes [8] are the first realization of a true digital fountain code. LT codes are rateless, i.e., the number of encoding symbols that can be generated from the data is potentially limitless, which enables encoding symbols to be generated on the fly, as few or as many as needed. LT codes properly designed on the basis of the robust soliton distribution have desired properties. The symbol for the code can be arbitrary, from one-bit binary symbols to a large w-bit symbols. If an original file can be split into k symbols, then each codeword can be independently generated, and the original file can be recovered with a vanishing overhead of $O(\sqrt{k}\ln^2(k/f))$ codewords for reconstruction probability $1 - f$. Thus LT codes can forgo costly and complicated retransmission protocols often needed to maintain a reliable data storage. Furthermore, LT codes have low encoding/decoding complexity of the order $O(k\ln k)$ exclusive-or operations [9].

3   AONT-LT Scheme

3.1   Architecture

We propose an AONT-LT scheme, an improved extension of Resch's with efficient rateless encoding and fast belief propagation decoding. The AONT-LT scheme can be considered similar to Resch's in that it employs Rivest's AONT as an encryption mode. But the AONT-LT incorporates a class of fountain codes instead of Reed-Solomon codes in Resch's.

At a sender, an integrity check of a data file is computed and appended to the file. The resultant file $F = (\mathbf{d}_1, \ldots, \mathbf{d}_N)$ is encrypted into the file $F' = (\mathbf{e}_1, \ldots, \mathbf{e}_N)$ by AES using a random key K. A hash value $H(\mathbf{e}_1, \ldots, \mathbf{e}_N)$ of the encrypted file F' is computed. The random key and hash value are then combined using bitwise exclusive-or to yield a signature $\mathbf{e}_{N+1}$, which is appended to the encrypted file F' to form the AONT package $F' = (\mathbf{e}_1, \ldots, \mathbf{e}_{N+1})$. Once processed by AONT, the package is split into slices $\mathbf{e}'_i$ of k symbols, and then fed into LT coding producing codeword slices $\mathbf{c}_i$ of n symbols for $1 \leq i \leq (N+1)w/k$. Each symbol of the slices are stored to storages in geographically separate locations.

At a receiver, the entire AONT package F' needs to be retrieved to recover the original file F. Once one retrieves at lease any k of the n symbols for the encrypted codeword $\mathbf{c}_i$, they can compute via a belief propagation (BP) algorithm the slice $\mathbf{e}'_i$ that is input to LT coding, thereby forming the AONT-LT package $F' = (\mathbf{e}_1, \ldots, \mathbf{e}_N)$. A random key K to AES encryption is retrieved by using bitwise exclusive-or of the signature $\mathbf{e}_{N+1}$ and the hash $H(\mathbf{e}_1, \ldots, \mathbf{e}_N)$ of the encrypted file. The random key is then used to decrypt the encrypted file into the original file $F = (\mathbf{d}_1, \ldots, \mathbf{d}_N)$, and the integrity check value is checked to detect corruption.

3.2   LT Codes

The basic property of a good degree distribution required for LT codes is that input symbols are added to the ripple at the same rate as they are processed [8].

The ideal soliton distribution ideally behaves in terms of the expected number of encoding symbols in order to recover the data. This distribution is $\Omega_I(1), \cdots, \Omega_I(k)$, where

$$\Omega_I(d) = \begin{cases} 1/k & d = 1 \\ 1/(d(d-1)) & d = 2, 3, \ldots, k \end{cases} \tag{19}$$

Unfortunately, this distribution is useless in practice. The problem with the ideal soliton distribution is that the expected number of degree one is too small. Any variation in the degree size is likely to make the ripple disappear and then the overall process fails.

The robust soliton distribution ensures that the expected size of the degree one is large enough at each point in the process so that it never disappears completely with high probability. This distribution is $\Omega_R(d)$ which is defined by

$$\Omega_R(d) = \frac{\Omega_I(d) + \tau(d)}{\sum_{j=1}^{k}(\Omega_I(j) + \tau(j))} \quad d = 1, 2, \cdots, k \tag{20}$$

Let $S = c\ln(k/f)\sqrt{k}$ for some suitable constant $c > 0$, then $\tau(d)$ is given by

$$\tau(d) = \begin{cases} S/(kd) & d = 1, 2, \ldots, (k/S) - 1 \\ S\ln(S/f)/k & d = k/S \\ 0 & d = k/S + 1, \cdots, k \end{cases} \tag{21}$$

3.3 Encoding Algorithm

Each time an output symbol is generated in an LT code, the robust soliton distribution is sampled which returns an integer d between 1 and the number k of input symbols. Then d random distinct input symbols are chosen, and their value is added to produce the output symbol.

3.4 Decoding Algorithm

In the case of an erasure channel, each output symbol represents a linear equation in the unknown k input symbols, thus the decoding can be viewed as solving a system of n linear equations in k unknown symbols. The cost of a typical algorithm is $O(nk^2)$, since the Maximum Likelihood (ML) decoding for this code is equivalent to Gaussian elimination, which is computationally expensive for a long LT code. However, LT codes properly designed on the basis of the robust soliton distribution have low encoding/decoding complexity of the order $O(k\ln k)$ exclusive-or operations [9].

4 Evaluation

We evaluate each of (n, k) threshold schemes in the case of the $(10, 5)$ configuration in terms of security, storage, and complexity criteria. Note that the principle of the proposed scheme differs from that of the existing schemes, we use $k = 10^2$, 10^4, and 10^6 depending on the file size for the proposed scheme.

Table 1. The required storage of (n, k) threshold scheme for a plain file of size M

Algorithm	Storage fomula (Byte)	$M = 10 \times 2^{10}$	$M = 10 \times 2^{20}$	$M = 10 \times 2^{30}$
Replication	$M(n - k + 1)$	60 (KB)	60 (MB)	60 (GB)
Shamir	Mn	100 (KB)	100 (MB)	100 (GB)
Rabin	Mn/k	20 (KB)	20 (MB)	20 (GB)
Krawczyk	Mn/k	20 (KB)	20 (MB)	20 (GB)
Resch	Mn/k	20 (KB)	20 (MB)	20 (GB)
AONT-LT	$M(k + O(\sqrt{k}\ln^2(k/f)))/k$	57.7 (KB)	23.3 (MB)	12.6 (GB)

4.1 Security Evaluation

We use the threat model where individual storage servers belong to different domains in terms of both administrative and physical manners. The security of servers may be compromised in such a way that a traitorous system administrator or outside anonymous attacker can steal data. Our assumption is that a judicious choice of k slices of any slices from physically dispersed storage servers in different domains is sufficient to protect the security of servers against malicious attackers. For each algorithm, we assume that the attacker knows how the slices were generated, except for the random numbers.

Shamir's scheme is information theoretic security. Attackers cannot get any information from fewer than k slices, irrespective of their computing power. Rabin's scheme has no security because of no randomness. Krawczyk's scheme has computational security because no one can break the data without the key that is dispersed using information theoretically secure secret sharing scheme. Like Krawczyk's, Resch's scheme is computational security because one needs all of the data to recover the key, and one cannot decode any of the data without the key. The security of the proposed scheme is equivalent to that of Krawczyk's and Resch's.

4.2 Storage Evaluation

We evaluate the storage capacity of each scheme for three different file sizes of 10 KB, 10 MB, and 10 GB. Table 1 shows the calculated storage capacity of each scheme. It is shown that the proposed scheme becomes effective, especially for a large number of k, and can reduce the total storage capacity.

4.3 Complexity Evaluation

The decoding complexity of the all existing schemes basically belongs to $O(k^3)$ because of Gaussian elimination. On the other hand, the proposed scheme can utilize a greedy graph pruning procedure of the BP decoding algorithm whose complexity is $O(k \ln k)$. Therefore, the proposed scheme is much more efficient than all the existing schemes.

5 Conclusions

This paper proposes the architecture of a securely distributed storage system. It maintains high security level by using packaging techniques that require no key management based on AES encryption. It scales out so that it is capable of storing a large amount of data safely and securely. It is shown that the proposed scheme has good storage efficiency and computational security.

References

1. Secure Hash Standard. National Institute of Standards and Technology, Washington, federal Information Processing Standard 180-2
2. Byers, J.W., Luby, M., Mitzenmacher, M., Rege, A.: A digital fountain approach to reliable distribution of bulk data. SIGCOMM Comput. Commun. Rev. 28, 56–67 (1998)
3. Chang, F., Dean, J., Ghemawat, S., Hsieh, W.C., Wallach, D.A., Burrows, M., Chandra, T., Fikes, A., Gruber, R.E.: Bigtable: A distributed storage system for structured data. In: Proceedings of The 7th Conference on Usenix Symposium on Operating Systems Design And Implementation, vol. 7, pp. 205–218 (2006)
4. Chervenak, A., Foster, I., Kesselman, C., Salisbury, C., Tuecke, S.: The data grid: Towards an architecture for the distributed management and analysis of large scientific datasets. Journal of Network and Computer Applications 23, 187–200 (1999)
5. Daemen, J., Rijmen, V.: The Design of Rijndael. Springer-Verlag New York, Inc., Secaucus (2002)
6. Gallager, R.G.: Low-Density Parity-Check Codes (1963)
7. Krawczyk, H.: Secret Sharing Made Short. In: Stinson, D.R. (ed.) CRYPTO 1993. LNCS, vol. 773, pp. 136–146. Springer, Heidelberg (1994)
8. Luby, M.: LT codes. In: Proceedings of the 43rd Symposium on Foundations of Computer Science, FOCS 2002, pp. 271–280 (2002)
9. Mackay, D.J.C.: Fountain codes. IEE Communications 152, 1062–1068 (2005)
10. MacWilliams, F.J., Sloane, N.J.A.: The Theory of Error-Correcting Code. North-Holland (1977)
11. MacWilliams, F., Sloane, N.: The Theory of Error-Correcting Codes, 2nd edn. North-holland Publishing Company (1978)
12. McEliece, R.J., Sarwate, D.V.: On sharing secrets and reed-solomon codes. Commun. ACM 24(9), 583–584 (1981)
13. Rabin, M.O.: Efficient dispersal of information for security, load balancing, and fault tolerance. J. ACM 36, 335–348 (1989)
14. Reed, I.S., Solomon, G.: Polynomial codes over certain finite fields. Journal of the Society of Industrial and Applied Mathematics 8, 300–304 (1960)
15. Resch, J.K., Plank, J.S.: AONT-RS: Blending security and performance in dispersed storage systems. In: 9th USENIX Conference on File and Storage Technologies, FAST 2011, pp. 191–202 (2011)
16. Rivest, R.L.: All-or-Nothing Encryption and the Package Transform. In: Biham, E. (ed.) FSE 1997. LNCS, vol. 1267, pp. 210–218. Springer, Heidelberg (1997)
17. Shamir, A.: How to share a secret. Commun. ACM 22, 612–613 (1979)

A Multi-modal Face
and Signature Biometric Authentication System
Using a Max-of-Scores Based Fusion

Youssef Elmir[1,*], Somaya Al-Maadeed[2], Abbes Amira[2,3], and Abdelǎali Hassaïne[2]

[1] Computer Science Department, The African University Ahmed Draya, Adrar, Algeria
y.elmir@univ-adrar.dz
[2] Qatar University, Doha, Qatar
{S_alali,abbes.amira,hassaine}@qu.edu.qa
[3] University of Ulster, NIBEC, Newtownabbey, United Kingdom
a.amira@ulster.ac.uk

Abstract. Face and signature based multimodal biometric systems are often required in various areas, such as banking biometric systems and secured mobile phone operating systems, among others. Our system combines these two biometric traits and provides better recognition performance compared with the systems based on a single biometric trait or modality. In multimodal biometric system, the most common fusion approach is integration at the matching score level because of the ease of combining and accessing the scores generated by different matchers. In this paper, we study the performance of a max-of-scores fusion technique based on the face and signature traits of a user. The experiments that were conducted on a database of 40 users indicate that the max-of-scores fusion-based method yields better authentication performance than single-face, single-signature, simple-sum or min-of-scores fusion-based biometric systems.

Keywords: Multimodal biometrics, Face, Signature, Score level fusion, SVM.

1   Introduction

Although there currently are highly reliable biometric techniques, such as retina or iris recognition, these available techniques are expensive and are generally not accepted by the general public; therefore, they can be used only for very high security applications. For other applications, and especially for banking applications, techniques such as signature or face authentication or face are very well accepted by users, but their performance is still too unsatisfactory for these systems to be used in real-world applications, which highlights the need for a robust and efficient fusion technique.

Multimodality is an alternative that improves the systematic performance of a biometric system [1]. Performance is characterized by both the precision of the system and its efficiency. Indeed, different classifiers have different errors, and it is possible to take advantage of this complementarity to improve the overall performance of the system.

---

* Corresponding author.

T. Huang et al. (Eds.): ICONIP 2012, Part V, LNCS 7667, pp. 576–583, 2012.
© Springer-Verlag Berlin Heidelberg 2012

J. Fierrez-Aguilar et al [2] experimentally compare some fusion strategies and use as a monomodal platform, their face verification system based on a global face appearance representation scheme, their minutiae-based fingerprint verification system, and their on-line signature verification system based on Hidden Markov Model (HMM) modeling of temporal functions; all models are applied to the MCYT multimodal database. A new strategy is also proposed and discussed to generate a multimodal combined score using Support Vector Machine (SVM) classifiers from which user-independent and user-dependent fusion schemes are derived and evaluated. In another study, Bernadette Dorizzi [3] has presented different types of score fusion methods discussed their complexity when used to model the systems-scores and proposed to a comparison of score fusion methods using a large multimodal database, BioSecure DS3. M.M. Kazi et al [4] have studied the performance of a single fast normalized cross-correlation matcher and a simple sum rule fusion technique based on face and signature traits of a user. Lorene Allano et al [5] have tested a number of score fusion methods for the purpose of multimodal biometric authentication. These tests were made for the SecurePhone project. The three biometrics, i.e., voice, face and signature were selected. All of the methods tested are based on fusion of the match scores output by each modality. All of the four fusion methods tested yield a significant performance increase. Luan Fang-jun and Lin Lan [6] have proposed a new fusion method based on principal component analysis (PCA). They have extracted features from the voice and face images and online handwritings of one person. Then, the three features are fused using the PCA method. Finally, the authorization is implemented through classification by minimizing the Euclidean distances of different people. Table 1 summarizes these works and their performance on well-known databases.

Table 1. Review of related works

Ref	Method	FAR		EER	Database	Population
[1]	Face + sign (sum rule based)	≈40	≈5		XM2VTS	50
[2]	SVM (face,sign) user-indep	≈30	≈10		XM2VTS	50
[3]	SVM classifier (face and sign)			5.54	Biosecure	500
[4]	Simple sum fusion (face and			2	BAMU	17
[5]	MinMax + GMM (face, voice			2.39,1.54,2.3	PDA	60
[6]	Fused feature with PCA (face,	9	2	1	Their own	20

2 Proposed Multimodal Authentication System

Our proposed system is based on fusion at the score level. The scores used are obtained from two verification systems (see Fig. 1). The first system is based on face verification using the combination of Gabor for feature extraction [8, 9] and SVM for classification. The second system is based on online signature verification using Nalwa's method [7]. We used various strategies to fuse face and online signature scores such as the simple sum, minimum and maximum of scores.

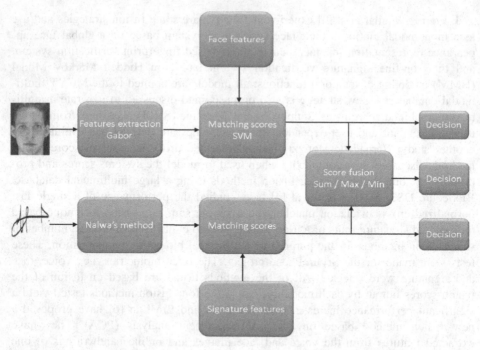

Fig. 1. Our proposed architecture for a score level fusion based multimodal biometric authentication system

2.1   Face Verification System

2.1.1   Gabor Filters

A Gabor filter bank has been used to construct the face vector code. The default parameters correspond to the most common parameters used in conjunction with face images of size 128 × 128 pixels. Optionally, the function returns a filter bank structure that contains the spatial and frequency representations of the constructed Gabor filter bank and some meta-data [8, 9].

The following steps are performed to create a 1D vector code for faces:

1. Sector-wise normalization followed by application of a bank of Gabor filters that has the general form in the spatial domain [10].

$$G(x, y; f, \theta) = \exp\left\{\frac{-1}{2}\left[\frac{x'^2}{\delta_{x'}^2} + \frac{y'^2}{\delta_{y'}^2}\right]\right\}\cos(2\pi f x') \qquad (1)$$

Where $x' = x\sin\theta + y\cos\theta$ and $y' = x\sin\theta - y\cos\theta$. f is the frequency of the sine plane wave along the direction θ from the x-axis, and δx' and δy' are the space constants of the Gaussian envelope along the X' and Y' axes, respectively.

2. Finally, the feature code is generated by obtaining the standard deviations of all the sectors [10].

2.1.2   Support Vector Machine

SVMs [11, 12] are a set of related supervised learning methods used for classification. They work by viewing input data as two sets of vectors in an n-dimensional space and constructing a hyper plane that separates the data with the greatest possible margin between the sets. This solution is achieved by calculating two parallel hyper planes, each of which is pushed up against" one of the data sets, the samples on the margin of the hyper planes are referred to as support vectors". The dividing hyper plane is then constructed halfway between the margins. The concept of SVMs has been extended to include the use of soft margins [11]. In cases where data cannot be cleanly split into two sets, a margin is chosen that splits the data as cleanly as possible, while still maximizing the margin between the cleanly split examples [13].

Although SVMs are linear, they are usually combined with kernel functions to produce a non-linear classifier. The kernel functions increase the dimensionality of the feature space and transform the distribution of data so that data sets that are not linearly separable in the original feature space become linearly separable in the transformed space. Kernel functions that have been used for this purpose include the polynomial, sigmoidal and radial basis functions (RBF) [13].

SVMs classify data into two classes. Because face recognition generally involves classifying data into one of many classes, a variety of approaches have been adopted by the authors of [11] to adapt SVMs to this task.

A SVM has been used as a classifier in a face verification system. It was trained on three 1D vectors for each subject. These 1D vectors were obtained using Gabor filters.

To test the robustness of the multi-class SVMs or face verification, a one-against-one approach was chosen. A Gaussian kernel with bandwidth $\sigma = 1$ was used for the discrimination. The penalty parameter C was set at a value high enough that the learning error remained low (C = 1000).

2.2   Signature Verification System

In our case, we have used the system developed by a Qatar University team to participate in a signature verification competition 2011 [14]. This system has two parts: the first one is for online verification and the second one is for offline verification. We have used the online tool to combine the most discriminant features described in Nalwa's method [7]. In particular, Nalwa's algorithm has three distinct components: normalization, description, and comparison. Normalization makes the algorithm largely independent of the orientation and aspect of a signature; the algorithm is inherently independent of the position and size of a signature. Description generates the characteristic functions of the signature. Comparison computes a net measure of error between the signature characteristics and their prototypes. This net error provides a measure of the discrepancy between the signature being verified and its model; comparison of the net error against a threshold determines whether signature being verified is accepted or rejected [7].

Additionally, this system was trained on each dataset of the competition using a logistic regression classifier. This tool uses the proposed z-calibration method. It was integrated in our proposed system as a black box and used scores obtained on five signatures as references and one more as a questioned signature. The obtained score is fused with the other score obtained from the face verification system.

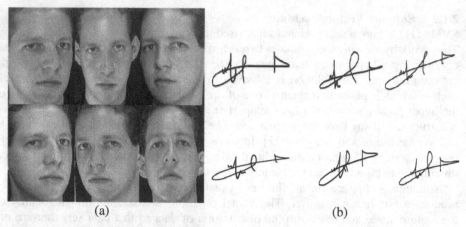

(a)   (b)

Fig. 2. Samples from the developed multimodal database; (a) six face samples for the first subject from the ORL database and (b) six signature samples for the 51st subject from the QU signature database

3   Experiments

3.1   Datasets

To evaluate our proposed fusion method, we have used the ORL face database and the QU signature database. The first database contains ten different images, each of 40 distinct subjects. For some subjects, the images were taken at different times, with varying lighting, facial expressions (e.g., open or closed eyes, smiling or not smiling) and facial details (glasses or no glasses). All the images were taken against a dark homogeneous background with the subjects in an upright, frontal position (with tolerance for some side movement) [15]. The second database used for the signature verification system is available from Qatar University; it contains 138 subjects with three reference signatures each, and some of the subjects have as many as six references.

Considering the number of subjects in the ORL face database, we have selected forty users from the QU-signature database including on-line signature samples. 40 different users from the ORL face database have been used. From both subsets, and by taking advantage of the independence of signature and face traits, we have created 40 virtual subjects who have signature and face traits (see Fig. 2).

The following training and testing process for monomodal systems has been established:

For training purposes, each signature has been modeled using 5 samples, and each face has been modeled using 3 samples.

For the testing, for each client 1 more sample of each trait (face and signature) was also selected for testing; we used the same 40 clients as impostors, except that each client claims an identity different from his own. Each client has been considered and, from each impostor, 1 sample has been selected.

Consequently, the sub-corpus for the experiments consists of 40 clients, and 39x40=1560 multimodal impostor attempts.

3.2   Results and Analysis

Our experiments demonstrate that fusion-based method yields an improvement in the efficiency of the multimodal authentication compared with the monomodal system. Table 2 demonstrates that the lowest equal error rate is obtained by using a fusion strategy based on max-of-monomodal systems scores. Additionally, we have studied the time spent for testing a subject with all existing subjects in the multimodal database; our results demonstrate that the online signature verification system is much faster than the other systems. Conversely, the multimodal system is penalized when the face verification system, which is much slower, is used.

Table 2. Results of the performance evaluation using a PC with a 2.4-GHz Core 2 Duo CPU

System	EER (%)	Testing Time (s)
Face	14	120
Signature	6	19
Sum of scores	6	139
Max of scores	3	139
Min of scores	8	139

Furthermore, as shown in Fig. 3, even when the higher rate of errors produced by the face authentication system compared with the others based on online signatures is considered, it is clear that the on max-of-scores based multimodal system can combine the advantages of both of the monomodal systems to improve the method's overall efficiency.

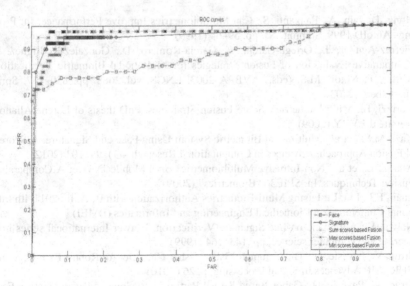

Fig. 3. Plots of ROC curves for each monomodal and multimodal system

4   Conclusions

Here, we explored different fusion techniques, such as the maximum, minimum and sum of scores, for face and online signature data. We devised a bimodal biometric system that merges evidence from both modalities. Our experiments highlight the benefits of using the max-of-scores strategy for face and online signature authentication. In fact, the combination of faces and online signatures based on a max-of-scores analysis outperformed the combination based on the min and sum-of-scores-based-strategies as well as the single modality-based authentication systems such as face and online signature biometric systems. Furthermore, the experimental results reveal an improvement of the monomodal biometric system compared with the best classifiers that use a single modality (online signature). Our future work consists of (i) improving the classifier combination techniques by weighting both classifiers differently, (ii) using a hierarchical fusion method that first requests an online signature because of the online signature method's high verification speed, and then facial verification, if there is a failure in the signature verification system, and (iii) using a novel generation of neural networks (spike neural networks) method as a face classifier to improve the method's performance.

Acknowledgements. This work has been funded by Qatar University (Internal Grant QUUG-CENG-DCS-11/12-7) and by Qatar National Research Fund (QNRF) grant through National Priority Research Program (NPRP) No. 09 - 864 - 1 - 128.

References

1. Hong, L., Jain, A., Pankanti, S.: Can Multibiometrics Improve Performance. In: Proceedings AutoID 1999, Summit, NJ, pp. 59–64 (1999)
2. Fierrez-Aguilar, J., Ortega-Garcia, J., Garcia-Romero, D., Gonzalez-Rodriguez, J.: A Comparative Evaluation of Fusion Strategies for Multimodal Biometric Verification. In: Kittler, J., Nixon, M.S. (eds.) AVBPA 2003. LNCS, vol. 2688, pp. 830–837. Springer, Heidelberg (2003)
3. Dorizzi, B.: Multi-biometrics: Score Fusion Strategies. PhD thesis of Lorene Allano, Université d'EVRY (2009)
4. Kazi, M.M., et al.: Multimodal Biometric System Using Face and Signature: A Score Level Fusion Approach. Advances in Computational Research 4(1), 99–103 (2012)
5. Allano, L., et al.: Non Intrusive Multibiometrics on a Mobile Device: A Comparaison of Fusion Techniques. In: SPIE36 "Biometrics" (2009)
6. Luan, F.J., Lin, L.: Fusing Multi-Biometrics Authorization with PCA. In: 2011 4th International Conference on Biomedical Engineering and Informatics (BMEI)
7. Nalwa, V.: Automatic on-line Signature Verification. Kluwer International series in engineering and computer science, pp. 143–164 (1999)
8. Struc, V., Pavesic, N.: The Complete Gabor-Fisher Classifier for Robust Face Recognition. EURASIP Advances in Signal Processing, p. 26 (2010)
9. Struc, V., Pavesic, N.: Gabor-Based Kernel Partial-Least-Squares Discrimination Features for Face Recognition. Informatica (Vilnius) 20(1), 115–138 (2009)

10. Jain, A.K., Prabhakar, S., Hong, L.: A Multichannel Approach to Fingerprint Classification. IEEE Transactions on Pattern Analysis and Machine Intelligence 21(4), 348–359 (1999)
11. Cortes, C., Vapnik, V.: Support-vector networks. Machine Learning 20(3), 273–297 (1995)
12. Osuna, E., Freund, R., Girosi, F.: Support vector machines: Training and applications. MIT A.I Lab, Tech. Rep. AIM-1602 (1997)
13. Nicholl, P.R.: Face Recognition using Multiresolution Statistical Approaches. PhD Thesis, Queen's University of Belfast (2008)
14. Liwicki, M., Malik, M.I., van den Heuvel, E.C., Chen, X.H., Berger, C., Stoel, R., Blumenstein, M., Found, B.: Signature Verification Competition for Online and Offline Skilled Forgeries (SigComp 2011), pp. 1480–1484 (2011)
15. Samaria, F., Harter, A.: Parameterization of a Stochastic Model for Human Face Identification. In: Proceedings of 2nd IEEE Workshop on Applications of Computer Vision, Sarasota FL (1994)

A Set of Geometrical Features
for Writer Identification

Abdelâali Hassaïne[1], Somaya Al-Maadeed[1], and Ahmed Bouridane[2]

[1] Computer Science and Engineering Department,
College of Engineering, Qatar University,
Doha, Qatar
{hassaine,s_alali}@qu.edu.qa
[2] CEIS, Northumbria University,
Newcastle upon Tyne, UK
ahmed.bouridane@northumbria.ac.uk

Abstract. Writer identification is an important field in the forensic document examination. We propose in this paper a set of geometrical features that makes it possible to characterize writers. They include directions, curvatures and tortuosities. We show how these features can be combined with edge based directional features as well as chain code based features. Evaluation of the method is performed on the IAM handwriting database.

Keywords: Forensic Document Examination, Writer identification, Handwriting curvatures, Handwriting directions, Handwriting tortuosities.

1 Introduction

Automatic writer identification is of a high importance in forensic document examination. Numerous cases over the years have dealt with evidence provided by handwritten documents such as wills and ransom notes [16]. Moreover, writer identification can be used in handwriting recognition when adapting the recognizers to a specific type of writers [13] and also in handwriting synthesis when generating a text as it was written by a specific writer [6].

Automatic methods for writer identification can be classified into two main categories: codebook-based and feature-based approaches.

In codebook-based approaches, the writer is assumed to act as a stochastic generator of graphemes. The probability distribution of grapheme usage is characteristic of each writer and can efficiently be used in order to distinguish between different writers. The methods of this category mainly differ in the way of segmenting the handwriting into graphemes and in the way of clustering these graphemes.

Feature-based approaches compares the handwritings according to some geometrical [1], structural [11] or textural features [5,14]. Feature-based approaches prove to be efficient and are generally preferred when there is a limited amount

T. Huang et al. (Eds.): ICONIP 2012, Part V, LNCS 7667, pp. 584–591, 2012.
© Springer-Verlag Berlin Heidelberg 2012

of available handwriting. Therefore, we are more interested in this paper in this category of approaches and we will give below an overview of such methods.

Srihari et al. propose a set of macro-features which are extracted from document / paragraph / word level and micro-features which are extracted at word / character level. This approach has been validated on a dataset of 1500 writers which have produced the same letter [16].

Said et al. describe a text-independent writer identification method based on Gabor filtering and gray scale co-occurrence matrices. The authors tested their method on a set of 20 different writers [14].

Marti et al. use text line based features for text-independent writer identification [11], the authors tested their method on a subset of 50 writers from the IAM database [10]. This database will also be used in this study.

Bulacu et al. use edge-based directional probability distribution functions (PDFs) as features for text-independent writer identification. The joint PDF of "hinged" edge-angle combinations outperformed all the other evaluated features [3].

Siddiqi and Vincent propose features extracted from chain-codes of the handwriting contour and achieve high identification rate on the IAM database [15].

In this paper, we propose new features based on directions, curvatures and tortuosities. We show that these features can efficiently be combined with edge-based directional features as well as the chain code based features.

The reminder of this paper is organized as follows: section 2 gives details of the method and the proposed features, in section 3, results are discussed. Section 4 concludes this work and draws some perspectives.

2 Method Description

Writer identification consists of two main steps: 1) feature extraction, and 2) feature matching and combination. These two steps are detailed below.

2.1 Feature Extraction

In this step, characterizing features are extracted from the handwriting. In order to have a pen independent system, images are first binarized using Otsu thresholding algorithm. The following features have been considered in this study.
Note that in writer identification, features do not correspond to a single value, but a probability distribution function (PDF) extracted from the handwriting images to characterize writer individuality [7].

Directions (f1) directions are known for well characterizing writers [3]. The method used in this study is very close to the one proposed by Matas et al. [12]: First, we compute the Zhang skeleton of the binarized image. This skeleton is well known for not producing parasitic branches in contrast to most skeletonization algorithms [17]. The skeleton is then segmented at its junction pixels as described in [8]. Then, we move along the pixels of the obtained segments of the skeleton

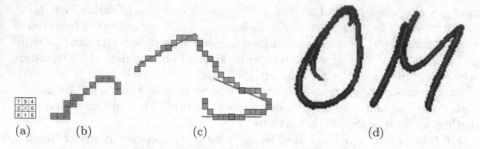

Fig. 1. Computing local directions. (a) The predefined order for traversing shapes. (b) Example of an ordered shape. (c) Estimating directions by computing linear regression of neighboring pixels. (d) Binary image and its corresponding Zhang skeleton: the red color corresponds to a $\frac{\pi}{2}$ tangent and the blue color corresponds to a zero tangent.

using the predefined order favoring the 4-connectivity neighbors as shown in figure 1(a). A result of such an ordering is shown in figure 1(b). For each pixel p, we consider the $2 \cdot N + 1$ neighboring pixels centered at p. The linear regression of these pixels gives a good estimation of the tangent at the pixel p (figure 1(c)). The value of N has empirically been set to 5 pixels. The PDF of the resulting directions is computed as a vector of probabilities for which the size has been empirically set to 10.

Curvatures (f2) it is commonly accepted in forensic document examination to consider the curvature as characterizing feature [9]. In order to estimate curvatures, we use the method introduced in [4] that we adapt to handwriting as follows: for each pixel p belonging to the contour, we consider a neighboring window which size is t. We compute inside this neighboring window, the number of pixels n_1 which belong to the background and the number of pixels n_2 which belong to the foreground (figure 2(a)). Obviously, the difference $n_1 - n_2$ increases with the local curvature of the contour. We estimate therefore the curvature as being $C = \frac{n_1 - n_2}{n_1 + n_2}$, this value is illustrated in figure 2(b) on a binary shape for which t has been empirically set to 5. The PDF of curvatures is computed in a vector which size has been empirically set to 100.

Tortuosity (f3) this feature makes it possible to distinguish between fast writers who produce smooth handwriting and slow writers who produce "tortuous"/twisted handwriting. In order to estimate tortuosity, we find for each pixel p of the text, the longuest line segment which traverse p and is completely included inside the foreground (figure 3(a)). An example of estimated tortuosities is shown in figure 3(b).

The PDF of the angles of the longuest traversing segments are produced in a vector which size has been set to 10.

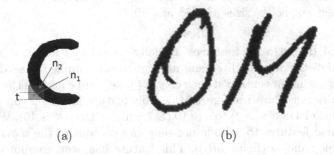

Fig. 2. (a) Computing curvatures. (b) Curvatures highlighted on binary image: red corresponds to the maximum curvature and blue corresponds to the minimum one.

Fig. 3. Computing tortuosity (a) Longuest traversing segment for 4 different pixels. (b) Length of maximum traversing segment: red corresponds to the maximum length, blue to the minimum one.

Chain code features Chain codes are generated by browsing the contour of the text and assigning a number to each pixel according to its location with respect to the previous pixel. Figure 4 shows a contour and its corresponding chain code.

Chain codes have been applied to writer identification in [15], those features makes it possible to characterize detailed distribution of curvatures in the handwriting. Chain code might be applied at different orders:

f4: PDF of i patterns in the chain code list such that $i \in 0, 1, ..., 7$. This PDF has a size of 8.

f5: PDF of (i, j) patterns in the chain code list such that $i, j \in 0, 1, ..., 7$. This PDF has a size of 64.

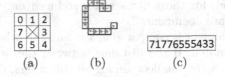

Fig. 4. (a) Order followed to generate chain code. (b) Example shape and, (c) its corresponding chain code.

Similarly **f6** and **f7** correspond to PDF of (i, j, k) and (i, j, k, l) in the chain code list, their respective sizes are 512 and 4096.

Edge based directional features initially introduced in [3], these features give detailed distribution of directions and can also be applied at several sizes by positionning a window centred at each contour pixel, and counting the occurences of each direction as shown in figure 5(a). This feature has been computed from size 1 (**f8** which PDF size is 4) to size 10 (**f17** which PDF size is 40). We have also extended these features to include not only the contour of the moving window but the whole window (figure 5(b)). This feature has been computed from size 2 (**f18** which PDF size is 12) to size 10 (**f26** which PDF size is 220).

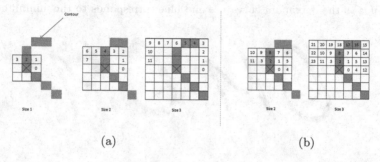

(a)   (b)

Fig. 5. Counting the edge based directional features when considering (a) the contour of the moving window, (b) the whole moving window

2.2 Matching

For comparing a query document q with any given document i, the difference between their features is computed. χ^2 distance is a natural choice for comparing two PDFs: $\chi^2 = \sum_{n=1}^{size} \frac{(F_q(n) - F_i(n))^2}{F_q(n) + F_i(n)}$.
The presented features have different discriminative level which will be shown in the next section.

3 Evaluation

The most widely used database for writer identification is the IAM database [10]. It contains handwritten documents by 657 different writers. Similarly to [3], we only kept the first two documents for the writers who produced more than 2 documents and, for the writers who produced only one document, we split it into two separate documents.

Therefore, each document is compared against all the others (1313) with only one possible correct match. If the distance between the questioned document and the correct match is the smallest among all the others, the document is said to be correctly identified. TOP 10 identification rate considers the matching as correct if the corresponding distance is among the 10 smallest distances. The identification rate of the presented features is given in table 1.

Table 1. Identification rates of the presented features

feature	TOP1	TOP10	feature	TOP1	TOP10
f1	40.56%	75.88%	f14	50.46%	78.31%
f2	36.15%	65.22%	f15	48.71%	77.09%
f3	42.24%	75.27%	f16	49.09%	76.03%
f4	32.19%	66.90%	f17	47.56%	74.66%
f5	59.74%	82.88%	f18	42.85%	71.99%
f6	70.40%	87.75%	f19	54.79%	81.43%
f7	71.46%	87.75%	f20	59.74%	84.32%
f8	16.74%	54.79%	f21	63.47%	85.92%
f9	41.55%	72.30%	f22	64.76%	86.45%
f10	49.32%	78.61%	f23	67.28%	87.21%
f11	52.28%	79.22%	f24	68.04%	88.13%
f12	53.50%	80.67%	f25	68.95%	89.27%
f13	51.29%	78.31%	f26	70.40%	89.57%

In order to combine features, we tried random forest classifier with 100,1000 and 2000 random trees [2]. The training/validation set is made of the third and fourth document of writers who produced at least four document from the IAM dataset. We also tested the sum of several features, the results are shown in table 2.

Table 2. Identification rates of the combination of features

Combination	TOP1	TOP10
Random forest 100	0.729833	0.847032
Random forest 1000	0.740487	0.869102
Random forest 2000	0.740487	0.871385
f1+f2+f3	0.69102	0.861492
f1+...+f26	0.76484	0.893455
Chain code features: f4+f5+f6+f7	0.687976	0.869863
Edge distribution features: f8+...+f17	0.670472	0.869102
Edge distribution features extended: f18+...+f26	0.655251	0.870624

Note that the sum of all features performs the random forest classifier.

4   Conclusion

We presented several new features for writer identification including curvatures, directions and tortuosities. We have shown to what extent these features might be used for writer identification and studied their combination with other state-of-the art features.

Future work includes the validation of the method on other languages as well as for offline signature verification.

Acknowledgments. This publication was made possible by a grant from the Qatar National Research Fund through National Priority Research Program (NPRP) No. 09 - 864 - 1 - 128. Its contents are solely the responsibility of the authors and do not necessarily represent the official views of the Qatar National Research Fund or Qatar University.

References

1. Al-Ma'adeed, S., Mohammed, E., Al Kassis, D.: Writer identification using edge-based directional probability distribution features for arabic words. In: IEEE/ACS International Conference on Computer Systems and Applications, Doha, pp. 582–590 (March 2008)
2. Breiman, L.: Random Forests. Machine Learning 45, 5–32 (2001)
3. Bulacu, M., Schomaker, L.: Text-Independent Writer Identification and Verification Using Textural and Allographic Features. IEEE Transactions on Pattern Analysis and Machine Intelligence 29(4), 701–717 (2007)
4. Enficiaud, R.: Algorithmes multidimensionnels et multispectraux en morphologie mathématique: Approche par méta-programmation. Ph.D. thesis, Ecole des Mines de Paris (2007)
5. Franke, K., Bünnemeyer, O., Sy, T.: Ink Texture Analysis for Writer Identification. In: International Workshop on Frontiers in Handwriting Recognition, Los Alamitos, CA, USA, p. 268 (2002)
6. Franke, K., Schomaker, L., Koppen, M.: Pen force emulating robotic writing device and its application. In: IEEE Workshop on Advanced Robotics and its Social Impacts, pp. 36–46 (2005)
7. Hassaïne, A., Al-Maadeed, S., Alja'am, J., Jaoua, A., Bouridane, A.: The ICDAR 2011 Arabic Writer Identification Contest. In: Proceedings of the Eleventh International Conference on Document Analysis and Recognition, Beijing, China (September 2011)
8. Hassaïne, A., Eglin, V., Bres, S.: Une méthode de compression sans perte pour les images de documents basée sur la séparation en couches. In: COmpression et REprésentation des Signaux Audiovisuels, Montpellier, France (November 2007)
9. Koppenhaver, K.: Forensic Document Examination: Principles and Practice. Humana Press (April 2007)
10. Marti, U.V., Bunke, H.: The IAM-database: an English sentence database for offline handwriting recognition. International Journal on Document Analysis and Recognition 5, 39–46 (2002)
11. Marti, U., Messerli, R., Bunke, H.: Writer Identification Using Text Line Based Features. In: Proceedings of the Sixth International Conference on Document Analysis and Recognition, Seattle, USA, pp. 101–105 (2001)
12. Matas, J., Shao, Z., Kittler, J.: Estimation of Curvature and Tangent Direction by Median Filtered Differencing. In: Braccini, C., Vernazza, G., DeFloriani, L. (eds.) ICIAP 1995. LNCS, vol. 974, pp. 83–88. Springer, Heidelberg (1995)

13. Nosary, A., Heutte, L., Paquet, T., Lecourtier, Y.: Defining writer's invariants to adapt the recognition task. In: Proceedings of the Fifth International Conference on Document Analysis and Recognition, Bangalore, India, pp. 765–768 (1999)
14. Said, H., Tan, T., Baker, K.: Personal Identification Based on Handwriting. Pattern Recognition 33(2), 149–160 (2000)
15. Siddiqi, I., Vincent, N.: Text independent writer recognition using redundant writing patterns with contour-based orientation and curvature features. Pattern Recogn. 43(11), 3853–3865 (2010)
16. Srihari, S., Cha, S.H., Arora, H., Lee, S.: Individuality of handwriting: a validation study. In: Proceedings of the Sixth International Conference on Document Analysis and Recognition, Seattle, USA, pp. 106–109 (2001)
17. Zhang, T.Y.: A fast parallel algorithm for thinning digital patterns. ACM 27(3), 236–239 (1984)

Comparative Analysis of Clustering Algorithms Applied to the Classification of Bugs

Anderson Santana, Jackson Silva, Patrícia Muniz,
Fabricio Araújo, and Renata Maria Cardoso R. de Souza

Fedederal Universitity of Pernambuco, Informatics Center, Recife, Pernambuco, Brazil
{ams8,jrfs,pfm,foa,rmcrs}@cin.ufpe.br

Abstract. This paper presents a study of clustering algorithms in bug classification for a company from a database that contains a description each bug. It is made a comparison these algorithms using a sample of the database of this company. Considering that the classification will encourage the decision process of the organization as the result of the efficiency and reliability increase, this study will conduct an investigation to identify, among the techniques employed, one that will produce satisfactory results for the company, so to provide a set of information that are relevant to strategic decision making.

Keywords: clustering algorithms, bug classification, knn, decision tree, Bayesian learning, Kolmogorov-Smirnov test, Kruskal-Wallis test, t-test.

1 Introduction

In the process of software development, all defects are human and, despite the best use of development methods, the defects remain in the product. Saying that a program has broken means that the program is not in accordance with the expected by the user. Therefore, it is necessary a control of the bugs that are found so that they can be managed from its identification to its correctness and availability.

In many software projects, bug tracking systems plays a central role in the management of defects [2]. This being the reality of many organizations, it would be of great interest to obtain more accurate information about the bugs reported in these databases.

In this paper, in front the need for a company in an automatic way to classify the bugs reported by customers on its systems will be made a comparison of classification algorithms using a sample of the database in which the same are described.

1.1 Overall Goal

This paper aims at providing a comparative analysis of classification algorithms kNN, Bayesian Learning and Decision Trees for classification of bugs through their textual descriptions.

T. Huang et al. (Eds.): ICONIP 2012, Part V, LNCS 7667, pp. 592–598, 2012.

1.2   Specific Goals

- Demonstrate the performance of each algorithm for categorization;
- Statistically compare the results obtained from the same database;
- Point which algorithm fits better in the categorization of bugs to the database presented;
- Check among the categories, the one with the largest amount of problems for the company.

2   Automatic Classification of Text

Text classification is to assign text documents to predefined thematic classes. For example, press reports can be associated with categories of politics, sports, entertainment, or even e-mail can be classified as spam or not spam, and so on [5].

One of the strategies used to assist in the organization and retrieval of documents in digital text format has been the use of automatic text classifiers, with the aim of linking digital documents to predefined classes, according to the content and structure of texts. According to [6], two distinct approaches of Artificial Intelligence (AI) have been adopted to develop automatic classifiers of text: Knowledge Engineering, in which rules are set manually by a knowledge engineer with the help of a specialist in field of application, and Machine Learning, in which a classifier is constructed automatically by a learning algorithm from a training set with pre-sorted documents.

The Machine Learning approach has some advantages in relation to Knowledge Engineering, as it reduces the need for interaction between knowledge engineers and domain experts (expensive and time consuming task), and still achieves good accuracy rate [6]. We emphasize in this context that different families of machine learning algorithms have been used in tasks of classification of text, such as learning algorithms based on analysis, decision trees and Bayesian Learning, which will be explained below.

2.1   k - Nearest Neighbor

A known method of classification is the kNN (k - Nearest Neighbor). It is the most basic method among the instance-based algorithms. It assumes all instances correspond to points in an n-dimensional space [1]. In the most common version of kNN to classify a new instance x, the algorithm assigns to x the most frequent class among the k nearest training instances, in other words, they are less distance x. The neighborhood of an instance is defined in terms of a distance function or a similarity function (such as Euclidean distance, among others).

It is an incremental technique, where the function is only approximated locally and all computation is deferred until classification. He has high computational cost to provide the ratings, does not generate explicit rules and requires definition of parameters (the value of k) for its implementation [5].

2.2   Decision Tree

Decision Tree [4] is a statistical model that uses a supervised training for classification and prediction data, whereby the learned function is represented by a decision tree, that is, can also be understood as a group rules "if-then".

The decision tree reaches its decision by performing a test sequence. Each internal node of the tree corresponds to a test of the value of properties, and the branches of the node are identified with the possible values of the test. Each tree leaf node specifies the return value if the leaf is reached. The classification of an instance starts at the root node of the tree, testing the attribute specified by this node, moving then to the branch corresponding to the attribute value in the given example. This process is then repeated for the new subtree fixed node [3].

Decision Trees using the training data to build a single set of rules that is used to classify all instances, has low computational cost, generates rules and does not require explicit definition of parameters [5].

2.3   Bayesian Learning

Naive Bayes is a classification algorithm based on the application of Bayes' theorem to determine the most likely class for each new instance to be classified. He believes that the attributes that describe instances are completely independent, which is rarely observed in real problems, so it is called naive. However, the algorithm has performed well in complex real-world situations, especially in text classification problems [5].

3   Development

3.1   Methodology

According to [5] the definition of a text classifier learning algorithms can be summarized in six steps: acquisition of documents (1), definition of a vocabulary of terms that will be used to represent these documents (2), defining the set attributes which describe the documents (3) selecting the most relevant attributes (4), induction of the binder from a training set (5) and performance evaluation from a test set (6).

All the above steps will be covered in the following sections.

3.2   Collection, Preprocessing and Definition of Relevant Attributes

The experiments were performed using the real data set from the database of a software development company in Recife-PE-Brazil. These data were obtained from the repository of the bug tracking tool used by the organization, the Mantis, where information about bugs are launched and monitored. Importantly, this foundation was built of course, the day-to-day organization, in the course of testing activities and use of applications by customers.

The sample taken for the experiment contains 713 records, which are in fact all elements of the population. Thus, besides ensuring the execution of the experiment using the algorithms chosen, it is expected to get an overview of the organization with regard to the occurrence of bugs reported.

The data set comprising the sample consists of texts that represent the descriptions of the bugs reported in Mantis. For such a database could be worked into the tool of statistical analysis and mining used - the Weka - data had to pass a pre-processing, which extracted from each record the relevant keywords (determined by an expert) as well as its classification according to categories established by the organization. They are:

- **cad**: a bug is found in a registration system;
- **fnc**: represents a bug found in a function, i.e., a component that is present in several system modules;
- **rel**: a bug is found in a report of the system;
- **utl**: represents a bug found utility in a system, for example, a switch or a filter parameters.

The attributes available for the research are: *id* (a unique numeric identifier for each bug), three keywords (identified by *kw1*, *kw2* and *kw3*) and the category that corresponds to one of four options explained in the previous paragraph. However, because it is semantically irrelevant information, the attribute "id" was excluded from analysis.

3.3   Algorithms Application

The same data were used for the implementation of the algorithms kNN, decision tree and NaiveBayes. In all we used cross-validation with 50 folds (or subsets), while it own tool to define subsets of training and testing. The specific parameters for each algorithm are varied so that it was also possible to include in these different evaluation results for the same algorithm.

Knn. In the execution of kNN, with k ranging from 1 to 26, the lowest error rate was found for k = 1, whose value was 20.48%. The confusion matrix generated by Weka shows that 567 instances were classified correctly, as outlined in the table.

Table 1. Confusion Matrix for Knn

	cad	fnc	rel	utl
cad	152	5	10	9
fnc	19	60	26	13
rel	10	5	263	2
utl	27	12	8	92

Decision Tree. In running with the Decision Tree, the experiment was performed in two ways: no pruning (i.e., all possible alternatives would be evaluated by the algorithm) or pruned (some of these alternatives would be disregarded in the analysis, according to the criteria of the tool itself). Based on the implementation it was observed that the best case was achieved without the use of pruning whose error rate was 24.54%. The confusion matrix generated by Weka shows that 538 instances were classified correctly, as outlined in the table.

Table 2. Confusion Matrix for Decision Tree

	cad	fnc	rel	utl
cad	123	12	23	18
fnc	25	64	11	18
rel	5	3	269	3
utl	28	18	11	82

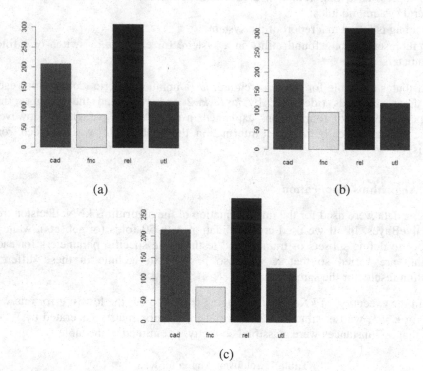

(a)   (b)

(c)

Fig. 1. Comparison between the algorithms knn (a), decision tree (b) and bayesian learning (c) according to the classification for the clusters cad, fnc, rel and utl (obtained by using the R statistical tool)

Naive Bayes. In one implementation of the Naive Bayes (since it requires no parameter that can vary its results), the percentage error achieved was 20.6%. The confusion matrix generated by Weka shows that 570 instances were classified correctly, as outlined in the table.

Figure 1 shows a comparative graph with classification numbers held by each of the algorithms.

Table 3. Confusion Matrix for Baysian Learning

	cad	fnc	rel	utl
cad	155	4	7	10
fnc	24	62	14	18
rel	9	6	260	5
utl	27	10	9	93

4   Analysis of Results

For each execution of the algorithms noted a set of values containing the number of errors of each fold. Once this amount is known, it was possible to calculate the error rate by folding. These values were included in the sample, with size 50, the rate of errors by the algorithm. From then on it was necessary to carry out an adhesion test to clarify whether these samples follow the same distribution, to be subsequently performed a hypothesis test, whose purpose is to define, with statistical reasoning, which of the three algorithms used has a lower rate errors.

4.1   Adhesion Test

We used the Kolmogorov-Smirnov test to detect if the three samples follow a normal distribution. Considering a significance level of 0.05, the p-value calculated for each algorithm is larger than this, we will not have to reject the null hypothesis, which states that the algorithm in question follows a normal distribution.

By applying these samples in R resulted the following values for the p-value: 0.3937 for knn, p-value is 0.424 for the decision tree, and 0.026 for Naive Bayes.

4.2   Hypothesis Test

In order to ensure that the error rate of the three algorithms is not equal, we applied the Kruskal-Wallis test. Considering a significance level of 0.05, the p-value provided from the tool R for the data in question resulted in 0.02, and thus cannot confirm that the error rate of the three algorithms is the same. At this point, it could be affirmed that at least one of the algorithms has an error rate lower than the error rate of the rest. To find out what, or which of these algorithms is responsible for rejecting the null hypothesis of the Kruskal-Wallis test was necessary to run three t-tests, where were found the values of p-value shown in the table below.

Table 4. p-value for t-test comparision between the algorithms

	knn x decision tree	knn x Bayes	Bayes x decision tree
p-value	0.007013	0.2297	0.009732

From the results, note that both knn and NaiveBayes have different error rates with respect to the decision tree. And as for the hypothesis testing performed and knn NaiveBayes reject the null hypothesis, or present as error rate equivalent. It is worth noting that during the test of adhesion to the error rate distribution algorithm NaiveBayes was given as a non-normal. To utilize the error rate of the t-test NaiveBayes had normalization of forcing the sample with the R function *rnorm* .

5   Conclusion

Perform automatic classification of texts is not a trivial task. However, the use of algorithms classification categories can provide support it, thus eliminating the arduous task of developing a purely manual.

This study conducted an analysis of algorithms kNN, Naive Bayes and Decision Tree using as a base set of data from a software company in the city of Recife. Through the experiments it was possible to observe the behavior of each of the algorithms for the categorization of classification of bugs through their textual descriptions, showing: best case situations, the rate of positivity in each situation developed. With respect to statistical analysis, it was possible to assess the feasibility raised on the efficiency of the algorithms applied at the base in question.

Two important fact is noted that part of the class of reports **rel** most bugs classified and that while the algorithm NaiveBayes have had a better performance, ie a lower error rate, hypothesis testing showed that the this error rate and the algorithm knn are equivalent and that they would be good classification algorithms in this case. This latter fact surprised the people involved in this work, the problem is present in the beginning, apparently easier to solve with decision tree.

References

1. Aha, D., Kibler, D.: Instance-based learning algorithms. Machine Learning 6(3), 37–66 (1991)
2. Breu, S., Premraj, R., Sillito, J., Zimmermann, T.: Information needs in bug reports: improving cooperation between developers and users. In: CSCW 2010, pp. 301–310 (2010)
3. Mitchell, T.: Machine Learning. McGraw Hill (1997)
4. Quinlan, J.R.: Programs for Machine Learning. Morgan Kaufmann, San Francisco (1993)
5. Rodrigues Pereira, J.: Sistemas Inteligentes Híbridos para Classificação de Texto. Dissertação (Mestrado em Ciência da Computação) - Centro de Informática, p. 119. Universidade Federal de Pernambuco, Pernambuco (2009)
6. Sebastiani, F.: Machine learning in automated text categorization. ACM Computing Surveys 34(1), 1–47 (2002)

DNS-Based Defense against IP Spoofing Attacks

Eimatsu Moriyama[1], Takeshi Takahashi[1], and Daisuke Miyamoto[2]

[1] National Institute of Information and Communications Technology, Tokyo Japan
{moriyama,takeshi_takahashi}@nict.go.jp
[2] University of Tokyo, Tokyo Japan
daisu-mi@nc.u-tokyo.ac.jp

Abstract. Many attacks on the Internet spoof the source IP addresses. Numerous techniques have been researched and developed thus far to cope with this, but they are not yet sufficient. This paper proposes a Domain Name System-based technique for handling the issue. An attacker needs the IP address of an application server, the target of attack, to access there. To obtain the address, the attacker queries the DNS full-service resolver to resolve the server's fully qualified domain name. While the attacker is inquiring about the address, it cannot spoof its address in the proposed scheme. The proposed scheme informs the application server-side gateway of the client's address, with which the gateway can ignore access by those other than the informed address.

Keywords: spoofing address; DNS log; SFP; TCP fallback.

1 Introduction

A denial-of-service (DoS) attack is an attempt to make resources unavailable to the expected users. To carry out the attack, the attacker's host sends a large amount of packets to the targeted victim destination server to deny the victim(s) access to a particular resource. Since the attacker makes heavy use of network resources during the attack, it is not only a serious threat to network service providers, but also a threat to application service providers. A distributed denial-of-service (DDoS) attack is one in which multiple hosts attack a single targeted victim server. DDoS attacks are often observed today on backbone networks and measures to counter such attacks at an appropriate cost remain a challenge. Moreover, on an Internet protocol, it is well known that one's IP address can easily be faked. It is extremely difficult to recognize the correct address from the spoofed one; that is, an attacker can easily hide its real address. This makes defending much more difficult for the victim side.

Moreover, an SYN flooding attack [3] with a spoofed source IP address of an attacker, a typical DDoS attack using a spoofed IP address, is a difficult situation to handle. The attack takes advantage of the state retention that Transmission Control Protocol (TCP) performs for some time after receiving a TCP SYN segment in a TCP port that has been put into the LISTEN state. The basic idea is to exploit this behavior by causing the host to retain enough state for

T. Huang et al. (Eds.): ICONIP 2012, Part V, LNCS 7667, pp. 599–609, 2012.

bogus half-connections that there are no resources left to establish new legitimate connections. The attacker creates a random source address for each TCP segment, and sets the TCP SYN flag in each segment as a request to the server to open a new connection from the spoofed IP address. When a victim server receives TCP SYN segments, it builds a Transmission Control Block (TCB) on the memory to store all the state information for an individual TCP connection. The victim server responds with the TCP SYN+ACK segment to the spoofed IP address, and then waits for ACK segments from the attacker host for connection confirmation. Since the attacker never sends back the ACK segment and continues to send a large number of SYN segments to the victim server, the TCB will eventually fill up. Once the TCB fills up, any new connections to the victim are denied. Access by legitimate users is also denied.

This paper proposes a technique that distinguishes the IP address of an incoming packet in terms of whether or not the IP address is spoofed. We paid attention to the fact that, before a host accesses the targeted server, the host requests the IP address of the object server from the Domain Name System (DNS) server. A host cannot spoof its address while it asks the DNS for the IP address of the object server. This is because if the host spoofs its address it cannot receive a response packet bearing the IP address of the object server from the DNS server. By informing the application server-side gateway of an address of the client that has been requested from the full-service resolver, the gateway can ignore access of those other than the informed address.

2   Related Works

Assorted countermeasures against DDoS and IP address spoofing attacks are researched until now.

2.1   Against IP Address Spoofing Attacks

Ingress filtering [2] inspects the source IP addresses of incoming datagrams at the edge routers and filters those with the source IP addresses that are not known to be reachable via the router interface, whereas egress filtering inspects the source IP addresses of outgoing datagrams at egress router and filters those with the source IP addresses that are not known to be allowed to leave. This filtering is, however, useless if a malicious host is spoofing its address that is within the range of the subnet address for preventing ingress filtering, within the range of network for preventing egress filtering.

Unicast Reverse Path Forwarding (RPF) [4], under its strict mode, filters packets as well; packets must be received on the interface that the router would use to forward the return packets. An inadequate side effect, however, arises from this mode in that Unicast RPF configured in strict mode may drop legitimate traffic that is received on an interface that was not the router's forwarding information base for sending return traffic.

IP traceback [6] finds the source of a packet without a help of a source address described in the datagram, but it is yet to be improved further so that it could be deployed over the Internet.

2.2 Against SYN Flooding Attack

SYN cache [3] system stores the hash of the data for TCP connection on the TCB to reduce its size. This reduction is effective for raising the immunity of the server against spoofed attacks, but the size is not reduced to zero. The effect of the SYN cache against an address-spoofed attack is thus limited.

SYN cookies [3] technique embeds the data as the initial sequence number of a TCP SYN+ACK segment instead of storing its hash on the TCB and then sends that back to the attacker's host. This technique causes absolutely zero state and no additional TCB is to be generated by a received SYN. However, it experiences several drawbacks. Since an ACK segment that echoes this sequence number plus one for a legitimate connection will be received, the basic TCB data can be regenerated and a full TCB can safely be instantiated by decompressing the Acknowledgment field. Some TCP options required for high performance might be disabled since not all TCB data can fit into the 32-bit Sequence Number field. Another problem is that TCP SYN+ACKs are not retransmitted since retransmission would require state. The server then has no means of detecting the drop of an ACK segment transmitted from a legitimate client host to the server for some reason. For a legitimate client, as the client sends the ACK segment, a connection seems to be accomplished. This situation may lead to a resource leak on a legitimate client. Activation of the SYN cookie at all times is therefore not recommended on the victim server.

2.3 Against e-mail Spoofing

E-mail on the Internet can be forged in a number of ways. Among them is the Sender Policy Framework (SPF) protocol [7], whereby a domain may explicitly authorize the hosts that are allowed to use its domain name, and a receiving host may check such authorization. An SPF record is a TXT record that is part of a domain's DNS zone file stored on an authoritative name server of mail sender domain. The TXT record specifies the list of authorized host names/IP addresses that mail can originate from for a given domain name. A mail receiver verifies that the IP address of the mail sender is matched to the address described on the SPF record of authoritative name server of sender's domain.

The proposed scheme applies SPF's concept to DNS full-service resolver.

3 Proposed Scheme

This section introduces the proposed scheme, including its motivation, design principle, and procedure.

3.1 Motivations

Many of techniques to cope with address spoofing inspect addresses when the mail or packet arrives at the victim. In the inspection, many procedures are needed on the victim server to distinguish whether the address is spoofed. Once a DoS attack begins, the targeted victim server receives a large number of packets whose IP addresses are spoofed and diversified. The targeted victim server cannot cope with the attack because the inspecting operation requires a large number of procedures. Therefore these cannot cope with address-spoofed DoS attacks as long as they inspect the packet after receiving it. For instance, IPSec provides security function by using encryption technology, but its ESPiEncapsulating Security Payload) mode with any transform is unable to handle source address spoofing [1]. Moreover, a receiver cannot notice the address was spoofed before they decrypt the payload. An attacker can exploit this weak point to DoS attack. Unlike above techniques, a method that judges whether arriving packets have spoofed their source IP address before they reach the victim server is needed.

3.2 Design Principle

The proposed scheme has the following three functions.

The proposed scheme informs the victim side, i.e., gateway, of this client's address correctly and protects application server from source address spoofing. We assume that a legitimate client queries the DNS full-service resolver about the address of the server before accessing the server. Based on this assumption, the DNS full-service resolver informs the gateway of the client's address only when the address is not spoofed, and the gateway registers the address and then allows only packets having an address registered with the list to be passed.

The proposed scheme implements a scheme that copes with source IP address spoofing, such as TCP fallback. TCP fallback enables DNS to use TCP, which prevents source IP address spoofing with its three-way handshake procedure, while UDP allows such spoofing. It enables DNS to use TCP instead of UDP when the DNS message size is above 512 bytes since UDP cannot accommodate the record on a single segment. Note that EDNS0 should not be activated to enable TCP fallback.

The resolver collaborates with authoritative servers in the proposed scheme to prevent a malicious client from spoofing a full-service resolver and sending faked client IP address to the server side. The resolver is given a legitimate right to rewrite the record of the authoritative server of the domain beforehand. As with the SPF scheme, which authenticates a mail from the sender-side mail server at the authoritative DNS on the mail-sender side, the proposed scheme authenticates the resolver at the authoritative DNS on the client side following the above procedure.

3.3 Procedures

Figure 1 describes the steps of the proposed scheme, which is elaborated as below.

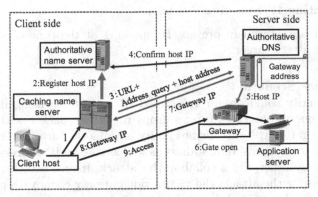

Fig. 1. Overview of the proposed approach

1. A client host asks the full-service resolver on the client side about the address of the gateway that was installed in the preceding section of the targeted victim application server. A scheme that copes with source IP address spoofing, such as TCP fallback, needs to be used in this step.
2. The resolver registers the IP addresses of itself and the client host with the authoritative name server on the client side.
3. The resolver reports the authoritative name server on the victim side of the IP address of the client host, fully qualified domain name (FQDN) of the authoritative name server on the client side and asks the authoritative name server on the victim side about the IP address of the gateway connected to the victim's server.
4. The authoritative name server on the victim side queries the one on the client side for the client host's IP address to confirm whether the query from the resolver is authorized. Since the resolver reports beforehand the IP address of the authoritative name server on the victim side, the one on the client side could drop packets except those from the one on the victim side.
5. The authoritative name server on the victim side notifies the gateway of the client host's IP address when the above items 3 and 4 are matched.
6. The gateway opens its gate to the reported IP address.
7. The authoritative name server on the victim side notifies the resolver of the gateway's IP address.
8. The resolver responds to the client host about the gateway's IP address.
9. The client host accesses the application server via the gateway.

With this protocol, the proposed scheme copes with IP spoofing attacks.

4   Analysis on the Proposed Protocol

This section provides initial analysis on the proposed scheme from the standpoint of protocol verification. Note that these discussions are based on the completeness of these servers and assume that an attacker compromises neither the DNS full-service resolver nor the authoritative DNS.

4.1   Preventing IP Spoofing Attacks

The proposed procedure can prevent IP spoofing at the time of obtaining IP address of the attack target.

Simple Attack Scenario

A malicious entity could obtain a targeted IP address by using its authentic address in a query, but the gateway drops its attack packet with a spoofed address at the gateway. If a malicious entity obtains the targeted IP address in the above way, it can reveal that address as well as its own IP address to attacker groups intending to perform a collaborative attack. If such attacker groups use these addresses for the attack, the victim-side gateway cannot drop the packet since the source address is already authenticated. In such a case, per-IP traffic shaping is preferable. Per-IP traffic shaping enables limiting of traffic to prevent one user from using all of the available bandwidth; it now is shared within a group equally.

Note that these discussions are based on the completeness of these servers and assume that an attacker compromises neither the DNS full-service resolver nor the authoritative DNS.

Complex Attack Scenario

A malicious entity could query the server without spoofing its source IP address and obtain the gateway information, with which it could attack the application server with a spoofed source IP address, but the gateway would block this attack because the spoofed source IP address is not registered on the gateway. If a malicious client using a spoofed IP address queries the DNS server to resolve a server name with an IP address, the DNS server returns the server's IP address to the spoofed address; thus, the malicious client cannot obtain the server's IP address. The malicious entity cannot obtain the gateway IP address, but could register the spoofed address on the gateway by a query using the spoofed source IP. This means there is a possibility of the following three stages of attack.

The first stage is the preliminary stage of the attack to obtain the target IP. To enable this, the malicious client does not spoof its address on the query. The second is the query from the malicious client by using a spoofed IP address to register the spoofed IP address on the gateway. The third is the real attack using the spoofed address that is registered in the second stage.

To avoid such an attack, the proposed protocol should prevent address spoofing by using the aforementioned TCP fallback or appropriate means while transaction between the full-service resolver and the client is carried on as a DNS query and response. This spoofing protection can be enabled by implementing more signaling than what is conventionally used between entities.

4.2   Protocol Specific Features

As described in section 3, to realize this system, collaboration between both the side of the DNS full-service resolver and the authoritative DNS is needed. This means that to introduce this system, modification is needed only for the full-service resolver and authoritative DNS for both sides. Moreover, for the client side, if the full-service resolver controls timeouts (e.g., as considered in Section 5.1) of the DNS record of the authoritative DNS server, the authoritative DNS server does not need to be modified. Conversely, once the client-side DNS full-service resolver obtains the IP address of the victim-side authoritative DNS, neither the root DNS nor TLD DNS are needed during this authentication procedure. This means that to introduce this, no modification is needed for the DNS server except for the full-service resolver on the client side and authoritative DNS on the victim side. This means that we never propose any modification of DNS servers, except for the client side name servers and the server side one.

For client-side full-service resolver modification, when the resolver issues a query to the server-side authoritative DNS server, it must send the FQDN of the authoritative DNS and host address. The full-service resolver must register the IP addresses of the client host and the full-service resolver with the authoritative name server. It must control timeouts of the DNS record of the authoritative DNS server. TCP fallback must be used to avoid spoofing.

For server-side authoritative DNS server modification, the DNS server must receive the FQDN and host address from the full-service resolver on the client side. After that, it issues a query to the authoritative DNS on the client side. If the reply from the authoritative DNS on the client side is consistent with the query on the client side full-service resolver, the authoritative DNS on the server side tells the gateway the client address, and responds to the full-service resolver of the gateway's address.

4.3 Security Consideration

The proposed scheme could suffer from various attacks. If the full-service resolver on the client side is a target of attacks, the attacker cannot resolve the name of the victim server, thus the victim server is free from attacks. If the authoritative name server on the client side is a target of attacks, the victim side cannot obtain the IP address of clients; so, all the packets sent from the client (attacker) are discarded, and the application server is kept safe. If the authoritative name server on the victim side is a target of attacks and a denial of service occurs, we have to consider the possibility of a legitimate client not being able to access the application server. No client can resolve the IP address from the FQDN of the application server in such a case. The attacker can attack the authoritative DNS server on the victim side, whether or not the proposed protocol is installed. Conversely, if the attacker can compromise the authoritative DNS on the victim side, all application services of the referring DNS are stopped. Therefore the DNS must be perfectly protected in all cases. If there is a case of targeting, the authoritative name server needs proper countermeasures, such as installation of DoS-resilient appliances. Using the proposed protocol, the attacker can attack the DNS, but its attacks on other servers (e.g., mail server or web server) are rendered ineffective by using the gateway. This enables easier protection of victim-side resources.

Even though the gateway will drop the attack packet, the attacker can fill up the access line to the gateway. This creates denial of service for legitimate clients. This is so-called bandwidth attack. To countermeasure this approach, the gateway should be installed at the backbone of a broadband network. Using such a configuration can improve the sustainability of services against network bandwidth DoS attacks with IP spoofing.

5   Consideration on DNS Operations

This paper introduced our initial research and proposed a scheme to cope with DDoS attacks. This prevents some spoofed DoS attacks, but suffers from various issues and must adhere to various assumptions. These are considered and discussed in this section.

5.1   Caching

As a DNS caching function, the name server performs various types of caching, but the caching function should be carefully made, if needed, for this approach.

By caching the IP address of the client based on the proposed protocol, the resolver can distinguish whether a client has up until this point queried the resolver about the IP address of the server. By using this cached client IP address, the resolver does not reply to a client about the cache if it was cached based on the query from the other client. This means that the cache is not diverted to the other client. In such a case, the full-service resolver starts the signaling procedure from step 2 in Section 3.

Cache data have their own TTL (time to live) value. If the TTL of the client does not match that of the gateway, the gateway may refuse access from the client. To avoid this, the TTL needs to be synchronized among the gateway, full-service resolver, server-side authoritative name server, and client.

Having a cache function on a client is preferable for a network, but some clients have no such system; e.g., Linux client. Such clients repeatedly issue a DNS query when needed. In these cases, the full-service resolver does not make further requests to the authoritative name server if the TTL of the cache has not expired. In reply to the client's query about the server's IP address, the full-service resolver responds with the cached data. This reduces useless traffic on the network. For TTL, by setting TTL long, it is preferable to reduce traffic between the full-service resolver and authoritative name server. On the other hand, long TTL is not adequate if the address is subject to dynamically changing, especially to avoid attacks. TTL is determined by considering these trade-off conditions.

5.2   DNS Record

Another issue is the size of the DNS record. The number of records inside the client-side authoritative DNS server increases following the number of clients issuing DNS queries. Since the record on the client-side authoritative DNS is prepared so as to be read by the victim-side authoritative DNS (step 4 of the signaling procedure described in Section 3), and once this takes place there is no longer need for it. To minimize the number of records, we can delete old records after they time out, or delete the old record after the client-side authoritative DNS server replies to the victim-side authoritative one. Apart from this deleting policy, the cache can be deleted after the full-service resolver receives the gateway IP address. More meticulous description design on the records needs to be considered with regard to the DNS record.

5.3 Anycast Support

Anycast is often used by the DNS, and the same IP address of the naming servers may identify differing naming servers. That may cause inconsistency of information registered inside the naming servers, and a mechanism to align the related information inside the naming servers is needed. To enable this we must consider a mechanism for both the full-service resolver and the authoritative name server of the DNS.

The inconsistency can be resolved by having multiple IP addresses of each full-service resolver for information transaction between entities. That is, of these addresses, the first are the same as each other for all full-service resolvers to receive the query from a client. This is called anycast. The other addresses of the full-service resolvers are different from one another so that the authoritative name server can distinguish each full-service resolver. Inconsistency can be resolved by communicating using different address among each name server.

Similar to the full-service resolver, the first IP address of the authoritative name server is used for reception of queries from the full-service resolver with anycast. Since the authoritative name server on the client-side and server-side gateway should distinguish each of the authoritative name servers on the server side, the other addresses of the authoritative name server are different from one another. They may in fact use the same IP address for all transactions by using different port numbers, but this is not preferable for security reasons since it may leave room for attack.

5.4 Subsidiary DNS Service

The client cannot access the server protected by this proposed system as long as the client-side default DNS full-service resolver and authoritative name server are conventional ones. It is preferable to introduce the proposed system as a new DNS service like as OpenDNS supplied by some kind of service provider to enable this research to be widely utilized.

Some network managers may prohibit clients from using an auxiliary DNS server operating outside their managed domain regardless of their intention. We expect that a client will start the TCP fallback sequence if a packet with a TC (truncated) bit flag set is received from the name server.

However, TCP fallback cannot be activated if TCP port 53 has been closed, or EDNS0 is activated instead of TCP fallback. In all such cases the proposed system is useless. To solve this difficulty, installation of a new stub resolver is essential for the client. Such a resolver may use TCP port 80 or 443 instead of port 53 to pass through the client side firewall or proxy. The client also needs to be notified about how to access the protected server if using this proposal.

5.5   Redundancy and Bandwidth Consumption

There are two types of authoritative name servers: primary and secondary. By using DNS NOTIFY transaction, a zone file modification on primary server of client side can be notified immediately to the secondary one. According to the passive monitor data measured with packet/s obtained by the CAIDA [5], DNS traffic data is about 100 times smaller than that of HTTP. Even if DNS traffic increases 10 times due to introduction of proposed system, the traffic is still about 10 times smaller than that of HTTP.

5.6   Other Considerations

The proposed scheme requires server side to implement several protocol specific features. To minimize the scalability hurdle of the server side, cookies could be used. Cookies releases servers from the burden of storing communication partner's state information, thus they are not required to keep any state information during the steps from 3 to 7 of the proposed scheme. This property is preferable to keep scalability of the system.

An application server may be co-operated by multiple servers to maintain system's scalability and availability. In this case, the authoritative name server may select the gateway of one of the servers based on certain criteria. The criteria could be server's average load, which is advertised by the servers intermittently, or it could be the number of hops to the destination, for instance.

If client access the Internet through the Network Address Translation (NAT), and if client address is not predictable beforehand, full-service resolver reports all candidate address of the client may given; e.g. 192.168.1.4/30; to the authoritative server on the victim side.

6   Conclusion and Future Works

This paper proposed a scheme to defend against IP-spoofed attacks by collaborating with the DNS. As with SPF and its successful deployment, we expect to develop this scheme to authenticate a unique client IP against an address-spoofed client. Indeed numerous modifications are needed to introduce this, but we believe that our consideration will provide some viewpoints for enhancing security by an aid of DNS infrastructure. Despite our careful consideration of this approach, it still contains issues that need to be considered, so the proposed scheme needs to be developed further to accommodate a practical environment. The efficiency and effectiveness of the proposed scheme will also be evaluated in our future work.

References

1. Aboba, B., Dixon, W.: IPsec-Network Address Translation (NAT) Compatibility Requirements. RFC 3715, Informational (March 2004), http://www.ietf.org/rfc/rfc3715.txt
2. Baker, F., Savola, P.: Ingress Filtering for Multihomed Networks. RFC 3704 (Best Current Practice) (March 2004), http://www.ietf.org/rfc/rfc3704.txt
3. Eddy, W.: TCP SYN Flooding Attacks and Common Mitigations. RFC 4987, Informational (August 2007), http://www.ietf.org/rfc/rfc4987.txt
4. Kumari, W., McPherson, D.: Remote Triggered Black Hole Filtering with Unicast Reverse Path Forwarding (uRPF). RFC 5635, Informational (August 2009), http://www.ietf.org/rfc/rfc5635.txt
5. Passive Network Monitors (August 2012), http://www.caida.org/data/realtime/passive/?monitor=equinix-chicago-dirA
6. Takahashi, T., Hazeyama, H., Miyamoto, D., Kadobayashi, Y.: Taxonomical approach to the deployment of traceback mechanisms. In: 2011 Baltic Congress on Future Internet Communications (BCFIC Riga), pp. 13–20 (February 2011)
7. Wong, M., Schlitt, W.: Sender Policy Framework (SPF) for Authorizing Use of Domains in E-Mail, Version 1. RFC 4408 (Experimental) updated by RFC 6652 (April 2006), http://www.ietf.org/rfc/rfc4408.txt

MOTIF-RE: Motif-Based Hypernym/Hyponym Relation Extraction from Wikipedia Links

Bifan Wei[1], Jun Liu[1], Jian Ma[1], Qinghua Zheng[1], Wei Zhang[2], and Boqin Feng[1]

[1] SPKLSTN Lab, Department of Computer Science, Xi'an Jiaotong University, Xi'an, China
[2] Amazon.com Inc, Seattle, WA 98109, USA
{Weibifan,xjtumajian}@gmail.com, {liukeen,qhzheng,
bqfeng}@mail.xjtu.edu.cn, wzhan@amazon.com

Abstract. Hypernym/hyponym relation extraction plays an essential role in taxonomy learning. The conventional methods based on lexico-syntactic patterns or machine learning usually make use of content-related features. In this paper, we find that the proportions of hyperlinks with different semantic type vary markedly in different network motifs. Based on this observation, we propose MOTIF-RE, an algorithm of extracting hypernym/hyponym relation from Wikipedia hyperlinks. The extraction process consists of three steps: 1) Build a directed graph from a set of domain-specific Wikipedia articles. 2) Count the occurrences of hyperlinks in every three-node network motif and create a feature vector for every hyperlink. 3) Train a classifier to identify semantic relation of hyperlinks. We created three domain-specific Wikipedia article sets to test MOTIF-RE. Experiments on individual dataset show that MOTIF-RE outperforms the baseline algorithm by about 30% in terms of F1-measure. Cross-domain experimental results show similar, which proves that MOTIF-RE has fairly good domain adaptation ability.

Keywords: hypernym/hyponym relation, Wikipedia link, network motif.

1 Introduction

Hypernym/hyponym relation is a type of fundamental semantic relation, which can be used to connect diverse concepts forming a semantic taxonomy[1]. Hypernym/hyponym relation extraction plays a crucial role in web information extraction, formal ontology learning, knowledge base construction, question answering, text entailment and other knowledge-rich problems. How to automatically extracting hypernym/hyponym relation with high quality from natural language text is a challenging issue.

In this paper, we focus on hypernym/hyponym relation extraction from Wikipedia links. Wikipedia can be viewed as a directed graph, with the articles being the nodes and the hyperlinks being the edges (directed links). By analyzing the directed graph formed by the hyperlinks in a set of domain-specific articles, we found that the graph contains a lot of significant recurring connectivity patterns, which is named as network motifs. Network motifs are connectivity-patterns that appear much more often

T. Huang et al. (Eds.): ICONIP 2012, Part V, LNCS 7667, pp. 610–619, 2012.

in a specific network than they do in randomized networks [2]. For example, the right hand side of Fig.1 is an instance of a motif showing on the left hand side, in which the arrows are the directed links between Wikipedia articles. In Fig.1, *Hyp(A,B)* indicates that A is superordinate word of B, while *Other(A,B)* indicates that A and B does not have a hypernym/hyponym relation.

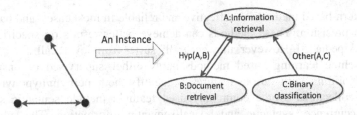

Fig. 1. Example of a network motif and its instance

After all instances of a motif have been enumerated for counting the different hyperlinks tagged as *hypernym/hyponym* or *other*, we found the proportion of hypernym/hyponym relation hyperlinks to all the hyperlinks in a motif varies in different types of motifs, and the hypernym/hyponym relation hyperlinks are more likely to appear in some types of motifs than in others.

The generalization of this observation is that motifs in which two Wikipedia articles appear can be used as indicators for discriminating different semantic relations of hyperlinks. That means that a hyperlink in different network motifs is tagged as hypernym/hyponym relation with different probability.

Moreover, after carefully analyzing different domains, we found that for a specific network motif, the proportion of hypernym/hyponym relation in each motif is similar cross different domain. It is possible the annotated hyperlinks from one domain can be used for learning the semantic relations of hyperlinks in other domains.

Based on these, we developed a novel extraction method called MOTIF-RE (MOTIF-based hypernym/hyponym Relation Extraction) to extract hypernym/hyponym relation from Wikipedia links. MOTIF-RE only relies on the hyperlink structure information for hypernym/hyponym relation extraction, without the information based on text content.

The rest of the paper is organized as follows. Section 2 discusses related work. Section 3 shows some three-node motifs and key statistics. Section 4 presents a novel method MOTIF-RE for extracting hypernym/hyponym relation from Wikipedia links and Section 5 discusses the extraction performance and the evaluation of the results. Finally, the conclusions and future work are presented in Section 6.

2   Related Work

Hypernym/hyponym relation extraction from text corpus has long been studied by the taxonomy learning, statistical relational learning, and ontology learning communities. Pattern-based and machine learning-based extraction methods are the two popular methods in the literatures.

The pattern-based methods usually employ lexico-syntactic patterns, either handcrafted or automatically induced from training data, to identify hyponymy. Hearst's research [3] took a great role in this subfield. She showed six basic patterns for hyponymy extraction and some rules for acquiring new patterns. Based on Hearst's seminal work, the researchers extended lexico-syntactic patterns in different aspects[4].

The pattern-based methods are effective and reliable in most cases, and have a relatively high precision. These methods can achieve better results on specific datasets such as Wikipedia [5]. However, they generally suffer from low recall.

The machine learning based methods apply either supervised or unsupervised learning algorithms with various features, to identify the hypernym/hyponym relation from text collections. Some commonly used features include semantic similarity, word co-occurrence, syntactic and lexical-semantic information. The subsumption algorithm [6] proposed by Sanderson and Croft can automatically derive hypernym/hyponym relation from a set of documents using term co-occurrence. Hasegawa et al. [7] exploited the contexts of named entity pairs for discovering relations by means of vector space model and cosine similarity based on complete-linkage clustering. Navigli and Velardi [8] applied the Word-Class Lattices, constructed from Wikipedia by means of sentence clustering, to the task of hypernym extraction.

The unsupervised learning-based methods do not need training data and hence can easily be applied to a wide range of fields; while the supervised learning-based methods depend on carefully selected training set which is harder to acquire. The performance of machine learning-based methods vary with algorithms and datasets, and the precision is low compared with pattern-based methods.

All these methods mainly rely on the features from the content of text for hypernym/hyponym relation extraction without employing topological structure of text in corpus.

3   Motifs Analysis Of Hypernym/Hyponym Relation Links

A lot of previous work has been done on studying the properties of the Wikipedia network. However, to our knowledge, all previous studies focused on the global statistical features of the whole Wikipedia, no attempts have been made to characterize the local connectivity-patterns of a domain-specific article set, and no works have been carried out for utilizing network motifs as main features for hypernym/hyponym relation discovery.

To investigate the motif properties of the hypernym/hyponym relation links, we developed three empirical datasets related to *data mining*, *data structure* and *computer network* from Wikipedia. We used the open source web search engine Nutch[1] to crawl the articles rooted at the Wikipedia entries of *data mining, computer network* and *data structure*, by respectively going up to a depth of 3. During the crawling process, we applied a set of URL regular expressions to remove irrelevant articles,

---

[1] http://nutch.apache.org/

such as External links, Languages, and View history; and then manually filtered out all articles that did not belong to the three domains. Each hyperlink was annotated with a semantic tag (*hypernym/hyponym* or *other*).

For a particular network, the statistical significance of a network motif is measured by Z-Score [2]. We proposed an additional parameter, Hyp-Prop(j), which is defined as Equation 1.

$$\text{Hyp-Prop}(j) = \frac{Hyp(j)}{Hyp(j) + Other(j)} \tag{1}$$

where $Hyp(j)$ and $Other(j)$ are the numbers of hyperlinks tagged with *hypernym/hyponym* and *other* in all the instances of network motif j respectively.

All the three-node motifs in *data mining*-related dataset and their normalized Z-Score and Hyp-Prop are plotted in Fig.2. The normalized Z-Score is formalized by Equation 2.

$$\text{Normalized Z-Score}(j) = \frac{|\text{Z-Score}(j)|}{\max_{k \in [1,13]}\{|\text{Z-Score}(j)|\}} \tag{2}$$

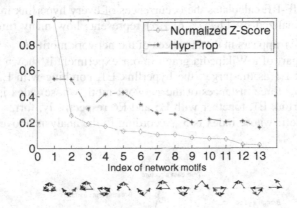

Fig. 2. The normalized Z-Score and Hyp-Prop of three-node motifs in data mining dataset

Two inferences can be drawn from Fig.2. First, there is a positive correlation between the Hyp-Prop and the normalized Z-Score of data mining. Second, hypernym/hyponym hyperlinks appear more frequently in the motifs with higher Z-Scores. Our experiments on two other datasets produce similar results. Different network motifs can be employed to build features to learn the hypernym/hyponym relation from the Wikipedia article links.

4   Hypernym/Hyponym Relation Extraction

Based on network motif analysis of Wikipedia links, we proposed a three-step method, named MOTIF-RE, for hypernym/hyponym relation extraction.

The preprocessing step generates a domain-specific Wikipedia graph. This includes scrawling Wikipedia articles and extracting links of these articles. The feature extraction step builds a motif-based feature vector for each hyperlink. The relation identification step identifies the semantic type of hyperlinks by a binary classifier that uses the above feature vectors.

4.1   Motif-Based Feature Extraction

Given a domain-specific Wikipedia article graph G having n hyperlinks, which is generated in the preprocessing step of MOTIF-RE, one hyperlink in G combined with other hyperlinks can form a set of network motifs; these network motifs can be used to build a feature vector for identifying the semantic type of this hyperlink. The feature extraction can be further divided into three sub-steps.

First, MOTIF-RE discoveries all instances of three-node motifs from a domain-specific Wikipedia graph. It then enumerates all hyperlinks in every motif instance automatically. A hyperlink may simultaneously appear in multiple instances of a network motif. Different occurrences imply different impact strength of this type of network motif.

Second, MOTIF-RE calculates the occurrences of every hyperlink in different network motifs, formalized by $O(i,j)$. $O(i,j)$ represents how many times the hyperlink i $(1 \leq i \leq n)$ appears in the instances of the network motif j.

For example, part of a Wikipedia graph in our experiment is shown in Fig.3. Taking the hyperlink E1 as the target, the hyperlink E1, combing with E2, E3, and E4 respectively, forms three instances of the network motif whose index is 13. Simultaneously the hyperlink E1, together with E5 and E6 respectively, forms two instances of the network motif whose index is 5. According to the analysis above, $O(1,5) = 3$ and $O(1,13) = 2$.

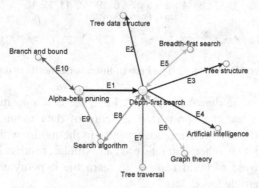

Fig. 3. Part of the Wikipedia graph

Third, MOTIF-RE builds a feature vector for every hyperlink in G based on the network motifs in which the hyperlink appears. The hyperlink occurrences of network motifs have a different significance, which can be weighted by the Z-Scores of corresponding network motifs; thus, the weighted vector component for hyperlink i and motif j can be formalized as follows.

$$F(i, j) = O(i, j) * \text{Z-Score}(j) \tag{3}$$

Moreover, the feature vector of hyperlink i is as follows:

$$FV(i) = (F(i, 1), F(i, 2), \cdots, F(i, 13)) \tag{4}$$

4.2   Relation Identification

By means of motif-based features of hyperlinks, MOTIF-RE transforms the hypernym/hyponym relation identification into a binary classification problem.

Furthermore, we find motifs of different domains are similar and it is possible to predict hypernym/hyponym relation of one domain by utilizing the annotated data of another domain. We proposed a weakly supervised method to bridge the gap between different domains. The domain having annotated data is called auxiliary domain and the domain that we intend to analyze is called target domain. First, MOTIF-RE automatically finds out the about 15% hyperlinks and these are annotated by domain experts. Finally all the annotated data of auxiliary domain and the annotated data of target domain are used as training data for predicting the unlabeled data of the target domain.

5   Experiments

We tested MOTIF-RE in two scenarios. The first one was to evaluate the effectiveness of MOTIF-RE on a single domain dataset; while the second one was to evaluate MOTIF-RE's cross-domain ability. The lexico-syntactic pattern based method was selected as the baseline.

5.1   Baseline Experiment

Variant extraction methods based on lexico-syntactic pattern are very popular and effective when processing web pages and better results can be achieved on the base of Wikipedia. So, in our experiment, the method based on lexico-syntactic patterns is selected for the baseline.

Context words of hyperlinks in Wikipedia articles are analyzed in-depth. The analyzing process is described as follows.

First, about one hundred of Wikipedia articles are selected to learn the lexico-syntactic patterns. In these articles the three sentences around every hyperlink are extracted automatically.

Second, after manual examination and detailed statistical analysis of the extracted sentences, twenty patterns with higher coverage are selected from the context sentences of hyperlinks. These patterns can be looked upon as features of hyperlinks and can be utilized for identifying the hypernym/hyponym relations.

Third, two sentences before hyperlinks and one sentence after hyperlinks are extracted automatically for POS-tag analyzing. Then pattern matching based on POS-tag is carried out to identify the semantic type of hyperlinks. If the words around a hyperlink are matched by these patterns, this hyperlink probably has a hypernym/hyponym relation.

By means of these patterns, the baseline method can achieve fair good results on the three datasets and the results are shown in Table 1 as follows.

Table 1. The Results Of Baseline On The Three Datasets

Datasets	Precision	Recall	F1
Data mining	0.698	0.259	0.378
Computer network	0.657	0.271	0.384
Data structure	0.683	0.264	0.381

5.2   Single-Domain Experiments

Some classic classification algorithms are adopted by MOTIF-RE for extracting hypernym/hyponym relation from Wikipedia article links, including Naïve Bayes, MP (MultilayerPerceptron), AdaBoost (voted perceptron as a weak learner), and SVM.

Three experiments are carried out using our three datasets respectively. The results based on 10-fold cross-validation are shown in Table 2 Every row of the table shows the results of different classification algorithms.

Table 2. Results of The Single-Domain Experiments

Datasets	Naive Bayes		MP		AdaBoost		SVM	
	Prec. Recall	F1	Prec. Recall	F1	Prec. Recall	F1	Prec. Recall	F1
Data mining	0.402 0.402	0.365	0.648 0.563	0.603	0.689 0.583	0.632	0.88 0.327	0.477
Computer network	0.684 0.316	0.433	0.681 0.553	0.610	0.657 0.561	0.605	0.977 0.513	0.673
Data structure	0.487 0.487	0.413	0.723 0.628	0.672	0.693 0.572	0.627	0.852 0.524	0.649

The following conclusions can be drawn from the statistical data in Table 2.

1) The F1 values of all the algorithms on the three datasets are all higher than that of the lexico-syntactic based method. From this evaluation, it can be concluded that the features based on network motifs is suitable for hypernym/hyponym relation extraction.

2) The F1 values vary by different classification algorithms. SVM can reach very high precision. The performances of Naive Bayes are the worst. This seems to be caused by the fact that the conditional independence assumption of features is not satisfied.

5.3 Cross-Domain Experiments

In this experiment every dataset is leveraged as an auxiliary domain for predicting hyperlinks of another domain. The auxiliary domain and target domain have same feature space and same task space. The TrAdaboost[9] algorithm is used by MOTIF-RE in this experiment, combined with different weaker learners including Naïve Bayes, MP (MultilayerPerceptron), and SVM.

Table 3. Results Of The Cross-Domain Experiments

Auxiliary Domain	Target Domain	Naive Bayes		MP		SVM	
		Prec. Recall	F1	Prec. Recall	F1	Prec. Recall	F1
Data mining	Data structure	0.391 0.348	0.368	0.681 0.532	0.597	0.741 0.519	0.61
Data structure	Data mining	0.423 0.373	0.396	0.664 0.529	0.589	0.588 0.479	0.528
Data mining	Computer network	0.401 0.351	0.374	0.675 0.513	0.583	0.93 0.422	0.581
Computer network	Data mining	0473 0. 337	0.394	0.655 0.498	0.566	0.676 0.482	0.562
Data structure	Computer network	0.419 0.353	0.383	0.673 0.536	0.597	0.87 0.437	0.581
Computer network	Data structure	0.456 0.364	0.405	0.633 0.485	0.549	0.848 0.502	0.631

Experimental results, shown in Table 3, can be briefly summarized as follows:

1) MOTIF-RE's performance in cross domain experiments is lower than that in the single domain experiments.
2) The performances of different weaker learners are similar for different domain used as auxiliary domain except Naïve Bayes. The precision of cross domain is fairly high and can match that of pattern-based methods.

6   Conclusion

Although hypernym/hyponym relation extraction has been developed for decades, there is not much work done in discovering the hypernym/hyponym relation from

Wikipedia by means of connectivity patterns. This paper proposes a principle solution. The main contributions of this paper can be summarized as follows.

1) Through the analysis of the topology of Wikipedia links in different domains, we found that three-node motifs are prevalent. These network motifs reflect the local connectivity-patterns of Wikipedia links.
2) We proposed a novel method for building feature vectors based on three-node motifs and extracting hypernym/hyponym relation from Wikipedia links by means of these feature vectors. Our method significantly outperforms the traditional methods based on lexico-syntactic patterns, but depends only on the motifs of Wikipedia links.

There are two potential future directions of this work. First, the features based on text content can be combined with local connectivity-patterns for improving the performance of extraction. Second, new algorithms should be developed for processing more complex connectivity patterns.

Acknowledgement. The research was supported in part by National Science Foundation of China under Grant Nos. 60825202, 61173112, 60921003; National High Technology Research and Development Program 863 of China under Grant No. 2012AA011003; Cheung Kong Scholar's Program; Key Projects in the National Science and Technology Pillar Program under Grant Nos. 2011BAK08B02, 2011BAK08B05, 2009BAH51B02.

References

1. Wu, W., Li, H., Wang, H., Zhu, K.Q.: Probase: a probabilistic taxonomy for text understanding. In: Proceedings of the 2012 International Conference on Management of Data, pp. 481–492. ACM, Scottsdale (2012)
2. Milo, R., Shen-Orr, S., Itzkovitz, S., Kashtan, N., Chklovskii, D., Alon, U.: Network motifs: simple building blocks of complex networks. Science 298, 824–827 (2002)
3. Hearst, M.A.: Automatic acquisition of hyponyms from large text corpora. In: Proceedings of the 14th Conference on Computational Linguistics, pp. 539–545. Association for Computational Linguistics, Nantes (1992)
4. Snow, R., Jurafsky, D., Ng, A.: Learning syntactic patterns for automatic hypernym discovery. In: Saul, L., Weiss, Y., Bottou, L. (eds.) Advances in Neural Information Processing Systems 17, pp. 1297–1304. MIT Press (2005)
5. Ponzetto, S.P., Strube, M.: Deriving a large scale taxonomy from Wikipedia. In: Proceedings of the 22nd International Conference on Artificial Intelligence, vol. 2, pp. 1440–1445. AAAI Press, Vancouver (2007)
6. Sanderson, M., Croft, B.: Deriving concept hierarchies from text. In: Proceedings of the 22nd Annual International ACM SIGIR Conference on Research and Development in Information Retrieval, pp. 206–213. ACM, Berkeley (1999)

7. Hasegawa, T., Sekine, S., Grishman, R.: Discovering relations among named entities from large corpora. In: Proceedings of the 42nd Annual Meeting on Association for Computational Linguistics, pp. 415–422. Association for Computational Linguistics, Barcelona (2004)

8. Navigli, R., Velardi, P.: Learning word-class lattices for definition and hypernym extraction. In: Proceedings of the 48th Annual Meeting of the Association for Computational Linguistics, pp. 1318–1327. Association for Computational Linguistics, Uppsala (2010)

9. Dai, W., Yang, Q., Xue, G.-R., Yu, Y.: Boosting for transfer learning. In: Proceedings of the 24th International Conference on Machine learning, pp. 193–200. ACM, Corvalis (2007)

Behavior Analysis of Long-term Cyber Attacks in the Darknet

Tao Ban, Lei Zhu, Junpei Shimamura, Shaoning Pang,
Daisuke Inoue, and Koji Nakao

National Institute of Information and Communications Technology
4-2-1 Nukui-Kitamachi, Tokyo, 184-8795, Japan
bantao@nict.go.jp
http://www.nict.go.jp

Abstract. Darknet monitoring provides us an effective way to counter-measure cyber attacks that pose a significant threat to network security and management. This paper aims to characterize the behavior of long term cyber attacks by mining the darknet traffic data collected by the nicter project. Machine learning techniques such as clustering, classification, function regression are applied to the study with promising results reported.

Keywords: Attack behavior analysis, cyber attacks, darknet, data mining, network monitoring.

1 Introduction

The malicious activities of a great variety of malware spread all over the Internet are leading to serious security incidents that cause significant damages to the Internet infrastructure and user assets as well. Basically, there are two main approaches to fight against these ever-evolving cyber-threats: The microscopic approach includes countermeasures that perform malware detection or quarantine base on system-level characteristics and behaviors obtained from close analysis on malware specimens. Commonly known commercial anti-malware products usually belong to this approach. The macroscopic approach usually refers to countermeasures that search for behavioral characteristics to reveal the existence of the malware in its communication style across a monitored network. Network based Intrusion Detection Systems (NIDS) [1] installed at the edges of large enterprise networks are examples of this approach. We have been developing the Network Incident analysis Center for Tactical Emergency Response (nicter) [2,3], which integrates both the macroscopic and microscopic approaches. The nicter binds the results of both macroscopic and microscopic analysis to obtain much richer information about malware activities. This paper describes the macroscopic component of the nicter, i.e., a large-scale *darknet* monitoring network, and its possible contributions to the Worldwide Observatory of Malicious Behavior and Attack Tools (WOMBAT).

T. Huang et al. (Eds.): ICONIP 2012, Part V, LNCS 7667, pp. 620–628, 2012.

A darknet is a portion of routed, allocated IP space where no active services or servers reside. The goals of the Darknet are to increase awareness of the malicious or mistaken 'activities and to ease the mitigation: Because of the absence of legitimate hosts within the darknet, any packet from Internet into the darknet is by its presence aberrant: it is either malicious or mis-configured. A Darknet can be used to host flow collectors, backscatter detectors, packet sniffers, etc., considerably cutting down the false positive for such device or technology.

The results reported in this paper are based on long term observation over darknet sensors installed in two class B networks. The setups of the sensors are shown in Fig. 1. For sensor I, all 65,535 IP addresses within the /16 network are allocated to the darknet sensor. For sensor II, the /16 IP address space is sparsely populated with darknet addresses interpreted with active IP addresses. Such a variant is also known as greynet [4]. Darknet monitoring relies on the presumption that most kinds of malware engage a random procedure in search for the next potential victims. Thus the more IP addresses encompassed by the darknet, the more essential information the darknet could gather. Of course, due to the limited scale of a darknet compared with the IPV4 space and its passive nature, only partial information is available from darknet monitoring.

The rest of the paper is organized as follows. Section II reports the ongoing trend of cyber threats as reflected in the long term darknet monitoring. Section III focuses on the analysis on long-term attacks. Section IV concludes the paper.

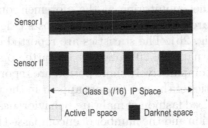

Fig. 1. Formulation of the darknet sensor

2   The Trend of Cyber Attacks

2.1   Basic Darknet Statistics

In the limited information observed in the darknet traffic, the source IP address field is of most practical value. First, all packets targeting the darknet are considered suspicious. These packets are often generated by malware or attackers when searching for the next potential target. Or it may caused by mis-configured machines which happen to direct their packets to the darknet. For any of the above reasons, there is a necessity to further inspect to the corresponding host. Second, due to the passive nature of the darknet, expect for some special cases, e.g., a mis-configured server which connects to a presumed printer in the dark-net space, there will be little persisting connections toward a darknet, especially toward a single IP address. Most of the observed hosts send only a couple of packets. This rendered the darknet connections more fragmented, lacking appli-cation level information. Therefore, the most important statistics for a darknet is the number of (distinct) hosts monitored during a period – which roughly implies the scale of the cyberspace in question.

The number of packets collected by the darknet sensors is also an important indicator of network incidents – assuming a perfect capturing rate. A majority part of the packets captured by a darknet sensor use TCP protocol. Since there will be no response from the darknet, a TCP_SYN packet sent by a host to a darknet IP cannot get a TCP_ACK reply. Therefore, there will be no established TCP session monitored in the darknet; nor will there be TCP packets containing substantial payload. Another majority of the darknet traffic are contributed by UDP protocol, which does not require a communication session established before actual payload is sent. For both of these protocols the number of packets scales well with the number of trials for vulnerability exploitation. Moreover, the number of packets and other packet statistics also carries discriminating information of the attack.

2.2   Trends in Statistics

In the following, we inspect more closely into the above statistics. Figure 2 shows the results of the darknet monitoring at the aforementioned sensors. Reported statistics in the figure are the number of unique hosts (Fig. 2a) and the number of observed packets (Fig. 2b). The statistics are reported on a daily basis, from October 1, 2006 to February 29, 2012. Sensor II started operation since July 10, 2009, thus results of a comparatively short period are reported. A strong correla-tion between the statistics of the two sensors appear in the two graphs. This is an evidence of the randomized fashion of malware behavior: despite of the differences in the network formulation and the number of encompassed IP addresses, the data collected by Sensor II seems to be sampled from the same distribution of that by Sensor I. Note that, due to scheduled maintenance of darknet sensors, there are a few cases that the numbers drop down suddenly, often to a zero value. These missing values shall not affect the results we reported hereinafter.

Number of Unique Hosts. As shown in Fig. 2a, the number of unique hosts shows a downward trend since the starting of the operation of Sensor I to late 2008. The downward trend was broken by a burst in late 2008, when the Con-ficker worm came into being and started to spread over the Internet[1]. During the

---

[1] Conficker, also known as Downup, Downadup and Kido, is a computer worm tar-geting the vulnerability of Microsoft Windows operating system (MS08-067). It was first detected in November 2008 and the estimated infected host is about seven mil-lion, making it the largest known computer worm infection since the 2003 Welchia [5]. The five major variants of Conficker, i.e., Conficker.A through Conficker.E was first reported in November 21, 2008, December 29, 2008, February 20, 2009, March 4, 2009, and April 8, 2009, respectively.

Conficker incident, the number of observed unique hosts appear to be 15 times more than before, which was an evidence of the large-scale contagion of Conficker. The Conficker incident remains as the most influential threat until late 2010, where the number of observed unique hosts was brought down to around 100 thousands – the spike during 2010 is verified to be a re-burst of the same worm.

Number of Packets. Before going into further detail of Fig. 2b, we have to explain more about the composition of darknet traffic. A considerable part of the TCP packets sent to the darknet are flagged with TCP_ACK. These packets are known as backscatters – a side-effect of a spoofed denial of service (DoS) attack. In this kind of attack, the attacker spoofs the source address (happened to be IP addresses in the darknet) in IP packets sent to the victim. The victim responds to the spoofed packets as it normally would. The number of these backscatters varies from the intensity of the DoS attack and the mechanism of source IP spoofing, and often forms aberrant spikes in packet number statistics. For example, the most significant spike in Fig. 2b occurred in January 2012 was formed by backscatters of a DoS attack. However, such an event will not be perceived in the unique-host number statistics: there are usually a handful servers attacked by DoS attacks at a time. Another kind of spikes in Fig. 2b associate with scans from a single host. It could be a port scan toward multiple ports of one darknet IP, an IP scan to multiple IP addresses in the darknet, or a combination of the two. Such events are also difficult to perceive in the number of unique-host statistics if it is only from one source IP address. Of course, there are events that are easily perceivable in both of the two graphs: a vast number of host burst out to send packets to the darknet.

3   Analysis Methods and Results

The following analysis is driven by the necessity to gain further understanding of the behavior of malware-infected hosts over time, to identify their temporal

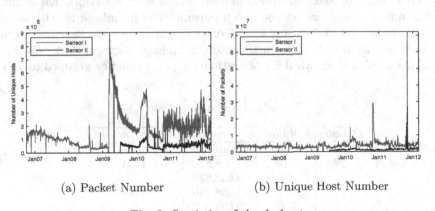

(a) Packet Number   (b) Unique Host Number

Fig. 2. Statistics of the darknet

regularities, and to predict their future activities based on their previous behavior. Towards this end, classification and regression modelling upon host behavior time series is considered reasonable for exploratory study [6,7,8,9]. Previous studies suggest that the temporal regularity of a host is usually malware-specific [10,11], i.e., the behavior of two hosts that are infected by two different types of malware are different from each other. Therefore, malware-specific behavioral models are generally more preferable, in terms of prediction performance, than an overall model learned from data composed of all kinds of malware behavior patterns. In the rest of this section, malware-specific behavior modelling is done as follows: In Section 3.1, hosts are clustered according to the destination ports they attack. Then, we take two approaches to quantify the temporal regularity of grouped hosts that are with similar behavior. In Section 3.2, a quantitative analysis, i.e., to predict the number of packets that will be sent to the darknet, is studied using regression model. In Section 3.3, a qualitative analysis, i.e., to prediction whether a host will continue attacking in the next time slot is done using a classification model.

3.1   Clustering Based on Attacked Destination Ports

It is well known that the information about destination port is closely related to the type of attack [11,12,13]. To find groups of hosts that have similar behavior, we first conduct destination port based clustering to categorize attacking hosts. To simplify the problem, we focus on which destination ports are attacked by the host and ignore the packet distribution among the destination ports [14]. Let the set of destination ports attacked by a host h_i be $P_i = \{p_{il}\}, p_{il} \in \{0, \ldots, 65535\}$, the dissimilarity between two hosts, h_i and h_j, are measured by their Jaccard distance,

$$D_\delta(h_i, h_j) = 1 - \frac{|P_i \cap P_j|}{|P_i \cup P_j|} \tag{1}$$

Then the *linkage* algorithm is applied to the set of destination ports to perform data clustering, To make hosts within each cluster have behaviors that conform with a simple model, we set the *cutoff* parameter c, in linkage to 0.01 so that clusters are formed when a host and all of its neighboring hosts have the Jaccard distance less than c. Such a small cutoff in linkage suggests that only hosts attacking an almost identical list of destination ports could be grouped together.

Table 1. Five largest clusters

Cluster	# of hosts	# of packets	Destination ports
1	25506	351,015,296	445
2	1108	87,008,630	1433
3	1049	40,628,838	22
4	651	21,408,210	3389
5	500	29,693,627	80

Consequently, the coincidence in temporal behavior of similar attacks will render the numerical modelling more accurate. Table 1 shows the top ranking clusters, which will be further investigated in the following experiments.

3.2 Regression and Analysis on Weekly Attack Volume Time Series

In this subsection, we try to exploit how a host's future behavior is effected by its previous behavior in a quantitative manner. To be more specific, we turn to the time series prediction scheme to formulate the task as to predict a host's attack behavior in terms of number of packets sent to the darknet based on its historical observation [15,16].

A host h_i is presented as a time series $x_{it}, t = 0, \cdots, 52$, measured from the first week to the last week of 2011, by counting the number of packets received from the host within each week. Such a time series is then normalized to be unit length to average out the difference between hosts that happen to direct most of their attacks on the darknet and hosts that only have a small portion of their attacks monitored in the darknet. On time series extracted for attacking hosts in a cluster, by setting a window size W, we formulate a W-dimensional regression problem where the number of packets send by a host at time $T + W + 1$ is predicted on the basis of its most recent statistics at $T, \cdots, T + W$.

The regression algorithm applied in this experiment is the Support Vector Regression (SVR) which was first introduced by Vapnik [17]. Unlike most of the traditional learning schemes that adopt the *empirical risk minimization* principle, SVR embodies the *structural risk minimization* principal, which seeks to minimize an upper bound of the generalization error. SVR shows better generalization performance than conventional techniques in many applications [17]. In the experiment, we employ the ν-SVR with Gaussian kernel function implemented in LibSVM [18] to solve the regression problems we have formulated. Five-fold cross validation is engaged to select parameters including the window size parameter W, the width parameter, γ, of the Gaussian kernel, and the penalty parameter ν for SVR.

Table 2. Crossover regression prediction on weekly attack volume time series

Model trained from	SCC tested on					MSE tested on				
	1	2	3	4	5	1	2	3	4	5
1	0.39	0.41	0.25	0.19	0.11	**3.61e-4**	2.17e-3	4.17e-3	8.04e-3	4.36e-3
2	0.47	0.29	0.28	0.2	0.11	4.69e-4	**3.18e-4**	4.35e-3	7.84e-3	4.80e-3
3	0.31	0.27	0.03	0.12	0.042	8.57e-4	2.44e-3	**2.00e-3**	8.16e-3	4.32e-3
4	0.4	0.41	0.26	0.19	0.13	6.31e-4	2.05e-3	3.74e-3	**3.77e-3**	4.03e-3
5	0.3	0.094	0.18	0.096	0.06	6.04e-4	3.20e-3	4.08e-3	8.64e-3	**1.28e-3**

Five regression models are trained, each of which on one of the above five clusters. The training sets and test sets are created as follows. First, a training set is composed of 20,000 patterns (W dimensional segments) randomly sampled from the host time series in each cluster. All the remaining patterns are assigned to the test set. Given the regression models built from the training sets, crossover test is performed by applying the models $M_i, (i = 1, \cdots, 5)$ on all five test sets. The objective of this crossover test is to discover whether the attack behavior patterns are different from cluster to cluster.

Result of the experiment is shown in Table 2, in which the prediction performance is measured by Squared Correlation Coefficient (SCC) and Mean Squared Error (MSE). As can be seen in the right half of the table, the MSE values on the diagonal appear to be the minimum of each row, which means that the regression model trained from a cluster best fits the test set from that cluster. These small MSE values along the diagonal indicate that a host future behavior is closely related to its past behavior and such relation could be learned in a quantitative sense. The comparatively large MSE values off the diagonal suggest that different type of attacks may conforms to different behavior models in terms of number of packets send to the darknet, which is in consistence with our intuition. For the SCC values on the left of Table 2, we cannot observe clear regularity. This may because of that the squared entry in SCC favors large output values which happen to be in accordance with the true value due to large noise ratio.

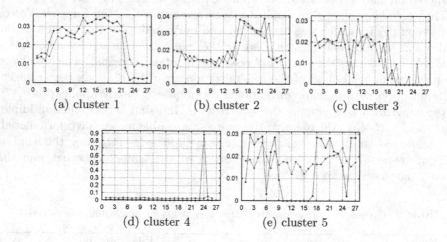

(a) cluster 1   (b) cluster 2   (c) cluster 3

(d) cluster 4   (e) cluster 5

Fig. 3. Examples of regression on weekly attack volume time series

Figure 3 shows some examples of the prediction result (red) compared with the ground truth (blue) of random selected host in each cluster. Despite of the perceptible *lag* in the figure, the predicted values align with the observed values for most of the cases.

3.3 Categorical Prediction on the Attack

In this subsection, we seek to answer the following question which is more qualitative about the host status: given the statistics of a host h_i in the past W time slots as $t_{i1}, \cdots, t_{iW}$, can we know whether it will continue its attack at time $t_{i(W+1)}$? This question is best modelled as a classification problem.

Based on the formulation in the previous subsection, we define a binary classification problem as follows: the input vectors for the classifier are kept the same as in the regression model while the output values are transformed into binary codes, where a host is labelled +1 if it no longer launches any attack at time $W + 1$, or -1 otherwise. This time, we apply the Support Vector Classifier (SVC) [17] to the so created dataset. The C-SVC implementation in LibSVM is adopted for numerical study, with the result shown in Table 3. Because classification problems formed from some of the clusters appeared to be imbalanced, i.e., samples from the one class overwhelm those from the other, we use G-mean and F-measure instead of accuracy to measure the generalization performance of the classifiers. As can be seen from the table, G-mean shows a similar pattern as the MSE in Table 2, which indicates that the hosts located in the same cluster behaves similarly. Despite of a little variation, the F-measure statistics also supports the above claim.

Table 3. Crossover classification prediction on weekly attack volume time series

Model trained from	G-mean tested on					F-measure tested on				
	1	2	3	4	5	1	2	3	4	5
1	**0.91**	0.94	0.88	0.77	0.79	0.94	0.92	0.8	0.73	0.60
2	0.87	**0.95**	0.86	0.75	0.82	0.92	0.92	0.8	0.71	0.69
3	0.89	0.92	**0.92**	0.73	0.79	0.92	0.90	**0.89**	0.69	0.66
4	0.78	0.94	0.92	**0.88**	0.85	0.91	0.90	0.78	**0.82**	0.60
5	0.76	0.91	0.88	0.77	**0.88**	**0.95**	**0.94**	0.85	0.76	**0.82**

4 Conclusion

In this paper, we have studied the behavior of malware-infected hosts via darknet monitoring. To form group of hosts that conform with simple behavioral patterns, hosts are firstly clustered by the destination ports they are attacking. Consequent numerical study based on function regression and classification verifies that there is strong predictability with regards to the attack behavior for hosts within the same cluster. The result of this study can be helpful in security enforcement such as adaptive blacklisting.

References

1. Herve, D., Marc, D., Andrea, W.: Towards a taxonomy of intrusion-detection systems. Computer Networks 31(8), 805–822 (1999)
2. Nakao, K., Yoshioka, K., Inoue, D., Eto, M.: A novel concept of network incident analysis based on multi-layer ovservation of malware activities. In: The 2nd Joint Workshop on Information Security (JWIS 2007), pp. 267–279 (2007)

3. Inoue, D., Yoshioka, K., Eto, M., Yamagata, M., Nishino, E., Takeuchi, J., Ohkouchi, K., Nakao, K.: An Incident Analysis System NICTER and Its Analysis Engines Based on Data Mining Techniques. In: Köppen, M., Kasabov, N., Coghill, G. (eds.) ICONIP 2008, Part I. LNCS, vol. 5506, pp. 579–586. Springer, Heidelberg (2009)

4. Harrop, W., Armitage, G.J.: Defining and evaluating greynets (sparse darknets). In: LCN 2005 (2005)

5. Markoff, J.: Worms infects millions of computers worldwide. New York Times (2009)

6. Bailey, M., Cooke, E., Jahanian, F., Provos, N., Rosaen, K., Watson, D.: Data reduction for the scalable automated analysis of distributed darknet traffic. In: Proceedings of the 5th ACM SIGCOMM Conference on Internet Measurement, IMC 2005, p. 21. USENIX Association, Berkeley (2005)

7. Bailey, M., Cooke, E., Jahanian, F., Myrick, A., Sinha, S.: Practical darknet measurement. In: 2006 40th Annual Conference on Information Sciences and Systems, pp. 1496–1501 (March 2006)

8. Song, J., Shimamura, J., Eto, M., Inoue, D., Nakao, K.: Correlation analysis between spamming botnets and malware infected hosts. In: 2011 IEEE/IPSJ 11th International Symposium on Applications and the Internet (SAINT), pp. 372–375 (July 2011)

9. Fukuda, K., Hirotsu, T., Akashi, O., Sugawara, T.: Correlation among piecewise unwanted traffic time series. In: IEEE Global Telecommunications Conference, GLOBECOM 2008, November 30-December 4, pp. 1–5 (2008)

10. Fukuda, K., Hirotsu, T., Akashi, O., Sugawara, T.: A pca analysis of daily unwanted traffic. In: 2010 24th IEEE International Conference on Advanced Information Networking and Applications (AINA), pp. 377–384 (April 2010)

11. Vinu, J., Theepak, T.: Realization of comprehensive botnet inquisitive actions. In: 2012 International Conference on Computing, Electronics and Electrical Technologies (ICCEET), pp. 915–921 (March 2012)

12. Limthong, K., Kensuke, F., Watanapongse, P.: Wavelet-based unwanted traffic time series analysis. In: International Conference on Computer and Electrical Engineering, ICCEE 2008, pp. 445–449 (December 2008)

13. McManamon, C., Mtenzi, F.: Defending privacy: The development and deployment of a darknet. In: 2010 International Conference for Internet Technology and Secured Transactions (ICITST), pp. 1–6 (Novemeber 2010)

14. Li, Z., Goyal, A., Chen, Y., Paxson, V.: Towards situational awareness of large-scale botnet probing events. IEEE Transactions on Information Forensics and Security 6, 175–188 (2011)

15. Ahmed, E., Clark, A., Mohay, G.: A novel sliding window based change detection algorithm for asymmetric traffic. In: IFIP International Conference on Network and Parallel Computing, NPC 2008, pp. 168–175 (October 2008)

16. Kalakota, P., Huang, C.-T.: On the benefits of early filtering of botnet unwanted traffic. In: Proceedings of 18th International Conference on Computer Communications and Networks, ICCCN 2009, pp. 1–6 (August 2009)

17. Vapnik, V.N.: The Nature of Statistical Learning Theory. Springer (1995)

18. Chang, C.-C., Lin, C.-J.: Libsvm: A library for support vector machines. ACM Transactions on Intelligent Systems and Technology 2, 27:1–27:27 (2011), Software available at http://www.csie.ntu.edu.tw/~cjlin/libsvm

Clock Synchronization Protocol
Using Resonate-and-Fire Type of Pulse-Coupled
Oscillators for Wireless Sensor Networks

Kazuki Nakada[1] and Keiji Miura[2]

[1] Advanced Electronics Research Division, INAMORI Frontier Research Center,
Kyushu University, 744 Motooka, Nishi-ku, Fukuoka 819-0395, Japan
k.nakada@ieee.org
[2] Graduate School of Information Sciences, Tohoku University,
Aramaki aza Aoba 6-3-09, Aoba-ku, Sendai 980-8579, Japan
miura@ecei.tohoku.ac.jp

Abstract. We present a system of pulse-coupled oscillators (PCOs) based on the resonate-and-fire neuron (RFN) model for an application to clock synchronization protocol for wireless sensor networks. Firstly, we show a novel type of PCO derived from the RFN model and its Type I/Type II phase response properties. Secondly, we demonstrate that global phase synchronization in a network of the RFNs as PCOs are robust against transmission delays. Finally, we propose a possible scheme for compensation of transmission delays.

Keywords: Coupled Phase Oscillator, Pulse-coupled Oscillator (PCO), Resonate-and-Fire Neuron (RFN), Wireless Sensor Network (WSN), Clock Synchronization Protocol.

1 Introduction

Collective synchronization phenomena are ubiquitously observed in nature. Such synchronization phenomena can be modeled as ensembles of nonlinear limit-cycle oscillators. On the basis of the phase reduction theory, an ensemble of coupled limit-cycle oscillators can be reduced into a system of coupled phase oscillators, and individual phase oscillators are characterized by the phase response curve (PRC) that represents the phase advanced and delay in response to a weak perturbation as an input. The PRC can be described as a function with respect to the phase of the input.

Coupled phase oscillators are classified into two types: the phase-coupled oscillators with continuous phase interactions and the pulse-coupled oscillators (PCOs) with discontinuous pulse interactions. Such interactions among phase oscillators are represented as a phase coupling function, which is derived from the convolution of the PRC and the input received from other oscillators. The phase dynamics of coupled phase oscillators can provide us with perspectives for understanding of collective dynamics of a general class of nonlinear oscillators and for the potential applications.

T. Huang et al. (Eds.): ICONIP 2012, Part V, LNCS 7667, pp. 629–636, 2012.

As one of the potential applications, the collective dynamics of PCOs have been applied to clock synchronization protocols for wireless sensor networks (WSNs) [3]-[9]. In the most of the conventional protocols, the PCOs are based on the type of the leaky integrate-and-fire (LIFN) model, which is modeled after synchronous fireflies [10]. These PCOs have the Type I PRC [10]. The performance and validity of such protocols are verified on practical platforms in real-world environments. However, the strong condition that transmission delays are enough small should be assumed in the most protocols. In the presence of strong transmission delays, the additional functional scheme, such as Reachback Firefly Algorithm (RFA) [3], should be needed for compensation of the effects of the delays. Recently, it has been reported that the PCO with the Type II PRC can increase the robustness in convergence to synchronization states in a large class of network graphs with delays in contrast to the case of the PCOs with the Type I PRC [11].

In this work, we propose to apply a system of novel type PCOs based on the resonate-and-fire neuron (RFN) model [12] to the clock synchronization protocol for WSNs. The RFN model is a complex extension of the LIFN model, and it has the second-order membrane dynamics and a firing threshold. Depending on the location of the reset value after firing, the RFN model can have both Type I and Type II PRCs [14] and can exhibit the transition among the two types of PRCs [17]. We demonstrate that global phase synchronization in a fully connected network are robust against slightly distributed delays because of the Type II PRC. Based on the results, we consider an adaptive compensation scheme for clock synchronization protocol in WSNs with delays.

2   Model Dynamics and Phase Response Curve

Let us explain the dynamics of the PCO derived from the RFN model for the clock synchronization protocol.

2.1   Resonate-and-Fire Neuron Model

We begin by considering the RFN model defined as a complex extension of the IFN model. The RFN model is a spiking neuron model that has second-order subthreshold membrane dynamics and a firing threshold [12]. The dynamics of the RFN model are described by:

$$\dot{z} = (-b + iw)z + I \qquad (1)$$

with

$$\text{if } \mathrm{Im}(z) > a_{th}, \quad \text{then } z \leftarrow z_o \qquad (2)$$

where $z = x + iy$ is a complex state variable. The real and imaginary parts, x and y, correspond to the current- and voltage-like variables, respectively. The constants b and w are parameters, and I is an external input. If the imaginary part of the

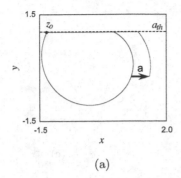
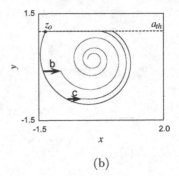

(a)   (b)

Fig. 1. Phase plane portraits of the RFN model in response to pules inputs at different timing

state variable y exceeds a certain threshold a_{th}, the variable z is reset to an arbitrary value z_o, which describes activity-dependent after-spike reset. Since this model has second-order membrane dynamics, and thus it exhibits the fast subthreshold oscillation of the membrane potential, resulting in the coincident detection, post-inhibitory-rebound, and frequency preference in response to the consequence of input pulses [12].

2.2   Phase Response Properties of the RFN Model

The PRC of the RFN model has been derived from the model dynamics as proven in Ref. [15]. For simplicity, the parameter b can be normalized to 1 by scaling t without loss of generality. The analytical solution of the PRC can be represented as follows: [15]

$$Z(\phi) = \frac{1}{T} \frac{e^{T\phi} \sin(\omega T(1 - \phi))}{\cos(\omega T) + I_0 \sin(\omega T) + \omega \sin(\omega T)} \tag{3}$$

where ϕ represents the phase variable, and T the repetitive firing period. We assume that the intensity of the perturbation input is normalized to 1.

The effect of the delay τ can be incorporated into the PRC as follows [17]:

$$Z(\phi) = \frac{1}{T} \frac{e^{T\phi} \sin(\omega(T - \tau - T\phi))}{\cos(\omega(T - \tau)) + I_0 \sin(\omega(T - \tau)) + \omega \sin(\omega(T - \tau))} \tag{4}$$

for $\phi < \frac{T-\tau}{T}$ and 0 for $\phi > \frac{T-\tau}{T}$. It should be noted that the phase variable ϕ advances in proportion to the time under no perturbation.

2.3   Pulse-coupled Oscillator Based on the RFN Model

In general, the dynamics of coupled phase oscillators are represented as follows:

$$\frac{d\phi}{dt} = \omega + Z(\phi)S(t) \tag{5}$$

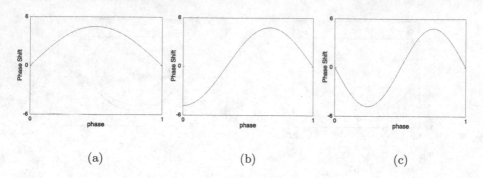

(a)   (b)   (c)

Fig. 2. Transition from Type I to Type II PRCs of the RFN model depending on T/T_o as (a) 0.5, (b) 0.75, and (c) 1.0, respectively

where $Z(\theta)$ represents an arbitrary PRC, $S(t)$ the input at time t, and ω the natural angular frequency. In the case of PCOs, the dynamics are rewritten as follows:

$$\frac{d\phi_i}{dt} = \omega_i + \frac{K}{N} \sum_{j}^{N} \sum_{k} Z(\phi)\delta(t - t_j^k) \tag{6}$$

where ϕ_i represents the phase variable of the i-th oscillator. The coupling strength determined by the input intensity K and the number of the oscillators N. The function $\delta(t)$ is the Dirac delta function, which represents the timing of input pulses at time t_j^k without the pulse width.

We can consider Eqs. (3) and (4) as a PRC of the PCO represented as in Eq. (6). It should be noted that $\omega_i = 2\pi/T \neq \omega$. In order to avoid confusion, we hereafter replace w in Eqs. (3) and (4) with $2\pi/T_o$, where T_o represents the maximum firing period near the grazing bifurcation point [18].

By changing the location of the reset value, the firing period T can be tuned as a control parameter. As a result, the PRC can be changed systematically, as shown in Fig. 1. In addition, the location of the reset value only change the scale of the PRC. The transition from the Type I PRC to the Type II PRC can be observed. Here, we set the parameters as $T = 0.2$ and the ratio of T/T_o as 0.5, 0.75, 1.0, corresponding to Figs. 1(a), (b), (c), respectively. Such phase response properties are suitable for clock synchronization protocol in WSNs with transmission delays.

3   Global Phase Synchronization in Network of the RFN type PCOs

Let us demonstrate global phase synchronization in a fully-connected network of the RFN type PCOs. In the following simulations, we set the number of the RFNs as PCOs as $N = 5$ and the reset value as $z_o = -2.0 - i$. Other parameters were set as: $\tau = 2.5, I = 2.5, a_{th} = 1.0$.

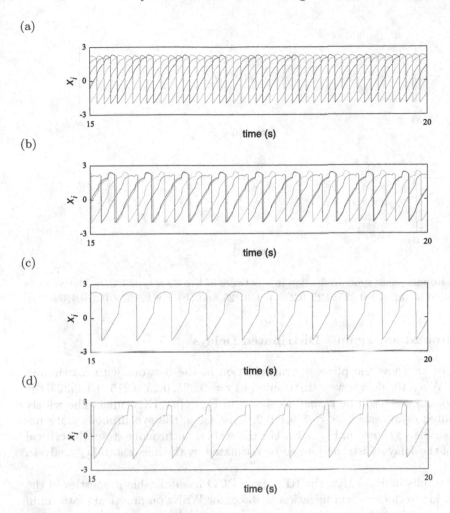

Fig. 3. Global phase synchronization in a network of the RFN type PCOs. (a) No delay, (b) $\tau = 0.1$, (c) $\tau = 0.25$, and (d) $\tau = 0.5$ for each PCO.

3.1   Global Phase Synchronization

Firstly, we compare the phase synchronization in the fully-connected network without/with delays. In the case of no delay, the network shows an asynchronous state, namely, the merry-go-round state, as shown in Fig. 3(a) [14]. If the small delays are distributed, the asynchronous state will be unstable, as shown in Fig. 3(b). However, the delays become large, a synchronous state becomes stable, as shown in Figs. 3(c) and (d). In both the cases, the delays were set as $\tau = 0.25$ for (c) and $\tau = 0.5$ for (d), respectively. The range of the delays for the stable synchronous states are relatively wide.

(a)

(b)

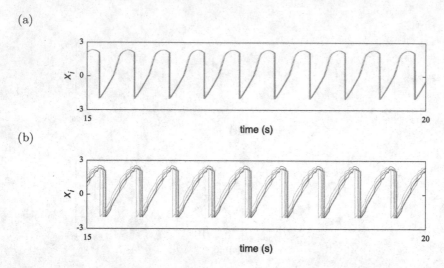

Fig. 4. Robust phase synchronization in a network of the RFN type PCOs. Delays were distributed as (a) $\tau = 0.225, 0.25, 0.275, 0.3, 0.325$ and (b) $\tau = 0.1, 0.2, 0.3, 0.4, 0.5$.

3.2   Robustness against Distributed Delays

Secondly, we show the phase synchronization in the network with distributed delays. When the delays were distributed as $\tau = 0.225, 0.25, 0.275, 0.3, 0.325$, the synchronous state will be stable, as shown in Fig. 4(a). Even under the widely distributed delays, such as $\tau = 0.1, 0.2, 0.3, 0.4, 0.5$, the synchronous state becomes stable, as shown in Fig. 4(b). For the stable synchronous state, the critical value of the delays exists, and it can be estimated by the linear stability analysis.

The results indicate that the RFN type PCO has desirable properties in the application to clock synchronization protocol for WSNs on practical platform in real-world environments suffered from transmission delays.

4   Compensation Protocol for Delay Effects

Let us propose a resetting control as a compensation scheme for transmission delays utilizing the phase response properties of the RFN type PCO. If delays are enough small, we set the reset value on the right side on the phase plane because the high natural angular frequency is preferable for accelerating the speed to convergence to a synchronous state [10]. If the phase differences between the PCOs are large, it indicates the existence of large delays. In the case, we change the reset value on the left side on the phase plane along the orbit of the PCOs, as shown in Fig. 5. Consequently, the synchronous state will be stable even in the presence of the delays. Since the resources of platforms for WSNs are restricted, the present simple scheme seems to be suitable for practical applications.

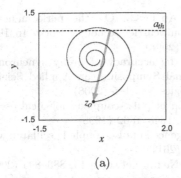

(a)

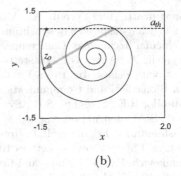

(b)

Fig. 5. Reseting control of the RFN model for delay effect compensation

5 Conclusion

In this work, we have presented a novel type PCO based on the RFN model for an application to clock synchronization protocol for WSNs. Firstly, we showed the origin of the PCO and the Type I/Type II phase response properties. Secondly, we demonstrated that the global phase synchronization in the network of the RFNs as PCOs can be robust against distributed delays in transmission. Finally, we proposed a possible scheme for compensation of the transmission delays utilizing the phase response properties of the RFN type PCO. In future work, we are going to implement the present protocol on a practical platform in the real-world environment.

References

1. Kuramoto, Y.: Chemical oscillations, waves, and turbulence. Springer, Berlin and New York (1984)
2. Izhikevich, E.M.: the geometry of excitability and bursting. Dynamical systems in neuroscience. The MIT press (2007)
3. Werner-Allen, G., Tewari, G., Patel, A., Welsh, M., Nagpal, R.: Firefly-inspired sensor network synchronicity with realistic radio effects. In: The 3rd International Conference on Embedded Networked Sensor Systems, pp. 142–153 (2005)
4. Hong, Y.W., Scaglione, A.: A scalable synchronization protocol for large scale sensor networks and its applications. IEEE Journal on Selected Areas in Communications 23, 1085–1099 (2005)
5. Wang, X.Y., Apsel, A.B.: Pulse coupled oscillator synchronization for low power UWB wireless transceivers. In: The 50th Midwest Symposium on Circuits and Systems, pp. 1524–1527 (2007)
6. Mutazono, A., Sugano, M., Murata, M.: Evaluation of robustness in time synchronization for sensor networks. Bio-Inspired Models of Network, Information and Computing Systems, 89–92 (2007)
7. Taniguchi, Y., Wakamiya, N., Murata, M.: A self-organizing communication mechanism using traveling wave phenomena for wireless sensor networks. In: International Symposium on Autonomous Decentralized Systems, pp. 562–570 (2007)

8. Sanguinetti, L., Tyrrell, A., Morelli, M., Auer, G.: On the performance of biologically-inspired slot synchronization in multicarrier ad hoc networks. In: IEEE Vehicular Technology Conference, pp. 21–25 (2008)
9. Tyrrell, A., Auer, G., Bettstetter, C.: On the accuracy of firefly synchronization with delays. In: Proc. First International Symposium on Applied Sciences on Biomedical and Communication Technologies, pp. 1–5 (2008)
10. Mirollo, R.E., Strogatz, S.H.: Synchronization of pulse-coupled biological oscillators. SIAM Journal on Applied Mathematics, 1645–1662 (1990)
11. Nishimura, J., Friedman, E.J.: Robust convergence in pulse-coupled oscillators with delays. Physical Review Letters 106, 194101 (2011)
12. Izhikevich, E.M.: Resonate-and-fire neurons. Neural networks 14, 883–894 (2001)
13. Izhikevich, E.M.: Dynamical systems in neuroscience: the geometry of excitability and bursting. The MIT press (2007)
14. Miura, K., Okada, M.: Pulse-coupled resonate-and-fire models. Physical Review E 70, 021914 (2004)
15. Miura, K., Okada, M.: Globally coupled resonate-and-fire models. Progress of Theoretical Physics Supplement 161, 255–259 (2006)
16. Nakada, K., Miura, K., Hayashi, H.: Burst synchronization and chaotic phenomena in two strongly coupled resonate-and-fire neurons. International Journal of Bifurcation and Chaos 18, 1249–1259 (2008)
17. Nakada, K., Miura, K.: Synchronization analysis of resonate-and-fire neuron models with delayed resets. SCIS-ISIS. Kobe, Japan (2012)
18. Nakada, K., Miura, K., Hayashi, H.: Noise-induced phenomena in a system of two strongly pulse-coupled spiking neurons. In: IUTAM Symposium on 50 years of Chaos, Kyoto, Japan (2011)

Neuro-Cryptanalysis of DES and Triple-DES

Mohammed M. Alani

Department of Computing, Middle-East College, Muscat, Sultanate of Oman
m.alani@d-crypt.org

Abstract. In this paper, we apply a new cryptanalytic attack on DES and Triple-DES. The implemented attack is a known-plaintext attack based on neural networks. In this attack we trained a neural network to retrieve plaintext from ciphertext without retrieving the key used in encryption.

The attack was practically, and successfully, applied on DES and Triple-DES. This attack required an average of 2^{11} plaintext-ciphertext pairs to perform cryptanalysis of DES in an average duration of 51 minutes. For the cryptanalysis of Triple-DES, an average of only 2^{12} plaintext-ciphertext pairs was required in an average duration of 72 minutes. As compared to other attacks, this attack is an improvement in terms of number of known-plaintexts required, as well as the time required to perform the complete attack.

Keywords: cryptanalysis, des, triple-des, 3des, neural, neuro-cryptanalysis.

1 Introduction

Cryptanalysis of block ciphers has witnessed rapid developments in the last few years. Many new cryptanalytic techniques were developed, especially after the adoption of the Advanced Encryption Standard (AES). Most of the recent developments in cryptanalysis focused on algebraic aspects of block ciphers [1].

In this paper we will focus on known-plaintext attacks which assume that an adversary knows some plaintexts-ciphertexts pairs. The objective of this attack is to find the associated key or to recover unknown parts of the plaintext.

An attack may not only provide information about the key but also about other kinds of data. As classified in [2], the attack applied in this paper is a Global Deduction; "an attacker finds an algorithm functionally equivalent to the original encryption/decryption ones but without knowing the secret key."

DES was adopted as a standard in 1977 [3]. Since then, this cipher was used in many applications worldwide. The high processing ability rendered the 56-bit key of the DES vulnerable to brute-force attacks. This attack, which is basically trying every possible key, was confronted by introducing the Triple-DES by the National Institute of Standards and Technology (NIST) as a more appropriate alternative algorithm, offering a higher level of security [4]. The complete description of DES can be found in [3].

Triple-DES is a repetition of DES algorithm for three times using three different keys, which sums the total key to be 168 bits, in a pattern called EDE. EDE is simply

T. Huang et al. (Eds.): ICONIP 2012, Part V, LNCS 7667, pp. 637–646, 2012.
© Springer-Verlag Berlin Heidelberg 2012

the process of Encryption-Decryption-Encryption. Each one of these processes is conducted using a different 56-bit key [5].

1.1   Cryptanalysis of DES and Triple DES

The DES has been a target for many attacks for a long time. Some of these attacks started by analyzing reduced-rounds DES and went up to the full-round DES. The most known were, differential cryptanalysis, and linear cryptanalysis [4,6].

Differential cryptanalysis is a chosen-plaintext technique developed by Biham and Shamir in [5] against reduced-round variants of the DES cipher, and later applied to the full 16-round DES [7]. The full-round DES was said to be broken by differential cryptanalysis using 2^{47} chosen-plaintexts with a time complexity of 2^{37}.

Linear cryptanalysis is a statistical, known-plaintext attack on block ciphers. This technique has been more extensively developed by Matsui in attacks on the DES cipher in [6]. The full linear cryptanalysis of DES required 2^{43} known-plaintexts. Many other cryptanalytic attacks were developed afterwards, such as differential-linear cryptanalysis [8], related-key attacks [9].

Triple-DES was a target for a slightly different type of attacks due to its repetition of DES algorithm. Repeated encryption was bounded by some limitations as mentioned in [10]. Meet-in-the-middle attacks rendered the effective key to be of 112-bits length [11]. Few attacks on Triple-DES were successful in terms of numbers of known/chosen ciphertexts/plaintexts needed to break the algorithm such as the attack introduced by Lucks which requires around 2^{32} known plaintexts, 2^{113} steps, 2^{90} single DES encryptions, and 2^{88} memory. Although it is the most successful attack on Triple-DES in terms of number of known-plaintexts needed, it was, and still, considered impractical by the NIST [12]. In 2005, another approach, relying on related-key and meet-in-the-middle attacks, was published [13]. However, many commercial applications still use Triple-DES as their encryption choice such as financial transactions, and email security.

1.2   Neural Networks and Cryptology

There have been few attempts to use neural networks in cryptography. In 1998, Clark and Blank introduced a cryptographic system based on neural networks [14]. A close encounter was made in [15] in 2005. Another application of neural networks in cryptography was published by Kinzel and Kanter which introduced a way of using neural networks in secret key exchange over a public channel [16]. Klimov, Mityagin, and Shamir introduced a method of cryptanalysis to the previous system in [17]. A scheme for remote password authentication was published in [18]. Significant work was done in [19] which suggested using neural networks in optimizing differential and linear cryptanalysis. Li, in [20], suggested a new cryptographic system which was later analyzed by [21].

Recently, [22] suggested the use of right-sigmoidal function as an activation function in a neural network used in cryptanalysis of Feistel-type ciphers. This paper did not achieve actual cryptanalytic results and contained many assumptions.

2   Neuro-Cryptanalysis

The attack implemented in this paper is a known-plaintext attack. This attack is based on training a neural network to do the decryption process without knowing the key. This attack was first introduced in [23] and was applied to DES. A schematic diagram of the proposed attack is shown in figure 1.

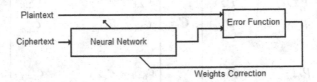

Fig. 1. A Schematic Diagram of the Neuro-Cryptanalysis System [23]

The neural network is fed with ciphertext as its input, and the plaintext is considered the reference output. After an adequate amount of training, with an adequate amount of plaintext-ciphertext pairs that are all encrypted with the same key, the neural network will be able to retrieve plaintext from ciphertext that has not been part of the training process, as long as this ciphertext is encrypted with the same key. As figure 2 shows, the key bits are not retrieved. Thus, the attack is considered a global deduction attack which was defined earlier as finding an algorithm functionally equivalent to the original decryption one but without knowing the secret key. To have a neural network trained to have a small, and acceptable, error rate, the network size need to be expanded such that longer time for each training cycle is required. The neural network has many parameters that need to be set. These parameters will be discussed in the next subsections along with the implementation specifications. The choice of neural networks to be used in this attack is that multilayer feedforward neural networks, which can be used as global approximators as proved in [24].

2.1   Implementation Environment

The software used in implementing the proposed attack was MATLAB R2008a with neural networks toolbox. The single computer used in the implementation had an AMD Athlon X2 processor with 1.9 Gigahertz clock frequency and 4 Gigabytes of memory.

2.2   Design Choices

a.   Data representation

Since DES and Triple-DES handle data as blocks of 64 bits, the data representation for both of their cryptanalytic network was the same. The 64-bit blocks were represented as simultaneous 64 inputs to the neural network and the output was also represented as 64 output lines. Obviously, the range of inputs and outputs was [0, 1].

The DES implementation was in Electronic Code Book (ECB) mode, while the triple-DES encryption was in Encryption-Decryption-Encryption (EDE) pattern.

Since the implementation was in MATLAB, the data was represented in the following matrix form:

$$P = \begin{bmatrix} p_{1,1} & p_{1,2} & & p_{1,k} \\ p_{2,1} & p_{2,2} & & p_{2,k} \\ p_{3,1} & p_{3,2} & \cdots & p_{3,k} \\ \vdots & \vdots & & \vdots \\ p_{64,1} & p_{64,2} & & p_{64,k} \end{bmatrix}, \; C = \begin{bmatrix} c_{1,1} & c_{1,2} & & c_{1,k} \\ c_{2,1} & c_{2,2} & & c_{2,k} \\ c_{3,1} & c_{3,2} & \cdots & c_{3,k} \\ \vdots & \vdots & & \vdots \\ c_{64,1} & c_{64,2} & & c_{64,k} \end{bmatrix}$$

where,

P is the plaintext matrix

C is the ciphertext matrix

$p_{i,j}$ is the ith plaintext bit in the jth plaintext block

$c_{i,j}$ is the ith ciphertext bit in the jth ciphertext block

k is the number of plaintext-ciphertext pairs used in the training

The first column in C is the ciphertext corresponding to the plaintext in the first column in P. The number of rows in the two matrices is chosen based on the number of inputs and outputs of the neural network. All ciphertext in the plaintext-ciphertext pairs are encrypted with the same key. The plaintext blocks and the ciphertext blocks that were encrypted using the same key are called a *dataset*. The average size of datasets used was about 2^{20} plaintext-ciphertext pairs.

b.   Network size and layout

Few different sizes and arrangements were tried and most of them were successful. In this paper, the network layout will be expressed as

n_1-n_2-n_3-...-n_m

where,

n_i is the number of neurons in the ith layer

m is the number of layers in the neural network

n_m is the number of neurons in the output layer

The input layer is not taken into account in this expression. For the DES and Triple-DES, the topology was chosen to be cascaded-forward network because other feedforward network was not successful in achieving the required cryptanalysis.

c.   Training algorithm

The choice was made to use scaled conjugate-gradient as it is a suitable algorithm for large networks. Scaled conjugate-gradient requires usually more training cycles than Levenberg-Marquardt but it requires less memory.

d.   Error function and initial weights

The error function chosen to correct the weights during the training was mean-squared error. The weights and biases are initialized in MATLAB using a function called `initlay`. This function generates randomized matrices for weights and biases.

e.   Number of training cycles (epochs)

The maximum numbers of training cycles, or epochs, is set to 10^4.

f.   Training stopping conditions

The first obvious condition is reaching the final number of training cycles. Another condition is a limit for an acceptable mean-squared error. In this paper we set the error stopping condition to be 10^{-4}. Maximum number of consecutive validation failures is also another stopping condition. In this paper, we set the maximum failures to be 10.

2.3   Methodology

The proposed cryptanalytic attack was implemented on DES using 100 different datasets to prove that this method does not rely on certain properties of the key used in encryption. Each dataset consisted of a certain number of plaintext-ciphertext pairs. For each dataset, all the plaintext in that dataset was encrypted with the same key to obtain the corresponding ciphertext. Each dataset used a key that is different from all other datasets (independent keys). Each dataset was divided into two parts; one part used in the training the neural network, and another part used for validation. The datasets were generated using a Pseudo-random Number (PN) generator and so were the keys. Despite that the data was generated using a PN generator, the data, as well as the key, was checked to assure their independence. The following subsections describe the processes used in the cryptanalysis.

Training Process

1. The training process starts by creating the neural network and selecting its layout and the stopping conditions for the training are then set.

2. The training starts by feeding the neural network with the ciphertext blocks selected from the dataset for training and using the corresponding plaintext blocks as reference output.

3. At the end of training it is decided whether this training trial is a failure or a success. Failure training is identified by reaching one of the stopping conditions with mean-squared-error more than 10^{-2}. On the other hand, successful training was identified by reaching the minimum possible mean-squared error with the minimum possible number of plaintext-ciphertext pairs.

4. If the training fails, the network is re-initialized with random weights and biases and the layout of the network is manually changed and the training is restarted at step 2. Most of the failures are caused by increasing slope of the mean-squared-error due to reaching local minima.

5. If the training was successful, we move to the validation process.

Validation Process

1. After the training stops, successfully, the same ciphertext matrix used in training is fed into the network to obtain neuro-decrypted-plaintext results; P'. Since the activation function used in all the neurons is non-linear, the output of the neural network

is a mostly floating-point numbers. Thus, the output of the neural network is rounded to 0 or 1 to produce data that is comparable to the reference output data.

2. P' is then compared to P, bit-by-bit and the percentage of bits in P' having a value different from the corresponding bit in P is calculated. This comparison is done using XOR operation. For the sake of simplicity, this error is called *inside-error*. The inside-error can be expressed in the following mathematical form:

$$inside\text{-}error = \frac{\sum_{i=1}^{m}\sum_{j=1}^{n} p'(i,j) \oplus p(i,j)}{m \times n} \tag{1}$$

where,

m is the number of bits in a block of training data

n is the number of blocks used in the training

$p'(i,j)$ is the jth bit in the ith block of the neural network output

$p(i,j)$ is the jth bit in the ith block of the plaintext used in training

For both DES and 3DES, $m=64$

3. Afterwards, more ciphertext is fed into the network; C_{new}. This ciphertext is not part of the ciphertext used in training, but encrypted with the same key. The output is also rounded to 0 or 1 in this step as in step 1.

4. The resulting neuro-decrypted-plaintext, P'_{new} is compared to the original plaintext P_{new} bit-by-bit, and the resulting error percentage is calculated. This error percentage is called *outside-error*. The outside-error can be expressed in the following mathematical formula:

$$outside\text{-}error = \frac{\sum_{i=1}^{m_{new}}\sum_{j=1}^{n_{new}} p'_{new}(i,j) \oplus p_{new}(i,j)}{m_{new} \times n_{new}} \tag{2}$$

where,

m_{new} is the number of bits in a block of validation data

n_{new} is the number of blocks used in the validation

p'_{new} is the jth bit in the ith block of output of the neural network in the validation

$p_{new}(i,j)$ is the jth bit in the ith block of plaintext used in the validation

For both DES and 3DES, $m_{new}=m=64$

5. As mentioned earlier, the ciphertext, C_{new}, and the plaintext P_{new} were chosen from the dataset such that they do not replicate any plaintext-ciphertext pair previously used in training, but the number of blocks in them ranged from 8-10 times the number of blocks used in the training.

Dataset Variation. The same training and validation procedures explained earlier are repeated for each dataset. These repetitions produce a separate neural network for each dataset (i.e. for each key used in encryption). This neural network is capable of retrieving plaintext from, virtually, any ciphertext that was encrypted with the same key used to encrypt the data used in training.

Encryption Algorithm Variation. The exact same processes mentioned previously were implemented on DES-encrypted datasets and Triple-DES-encrypted datasets using 100 different dataset for each algorithm.

3   Results of Implementation

After reaching a successful training trial on each one of the 100 datasets (i.e. producing a trained neural network for each dataset) for both DES and Triple-DES, some calculations were made to get a general tabulation of results. Table 1 shows the general results of implementing the cryptanalytic attack on 100 datasets for DES and 100 data sets for Triple-DES.

Table 1. Results of Implementing Neuro-cryptanalysis on DES and Triple-DES

Calculation	DES	Triple-DES
Total Number of trials	833	1093
Number of successful trials	100	100
Number of failure trials	733	993
Average number of plaintext-ciphertext pairs needed for training	2^{11}	2^{12}
Average time of successful training (min.)	51	72
Average time of failure trial (min.)	3	5
Average time until success (min.)	21	52
Average number of trials required to reach success	7.33	9.93
Average inside-error	0.022	0.028
Average outside-error	0.083	0.114

Table 2. Results of most successful 10 trials on DES

Net. Layout	text pairs needed	Inside error	Outside error	MSE reached	Training time (min.)	Epochs
128-256-256-128	2048	0.0279	0.0859	0.0133	39	308
128-256-256-128	2048	0.0273	0.0997	0.0104	39	298
128-256-256-128	2048	0.0298	0.1116	0.0274	41	334
128-256-512-256	2048	0.0341	0.1310	0.0151	42	356
128-256-256-128	2048	0.0283	0.0999	0.0049	42	341
128-512-256-256	2048	0.0378	0.1239	0.0474	42	351
128-256-256-128	2048	0.0344	0.1352	0.0359	42	355
128-256-256-128	2048	0.0295	0.0858	0.0006	43	388
128-256-512-128	2048	0.0415	0.1348	0.0768	43	379
128-256-256-128	2048	0.0267	0.1017	0.0759	43	411

For more detailed results, specific details of the most successful 10 trials in DES and Triple-DES cryptanalysis were provided in Tables 2 and 3. The attack with the least time required for training is considered most successful.

Table 3. Results of most successful 10 trials on Triple-DES

Net. Layout	text pairs needed	Inside error (%)	Outside error (%)	MSE reached	Training time (min.)	Epochs
128-512-512-128	4096	0.0303	0.1170	0.0437	62	242
128-512-512-128	4096	0.0495	0.2103	0.0590	63	248
128-512-512-128	4096	0.0473	0.1259	0.0309	63	239
64-128-256-512-1024	4096	0.0407	0.1639	0.0373	63	218
128-512-512-128	4096	0.0408	0.1748	0.0455	63	265
128-512-512-128	4096	0.0538	0.2171	0.0575	64	233
128-256-512-256	4096	0.0319	0.1507	0.0236	64	253
128-512-512-128	4096	0.0459	0.1947	0.0155	64	235
128-512-512-128	4096	0.0334	0.1313	0.0242	64	229
128-256-512-512	4096	0.0362	0.1724	0.0350	64	230

4   Discussions

This attack, as shown in the results tables, has succeeded in getting global deduction for DES with an average less than 2^{11} known-plaintexts. As compared to other attacks, this is the best result reached for a known-plaintext attack on DES. The fact that less than 60 minutes and one computer is enough to break the cipher renders the attack a practically implementable one.

As regarding Triple-DES, the results achieved by this attack are interesting. The need of an average of about 2^{12} known-plaintexts has never been achieved by any other cryptanalytic attack on Triple-DES. Table 4 compares the neuro-cryptanalysis results to other attacks on DES in terms of required number of plaintext-ciphertext pairs.

Table 4. Number of plaintext-ciphertext pairs required for different attacks on DES

Attack Type	Neuro-Cryptanalysis	Differential Cryptanalysis	Linear Cryptanalysis
Number of pairs required	2^{11}	2^{47}	2^{43}

A downside of this cryptanalytic method is the outer error. It is clear from the tables 2 and 3 that having less plaintext-ciphertext pairs can increase the outer error percentage. Having a time-until-success average of about 21 minutes for DES and 52 minutes for Triple-DES is a noticeable advancement in cryptanalysis of both of these standards. Hardware implementation of the proposed system can be done through dedicated neural processors or through other generalized hardware like Field-Programmable Gate Array (FPGA).

5   Conclusions

This paper focuses on implementing a new cryptanalytic attack on DES and Triple-DES based on neural networks. This attack was implemented by training a neural network to retrieve plaintext of the ciphertext fed into it. The attack presented here is a known-plaintext attack that retrieves most of the plaintext of a given ciphertext without retrieving the key.

The attack was successfully implemented using MATLAB. The successful cryptanalysis of DES required an average of 2^{11} plaintext-ciphertext pairs, while the cryptanalysis of Triple-DES required and average of 2^{12} plaintext-ciphertext pairs. The average time required to train the neural network to perform cryptanalysis of DES was 51 minutes, while this average training time was 72 minutes to achieve Triple-DES cryptanalysis. These results are considered an important achievement regarding the cryptanalysis of DES and Triple-DES. The results obtained for Triple-DES cryptanalysis are particularly important because Triple-DES is still widely used in many commercial applications. The proposed cryptanalytic attack can be further implemented in hardware to improve its performance. The scope of this attack can also be expanded to other block ciphers and other cryptanalytic techniques.

References

1. Albrecht, M., Cid, C.: Algebraic Techniques in Differential Cryptanalysis. In: Proceedings of the First International Conference on Symbolic Computation and Cryptography (2008)
2. Knudsen, L.R.: Block Ciphers - A Survey. In: Preneel, B., Rijmen, V. (eds.) State of the Art in Applied Cryptography. LNCS, vol. 1528, pp. 18–48. Springer, Heidelberg (1998)
3. NIST, Data Encryption Standard (DES), FIPS PUB 46, Federal Information Processing Standards Publication 46, U.S. Department of Commerce (1977)
4. Biham, E., Shamir, A.: Differential Cryptanalysis of DES-like Cryptosystems. Journal of Cryptology 4, 3–72 (1991)
5. NIST, Data Encryption Standard (DES), FIPS PUB 46-3, Federal Information Processing Standards Publication 46-3 (1999)
6. Matsui, M.: Linear Cryptanalysis Method for DES Cipher. In: Helleseth, T. (ed.) EUROCRYPT 1993. LNCS, vol. 765, pp. 386–397. Springer, Heidelberg (1994)
7. Biham, E., Shamir, A.: Differential Cryptanalysis of the Full 16-Round DES. In: Brickell, E.F. (ed.) CRYPTO 1992. LNCS, vol. 740, pp. 487–496. Springer, Heidelberg (1993)
8. Langford, S.K.: Differential-Linear Cryptanalysis and Threshold Signatures. PhD thesis, Stanford University, USA (1995)

9. Biham, E.: New Types of Cryptanalytic Attacks Using Related Keys. In: Helleseth, T. (ed.) EUROCRYPT 1993. LNCS, vol. 765, pp. 398–409. Springer, Heidelberg (1994)
10. Merkle, R., Hellman, M.: On the Security of Multiple Encryption. Communications of the ACM 24, 465–467 (1981)
11. Diffie, W., Hellman, M.E.: Exhaustive Cryptanalysis of the NBS Data Encryption Standard. Computer 10, 4–84 (1977)
12. Lucks, S.: Attacking Triple Encryption. In: Vaudenay, S. (ed.) FSE 1998. LNCS, vol. 1372, pp. 239–253. Springer, Heidelberg (1998)
13. Choi, J., Kim, J.-S., Sung, J., Lee, S.-J., Lim, J.-I.: Related-Key and Meet-in-the-Middle Attacks on Triple-DES and DES-EXE. In: Gervasi, O., Gavrilova, M.L., Kumar, V., Laganá, A., Lee, H.P., Mun, Y., Taniar, D., Tan, C.J.K. (eds.) ICCSA 2005. LNCS, vol. 3481, pp. 567–576. Springer, Heidelberg (2005)
14. Clark, M., Blank, D.: A Neural-Network Based Cryptographic System. In: Procceedings of the 9th Midwest Artificial Intelligence and Cognitive Science Conference (MAICS 1998), pp. 91–94 (1998)
15. Godhavari, T., Alamelu, N.R., Soundararajan, R.: Cryptography Using Neural Network. In: Proceedings of 2005 Annual IEEE INDICON, pp. 258–261 (2005)
16. Kanter, K.W., Kanter, E.: Secure exchange of information by synchronization of neural networks. Europhysics Letters 57, 141 (2002)
17. Klimov, A.B., Mityagin, A., Shamir, A.: Analysis of Neural Cryptography. In: Zheng, Y. (ed.) ASIACRYPT 2002. LNCS, vol. 2501, pp. 288–298. Springer, Heidelberg (2002)
18. Li, L., Lin, L., Hwang, M.: A remote password authentication scheme for multiserver architecture using neural networks. IEEE Transactions on Neural Networks 12, 1498–1504 (2001)
19. Dourlens, S.: Neuro-Cryptography. MSc Thesis, Dept. of Microcomputers and Microelectronics, University of Paris, France (1995)
20. Li, S.: Analyses and New Designs of Digital Chaotic Ciphers. PhD thesis, School of Electronics and Information Engineering, Xi'an Jiaotong University, Xi'an, China (2003)
21. Li, C., Li, S., Zhang, D., Chen, G.: Chosen-Plaintext Cryptanalysis of a Clipped-Neural-Network-Based Chaotic Cipher. In: Proceedings of International Symposium on Neural Networks (2005)
22. Rao, K., Krishna, M., Babu, D.: Cryptanalysis of a Feistel Type Block Cipher by Feed Forward Neural Network Using Right Sigmoidal Signal. International Journal of Soft Computing 4, 131–135 (2009)
23. Alani, M.M.: Neurocryptanalysis of DES. In: Proceedings of World Congress on Internet Security. University of Guelph, Guelph (2012)
24. Hornik, K., Stinchcombe, M., White, H.: Multilayer Feedforward Neural Networks are Universal Approximators. Neural Networks 2, 359–366 (1989)

An Expanded HP Memristor Model
for Memristive Neural Network

Yu Dai[1,2] and Chuandong Li[1]

[1] College of Computer Science, Chongqing University, Chongqing 400044, China
licd@cqu.edu.cn
[2] College of Computer Science, Chongqing University of Post and Telecommunications,
Chongqing 400065, China
daiyu@cqupt.edu.cn

Abstract. In this paper, based on classical HP memristor we present an expanded model that fully considers the influence of R_{on}. We demonstrate the hysteresis effect of the expanded model, and then make a comparison with the HP model under certain voltage load. Simulations show the Expanded HP memristor is a good candidate for memristive neural networks.

Keywords: Expanded HP memristor, R_{on}, memristance, memristive neural network.

1 Introduction

The memristor is considered as the fourth fundamental circuit element, which is first proposed by Chua in 1971 [1]. After 37 years in 2008, HP Lab announced that they had established the existence of what Chua has postulated, and presented a physical model [2], which usually called as HP memristor. Memristor itself acts like a passive [1], non-volatile [3-5], nonlinear [6] resistor, and has some astonishing properties, attracting lots of researchers' interests [7-11]. One of these researches is to find a proper use of memristor in neural network, which was named as Memristive Neural Network (MNN) [12]. Snider [13] describes a solution of instar learning in MNN that utilizing a hypothetical curve, but not based on the classical HP memristor. The reason is that it is hard to use HP memristor to act as an analogical 'Synapse' [14].

In this paper, our main aim is to extend HP memristor model by making a meaningful change in its classical definition in order to make the expanded memristor model easier to act as an analogical synapse. Simulation results in Matlab demonstrate the memristive characteristics in expended memristor model, and therefore, the Expanded HP memristor will be a good candidate for MNN.

2 HP Memristor

First, we take a review of this classical model. The memristor is a two-terminal device that by definition it obeys equations of the form given by Chua and Kang [15]:

T. Huang et al. (Eds.): ICONIP 2012, Part V, LNCS 7667, pp. 647–653, 2012.

$$i_M = G(x)v_M \tag{1}$$

$$\frac{dx}{dt} = f(x, v) \tag{2}$$

where v_M is the voltage driven on the memristor, i_M is the current; x denotes some physical states which could often change with time, so a nonlinear function $G(x)$ become the memductance of the memristor. As we know, it is a voltage-controlled (or flux-controlled) memristor [1]. Another kind of these devices, called current-controlled (or charge-controlled) memristor [1] obeys the follows:

$$v_M = R(x)i_M \tag{3}$$

$$\frac{dx}{dt} = g(x, i) \tag{4}$$

where $R(x)$ is the memristance of the memristor.

Fig.1 shows the physical model of HP memristor. A semiconductor film sandwiched between two metal contacts [2]. The inside film has two parts, of which contains dopants and the other not. The total memductance is determined by the position of the boundary between two parts. The HP Lab assumed that the Doped side filled with positive ions. Therefore, when a bias voltage drives on this device, the boundary will move because of the drift of the dopants. Especially, when a positive polar connects to the left side of the model in Fig.1, the Doped part's width increases and meanwhile the Undoped region decreases.

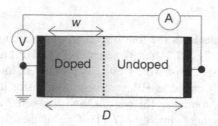

Fig. 1. HP memristor

Because the doped region has high concentration of ions, when the width of the Doped increased, the total resistance will decrease, which means the memductance grows to a high level. The Doped and the Undoped have resistance R_{on} and R_{off} when each of them reaches the full length D, respectively. According to the Kirchhoff's circuit laws, the total resistance (memristance) is given by the sum of inseries connected parts, means:

$$R_{mem} = R_d + R_{ud} \tag{5}$$

$$R_d \triangleq R_{on} \frac{w(t)}{D} \tag{6}$$

$$R_{ud} \triangleq R_{off}(1 - \frac{w(t)}{D}) \tag{7}$$

where R_{mem} is the resistance (memristance) of the HP memristor, R_d and R_{ud} denote the resistance of the Doped region and the Undoped region, respectively. D is the thickness of the film.

When the boundary reaches either side of the memristor, namely, $w = 0$ or $w = D$, R_{mem} equals to the R_{off} or R_{on}, respectively. Then, the memristor reduce to act like an ordinary resistor, which is not our interest. So subsequent discussion is based on $w \in (0, D)$, which means neither R_d nor R_{ud} could equal to 0 [6].

Compared R_d with R_{ud}, and noting that R_{on}, R_{off} and D are always constants in certain situation, we obtain

$$\frac{R_d}{R_{ud}} = \frac{R_{on}}{R_{off}} \cdot \frac{\frac{w(t)}{D}}{\left(1 - \frac{w(t)}{D}\right)}$$

$$= \alpha \cdot \frac{X(t)}{(1 - X(t))}$$

$$= \alpha \cdot \left(\frac{1}{1 - X(t)} - 1\right) \tag{8}$$

where $\alpha = R_{on}/R_{off}$ is a constant too, and $X(t) = w(t)/D$ is a nonlinear function with $X(t) \in (0, 1)$.

When the bias voltage crossed on the memristor does not change its polarity, the curve of $X(t)$ is always monotonically increasing (or decreasing). For simplification, we only consider the 'increasing' case, the 'decreasing' case is easy to obtain based on our discussion.

As mentioned above, it is assumed that $X(t)$ is monotonically increasing. According to Kirchhoff's Voltage Law, we can obtain

$$\frac{v_{R_d}(t)}{v_{R_{ud}}(t)} = \frac{R_d}{R_{ud}} = \alpha \cdot \left(\frac{1}{1 - X(t)} - 1\right)$$

$$= \alpha \cdot h(X(t)) \tag{9}$$

Making derivation of $h(X(t))$ by X yields:

$$\frac{dh(X)}{dX} = \frac{1}{(1 - X)^2}, X \in (0, 1) \tag{10}$$

We can learn from Eq.(10) that in region (0, 1), $h(X)$ is also monotonously increasing. Especially in (0.8, 1), the change is more rapid than that in (0, 0.8), as shown in Fig.2. This implies that the most influential drift takes place in (0.8, 1).

3   Expanded HP Memristor

As Strukov etc suggested in their great discovery [2], R_{on} is always far less than R_{off}. This conclusion allows us to simplify Eq.(5) to $R_{mem} = R_{ud}$. Therefore, the influence of R_{on} is totally eliminated.

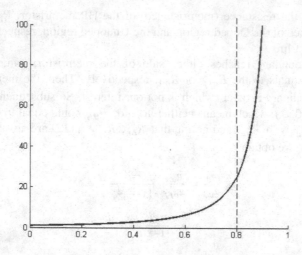

Fig. 2. Curve of derivation of $h(X)$

In our research, we suppose a kind of HP-like memristors, but its R_{on} is the similar order of magnitude of R_{off}. This may lead R_{on} to play an important role in the change of the total memristance.

The first thing we care about is the hysteresis effect in such situation. We choose α^{-1} equal to 2 as an example to examine our thought. When an AC voltage source connect to the emulational model, the v-i curve we obtain in Matlab looks almost like a straight line, as shown in Fig.3(a). The parameters we set refer to [2]; namely, v_0 is 1, w_0 is 1. Then we attempt to change the AC source's amplitude v_0. Accidentally when v_0 set to 0.02, we observe an expected ideal hysteresis pinched loop as shown in Fig.3(b). We also notice that no matter what values of R_{on} and R_{off} are chosen, the v-i curves are the same when v_0 is equal to 0.02, α^{-1} and w_{init} are always 2 and 0.5, respectively. The only difference is the region of i changes according to the magnitude of computed memristance. Obviously, this result meets the Ohm's Law.

Although we do not confirm the mathematical relationship among v_0, α and w_{init}, even including w_0 yet, we are well sure that even when R_{on} and R_{off} are the similar order of magnitude, the memristor is working well in some given situations. We call these kinds of memristors are Expanded HP memristors. Contents beneath this paragraph will compare Original and Expanded HP memristor to show the latter is much proper for MNN.

Fig.4(a) shows a schema to compare an Expanded HP memristor (denoted by M_{E-HP} in the sequel) to a standard HP memristor (denoted by M_{HP}) in the same direct bias voltage load. The curves are shown in Fig.4(b), where the value of DC is 1V, $\alpha_{M_{E-HP}}^{-1}$ is 2, $\alpha_{M_{HP}}^{-1}$ is 160, w_{init} are both 0.5.

From Fig.4, we realize that while α^{-1} is getting bigger, the more time is needed to change the memristance.

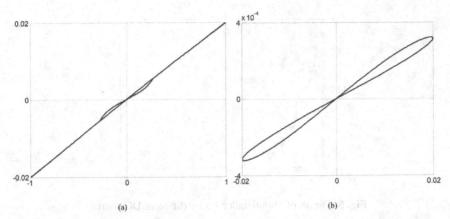

(a)   (b)

Fig. 3. v-i curve of an Expanded HP memristor ($\alpha^{-1} = 2$) under different AC voltage source ($w_0 = 1$). (a) v_0 equals to 1V. (b) v_0 equals to 0.02V

Fig.5 displays the change curves of memristance of same M_{E-HP} under different DC source. Curve 1 is under the larger voltage of 2 times of Curve 2, which memristance drops nearly 2 times faster than the latter.

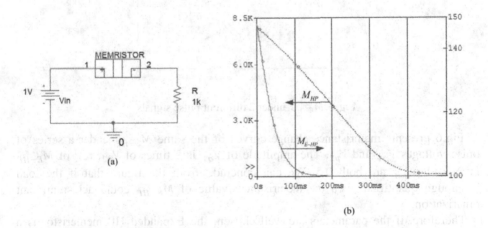

(b)

Fig. 4. Comparison between M_{E-HP} and M_{HP} under DC Voltage Source

We also noticed, in HP memristor, $w(t)$ is given by

$$\frac{dw(t)}{dt} = \mu_v \frac{R_{on}}{D} i(t) \tag{11}$$

where μ_v is the average ion mobility. We conjecture that different material used as dopants has different μ_v. Therefore, it would be lots of HP memristor-like models, which have different characteristics curves. Even the μ_v is fixed, we can get different curves by modifying the HP memristor's manufactured parameters, or changing the load on it.

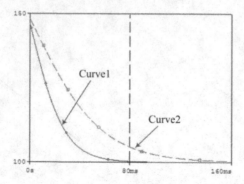

Fig. 5. Curves of Memristance under different DC source

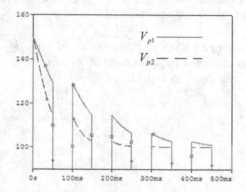

Fig. 6. M_{E-HP} under 2 different pulse signals

Fig.6 presents memristance change curves of the same M_{E-HP} under a series of pulse voltages V_{p1} and V_{p2}. The amplitude of V_{p1} is 2 times of V_{p2}, α^{-1} of M_{E-HP} is 2, and w_{init} are both 0.5. We can conclude from the figure that if the load is enough quantified, then the memristance value of M_{E-HP} could act significant binarization.

Therefore, if the parameters are well chosen, the Expanded HP memeristor is a good candidate for MNN, especially in binary computation. Furthermore, because the memristance of Expanded HP memristor is much easier to change than the standard one; for example, a voltage-controlled Expanded HP memristor with $\alpha^{-1} = 2$ only needs 2 times of given voltage to significantly change its memristance. Therefore, MNN using the expanded model can obviously reduce the energy dissipation [14]. That would be constructive to build high density of neuromorphic circuits.

4   Conclusions

We have proposed an Expanded HP memristor which memristance is easy to change in certain voltage load (for voltage-controlled memristor, and current-controlled memristor is easy to prove). This kind of memristor model is a good candidate for MNN, especially in binary computation. Our future work mainly includes two parts. One is to discover the mathematical relationship among v_0, α, w_{init}, and w_0. The other is to apply the Expanded HP memristor to large-scale memristive neural networks.

References

1. Chua, L.O.: Memristor - The Missing Circuit Element. IEEE Transactions on Circuit Theory 18, 507–519 (1971)
2. Strukov, D.B., Snider, G.S., Stewart, D.R., Williams, R.S.: The missing memristor found. Nature 453, 80–83 (2008)
3. Rozenberg, M.J., Inoue, I.H., Sánchez, M.J.: Nonvolatile Memory with Multilevel Switching: A Basic Model. Physical Review Letters 92, 178302 (2004)
4. Jo, S.H., Lu, W.: CMOS Compatible Nanoscale Nonvolatile Resistance Switching Memory. Nano Letters 8, 392–397 (2008)
5. Ho, Y., Huang, G.M., Li, P.: Nonvolatile Memristor Memory: Device Characteristics and Design Implications. In: 2009 IEEE/ACM International Conference on Computer-Aided Design Digest of Technical Papers, pp. 485–490. ACM, New York (2009)
6. Biolek, Z., Biolek, D., Biolková, V.: SPICE Model of Memristor with Nonlinear Dopant Drift. Radioengineering 18, 210–214 (2009)
7. Snider, G.: Self-organized computation with unreliable, memristive nanodevices. Nanotechnology 18, 365202 (2007)
8. Biolek, D., Biolek, Z., Biolkova, V.: SPICE modelling of memcapacitor. Electronics letters 46, 520–522 (2010)
9. Pershin, Y.V., Ventra, M.D.: Practical Approach to Programmable Analog Circuits With Memristors. IEEE Transaction on Circuits and Systems 57, 1857–1864 (2010)
10. Batas, D., Fiedler, H.: A Memristor SPICE Implementation and a New Approach for Magnetic Flux-Controlled Memristor Modeling. IEEE Transactions on Nanotechnology 10, 250–255 (2011)
11. Lin, Y.J., Hou, C.L., Su, T.J.: Cellular Neural Networks for Noise Cancellation of Gray Image Based on Hybrid Linear Matrix Inequality and Particle Swarm Optimization. In: 2009 International Conference on New Trends in Information and Service Science, pp. 613–617. IEEE, New York (2009)
12. Pershin, Y.V., Ventra, M.D.: Experimental demonstration of associative memory with memristive neural networks. Neural Networks 23, 881–886 (2010)
13. Snider, G.: Instar and outstar learning with memristive nanodevices. Nanotechnology 22, 015201 (2011)
14. Jo, S.H., Chang, T., Ebong, I., Bhadviya, B.B., Mazumder, P., Lu, W.: Nanoscale Memristor Device as Synapse in Neuromorphic Systems. Nano Letters 10, 1297–1301 (2010)
15. Chua, L.O., Kang, S.M.: Memristive Devices and Systems. Proceedings of the IEEE 64, 209–223 (1976)

Aimbot Detection in Online FPS Games Using a Heuristic Method Based on Distribution Comparison Matrix

Su-Yang Yu*, Nils Hammerla, Jeff Yan, and Peter Andras

School of Computing Science, Newcastle University,
Newcastle upon Tyne, NE1 7RU, United Kingdom
{s.y.yu,nils.hammerla,jeff.yan,peter.andras} @ncl.ac.uk

Abstract. Online gaming is very popular and has gained some recognition as the so called *e-sport* over the last decade. However, in particular First Person Shooter (FPS) games suffer from the development of sophisticated cheating methods such as aiming robots (aimbot), which can boost the players ability to acquire and track targets by the illicit use of internal game states. This not only gives an obvious unfair advantage to the cheater, but has negative impact on the gaming experience of honest players.

In this paper we present a novel supervised method based on distribution comparison matrices that shows very promising performance in the identification of players that use such aimbots. It extends our previous work in which two features were identified and shown to have good predictive performance. The proposed method is further compared with other classification techniques such as Support Vector Machines (SVM). Overall we achieve true positive and true negatives rates well above 98% with low computational requirements.

Keywords: Cheating Detection, Distribution Comparison, Computational Intelligence, Computer Games, Game Bots, First Person Shooters.

1 Introduction

First Person Shooter (FPS) is a popular genre in multiplayer online games. It is viewed from the first person perspective where the player can shoot at targets using a selection of weapons. In the multiplayer variety, the players compete with humans instead of scripted opponents. Reaching high levels of performance in this very competitive type of game requires years of extensive training, where some players naturally seek to boost their performance with other, unauthorised ways.

Aiming robots (*aimbots*) support the player in acquisition and tracking of targets (i.e. their aim) and represent the most common and effective way to cheat

---

* This work was supported by Microsoft Research through its PhD Scholarship Programme.

T. Huang et al. (Eds.): ICONIP 2012, Part V, LNCS 7667, pp. 654–661, 2012.

in this genre. They access internal game states that are otherwise unavailable to honest players [7] and easily outperform the human's ability in these crucial aspects of performance. Beyond the obvious unfair advantage for the individual, this type of *bot* has a severe negative impact on the enjoyment of other honest players.

In our previous work [8] we identified two features, and proposed a voting scheme using the Kolmogorov-Smirnov test (K-S test). Although showing promising performance this approach had its limitations such as lack of scalability, as (multidimensional) K-S tests are computationally expensive.

In this paper, we present a novel heuristic scheme based on well-known statistical tests (T-test, F-test, and K-S test) that form the basis for classification experiments. Fundamentally, we train a binary classifier to perform labelling by finding the category of distributions that the sample distribution is closest to. This differs from a 2-sample statistical test where the comparison can only be made between the sample and a single target distribution. With it we address the aforementioned limitations of the K-S approach employed in our previous work, and improve computational performance as well as overall accuracy. Comparisons to other classification methods justify our approach, as significantly better performance can be achieved.

2   Background and Related Work

In the current literature, works on anti-cheating can be divided into prevention and detection. The former focuses on techniques that are designed to prevent hacking such as hiding or encrypting the game states, and the latter focuses on collecting evidence that can then be used to identify the cheaters. Our work belongs to a subgroup of the latter which evaluates the player's in game behaviour. In this subgroup different machine learning techniques with their associated features have been proposed.

Kim et al.[4], and Gianvecchio et al.[2] looked into identifying automatic robots (auto-bots) in Massive Multiplayer Online RPGs (MMORPGs). Kim et al.'s work compared different machine learning techniques using 26 windows event based features [4], and Gianvecchio et al.'s work used neural network and evaluate the reaction times of the players' input controls [2].

In the FPS genre, Chen et al.[1] have looked into identifying auto-bots in FPS by evaluating the avatar's trajectories, they compared techniques using k-NN, SVM, both with and without dimensionality reduction. Yeung et al. proposed the only other work specific to aimbot detection in FPS known to us, using a dynamic Bayesian network [7]. Yan discussed the challenge of differentiating aimbot users and super human players in [6].

In our previous work [8], we introduced two features and have shown with the 2-sample Kolmogorov-Smirnov tests (K-S test) [3] that statistical differences between the distributions of honest players and aimbots users are present in both features. We also proposed a statistical voting scheme based on K-S tests and have achieved good results. Which will be briefly overviewed in section 4.

3 Data Collection

We collected data using a modified version of the open source game Cube [5]. We modified the client to include 3 modes of aimbots from [7]: i) always on, ii) randomly turns on and off, and iii) randomly misses. We also modified the server to include a logging facility, and the means to control the weapon selection and aimbot activation of its clients[1]. We logged all the data packets the server receives. As all packets are necessary for the game synchronisation, we guarantee that any features we then compute can be performed by the server alone (i.e. can be performed in a real world scenario without having to trust the clients).

In total, 21 players with a wide spread of skill levels, ranging from beginner to veteran with 5+ years of experience, participated in our study. We randomly allocated participants into 4 game sessions. Each session contains 6 rounds (10 minutes each), where in each round the game is restricted to a single weapon-choice for all participants. All sessions are run using a fixed protocol with a 15 minute break after the first 3 rounds, and a warm up period[2] (up to 10 minutes) before the session and after the break. An examiner is present for briefing, debriefing and general co-ordination.

We allocated a 'no aimbot' round for each of the 4 weapons available (chain gun, shotgun, rocket launcher, and rifle), and an 'aimbot' round[3] for only the chain gun, and rifle rounds. This decision is made to limit the duration of the sessions. The above two weapons chosen are typical weapons of choice for aimbot users, thus we believe the data we collect is sufficient.

4 Kolmogorov-Smirnov Test Based Evaluations

4.1 Feature Selection and Analysis

In our previous work [8] we introduced the following features, Cursor Acceleration when acquiring Aim (AccA) and Time on Target (ToT). Both features occur exactly once in each targeting event, where a targeting event is defined as the period starting from when the target is acquired up until when the target has changed or is lost.

1. Cursor Acceleration when acquiring Aim (AccA). This is a measurement of the change in acceleration when the targeting event starts. We calculate this by taking the 2nd derivative of $\sqrt{(\varDelta\text{Yaw})^2 + (\varDelta\text{Pitch})^2}$ where $\varDelta$ is the difference of the angles between time t and $t-1$. Yaw and Pitch are used by the game to model view orientation. The unit for this feature is

---

[1] The server controls are put in place to ensure sufficient data can be collected for all weapons, as in earlier pilot studies most players neglected the weaker weapons, and did not use the 2nd or 3rd aimbot modes.

[2] The warm up period is to ensure all participants are present, and for them to readjust to the virtual environment.

[3] In the 'aimbot' rounds, we set the aimbot modes to automatically toggle after every 3 kills, and the participants were informed of this during the briefing.

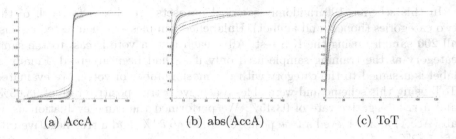

| (a) AccA | (b) abs(AccA) | (c) ToT |

Fig. 1. The Empirical Cumulative Distribution Functions (ECDF) for the 3 primary features The ECDFs are evaluated for a single player for the duration of a single round. This shows the mean (thicker lines) and the std devs (dotted lines) of the two categories (honest-solid, aimbot-dashed). Horizontal axes are the units of the feature. Vertical axes are the cumulative probability. The unit in (a) & (b) are $degree/(frame \times frame)$. The unit in (c) is $frame$ (approx.50 ms).

$degree/(frame \times frame)$, where each frame is approximately 50 milliseconds. In this paper, we investigate both the signed and the absolute value variant of this feature, denoted by AccA and abs(AccA) respectively.

2. Time on Target (ToT). This is a duration of the targeting event, measured in the number of frames.

For the rest of this paper, we will refer to AccA, abs(AccA) and ToT as our 3 primary features. Fig. 1 shows the Empirical Cumulative Distribution Functions (ECDF) for each feature. With each ECDF plotted over the data of a single player, in a single round.

Analysing the abs(AccA) data, we find that only 9.7% of the overall honest players' targeting events had absolute accelerations that are greater than 4.5degrees/(frame × frame), compared to the 24.8% by the aimbot users (Fig. 1(b)). By looking at the AccA data, we see that this spread is wider in the decelerations than in the accelerations (Fig. 1(a)).

Analysing the ToT data, we find that only 9.9% of the overall honest players' targeting events lasted longer than 8 frames, whereas this is the case for 27.4% of the aimbot users' events (Fig. 1(c)).

These results are intuitive as we expect the aimbots to be able to perform superior functions more frequently than honest players.

4.2 Limitations of the Kolmogorov-Smirnov Test

In [8] we used 2-sample Kolmogorov-Smirnov tests (K-S test) [3] to show there are statistical differences between the distributions of honest players and aimbots users. However, given the intra-category distribution spread shown in Fig. 1, it appears that the distributions within each category can also be significantly different from each other. A single 2-sample K-S test is therefore not reliable enough, as it relies on high intra-category similarity. To address this we introduced a voting scheme.

In this scheme, 100 randomly sampled datasets are chosen for each of the two categories (honest and aimbot). Unlabelled samples are then tested against all 200 samples using the K-S test. After each test, a vote is cast to the same category as the training sample used only if h_0 had been accepted. Finally, a label is assigned to the category with the most number of votes. We evaluated ToT using this scheme and were able to achieve a true positive rate of 93.65%, and a true negative rate of 93.45%. We performed the same evaluation using abs(AccA) and achieved a true positive rate of 46.83%, and a true negative rate of 97.82%.

There are limitations to this approach, including the following: 1) only a single feature can be used (computing the K-S statistics in more than one dimension using this scheme is infeasible), and 2) we could not make effective use of 1-tailed tests because it is possible for the test statistics of two different distributions to be significantly larger and smaller than the other at the same time.

5   A Heuristic Approach

5.1   Overview

We addressed the limitations of the K-S test with a novel method. The aim is to learn which category the sample dataset is closest to, given the deviation between the distribution of the training and sample datasets. First, we define a *heuristic* as a set of values that can describe a distribution, and a *comparison function* as a function which computes the deviation of two heuristics.

In the training phase, we start by sampling 2 sets of random sample datasets, one for each category and with equal length. Where each dataset is the targeting events of a single category (honest or aimbot) randomly sampled from the training data. For each primary feature we retain some heuristic to describe the overall distribution of the training dataset (*training heuristic*) and the random samples (*sample heuristic*). Using a comparison function we compute the differences between the training and sample heuristics which we denote as *secondary features*.

Once this is complete, we train a classifier using an augmented input, whereby each row of the input is a set union of the secondary features from the 3 primary features. Only the training heuristics and the trained classifier are kept for the labelling phase.

In the labelling phase, for each primary feature we compute the secondary feature using sample heuristic of the unlabelled dataset and the relevant training heuristic. Once complete, we use the trained classifier to assign a label to the dataset, using augmented secondary features (in the same manner as the training phase) as input.

Note, we use random sample datasets because it is possible for the same targeting event to reoccur, and in different ordering, therefore these are valid theoretical scenarios. As random samples are likely to deviate from the original distribution they were sampled from, using a sufficiently large number of random samples we can provide a good coverage for possible intra-category deviations.

5.2   Heuristics Sets

We derived 2 sets of heuristics to be used by our method, the first from the K-S test, and the second from the T-test & the F-test. Each heuristic set consists of training heuristic, sample heuristic, comparison functions and their resultant secondary feature.

1. The K-S heuristic set. These calculations are derived from the K-S test, the idea comes from performing 1-tailed tests. With a Middle Cumulative Distribution Function (MCDF) whereby the CDFs from one category is vertically above it and the other below it, we can use 1-tailed tests to label datasets by evaluating the vertical difference between its CDF and the MCDF.
   In this set the training heuristic is {MCDF}, the secondary feature {$Max\Delta$, $Min\Delta$, and $MaxABS\Delta$} are computed using the sample heuristic {Sample's CDF (SCDF)} in the following comparison functions:

$$Max\Delta = max(SCDF, MCDF), Min\Delta = min(SCDF, MCDF)$$

$$MaxABS\Delta = \begin{cases} Max\Delta & \text{if } abs(Max\Delta) > abs(Min\Delta) \\ 0 & \text{if } abs(Max\Delta) = abs(Min\Delta) \\ Min\Delta & \text{if } abs(Max\Delta) < abs(Min\Delta) \end{cases}$$

   Where $max()$ and $min()$ respectively return the maximum and minimum vertical difference between the input CDFs. We use this heuristic set for ToT and abs(AccA) but not AccA, because in the AccA the vertical ordering of the two categories changes around 0 (see Fig.1(a)).

2. The Mean Var heuristic set. These calculations are derived from the T-Test and the F-Test, and we use it for AccA because its distribution is similar to normal. The training heuristic are the mean and variance for the honest and aimbot distributions {$\bar{x}_H$, s_H^2, $\bar{x}_A$, s_A^2}. We compute the secondary feature {$\Delta\bar{x}_H$, Δs_H^2, $\Delta\bar{x}_A$, and Δs_A^2} for sample heuristic {$\bar{x}$, s^2} using the following comparison functions:

$$\Delta\bar{x}_H = (\bar{x} - \bar{x}_H)^2, \Delta s_H^2 = (s^2 - s_H^2)^2, \Delta\bar{x}_A = (\bar{x} - \bar{x}_A)^2, \Delta s_A^2 = (s^2 - s_A^2)^2$$

5.3   Evaluation

Using linear regression and 7-fold cross validation, we performed two evaluations using different datasets for validation: 1) random sample datasets, 2) empirical datasets from the collection process. The former is a theoretical evaluation to see how sample sizes affect our method (results are shown in Table 1), and the latter is to see how effective our method would have been if it were deployed during the data collection process. We used 1,000 as the number of random sample datasets for each category.

In the second evaluation, we set the training sample size to 100 because based on the results of the first evaluation, further increase in size did not yield significant improvements. The empirical datasets used in the evaluation each contains exactly one player's data from a single round (due to the format of the collection process these were either all honest or all aimbot). For this, we achieved a true positive rate of 98.21% and true negative rate of 98.81%.

Table 1. Random Sample Evaluation[4] This table shows the True positive (TP) and True negative (TN) rates of our method using randomly sampled datasets for validation. The input is all 3 primary features, and the classifier is trained using Linear Regression.

		Sample size of the random sample datasets used for validation								
		10		30		50		**100**		150	
		TN	TP	TN	TP	TN	TP	TN	TP	TN	TP
Training size	10	81.57%	88.82%	94.71%	95.94%	97.14%	97.69%	98.94%	99.32%	99.46%	99.71%
	30	74.50%	89.11%	94.28%	96.44%	97.61%	98.28%	99.60%	99.70%	99.91%	99.90%
	50	69.87%	89.65%	93.03%	96.70%	97.24%	98.39%	99.49%	99.73%	99.85%	99.91%
	100	63.17%	89.70%	89.55%	95.98%	95.89%	98.05%	**99.33%**	**99.62%**	99.81%	99.91%
	150	62.03%	88.73%	87.87%	95.66%	94.73%	97.88%	99.00%	99.55%	99.70%	99.87%
	200	61.55%	88.35%	87.16%	95.69%	94.55%	98.07%	99.00%	99.61%	99.75%	99.89%

6   Comparison to Support Vector Machines

We compared our methods to Support Vector Machines (SVM). In this evaluation, we used augmented inputs where each row is the set union of all 3 primary features for a sequence of targeting events from event T to $T - x$ where T is the latest event in the sequence. We repeated the tests for different sequence lengths at 10, 30, 50 and 100. The experiment was carried out using four different kernels (linear, polynomial, sigmoid, and radial basis functions). Linear and polynomial kernels performed well (e.g. with the sequence length of 100, linear: TN 95.99%, TP 97.82%, polynomial: TN 99.05%, TP 93.91%), but the other two did not for this dataset. We compared our proposed method against these using the 2-sample T-Test. The comparisons were performed between the results of our method using the same sample size for training and validation as the sequence length used for the input of the SVM. We found our method to be statistically significantly better for all sample sizes.

7   Conclusion and Future Work

In this paper, we presented a novel heuristic method to differentiate aimbot users from honest players in online FPS games. The method utilises ideas and calculations taken from 2-sample statistical tests to perform feature extraction. The new features describe the deviations of a sample distribution from that of the training data. Using a linear classifier with the new features, we then determine the category of the sample. With this, we were able to achieve a true positive rate of 98.21% and a true negative rate of 98.81% on a realistic dataset.

This approach requires far less computation allowing it to be usable in real time, and has achieved better overall accuracy in comparison to our previous work [8]. It addresses several limitations, such as being able to perform feasible

---

[4] The values in bold are used to compare against the SVMs with the length of the input sequence being set to 100 as described in section 6.

evaluation using multi-dimensional inputs, and compare new datasets against a category of distributions rather than only a single distribution. For this, we believe our method is worth further investigation.

In our future work, we would like to perform more extensive user studies, test different sub-genres of FPS (e.g. tactical, role playing etc.) and platforms (e.g. Xbox 360). This may help us to understand further the limitations of our methods and how we could generalise it.

Acknowledgements. We thank Siu Fung Yeung for providing us with the implementation of their aimbots, thank Carl Gamble, Richard Payne and anonymous reviewers for their useful comments, and to all the participants of our experiment.

References

1. Chen, K.T., Pao, H.K.K., Chang, H.C.: Game bot identification based on manifold learning. In: Proceedings of the 7th ACM SIGCOMM Workshop on Network and System Support for Games, pp. 21–26 (2008)
2. Gianvecchio, S., Xie, M., Wu, Z., Wang, H.: Battle of Botcraft: fighting bots in online games with human observational proofs. In: Proceedings of the 16th ACM Conference on Computer and Communications Security, pp. 256–268 (2009)
3. Frank, J., Massey Jr.: The Kolmogorov-Smirnov Test for Goodness of Fit. Journal of the American Statistical Association 46, 68–78 (1951)
4. Kim, H., Hong, S., Kim, J.: Detection of Auto Programs for MMORPGs. In: Zhang, S., Jarvis, R.A. (eds.) AI 2005. LNCS (LNAI), vol. 3809, pp. 1281–1284. Springer, Heidelberg (2005)
5. Van Oortmerssen, W.: Cube (2005), http://www.cubeengine.com
6. Yan, J.: Bot, Cyborg and Automated Turing Test. In: Security Protocols Workshop, pp. 190–197 (2006)
7. Yeung, S.F., Lui, J.C.S., Liu, J., Yan, J.: Detecting cheaters for multiplayer games: theory, design and implementation. In: Consumer Communications and Networking Conference, pp. 1178–1182 (2006)
8. Yu, S.-Y., Hammerla, N.Y., Yan, J., Andras, P.: A statistical aimbot detection method for online FPS games. In: The Preceedings of International Joint Conference on Neural Networks (2012)

An Improved NN Training Scheme Using Two-Stage LDA Features for Face Recognition

Behzad Bozorgtabar[1] and Roland Goecke[1,2]

[1] Human-Centred Computing Lab, University of Canberra, Australia
[2] Research School of Computer Science, Australian National University, Canberra
behzad.bozorgtabar@canberra.edu.au, roland.goecke@ieee.org

Abstract. This paper presents a new approach based on a Two-Stage Linear Discriminant Analysis (Two-Stage LDA) and Conjugate Gradient Algorithms (CGAs) for face recognition. A Two-Stage LDA technique is proposed that utilises the null space of the sample covariance matrix as well as using the range space of the between-class scatter matrix to extract discriminant information. Classic *Back Propagation* (BP) is a widely used *Neural Network* (NN) training algorithm in many detectors and classifiers. However, it is both too slow for many practical problems and its performance is not satisfactory in many application areas, including face recognition. To overcome these problems, four CGA algorithms (Fletcher-Reeves CGA, Polak-Ribiere CGA, Powell-Beale CGA, scaled CGA) have been proposed, the utility of which we investigate here in combination with Two-Stage LDA features. To further improve the accuracy, a modified AdaBoost.M1 approach was employed, which combines results of several NN classifiers as a single strong classifier. Experiments are performed on the ORL, FERET and AR face databases. The results show that all of the proposed methods lead to increased recognition rates and shorter training times compared to the classic BP.

Keywords: Neural network, improved learning, face recognition.

1 Introduction

Face recognition has had many effects on daily life, e.g. security access control systems, content-based indexing, and bank teller machines [1]. Recent developments have facilitated faster processing and higher accuracy. However, difficulties exist in the development of efficient visual feature extraction algorithms and the high computational cost for retrieval from huge image databases.

LDA is a popular technique for dimensionality reduction and feature extraction, which has been employed in face recognition (e.g. [1,2]). LDA can suffer from the small sample size (S3) problem, meaning the dimension of the feature space is larger than the number of training examples per class. The discriminant vectors are computed from the eigenvalue decomposition of $S_W^{-1}S_B$, where S_W is the within-class scatter matrix and S_B is the between-class scatter matrix [3]. The problem is that S_W can become singular and the computation of its inverse

T. Huang et al. (Eds.): ICONIP 2012, Part V, LNCS 7667, pp. 662–671, 2012.

impossible. Several approaches have been proposed to overcome this drawback of LDA, such as the pseudo inverse [4] and regularized [2] LDA techniques. Our algorithm is a two-stage process. First, it selects those bases of the sample covariance matrix that are most correlated to the basis vectors of S_B. Then, it uses the Fisher criterion to select the most discriminative vectors from $\mathcal{R}^p$, where p is the dimensionality of the original feature vectors.

Many previous face recognition methods use classic BP to train a NN. Despite the widespread use, the BP algorithm has many deficiencies, e.g. a low convergence speed, convergence to local minima, and easily being affected by outliers in the signal during the learning process. To overcome these problems, in this paper, four kinds of *Conjugate Gradient Algorithms* – Fletcher-Reeves CGA, Polak-Ribiere CGA, Powell-Beale CGA, and scaled CGA – are used for training a feed-forward NN, which we evaluate on a face recognition task. In the CGAs, a search is done along conjugate directions, which normally generates faster convergence than that of the steepest descent directions [5]. Finally, the improved AdaBoost.M1 [6] approach, useful for classification of multiple classes, is used. Here, subsequent classifiers are tweaked in favour of those instances misclassified by previous classifiers.

The remainder of this paper is organised as follows. The Two-Stage LDA as a feature extraction method is briefly described in Section 2. Section 3 presents the four CGA types investigated here. Section 4 describes the experiments on the ORL and FERET face databases. The modified AdaBoost.M1 is presented in Section 5 and the conclusions given in Section 6.

2   A Two-Stage LDA

In the LDA, we can find the discriminant vectors by defining

$$(S_T^{-1} S_B)_S = \sum_{j \in S} \sum_{i=1}^{q} \frac{\lambda_{B_i}}{\lambda_{X_j}} u_j^T w_i u_j w_i^T \tag{1}$$

where S_T is the sample covariance matrix, $U = \{u_1, \ldots, u_p\}$ the eigenvectors of S_T, and $\{\lambda_{X_1}, \ldots, \lambda_{X_p}\}$ the corresponding eigenvalues. Decomposing Eq. 1, $u_j^T w_i$ acts as an important factor, showing u_j has to be correlated to the $span(w_i)$. This result can be described as a correlation-based criterion [7]

$$I_j = \sum_{i=1}^{q} \left(u_j^T w_i \right)^2 (1 \leq j \leq p) \tag{2}$$

We used this criterion for experiments on the ORL face database, with 5 training images per class. Using the classic approach for selecting the eigenvectors of S_T, we sort the eigenvectors u_i in descending order of eigenvalues (Figure 1 (left)). This figure shows that several eigenvectors associated to large eigenvalues are not strongly correlated to the $span(S_B)$, while some eigenvectors associated with

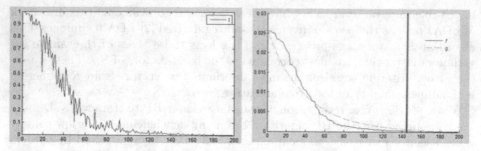

Fig. 1. (Left) Correlation values for the ORL face database. (Right) Exponential function $g_x = \lambda e^{-\lambda x}$. Red line: Cut where 80% of the total variance is kept.

small eigenvalues are. To solve this problem, we re-order the eigenvectors of S_T, with the new order given by the correlation value I_j (i.e. $I_1, \ldots, I_p$). Since

$$\frac{1}{q} \sum_{j=1}^{p} I_j = \frac{1}{q} \sum_{j=1}^{p} \sum_{i=1}^{q} \left(u_j^T w_i \right)^2 = 1 \tag{3}$$

the normalised result will be

$$f_j = \frac{1}{q} I_j = \frac{1}{q} \sum_{i=1}^{q} \left(u_j^T w_i \right)^2 \tag{4}$$

This function is monotonically decreasing and is guaranteed to start at $f_1 \approx \frac{1}{q}$ and end at $f_k = 0$ for some $k > 1$. Considering that when the number of classes q is large, the curve defined by f_j is less steep (as $span\,(w_i)$ is a subspace of high dimensionality) and that when q is small, the curve approaches zero much faster (as $span\,(w_i)$ represents a subspace of low dimensionality), allowing us to approximate the curve defined by $\{f_1, \ldots, f_p\}$ with an exponential function $g_x = \lambda e^{-\lambda x}$ described by discrete instances. I_j is considered when selecting the eigenvectors of S_T that are most correlated to the $span\,(S_B)$ [7]. Figure 1(right) shows how g_x can approximate the shape of the original curve defined by the f_i. The line represents the cut where 80% of the total variance is kept.

3   The Neural Network Model

In this section, we first describe the architecture of the proposed NN and then present a detailed model of the improved network.

3.1   Network Architecture

The *Feed-Forward Neural Network* (FFNN) is a suitable structure for nonlinear separable input data. After calculating the feature vectors using a two-stage LDA, the projection vectors are calculated for the training set and then used to

Table 1. Main steps of the Conjugate Gradient Algorithms

Conjugate Gradient	Weights are updated along the search direction: $\Delta w = \alpha_k p_k$, where search direction defined as: $p_k = -g_k + \beta_k p_{k-1}$, where g_k is the gradient in the kth iteration. Learning step α_k is found through the line search. β_k is a positive scalar that varies in different versions of CGAs. For example, β_k in Fletcher-Reeves is updated as follows: $\beta_k = \dfrac{g_k^T g_k}{g_{k-1}^T g_{k-1}}$

train the NN. When a new test image is to be recognised, the image is mapped to the feature space and assigned a feature vector. Each feature vector is fed to its respective NN and the network outputs are compared.

3.2 Network Training

The aim of NN training is to iteratively reduce an error function. Here, we use the *Mean-Square-Error* (MSE) function [8], which uses the network to represent the discriminant function directly. The overall error is defined as the MSE between the network outputs and the desired outputs

$$E_{MSE} = \frac{1}{K \times N} \sum_{k=1}^{K} \sum_{n=1}^{N} |y_n^k - d_n^k|^2, \quad n = 1 : N, \quad k = 1 : K \qquad (5)$$

For each network output y_n^k, vector d_n^k is the corresponding desired output, where n and k are the number of outputs and feature vectors, respectively.

3.3 Conjugate Gradient Algorithms

In the classic BP algorithm, the weights are adjusted in the steepest descent direction. Although the performance function decreases most quickly along the negative slope of the gradient, it does not necessarily generate the fastest convergence [5,9]. In contrast in CGAs, the weights and biases are not adjusted in the steepest descent direction but along the conjugate directions, resulting in faster convergence. In this paper, we propose to use CGAs (see Table 1) for the training of a FFNN and investigate the performance of four different types of CGAs – Fletcher-Reeves CGA, Polak-Ribiere CGA, Powell-Beale CGA, and scaled CGA – on the example of a face recognition task [10].

4   Experimental Results and Discussion

Three widely used face datasets (ORL, FERET, AR) are utilised in the experiments. For a fair comparison, the performance of the different conjugate gradient algorithms is evaluated via the recognition rates and variables such as learning rate and gradient according to the number of epochs for the databases.

4.1   Experiments on the ORL Face Database

The ORL face database contains 400 images of 40 individuals (10 images per person) with various facial expressions and lighting conditions. To extract visual features, we first implemented a *Two-Stage LDA* feature extractor. For all methods, the images were normalised and cropped to 70×70 pixels. Then, we trained a FFNN with 40% of the images in the database (images 1-4 for each person) and used the other 60% for testing (images 6-10 for each person).

Deciding how many neurons to use in the hidden layer is one of the most important characteristics in an NN. When the number of neurons is too low, the NN cannot model complex data and the result may be unacceptable. If too many neurons are used, it not only increases the training time, but may also reduce the performance. We investigated a range from 10 to 200 neurons in the hidden layer and tested the performance. We empirically selected a tangent sigmoid function for the NN with training by all of the CGAs. When we increased the number of iterations to more than 700, the accuracies of the proposed methods do not change significantly, but the training time increases considerably. Thus, we suggest using 700 iterations for training a FFNN by using the improved BP training algorithms. Due to the uncertain behaviour of NNs, we ran all algorithms 40 times and the average results are presented here. Figure 2 shows recognition rates for the experiments on the ORL dataset, the number of features and hidden layer neurons for the classic BP and the four CGA methods.

4.2   Experiments on the FERET Face Database

The FERET face database is a widely used database for evaluating face recognition algorithms. The proposed CGA methods were tested on a subset of the FERET database, comprising 1050 images of 150 individuals (7 images per individual). This subset involves variations in facial expression, illumination and pose. The facial portion of each original image is first cropped to 65×65 pixels, i.e. the dimensionality is 4225. In the experiments, we used the first $(l = 4, 5)$ images per class for training and the remaining images for testing. The results indicate that for all proposed training methods, 70 neurons are sufficient to reach a classification accuracy close to the maximum. Table 2 compares the highest obtained recognition rates for the different NN training algorithms with 700 iterations and with a different number of samples $(l = 4, 5)$ for the ORL and FERET databases, respectively.

As per Table 2, all proposed CGA methods result in higher correct recognition rates (CRR) than the classic BP algorithm, with the Powell-Beale CGA achieving the highest recognition rate. Figure 3 shows the Powell-Beale CGA performance in terms of the number of training epochs versus step size and gradient evaluations, and the MSE via training regression model for the ORL and FERET databases, respectively. In these experiments, we trained a FFNN with half of the images in this database (images 1-5 for each person) and then used the other half for testing (images 5-10 for each person).

Training times for all proposed methods were approximately equal. The advantages of the proposed methods are the higher accuracy and lower CPU time.

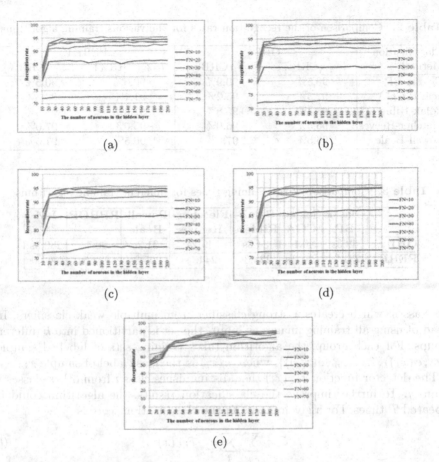

Fig. 2. Recognition rates of the ORL dataset classification using the (a) Scaled CGA, (b) Powell-CGA, (c) Polak-Ribiere, (d) Fletcher-Reeves CGA, and (e) classic BP for different numbers of features (FN) and hidden layer neurons

In our experiments, with a configuration of 2.8GHz CPU and 1GB RAM, the proposed algorithms were compared with Resilient Backpropagation (RPROP) [11] and Levenberg-Marquardt (LM) [12,11]. The results are given in Table 3.

5   Boosting the Results

To further increase accuracy, especially for large subject numbers (we used the AR and FERET databases), a modified AdaBoost.M1 method was implemented.

5.1   Modified AdaBoost.M1 Algorithm for Large Face Databases

AdaBoost.M1 is a popular sequential learning boosting algorithm for binary classification [6]. We used a modified Adaboost.M1 method suitable for multi-

Table 2. Comparison of the recognition rates for the various training algorithms

Classification	ORL		FERET	
Method	$l=4$ (CRR%)	$l=5$ (CRR%)	$l=4$ (CRR%)	$l=5$ (CRR%)
Classic BP	90.5%	90.9%	90.0%	90.5%
Scaled CGA	92.5%	95.2%	91.1%	93.3%
Polak-Ribiere	93.3%	95.8%	92.6%	95.5%
Fletcher-Reeves	95.5%	96.0%	95.0%	97.0%
Powell-Beale	96.5%	97.0%	96.5%	97.5%

Table 3. Comparison of the training times for different training algorithms

	Classic BP	Scaled CGA	Polak-Ribiere	Fletcher-Reeves	Powell-Beale	RPROP	LM
ORL	42.3s	40.1s	34.3s	49.2s	31.2s	80.4s	230.9s
FERET	231.2s	238.5s	223.3s	244.5s	205.3s	922.5s	2062.0s

ple classes, which creates a strong classifier from multiple weak classifiers. Instead of using all training images as input, the set is partitioned into k different groups. For each group, the algorithm takes training sets of labelled samples $\{(x_1, c(x_1)), \ldots, (x_m, c(x_m))\}$, where $c(x_i)$ is the class label of sample x_i.

The detector function $f_{j,t}(x_i)$ then distinguishes class i from other classes in group j. To further improve the classification results, the algorithm could be repeated T times. For a given input X, the classification score is

$$C = \sum_j \sum_t a_{j,t} f_{j,t}(x) \tag{6}$$

where $a_{j,t} = \frac{1}{2} \log \frac{1-\varepsilon_{j,t}}{\varepsilon_{j,t}}$ are the weights of the weak classifiers. $\varepsilon_{j,t}$ is the average error of iteration t in the jth group, which can be computed as

$$\varepsilon_{j,t} = \sum_{i=1}^{m} P_{i,j}^t |f_{j,t}(x_i) - c_j(x_i)| \tag{7}$$

Here, $c_j(x_i)$ is the class label of the ith sample of the jth group. $P_{i,j}^t$ is the weight of observation at step t and m is the number of samples in the jth group.

The AdaBoost method then increases the weights for observations misclassified by learner t and reduces weights for observations correctly classified [6]. The next learner $t+1$ is then trained on the data with updated weights

$$P_{i,j}^{t+1} = P_{i,j}^t \beta_t, \qquad \text{where } P^1 = \frac{1}{N}, \ \beta_t = \frac{\varepsilon_{j,t}}{1 - \varepsilon_{j,t}} \tag{8}$$

To classify a new face image, all scores C (Eq. (6)) are compared. The class, which yields the greatest C, is nominated as the new candidate class. It should

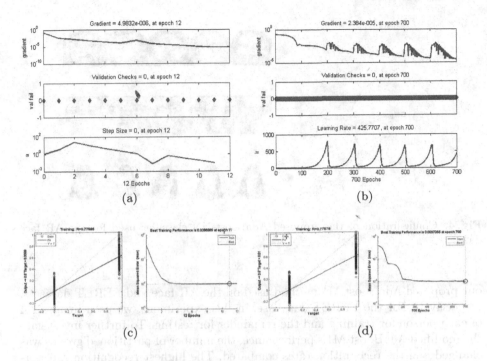

Fig. 3. (Top) Powell-Beale CGA performance in terms of the number of training epochs versus step size and gradient evaluations for the (a) ORL and (b) FERET datasets. (Bottom) Powell-Beale CGA regression model (left) and its performance in terms of the number of training epochs versus MSE (right) for the (c) ORL and (d) FERET datasets.

be noted that a threshold could be defined, as if the results of all classifiers for an image yield lower than a certain value, the image is certainly misclassified. Figure 4 shows the configuration of the proposed AdaBoost-NN classifier, which was tested on the AR face dataset.

5.2   Experiments on the AR Face Database

This face database contains over 4000 colour images of 126 people [13]. Images feature frontal view faces with different facial expressions, illumination conditions, and occlusions (sun glasses and scarf). We randomly selected 1040 images of 40 males and 40 females (13 images per person) to test our method. The original resolution is 768 768 × 576 pixels normalised to 70 × 70 pixels. We selected 6 samples from each class for the training and the rest for the testing. The modified AdaBoost.M1 algorithm was executed 8 times with the same parameters and the results averaged. The whole database was randomly divided into separate groups and the NN output of each group was fed to the modified AdaBoost.M1. Table 4 shows a comparison of recognition rates for different CGAs with and without

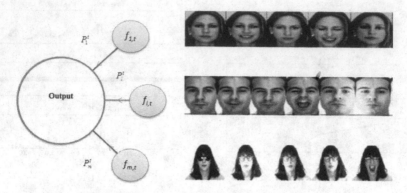

Fig. 4. Configuration of the proposed AdaBoost-NN classifier used for the AR face dataset

our proposed AdaBoost.M1 method used on the AR face and FERET datasets, respectively. For the FERET database, in this experiment, we used images 1-4 of each person for training and the remainder for testing. To further investigate the modified AdaBoost.M1's performance, the number of partitioned groups was changed and the recognition rates compared. The highest recognition rate was obtained using 3 partitions.

Table 4. Comparison of the recognition rates for various training algorithms

Classification Method	AR		FERET	
	NN	Boosting-NN	NN	Boosting-NN
Simple BP	85.0%	90.5%	90.0%	91.2%
Scaled CGA	86.5%	92.5%	91.1%	92.5%
Polak-Ribiere	88.0%	95.0%	92.6%	94.4%
Fletcher-Reeves	89.2%	95.8%	95.0%	96.5%
Powell-Beale	90.5%	96.5%	96.5%	98.0%

6   Conclusions

To enhance the performance of NN-based face recognition systems, this study investigated the utility of four types of conjugate gradient algorithms – Fletcher-Reeves CGA, Polak-Ribiere CGA, Powell-Beale CGA, and scaled CGA. After extracting discriminant features via a two-stage LDA by eliminating feature vectors that are less correlated, various experiments were carried out on the ORL, FERET and AR face datasets using these training methods. Among these methods, the Powell-Beale CGA gave the best results on the ORL face dataset with 110 neurons in the hidden layer and 70 features. It improved the recognition rate from 90.9% for the basic BP algorithm to 97.0% without any further pre-processing step. Furthermore, when faced with many classes, we proposed a modified AdaBoost.M1 to create a strong classifier from multiple weak classifiers, which resulted in a 6.0% and 1.5% absolute improvement in the recognition rate for the AR and FERET databsets, respectively.

References

1. Belhumeur, P.N., Hespanha, J.P., Kriegman, D.J.: Eigenfaces vs. Fisherfaces: Recognition using class specific linear projection. IEEE Transactions on Pattern Analysis & Machine Intelligence 19(7), 711–720 (1997)
2. Zhao, W., Chellappa, R., Rosenfeld, A., Phillips, P.J.: Face recognition: A literature survey. ACM Computing Surveys 35(4), 399–458 (2003)
3. Duda, R., Hart, P., Stork, D. (eds.): Pattern Classification, 2nd edn. Wiley, New York (2000)
4. Tian, Q., Barbero, M., Lee, Z.: Image Classification by the Foley Sammon Transform. Journal of Optical Engineering 25(7), 834–840 (1986)
5. Azami, H., Sanei, S., Mohammadi, K.: Improving the Neural Network Training for Face Recognition using Adaptive Learning Rate, Resilient Back Propagation and Conjugate Gradient Algorithm. International Journal of Computer Applications 34(2), 22–36 (2011)
6. Friedman, J., Hastie, T., Tibshirani, R.: Additive Logistic Regression: A Statistical View of Boosting. Annals of Statistics 28(2), 337–407 (2000)
7. Zho, M., Martinez, A.: Selecting Principal Components in a Two-Stage LDA Algorithm, pp. 132–137 (2006)
8. Bishop, C.M. (ed.): Neural Networks for Pattern Recognition. Clarendon, Oxford (1996)
9. Paulin, F., Santhakumaran, A.: Classification of Breast Cancer by Comparing Back Propagation Training Algorithms. International Journal on Computer Science and Engineering 3(1), 327–332 (2011)
10. Shaheed, M.: Performance analysis of 4 types of conjugate gradient algorithms in the nonlinear dynamic modelling of a TRMS using feedforward neural networks. In: Proc. 2004 IEEE Int. Conf. Systems, Man and Cybernetics, vol. 6, pp. 5985–5990 (2004)
11. Phung, S., Bouzerdoum, A.: A Pyramidal Neural Network for Visual Pattern Recognition. IEEE Transactions on Neural Networks 18(2), 329–343 (2007)
12. Hagan, M., Menhaj, M.: Training FeedForward Networks with the Marquardt Algorithm. IEEE Transactions on Neural Networks 5(6), 989–993 (1994)
13. Martinez, A.: Recognizing Imprecisely Localized, Partially Occluded, and Expression Variant Faces from a Single Sample per Class. IEEE Transactions on Pattern Analysis & Machine Intelligence 24(6), 748–763 (2002)

Medical Image Thresholding
Using Online Trained Neural Networks

Ahmed A. Othman

The Department of Systems Design Engineering
University of Waterloo, Ontario, Canada
ahmed.othman@uwaterloo.ca

Abstract. Medical images are used mainly in the diagnosing process
and as an aid in determining correct treatment. Therefore, the process
of segmenting different regions of interests (ROIs) within the medical
images is considered a critical one. When provided with a segment with
high segmentation accuracy, the physician can easily detect the problem and determine the best treatment. In this paper, a neural network
retrained on-line is proposed to automatically segment medical images
using a global threshold. The network is initially trained off-line using a
set of features extracted from a set of randomly selected training images,
along with their best thresholds, as targets for the neural network. The
features are extracted using Seeded Up Robust Feature (SURF) technique from a rectangle around the ROI. This network continues training
on-line as new images arrive, based on a feedback correction done by the
clinician to the segmented image. This process is repeated multiple times
to verify the generalization ability of the network.

Keywords: Neural network, Seeded Up Robust Feature (SURF).

1 Introduction

Separating an object from its background based on the gray levels is performed
using thresholding [1]. Global threshold is a value T between 0 and 255 which
is used to separate the image into two parts, one part greater than T and the
other part less than or equal to T. Therefore, image thresholding is considered
the easiest and the fastest segmentation technique. Fast segmentation is still
required in many applications, especially when processing a large number of
images. Aside from fast segmentation, segmentation accuracy is very important.
Therefore, a segmentation approach that generates accurate results in a short
time is considered an optimum approach. However, using a fixed threshold for all
images may generate accurate results for some images - but not for all. Therefore,
the threshold must be tuned for each different image in order to achieve a high
segmentation accuracy for all images. This process is very difficult, especially
when there are a lot of images that need to be processed. There is still a need
for a simple and efficient thresholding approach that can assign threshold for
different images based on image features. This approach would combine the

T. Huang et al. (Eds.): ICONIP 2012, Part V, LNCS 7667, pp. 672–680, 2012.

fast capability of the global thresholding with the high segmentation accuracy. This paper is organized as follows: In sections 2 and 3 a brief survey of image thresholding and segmentation using neural networks is provided. In section 4, the proposed technique is presented. In section 5, the experiment and the results are presented. In section 6, the paper is concluded.

2 Image Thresholding

Sezgin et al. [1] provide a survey of the most popular thresholding methods and categorize them (based on the information used) into six different classes: histogram shape-based, clustering-based, entropy-based, object attribute-based, spatial methods, and local methods. Most thresholding techniques are based mainly on the histogram of the image. Because the shape of the histogram is not the same for all images, most of these methods work well when the histogram is bi- or multi-modal. Other methods use a static scheme for all images, that is, one without any ability to adjust to the features of the image, or are based on several parameters that must be tuned separately for each image in order to obtain maximum accuracy. Many studies have been conducted with the goal of overcoming the limitations of thresholding techniques, using type II fuzzy sets [2], the entropy approach [3], expectation-maximization algorithm [4], multi-objective optimization [5], genetic algorithm [6], Parzen window technique [7], digital fractional differentiation of the image histogram [8], the sum of the variance and the discrepancies between the variances [9], and median-based criteria [10]. Although a great deal of effort has been dedicated to solving the problem of thresholding, many challenges remain. Most methods still depend on the shape of the image histogram, which prevents the generalization of the methods [3,4,10]. another method has a complicated process [8]. Other methods have lengthy computational time [4, ?? ,7,10], are parameter-dependent [5], assume previous or customized assumptions [8,3] or face difficulty in selecting parameters [9]. Generally, no thresholding method works well for all types of images because they are not based on a consideration of the image features and are disconnected from the visual perception of the user. There is therefore a need for a thresholding method that can assigns a threshold for an image based on its features and user feedback.

3 Segmentation Using Neural Networks

Neural network has been widely used in medical image segmentation such as segmentation of swallow [11]. The problem with the methods that depend on a self-organizing map is the large number of parameters that must be tuned before training. Also, they depend mainly on prior information about the image, which renders the generalization of these methods very difficult. A discrete Fourier transform, discrete cosine transform (DCT) [12], DCT and vector quantization [13], wavelet transform, and the moments of the gray-level histogram [14] are used to train incremental neural network. These incremental methods depend on tuning parameters, such as the threshold value that controls the number of nodes and the gain constraint of the network when using DCT. moreover, the region of interest selection is performed manually. Fu et al. [15] used a pulse-coupled neural network along with a statistical expectation maximization applied to the image histogram to estimate initial thresholds for segmenting different regions of the brain, and for adjusting the parameters of the neural network. This method is parameter-dependent and also depends on the initial threshold generated from the histogram which can not be trusted when the shape of the histogram is unclear. Texture features extracted from a rectangle around the region of interest are used with some neural networks to segment different medical images [18]. The problem with this method is that prior information about the region of interest is required, otherwise manual user interaction is needed.

4   Proposed Approach

The proposed approach starts with extracting a set of features from a set of randomly selected training images with their best thresholds as target outputs to train a supervised neural network. In on-line mode, newly arriving images are not only segmented using the trained neural network but also help in retraining the network, which increases its capability to threshold images. Therefore, the results generated by the retrained neural network are better than those generated by a static network. The neural network segments test images directly by assigning a threshold $T \in \{0, 1, \ldots, 255\}$. The proposed approach consists of two major stages: off-line and on-line modes Fig. 1.

4.1   Off-line Mode

First, the best possible threshold for every available image is calculated by trying all thresholds between $\{0, \ldots, 255\}$ and the best threshold for the image is the threshold that generates the maximum segmentation accuracy compared to the gold standard image. The best thresholds for the available images are stored in a matrix T .

Detection Algorithm. This algorithm finds the position of the first seed point (x_s, y_s) inside the ROI and constructs a rectangle R around the ROI with the point (x_s, y_s) as the center of the rectangle R. In previous work [17], we described the detection algorithm; there is a difference here that the maximum size of the initial region around (x_s, y_s) is 120×120 and the limit for stopping the enlarging process is 2%. This change in the initial region and the enlarging limit is done to ensure that R will return sufficient SURF points. The initial size of 120×120 may not be achieved if the point (x_s, y_s) is located close to any border of the image. In most images, the detection algorithm detects a correct point inside the ROI and failed to detect a correct point in some images, but we proceeded with the rectangles generated by the algorithm to verify the overall performance.

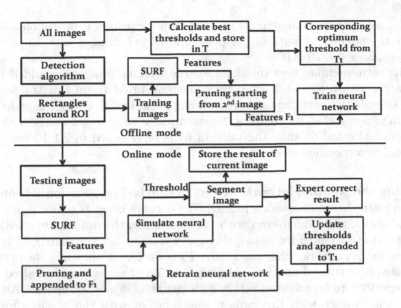

Fig. 1. The proposed approach

All available images are processed by the detection algorithm and the resulting rectangles around the ROI $R_i, i = \{1, \ldots, v\}$(where v is the number of available images in this paper $v = 26$) are stored in one folder to be used for extracting features in off-line and on-line modes.

Feature Extraction. Six training images are randomly selected to be the training images. The process of feature extraction is performed as follows:

– For every training image I_1,
  1. The rectangle R_1 generated from I_1 around the ROI is passed to the SURF technique.
  2. SURF is updated to first apply 5×5 median filter on R_1 and to adjust R_1 using the matlab functions "*medfilt2*" and "*imadjust*" respectively.
  3. The SURF detect n_1 key points inside R_1.
  4. Selecting 10 points from the n_1 points, a rectangle with a maximum size of 40×40 (may be less if the point is close to the border of R_1) is generated around each point $p \in n_1$, and the mean of each rectangle is calculated. The 10 points that have the maximum means values are the selected 10 points.
  5. From the selected 10 points, the SURF features are recorded. For every point, a feature vector consisting of the scale, the orientation and 64 SURF descriptors [16].
  6. The features of every point are stored in a column that consists of 66 rows (scale + orientation + 64 descriptors).
  7. The size of the feature matrix $Fea1$ for every image ($Fea1_i$, where $i = \{1, \ldots, 6\}$) is 66×10 for the 10 points.

- The general feature matrix M of the 6 training images is of size 66×60 (66 are the number of features per point and 60 is 6 images with 10 points in each image $6 \times 10 = 60$).
- The corresponding best threshold for the training images are added to $T1$ matrix with a size of 1×60. Every column in M (represent the features of one SURF point) has a corresponding value in $T1$. Every image has the same value for all its columns, for example, if there is an image with best threshold equal 64 then the value of 64 is represented in $T1$ 10 times one time corresponding to one columns.

Training the Neural Network. A feed-forward backpropagation network was used to learn the set of training images. The network consists of one input layer with 60 nodes, one hidden layer with 100 nodes, and the output layer with one output (=the estimated threshold). For every training set, the matrix M is used as input to the network, while the matrix $T1$ is the target output of the network. The network is trained using the Matlab function "*trainscg*" with desired error set to $\epsilon = 10^{-7}$ to be achieved within a maximum of $N_E = 10^5$ epochs. Another experiment is done with the same parameters but with the Matlab function "*trainrp*".

4.2   On-line Mode

In on-line mode, the algorithm procees for every incoming image I_2 as follow:

1. A feature matrix $Fea2$ (66×10) is extracted just as was $Fea1$ in 4.1.
2. The trained neural network assigns a threshold for every columns in $Fea2$ when simulated by $Fea2$.
3. The generated thresholds from the network is a vector $Thresholds$ (1×10) and the final threshold is the mean of $Thresholds$. $T^* = \frac{1}{10} \sum_{i=1}^{10} Thresholds_i$.
4. The threshold T^* is applied to the original image I_2 (not the rectangle around the ROI image R_2). I_2 is first filtered and adjusted just as was R_1 in 4.1. A binary segment B is generated from I_2 using T^* by a user click inside the ROI. The holes in B are filled using the Matlab function "*imfil*" and an open morphological operation is performed on B using the Matlab function "*bwmorph*". B is stored for calculating statistics.
5. B is provided to the expert to correct it to B^* (the gold standard images are assumed to be the output of the expert B^*).
6. B^* is used to recalculate the best threshold t for I_2 using a detailed algorithm consisting of many functions; to save space a very brief description will be presented in the following steps:
   (a) Keep T^* if the Jaccard accuracy (see section 5) of B is greater than 86%.
   (b) Otherwise, the decision as to whether T^* should be increased or decreased is made and applied to T^*. For example, if T^* should be increased, different segments and the corresponding Jaccard accuracies using values greater than T^* are calculated and t (the updated threshold)

that generate the maximum accuracy before accuracy start decreasing is returned.

7. A pruning process that prevents redundancy and similar features from re-adding in M is as follows:

   (a) The Euclidean distance E between every column in $Fea2$ and every column in M is calculated. $E(k,f) = \sqrt{\sum((Fea2(1:66,k) - M(1:66,f)).^2)}$, where $k = \{1,\ldots,10\}$, $f = \{1,\ldots,l\}$, and l is the number of columns in M.

   (b) Every row in E represent the Euclidean distance between one column in $Fea2$, and every column in M.

   (c) If the Euclidean distance between any column in $Fea2$ and any column in M is less than 0.5, this column is discarded and not added to M.

8. t is added to $T1$ matrix a number of times equal to the number of columns in $Fea2$ after pruning.

9. The modified M and $T1$ are used to retrain the neural network with the same network parameters used in off-line mode.

10. The next imcoming image is segmented using the retrained neural network.

11. This process is repeated as long as there are new images.

5 Experiments and Results

In this section, a set of 26 different-sized breast ultrasound images are employed to train and test the proposed technique. Six of the images are used as a training set and the remaining 20 images are used for testing. This process is repeated 10 times to generate different training sets. For every training set, the process is repeated five times and the average of the results was taken. The results of the proposed techniques (Retrained Neural Network RNN) are compared with the results generated from the static neural network SNN (images simulate the neural network that trained off-line and no online retraining is done) via accuracy calculation using the gold standard images. The purpose of this experiment is to prove that the results generated from a static neural network (SNN) could be improved if the network is retrained online (RNN). The following accuracy measures have been employed to verify the performance of the technique(A is the resulted binary segmented image and B is the gold standard image): 1) The average of segmentation accuracy J is calculated using the area overlap (also called Jaccard Index): $J(A,B) = \frac{|A \cap B|}{|A \cup B|}$, 2) The standard deviation σ of J, and 3) The 95% confidence interval (CI) of J. 4) The Dice coefficient D: $D(A,B) = \frac{2*J(A,B)}{1+J(A,B)}$, 5) The standard deviation σ_D of D, 6) The 95% confidence interval (CI) of D, 7) The null hypothesis (t-test) between the five average rsults of SNN and RNN to ensure that they are statistically different.

Table 1 present a complete set of results of SNN and RNN where the proposed algorithm, RNN has the highest statistics for all training sets except training sets five and six. For example, the average Jaccard index of the third training set is increased from 45% SNN to 66% in RNN. On the other hand, Table 2 present the results of the second experiment that performed with a training Matlab function "*trainrp*" in terms of average Jaccard index only. The results of the second experiment are not good as the average Jaccard index in RNN is higher than SNN in five training set and worst in the other five.

Table 1. Results of SNN and RNN (all numbers are rounded)

Training	Method	J	σ_J	CI_J	D	σ_D	CI_D	$Null.$
1st set	SNN	55%	28%	42%-69%	67%	26%	55%-79%	
	RNN	67%	25%	56%-79%	77%	21%	67%-87%	1
2nd set	SNN	63%	24%	52%-75%	74%	22%	64%-85%	
	RNN	66%	23%	55%-77%	77%	20%	67%-86%	0
3rd set	SNN	45%	30%	31%-59%	57%	29%	43%-70%	
	RNN	66%	23%	55%-77%	77%	20%	67%-86%	1
4th set	SNN	74%	18%	65%-82%	83%	14%	77%-90%	
	RNN	75%	18%	67%-83%	84%	14%	78%-91%	1
5th set	SNN	72%	18%	64%-80%	82%	14%	76%-89%	
	RNN	70%	21%	60%-80%	80%	18%	72%-89%	0
6th set	SNN	68%	20%	59%-79%	79%	17%	71%-87%	
	RNN	68%	22%	58%-79%	79%	19%	70%-88%	0
7th set	SNN	63%	23%	52%-74%	75%	21%	65%-84%	
	RNN	65%	22%	55%-75%	76%	19%	68%-85%	1
8th set	SNN	68%	22%	58%-78%	78%	19%	69%-87%	
	RNN	71%	20%	62%-80%	81%	16%	73%-89%	0
9th set	SNN	67%	22%	57%-77%	78%	19%	69%-87%	
	RNN	68%	22%	57%-78%	78%	20%	69%-88%	0
10th set	SNN	55%	26%	43%-67%	67%	24%	56%-78%	
	RNN	60%	25%	48%-72%	72%	23%	61%-82%	0

Table 2. Results of SNN and RNN: Jaccard Index J using Matlab function "*trainrp*"

Training Set	Method	1	2	3	4	5	6	7	8	9	10
	SNN	57%	65%	44%	72%	70%	66%	60%	66%	65%	50%
	RNN	63%	62%	57%	69%	68%	66%	60%	69%	66%	51%

6   Conclusions

Using a static threshold for all images may generate good results in some images and bad results for other images. the threshold must be tuned for every single image to achieve the highest segmentation accuracy for all images. This process is very difficult and time consuming, especially when processing a large number of images. Using a neural network that could update its parameters and be retrained on-line to threshold images seems to be a good solution. In off-line mode, the network is trained using a set of features extracted using SURF technique from a rectangle around the ROI. In on-line mode, every incoming image is thresholded by a threshold assigned by the neural network and the result is provided to an expert for verification and correction. The corrected result is used to calculate the best threshold of the image to be used along with the image features to retrain the neural network. The results prove that the output of the proposed algorithm RNN is better than SNN in most training sets.

References

1. Sezgin, M., Sankur, B.: Survey over image thresholding techniques and quantitative performance evaluation. Journal of Electronic Imaging 13, 146–165 (2004)
2. Tizhoosh, H.R.: Image thresholding using type II fuzzy sets. Pattern Recognition 38, 2363–2372 (2005)
3. Liu, D., Jiang, Z., Feng, H.: A novel fuzzy classification entropy approach to image thresholding. Pattern Recognition Letters 27, 1968–1975 (2006)
4. Bazi, Y., Bruzzone, L., Melgani, F.: Image thresholding based on the EM algorithm and the generalized Gaussian distribution. Pattern Recognition 40, 619 634 (2007)
5. Nakib, A., Oulhadj, H., Siarry, P.: Image histogram thresholding based on multiobjective optimization. Signal Processing 87, 2516–2534 (2007)
6. Hammouche, K., Diaf, M., Siarry, P.: A multilevel automatic thresholding method based on a genetic algorithm for a fast image segmentation. Computer Vision and Image Understanding 109, 163–175 (2008)
7. Wang, S., Chung, F., Xiong, F.: A novel image thresholding method based on Parzen window estimate. Pattern Recognition 41, 117–129 (2008)
8. Nakib, A., Oulhadj, H., Siarry, P.: Fractional differentiation and non-Pareto multiobjective optimization for image thresholding. Engineering Applications of Artificial Intelligence 22, 236–249 (2009)
9. Li, Z., Liu, C., Liu, G., Cheng, Y., Yang, X., Zhao, C.: A novel statistical image thresholding method. International Journal of Electronics and Communications 64, 1137–1147 (2010)
10. Xue, J.H., Titterington, D.M.: Median-based image thresholding. Image and Vision Computing 29, 631–637 (2011)
11. Lee, J., Steele, C.M., Chau, T.: Swallow segmentation with artificial neural networks and multi-sensor fusion. Medical Engineering & Physics 31, 1049–1055 (2009)
12. Kurnaz, M.N., Dokur, Z., Ölmez, T.: An incremental neural network for tissue segmentation in ultrasound images. Computer Methods and Programs in Biomedicine 85, 187–195 (2007)
13. Dokur, Z.: A unified framework for image compression and segmentation by using an incremental neural network. Expert Systems with Applications 34, 611–619 (2008)
14. Iscan, Z., Yüksel, A., Dokur, Z., Korürek, M., Ölmez, T.: Medical image segmentation with transform and moment based features and incremental supervised neural network. Digital Signal Processing 19, 890–901 (2009)
15. Fu, J.C., Chen, C.C., Chai, J.W., Wong, S.T.C., Li, I.C.: Image segmentation by EM-based adaptive pulse coupled neural networks in brain magnetic resonance imaging. Computerized Medical Imaging and Graphics 34, 308–320 (2010)

16. Bay, H., Tuytelaars, T., Van Gool, L.: SURF: Speeded Up Robust Features. In: Leonardis, A., Bischof, H., Pinz, A. (eds.) ECCV 2006. LNCS, vol. 3951, pp. 404–417. Springer, Heidelberg (2006)
17. Othman, A.A., Tizhoosh, H.R.: Segmentation of Breast Ultrasound Images Using Neural Networks. In: Iliadis, L., Jayne, C. (eds.) Engineering Applications of Neural Networks. IFIP AICT, vol. 363, pp. 260–269. Springer, Heidelberg (2011)
18. Sharma, N., Ray, A.K., Sharma, S., Shukla, K.K., Pradhan, S., Aggarwal, L.M.: Segmentation and classification of medical images using texture-primitive features. Application of BAM-Type Artificial Neural Network 33, 119–126 (2008)

Subspace Echo State Network for Multivariate Time Series Prediction

Min Han and Meiling Xu

Dalian University of Technology, Dalian, China
minhan@dlut.edu.cn, xuml@mail.dlut.edu.cn

Abstract. Echo state network is a novel recurrent neural network, with a fixed reservoir structure and an adaptable linear readout layer, facilitating the application of RNNs. Often the network works beautifully. But sometimes it works poorly because of ill-posed problem. To solve it, we introduce a new approach towards ESNs, termed FSDESN, herein. It combines the merits of ESNs and fast subspace decomposition algorithm to provide a more precise alternative to conventional ESNs. The basic idea is to extract the subspace of the redundant large-scale reservoir state matrix by Krylov subspace decomposition algorithm, subsequently, calculate the readout weights using the subspace to replace original reservoir matrix. Hence, it can eliminate approximate collinear components and overcome the ill-posed problems so as to improve generalization performance. We exhibit the merits of our model in two multivariate benchmark datasets. Experimental results substantiate the effectiveness and characteristics of FSDESN.

Keywords: Echo state network, fast subspace decomposition, reservoir, multivariate time series.

1 Introduction

Recently, an artificial recurrent neural network, called Echo State Network (ESN), has become an active research area [1-3]. It consists of a non-trainable sparse recurrent part, i.e., reservoir, and a linear readout part. The reservoir can maintain active even in the absence of input, exhibiting dynamic memory. Connection weights inside the reservoir, as well as input weights, are generated randomly. Only the synaptic connections from the recurrent reservoir to output readout neurons are adaptable via supervised learning. Training an ESN network becomes a simple linear regression task, which expedites convergence [4].

As a result of the above merits, ESNs have been successfully applied in time-series prediction. For example, in [5], to modeling time series, the authors propose a simple cycle reservoir topology with its memory capacity arbitrarily close to the proved optimal value. In [3], the authors propose a support vector echo state machine (SVESM) for predicting the chaotic time series. The basic idea is to replace the "kernel trick" with "reservoir trick" in nonlinear system modeling. But, when we use least square method to estimate the readout weights, the reservoir state matrix presents the ill-

T. Huang et al. (Eds.): ICONIP 2012, Part V, LNCS 7667, pp. 681–688, 2012.

posed nature, which eases the training process [6]. Fast subspace decomposition (FSD) is a computational efficient reduction method, which could avoid the ill-posed problem [7]. We dub the fast subspace decomposition algorithm toward ESNs, termed FSDESN, to improve time series forecasting accuracy herein.

This paper is organized as follows. In Section 2, we give a brief review of conventional ESNs. In Section 3, we concisely introduce the fast subspace decomposition algorithm and propose the FSDESN model. Afterwards, we analyze the computational complexity of the proposed model. In Section 4, experiments are conducted. Finally, in Section 5, discussions and conclusions are given.

2   Brief Introduction to Echo State network

An ESN is a novel recurrent discrete-time neural network, which consists of a feed-forward input layer, a recurrently connected hidden layer, and a linear output layer. Denote $u(t)$ and $y(t)$ as the input and output units at time step t, respectively. The echo state $x(t)$ and output $y(t)$ of the reservoir are generated as follows:

$$x(t) = f(W_{in}u(t) + W_x x(t-1)); \quad y(t) = W_{out}x(t) \tag{1}$$

where f denotes the reservoir activation function, with tanh transfer function [4] herein. W_{in} and W_x are time-invariant and known, whereas W_{out} is time-varying and unknown, i.e., only W_{out} is modified via supervised learning. We collect echo states $x(t)$ row-wise into a matrix X, and make the corresponding target output value $y(t)$ row-wise into a vector Y. Then we use least square estimation to calculate W_{out}.

$$\hat{W}_{out} = Y(X^T X)^{-1} X^T \tag{2}$$

In practice, when the size of reservoir becomes large, some features are redundant, the matrix X is ill conditioned [3]. In such way the solution $\hat{W}_{out}$ is very sensitive to the perturbation of X and Y, which leads to bad numerical stability and unsatisfied generalization ability of ESNs. Modified ESN methods are required in order to compute a stabilized solution to discrete ill-posed problems.

3   Fast Subspace Decomposition of Echo State Network Prediction

In this section, we propose a novel echo state network model combining fast subspace decomposition algorithm (FSD) with ESNs, termed FSDESN, to solve the large-scale matrix ill-posed problem, and bring a jump on computational complexity compared with conventional eigenvalue decomposition algorithm [8-9].

3.1   Fast Subspace Decomposition Algorithm

In this part, we introduce the fast subspace decomposition algorithm [8] for extracting a subspace from the large-scale state matrix X:

1) Compute the covariance matrix $\hat{A}_{N \times N}$ of reservoir state matrix X. r_0 is randomly chosen as a unit-norm vector. Give $\beta_0 = 1$, $j = 0$, $m = 1$, and $\hat{d} = 1$.

2) Do the mth step of the Lanczos algorithm.

$while\left(\beta_j \neq 0\right) q_{j+1} = r_j / \beta_j; j = j+1; \alpha_j = q_j^H A q_j; r_j = A q_j - \alpha_j q_j - \beta_{j-1} q_{j-1}; \beta_j = \left\| r_j \right\|_2; end.$

At the mth step of the Lanczos algorithm, we get $Q_m == \{q_1, q_2, \cdots, q_m\}$. In the process, we have $\{a_i, \beta_i\}_{i=1}^m$. Then we can find an $m \times m$ tridiagonal matrix.

$$T_m = Q_m^H \hat{A} Q_m = \begin{bmatrix} \alpha_1 & \beta_1 & & & \\ \beta_1 & \alpha_2 & \beta_2 & & \\ & \ddots & \ddots & \ddots & \\ & & \ddots & \alpha_{m-1} & \beta_{m-1} \\ & & & \beta_{m-1} & \alpha_m \end{bmatrix} \tag{3}$$

3) Calculate the eigenvalues $\theta_i^{(m)}$ of T_m. Assume the ith eigenvalue of T_m is $v_i, i = 1, 2, \cdots, m$, thus we have $\theta_i^{(m)} = v_i$.

4) Form the test statistic $\phi_{\hat{d}}$ as to $\hat{d} = 1, 2, \cdots, m$,

$$\phi_{\hat{d}} = p(N - \hat{d}) \log \left[\sqrt{\left(\left\| \hat{A} \right\|_F^2 - \sum_{k=1}^{\hat{d}} \theta_k^{(m)^2} \right) / N - \hat{d}} \middle/ \left(tr(\hat{A}) - \sum_{k=1}^{\hat{d}} \theta_k^{(m)} \right) / N - \hat{d} \right].$$

where p is the number of samples, N and d are the column numbers of A and its Krylov subspace respectively, and $\hat{d}$ is the estimate of the d.

If $\phi_{\hat{d}} \leq r_{\hat{d}} c(N)$, set $\hat{d} = d$, go to step 5), else take $m = m+1$, go back to step 2), where $\gamma_{\hat{d}}$ is a precalculated threshold based on the tail area of a certain χ^2 distribution with $[(N-d)(N-d+1)-1]/2$ degrees of freedom, and we set confidence level as 99% herein. $c(N)$ is taken as $\sqrt{\log N}$ [8].

5) Compute the d principal RR vector $g_k^{(m)}$. Assume the ith eigenvector of T_m is $s_i, i = 1, 2, \cdots, m$, thus we have $g_i^{(m)} = Q s_i$.

Denote $G \triangleq span\{g_1^{(m)}, g_2^{(m)}, \cdots, g_{\hat{d}}^{(m)}\}$, then $\hat{X} = X \times G$ form the final subspace estimate.

3.2 Fast Subspace Decomposition for Echo State Network Prediction Model

Fast subspace decomposition algorithm extracts a low full column rank subspace $\hat{X}_{p \times d}$ out of a large reservoir state matrix $X_{p \times N}$, with p being the number of training sample sets, and d denoting the number of the principal columns of $X_{p \times N}$. The Moore-Penrose generalized inverse matrix of $\hat{X}$ can be stated in the form $\hat{X}^{\dagger} = (\hat{X}^T \hat{X})^{-1} \hat{X}^T$. So the readout weights are calculated in the following form instead of (2)

$$\hat{W}_{out} = Y \cdot \hat{X}^{\dagger} = Y \cdot (\hat{X}^T \hat{X})^{-1} \hat{X}^T \qquad (4)$$

Compared with $X^T X$, $\hat{X}^T \hat{X}$ has a compact eigenvalue distribution and low condition number. Using subspace rather than original reservoir matrix to compute readout weights can solve the ill-posed problem.

Based on the above analysis, the proposed FSDESN model can be shown in the following five steps and the model's structure is shown in Figure 1.

1) Initialize the reservoir. And preprocess the input data.

2) Train the model. Use input data to excite the internal neurons, generating reservoir state matrix X.

3) Extract the low-dimensional subspace $\hat{X}$ from X by FSD algorithm detailed presented in Part 3.1.

4) Calculate the readout weights based on the Moore-Penrose generalized inverse matrix of $\hat{X}$ as (4).

5) With the trained $\hat{W}_{out}$ in place, the FSDESN disconnects from the training data and runs freely on the test data as (1).

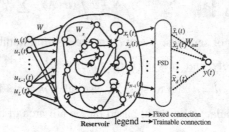

Fig. 1. FSDESN structure

3.3   Computational Complexity

Since the $d(d \ll N)$ RR vectors span the subspace of the covariance matrix A, we can obtain them by Lanczos iteration one by one, until all the d RR vectors are calculated. The operation of each Lanczos step is $O(N^2)$ flops, for d Lanczos steps are required, the overall cost is around $O(N^2 d)$ flops [8]. Whereas a standard matrix eigenvalue decomposition usually requires $O(N^3)$ flops for an $N \times N$ matrix, which may be a barrier to implement in real time. Both TSVD and PCA method or their modifications still need $O(N^3)$ step. So FSD is a computational efficient reduction algorithm.

4   Experimental Results

In this section, we provide a thorough experimental analysis of the FSDESN model based on Henon map and Rossler time series. Source codes are executed in MATLAB on the Windows XP operating system, Pentium(R) 4 CPU 2.66GHz, 1.5GB RAM. The obtained results provided in the ESN-based models [2-3] are averages of 50 times of random reservoir initialization and means of 50 test sequences. In order to compare the performance of different models, the normalized mean square error (NMSE) is used as a prediction measure given by

$$NMSE = \sum_{t=i}^{S}[\hat{y}(t) - y(t)]^2 \bigg/ \sum_{t=i}^{S}[y(t) - \bar{y}]^2 \tag{5}$$

where $\hat{y}(t)$ is the predicted value, $y(t)$ is the target value, and $\bar{y}$ is the mean value of $y(t)$.

Henon map is a discrete-time dynamical system that exhibits chaotic behavior. It maps a point $[x(t), y(t)]$ to a point $[x(t+1), y(t+1)]$ given by

$$x(t+1) = y(t)+1-ax(t)^2, \quad y(t+1) = bx(t) \tag{6}$$

where a=1.4 and b=0.3 for a canonical Henon map. In our experiment, we used a sample comprising 1000 time points to train the ESN-based models. White noise with standard deviation of 0.05 was added to the training sets of each coordinates. Finally, evaluation was conducted by using the trained models to predict the following 250 time steps of $x(t)$. The spectral radius and sparse connectivity of W_x are 0.9 and 0.05 respectively. The performance curves are depicted in Fig.2. Graph (a) presents the point-to point comparison between target and predicted values, and Graph (b) depicts the prediction absolute error curves. In addition, the predicted NMSEs and computational times of the evaluated models for reservoir size ranging from 100 to 800 are provided in Table 1 and Table 2. Subspace time denotes the runtime of extracting subspace from the reservoir state matrix required by the three dimensionality-reduction methods, and total time presents the whole runtime for prediction required by all the evaluated models.

As we can observe in Table1, the proposed model FSDESN performs much better than other considered alternatives, being capable of obtaining much lower NMSEs value. We note that conventional ESNs perform worse when reservoir dimension increases because of ill-posed problems. This sensitivity to reservoir dimensions makes conventional ESNs suffer a lack of robustness towards large scale reservoir. However, the FSDESN can automatically remove the approximate collinear components of the reservoir state matrix, and extract a well-conditioned subspace.

As to the computational cost, when reservoir dimension is small, all the three dimensionality-reduction methods run fast, difference among computational time is very small. However, when reservoir dimension is large, it becomes apparent that FSDESN outweighs other models significantly on both the subspace extracted runtime and total prediction runtime. For example, when reservoir size is 800, the subspace runtime and total runtime of FSDESN are 8.56 seconds and 48.50 seconds and that for PCAESN are 29.64 seconds and 61.63 seconds. Moreover, SVESM spends 126.57seconds, almost three times longer of FSDESN.

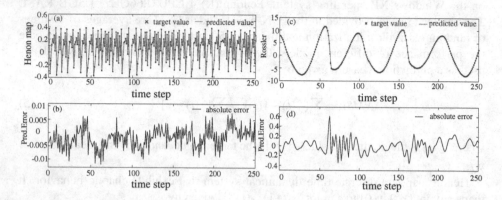

Fig. 2. (a) and (b) Prediction of Henon map $x(t)$,   (c) and (d) Prediction of Rossler $x(t)$

Table 1. Henon map: Performance of the evaluated models (NMSEs: 10^{-4})

Reservoir size	FSDESN	PCAESN	T-ESN	ESN	SVESM
100	3.20	5.87	3.66	6.80	10.00
300	3.05	6.22	5.86	21	9.64
500	3.71	7.24	6.71	49	8.63
800	3.50	7.21	7.13	339	9.91

Table 2. Henon map: Runtime of the evaluated models for reservoir size ranging from 100 to 800 (seconds)

Reservoir size		FSDESN	PCAESN	T-ESN	ESN	SVESM
100	Subspace	0.09	0.86	0.09	-	-
	total	4.13	4.90	5.32	4.13	8.61
300	Subspace	0.54	1.52	0.96	-	-
	total	14.19	14.50	19.37	14.29	31.69
500	Subspace	2.27	4.06	4.01	-	-
	total	24.07	25.82	33.29	25.28	68.48
800	Subspace	8.56	29.64	30.66	-	-
	total	48.50	61.63	85.19	61.02	126.57

Table 3. Rossler: Performance of the evaluated models (NMSEs: 10^{-4})

Reservoir size	FSDESN	PCAESN	T-ESN	ESN	SVESM
100	0.93	1.94	1.71	1.04	0.93
300	0.76	0.83	1.02	2.71	0.76
500	0.93	1.10	0.91	7.73	0.73
800	1.01	1.17	1.01	110	1.97

Table 4. Rossler: Runtime of the evaluated models for reservoir size ranging from 100 to 800 (seconds)

Reservoir size		FSDESN	PCAESN	T-ESN	ESN	SVESM
100	subspace	0.09	0.77	0.07	-	-
	total	3.76	4.35	3.81	4.05	6.69
300	subspace	0.51	1.28	0.72	-	-
	total	12.01	12.66	12.57	13.33	29.92
500	subspace	2.18	3.56	3.11	-	-
	total	21.66	22.84	22.83	23.39	70.85
800	subspace	8.56	24.62	21.42	-	-
	total	40.08	56.51	53.05	56.82	148.45

Another experiment is about the Rossler system, which is defined by

$$dx/dy = -y - z, \, dy/dz = x + ay, dz/dt = b + z(x - c) \tag{7}$$

where the parameters are set at a=0.2, b=0.4, and c=5.7. We also used a set of 1000 time series for training model, and a set of 250 time series for evaluating. In addition, white noise with standard deviation of 5% was added in the training data. We provide the obtained NMSEs of the evaluated models in Table 3, the computational runtime of the evaluated models in Table 4. In Fig.2(c) and (d), we depict the trajectories produced by the FSDESN model.

As we observe, FSDESN is a favorable alternative in large scale time series prediction, which performs more robustness to perturbation than other evaluated models. The performance of conventional ESNs is getting worse but FSDESN keeps high accuracy when reservoir size ranges from 100 to 800. It is noteworthy that in some conditions, SVESM performs higher accuracy than FSDESN, but SVESM is less computationally efficient. So the FSDESN offers a very good tradeoff between computational complexity and forecast accuracy.

5   Conclusions

In this paper, we propose a new model for forecasting multivariate time series, namely, the FSDESN. It uses fast subspace decomposition method to extract a low-dimensional subspace from the reservoir state matrix, which is a large-scale ill-posed matrix with some column vectors collinear or approximately collinear. Subsequently, the model computes the readout weights using the subspace instead of the original reservoir matrix. By this way, computation complexity is reduced, and ill-condition solutions are avoided. This is the main reason for the jump in forecast performance. It has significant robustness to perturbation. Experiment results of two time-series benchmark datasets have shown our proposed model's promise in multivariate time series forecasting application.

References

1. Jaeger, H.: The" Echo State" Approach to Analysing and Training Recurrent Neural Networks, GMD Report 148, German National Research Center for Information Technology (2001)
2. Jaeger, H., Haas, H.: Harnessing Nonlinearity: Predicting Chaotic Systems and Saving Energy in Wireless Communication. Science 304, 78–80 (2004)
3. Shi, Z., Han, M.: Support Vector Echo-State Machine for Chaotic Time-Series Prediction. IEEE Transactions on Neural Networks 18, 359–372 (2007)
4. Lukosevicius, M., Jaeger, H.: Reservoir Computing Approaches to Recurrent Neural Network Training. Computer Science Review 3, 127–149 (2009)
5. Rodan, A., Tiňo, P.: Minimum Complexity Echo State Network. IEEE Transactions on Neural Networks 22, 131–144 (2011)
6. Chatzis, S.P., Demiris, Y.: Echo State Gaussian Process. IEEE Transactions on Neural Networks 22, 1435–1445 (2011)
7. Tang, X., Han, M.: Partial Lanczos Extreme Learning Machine for Single-Output Regression Problems. Neurocomputing 72, 3066–3076 (2009)
8. Xu, G., Kailath, T.: Fast Subspace Decomposition. IEEE Transactions on Signal Processing 42, 539–551 (1994)
9. Reichel, L., Sgallari, F., Ye, Q.: Tikhonov Regularization based on Generalized Krylov Subspace Methods. Applied Numerical Mathematics 62, 1215–1228 (2012)

A Framework of a Route Optimization Scheme for Nested Mobile Network

Shayma Senan[1], AishaHassan A. Hashim[1], Akram M. Zeki[2], Rashid A. Saeed[1], Shihab A. Hameed[1], and Jamal I. Daoud[1]

[1] Faculty of Engineering, IIUM, Jalan Gombak, 53100 KL, Selangor, Malaysia
shay_sinan@yahoo.co.uk,
{aisha,rashid,shihab,jamal58}@iium.edu.my
[2] Faculty of ICT, IIUM, Jalan Gombak, 53100 Kuala Lumpur, Selangor, Malaysia
akramzeki@iium.edu.my

Abstract. Network mobility technology is now being accomplished with the foundation of NEMO (NEtwork MObility), developed by Internet Engineering Task Force (IETF). Although, it achieves optimal and continuous communication, it still suffers from some limitations, especially when the level of nesting increases. To overcome these drawbacks, this paper will present a route optimization framework for nested mobile network using hierarchical structure with Binding Update Tree (BUT). This framework should reduce packet overhead, handoff latency, packet transmission delay, and achieve optimal routing. At last, a comparison will be done with bi-directional tunneling used by NEMO Basic Support to evaluate the performance of the proposed framework.

Keywords: Mobile IPv6, Network Mobility, nested NEMO, nested Mobile Networks, NEMO BS, Mobile Router.

1 Introduction

Today wireless devices and mobile technologies are widely used in IPv6 communication. However, as users expect to be accessible to the Internet "anywhere" and "anytime", Mobile IP and Mobile IPv6 aim at maintaining Internet connectivity while a host is roaming. Any mobile node, in Mobile IPv6, is identified by its home address regardless of its point of attachment to the Internet [1].

2 An Overview of NEMO

Extending from MIPv6 [1], Internet Engineering Task Force (IETF) proposed Network mobility (NEMO) [2] to support network mobility management. A mobile network includes one or more mobile routers (MRs) that provides access to the Internet. The MR transmits the packet to mobile network nodes (MNNs). Moreover, the MR performs the binding update (BU) to the home agent (HA) without additional registration such that NEMO can reduce the signaling overhead.

T. Huang et al. (Eds.): ICONIP 2012, Part V, LNCS 7667, pp. 689–696, 2012.

Although several advantages could be achieved by using NEMO Basic Support Protocol, such as reducing handoff overhead and power consuming, it still has several drawbacks. NEMO Basic Support suffers the pinball routing problem and non-optimal transmission path [3] in the nested NEMO [2], which further introducing significant delays and packet overheads.

3   Related Works

Many studies and efforts have been done to solve the problems encountered by Mobile Networks and to provide a more efficient and secure NEMO protocol. An optimization mechanism for MIPv6 enabled nodes is proposed in [4]. Although this technique does not require any support from the infrastructure, it increases the signaling load on the wireless link.

The Care-of Prefix (CoP) [5] proposed a routing mechanism using hierarchical mobile network prefix assignment and hierarchical re-routing to optimize the routing and to reduce handoff signal overheads. CoP resolves the pinball routing problem but it spends more time to delegate CoPs for MRs of each level, which in turn raises total handoff latency. Second, the AGR cannot be placed at an optimal location for all CNs, which increases transmission delay and consumed bandwidth.

HCoP-B [6] focused on resolving pinball routing and RO storm problems for the nested NEMO by proposing a novel hierarchical care-of prefix (CoP) with the binding update tree (BUT) scheme. HCoP-B achieves shorter playback disruption time and buffering time in the nested NEMO, but it suffers from Long handoff latency and packet losses upon handoff.

Based on NEMO BS, ROTIO [7] proposed a routing optimization scheme with the extended tree information option (xTIO) [8]. ROTIO guarantees location privacy and mobility transparency. This scheme also achieves intra-NEMO route optimization and seamless handoff support but it suffers only two levels of nested tunnels.

As a conclusion, extensions for NEMO BS solve some problems but today none of them provides a complete, coherent, and integrated framework to address all the issues of mobile networks.

4   The Framework Architecture

Nested Mobile Network is established when the mobile network having a hierarchical structure, i.e. a mobile network is attached to another mobile network through a MR. To localizing handoff signaling and optimize routing of the nested NEMO, the MAP of the hierarchical approach, like HMIPv6, will be proposed into ROTIO [7]. However, the TLMR of the nested NEMO is proposed to be as a MAP in HMIPv6 for the nested NEMO. Thus, the entire nested NEMO becomes a local MAP domain. The MAP will maintain the binding cache for all MRs/MNNs, and achieves optimal routing from the CN to the MNN in the nested NEMO.

4.1   Binding Update

After handoff, each MR and each underlying LMN/VMN configure their new CoAs layer by layer in the nested NEMO. Because the hierarchical structure has been proposed, MRs and MNs in the nested mobile network will configure two CoAs: RCoA (Regional CoA) based on the mobile network prefix of the TLMR (MAP) and LCoA (On-link CoA) based on the mobile network prefix of its access router (MR or fixed router) as described in [9].

After each MR has configured its CoAs, it sends only one local binding update (LBU) to the MAP to update the binding cache, as well as to build the nested NEMO topology in BUT. Finally, the MAP will send a normal BU to each MR's HA for building an entry in the MR-HA's binding cache. The normal BU message contains the HoA of the TLMR (MAP). These steps are illustrated on Fig. 1.

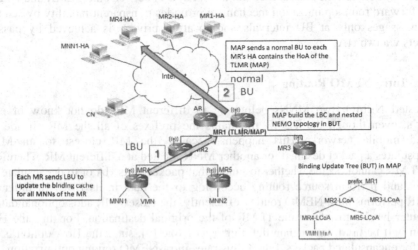

Fig. 1. Binging Update in the proposed framework

4.2 Route Optimization

Fig. 2 illustrates forward route optimization. A packet sent from the CN toward the MNN1 is routed to the closest MR's HA. Since MR4's HA already has the binding

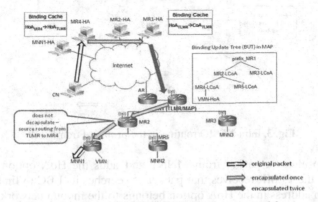

Fig. 2. forward routing path (CN-MNN1) in the proposed framework

information that MR4 (the closest MR to the MNN1) is located below the TLMR, the packet is encapsulated and sent to the HA of the TLMR.

When the packet is delivered to the HA of the TLMR, it is encapsulated again and sent to the current location (CoA) of the TLMR. After receiving the packet, the TLMR decapsulates it and searches its LBC and BUT to find the route to the MNN1.

By searching the Local Binding Cache (LBC), the TLMR discovers that the MNN1 is under MR4, which can be reachable via MR2. The TLMR forwards the packet using source routing to MR2. When the packet arrives at MR4, it decapsulates the packet and forwards it towards the MNN. There are only two levels of nested tunnels: 1) between the closest MR and its HA and 2) between the TLMR and its HA. This forward route optimization mechanism provides transparent mobility by sending BU messages only at BU intervals and location privacy is achieved by passing packets via two HAs.

4.3   Intra-NEMO Routing

In nested NEMO, two MNNs belonging to different MRs do not know of each other's, even though the TLMR maintains the prefixes of all the MRs inside the nested mobile network. This happens because the MR closest to an MNN encapsulates a packet destined for another MNN located at a different MR. Therefore, the TLMR cannot decide whether to tunnel that packet to its HA or to decapsulate the packet and perform source routing according to the data in its LBC. In order the TLMR performs intra-NEMO routing efficiently, the closest MR adds optional data to the outer header to inform the TLMR of the original destination. For this, the HoA option will be used, rather than defining a new one [9], since the HoA option is not used for encapsulated packets. Fig. 3 illustrates intra-NEMO routing optimization.

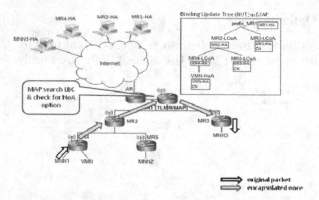

Fig. 3. Intra-NEMO routing in the proposed framework

MR4 encapsulates a packet from MNN1 and adds the HoA option to the outer header. When the TLMR receives that packet, it searches its LBC to find out whether the destination address in the HoA option belongs to the mobile network itself or not. If there is a match in the LBC, the TLMR decapsulates and forwards the packet using source routing. By using the intra-NEMO route optimization feature, the delay in communication significantly will be reduced.

The signaling flow of the proposed framework's messages is shown in Fig. 4.

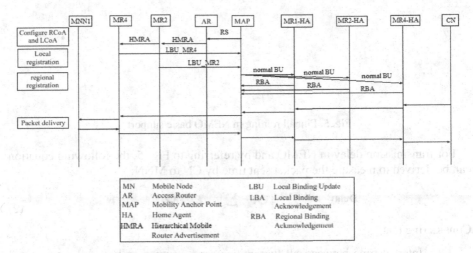

Fig. 4. Signaling flow of the proposed framework

5   Analytical Evaluation of the Proposed Solution

NEMO basic support protocol [2] uses Bi-directional tunneling to support mobile networks. Thus, it doesn't take route optimization into consideration. The CN has no binding information about the MNN, and packets heading toward the MNN have to go through all of the HAs, which is known as pinball routing.

Route optimization is essential for longer routing distance [10], since HAs can be located any place all over the world. So to evaluate the network performance, routing cost (which represents in this research link delay) will be considered as well as transmission delay as the following:

As shown in Fig. 5, Routing cost in NEMO (using bi-directional tunneling) may be measured by:

$$\text{Routing cost} = c_1 + \sum_{i=1}^{N} Ci \tag{1}$$

Where,

N: degrees of nesting

C_i: routing costs of link delay from the HA of the MR_i to the HA of the $MR_{(i+1)}$

C_N: routing cost of link delay from the TLMR's HA to the TLMR

c_i: routing cost of link delay from the CN to the HA of the MRi (MR_N is TLMR)

Also, it's assumed that the routing cost from the MR_N (TLMR) to the MNN is trivial in comparison with the routing cost from the CN to the MR_N.

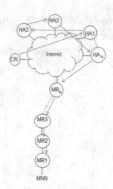

Fig. 5. Pinball routing in NEMO basic support

For transmission delay in NEMO and by referring to Fig. 5, the following equation can be derived to measure the packet sent time by CN to MNN:

$$\textbf{Delay} = (\sum_{i=1}^{N} Di + \sum_{i=1}^{N} Ti)\text{X}2 \tag{2}$$

Considering that,

- interval time between all the entities are assumed to be similar and equal to (T)
- Time spent in encapsulation and decapsulation are assumed to be identical and equal to (D)

Similarly, using Fig. 6, routing cost in the proposed framework can be measured as follows:

$$\text{Routing cost} = c_1 + C_N \tag{3}$$

Where,

N: degrees of nesting

C_i: routing costs of link delay from the HA of the MR_i to the HA of the $MR_{(i+1)}$

C_N: routing cost of link delay from the TLMR's HA to the TLMR

c_i: routing cost of link delay from the CN to the HA of the MR_i (MR_N is TLMR)

Assuming that the routing cost from MR_N to MNN is trivial compared to routing cost from CN to MR_N.

To calculate transmission delay from CN to MNN, with the following assumption:

- interval time between all the entities are assumed to be similar and equal to (T)
- Time spent in encapsulation and decapsulation are assumed to be identical and equal to (D)

So, the packet sent by CN to MNN will take time of

$$\textbf{Delay} = \text{x}D + \text{y}T + \sum_{i=1}^{N} Ti \tag{4}$$

Where, x = 4 and y = 3 in the proposed framework, regardless the nesting degree.

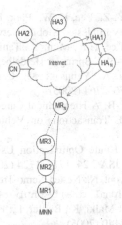

Fig. 6. Route optimization in the proposed framework

As shown in above equations, for higher degrees of nesting, the routing cost for bidirectional tunneling increases linearly. However, for the proposed framework, the routing cost remains constant because there will always be 2 levels of tunnelling regardless the degree of nesting.

6   Conclusion

In this paper, a framework of new scheme has been proposed which is based on route optimization using hierarchical structure with binding update tree (BUT). This proposed scheme aims to solve the pinball routing problem (non-optimal routing problem), binding update storm (signaling problem) and achieve seamless handoff. This will lead to reduce packet overhead, handoff latency, packet transmission delay, and achieve optimal routing. Hence, it solves the problems face NEMO Basic Support, especially in nested mobile network.

In addition, this paper evaluated a comparison between the rout optimization of the proposed scheme and bi-directional tunneling of NEMO. It shows that in the proposed scheme the communication will not be interrupted by increasing the degree of nesting, while NEMO will be extremely affected with higher degree of nesting. As future work, this framework will be simulated using QualNet Developer 5.0.1 [11] to evaluate its performance.

References

1. Johnson, D., Perkins, C., Arkko, J.: Mobility Support in IPv6 (2005), http://www.ietf.org/rfc/rfc3775.txt
2. Devarapalli, V., Wakikawa, R., Petrescu, A., Thubert, P.: NEMO basic support protocol (2005), http://www.ietf.org/rfc/rfc3963.txt.3963
3. Ng, C., Thubert, P., Watari, M., Zhao, F.: Network Mobility Route Optimization Problem Statement (2007), http://tools.ietf.org/html/rfc4888

4. Ryu, H.-K., Kim, D.-H., Cho, Y.-Z., Lee, K.-W., Park, H.-D.: Improved Handoff Scheme for Supporting Network Mobility in Nested Mobile Networks. In: Gervasi, O., Gavrilova, M.L., Kumar, V., Laganá, A., Lee, H.P., Mun, Y., Taniar, D., Tan, C.J.K. (eds.) ICCSA 2005. LNCS, vol. 3480, pp. 378–387. Springer, Heidelberg (2005)
5. Suzuki, T., Igarashi, K., Miura, A., Yabusaki, M.: Care-of Prefix Routing for Moving Networks. IEICE Trans. 88, 2756–2764 (2005)
6. Chang, I.C., Chou, C.H.: HCoP-B: A Hierarchical Care-of Prefix With BUT Scheme for Nested Mobile Networks. IEEE Transactions on Vehicular Technology 58, 2942–2965 (2009)
7. Cho, H., Kwon, T., Choi, Y.: Route Optimization Using Tree Information Option for Nested Mobile Networks. IEEE JSAC 24, 1717–1724 (2006)
8. Thubert, P., Bontoux, C., Montavont, N.: Nested Nemo Tree Discovery. Internet Draft (2009), http://tools.ietf.org/html/draft-thubert-tree-discovery-08
9. Soliman, H., Castelluccia, C., El-Malki, K., Bellier, L.: Hierarchical Mobile IPv6 Mobility Management (HMIPv6). IETF 4140 (2005)
10. Cho, H., Paik, E.K., Choi, Y.: RBU+: Recursive Binding Update for End-to-End Route Optimization in Nested Mobile Networks. In: Mammeri, Z., Lorenz, P. (eds.) HSNMC 2004. LNCS, vol. 3079, pp. 468–478. Springer, Heidelberg (2004)
11. QualNet Developer, Scalable Networks. Qualnet user manual (2004), http://www.scalable-networks.com/products/qualnet/

Pathway-Based Multi-class Classification of Lung Cancer

Worrawat Engchuan and Jonathan H. Chan[*]

Data and Knowledge Engineering Laboratory (D-Lab), School of Information Technology,
King Mongkut's University of Technology Thonburi, Bangkok, Thailand
worrawat.bank@gmail.com, jonathan@sit.kmutt.ac.th

Abstract. The advances in high throughput microarray technology have enabled genome-wide expression analysis to identify diagnostic biomarkers of various disease states. In this work, muti-class classification of lung cancer data is developed based on our previous accurate and robust binary-class classification using pathway activity data. In particular, the pathway activity of each pathway was inferred using a Negatively Correlated Feature Set (NCFS) method based on curated pathway data from MSigDB, which combines pathway data of many public databases such as KEGG, PubMed, BioCarta, etc. The developed technique was tested on three independent datasets as well as a merged dataset. The results show that using a two-stage binary classification process on independent datasets provided the best performance. Nonetheless, the multi-class SVM technique also yielded acceptable results.

Keywords: Pathway activity, multi-class classification, lung cancer, gene expression analysis, SVM.

1   Introduction

Among the cancer types, lung cancer is one of the top killer diseases, which caused approximately 157,000 Americans to die in 2011 [1]. It also has the highest incidence rates in East Asia [2]. Lung cancer can be divided into two types: non-small cell lung cancer (NSCLC) and small cell lung cancer (SCLC). Roughly 80% of lung cancer is NSCLC type, which can also be sub-categorized as Squamous cell carcinoma (SCC) and Adenocarcinoma (AC) [3]. In the past few decades, lung cancer diagnosis and prognosis were done based on the physical analysis of tissues like Computed Tomography (CT) scan [4-5]. However, the physical-based diagnosis can only detect the malignant in late stage of lung cancer and resulted in low survival rates (approximately 16% for NSCLC and 6% for SCLC) [6]. With the advances in molecular biology, information about DNA, RNA and proteins can be applied for these medical tasks in order to detect the tumor in earlier stages, which should result in an increase in the survival rate. In 1995, Plebani et al. evaluated seven tumour markers for lung cancer diagnosis by using protein assays. However, it needed some expert knowledge in order to yield an accurate diagnosis [7]. Then with the advent of the microarray

---

[*] Corresponding author.

T. Huang et al. (Eds.): ICONIP 2012, Part V, LNCS 7667, pp. 697–702, 2012.

technique, which can simultaneously measure the expression of thousands of genes, it was utilized for this medical purpose. Arindam et al. used hierarchical clustering and feature selection on microarray data [8]. Gavin et al. calculated gene expression ratio of pairs of genes which expressed differently in two groups and used that ratio for diagnosis [9]. However using only a single gene is usually inadequate for understanding the complex traits like lung cancer. So, multiple-genes analysis based on biological knowledge have been developed to improve understanding and increase the accuracy of lung cancer diagnosis. Hosgood et al. analyzed lung cancer data using pathway-based tool, GoMiner and can found that the role of cell cycle pathway is relevance to lung cancer development [10]. Sootanan et al. develop a pathway-based approach named Negatively Correlated Feature Sets (NCFS-i), which identifies subset of cancer-related gene set of each pathway called phenotype-correlated genes (PCOGS) and then calculates pathway activity based on those sets of genes [11]. The classification using pathway activity data obtained from their method shows higher classification power and robustness. Chan et al. evaluated the feature selection approaches on pathway activity data to increase the accuracy and robustness of classification [12].

Much works have been done on binary-class classification in the more traditional studies. However, many medical works now need to diagnose more than two classes of disease, like staging. So, multi-class classification is required. In 2001, Sridhar et al. employed Support Vector Machine (SVM) approach and One vs. All (OVA)-based approach on microarray data and yield approximately 78% overall accuracy [13]. Jane et al. combined SVM and GA-algorithm for this multi-class classification tasks and show 68-85% on leave-one-out cross-validation [14]. In early 2012, a competition for improving the performance of complex disease diagnosis called SBVImprover was launched. One of challenges in the competition aims to develop classifier for classifying lung cancer subtype and staging simultaneously [15]. In this paper, we aim to utilize the pathway activity data, which show good performance for binary-class classification to tackle this multi-class classification problem using a SVM-based approach.

2   Materials and Method

2.1   Microarray Datasets

The purpose of this study is to build a classifying model to classify lung cancer subtypes: Adenocarcinoma (AC) and Squamous cell carcinoma (SCC) together with staging of each subtype (stage 1-2). Since the pathway activity inferring method developed by Sootanan et al. has a limitation on microarray platform, which all dataset must be from the same platform [11]. So, it is necessary to pick only the datasets from the same platform. There are 3 microarray datasets generated by GPL570 microarray platform, providing all those information about subject's subtypes and stages. Those datasets are available at GEO databases with following accession number: GSE10245 (51 samples), GSE18842 (42 samples) and GSE2109 (52 samples).

2.2   Pathway-Based Multi-class Classification

The muti-class classification of lung cancer data is developed based on the previous accurate and robust binary-class classification using pathway activity data [11]. The pathway activity of each pathway was inferred using NCFS-i method based on curated pathway data from MSigDB, which combines pathway data of many public databases such as KEGG, PubMed, BioCarta, etc [16].   Fig. 1 shows the algorithm of NCFS-i method, in each pathway, all the members will be ranked by t-score, which calculated by Student t-test based on the difference of mean between two classes. In each iteration, pathway activity was calculated by subtraction of the top and bottom genes, then discriminative score of that obtained pathway activity, which denote the ability of distinguish two classes is calculated. In next iteration, another top and bottom genes are added in subtraction for inferring the new pathway activity. The process will be repeated until the discriminative score of new pathway activity is less than the old one, or in other words, when the ability to distinguish two classes based on new pathway activity is lower than using the previous pathway activity.

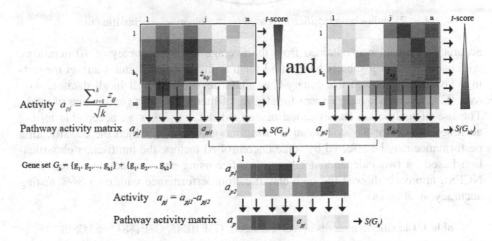

Fig. 1. NCFS-i pathway inferring method

Since, NCFS-i pathway activity inferring method is based on binary class, so to use pathway activity data on multi-class classification, it is necessary to build classifying model based on the binary-class classification. First two independent binary classifying models based on SVM approach were constructed to deal with subtype diagnosis and staging. The results of two models were then combined to be the final output of the classification. A schematic of the proposed multi-class classification approach based on pathway activities is shown in Fig. 2.

3   Results and Discussion

The binary-class and multi-class classifying models were validated using 10-fold cross-validation, which would randomly divide data into 10 equal subsets. The

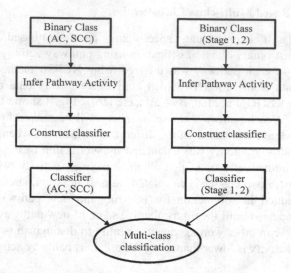

Fig. 2. Multi-class classification based on two binary-class classificaltions

accuracy of model was obtained from the average testing accuracy of 10 iterations, with each iteration 9 subsets being used for training the model and 1 subset for testing. The results showed that binary-class classifier worked well in all datasets with over 92% testing accuracy except for staging of merged dataset (with value of 0.83). The use of multi-class classification of merged dataset shows an acceptable testing accuracy of 0.81 or 81%. Instead of using merged dataset, which the classification performance may be affected by some uncontrolled factors, the multi-class classification based on two independent binary classifier using pathway activity inferred by NCFS-i approach showed the high classification performance with over 86% testing accuracy in all cases.

Table 1. Classification results of merge datasets, GSE10245, GSE18842 and GSE2109

Dataset	Classification accuracy		
	Subtype diagnosis	Staging	Multi-class
Merged	0.97	0.83	0.81
GSE10245	0.96	0.92	0.86
GSE18842	0.98	1.0	0.98
GSE2109	0.98	0.96	0.94

The results from Table 1 show overall testing accuracy of better than 80% in all cases. Referring to each binary-class classification, subtype diagnosis provided higher accuracy than staging. One possible reason may be because staging of each lung cancer subtype may have different characteristics. So if the staging is divided into two staging for two subtypes, the classification performance may be improved. Thus, instead of constructing a multi-class classifier based on two independent binary-class classifiers, it would be better if the classifier is constructed using a two-step or two-stage approach. First, the data are classified according to the subtype and then try to identify the stage of cancer. Another issue of this work is about the merging of datasets. Since, the datasets are from different experiments, so some uncontrolled factors could affect the expression data. Some typical factors include the skill of researcher who did the experiment, the RNA quality or RNA integrity and also the contamination of RNA sample [17]. Consequently, the development of a microarray data combining approach that addresses the effect of uncontrolled factors is required in order to improve classification performance.

4   Conclusions

This work demonstrated an effective extension to our previous pathway-based microarray technique for disease classification from binary class to multi-class. The developed technique was tested on three independent datasets as well as a merged dataset. The results show that using a two-stage binary classification process on independent datasets provided the best performance. Nevertheless, the use of a SVM multi-class classifier also showed promising results. Further refinement to the NCFS method is being made to address the biased approach taken previously. In addition, more disease datasets will be used to further test and validate our pathway-based approach.

References

1. American Cancer Society: Cancer Facts & Figures 2011. American Cancer Society, Atlanta (2011)
2. Wang, L., Cher, G.B.: An overview of Cancer trends in Asia. Innovationmagazine.com (2012)
3. Stöppler, M.C.: LungCancer. Medicine.net. (2011)
4. Mountain, C.F., Dresler, C.M.: Regional Lymph Node Classification for Lung Cancer Staging. CHEST 111, 1718–1723 (1997)
5. Mountain, C.F.: Revisions in the international System for Staging Lung Cancer. CHEST 111, 1710–1717 (1997)
6. Tsou, J.A., et al.: DNA methylation analysis: a powerful new tool for lung cancer diagnosis. Oncogene 21, 5450–5461 (2002)
7. Plebani, M., et al.: Clinical evaluation of seven tomour markers in lung cancer diagnosis: can any combination improve the results? British Journal of Cancer 72, 170–173 (1995)
8. Arindam, B., et al.: Classification of human lung cancer carcinoma by mRNA expression profiling reveals distinct adenocarcinoma subclasses. PNAS 98, 13790–13795 (2001)
9. Gavin, J., et al.: Translation of Microarray Data into Clinically Relevant Cancer Dianostic Test using Gene Expression Ratios in Lung Cancer and Mesothelioma. Cancer Research 62, 4963–4967 (2002)

10. Hosgood, H.D., et al.: Pathway-based evaluation of 380 candidate genes and lung cancer susceptibility suggests the importance of the cell cycle pathway. Carcinogenesis 10, 1938–1943 (2008)
11. Sootanan, P., et al.: Pathway-based microarray analysis for robust disease classification. Neural Computing & Applications 21, 649–660 (2012)
12. Chan, J.H., et al.: Feature selection of pathway markers for microarray-based disease classification using negatively correlated feature sets. In: International Joint Conference on Neural Networks (IJCNN 2011), pp. 3293–3299. IEEE Press, New York (2011)
13. Sridhar, R., et al.: Multiclass cancer diagnosis using tumor gene expression signatures. PNAS 98, 15149–15154 (2001)
14. Jane, J.L., et al.: Muticlass cancer classification and biomarker discovery using GA-based algorithms. Bioinformatics 21, 2691–2697 (2004)
15. SBVImprover, http://www.sbvimprover.com/
16. Aravind, S., et al.: Gene set enrichment analysis: A knowledge-based approach for interpreting genome-wide expression profiles. PNAS 102, 15545–15550 (2005)
17. Fleige, S., Pfaffl, M.W.: RNA Integrity and the effect on the real-time qRT-PCR performance. Molecular Aspects of Medicine 27, 126–139 (2006)

Spontaneous Emergence of the Intelligence in an Artificial World

Istvan Elek, Janos Roden, and Thai Binh Nguyen

Eotvos Lorand University Faculty of Computer Science, Budapest, Hungary
elek@map.elte.hu
http://mapw.elte.hu/elek/

Abstract. The aim of this paper is to introduce the principles of the evolution units, which are named the digital evolutionary machines. The entities will be constructed based on these principles. Their properties and abilities will be shown with some experimental results. The knowledge base of DEMs was stored in a database, which is the source of the individual knowledge graph. It helps to make decisions of each entity. Their life was observed and analyzed. Some interesting analysis and charts will be shown in order to understand how DEMs work in the artificial world. Finally the DEMs geographical extension will be shown.

Keywords: Emergence of intelligence, Artificial world, Knowledge base.

1 Introduction

The evolution modeling projects focus on multiplication and the competitive exclusion theory, which is one of the most famous low after Darwin in the evolution biology. The competitiv exclusion theory comes from Georgii Frantsevich Gause, 1932 [12], who was a Russian biologist in the 20th century. An important low of the evolution process, that is irreversible, which was revealed by Dollo [6], and expounded by Gould, 1970 [13]. The key moment of the evolution is the emergence of the intelligence. This early intelligence was really low level, but it can help to survive the fluctuating environment. The changing Earth and its climate were the most remarkable challenge for the ancient organisms.

Many papers (Adami, 2002 [1], 2003 [2], Ofria at al, 2004 [4], Ostrowski, 2007 [15]), dealing with artificial organisms emphasized the importance of complexity. Not only life shows serious complexity. There are human made constructions as well, such as the topology of Internet, that have increasing complex structures with fractal properties (Barabasi, 2002 [5], 2003 [3], Barabasi–Newman, 2006 [14]).

The multiplication and the competition are in the center of works mentioned above. The evolution of organisms' body and their intelligence can not be independent. While we are going to model the organisms' evolution, we need to model the development of their intelligence too. There is no right definition of intelligence of the organisms (Turing 's definition does not serve path). Instead of making useless definitions, it is more promising to set up some essential principles

T. Huang et al. (Eds.): ICONIP 2012, Part V, LNCS 7667, pp. 703–712, 2012.

that regulate the process of collecting and interpreting data from the surrounding world. There are many research papers and books that describe machines which collect data from the surroundings and they have some kind of remembrance (Russel, 2002 [16]). Their ability of interpretation of the environment is restricted. The limits of these constructions are obvious: their intelligence never becomes similar to that of mammals or octopuses.

This paper is about a computerized approach of the digital evolutionary machines as the elements of the evolution process. The principles of them were established in Elek, 2008 [7], 2009 [8], 2010 [9], [10] and 2011 [11].

2   Knowledge and Evolution

Paleontology and geology serve many exciting examples of the one way evolution. The time flows in one direction, forward. Life always tries to adapt itself to the circumstances, mainly to the weather (temperature, seasons, climate). If the climate has changed, the adaptive organisms also change their right properties, skills and manners. If one million years later the climate changed back, the evolution did not remember the previous stage of the environment. The adaptivity also produces new properties that help to survive the changes. The evolution seems to be recursive. It means the regulation is a kind of recursion, like a feedback. The current stage is the basis of the next step. Evolution never gets back. There are no general physical laws in the background where the processes can be computed from.

If an organism did not recognize the enemy, the food and the environment, it had died soon or immediately. Consequently, organisms had to collect data from their surroundings (perception) and interpret them (understanding). The remembrance is probably one of the most important components of understanding the world. The irreversible time, i.e. the serial recording of events and data of the environment produces a huge database. It contains everything that happened to an organism in its lifetime. The key of surviving is the interpretation of the surrounding world and the action.

The weather and climate were the most effective factors of the evolution of organisms. Every organism needs to interpret the measured and stored data of the environment if they wanted to survive. This ability required a huge database containing everything that ever happened to the organism. This is the experience based knowledge base.

3   Principles

Biologists say there is a kind of self organization among protein molecules in vitro. Even if we assume it was the first step toward a real organism, it is evident there is no huge knowledge database and complicated interpretation logic in it. Protein molecules seem to be inclined to form combinations. Their data collection logic also has to be simple. Consequently a knowledge database can be constructed from simple steps of data collection. The algorithm of the knowledge graph search has to be simple.

Let us construct the knowledge-graph. It consists of atoms and their connections. Atoms are the nodes and connections are the edges of knowledge-graph. Let us define a context that includes arbitrary atoms of the tree; consequently, the structure of the graph is not predefined. It depends on the atomic connections that depend on a time series, when events took place in time order one after the other. The general principles declared the equality of atoms and contexts. In other worlds a context can contain simple atoms or other contexts as well.

Let a_{ij} denote the i-th atom in the j-th context which contains N atoms. Let the quantity of the knowledge of a context (k^j) be the sum of the quantity of the knowledge of every atom in it:

$$k^j = \sum_{i=1}^{N} a_i^j \qquad (1)$$

The knowledge-base has two basic functions: receiving a question and answering. The key of the problem is to find the path from the questioning node to the answering node in the knowledge-base.

Let l_{ij} denote the strength of the connection between a_i and a_j . Let $l_{ij} = l_{ij} + 1$ if the tour produces good result and $l_{ij} = l_{ij} - 1$ if the result is bad. This logic makes good paths stronger and bad ones weaker. Look at the u and f nodes in the knowledge-graph. The right distance definition depends on the length of the path along branches, so it is graph tour based.

Let a_i and a_j be two nodes of the knowledge-graph where the path includes $m = j - i$ atoms between them. Let d_{ij} be the distance of these two nodes, let the strength of their connection be denoted by l_{ij} , which is the reciprocal of the sum of the strength of the connections between a_i and a_j. The stronger the connection between two nodes, the closer they are.

The goal of the knowledge-graph is to answer a question arising from the circumstance. How to find the right path from the questioning node to the answering one? The fastest is the right path, probably. This logic produces very fast reaction in well known problems and may result fail in unknown cases. The fail means unsuccessful escape, capture or something important for the organism. If it survived the situation, i.e. the reaction was successful, and the path that produced the success became stronger. If it did not survive the situation or the result of the action was failed, i.e. the result of the action was unsuccessful, the organism was knocked out or a path in the knowledge-graph became weaker.

3.1   The Knowledge Representation

Let $\mathbf{K}$ denote the knowledge-base which consists of n atoms a_i, $a_j \in \mathbf{K}$. Let us name it the knowledge-matrix.

$$\mathbf{K} = \begin{pmatrix} a_{11} & a_{12} & \ldots & a_{1n} \\ a_{21} & a_{22} & \ldots & a_{2n} \\ \vdots & \vdots & \ddots & \\ a_{n1} & a_{n2} & \ldots & a_{nn} \end{pmatrix} \tag{2}$$

Some of the atoms are in touch with other atoms in $\mathbf{K}$. Let us describe the links of a_i and a_j atoms with l_{ij}, where

$$l_{ij} = \begin{cases} 1 & \text{if } i = j \\ 0 & \text{if } i \neq j \text{ and no link between them} \\ u - v & \text{else where } u \text{ succeeful and } v \text{ unsuccessful} \end{cases}$$

Let us organize the atomic links into a matrix form, and name it a link matrix and denote it by $\mathbf{L}$. The elements of the link matrix are l_{ij} that describe the link of the atomic pointpairs. $\mathbf{L}$ is diagonal ($l_{ii} = 1$) and describes the relationships of the atoms in the knowledge-matrix:

$$\mathbf{L} = \begin{pmatrix} 1 & l_{12} & \ldots & l_{1n} \\ l_{21} & 1 & \ldots & l_{2n} \\ \vdots & \vdots & \ddots & \\ l_{n1} & l_{n2} & \ldots & 1 \end{pmatrix} \tag{3}$$

1. Every atom has one link at least.
2. Let $\mathbf{C}$ be a context. $\mathbf{C} \subseteq \mathbf{K}$, i.e. $\mathbf{C}$ consist of any atoms of $\mathbf{K}$.
3. Any atoms can be the member of any context.
4. Any contexts can be a member of any contexts. In this case, the knowledge-matrix ($\mathbf{K}$) is a hyper matrix where matrix elements can be matrices.
5. In summary an atom can be a
   (a) Simple atom that is really elementary and belongs to one context.
   (b) Multi-member atom that is also elementary but belongs to more than one context.
   (c) Aggregated atom that is a kind of simple atom, but its value is a representation of a context. In other words, its value is a determinant or a spur of the knowledge-matrix of a certain context.
   (d) A complex atom that is a context that includes any kinds of atoms above.

3.2 Properties of the Knowledge Base

1. The knowledge-base is a continuously increasing database which stores everything that happened to it. This is a one way process. The knowledge-base is different from entity to entity.
2. There is no changing or erasing function in the knowledge-base.
3. Let the data collection be extensive if a DEM-entity perceives the circumstance and stores data. There are some important consequences of the extensive mode:

(a) Every individual knowledge-base is different. It depends on the career of a certain DEM-entity. There are as many DEM-entities as many kinds of knowledge-base exist.

(b) If some changes happen that certainly produces the same circumstance in the past, the previously recorded atoms has not been changed, simply a new atom has appended the end of the knowledge-base.

4. Let us have another mode that was named intensive, when there is no perception. It is a meditative stage when the knowledge-base acquires the data came from the extensive stage. The result of the data acquisition in intensive mode may produce new contexts, faster path, more reliable work. This stage is extremely important in learning the circumstance.

5. The feedback is a process when a DEM-entity is informed about the result of its reaction. The result of this process is success or fail. As mentioned previously, the success/fail makes stronger/weaker a certain path in the knowledge-graph of the knowledge-base. The DEM-entity's knowledge-base becomes much stronger/ weaker if the feedback such as rewarding or punishment comes from an external intelligent entity, because its knowledge-base can be considered as an included context.

6. In the history of the Earth there was never only one organism. There were always ensembles. Ensembles make organisms competitive. Competition results in different skills i.e. different knowledge-bases.

7. Different circumstances cause different experiences for the organisms.

The evolution is a process in time and in space dimensions. Regarding the results of paleontology it is known that evolution is recursive, i.e. an evolution step depends on the previous step only and it is a one way process, it can not be turned back. This is known as Dollo's law in paleontology [6]. This law was first stated by Dollo in this way: *An organism is unable to return, even partially, to a previous stage already realized in the ranks of its ancestors.* According to this hypothesis, a structure or organ that has been lost or discarded through the process of evolution will not reappear in that line of organisms.

The development means not only physical but mental changes as well in an organism. Mental means intelligence in this context. The adaptivity comes from its knowledge base. The quality of the knowledge base has to influence the physical properties also. Consequently the physical and mental evolution work collaterally, since the organisms live in various climate, in many challenges, having different experiences with different chances with various knowledge but in many, different places at the same time.

4 Internet Is the Artificial World

If we are willing to create experienced DEM-entities there is no wizard unfortunately. There is no a recipe to install them from little pieces. How to construct a DEM-entity? Probably, it is impossible to construct only one of it. First we should create an ensemble from many initial DEM-entities and leave them to live in their own world.

We have two access points to this problem. The first one is to construct an artificial circumstance where DEM-entities live in. The second one is to construct many initial DEM entities with simple perception and interpretation functions. Let us look at some of the details:

1. The artificial world (AW) has some essential properties that define the frame of this world.
   (a) Let AW be huge where circumstances have spatial dependencies. Regarding the size of this world, environmental parameters are obviously different. If we leave DEM-entities alone in this world they will have different experiences and different knowledge-bases because of climatic differences.
   (b) If the AW is huge enough, the survival strategies will be different. One of the DEM-entities escapes from the unfriendly circumstances but others try to adapt. Different strategies result different knowledge-bases.
2. If there are many DEM-entities on the same territories, what is the consequence?
   (a) There are many DEM-entities who try to get better strategy in order to be more successful than others. Someone gets advantages but someone gets disadvantages since it fails to answer a certain question.
   (b) Regarding the different DEM-entities and unique knowledge-bases, many different strategies can coexist in the AW. Consequently, many different strategies can be successful at the same time. Someone prefers the escape, but someone the competition.
3. Many DEM-entities will have many different knowledge-bases.

The question is how to construct the prototype of a DEM-entity? Before the construction of the prototype, let us create the artificial world that will be the space for DEM-entities. If AW has been created already many DEM-entities should be available to start up in it. Properties and abilities of DEM-entities were introduced previously, so the task is to make their software representation. Regarding the quantity of DEMs and the huge sized world with different spatial properties may result many formed DEMs, and these entities have different knowledge-bases. An intelligent DEM-entity can exist in ensemble only. There is no lonely intelligence because it is a collective product that is realized in some successful entities' knowledge bases.

4.1   The Computerized Approach

This paper focused on the appearance of the intelligence and its properties in a digital representation. Digital evolution machines and the Internet, which is the artificial world, where they live in , are in interaction. The purpose of this work is to model this interaction. The task is to store everything sequentially that happened to DEMs, and to construct the knowledge graph, which nodes are the pointers for sequentially stored events.

Briefly the system works as follows:

1. There are a lot of DEMs in the Internet. An individual DEM starts his carrier in a starting URL.
2. This DEM investigates the certain URL, what kind of files are linked to this. Some tags in this file point to another URL-s ($<$ a ref $=$ http://...$>$). Some of the pointed files are considered as resources which are needed to DEM's life (for instance *doc, pdf, txt, jpg, gif, tif* files are considered resources with various energy content, but *exe, bat* files are considered poisons, and so on).
3. A successful step, when the reached resource produces energy input for a certain DEM, makes path stronger. If the reached node (file) is neutral or contains poison, the path becomes weaker.
4. if a DEM visited a resource, the node (file) runs out for a short period (it needs time for recover itself), so DEM has to move to another URL (for instances, it follows an $<$ a ref $=$ http://...$>$ html tag) for further resources.
5. go to 2

Last but not least, in order to find the rigth URL for the next move, we need an *adviser*, which is one of the most remarkable component of the system. *Adviser* uses the knowledge graph, and give the most promising URL for a DEM. Summarily sequentially stored events and *adviser* functionality figure out a certain DEM's intelligence.

For the technical implementation, there is a MS Sql server, which stores sequentially the every day events, (the table name is *Logbook*). The next table *DemWorkers* stores data for the identification of individual DEMs. This table contains the resource requirements also of a certain DEM. The visited web nodes are stored in the table *ArtiworldNodes*.

A related table, named *UrlsContent* stores the link of an URL (resource files, and potential URLs for further moves). The table *ContentTypes* contains the possible energy sources (food) with energy content units. You can change properties of the artificial world via table *ContentTypes* if you want the energy files to have different energy content. For example a pdf file has higher energy content than a simple txt file. You can adjust the world more, if the energy content depends on the file size too. Do not forget, that the world structure, its topology

DEMs will affect the artificial world, and AW influences DEMs life. This interaction is the target of our research. The system is working, but not finished yet, of course, because this is a changing database. The project info page can be reached on *http://dem.elte.hu* site. Here sooner or later you can reach the database and observe what DEMs do.

4.2   Results

In this section some figures will be shown, which were made from a program package. The purpose of this system is to handle DEMs' life, such as movements, destiny, status and so on. The other task of the system is to register the circumstance, URLs and their contents. This is a kind of projection of the known artificial world, where DEMs have already visited.

Look at the figure series which show the DEMs' life. First, 1000 DEMs were started. The properties of DEMs can be set. In addition we can set up any domain disabled if it is required. This set is obvious, regarding the early evolution stage, when the chemical evolution was typical. It is extremely interesting whether they have any chance to survive.

Look at the dem workers' status in the following figures (fig. 1, fig. 2 and fig. 3), which show those workers who alive certain time intervals.

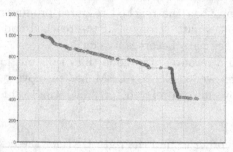

Fig. 1. Alive DEM workers after 5 minutes from start. X, Y axes means time and population (started with 1000 DEMs).

It is surprising that at the beginning DEMs die in large quantities. The most of URLs are not green grass for hunger DEMs, so they have no foods. They obviously perished.

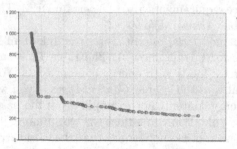

Fig. 2. Alive DEM workers after 120 minutes from start

If you study the figures (fig. 1, fig. 2 and fig. 3) you will find, there is a small population which survived the artificial deserts and all troubles. There were 5 runs with 1000 DEMs, and the results were the same. The consequence is really interesting. Generally the 85 % of DEMs perished after some days, but there remains a little group with 15 % of the starting population, who survived the troubles, and they will be alive for a long time, if the circumstances are steady.

4.3   Geographical Extension of DEMs

The geographical extension of DEMs is interesting. They extend almost all over the world, which can be seen on the figures and 4.

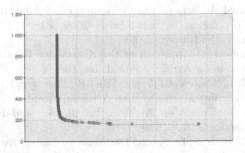

Fig. 3. Alive DEM workers after 4 days from start

Fig. 4. Cities in the world which are visited by DEMs

References

1. Adami, C.:Ab Initio of Ecosystems with Artificial Life, arXiv:physics0209081, 22 (2002)
2. Adami, C.: Sequence Complexity in Darwinian Evolution, Comlexity, vol. 8(2). Wiley Periodicals (2003)
3. Albert, B.: Statistical mechanics of complex networks. Reviews of Modern Physics 74 (2002)
4. Chow, S.S., Wilke, C.O., Ofria, C., Lenski, R.E., Adami, C.: Adaptive radiation from resource competition in digital organisms. Science 305, 84–86 (2004)
5. Barabsi, A., Albert, R., Jeong, H.: Mean-field theory for scale-free random networks. Preprint submitted to Elsevier Preprint, 5 (2002)
6. Dollo, L.: http://en.wikipedia.org/wiki/Dollo's_law
7. Elek, I.: Principles of Digital Evolution Machines. In: International Conference of Artificial Intelligence and Pattern Recognition, Orlando, FL (2008)

8. Elek, I.: Evolutional Aspects of the Construction of Adaptive Knowledge Base. In: Yu, W., He, H., Zhang, N. (eds.) ISNN 2009, Part I. LNCS, vol. 5551, pp. 1053–1061. Springer, Heidelberg (2009)
9. Elek, I.: A Computerized Approach of the Knowledge Representation of Digital Evolution Machines in an Artificial World. In: Tan, Y., Shi, Y., Tan, K.C. (eds.) ICSI 2010, Part I. LNCS, vol. 6145, pp. 533–540. Springer, Heidelberg (2010)
10. Elek, I.: Digital Evolution Machines in an Artificial World: a Computerized Model. In: International Conference on Artificial Intelligence and Pattern Recognition (AIPR 2010), Orlando, USA, p. 169 (2010)
11. Elek, I., Roden, J., Binh, N.T.: Spatial Database for Digital Evolutionary Machines in an Artificial World without Knowledge-Base. In: Fourth International Workshop on Advanced Computational Intelligence (IWACI 2011), Wuhan, China (2011)
12. Gause, G.F.: Experimental studies on the struggle for existence. Journal of Experimental Biology 9, 389–402 (1932)
13. Gould, S.J.: Dollo on Dollo's law: irreversibility and the status of evolutionary laws. Journal of the History of Biology (Netherlands) 3, 189–212 (1970)
14. Newman, M., Barabasi, A.L., Watts, D.J.: The structure and Dynamics of Networks. Princeton University Press (2006)
15. Ostrowski, E.A., Ofria, C., Lenski, R.E.: Ecological specialization and adaptive decay in digital organisms. The American Naturalist 169(1), E1–E20 (2007)
16. Russel, S., Norvig, P.: Artificial Intelligence: A Modern Approach. Prentice Hall (2002)

Adaptive Backstepping Neural Control for Switched Nonlinear Stochastic System with Time-Delay Based on Extreme Learning Machine

Yang Xiao[1], Fei Long[1,*], and Zhigang Zeng[2]

[1] College of Computer Science and Information, Guizhou University, Guiyang 550025, China
flong1973@yahoo.com.cn
[2] Department of Control Science and Engineering,
Huazhong University of Science and Technology, Wuhan 430074, China
zgzeng@mail.hust.edu.cn

Abstract .n this paper, for a class of switched stochastic nonlinear systems with time-varying delays, the output feedback stabilization problem is addressed based on single hidden layer feed-forward network (SLFN) and backstepping technique. Furthermore, an adaptive backstepping neural switching control scheme is presented for the above problem. In the scheme, only a SLFN is employed to compensate for all known system nonlinear terms depending on the delayed output. The output weights and control laws are updated based on the Lyapunov synthesis approach and backstepping technique to guarantee the stability of the overall system. Then a special switching law is given based on attenuation speed of each subsystem. Different from the existing techniques, the parameters of the SLFN are adjusted based on a new neural networks learning algorithm named as extreme learning machine (ELM), where all the hidden node parameters randomly be generated. Finally, the proposed control scheme is applied to an example and the simulation results demonstrate good performance.

Keywords: Switched Nonlinear Stochastic Systems, Neural Networks Control, Backstepping, Extreme Learning Machine.

1 Introduction

The switched systems are a class of hybrid dynamical systems consisting of a finite number of continuous-time (or discrete-time) subsystems and a particular type switching law. Many real world processes can be modeled as switched system, for example, networked control systems, chemical process systems, computer controlled systems and communication systems. Moreover, intelligent control strategies based on the idea of switching controllers can overcome the shortcoming of using a single controller and improve overall system performance.

---

* Corresponding author.

T. Huang et al. (Eds.): ICONIP 2012, Part V, LNCS 7667, pp. 713–721, 2012.

Recently, the investigations on stochastic systems have received considerable attention because stochastic disturbances phenomenon widely exists in science theory research and engineering applications field. Although many nonlinear control theories were extended to the case of stochastic nonlinear field, the technical obstacle of stochastic Lyapunov analysis which It $\hat{o}$ differentiation involves not only the gradient but also the higher order Hessian term was a troublesome question. Pan and Basar [1] first solved the stabilization problem for a class of strict-feedback stochastic nonlinear systems by using a quadratic Lyapunov function and a risk-sensitive cost criterion. By employing the quartic Lyapunov functions instead of the traditional quadratic functions, Deng and Krstic [2] presented an adaptive backstepping design algorithm which can be coded. Several works [3, 4] for the stochastic nonlinear stabilization problem of switched systems had been proposed.

Due to neural networks can represent the nonlinear relationship between the input and output data by its inherent structure and parameters, adaptive neural networks control method have been carried out for the nonlinear systems [5-7]. In recent years, a new neural network algorithm named extreme learning machine (ELM) was proposed [8-10]. Owing to that ELM have better performance than the existing NN algorithm; we will try to employ ELM as a new approximate means in adaptive backstepping control method for stochastic nonlinear stabilization problem of switched systems.

In this paper, we give an adaptive neural switching control scheme to solve the output feedback stabilization problem for such system. The main characteristic of this control scheme include: Firstly, all nonlinear functions of system are lumped into a suitable function that is approximated by SLFN whose parameters are adjusted by ELM with additive hidden nodes or RBF hidden nodes. Moreover, all the hidden node parameters in ELM algorithm can be generated randomly according to any given continuous probability distribution without any prior knowledge about the target function. Secondly, a switching control law is given based on attenuation speed of each subsystem, multiple Lyapunov function and backstepping techniques. The validity of the proposed NN switching control scheme is finally briefly clarified by an example.

2   Problem Statements and Preliminaries

2.1   Extreme learning machine (ELM)

The output of SLFN with L hidden nodes can be represented by $f(x) = \sum_{i=1}^{L} \theta_i F(x, a_i, b_i); x, a_i \in \mathbb{R}^n$, where a_i and b_i are the learning parameters of hidden nodes, $\theta_i \in \mathbb{R}^m$ is the weight vector connecting the i th hidden node to the output nodes and $F(x, a_i, b_i)$ is the output of the i th hidden node with respect to the input x. For additive hidden nodes with activation function is defined as $F(x, a_i, b_i) = F(a_i \cdot x + b_i)$, $b_i \in \mathbb{R}$. For RBF hidden nodes with activation function is defined as $F(x, a_i, b_i) = F(|x - a_i| b_i^{-1}), b_i \in \mathbb{R}^+$. For l arbitrary distinct samples (x_j, t_j), if the standard SLFN with L hidden nodes can approximate these l samples with zero error, there exists θ_i, a_i and b_i such that $\sum_{i=1}^{L} \theta_i F(x_j, a_i, b_i) = t_j$. This equation can be written compactly as $F(x, a, b)\theta = T$, where

$$F(x_1, \ldots, x_l, a_1, \ldots, a_L, b_1, \ldots b_L) = \begin{pmatrix} F(x_1, a_1, b_1) & \cdots & F(x_1, a_L, b_L) \\ \vdots & \cdots & \vdots \\ F(x_l, a_1, b_1) & \cdots & F(x_l, a_L, b_L) \end{pmatrix}_{l \times L}, \Theta = \begin{pmatrix} \theta_1 \\ \vdots \\ \theta_L \end{pmatrix}, T = \begin{pmatrix} t_1 \\ \vdots \\ t_L \end{pmatrix}^T$$

In ELM algorithm, all the hidden node parameters can randomly be generated according to any given continuous probability distribution without any prior knowledge about the target function. The ELM algorithm has the following property [19]: For a SLFNN with RBF hidden nodes and an activation function from $\mathbb{R}$ to $\mathbb{R}$, which is infinitely differentiable in any interval, there exists $L \leq l$ and a small positive value ε such that for arbitrary distinct input vectors x and (a_i, b_i), $P\{\|F(x, a, b)\theta - T\| < \varepsilon\} = 1$.

2.2 Problem Statements

Consider the following switched nonlinear stochastic system with time-delay

$$\begin{cases} dx_i = x_{i+1}dt, \ 1 \leq i \leq n-1 \\ dx_n = \left[f_{\sigma(t)}(y) + m_{\sigma(t)}(y, y(t-d(t))) + \varphi_{\sigma(t)} u_{\sigma(t)} \right]dt + \left[g_{\sigma(t)}(y) + n_{\sigma(t)}(y, y(t-d(t))) \right]d\omega \\ y = x_1, \\ y(t) = \varphi(t), \ -\eta \leq t \leq 0 \end{cases} \quad (1)$$

where $x_i \in \mathbb{R}, i = 1 \sim n$ is the system state. $y = x_1 \in \mathbb{R}$ is the system output. The right continuous function $\sigma(\cdot) : [0, +\infty) \to \overline{\mathbb{N}} = \{1, 2, \cdots, l\}$ stands for the piecewise constant switching signal to be designed and $\sigma(t) = k$ implies that the kth subsystem is active at the instantaneous t. $u_k \in \mathbb{R}$ is control input. φ_k is a positive real constant. The stochastic variable ω is an r-dimensional independent standard Wiener process defined on the complete probability space (Ω, F, P). The function $f_k(\cdot), g_k(\cdot) : \mathbb{R} \to \mathbb{R}$ and $m_k(\cdot), n_k(\cdot) : \mathbb{R}^2 \to \mathbb{R}$ is continuous. The time delay $d(t) : \mathbb{R}_+ \to [0, \eta], d'(t) \leq d < 1$. Here assume that $m_k(y(0), y(-d(t))) = \mu_k$, $n_k(y(0), y(-d(t))) = v_k, \mu_k, v_k \in \mathbb{R}$.

The control aim of this paper is to design the adaptive backstepping neural switching control scheme by employing SLFN which are trained by ELM such that the SISO closed-loop system is asymptotically stable in probability.

The following assumption is made for accomplishment of controller design.

Assumption 2.1: $f_k(\cdot), g_k(\cdot), m_k(\cdot)$ and $n_k(\cdot), k \in \overline{\mathbb{N}}$ is locally Lipschitz continuous, respectively. Furthermore $f_k(0) = 0, g_k(0) = 0, m_k(y,0) = 0, n_k(y,0) = 0$.

Remark 2.1: According to the Assumption 2.1 and Mean Value Theorem, the following equalities hold

$$\begin{cases} f_k(y) = y\overline{f}_k(y), \ g_k(y) = y\overline{g}_k(y) \\ m_k(y, y(t-d(t))) = y^4(t-d(t))\overline{m}_k(y, y(t-d(t))) \\ n_k(y, y(t-d(t))) = y^4(t-d(t))\overline{n}_k(y, y(t-d(t))) \end{cases} \tag{2}$$

where $\overline{f}_k(\cdot), \overline{g}_k(\cdot), \overline{m}_k(\cdot)$ and $\overline{n}_k(\cdot)$ are known nonlinear functions. All these known nonlinear functions will be lumped into a suitable function that is compensated only by a SLFN with additive nodes or RBF nodes in this paper.

In accordance with the adaptive backstepping design idea, define the following coordinate transformation

$$z_1 = y, \ z_i = x_i - \alpha_{(i-1)\sigma(t)}\left(y, x_2, \cdots, x_{i-1}, \hat{\theta}_{\sigma(t)}, \hat{h}_{\sigma(t)}\right), 2 \leq i \leq n \tag{3}$$

where the smooth function $\alpha_{(i-1)k}\left(y, x_2, \cdots, x_{i-1}, \hat{\theta}_k, \hat{h}_k\right), 2 \leq i \leq n, k \in \overline{\mathbb{N}}$ stands for dummy control to be designed.

According to Itô differentiation rule and coordinate transformation (3), system (1) can be rewritten as

$$\begin{cases} dz_1 = x_2 dt \\ dz_i = \left[x_{i+1} - \sum_{l=1}^{i-1} \frac{\partial \alpha_{(i-1)\sigma(t)}}{\partial x_l} x_{l+1} - \frac{\partial \alpha_{(i-1)\sigma(t)}}{\partial \hat{\theta}_{\sigma(t)}} \dot{\hat{\theta}}_{\sigma(t)} - \frac{\partial \alpha_{(i-1)\sigma(t)}}{\partial \hat{h}_{\sigma(t)}} \dot{\hat{h}}_{\sigma(t)} \right] dt \\ dz_n = \left[\begin{array}{l} f_{\sigma(t)}(y) + m_{\sigma(t)}(y, y(t-d(t))) + \varphi_{\sigma(t)} u_{\sigma(t)} \\ -\sum_{l=1}^{n-1} \frac{\partial \alpha_{(n-1)\sigma(t)}}{\partial x_l} x_{l+1} - \frac{\partial \alpha_{(n-1)\sigma(t)}}{\partial \hat{\theta}_{\sigma(t)}} \dot{\hat{\theta}}_{\sigma(t)} - \frac{\partial \alpha_{(n-1)\sigma(t)}}{\partial \hat{h}_{\sigma(t)}} \dot{\hat{h}}_{\sigma(t)} \end{array} \right] dt \\ \quad + \left[g_{\sigma(t)}(y) + n_{\sigma(t)}(y, y(t-d(t))) \right] d\omega \end{cases} \tag{4}$$

3   Neural Switching Controller Design

3.1   Design of Adaptive Neural Sub-Controller

In order to derive the tuning law of the parameters $\hat{\theta}_k, \hat{h}_k$ and the control scheme u_k for the kth subsystem of switched model (4), consider the Lyapunov function candidate $V_k = 4^{-1}\sum_{i=1}^{n} z_i^4 + (1-d)^{-1} \int_{t-d(t)}^{t} y^4(s)\beta(y, y(s))ds + 2^{-1}\tilde{\theta}_k^T \Gamma_k^{-1}\tilde{\theta}_k + (2\rho_k)^{-1}\tilde{h}_k^2$, where

$\tilde{\theta}_k = \hat{\theta}_k - \theta_k$, $\tilde{h}_k = \hat{h}_k - h_k$, Γ_k is symmetric positive definite matrix, θ_k is the output weight, h_k is the bounded approximate error, ρ_k is a known constant, $\beta(y, y(s))$ is positive function to be determined.

Note that $\dot{\tilde{h}}_k = \dot{\hat{h}}_k$ and $\dot{\tilde{\theta}}_k = \dot{\hat{\theta}}_k$, the infinitesimal generator of Lyapunov function alone with the trajectory of the kth subsystem of switched model (4) is given by

$$LV_k = z_n^3 \left[f_k(y) + m_k(y, y(t-d(t))) + \varphi_k \mu_k - \sum_{i=1}^{n-1} \frac{\partial \alpha_{(n-1)k}}{\partial x_i} x_{i+1} - \frac{\partial \alpha_{(n-1)k}}{\partial \hat{\theta}_k} \dot{\hat{\theta}}_k - \frac{\partial \alpha_{(n-1)k}}{\partial \hat{h}_k} \dot{\hat{h}}_k \right]$$

$$+ \sum_{i=2}^{n-1} z_i^3 \left[x_{i+1} - \sum_{l=1}^{i-1} \frac{\partial \alpha_{(i-1)k}}{\partial x_l} x_{l+1} - \frac{\partial \alpha_{(i-1)k}}{\partial \hat{\theta}_k} \dot{\hat{\theta}}_k - \frac{\partial \alpha_{(i-1)k}}{\partial \hat{h}_k} \dot{\hat{h}}_k \right] + z_1^3 x_2$$

$$+ 3 \times 2^{-1} z_n^2 \left[g_k(y) + n_k(y, y(t-d(t))) \right] \left[g_k(y) + n_k(y, y(t-d(t))) \right]^T$$

$$+ (1-d)^{-1} y^4 \beta(y, y) - (1 - d'(t))(1-d)^{-1} y^4(t-d(t)) \beta(y, y(t-d(t))) + \hat{\theta}_k^T \Gamma_k^{-1} \tilde{\theta}_k + \rho_k^{-1} \tilde{h}_k \dot{\hat{h}}_k$$

Using the Young's inequality and (2-3), the above equality can be given by

$$LV_k \leq z_n^3 \left[\begin{array}{l} \varphi_k \mu_k - \sum_{l=1}^{n-1} \frac{\partial \alpha_{(n-1)k}}{\partial x_l} x_{l+1} - \frac{\partial \alpha_{(n-1)k}}{\partial \hat{\theta}_k} \dot{\hat{\theta}}_k - \frac{\partial \alpha_{(n-1)k}}{\partial \hat{h}_k} \dot{\hat{h}}_k \\ + \left(\left(4\varepsilon_{k1(n-1)}^4\right)^{-1} + 3\left(4\varepsilon_{k2n}^{\frac{4}{3}}\right)^{-1} + 3\left(4\varepsilon_{k3n}^{\frac{4}{3}}\right)^{-1} + 3\left(4\varepsilon_{k4n}^2\right)^{-1} + 3\left(4\varepsilon_{k5n}^2 \varepsilon_{k6n}^2\right)^{-1} \right) z_n \end{array} \right]$$

$$+ \sum_{i=2}^{n-1} z_i^3 \left[\alpha_{ik} - \sum_{l=1}^{i-1} \frac{\partial \alpha_{(i-1)k}}{\partial x_l} x_{l+1} - \frac{\partial \alpha_{(i-1)k}}{\partial \hat{\theta}_k} \dot{\hat{\theta}}_k - \frac{\partial \alpha_{(i-1)k}}{\partial \hat{h}_k} \dot{\hat{h}}_k + \left(\frac{3}{4}\varepsilon_{k1i}^{\frac{4}{3}} + \frac{1}{4\varepsilon_{k1(i-1)}^4} \right) z_i \right]$$

$$+ z_1^3 \left(\alpha_{1k} + 3 \times 4^{-1} \varepsilon_{k11}^{4/3} y \right) + \hat{\theta}_k^T \Gamma_k^{-1} \tilde{\theta}_k + \rho_k^{-1} \tilde{h}_k \dot{\hat{h}}_k + 3\varepsilon_{k5n}^2 \left(4\varepsilon_{k7n}^2 \right)^{-1} z_n^4 + 3\left(4\varepsilon_{k8n}^2\right)^{-1} z_n^4$$

$$+ y^4 \left(4^{-1} \varepsilon_{k2n}^4 \left| \overline{f}_k(y) \right|^4 + 3\varepsilon_{k4n}^2 4^{-1} \left| \overline{g}_k(y) \overline{g}_k^T(y) \right|^2 + 3\varepsilon_{k5n}^2 \varepsilon_{k7n}^2 4^{-1} \left| \overline{g}_k(y) \right|^4 + (1-d)^{-1} \beta(y, y) \right)$$

where

$$\beta(y, y(t-d(t))) = \varepsilon_{k3n}^4 4^{-1} \left| \overline{m}_k(y, y(t-d(t))) \right|^4 + \left(4\varepsilon_{k5n}^2\right)^{-1} 3\varepsilon_{k6n}^2 \left| \overline{n}_k(y, y(t-d(t))) \right|^4$$

$$+ 3\varepsilon_{k8n}^2 4^{-1} \left| \overline{n}_k(y, y(t-d(t))) \overline{n}_k^T(y, y(t-d(t))) \right|^2$$

For convenience, define the following function

$$e_k(y) = (1-d)^{-1} \beta(y, y) + 4^{-1} \varepsilon_{k2n}^4 \left| \overline{f}_k(y) \right|^4 + 3\varepsilon_{k4n}^2 4^{-1} \left| \overline{g}_k(y) \overline{g}_k^T(y) \right|^2 + 3\varepsilon_{k5n}^2 \varepsilon_{k7n}^2 4^{-1} \left| \overline{g}_k(y) \right|^4$$

Now, all system nonlinear functions are lumped into a suitable function $e_k(y)$ that will be compensated by SLFN which djusted based on ELM. Set the network approximate of $e_k(y)$ is given by $e_k(y) = E_k(y, a_k, b_k)\theta_k + h_k$, then we have

$$LV_k \leq z_n^3 \left[\varphi_k \mu_k - \sum_{l=1}^{n-1} \frac{\partial \alpha_{(n-1)k}}{\partial x_l} x_{l+1} - \frac{\partial \alpha_{(n-1)k}}{\partial \hat{\theta}_k} \dot{\hat{\theta}}_k - \frac{\partial \alpha_{(n-1)k}}{\partial \hat{h}_k} \dot{\hat{h}}_k \right.$$
$$\left. + \left(\left(4\varepsilon_{k1(n-1)}^4 \right)^{-1} + 3 \left(4\varepsilon_{k2n}^{\frac{4}{3}} \right)^{-1} + 3 \left(4\varepsilon_{k3n}^{\frac{4}{3}} \right)^{-1} + 3 \left(4\varepsilon_{k4n}^2 \right)^{-1} + 3 \left(4\varepsilon_{k5n}^2 \varepsilon_{k6n}^2 \right)^{-1} \right) z_n \right]$$

$$+ 3 \left(4\varepsilon_{k8n}^2 \right)^{-1} z_n^4 + \sum_{i=2}^{n-1} z_i^3 \left[\alpha_{ik} - \sum_{l=1}^{i-1} \frac{\partial \alpha_{(i-1)k}}{\partial x_l} x_{l+1} - \frac{\partial \alpha_{(i-1)k}}{\partial \hat{\theta}_k} \dot{\hat{\theta}}_k - \frac{\partial \alpha_{(i-1)k}}{\partial \hat{h}_k} \dot{\hat{h}}_k + \left(\frac{3}{4} \varepsilon_{k1i}^{\frac{4}{3}} + \frac{1}{4\varepsilon_{k1(i-1)}^4} \right) z_i \right]$$

$$+ 3\varepsilon_{k5n}^2 \left(4\varepsilon_{k7n}^2 \right)^{-1} z_n^4 + z_1^3 \left[\alpha_{1k} + 3 \times 4^{-1} \varepsilon_{k11}^{4/3} y + y \left(E_k(y, a_k, b_k)\theta_k + h_k \right) \right] + \dot{\hat{\theta}}_k^T \Gamma_k^{-1} \tilde{\theta}_k + \rho_k^{-1} \tilde{h}_k \dot{\hat{h}}_k$$

If $\alpha_{ik}, u_k, \hat{h}_k, \hat{\theta}_k$ are selected as

$$\begin{cases} \alpha_{1k} = -c_{k1} y - 3 \times 4^{-1} \varepsilon_{k11}^{4/3} y - y \left(E_k(y, a_k, b_k)\hat{\theta}_k + \hat{h}_k \right) \\ \alpha_{ik} = -c_{ki} z_i + \sum_{l=1}^{i-1} \frac{\partial \alpha_{(i-1)k}}{\partial x_l} x_{l+1} + \frac{\partial \alpha_{(i-1)k}}{\partial \hat{\theta}_k} \dot{\hat{\theta}}_k + \frac{\partial \alpha_{(i-1)k}}{\partial \hat{h}_k} \dot{\hat{h}}_k - \left(\frac{3}{4} \varepsilon_{k1i}^{4/3} + \frac{1}{4\varepsilon_{k1(i-1)}^4} \right) z_i, 2 \leq i \leq n-1 \end{cases} \quad (5)$$

$$\begin{cases} \mu_k = \frac{1}{\varphi_k} \left[\begin{array}{l} -c_{kn} z_n + \sum_{l=1}^{n-1} \frac{\partial \alpha_{(n-1)k}}{\partial x_l} x_{l+1} + \frac{\partial \alpha_{(n-1)k}}{\partial \hat{\theta}_k} \dot{\hat{\theta}}_k + \frac{\partial \alpha_{(n-1)k}}{\partial \hat{h}_k} \dot{\hat{h}}_k \\ -3 \left(\left(12\varepsilon_{k1(n-1)}^4 \right)^{-1} + \left(4\varepsilon_{k2n}^{4/3} \right)^{-1} + \left(4\varepsilon_{k3n}^{4/3} \right)^{-1} + \left(4\varepsilon_{k4n}^2 \right)^{-1} \\ + \left(4\varepsilon_{k5n}^2 \varepsilon_{k6n}^2 \right)^{-1} + \varepsilon_{k5n}^2 \left(4\varepsilon_{k7n}^2 \right)^{-1} + \left(4\varepsilon_{k8n}^2 \right)^{-1} \right) z_n \end{array} \right] \\ \dot{\hat{h}}_k = \rho_k y^4 \\ \dot{\hat{\theta}}_k = \Gamma_k^T E_k^T (y, a_k, b_k) y^4 \end{cases} \quad (6)$$

then the infinitesimal generator $LV_k \leq -\sum_{i=1}^{n} c_{ki} z_i^4 < 0$, where $c_{ki} > 0, i = 1, 2, \cdots, n$.

3.2   Switching Laws Design

In this part, we will adopt energy attenuation speed of each subsystem to design switching law for system (1). According to the multiple Lyapunov functional method, we can orchestrate the switching in such a way.

For initial condition $x(t_0) = (x_1(t_0), x_2(t_0), \cdots, x_n(t_0))^T$, set $\sigma(t_0) = \arg \min_{1 \leq k \leq N} \{LV_k(t_0)\}$. The first switching time instant determined by

$$t_1 = \inf \left\{ t > t_0 \middle| \text{there exists a } i \in \overline{\mathbb{N}}, i \neq \sigma(t_0) \text{ such that } LV_i(t_1) < LV_{\sigma(t_0)}(t_1) \right\}$$

The corresponding switching index is chosen as $\sigma(t_1) = \arg \min\limits_{1 \leq k \leq N} \{LV_k(t_1)\}$. Finally, we define the switching time sequences recursively by

$$t_j = \inf\left\{t > t_{j-1} \middle| there\ exists\ a\ i \in \overline{\mathbb{N}}, i \neq \sigma(t_{j-1})\ such\ that\ LV_i(t_j) < LV_{\sigma(t_{j-1})}(t_j)\right\}, j \geq 2$$

The corresponding switching index is chosen as $\sigma(t_j) = \arg \min\limits_{1 \leq k \leq N} \{LV_k(t_j)\}, j \geq 2$.

According to the analysis of Subsection 3.1 and 3.2, we can obtain the main results in this paper.

Theorem: The switched nonlinear stochastic system (1) is asymptotically stable in probability if the adaptive backstepping neural networks switching control scheme is selected as (5-6) and the switching law in subsection 3.2.

4   Simulation Example

Example: Consider system (1) with three subsystems are given by

$$\Sigma_1 \begin{cases} dx_1 = x_2 dt \\ dx_2 = [2y + y(t - d(t))\cos y + u_1] dt + [\sin y + yy(t - d(t))] d\omega \\ y = x_1 \end{cases}$$

$$\Sigma_2 \begin{cases} dx_1 = x_2 dt \\ dx_2 = [y^2 + 2y\sin(y(t - d(t))) + u_2] dt + [4y + yy(t - d(t))] d\omega \\ y = x_1 \end{cases}$$

$$\Sigma_3 \begin{cases} dx_1 = x_2 dt \\ dx_2 = [y + y(t - d(t))e^{-y} + u_2] dt + [4y + yy(t - d(t))] d\omega \\ y = x_1 \end{cases}$$

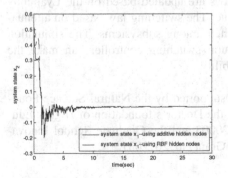

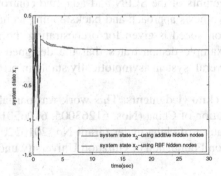

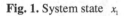

Fig. 1. System state x_1 **Fig. 2.** System state x_2

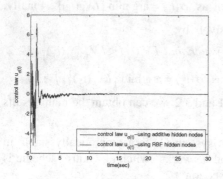

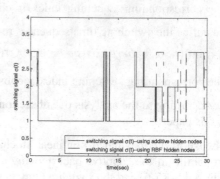

Fig. 3. Control law   **Fig. 4.** Switching signal

Set $x(0) = (0.5, -0.5)^T$, $0 \le d(t) = 1 - 0.6\cos t \le \eta = 1.6$, $\varepsilon_{\sigma(t)52} = \varepsilon_{\sigma(t)62} = \varepsilon_{\sigma(t)72} = \varepsilon_{\sigma(t)82} = 1$,

$\rho_{\sigma(t)} = \varphi_{\sigma(t)} = \varepsilon_{\sigma(t)11} = \varepsilon_{\sigma(t)22} = \varepsilon_{\sigma(t)32} = \varepsilon_{\sigma(t)42} = \varepsilon_{\sigma(t)52} = \varepsilon_{\sigma(t)62} = \varepsilon_{\sigma(t)72} = \varepsilon_{\sigma(t)82} = 1$, $\Gamma_{\sigma(t)} = I$

and $c_{\sigma(t)1} = c_{\sigma(t)2} = 0.25$. $L=10$. $E_{\sigma(t)}(y, a_{\sigma(t)}, b_{\sigma(t)}) = \exp(-(a_{\sigma(t)} \cdot y + b_{\sigma(t)})^2)$ for addi-

tive nodes and the activation function, $E_{\sigma(t)}(y, a_{\sigma(t)}, b_{\sigma(t)}) = exp(-b_{\sigma(t)}^{-2} |y - a_{\sigma(t)}|^2)$ for

RBF nodes. The hidden node parameters $(a_{\sigma(t)}, b_{\sigma(t)})$ are assigned randomly in the in-

tervals [-1,1] and [0,1] respectively. The simulation results are shown in Figs. 1-4.

5   Conclusions

In this paper, a adaptive backstepping neural switching control scheme for switched
nonlinear stochastic systems with time-delay is proposed. All system nonlinear terms
are lumped into a suitable function that is approximate by SLFN. The hidden node
parameters of the SLFN randomly be generated based on ELM algorithm. The output
weights of the SLFN and adaptive control laws are updated based on the Lyapunov
synthesis approach and backstepping technique. The switching law based on attenuation speed is given for orchestrating the switch among subsystems. The simulation
example demonstrates that the designed neural switching controller can make the
overall system asymptotically stable in probability.

Acknowledgments. This work was partially supported by the Natural Science Foundation of China (Nos. 61263005, 61065010), the Doctor's foundation of Higher Education of china under Grant No. 20105201120003 and the Graduate School Innovation Foundation of Guizhou University under Grant No. 2012031.

References

1. Pan, Z., Basar, T.: Backstepping controller design for nonlinear stochastic systems under a risk-sensitive cost criterion. SIAM J. Control Optim. 37(3), 957–995 (1999)
2. Deng, H., Krstic, M.: Stochastic nonlinear stabilization–Part I: A backstepping design. System Control Letter 32(3), 143–150 (1997)
3. Wu, L., Ho, D.W.C., Li, C.W.: Stabilization and performance synthesis for switched stochastic systems. IET Control Theory Applic. 4(10), 1877–1888 (2010)
4. Wu, Z.J., Yang, J., Shi, P.: Adaptive tracking for stochastic nonlinear systems with markovian switching. IEEE Trans. Automatic Control 55(9), 2135–2141 (2010)
5. Wang, D., Huang, J.: Neural network-based adaptive dynamic surface control for a class of uncertain nonlinear systems in strict-feedback form. IEEE Trans. Neural Networks 16(1), 195–202 (2005)
6. Chen, W.S., Jiao, L.C., Li, J., Li, R.H.: Adaptive NN backstepping output feedback control for stochastic nonlinear strict feedback systems with time varying delays. IEEE Trans. Systems Man Cyber. B 40(3), 939–950 (2010)
7. Rong, H.J., Suresh, S., Zhao, G.S.: Stable indirect adaptive neural controller for a class of nonlinear system. Neurocomputing 74, 2582–2590 (2011)
8. Huang, G.B., Zhu, Q.Y., Siew, C.K.: Extreme learning machine: theory and applications. Neurocomputing 70(1-3), 489–501 (2006)
9. Huang, G.B., Wang, D.H., Lan, Y.: Extreme Learning Machines: A Survey. Int. J. Machine Leaning and Cybernetics 2(2), 107–122 (2011)
10. Huang, G.B., Zhou, H., Ding, X., Zhang, R.: Extreme learning machine for regression and multiclass classification. IEEE Trans. Systems Man Cyber. B 42(2), 513–529 (2011)

Clustering Based on Rank Distance
with Applications on DNA

Liviu Petrisor Dinu and Radu-Tudor Ionescu

Faculty of Mathematics and Computer Science,
University of Bucharest, No. 14 Academiei Street, Bucharest, Romania
{liviu.p.dinu,raducu.ionescu}@gmail.com

Abstract. This paper aims to present two clustering methods based on rank distance. The K-means algorithm represents each cluster by a single mean vector. The mean vector is computed with respect to a distance measure. A new K-means algorithm based on rank distance is described in this paper. Hierarchical clustering builds models based on distance connectivity. Our paper introduces a new hierarchical clustering technique that uses rank distance. Experiments using mitochondrial DNA sequences extracted from several mammals demonstrate the clustering performance and the utility of the two algorithms.

Keywords: k-means, hierarchical clustering, rank distance.

1 Introduction

Clustering has long played an important role in a wide variety of fields: biology, statistics, pattern recognition, information retrieval, machine learning, data mining, psychology and other social sciences. There are various clustering algorithms that differ significantly in their notion of what constitutes a cluster. The appropriate clustering algorithm and parameter settings depend on the individual data set. But not all clustering algorithms can be applied on a particular data set. For example, clustering methods that depend on a standard distance function (such as K-means) cannot be applied on a data set of objects (such as strings) for which a standard distance function cannot be computed.

The goal of this work is to introduce two clustering algorithms for strings, or more precisely, that are based on a distance measure for strings. A few distance measures for strings can be considered, but we focus on using a single distance that has very good results in terms of accuracy and time for many important problems. The distance we focus on is termed *rank distance* [2] and it has applications in biology [3, 6], natural language processing and many other fields. The first clustering algorithm presented here is a centroid model that represents each cluster by a single median string. The second one is a connectivity model that builds a hierarchy of clusters based on distance connectivity. Both use rank distance.

In section 2 we introduce notation and mathematical preliminaries. Section 3 gives an overview of related work regarding sequencing and comparing DNA,

T. Huang et al. (Eds.): ICONIP 2012, Part V, LNCS 7667, pp. 722–729, 2012.

rank distance and clustering methods. The clustering algorithms are described in section 4. The experiments using mitochondrial DNA sequences from mammals are presented in section 5. Finally, we draw our conclusion and talk about further work in section 6.

2   Preliminaries

A ranking is an ordered list and is the result of applying an ordering criterion to a set of objects. A ranking defines a partial function on $\mathcal{U}$ where for each object $i \in \mathcal{U}$, $\tau(i)$ represents the position of the object i in the ranking τ.

We define the order of an object $x \in \mathcal{U}$ in a ranking σ of length d, by $ord(\sigma, x) = |d + 1 - \sigma(x)|$. By convention, if $x \in \mathcal{U} \setminus \sigma$, we have $ord(\sigma, x) = 0$.

Definition 1. *Given two partial rankings σ and τ over the same universe $\mathcal{U}$, we define the rank distance between them as:*

$$\Delta(\sigma, \tau) = \sum_{x \in \sigma \cup \tau} |ord(\sigma, x) - ord(\tau, x)|.$$

Rank distance is an extension of the Spearman footrule distance [1].

The next definition formalizes the transformation of strings into rankings.

Definition 2. *Let n be an integer and let $w = a_1 \ldots a_n$ be a finite word of length n over an alphabet Σ. We define the extension to rankings of w, $\bar{w} = a_{1,i(1)} \ldots a_{n,i(n)}$, where $i(j) = |a_1 \ldots a_j|_{a_j}$ for all $j = 1, \ldots n$ (i.e. the number of occurrences of a_j in the string $u_1 a_2 \ldots a_j$).*

We extend the rank distance to arbitrary strings as follows:

Definition 3. *Given $w_1, w_2 \in \Sigma^*$, we define $\Delta(w_1, w_2) = \Delta(\bar{w}_1, \bar{w}_2)$.*

Note that the transformation of a string into a ranking can be done in linear time (by memorizing for each symbol, in an array, how many times it appears in the string). Also, the computation of the rank distance between two rankings can be done in linear time in the cardinality of the universe [6].

Let χ_n be the space of all strings of size n over an alphabet Σ and let $P = \{p_1, p_2, \ldots, p_k\}$ be k strings from χ_n. The closest string problem (CSP) is to find the center of the sphere of minimum radius that includes all the k strings.

Problem 1. The *closest string problem via rank distance (CSRD)* is to find a minimal integer d (and a corresponding string t of length n) such that the maximum rank distance from t to any string in P is at most d. We say that t is the closest string to P and we name d the radius. Formally, the goal is to compute:

$$\min_{x \in \chi_n} \max_{i=1..k} \Delta(x, p_i).$$

The median string problem (MSP) is similar to the closest string problem, only that the goal is to minimize the average distance to all the input strings.

Problem 2. The *median string problem via rank distance (MSRD)* is to find a minimal integer d (and a corresponding string t of length n) such that the average rank distance from t to any string in P is at most d. We say that t is the median string of P. Formally, the goal is to compute:

$$\min_{x \in \chi_n} avg_{i=1..k} \Delta(x, p_i).$$

3   Related Work

3.1   Sequencing and Comparing DNA

In many important problems in computational biology a common task is to compare a new DNA sequence with sequences that are already well studied and annotated. The standard method used in computational biology for sequence comparison is by sequence alignment and it is based on dynamic programming [12]. Although dynamic programming for sequence alignment is mathematically optimal, it is far too slow for comparing a large number of bases, and too slow to be performed in a reasonable time. The standard distances with respect to the alignment principle are edit (Levenshtein) distance or its ad-hoc variants. The study of genome rearrangement [9] was investigated also under Kendall tau distance.

3.2   Rank Distance

To measure the distance between two strings with RD we scan (from left to right) both strings and for each letter from the first string we count the number of elements between its position in the first string and the position of its first occurrence in the second string. Finally, we sum up all these scores and obtain the rank distance. Note that in the Hamming and rank distance case the median string problem is tractable [4], while in the edit distance case it is NP-hard. In [5] it is shown that the CSRD is NP-hard. On the other hand, a solution for MSRD can be found in polynomial time. In the clustering algorithms that are about to be presented in this paper, CSRD and MSRD are important because steps in both algorithms involve finding the closest or median string for a cluster (subset) of strings.

3.3   Clustering Methods

In recent years considerable effort has been made for improving algorithm performance of the existing clustering algorithms. The authors of [7] propose an extensions to the K-means algorithm for clustering large data sets with categorical values. An unsupervised data mining algorithm used to perform hierarchical clustering over particularly large data-sets is presented in [13]. With the recent need to process larger and larger data sets (also known as big data), the willingness to treat semantic meaning of the generated clusters for performance has been increasing. This led to the development of pre-clustering methods such as canopy clustering [8].

4   Clustering Methods Based on Rank Distance

Cluster analysis groups data objects based only on information found in the data that describes the objects and their relationships. The goal is that the objects within a group (or cluster) are similar (or related) to one another and different from the objects in other groups. The greater the similarity within a group and the greater the difference between groups, the better or more distinct the clustering. The clusters should capture the natural structure of the data. There are various clustering algorithms that differ significantly in their notion of what constitutes a cluster. We introduce two clustering algorithms that are based on rank distance. These algorithms are to be applied on data sets that contain objects represented as strings, such as text, DNA sequences, etc.

4.1   K-Means Based on Rank Distance

The K-means clustering technique is a simple method of cluster analysis which aims to partition a set of objects into K clusters in which each object belongs to the cluster with the nearest mean. The algorithm begins with choosing K initial centroids, where K is an *a priori* parameter, namely, the number of clusters desired. Each object is then assigned to the nearest centroid, and each group of objects assigned to a centroid is a cluster. The centroid of each cluster is then updated based on the objects assigned to that cluster. We repeat the assignment and update steps until no point changes clusters or until a maximum number of iterations is reached. If string objects are considered for clustering, then we need a way to determine the centroid string for a certain cluster of strings. We propose the use of the median string computed with rank distance. Note that computing the median string for a cluster of strings is equivalent to solving Problem 2. We also use rank distance to assign strings to the nearest median string. Thus, the algorithm that we propose is entirely based on rank distance.

Algorithm 1 partitions a set of strings into K clusters using rank distance. It aims at minimizing an objective function, given by $J = \sum_{j=1}^{k} \sum_{i=1}^{n} \Delta(x_i^{(j)}, c_j)$, where $\Delta(x_i^{(j)}, c_j)$ is the rank distance between a string $x_i^{(j)}$ in cluster j and the cluster centroid c_j. The objective function is simply an indicator of the distance of the input strings from their respective cluster centers.

Algorithm 1. *K-means based on rank distance*
1. *Initialization: Randomly select K strings as initial centroids.*
2. *Loop: Repeat until centroids do not change or until a maximum number of iterations is reached*
    a. Form K clusters by assigning each string to its nearest median string.
    b. Recompute the centroid (median string) of each cluster using rank distance.

Theorem 1. *Algorithm 1 converges to a solution in a finite number of iterations for any K strings as initial centroids.*

Proof. The demonstration of the finite convergence of K-means algorithm for any metric is given in [11]. The rank distance is a metric function [2].

Although Algorithm 1 will always terminate, it does not necessarily find the most optimal configuration, corresponding to the global objective function minimum. The algorithm is also significantly sensitive to the initial randomly selected centroids. However, it can be run multiple times to reduce this effect.

Regarding computational complexity, the K-means algorithm proposed here depends on the complexity of the algorithm that computes the median string. We used the algorithm described in [4] that computes the median string in $O(n^3)$ time, where n is the string length. If we denote the maximum number of iterations by m, then the complexity of Algorithm 1 is $O(m \times K \times n^3)$.

4.2   Hierarchical Clustering Based on Rank Distance

Many hierarchical clustering techniques are variations on a single algorithm: starting with individual objects as clusters, successively join the two nearest clusters until only one cluster remains. These techniques connect objects to form clusters based on their distance. Note that these algorithms do not provide a single partitioning of the data set, but instead provide an extensive hierarchy of clusters that merge with each other at certain distances. Apart from the choice of a distance function, another decision is needed for the linkage criterion to be used. Popular choices are single-linkage, complete-linkage, or average-linkage. A standard method of hierarchical clustering that uses rank distance is presented in [6]. It presents a phylogenetic tree of several mammals comparable to the structure of other trees reported by other researches [10]. It is our belief that a hierarchical clustering method designed to deeply integrate rank distance would perform better. The hierarchical clustering method presented here works only for strings. It uses rank distance, but instead of a linkage criterion we propose something different, that is to determine a centroid string for each cluster and join clusters based on the rank distance between their centroid strings. In the implementation of Algorithm 2 we choose the median string as cluster centroid. To compute the median string is equivalent to solving Problem 2. Another implementation is to use the closest string which is equivalent to solving Problem 1.

Algorithm 2. *Hierarchical clustering based on rank distance*
1. *Initialization: Compute rank distances between all initial strings.*
2. *Loop: Repeat until only one cluster remains*
   a. *Join the nearest two clusters to form a new cluster.*
   b. *Determine the median string of the new cluster.*
   c. *Compute rank distances from this new median string to existing median strings.*

The analysis of the computational complexity of Algorithm 2 is straightforward. If we consider m to be the number of the initial strings, then $O(m^2)$ time is required to compute rank distances between them. The algorithm builds a binary tree structure where the leaves are the initial m strings. Thus, it creates $m - 1$ intermediate clusters until only one cluster remains. The most heavy computational step is to determine the median string of a cluster which takes $O(n^3)$ time, where n is the string length. Usually n is much greater than m and the algorithm complexity in this case is $O(m \times n^3)$.

5   Experiments

Our clustering methods are tested on a classical problem in bioinformatics: the phylogenetic analysis of the mammals. In the experiments presented in this paper we use mitochondrial DNA sequence genome of 22 mammals available in the EMBL database. The DNA sequence of mtDNA has been determined from a large number of organisms and individuals, and the comparison of those DNA sequences represents a mainstay of phylogenetics, in that it allows biologists to elucidate the evolutionary relationships among species. In mammals, each double-stranded circular mtDNA molecule consists of 15,000-17,000 base pairs.

In our experiments each mammal is represented by a single mtDNA sequence that comes from a single individual. Note that DNA from two individuals of the same species differs by only 0,1%.

5.1   K-Means Experiment

The first experiment is to cluster the 22 mammalian DNA sequences using K-means clustering into 6 clusters. The clustering is relevant if the 6 clusters match the 7 families of mammals available in our data set: Primates, Perissodactylae, Cetartiodactylae, Rodentia, Carnivora, Metatheria, Monotremata. Note that we have only one member of the Metatheria family, that is the North American opossum, and only one member of the Monotremata family, that is the platypus. The opossum and the platypus should be clustered together or with members of the Rodentia family. In the experiment we use only the first 1000 nucleotides extracted from each DNA sequence. Here are the clustering results: cow (Cetartiodactylae) - 2, sheep (Cetartiodactylae) - 0, big whale (Cetartiodactylae) - 2, fin whale (Cetartiodactylae) - 2, cat (Carnivora) - 4, gray seal (Carnivora) - 4, harbour seal (Carnivora) - 4, opossum (Metatheria) - 3, platypus (Monotremata) - 0, human (Primates) - 5, gibbon (Primates) - 5, gorilla (Primates) - 5, pygmy chimpanzee (Primates) - 5, orangutan (Primates) - 5, chimpanzee (Primates) - 5, Sumatran orangutan (Primates) - 5, horse (Perissodactylae) - 0, donkey (Perissodactylae) - 0, Indian rhinoceros (Perissodactylae) - 1, white rhinoceros (Perissodactylae) - 1, house mouse (Rodentia) - 3, rat (Rodentia) - 3. Note that numbers from 0 to 5 represent cluster labels. Our K-means algorithm is able to clearly distinguish the Carnivora and Primates families. It only leaves the sheep out of the Cetartiodactylae family. The Rodentia and Metatheria families are joined together in a single cluster. It seems that our clustering method was able to find that the opossum is very similar (at least in appearance) to the rat, even if they come from different families. On the other hand, the Perissodactylae family is separated in two clusters. One cluster joins the Indian and white rhinoceroses, while the other one joins the horse and the donkey with the platypus and sheep. Overall, using Algorithm 1 we are able to create relevant groups for most of the families, even if we only use the first 1000 nucleotides.

5.2   Hierarchical Clustering Experiment

The second experiment is to apply the hierarchical clustering proposed in section 4 on the 22 DNA sequences. The resulted phylogenetic tree is going to be compared with phylogenetic trees obtained with standard hierarchical clustering methods, such as the one presented in [6]. The phylogenetic tree obtained with Algorithm 2 (using the first 3000 nucleotides) is presented in Figure 1.

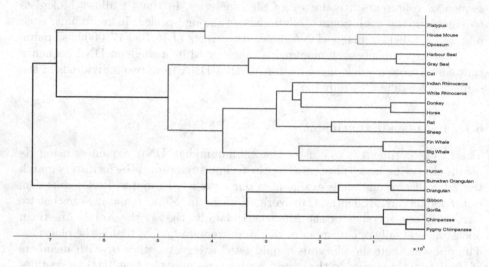

Fig. 1. Phylogenetic tree obtained from mammalian mtDNA sequences using rank distance

Analyzing the phylogenetic tree in Figure 1 we observe that our hierarchical clustering method is a able to separate the Primates, Perissodactylae, Carnivora and Cetartiodactylae families. The house mouse is joined together with the opossum and the platypus in a cluster that is relevant from a biological point of view, since they are related mammals. The only confusion of our clustering method is that the rat is clustered with the sheep instead of the house mouse. By contrast to our result, in the phylogenetic tree presented in [6] the Perissodactylae and Carnivora families are mixed together. The standard hierarchical clustering method used in [6] is also unable to join the rat with the house mouse. Therefore, we state that the hierarchical clustering method presented in this paper performs better, despite that it uses only the first 3000 nucleotides instead of the entire mtDNA sequences.

6   Conclusion and Further Work

In this paper we exibit two clustering methods based on rank distance. Both compute the median string of a cluster using rank distance. Our experiments on the phylogenies of mammals produced results which are similar or sometimes better to those reported in the literature [6, 10]. The experiments also demonstrate the utility of the two algorithms proposed in this paper. Note that the hierarchical clustering method presented here uses rank distance, but instead of a linkage criterion we proposed to determine the median string for each cluster and join clusters based on the rank distance between their median strings. In the future we want to try another implementation that uses the closest string instead of the median string, or both.

Acknowledgment. The contribution of the authors to this paper is equal. The research of Liviu P. Dinu was supported by the CNCS-PCE Idei grant 311/2011.

References

1. Diaconis, P., Graham, R.L.: Spearman footrule as a measure of disarray. Journal of Royal Statistical Society. Series B (Methodological) 39(2), 262–268 (1977)
2. Dinu, L.P.: On the classification and aggregation of hierarchies with different constitutive elements. Fundamenta Informaticae 55(1), 39–50 (2003)
3. Dinu, L.P., Ionescu, R.T.: An efficient rank based approach for closest string and closest substring. PLoS ONE 7(6), e37576 (June 2012)
4. Dinu, L.P., Manea, F.: An efficient approach for the rank aggregation problem. Theoretical Computer Science 359(1-3), 455–461 (2006)
5. Dinu, L.P., Popa, A.: On the Closest String via Rank Distance. In: Kärkkäinen, J., Stoye, J. (eds.) CPM 2012. LNCS, vol. 7354, pp. 413–426. Springer, Heidelberg (2012)
6. Dinu, L.P., Sgarro, A.: A Low-complexity Distance for DNA Strings. Fundamenta Informaticae 73(3), 361–372 (2006)
7. Huang, Z.: Extensions to the k-means algorithm for clustering large data sets with categorical values. Data Mining Knowledge Discovery 2(3), 283–304 (1998)
8. McCallum, A., Nigam, K., Ungar, L.H.: Efficient clustering of high-dimensional data sets with application to reference matching. In: Proceedings of ACM SIGKDD, pp. 169–178 (2000)
9. Palmer, J., Herbon, L.: Plant mitochondrial DNA evolves rapidly in structure, but slowly in sequence. Journal of Molecular Evolution 28, 87–89 (1988)
10. Reyes, A., Gissi, C., Pesole, G., Catzeflis, F.M., Saccone, C.: Where do rodents fit? evidence from the complete mitochondrial genome of sciurus vulgaris. Mol. Biol. Evol. 17(6), 979–983 (2000)
11. Selim, S.Z., Ismail, M.A.: K-means-type algorithms: A generalized convergence theorem and characterization of local optimality. IEEE Transactions on Pattern Analysis and Machine Intelligence PAMI-6(1), 81–87 (1984)
12. Smith, T., Waterman, M.: Comparison of biosequences. Advances in Applied Mathematics 2(4), 482–489 (1981)
13. Tian, T.Z., Ramakrishnan, R., Livny, M.: Birch: an efficient data clustering method for very large databases. SIGMOD Rec. 25(2), 103–114 (1996)

Petrophysical Parameters Estimation
from Well-Logs Data Using Multilayer Perceptron
and Radial Basis Function Neural Networks

Leila Aliouane, Sid-Ali Ouadfeul, Noureddine Djarfour, and Amar Boudella

LABOPHYT, Faculté des Hydrocarbures et de la Chimie,
Université M'hamad Bougara de Boumerdes,
Avenue de l'indépendance, Boumerdes, Algeria,
Algerian Petroleum Institute, IAP.,
Boumerdes, Algeria,
Geophysics Department, FSTGAT, USTHB,
Bab Ezzouar, Algiers, Algeria
Lil_aldz@yahoo.fr, Souadfeul@ymail.com

Abstract. The main objective of this work consists to use the two neural network models to estimate petrophysical parameters from well-logs data. Parameters to be estimated are: Porosity, Permeability and Water saturation. The neural network machines used consist of the Multilayer perceptron (MLP) and the Radial Basis Function (RBF). The main input used to train these neural models is the raw well-logs data recorded in a borehole located in the Algerian Sahara. Comparison between the two neural machines and conventional method shows that the RBF is the most suitable for petrophysical parameters prediction.

Keywords: estimation, MLP, radial basis function, comparison.

1   Introduction

Recently, Neural Networks became a popular method in geosciences in the context of pattern recognition problems. They can be used for classification [1] and approximation [2],[3],[4] in reservoir characterization by well logs-data. Approximation consists in estimating of petrophysical parameters such as porosity, permeability and water saturation. These parameters are the three fundamental rock properties which relate to the amount of fluid contained in a reservoir and its ability to flow when subjected to applied pressure gradients. These properties have a significant impact on petroleum fields operations and reservoir management. In un-cored intervals and well, the reservoir description and characterization methods utilizing well logs represent a significant technical as well as economic advantage because well logs can provide a continuous record over the entire well where coring is impossible.

However, porosity, permeability and water saturation estimation from conventional well logs in heterogeneous formation has a difficult and complex problem to solve by conventional statistical method and generally are affected by the methods used. Most

T. Huang et al. (Eds.): ICONIP 2012, Part V, LNCS 7667, pp. 730–736, 2012.

of these methods use theoretical and empirical formulae which have been developed to the specific area due to the geological variability such as facies changes.

In this work, we propose two neural network models to predict petrophysical parameters from wireline logs. First one is the Multilayer Percepron (MLP) using a supervised learning such a back propagation gradient algorithm. The second one is the Radial Basis Function (RBF) using both a supervised and unsupervised learing.

Neural network as a nonlinear and non-parametric tool is becoming increasingly popular in well log analysis. Neural network is a computer model that attempts to mimic simple biological learning processes and simulate specific functions of human nervous system. Neural network can be used as a nonlinear regression method to develop transformation between the selected well logs and core analysis data. Data from Algerian Sahara of two oilfields holes are used to test this technique. The data of one well is exploited to train neural network machine for each architecture to extract the weigts connection corresponding to minimum root mean quare between computed and target inputs. The other well is used for generalization using determined weights.

2   Data analysis and Reservoir Parameters Estimation Problem

The overall objective of our research is to demonstrate the effectiveness ANN computing techniques in providing an accurate estimation of petrophysical parameters that describe reservoir properties.

Several recordings of petrophysical parameters are exploited of two boreholes from Algerian Sahara (Figure.1). These measurements are total radioactivity (GR), deep and shallow resistivities (Rlld and Rmsfl), bulk density (Rhob), neutron porosity (Nphi) and sonic log (DT). These lasts allow the computation of reservoir parameters such as porosity, permeability and water saturation to characterize such a reservoir. In addition core data are also exploited to characterize reservoir.

A large number of data points were compiled over a wide range of reservoir water saturation, porosity and permeability. The data have been gathered from a typical Triassic petroleum reservoir of Hassi R'mel field in south eastern of Algeria. While most of the porosity and permeability data are obtained from core analysis under reservoir conditions in the laboratory, data. The water saturation is calculated from empirical formulae specified for this field where the Simandoux modified equation has been used [6]. It has been a fairly common practice to present permeability data vs porosity of the reservoir in order to generate a correlation between permeability and these variables. The variations of permeability and porosity of the studied reservoir divided in two zones are shown in Figure.2 and 3, respectively. The scatteredness of this plot indicates that this functionality is very complex and exhibits strong nonlinearity. A regression approach may be adopted; however, the selection of an appropriate general regression would be problematic

Data of well B can be used to train the two neural network machine which are MLP and RBF while the data of well A will be used for the generalization.

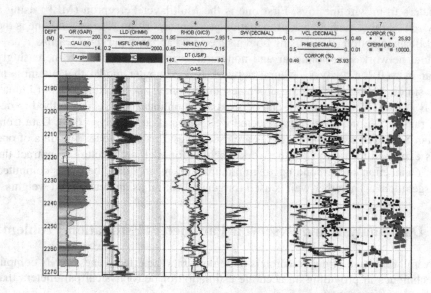

Fig. 1. Petrophysical parameters of a reservoir of wellA

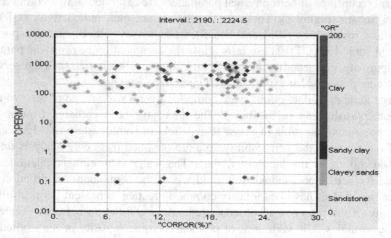

Fig. 2. Variation of permeability with porosity in a reservoir –zone1 for a well A

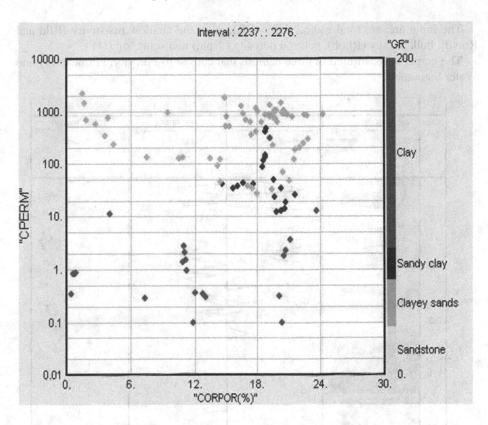

Fig. 3. Variation of permeability with porosity in a reservoir –zone2 for a well A

3   Multilayer Perceptron

The employed Neural Network type is a standard layered Neural Network type with a linear accumulation and a sigmoid transfer function, called multi-layer perceptron. Usually the network consists of an input layer, receiving the measurement vector x, a hidden layer and an output layer of units (neurons) (Figure 4). In this configuration each unit of the hidden layer realizes a hyperplane dividing the input space into two semi-spaces. By combining such semispaces the units of the output layer are able to construct any polygonal partition of the input space. For that reason it is theoretically possible to design for each (consistent) fixed sample a correct Neural Network classifier by constructing a sufficiently fine partition of the input space. This may necessitate a large number of neurons in the hidden layer. The model parameters consist of the weights connecting two units of successive layers. In the training phase the sample is used to evaluate an error measure and a gradient descent algorithm can be employed to minimize this net error. The problem of getting stuck in local minima is called training problem.

The input are: are total radioactivity (GR), deep and shallow resistivity (Rlld and Rmsfl), bulk density (Rhob), neutron porosity (Nphi) and sonic log (DT).

The output is constituted of three neurons that consist to Porosity, Permeability and Water Saturation.

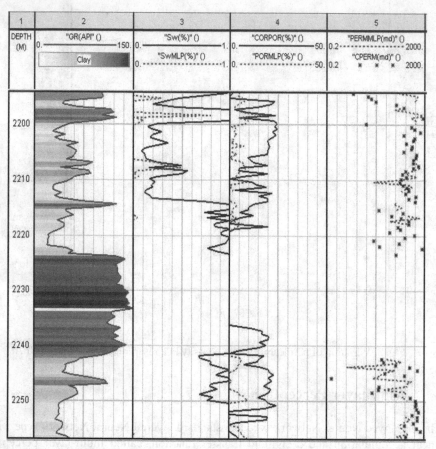

Fig. 4. Reservoir parameters estimated by MLP compared with conventional methods of wellA

4   Radial Basis Function

Powell (1985) surveyed the early work on RBF neural networks, which presently is one of the main fields of research in numerical analysis. With respect to this network, learning is equivalent to finding a surface in a multidimensional space that provides a best fit to the Learning data. Correspondingly, generalization is equivalent to the use of this multidimensional surface to interpolate the test data. The construction of a RBF network in its most basic form involves three entirely different layers. The input layer is made up of input nodes. The second layer is a hidden layer of high enough dimensions, which serves a different purpose from that in the multilayer perceptron. The output layer supplies the response of the network to the activation patterns applied to the input layer. In contrast to the multilayer perceptron, the transformation from the input space to the hidden layer space is non-linear, whereas the transformation from the hidden layer space to the output space is linear (Figure 5).

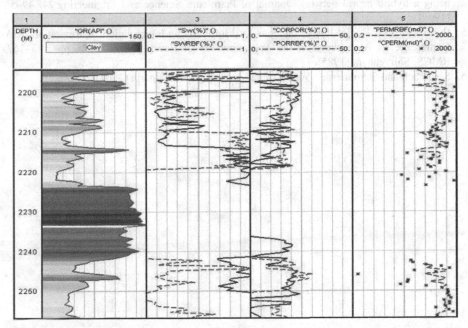

Fig. 5. Reservoir parameters estimated by RBF compared with conventional methods of wellA

5   Results Discussion and Conclusion

Obtained results show that the RBF neural network architecture is able to predict formation permeability, porosity and water saturation using laboratory measurement of the cores and well-logs data. Availability of core data for training process was proven to be essential. A comparison between the Multilayer perceptron and the Radial basis function shows clearly that the RBF gives better results than the MLP. By consequence we recommend the use of this last in the petrophysical parameters prediction from well-logs. Our results can help for reservoir characterization and petroleum exploration.

References

1. Aliouane, L., Ouadfeul, S., Boudella, A.: Fractal analysis based on the continuous wavelet transform and lithofacies classification from well-logs data using the self-organizing map neural network. Arab. J. Geosci (2011), doi:10.1007/s12517-011-0459-4
2. Lim, S.L.: Reservoir properties determination using fuzzy logic and neural networks from welldata in offshore Korea. Journal of Petroleum Science and Engineering 49, 182–192 (2005)

3. Aminian, K., Ameri, S.: Application of artificial neural networks for reservoir characterization with limited data. Journal of Petroleum Science and Engineering 49, 212–222 (2005)
4. Aminzadeh, F., Barhen, J., Glover, C.W., Toomarian, N.: Estimation of reservoir parameter using a hybrid neural network. Journal of Petroleum Science and Engineering 24, 49–56 (1999)
5. Powell, M.J.D.: Radial Basis Functions for Multivariable Interpolation: A Review in IMA Conference on Algorithms for the Approximation of Functions and Data, pp. 143–167. RMCS, Shirvenham (1985)
6. Sonatrach and Shlumberger: Well Evaluation Conference, Algeria (2007)

Lithofacies Classification Using the Multilayer Perceptron and the Self-organizing Neural Networks

Sid-Ali Ouadfeul and Leila Aliouane

Algerian Petroleum Institute, IAP. Boumerdes, Algeria,
Labophyt, FHC, UMBB, Algeria
Souadfeul@ymail.com, lil_aldz@yahoo.fr

Abstract. In this paper, we combine between the Self-Organizing Map (SOM) neural network model and the Multilayer Perceptron (MLP) for lithofacies classification from well-logs data. Firstly, the self organizing map is trained in an unsupervised learning; the input is the raw well-logs data. The SOM will give a set of classes of lithology as an output. After that the core rocks data are used for the map indexation. The set of lithology classes are generalized for the full depth interval, including depths where core rock analysis doesn't exist. This last will be used as an input to train an MLP model. Obtained results show that the coupled neural network models can give a more precise classification than the SOM or the MLP.

Keywords: Well-logs data, SOM, Supervised, Unsupervised.

1   Introduction

The artificial neural networks (ANNs) have been widely used in geophysics. In seismic data processing the ANNs are used for data inversion and facies segmentation. In seismology the artificial neural networks are used for earthquake prediction and characterization. In gravity and magnetism they are used for causative sources characterization and structural boundaries delimitation. In petrophysics the artificial neural networks are used petrophysical parameter predication and estimation. In fact my scientific researches have been published in the topic.

Ouadfeul et al [1] have published a paper that uses the ANNs in combination with the fractal analysis for lithofacies segmentation and for segmentation of thin bed. The proposedidea is applied at two boreholes located in the Algerian Sahara.

In this paper, we combine between the Self-Organizing Map neural network model proposed by Kohonen [2] and the Multilayer Perception (MLP) for better lithofacies segmentation. We start the paper by describing the principle of the SOM and the MLP techniques, after that we give the geological context of an area called Berkine basin located in the Algerian Sahara. Where the idea is tested on the data of two boreholes Sif-Fatima2 and Sif-Fatima3. We finalize the paper by the results discussion and a conclusion.

T. Huang et al. (Eds.): ICONIP 2012, Part V, LNCS 7667, pp. 737–744, 2012.

2   The Multilayer Perceptron (MLP)

Multilayer feed-forward networks form an important class of neural networks. Typically the network consists of a set of sensory units or input nodes, that constitute the input layer, one or more hidden layers of neurons or computation nodes, and an output layer. Multi-layer Perceptron (MLP) neural networks with sufficiently many nonlinear units in a single hidden unit layer have been established as universal function approximators. The advantages of the MLP are:

Hidden unit outputs (basis functions) change adaptively during training, making it unnecessary for the user to choose them beforehand.

The number of free parameters in the MLP can be unambiguously increased in small increments by simply increasing the number of hidden units.

The basic functions are bounded making overflow errors and round-off errors unlikely. The MLP is a feed-forward network consisting of units arranged in layers with only forward connections to units in subsequent layers. The connections have weights associated with them. Each signal traveling along a link is multiplied by its weight. The input layer, being the first layer, has input units that distribute the inputs to units in subsequent layers. In the following (hidden) layer, each unit sums its inputs and adds a threshold to it and nonlinearly transforms the sum (called the net function) to produce the unit output (called the activation). The output layer units often have linear activations, so that output activations equal net function values.

The layers sandwiched between the input and the output layers are called hidden layers, and the units in the hidden layers are called hidden units. The network shown below has one hidden layer.

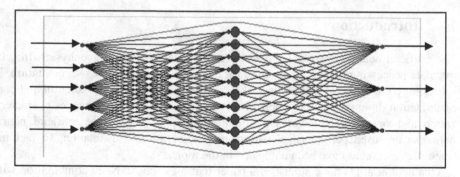

Fig. 1. An example of a Multilayer Perception neural network model with one hidden layer

3   The Self-Organizing Map (SOM)

A Self Organizing neural network, or SOM, is a collection of n reference vectors organised in a neighbourhood network, and they have the same dimension as the input vectors [3]. Neighbourhood function is usually given in terms of a two-dimensional neighbourhood matrix $\{W(i,j)\}$. In a two-dimensional map, each node has the same neighbourhood radius, which decreases linearly to zero during the self-organizing process. The conventional Euclidian distance is used to determine the best-matching unit (so called 'winner') {W(iw, jw)} on a map for the input vector {X}. Kohonen's SOMs are a type of unsupervised learning. The goal is to discover some underlying structure of the data. Kohonen's SOM is called a topology-preserving map because there is a topological structure imposed on the nodes in the network. A topological map is simply a mapping that preserves neighbourhood relations. In the nets we have studied so far, we have ignored the geometrical arrangements of output nodes. Each node in a given layer has been identical in that each is connected with all of the nodes in the upper and/or lower layer. In the brain, neurons tend to cluster in groups. The connections within the group are much greater than the connections with the neurons outside of the group. Kohonen's network tries to mimic this in a simple way. The algorithm for SOM can be summarized as follows :

- Assume output nodes are connected in an array (usually 1 or 2 dimensional)
- Assume that the network is fully connected (i.e. all nodes in the input layer are connected to all nodes in the output layer). Use the competitive learning algorithm as
  follows:
- Randomly choose an input vector x
- Determine the "winning" output node i, where W_i is the weight vector connecting the
  inputs to output node i. Note the above equation is equivalent to W_i x $\geq W_k$ x only if
  the weights are normalized.

$$|W_i - X| \leq |W_k - X|.....................\forall k$$

- Given the winning node i, the weight update is

$$W_k\,(new) = W_k\,(old) + X\,(i,k) \times (X - W_k)$$

Where $X(i,k)$ is called the neighborhood function that has value 1 when i=k and falls off with the distance $|r_k$ - r_i | between units i and k in the output array. Thus, units close to the winner as well as the winner itself, have their weights updated apprecia-bly. Weights associated with far away output nodes do not change significantly. It is here that the topological information is supplied. Nearby units receive similar updates and thus end up responding to nearby input patterns. The above rule drags the weight vector W_i and the weights of nearby units towards the input x.

Example of the neighbourhood function is given by the following relation

$$X\,(i,k) = e^{\left(-|r_k - r_i|^2\right)/(\sigma)^2}$$

where σ^2 is the width parameter that can gradually be decreased as a function of time.

4   The Processing Algorithm

The processing algorithm consists to combine between the Self-Organizing Map (SOM) and the Multilayer perceptron (MLP) for a better lithofacies prediction. The basic idea is to use the SOM as a lithofacies classifier form well-logs data, note that the Kohonen's map is based on the unsupervised learning. So at this step the Geological core rocks are used for the map indexation. The next step is to use the Multilayer perceptron (MLP) for lthofacies classification, the output of the SOM obtained by propagating the map on the full depth are used. The MLP is based on the supervised learning. The final lithofacies model is the obtained one by propagating the MLP on the full depth.

5   Application on Real Data

The proposed idea is applied at the data of two boreholes named Sif-Fatima2 and Sif-Fatima3 located in the Berkine basin (Algerian Sahara).   The first borehole is used as a pilot and the second one is used for data propagation. This section   is constructed of three part. Firstly we give the geological context of the area; secondly the used well-logs data are described, and finally an application on real data are detailed. Lets us start by the geological context of the area.

5.1   Geological Setting

South of the Alpine Algeria, the Saharan Platform is extended at vast area of over 2 million km2.   On a Panafrican substratum, most of its history is Hercynian with basins separated by high zones, synclines, among them that of Illizi-Berkine (See fig 2). The Syncline of Illizi-Berkine occupies the northeastern part over more than 400 km2. In this basin, ''another basin'', called the Triassic Province, was superimposed during the Mesozoic. Some structural features delimit the Berkine area: -To the North, the Djeffara ridge and the Sidi Touil ridge

-To the West, The bottom-up El Biod - Hassi Messaoud
-To the South, the Ahara ridge and the Zeghar-Gargaf ridge
-To the East, the Syrte Basin in Libya

The filling has Paleozoic sediment accumulation (4000 to 5000 m) mainly sandy clay of lower Paleozoic subsequently admitting some levels of carbonates and evaporites. The Mesozoic (Triassic-Upper Cretaceous) is not very thick and uncomformably overlaying the previous deposits; it is more or less eroded, according to regions. The Triassic outcrops very little (In Aménas region) but it is very thoroughly developed in depth and is recognized by many petroleum surveys. It is not very thick (50 to 100 m) and is characterized by continental and varied sedimentation (fluvial, floodplain, aeolian, sabkha, playa, volcanism at various epochs). The sedimentary bodies thus generated are discontinuous with lateral passages of facies. This sedimentation is interrupted by quiet periods during which paleosoils (complex and varied crusts) develop at the origin of discontinuities In the Triassic, petroleum geologists distinguish a number of lithological units:

- Lower Triassic shaly sandstones (TAGI)
  - Triassic carbonates and their equivalents
  - Upper Triassic shaly sandstones (TAGS)

Ages given on the basis of palynology are of the Upper Triassic.

At Berkine regional level, three sets are defined: the lower series, the intermediate series and the upper Series (T1, T2)

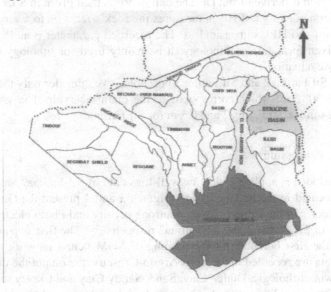

Fig. 2. Geographic location of the Berkine basin

5.2   Data Description

The physical properties that we analyze were recorded in situ in both Sif Fatima2 and Sif Fatima3 boreholes. They include Gamma ray (GR), Bulk density (RHOB), Photoelectric cross section (PEF), Neutron porosity (NPHI),   and Sonic P wave velocity (Vp).

Let us briefly describe the different measured parameters:

The Gamma ray log (GR) : The unit is usually given in API (American Petroleum Institute). It measures the radioactivity of rocks and it is used mainly for (1) differentiation between clean zones and clayey zones (2) estimation of the percentage of clay in the rocks.

The Bulk density (RHOB) : The unit is g/cc (Gramm per cubic centimetre). It measures the bulk density of rocks, by measuring the scattered rays following the bombardment of medium with medium to high energy gamma rays. The most frequently measured densities vary between 2 and 3 g/cc. It is mainly used for determination of porosity in zones of hydrocarbons or in geological formations containing the clay and, in combination with Neutron; it is used for differentiation between liquids and gases.

The photoelectric cross section (PEF): When the energy of the Gamma ray bombardment is smaller than 0.1 Mev , the measured interaction is named the photoelectric cross section , the unit   is (B/E).

The Sonic log (Dt) : The unit is µs/ft (micro-second/foot). It measures the variation of the slowness of acoustic wave propagation according to the depth. It must be done in "open" hole that means before the pose of the protective intubation. It is mainly used for determination of porosity in non-clayey formation and for the identification of lithology. Note that in this paper we used the velocity of the propagation of the P wave(Vp) which is derived from Dt. The unit of Vp is then given in S.I.   (m/s)..

The Neutron porosity (NPHI): it measures the rock's reaction to a very fast neutron bombardment.   NPHI is dimensionless. The recorded parameter is an index of hydrogen for a given formation of lithology. It is mainly used for lithology identification and porosity evaluation.

For both Sif Fatima2 and Sif Fatima3 boreholes, we consider only the depth interval [2838.5, 3082 m] which corresponds to the main reservoir. The sampling depth rate is 0.1524 m for Sif Fatima2 and 0.5 m for Sif Fatima3.

5.3   Data Processing

The proposed idea is applied at the raw well-logs data of Sif-Fatima2 and Sif-Fatima3 boreholes, located in the berkine basin.   Figures 3 and 4 present the Gama ray, Bulk density, Velocity of the Primary wave, neutron porosity and Photo-electric absorption coefficient, for Sif-Fatima2 and Sif-Fatima3 respectively. The first step consists to use the data of the first borehole for the training of SOM neural network machine. Because, the data are recorded in the Trias Argilo-Gréseux, the output the the map is one of the following lithological units: Clay, Sand, Sandy Clay and Clayey sand.

A preliminary raw lithofacies classification based on the natural gamma radioactivity well-log data was made. First, recall that the maximum value GRmax of the data is considered as a full clay concentration while the minimum value GRmin represents the full sandstone concentration. The mean value (GRmax+GRmin)/2 will then represent the threshold that will be used as a decision factor within the interval studied:

- Geological formationsbearing a natural GR activity characterized by:

$GR_{Threshold} < GR < GR_{max}$
are considered as Sandy Clay.

- Geological formations with a natural GR activity characterized by:

$GR_{min} < GR < GR_{Threshold}$are considered as a Clayey Sandstone.

The results for Sif Fatima2 and Sif Fatima3 boreholes are illustrated in Figure 5.1 and Figure 6.1that shedlighton the obtained segmentation.

The preliminary segmentation based on the Gama ray is used for the SOM indexation of the training data of Sif-Fatima2 borehole. The SOM output of this borehole is presented in figure 5.3. This output is used to train a MLP neural network machine based on the Levenberg Marquardt Algorithm. The MLP is constituted of three layers, an input layer with 05 neurons, a hidden layer with 08 neurons and an output layer with one neuron. The MLP is based on a supervised learning; the output of the SOM is used to train this MLP. The output of this machine after propagation is presented in figure 5.4. The weights of connection calculated by the training of the SOM and the MLP are used for lithofacies prediction on Sif-Fatima3 borehole (See figure 6).

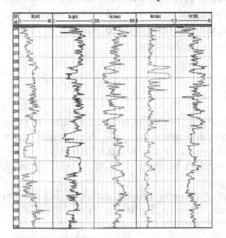

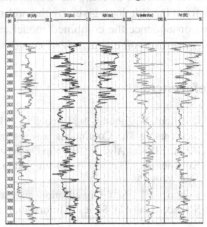

Fig. 3. Raw well-logs data of Sif-Fatima2

Fig. 4. Raw well-logs data of Sif-Fatima3 borehole

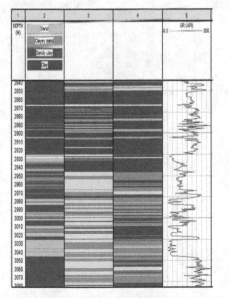

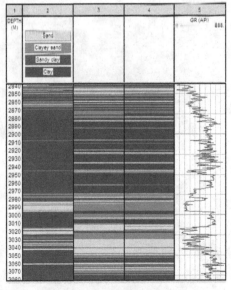

Fig. 5. Lithfacies classification in Sif-Fatima2 Using: (1) Gama ray (2) SOM (3) SOM and MLP

Fig. 6. Lithfacies classification in Sif-Fatima3using: (1) Gama ray (2) SOM (3) SOM and MLP

6   Results Interpretation and Conclusion

Application on real data shows that the Self-Organizing MAP combined with the Multilayer Perception MLP is able to give more precise lithology than the Self-Organizing MAP for example for Sif-Fatima3 borehole (figure 6) and in the depth interval [3000m, 3070m] the Combined learning gives lithological units that are not far from the lithology given by the preliminary segmentation based on the gamma ray. By consequence the combined model can be used to train a large number of data to enhance reservoir characterization and petroleum exploration

References

1. Ouadfeul, S., Zaourar, N., Boudella, A., Hamoudi, M.: Modeling and classification of litho-facies using The Continuous wavelet transform and neural network: A case study from Berkine Basin (Algeria). Buletin Du Service Géologique d'Algérie 22(1) (2011)
2. Kohonen, T.: The self organazing Map. Information Sciences, vol. 30, p. 312. Springer, New York (2010)
3. Kohonen, T.: Self Organization and associative memory pringer Series in Information Sciences, 2nd edn. Series in Information Sciences. Springer, Berlin (1998)

Author Index